29.50
C0-AQZ-979

Critical Materials Problems in Energy Production

ACADEMIC PRESS RAPID MANUSCRIPT REPRODUCTION

Critical Materials Problems in Energy Production

Edited by

Charles Stein

Air Force Weapons Laboratory
Albuquerque, New Mexico
and
New Mexico Institute of Technology
Socorro, New Mexico

ACADEMIC PRESS, INC. New York San Francisco London
1976

A Subsidiary of Harcourt Brace Jovanovich, Publishers

ACADEMIC PRESS, INC.
111 Fifth Avenue, New York, New York 10003

United Kingdom Edition published by
ACADEMIC PRESS, INC. (LONDON) LTD.
24/28 Oval Road, London NW1

Library of Congress Cataloging in Publication Data

Main entry under title:

Critical materials problems in energy production.

"This volume is the result of a series of distinguished lectures sponsored by the Joint Center for Materials Science in New Mexico."
Includes index.
1. Materials. 2. Power (Mechanics)
I. Stein, Charles, Date II. Joint Center for Materials Science.
TA403.C74 621.4′0028 76-47484
ISBN 0-12-665050-0

PRINTED IN THE UNITED STATES OF AMERICA

CONTENTS

CONTRIBUTORS

W. T. BAKKER, Chief, Materials Branch, Materials and Power Generation, Fossil Energy Research, ERDA

JAMES E. BATTLES, Chemical Engineering Division, Argonne National Laboratory, Argonne, Illinois

R. BEHRISCH, Max-Planck-Institut für Plasmaphysik, EURATOM Association, D-8046 Garching/München, Germany

M. BOUDART, Department of Chemical Engineering, Stanford University, Stanford, California

H. K. BOWEN, Department of Materials Science and Engineering, Massachusetts Institute of Technology, Cambridge, Massachusetts

RICHARD H. BUBE, Department of Materials Science and Engineering, Stanford University, Stanford, California

ELTON J. CAIRNS, Research Laboratories, General Motors Corporation, Warren, Michigan

R. S. CLAASSEN, Director of Materials and Processes, Sandia Laboratories, Albuquerque, New Mexico

FRANK W. CLINARD, JR., University of California, Los Alamos Scientific Laboratory, Los Alamos, N.M.

JAMES A. CUSUMANO, Catalytica Associates, Inc., 2 Palo Alto Square, Palo Alto, California

H. E. FRANKEL, Assistant Director, Materials and Power Generation, Fossil Energy Research, ERDA

T. H. GEBALLE, Department of Applied Physics, Stanford University, Stanford, California

ALEXANDER J. GLASS, Head, Theoretical Studies, Y Division, Lawrence Livermore Laboratory, Livermore, California

NICHOLAS J. GRANT, Department of Materials Science and Engineering, Massachusetts Institute of Technology, Cambridge, Massachusetts

ARTHUR H. GUENTHER, Chief Scientist, Air Force Weapons Laboratory, Kirtland AFB, New Mexico

R. G. HICKMAN, Lawrence Livermore Laboratory, University of California, Livermore, California

ROBERT A. HUGGINS, Department of Materials Science and Engineering, Stanford University, Stanford, California

G. G. LIBOWITZ, Materials Research Center, Allied Chemical Corporation, Morristown, New Jersey

RICARDO B. LEVY, Catalytica Associates, Inc., 2 Palo Alto Square, Palo Alto, California

J. J. LOFERSKI, Division of Engineering, Brown University, Providence, R. I.

BERND T. MATTHIAS, Institute for Pure and Applied Physical Sciences, University of California, San Diego, La Jolla, California, and Bell Laboratories, Murray Hill, New Jersey

G. A. MILLS, Director, Fossil Energy Research, ERDA

L. E. MURR, Department of Metallurgical and Materials Engineering, New Mexico Institute of Mining and Technology, Socorro, New Mexico

S. F. PUGH, Head of Metallurgy Division, A.E.R.E. Harwell, Oxfordshire, England OX11 ORA

DAVID W. RABENHORST, The Johns Hopkins University, Applied Physics Laboratory, Laurel, Maryland

B. R. ROSSING, Materials Science Division, Westinghouse Research Laboratories, Pittsburgh, Pennsylvania

R. W. STAEHLE, Department of Metallurgical Engineering, The Ohio State University, Columbus, Ohio

B.C.H. STEELE, Department of Metallurgy and Materials Science, Imperial College, London SW 7, England

D. STEINER, Thermonuclear Division, Oak Ridge National Laboratory, Oak Ridge, Tennessee

MARIA TELKES, Institute of Energy Conversion, University of Delaware, Newark, Delaware

W. van GOOL, Department of Inorganic Chemistry, State University, Utrecht, Netherlands

M. J. WEBER, Lawrence Livermore Laboratory, University of California, Livermore, California

F. W. WIFFEN, Metals and Ceramics Division, Oak Ridge National Laboratory, Oak Ridge, Tennessee

GENE A. ZERLAUT, Desert Sunshine Exposure Tests, Inc., Phoenix, Arizona

PREFACE

As far as we know, man is the only animal both cognizant of and endowed with sufficient intellect to change his environment. However, man's ability to do so is heavily predicated on his ability to sustain his continued scientific and technological advance. In the main fields in which he conducts this struggle—agriculture, urbanization, medicine, and communications—the availability of huge supplies of energy has determined the rate at which man has been able to achieve his goals.

The problem of providing energy in such vast quantities, distributing it, and storing portions of it for later use during peak demand periods consists of several different but closely related aspects: sociological (priorities—who gets it and how much), ecological (at what price to our environment), economical (how much are we willing to pay for it), and technological (can we produce large amounts of energy, in a wide variety of forms, at an economically acceptable price). While there have recently been several excellent books and articles that address each of these problems, this volume is concerned with those technological phenomena which are limiting progress in energy production due to materials related inadequacies.

It is interesting to note that while progress in other fields depends on the close synergism between scientific discoveries and breakthroughs and the transformation of this knowledge into technologically useful systems, the production of new sources of large quantities of energy is more dependent on the extraction of technical knowledge from several traditional fields of engineering. For the most part, the need is to translate, advance, and utilize knowledge already on hand in the chemical engineering, metallurgical engineering, high-temperature ceramic engineering, corrosion engineering, and electrochemical engineering industries into solutions of problems in energy production.

Out of many technical problems encountered in the production of energy, this volume identifies and concentrates on the most challenging of the materials problems in the three areas of production, distribution, and storage of energy and treats them in seven sections: (1) nuclear power, (2) materials for high-temperature applications, (3) solar energy, (4) direct solar conversion, (5) coal and other fossil fuels, (6) superconducting materials, and (7) energy storage devices.

This volume is the result of a series of Distinguished Lectures sponsored by the Joint Center for Materials Science in New Mexico. The Center is a manifestation of an endeavor on the part of the Air Force Weapons Laboratory, the Los Alamos Scientific Laboratory, and Sandia Laboratories to cooperate with the New Mexico Institute of Mining and Technology, New Mexico State University, and the University of New Mexico to provide a program of continuing education in the field of materials science for the New Mexico scientific community. It was appropriate, therefore, that the list of guest lecturers be compiled from suggestions made by researchers working directly in the field of energy production. Consequently, the individuals invited to participate in this symposium represent the most distinguished scientists from both the United States and Western Europe, who are contributing to our present understanding of the fundamental materials problems that are limiting energy production.

As a part of the Joint Center's continuing educational program, the collected papers of the Distinguished Lecture Series on "Critical Materials Problems in Energy Production" have been edited and produced in a format suitable to be used as a text by both senior students and first-year graduate students in materials science. Each of the 28 chapters has sufficient introductory information to provide the student with the background and the context from which the materials problems in specific topics of energy production can be understood. In addition, a set of problems and recommended readings that are related to important technical points in each chapter is appended to the final section of the book. They are intended to help the instructor amplify materials science concepts and to offer the student an opportunity to augment the material presented with further, in-depth background information.

The Joint Center for Materials Science, which sponsored these lectures, gratefully wishes to thank and acknowledge the Air Force Weapons Laboratory; the European Office of Aerospace Research and Development, U.S. Air Force, London, England; the Los Alamos Scientific Laboratory; and Sandia Laboratories for their financial support of the Distinguished Lecture Series, and the University of New Mexico for their hospitality in providing lecture hall space.

The editor wishes to thank those who participated in the Distinguished Lecture Series and who contributed to this volume; and Mr. L. DeJohn, who provided the original graphic designs used to introduce each energy section.

Charles Stein

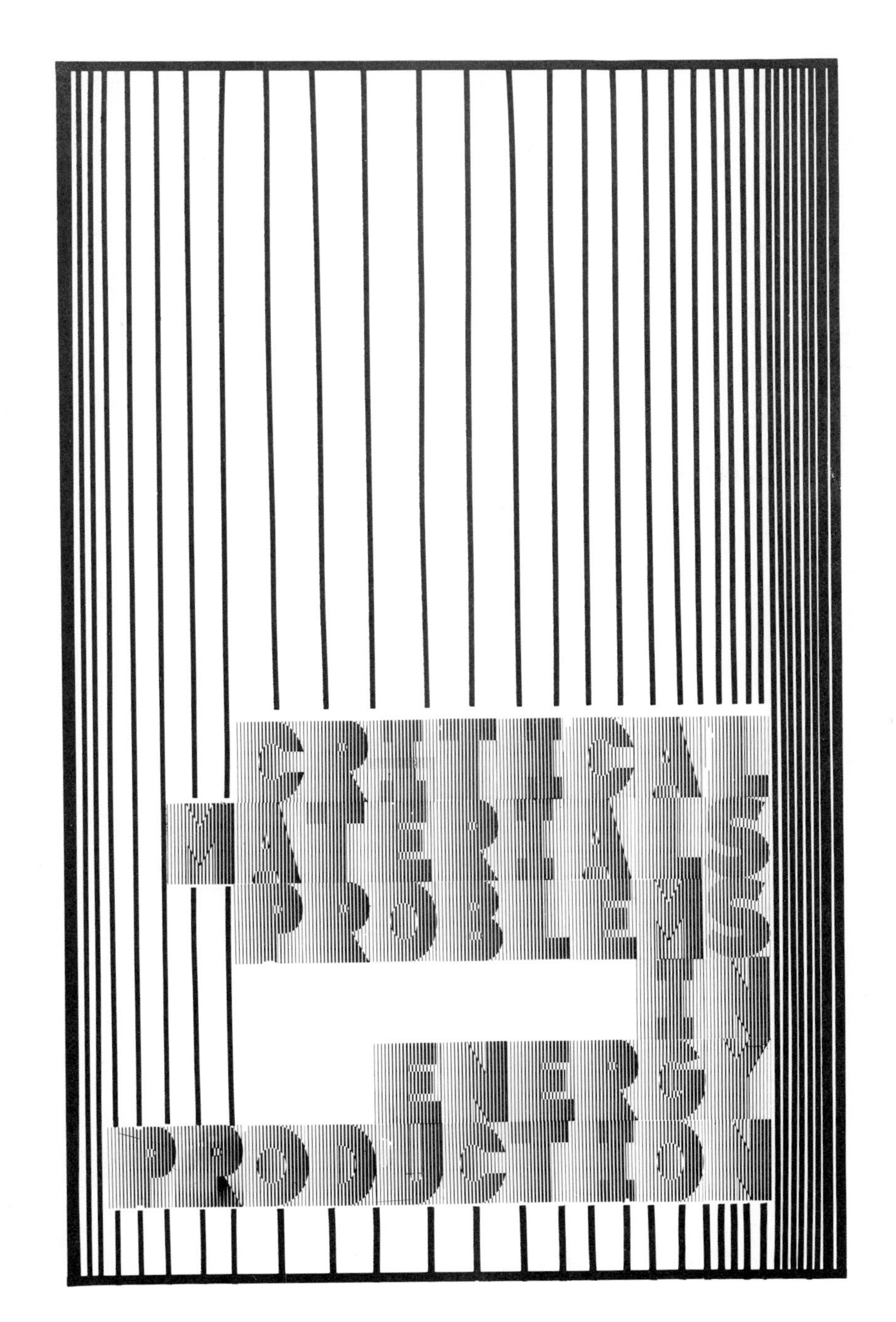
CRITICAL
MATERIALS
PROBLEMS
IN
ENERGY
PRODUCTION

CHAPTER 1

A PERSPECTIVE ON MATERIALS IN THE ENERGY PROGRAM

R. S. Claassen

Director of Materials and Processes
Sandia Laboratories
Albuquerque, New Mexico 87115

The production and consumption of energy, generally recognized by the technical community as a crucial program for the next few decades, will require significant advances in many areas of practice and knowledge. The Committee for the Joint Center for Material Science in the State of New Mexico feels that it is timely to address the question of contributions which the materials community may make to the energy program. Thus, we are embarking on this series of presentations entitled "Critical Material Problems in Energy Production." This introductory paper will serve to provide an overall perspective of the energy program within which the specific materials problems can be considered.

To many of us, the energy problem is most apparent through the rapid rise in fuel costs—gasoline for the car or fuel to heat the home. Therefore, people decry every day that energy is getting more and more expensive. The interesting point is not so much that energy is getting more expensive but that energy has always been so cheap. This situation has given rise to a number of rather inconsistent but understandable behavior patterns.

For example, in Albuquerque the average homeowner's lot receives about 8800 million BTUs per year from the sun. For heating, hot water and cooking by natural gas, the average consumption is only 130 million BTUs a year(1)—considerably less than the energy received from the sun. And yet each Albuquerque homeowner will pay $177 on the average this year to have gas piped in from a well someplace far away. In our forests, enough timber matures to furnish nearly all the energy we need, but we drill holes four miles deep in search of more oil. We are planning construction of a whole series of incredibly complex and expensive nuclear power plants to provide even cheaper energy than these traditionally available forms.

We could propel our ships across the ocean with sails as we once did. The energy is free; but instead we fuel our tankers with inexpensive oil out of our own deposits so we can go overseas and import expensive oil from the Near East. The University of New Mexico closed down for three extra days during the Christmas holiday season to save energy; the result was a saving of 26 cents per day for each person who remained away from work.(2) It was not a favorable trade-off, precisely because energy is so cheap. Consequently, we may decide that we are willing to pay considerably higher prices for energy.

Energy and materials are closely linked in many ways. For example, the production of materials, which requires their recovery from natural resources, and the processing of them until they are ready for manufacturing, consume something like 17% of all energy used in this country. Conservation of energy in material processing is therefore an important problem for the materials community but there is not time in this Series to cover energy conservation. Similarly, in other aspects of the energy situation, there is the question of source materials and substitution, which is also an appropriate problem for the materials community to address. We need certain materials, or at least we need the performance characteristics which they presently yield (for example, stainless properties of chromium alloys). And if chromium should become unavailable, the materials community will have to provide the same performance from other elements through materials development. But materials substitution is also beyond the scope of this volume.(3)

To fully understand how materials affect our evolving energy program, one needs first to understand the interconnections with other aspects of the situation. Thus, in Section I of this chapter, some historical perspective will reveal what energy consumption has been thus far and what it is likely to be in the future. Section II provides the units and conversion factors most often used in energy discussions. The uses of energy and the forms that energy must take to be consumable are summarized in Section III. Financial aspects, particularly capitalization problems and fuel expenses, are covered in Section IV. The final section then provides a brief description of the materials problems to be discussed by the other authors in this volume.

I. HISTORICAL PERSPECTIVE

An historical review of energy reveals the overwhelming point that energy has always been and remains a growth business. Energy used in this country as measured in quadrillions of BTUs (Quads) is plotted in Figure 1 as a function of year since 1947. Note that for nearly two decades we were on about a four-percent

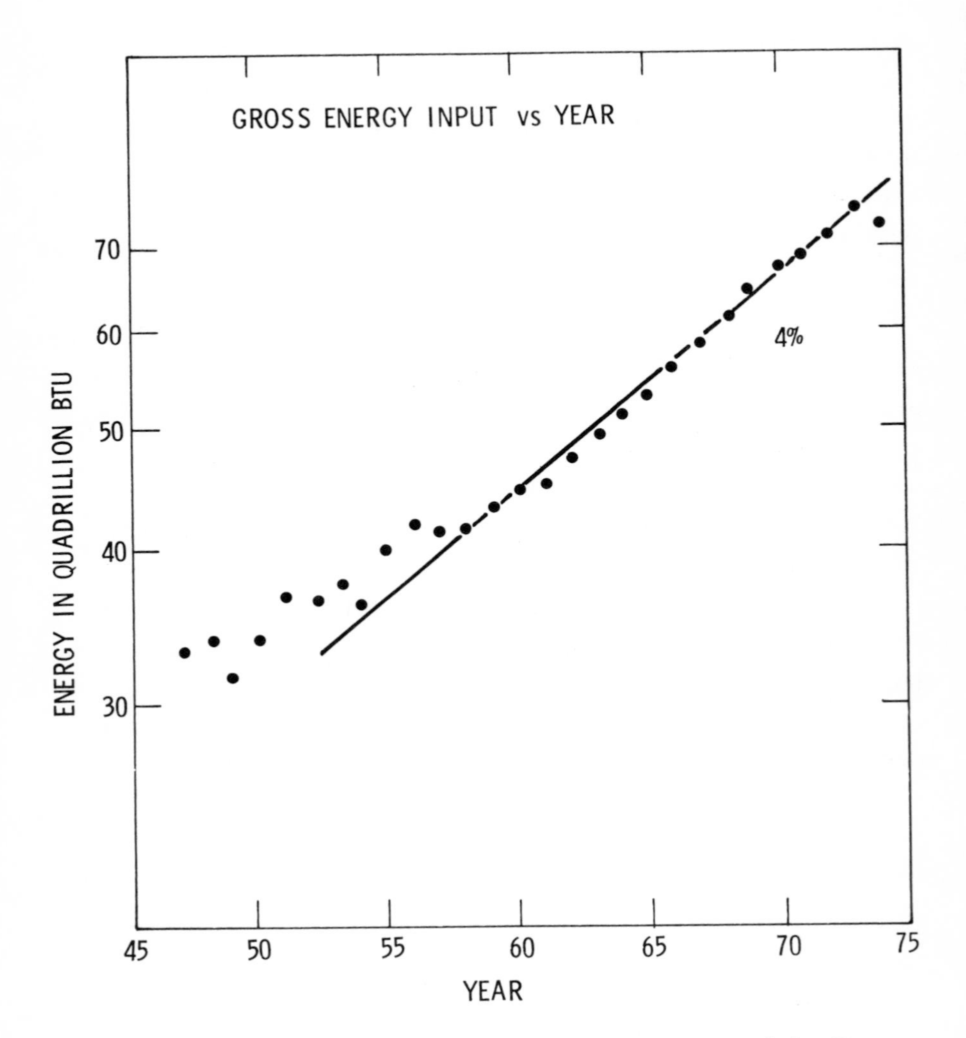

FIG. 1. Energy of all forms consumed in the United States in the years 1947-1974. The straight line represents an exponential of 4% per year.(4)

growth curve for total energy used in all forms. Only in the last couple of years have we fallen below that curve, and that drop is probably attributable mainly to the economic recession.

Use of electricity has increased even more dramatically. Figure 2 is a semilog plot of electricity generated in this country (in terawatt hours) as a function of year. For the entire period we had very close to a seven percent exponential growth, or a doubling every ten years. Even more remarkably, this rapid growth rate extends back to at least 1900, reflecting a very long-term trend indeed. Again, although the production of energy in 1974 and 1975 was nearly level compared with the previous years, most of the lack of increase is attributable to the slowing down of the industrial sector. Since both home and commercial consumption (that is, buildings, hotels, and so forth) decreased growth only slightly, perhaps this also suggests that stagnation is attributable to the industrial slowdown.

What has been happening in the electric generation business with regard to fuel type? Figure 3 shows the same overall growth rate. Although coal does not follow the exponential curve, the other energy forms do. The use of gas has increased until just the last couple of years. Nuclear energy in terms of fuel availability has increased at a rate of 47% per year, but of course is only at the beginning of its career and, under the pressure of environmental control, is not yet contributing a large fraction of energy. (Hydroelectric and other minor sources are not included in Figure 3.)

Something that is highly controversial needs to be considered: the relationship between consumption of energy and the level of affluence or standard of living. While I believe that standard of living is closely tied to energy consumption, others feel strongly that it need not be. The only quantitative measure we have for standard of living is gross national product per capita; Figure 4 compares this measure among countries. On

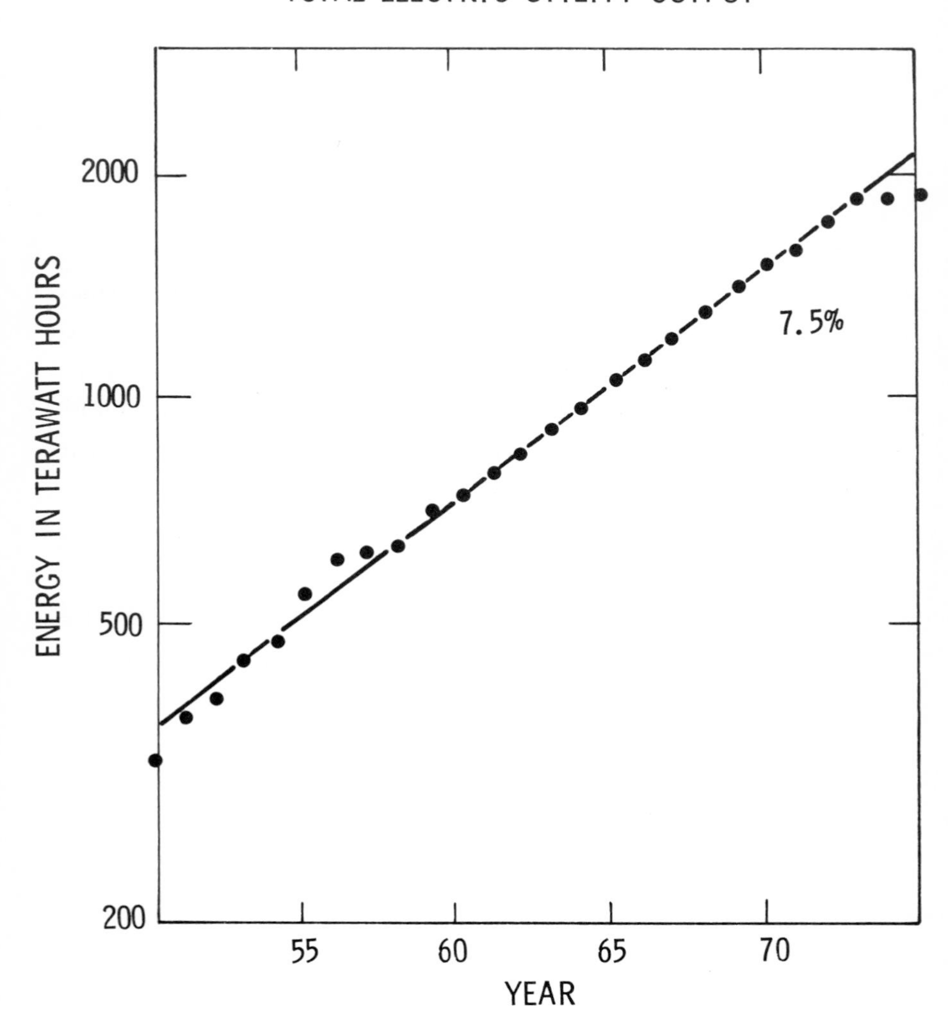

FIG. 2. Total electric utility output in the United States for the years 1950 through 1975. The straight line is an exponential with a slope of 7.5% per year.(5)

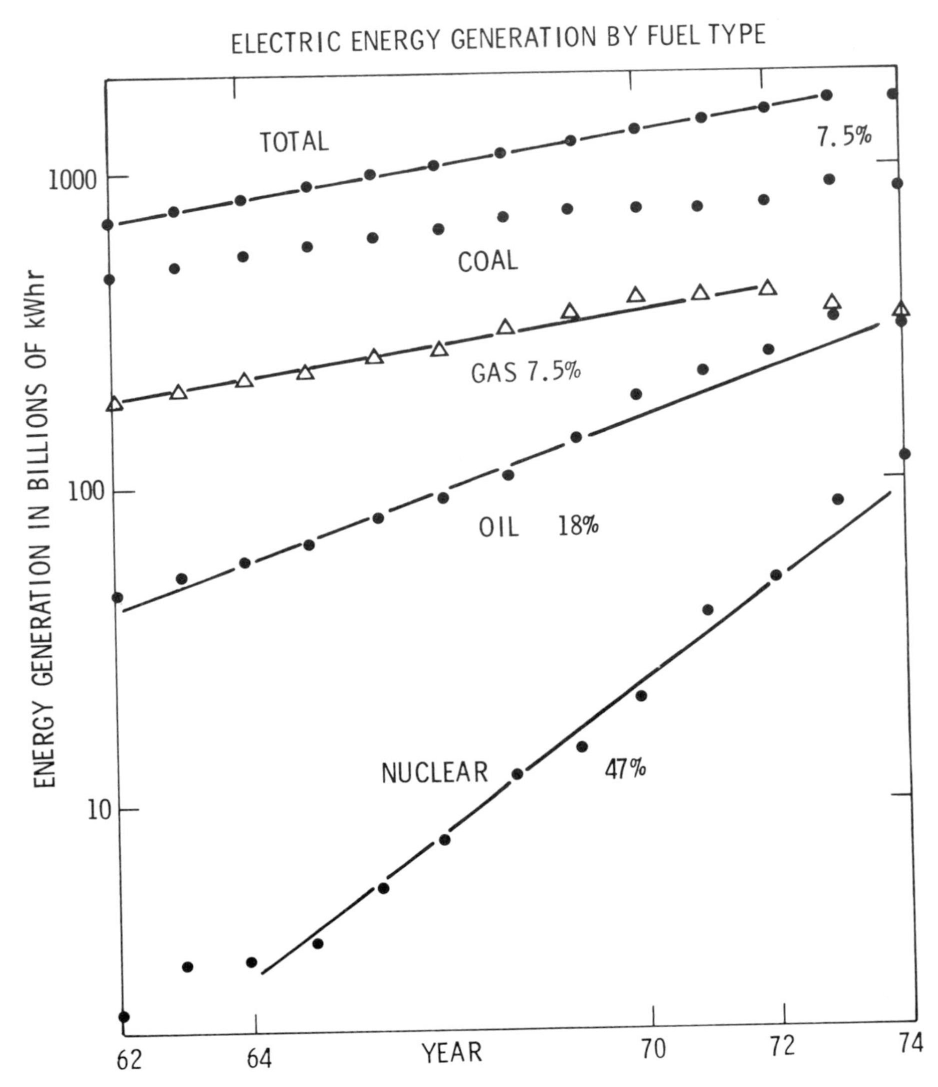

FIG. 3. Electric generation by fuel types for the period 1962-1974. The straight lines indicate the approximately exponential growth for natural gas, oil and nuclear. Hydroelectric and minor sources are not included.(6)

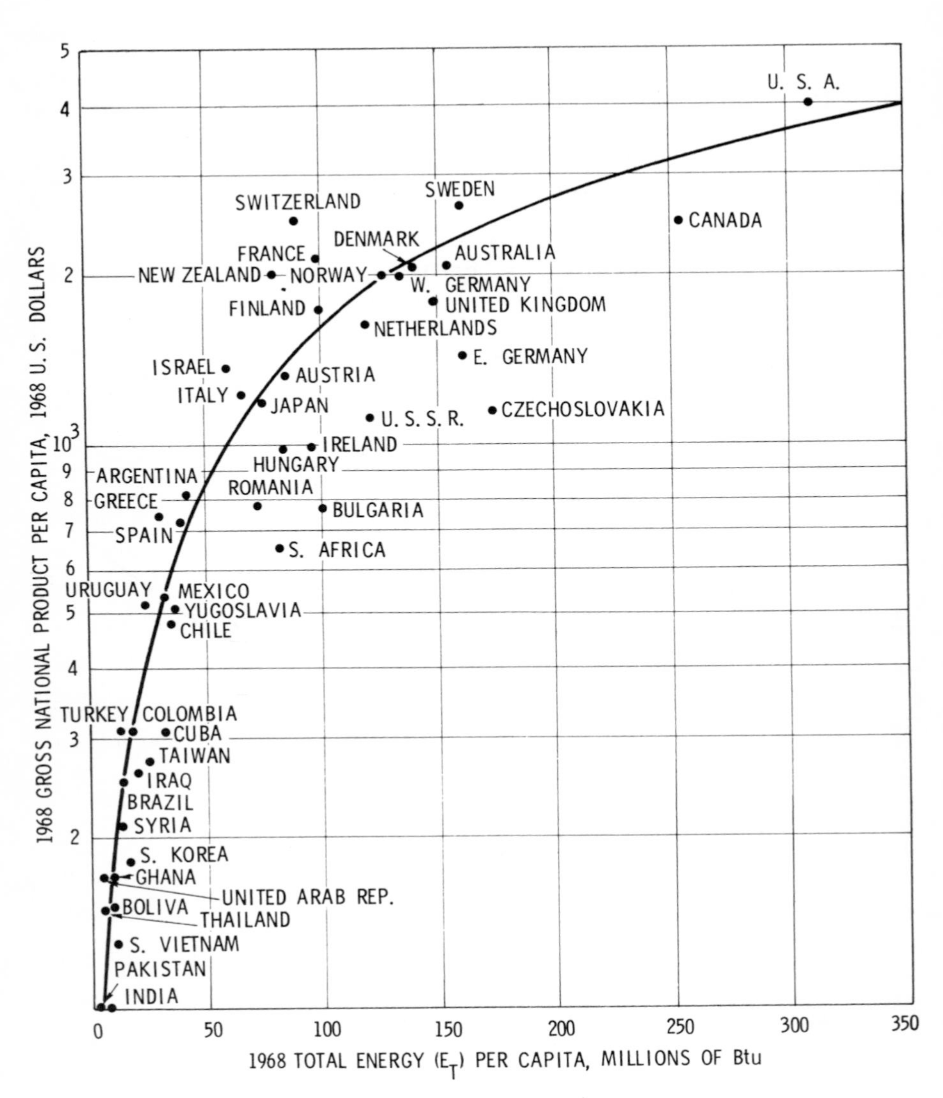

FIG. 4. Gross national product per capita vs total energy per capita indicating the general correlation between the two quantities.(7)

the ordinate is the gross national product per capita, measured in 1968 U.S. dollars. Along the abcissa is the total energy used per capita in millions of BTUs. When gross national product per individual is plotted in this way, the United States is highest, followed by other countries that are generally thought to be well-developed and prosperous, such as Canada and Sweden. Down in the left-hand corner are countries such as India, Pakistan, Vietnam, Thailand, Bolivia, Ghana, Korea, and others generally considered underdeveloped. The correlation is not perfect, but there seems to be much evidence that gross national product is closely correlated with energy consumption. If, referring back to the growth curves in Figure 3, one says, "Let's stop the growth in energy use and hold at a constant level from now," it would imply that we will also stop the growth of gross national product. We are heavily geared in this country to growth of the GNP and, if we halt the growth, we will probably produce a very substantial effect on all aspects of our lives. The country may make that choice but, in case it does not, the technologists should be prepared to offer alternatives. A capability for greater energy production in the future will require many contributions from the materials technologists, as will become evident throughout this volume.

Many different projections have been made regarding the future of energy in this country. ERDA, the Energy Research and Development Administration, has of course made projections.(8) Last year they projected that the 73 quads we now consume each year would increase to about 107 quads by 1985. In other words, we will stay on the 4% growth rate shown in Figure 1. By the year 2000, ERDA projects that we will drop to a 2½% annual growth rate, reaching 150 quads by that year; that is, consumption will double between now and the end of the century.

Figure 5 shows ERDA's projection of how electrical energy will grow and how it will be fueled. Two salient points need to

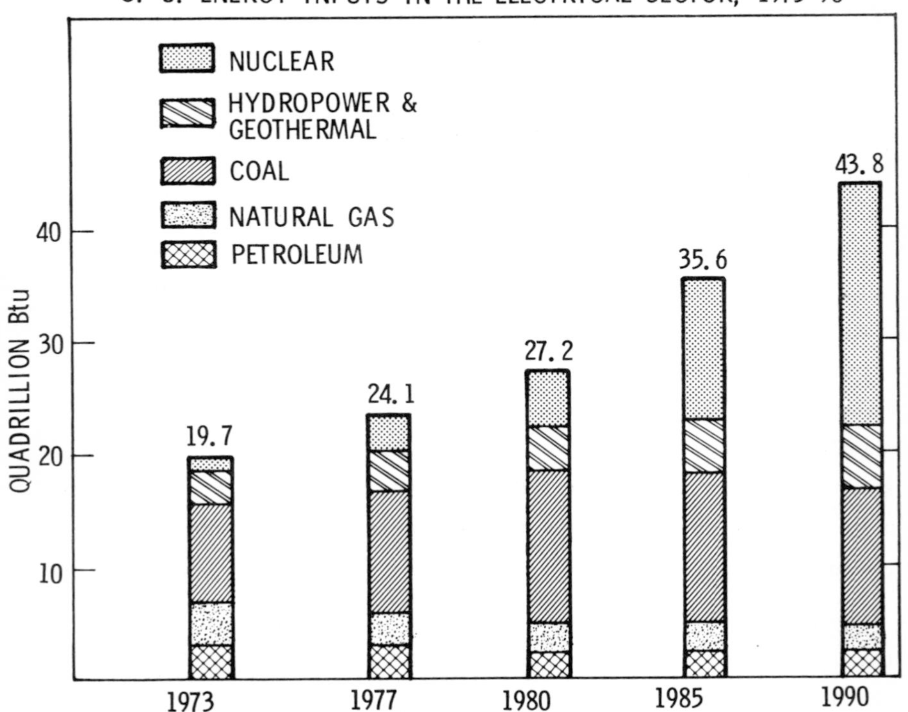

FIG. 5. Forecast of energy inputs for electrical generation in the United States, "business as usual," $11 per barrel oil, with conservation scenario.(9)

be noted: First, strong, continued growth is predicted (about 5% per year). More important, though, is that most of the growth is expected to be made up by nuclear. Note that coal is not shown as continuing to increase substantially, even though we have great reserves. The most discouraging fact is that natural gas is still projected to be a heavy contributor through 1990. The next projection, Figure 6, is for coal converted to two forms of synthetic fuel. The upper bars project synthetic gas, that is, high-BTU methane derived from coal, for pipeline use. Multiplying the billions of cubic feet per year by one thousand BTU per cubic foot, one gets BTUs amounting to 2.8 quads in 1990. Note that the far right bar represents exactly what ERDA projected in Figure 5 for the gas consumption needed to make electric power in the same year. That seems a terrible waste of gas—the equivalent of all high-grade gas from coal being burned to make electricity, when the coal could be burned directly with some less expensive but environmentally acceptable process.

We are going to change, but how fast can we? Let's look at past time constants for changing fuels in this country. They may not be inevitable constants, but they are interesting to consider. Figure 7 plots energy consumption patterns since 1850, showing the fuel sources of this country by type. A heavy dependence on wood gave way to coal, and coal dropped off as petroleum and natural gas came into use. Hydropower, a low fraction, will always be limited. Nuclear energy is just now beginning to demonstrate usefulness. If you study this data and ask the question, "Starting from 10% of its ultimate use, how long did it take coal, petroleum or natural gas to arrive at 50% of ultimate use?," the answer is about the same for all three. Each took about 30 years to rise from 10% of ultimate to 50% of ultimate use. Even though that period might be shortened somewhat by a crash program, three decades may be a good measure of the time it is going to take for some new fuel technology to make a real market penetration.

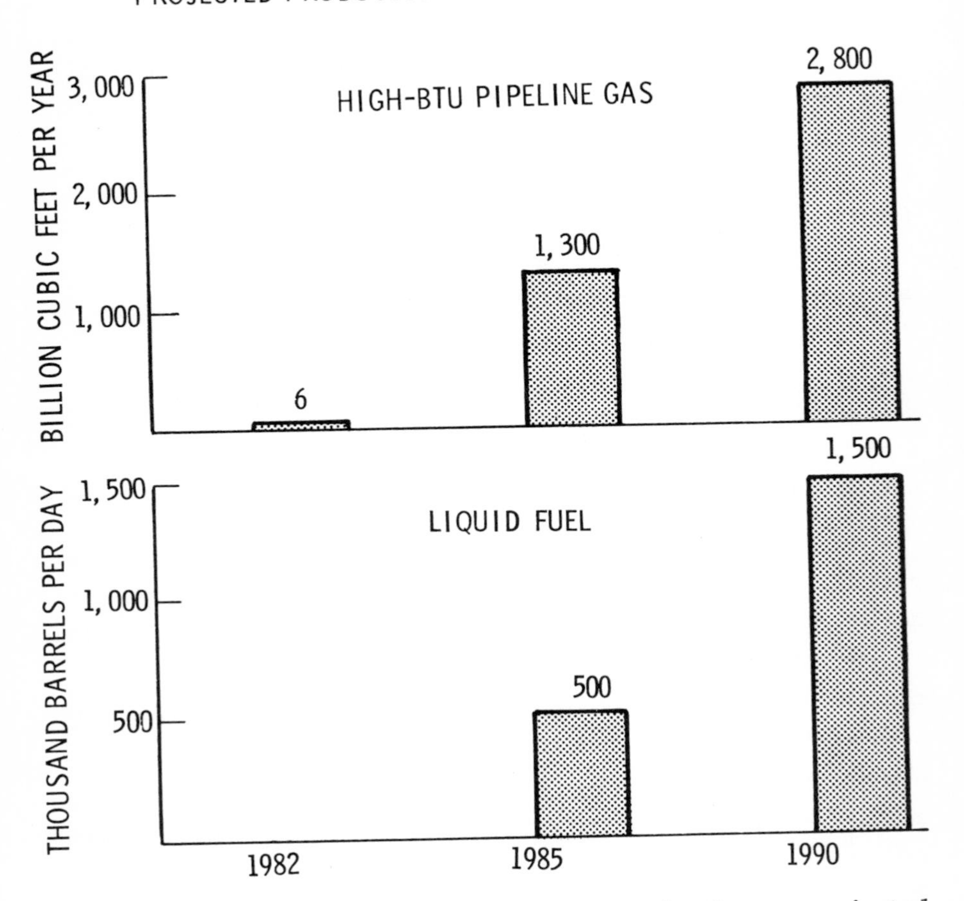

FIG. 6. Synthetic fuel production as projected by the Federal Energy Administration. The pipeline gas quantity for 1990 is similar to that for natural gas shown in Fig. 5 for electric production that year.(10)

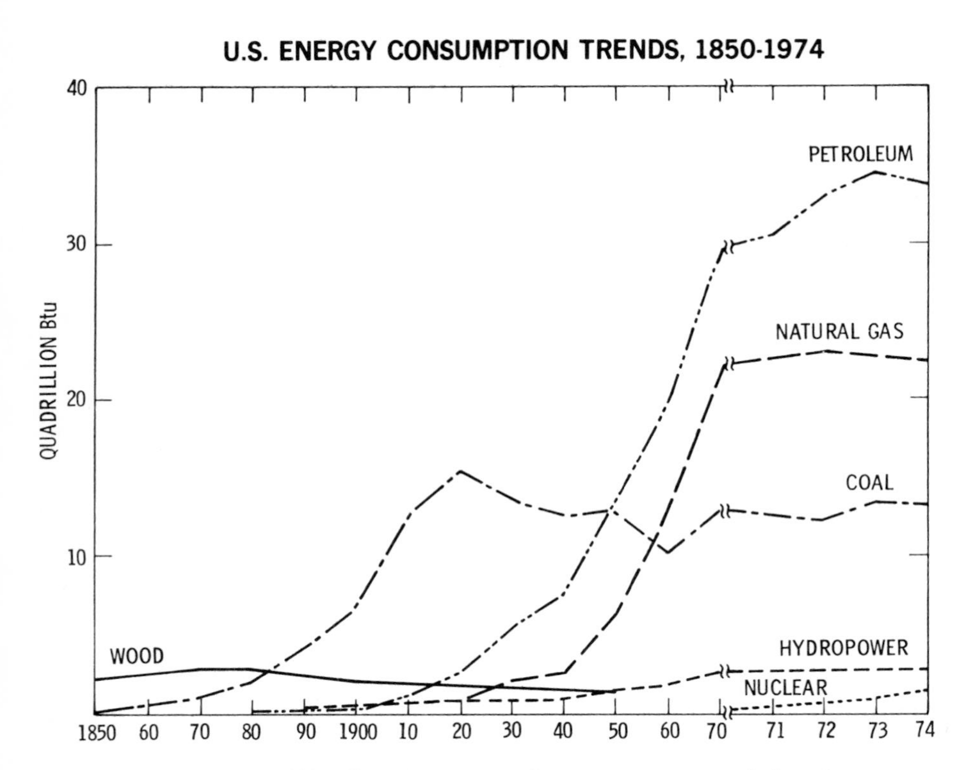

FIG. 7. Sources of energy consumed in the U.S. Coal reached its maximum production rate in 1920. The time to go from 10% of ultimate (1873) to 50% (1902) was 29 years. The growth rates for oil (32 years) and gas (27 years) were similar.(11)

Another problem that enters into consideration is the time that it takes to construct or obtain new physical plants. For different kinds of plants, Table I shows time from go-ahead to production. A new coal-fired power plant of very standard technology, ordered out of the "catalog," takes five to eight years. Nuclear power is now taking 9 to 10 years from the decision to commit the capital until the time the plant is available to contribute to on-line power production. Nuclear plants experience very long times because of inevitably long construction times plus delays in licensing and other factors. Coal mines do not take as long to start up; but in addition to the start-up times shown, one has to add as many as four years for delivery of major equipment because that particular sector of manufacturing is backordered. If one does get the equipment and the coal mine opened, one then finds that there are not sufficient transportation units to transport the coal economically to its conversion point. Changes in consumer habits take time, as does the creation of wholesale and retail distribution networks. So it is understandable that a new or changing technology of fuel requires considerable buildup time.

In terms of materials technology, the point I want to stress is that we are dealing in decades, and we ought properly to think in those time scales. It is fortunate that we have the latitude to deal in decades, because we need that kind of time. The normal engineering development period approaches a decade. In addition, if a new field is involved, one needs another decade prior to development to conduct research. After development is completed and the demonstration plant is operating, one needs yet another decade or so to commit the capital and to build full-production plants.

Let me use nuclear power as one more example illustrating this time consideration. Fission was first demonstrated in Chicago in 1942-3. It was not until 30 years later that nuclear

TABLE I

TYPICAL 1973 OVERALL PROJECT TIMES

(FROM GO-AHEAD TO PRODUCTION) *(12)*

TYPE OF FACILITY	YEARS
COAL-FIRED POWER PLANT	5-8
SURFACE COAL MINE	2-4
UNDERGROUND COAL MINE	3-5
URANIUM EXPLORATION AND MINE	7-10
NUCLEAR POWER PLANT	9-10
HYDROELECTRIC DAM	5-8
PRODUCE OIL AND GAS FROM NEW FIELDS	3-10
PRODUCE OIL AND GAS FROM OLD FIELDS	1-3

power for the first time contributed even as much as 1% of our energy base. If we look ahead to a new technology like fusion and say, "One of these days we'll have scientific breakeven," (which will be comparable to the Chicago fission demonstration), then 30 years more is an extremely short time to move from laboratory experiment to the point at which enough power is being produced to have any impact on the nation's energy situation.

When you attack a materials problem related to energy, there must be time to think the problem through, to plan and execute a systematic research and development program.

II. UNITS OF ENERGY

Table II will provide some common ground for understanding the conversion units of energy. A handy unit is the quad = 10^{15} BTU. (You may see reference to the unit Q = 10^{18} BTU.) It turns out that we are using about 73 quads this year. Since people in the government like to talk about number of barrels of oil a day equivalent, there is a handy, rough conversion: one million barrels per day $\cong$ 2 quads per year. A standard cubic foot of natural gas contains about 1000 BTUs. It takes about 10,000 BTUs to make one kilowatt of electricity through normal steam conversion, called the heat rate.* For those adopting the SI metric system, simply remember that a gigajoule is about the same as one billion BTUs. And a nice number to know is that your yard collects about one horsepower per square yard of surface from solar input at noon on a clear day.

---

*The average heat rate for the total electric utility industry in 1973 was 10,429 BTU/kWhr.[13]

TABLE II

UNITS AND CONVERSIONS

1 QUAD = 10^{15} BTU = 472,000 BBLS. OIL DAILY EQUIVALENT

1 BARREL OIL = 5.8 MBTU = 5.8×10^{6} BTU

1 STANDARD CUBIC FOOT (SCF) NATURAL GAS = 1030 BTU

1 KILOWATT HOUR ELECTRICAL = 3413 BTU

1 kWhr = 10,000 BTU RULE OF THUMB CONVERSION RATE FOR STEAM TURBINES

1 GIGAJOULE (GJ) = 0.96 MBTU

1 KILOCALORIE = 4187 JOULES

SOLAR INSOLATION (GOOD WEATHER) $\sim$ 1 kW/m^2 OR 1 hp/yd^2 BUT AVERAGE ABOUT 1/5 OF THIS

III. USE AND FORMS OF ENERGY

How is our fuel used? Table III summarizes all forms of energy used in the U. S. in 1973. First, note that one-fourth of all our fuel is used for transportation. Next, space heating, process steam, and direct heat, all three representing the simplest kind of heating, use 45% of our fuel. Electric drive, lighting, and electrolytic processes use 14%.

These three large user groupings represent three distinct forms of energy. The first, for transportation, is liquid hydrocarbon which is so convenient for carrying along with us as we do in automobiles, buses, airplanes and ships. Unless we achieve a major technological revolution in motive power, we will continue to depend on liquid hydrocarbons as fuel for transportation. The second form of fuel, for space or process heating, can be one of many types since the heat required is relatively low-grade. Much of the process steam produced in industry is below 200°C and, of course, space heating is at even lower temperatures. Combustion of almost any fuel will meet these requirements, but simple solar panels, for example, could also supply this demand. Electricity is the third form of energy and, in addition to the functions listed in the previous paragraph, is the natural energy supply for information processing and for much of our high technology. To some extent these three fuel forms are interchangeable, but not entirely.

Some fuels are not suited for some uses. For example, if you simply burn oil, you don't get a very efficient light source. If you burn oil to first make electricity, and then run that electricity through a resistor, you do not have an efficient way of heating things. Because fuel forms lack this easy interchangeability, we have difficulty even talking about our energy programs and priorities; and we have to be very careful about suggesting substitutions. In this country, our basic problems

TABLE III

ENERGY CONSUMPTION IN THE UNITED STATES
BY END-USE IN 1973 (ESTIMATED)*(14)*

END USE	ENERGY CONSUMPTION QUADS	% OF USE
TRANSPORTATION	18.7	25
SPACE HEATING	13.5	18
PROCESS STEAM	12.2	16
DIRECT HEAT	8.2	11
ELECTRIC DRIVE	5.8	8
LIGHTING	4.2	5
WATER HEATING	3.0	4
FEED STOCKS	2.8	4
AIR CONDITIONING	2.3	3
REFRIGERATION	1.5	2
COOKING	0.7	1
ELECTROLYTIC PROCESS	1.0	1
OTHER	1.8	2
TOTAL	75.6	100

center around shortages of petroleum and natural gas, so a program which merely substitutes coal for hydropower or nuclear energy for coal does not directly resolve our major problems. Since it is very difficult to compare all these fuels against one reference of quality, I have been unable to find an easy way to address interchangeability.

IV. FINANCIAL ASPECTS

One cannot understand the energy program without understanding the financial aspects of it at the simplest level. Since energy programs are huge consumers of capital, they are called "capital intensive." Several groups have studied what capital investments will be required over the next decade to support the various energy programs, and they are in substantial agreement. Table IV summarizes the findings of three groups who looked at the 1975-85 decade and were asked to project the capital required to sustain the level of our energy programs. They all found that about 400 billion (1973) dollars would be needed. The National Academy of Engineering's study of a year and a half ago suggested that more like $600-800 billions would be required.(15) So, at least $40 billions are needed each year for just the capital requirements of our energy programs. Perhaps that number no longer sounds like much to you because of the federal deficit of $70 billions this year. A better comparison might be the number of dollars invested by all the manufacturing industries for plant and equipment. In 1973 all manufacturing invested only $32 billions. So $40 billions a year represents a major impact on capital flow in our economy. (Public utilities plus the petroleum industry spent $24 billions in 1973.)(16)

TABLE IV

ENERGY FACILITY
CUMULATIVE CAPITAL REQUIREMENTS,
1975-85
(BILLIONS OF 1973 DOLLARS)(17)

ITEM	NATIONAL PETROLEUM COUNCIL	ARTHUR LITTLE CO.	FEA ACCELERATED SUPPLY
OIL AND GAS (INCLUDING REFINING)	133	122	98.4
COAL	8	6	11.9
SYNTHETIC FUELS	10	6	0.6
NUCLEAR	7	84	138.5
ELECTRIC POWER PLANTS (EXCLUDING NUCLEAR)	137	43	60.3
ELECTRIC TRANSMISSION	42	90	116.2
TRANSPORTATION	43	43	25.5
OTHER	--	8	2.2
TOTAL	380	402	453.6

Let's look specifically at coal gasification. The conversion of coal to methane as a substitute for natural gas, as noted earlier, would reach 2.8 quads in 1990. When we did a summer materials study(18) a year and a half ago, we compared that goal with what has been done in the nuclear energy program in terms of installed annual heat capacity in quads. We wanted to address the question of how much heat the nuclear system could produce for conversion to electricity through steam cycles. Figure 8 shows that the nuclear industry has supported a very strong rising exponential of capacity. The dotted curve shows the growth required to meet the targets written by the government for coal-derived methane. What the curve indicates is that the rate of increase of the coal gasification program would have to be just as great as the nuclear has been so that in only 12 to 14 years it equals today's nuclear heat capacity. One can observe that and say, "That's fine, but of course nuclear is very expensive, and the coal conversion won't require as much capital." To address that comment, we plotted the investment dollars in Figure 9. On the left are 1973 constant dollars of capital investment in billions. The nuclear line again shows the rapid buildup. The dashed line on the right is what would have to be invested to meet coal gasification targets. (Since 1973, capital estimates for synthetic gas have increased by a factor of two and a half.) What is startling is that by 1985 or 1990 we will have built an industry and have spent as much capital as we have already committed to the nuclear industry, which has been a mammoth undertaking with very substantial government support.

What about fuel costs? In discussing energy, you may find the data in Table V interesting and useful. Back in 1970, the cost per million BTU at the source for coal, oil, gas and nuclear was quite low and, although somehwat higher when resold to utilities, still very low. The higher cost to the final customer includes the distribution and retail costs. Note that electric

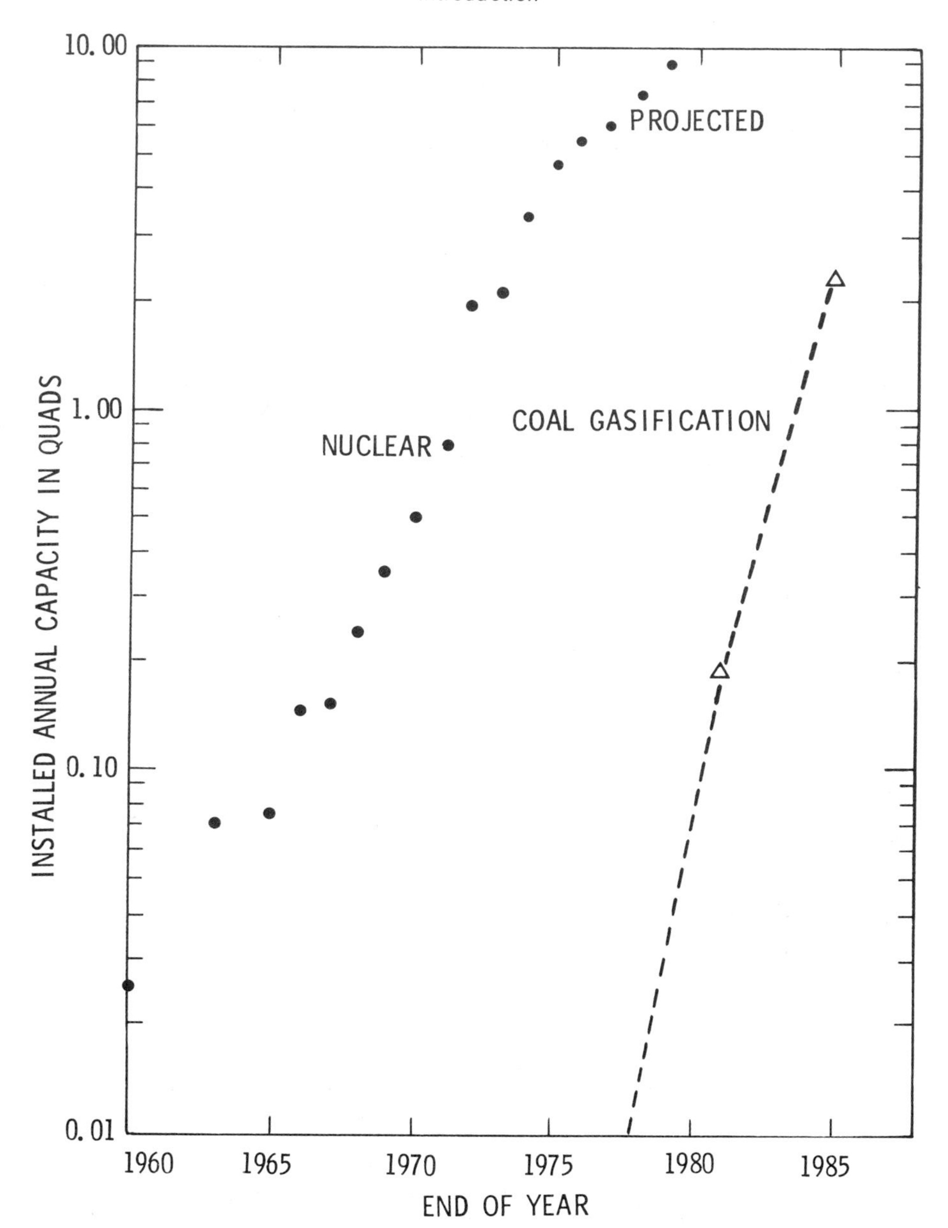

FIG. 8. The cumulative nominal heat generating capacity of all nuclear power plants in the U.S. (actual and projected) and the nominal heat content of the high BTU coal gas from plants projected to be in operation as a function of year. The nuclear figures assume 10,000 BTU/kWh; the gas contains 1000 BTU per standard cubic foot.(19)

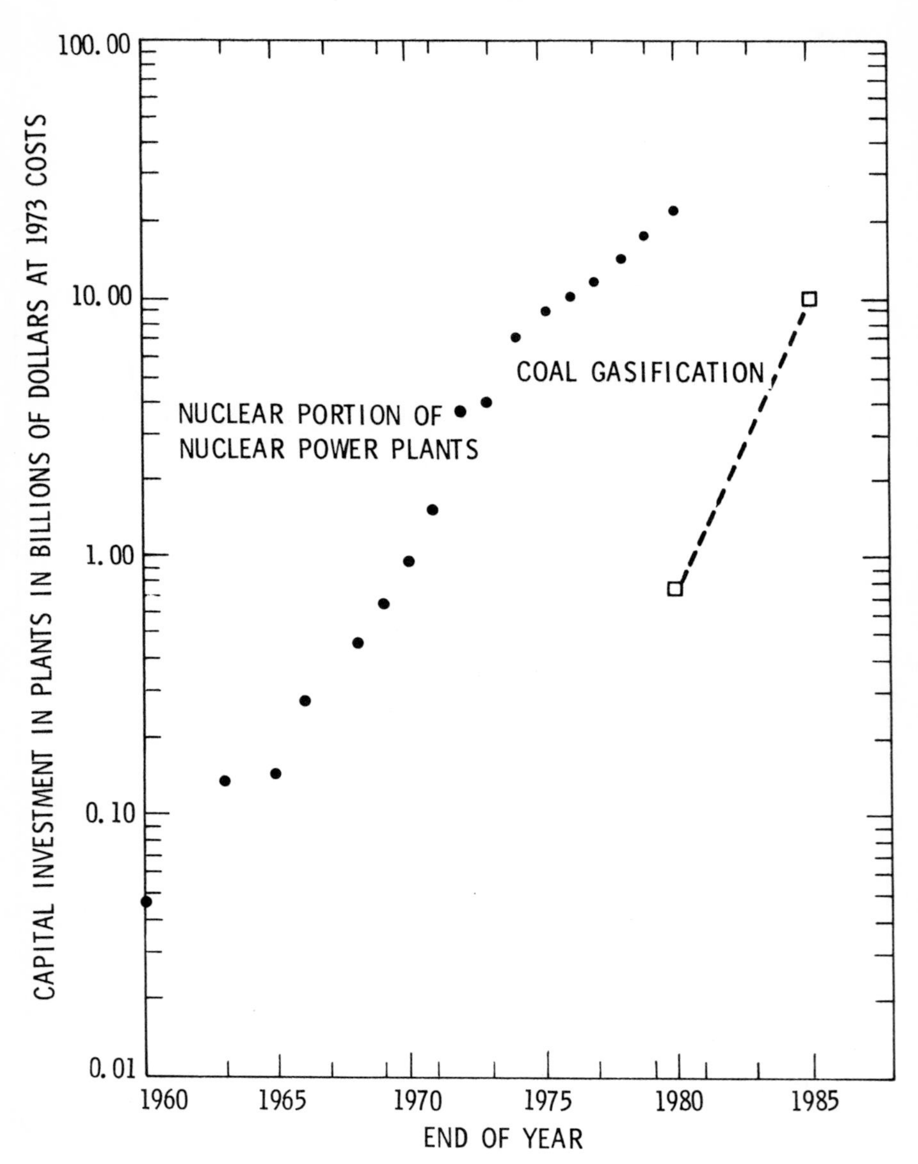

FIG. 9. The cumulative capital investment (in 1973 dollars) in the U.S. in nuclear power plants and in high BTU coal gasification plants versus year. The nuclear figures are based on a cost of $167 per kilowatt electrical for the nuclear or heat generating portion and 10,000 BTU per kWh. Comparable cost for the entire electric generating plant is $440 per kilowatt. The coal gasification figures are based on a 1973 estimate of $390 million for a 250 million SCF per day plant.(20)

TABLE V

ENERGY COSTS

CURRENT DOLLARS/MILLION BTU

	SOURCE	UTILITIES		FINAL CONSUMER	
	1970	1969	1975	1969	1974
COAL	0.24[21]	0.27[24]	0.82[27]	0.31[31]	---
OIL	0.57[22]	0.32[25]	2.02[28]	---	2.51[33]
GAS	0.17[23]	0.25[26]	0.65[29]	0.62[32]	1.22[34]
NUCLEAR	---	---	0.07[30]	---	---
WEIGHTED AVERAGE	0.36	0.27	0.97	---	---

ELECTRICAL ENERGY

CURRENT DOLLARS/kWh_E

	1969	1975
RESIDENTIAL	0.021[35]	0.032[37]
COMMERCIAL	0.013[36]	0.033[38]
INDUSTRIAL		0.020[39]
WEIGHTED AVERAGE	0.015	0.028

energy has not gone up as much during this period as oil and gas have. One of the facts I find fascinating is that in 1975 natural gas cost the electric utilities one-third of what oil did. So if they had any choice, it is pretty obvious which fuel they would use; yet natural gas has qualities for home distribution that cannot be matched by any other fuel now available, and we believe it will be the first to be exhausted.

V. THE ROLE OF MATERIALS IN THE ENERGY PROGRAM

Now let me discuss the role of materials in this emerging energy program. To do that, let us refer to the rather complicated Figure 10. The amount of detail displayed there is necessary so that you will have a relatively complete picture of what is happening in the energy program. On the left are the different fuel sources (uranium, gas, coal, oil, and so forth), and the energy conversion or processing that is normally appropriate to produce electricity. In addition one has to distribute or transmit this energy. We don't think so much of storing liquid fuels, because that is easy; but in the case of electricity, storage is difficult and the technology is only beginning to emerge. We eventually produce a final fuel form suitable for use by some consumer. The reason that materials are so involved in the energy picture is that processing is becoming such an important part of the system illustrated in Figure 10. In simpler times, we took coal out of the mine and burned it directly without changing it. Similarly, we take natural gas out of the wells and with only very minor chemical processing burn it as is. For the new technologies, only solar heating of homes and of hot water will pass directly to the user. All other forms will

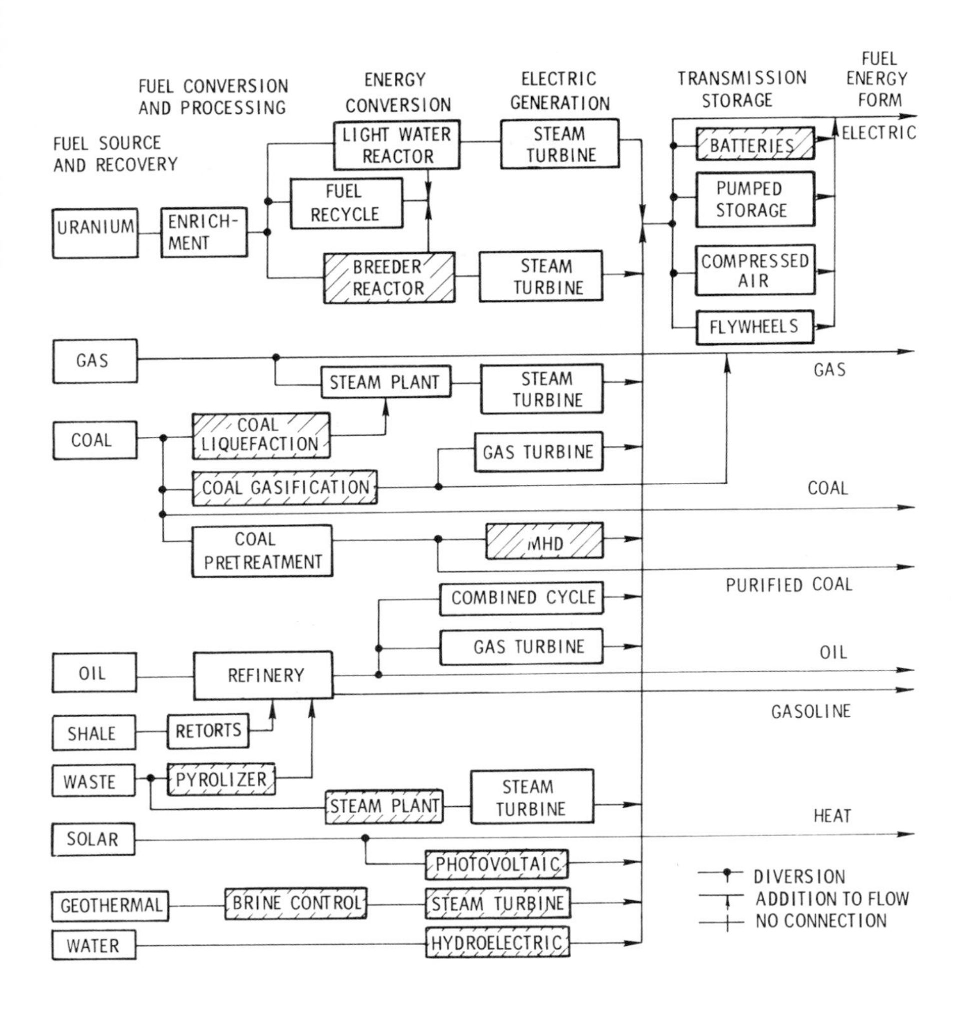

FIG. 10. Partial schematic representation of the energy flow as it may exist in the United States in 1990 when the first four fuel sources will probably still contribute more than 95 percent of the total. Note that the energy sources under development for the future depend on processing or conversion. Boxes with hatching show steps for which new materials are essential.

require processing. For example, we are going to have to process coal in some way to make it into a liquid or a gas—or at least we are going to clean it up. Uranium has first of all to be enriched; geothermal energy from the hot brines has to be processed and treated; and so on. As soon as one starts this kind of processing, one has to use materials; and then one immediately runs into their limitations. Let me review some materials limitations to identify the problems one faces.

First, we have to distinguish among three kinds of problems: efficiency, critical materials, and availability. Almost any box in Figure 10 can be improved to some extent by materials improvements. The designer normally designs the "optimum" within the limitations of the material he specifies. Given better materials, he designs more efficient equipment and systems. Thus, the first kind of problem is improving materials for greater efficiency. But let's concentrate on the second kind of problem, the critical materials. In many of the different boxes of Figure 10, roadblocks get erected because of materials limitations. We cannot always have what we want because of two types of limitations: one technical or technological; the other economic. To illustrate the first, consider the box "MHD" (magnetohydrodynamics). In this process, when coal is combusted at very high temperatures, the resulting plasma travels past electrodes in a magnetic field so that electricity is generated directly. This process will work only at very high temperatures.

Figure 11 presents all the data regarding experimental runs of MHD systems gathered through the summer of 1975. Plotted vertically is the electrical output in megawatts and horizontally the duration in seconds (as well as in more familiar units). In almost every case, the run was terminated by materials failure—and there is no known material to allow us to exceed these performances. The black dots represent the existing "envelope." The target point on the graph shows where economic performance

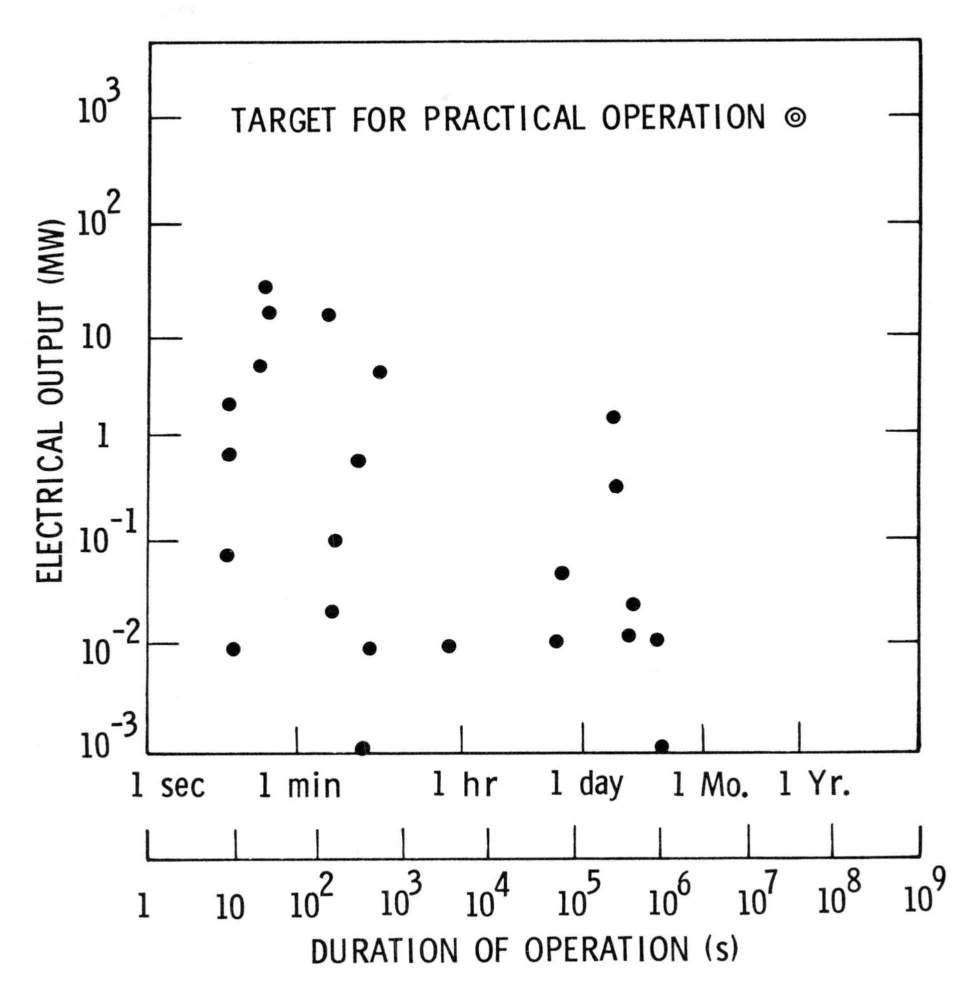

FIG. 11. Output electric power as a function of test duration for open-cycle MHD units using clean fuel.(40) *Operating time was limited principally by materials properties.*

would start. The difference between these points can be measured in hundreds of degrees Kelvin. The whole industry of aircraft turbine development has managed an improvement of only ten degrees per year with all the effort they have expended. Here, because we want to achieve a rise of hundreds of degrees, we face a technological limitation.

To illustrate the second type of limitation, economic, let me cite another problem. We would like to use the sun to make electricity through photovoltaic conversion. We know how to do that perfectly. All the solar system satellites use solar power. Though reliable and effective, they cost too much. Those systems, even if purchased for ground installation, now run about $50,000 a kilowatt (Peak power). If they are to compete economically with our present sources of electricity, they should be in the $200-a-kilowatt range. The difference between $50,000 and $200 may or may not be achievable; but if it is, one thing that will be required is a breakthrough in our understanding of how we could produce large sheets of converter materials at very low cost.

Finally, any material developed for the energy program must be based on adequate supplies at reasonable cost.

A new problem is also affecting our decisions about energy—an increased public awareness of and requirement for reliability. We will no longer forgive engineering mistakes. And so, when any new system is introduced, not only do we have to have one that passes initial tests, but we also have to have a high confidence that we have introduced a system which will operate for its lifetime with negligible chance of creating any major disaster. The only way that we can guarantee a negligible chance of failure is to furnish materials which we thoroughly understand so that we can predict life performance in demanding environments on the basis of our understanding of degradation mechanisms.

Illustrated in Figure 10 are examples of other areas requiring solution to a materials problem. At the top, right, is transmission. In many cases, as we know, it is convenient to produce electricity in one location and transmit it as electricity across long distances to the point of consumption. However, there is a transmission loss. One alternative would be to use superconductors for the transmitting line to avoid the inevitable I^2R loss in normal conducting materials. Two of the chapters of this book address the problems inherent in the development of these new superconducting materials.

It would be wonderful if we had batteries which would allow the economic operation of automobiles without pollution or noise; or which would allow storage of electrical energy generated by intermittent solar and wind power. Our lead-acid batteries simply have neither the specific energy density nor the cyclic life that is needed. If we are to have the batteries we want, we will have to depend upon entirely new materials systems. Five chapters in this volume are concerned with the various approaches that are being explored to solve battery problems along with two chapters on other methods of energy storage.

Earlier, I mentioned the kind of costs that are implicit in coal gasification, and in three of the chapters of this book you will read about the details of the serious material problems such as erosion-corrosion connected with this form of energy production. Similarly in magnetohydrodynamics, the very feasibility of the process depends on developing materials that we do not now have. Two chapters discuss materials for magnetohydrodynamics.

One of the limitations of solar-based photovoltaic-generated energy is cost of materials. Even when we consider just solar heating, involving panels to collect heat for the home, we have a problem of lifetime. The capital costs are sufficiently high that these devices must have a lifetime measured in decades. Almost inevitably one finds that a considerable amount of materials development is necessary, both to make the material available, and to assure long lifetimes of service. Two chapters treat this subject.

In the geothermal area, the only major reserves that are known in this country are exceedingly corrosive waters of 23% salinity (compared with a few percent for the oceans). Again, we will have to develop materials which can operate at high temperatures while in a severely corrosive environment.

I have not shown fusion as a source of energy production because Figure 10 indicates only what things might be like in 1990, and it is not anticipated that fusion will be contributing to our energy needs at that time. Fusion by whatever scheme—magnetic confinement, laser, electron beam, ion beam, whatever—has created problems that are far beyond the scale of any the materials community has solved in the past—a truly formidable set of new problems requiring a depth of understanding we've not even approached. For example, we don't even have the simulation sources needed to start evaluating materials against the kind of environment produced in a fusion reactor. Here is a program that clearly illustrates my earlier contention that we need a decade right now of intensive research to lay the basis for understanding so that we can be prepared for the next cycle of development of that program. Seven chapters provide insight on the magnitude of the materials problem in fusion.

In conclusion, this country has a very large energy problem and a program designed to solve it. It is *the* technological challenge of our time. As I've tried to illustrate, materials will play a central role in that program; often they will play the determining role.

REFERENCES

1. Southern Union Gas Company, Albuquerque, New Mexico.
2. Albuquerque Journal, January 18, 1976.
3. For a brief discussion of nine elements that may be in short supply for the energy program, see Ad Hoc Committee on Critical Materials Technology in the Energy Program, Materials Technology in the Near-Term Energy Program (National Academy of Sciences, Washington, D.C., 1974), pp. 115-125.
4. U.S. Energy through the Year 2000 (U.S. Dept. of Interior, Washington, D.C., 1972).
5. U.S. Energy through the Year 2000 (U.S. Dept. of Interior, Washington, D.C., 1972).
6. Electrical World, March 15, 1976.
7. The U.S. Energy Problem (Inter-Technology Corporation, Warrenton, Virginia, 1971), NSF-RANN 71-1-1.
8. National Plan for Energy Research - Creating Energy Choices for the Future (ERDA, 1975).
9. Energy Perspectives (U.S. Dept. of Interior, Washington, D.C., 1975).
10. Energy Perspectives (U.S. Dept. of Interior, Washington, D.C., 1975).
11. Energy Perspectives (U.S. Dept. of Interior, Washington, D.C., 1975).
12. U.S. Energy Prospects: An Engineering Viewpoint (National Academy of Engineering, Washington, D.C., 1974).
13. Statistical Yearbook (Edison Electric Institute, New York, New York, 1974).
14. U.S. Energy Prospects: An Engineering Viewpoint (National Academy of Engineering, Washington, D.C., 1974).
15. U.S. Energy Prospects: An Engineering Viewpoint (National Academy of Engineering, Washington, D.C., 1974).
16. Statistical Abstract of the United States: 1975, 96th edition (U.S. Bureau of the Census, Washington, D.C., 1975), Table 811.
17. Energy Perspectives (U.S. Dept. of Interior, Washington, D.C., 1975).
18. Materials Technology in the Near-Term Energy Program (National Academy of Sciences, Washington, D.C., 1974).
19. Materials Technology in the Near-Term Energy Program (National Academy of Sciences, Washington, D.C., 1974).

20. Materials Technology in the Near-Term Energy Program (National Academy of Sciences, Washington, D.C., 1974).

21. Statistical Abstract of the United States: 1975, 96th edition (U.S. Bureau of the Census, Washington, D.C., 1975).

22. Annual Statistical Review, Petroleum Industry Statistics 1965-1974 (American Petroleum Institute, Washington, D.C., 1975).

23. Annual Statistical Review, Petroleum Industry Statistics 1965-1974 (American Petroleum Institute, Washington, D.C., 1975).

24. Bituminous Coal Facts 1972 (National Coal Association, Washington, D.C.).

25. Bituminous Coal Facts 1972 (National Coal Association, Washington, D.C.).

26. Bituminous Coal Facts 1972 (National Coal Association, Washington, D.C.).

27. Electrical World, July 15, 1975.

28. Electrical World, July 15, 1975.

29. Electrical World, July 15, 1975.

30. Ref. Based on 33 lbs. U_3O_8 to produce 10^6 kWhr at 10,000 BTU/kWhr heat rate, $8/lb. and $36 per separative work unit, William I. Neef, private communication.

31. The 1970 National Power Survey (Federal Power Commission, Washington, D.C., 1971).

32. Statistical Abstract of the United States: 1975, 96th edition (U.S. Bureau of Census, Washington, D.C., 1975).

33. Monthly Energy Review (Federal Energy Administration, Washington, D.C., January, 1976).

34. Monthly Energy Review (Federal Energy Administration, Washington, D.C., January, 1976).

35. Statistical Yearbook (Edison Electric Institute, New York, New York, 1974), Tables 33S and 45S.

36. Statistical Yearbook (Edison Electric Institute, New York, New York, 1974), Tables 33S and 45S.

37. Electrical World, March 15, 1976.

38. Electrical World, March 15, 1976.

39. Electrical World, March 15, 1976.

40. W. K. Jackson, et al., in Sixth International Conference on Magnetohydrodynamics Electrical Power Generation, ed. J. E. Keteis (National Technical Information Service, Springfield, Virginia, 1975), V. 5, p. 40, Fig. 8; see also P. C. Strangeby, University of Toronto Institute Aerospace Study, Rev. No. 39 (1974), p. 35, Fig. 7. (Cited in Science, V. 191 (2/76).

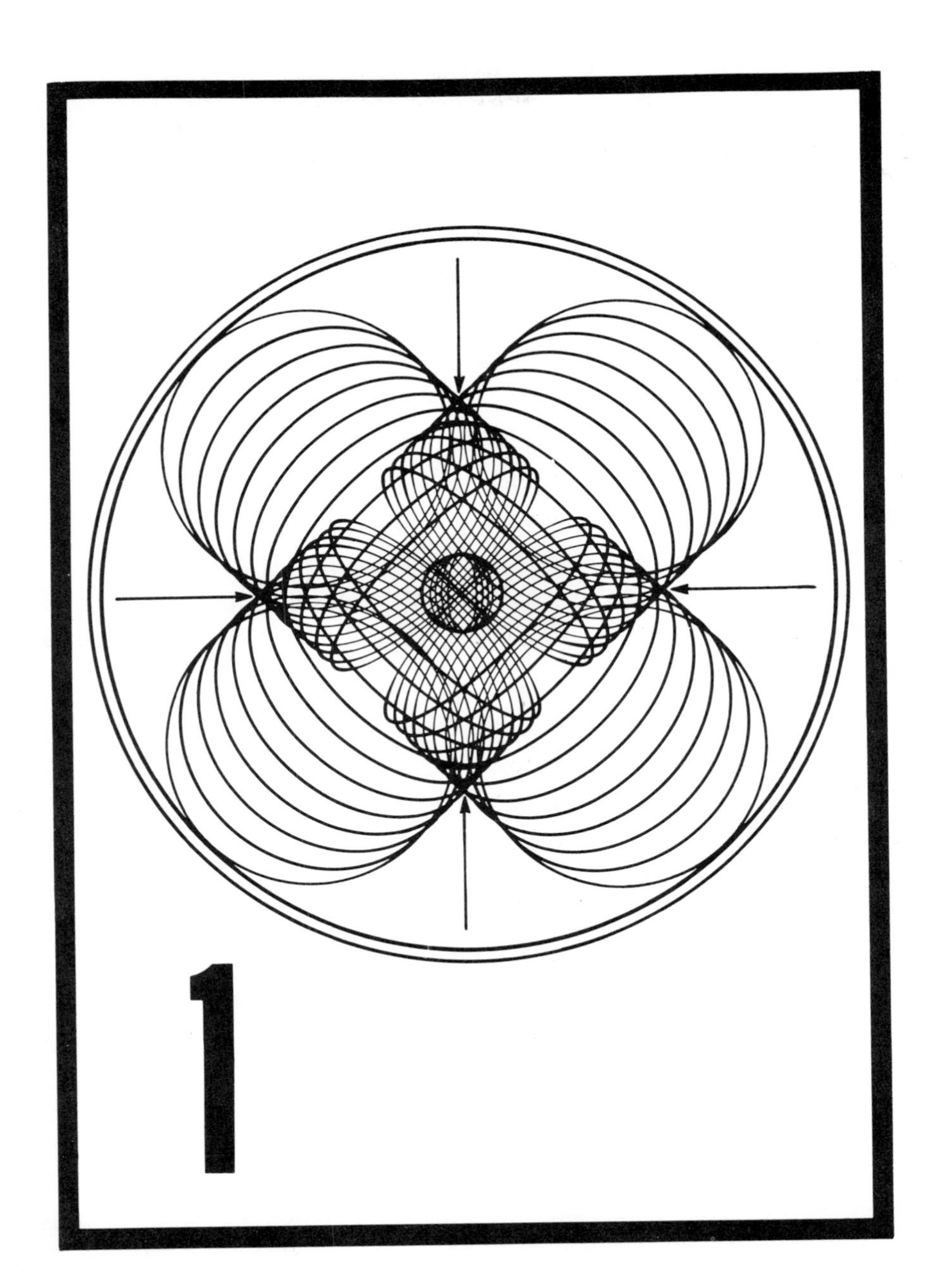
1

I. MATERIALS PROBLEMS IN NUCLEAR POWER GENERATORS

Of the seven topics relating to critical materials problems in energy production which are treated in this volume, none seems more challenging than those encountered in nuclear power generation. The number of problems in this area, the complication of their interdependence, the severity of the service parameters and the need, above all, for safety and reliability requires a monumental effort embracing the entire spectrum of materials science.

Such formidable problems as: plasma containment; surface erosion; radiation induced blistering, radiation damage to structural members, liquid metal corrosion attack, tritium permeability with the steam system and the materials problems involved in reactor safety and radioactive waste disposal are so demanding as to test the ability of the most astute and assiduous materials scientist.

In order to explore these problem areas in some depth, the topic of nuclear power generation has been subdivided into several different material categories which are discussed sequentially, from the confinement core outward, in the eight chapters which comprise this section: first wall problems; insulators for magnetic confinement of the plasma; radiation effects in structural materials; materials problems related to the tritium cycle; laser-fusion materials problems and critical materials problems in fission reactors.

CHAPTER 2

MATERIALS REQUIREMENTS FOR FUSION POWER

D. Steiner

Thermonuclear Division
Oak Ridge National Laboratory
Oak Ridge, Tennessee 37830

I. INTRODUCTION

During recent years significant progress in plasma performance has generated a renewed enthusiasm in the prospects for fusion power. This enthusiasm has manifested itself in a number of ways, but perhaps most significantly in that today fusion power technology is being pursued actively by groups in all countries with fusion programs. Within the past few years a number of groups have completed rather comprehensive design studies of fusion power plants. It is emphasized that these design studies are neither definite nor exhaustive; they are, rather, descriptions of approaches to fusion power based on extrapolations and assumptions in the areas of plasma physics and technology.

The purpose of this chapter is to elucidate and assess the major material requirements associated with fusion power as implied by current studies of fusion power plants. The chapter is organized as follows: Section II is a general discussion of the characteristics of fusion reactors; Section III describes the fusion reactor concepts that have been most thoroughly examined in the literature and summarizes the functional requirements that materials must satisfy in a fusion power plant; Section IV examines the implications of surface erosion associated with plasma-particle bombardment; Section V discusses neutron-induced phenomena and their anticipated radiation effects in fusion reactors; Section VI considers design requirements and limitations arising from corrosion processes in fusion reactors; Section VII illustrates some environmental implications associated with materials in fusion reactors. Finally, Section VIII summarizes the major conclusions of this chapter.

II. GENERAL CONSIDERATIONS

To derive useful amounts of power from nuclear fusion, it will be necessary to confine a suitably dense plasma at fusion temperatures ($\sim 10^8$K) for a specific length of time. This fundamental aspect of fusion power was first quantified by the British scientist, J. D. Lawson (1,2), who formulated a convenient expression for the product of the plasma density, n, and the plasma energy-confinement time, τ, required for fusion power breakeven (that is, the condition at which the fusion power release equals the power input necessary to heat and confine the plasma). Lawson showed that the required product, $n\tau$, depends on the fusion fuel and is primarily a function of the plasma temperature. Of all the fusion fuels under current consideration the deuterium-tritium fuel mixture requires the lowest value of $n\tau$ by at least an order of magnitude and the lowest fusion temperatures by at least a factor of five. When the plasma requirements for significant power generation are compared with the anticipated plasma performance of current approaches to fusion power, it is apparent that fusion power must initially be based on a deuterium-tritium fuel economy. Therefore, the emphasis of this discussion will be on deuterium-tritium fusion reactors.

Plasma confinement schemes can be categorized as electromagnetic or inertial. The objective of electromagnetic confinement is to provide a magnetic pressure that restrains the kinetic pressure of the plasma. Plasma heating can be intrinstic to the magnetic field system or can be provided by an auxiliary system. The objective of inertial confinement is to heat a fuel pellet to fusion temperatures in a time scale such that a significant fraction burns before thermal disassembly of the pellet terminates the reaction. Both lasers (3) and high-current electron beams (4) have been proposed as triggers for inertial confinement. Electromagnetic confinement schemes can be divided into two general classes on the basis of geometrical configurations, i.e., open and closed geometry. In a open configuration the magnetic field lines leave the confinement region and the plasma is able to escape through the ends of the system along field lines. In a closed configuration the field lines are contained in a toroidal volume and the plasma can escape only by moving across field lines.

The fusion concepts under current consideration result in both steady-state and pulsed reactor systems. In a steady-state reactor, fuel would be fed continuously into the fusion plasma and spent fuel (that is, unreacted fuel and reaction products) would be removed continuously. In a pulsed reactor, an initial charge of fuel would be ignited and would burn for a length of time determined by the confinement scheme. Unreacted fuel and reaction products would be removed at the end of the pulse and a fresh charge of fuel would be introduced into the reaction chamber and ignited. A quasi-steady-state mode of operation has also been visualized. In such a scheme the plasma burn-time would be long compared with the particle confinement time and plasma fueling, and spent-fuel removal would be required during the burning phase. The quasi-steady-state reactor is limited to intermittent plasma burning and, therefore, is characterized by a cyclic mode of operation.

The fusion of deuterium and tritum yields an alpha particle having an energy of about 3.5 MeV and a neutron having an energy of about 14.1 MeV. The natural abundance of deuterium is so great that it poses no practical fuel resource limitations in a fusion power economy. However, the natural abundance of tritium is not sufficient to support such a power economy and, therefore, it will be necessary to breed tritium in D-T fusion reactors. This is accomplished by neutron-induced reactions in a lithium-bearing blanket which surrounds the plasma. Thus, the primary fuels in a D-T fusion economy are deuterium and lithium, and tritium is only an intermediate product. In addition to tritium breeding, the fusion reactor blanket also serves the function of converting the kinetic energy of the fusion neutrons into recoverable heat. Nuclear power densities in fusion reactor blankets, limited by both plasma physics and materials considerations, will be 1 to 2 orders of magnitude lower than those in fission reactor cores.

The energy released in fusion can appear as neutron energy and as charged-particle energy. The neutron energy eventually appears as heat within the blanket and would be recovered by a thermal energy-conversion system. The charged-particle energy could also be recovered as heat via a thermal energy-conversion system. However, direct recovery of this portion of the energy may also be possible. Post (5) has proposed an electrostatic scheme for converting the kinetic energy of charged particles directly into electricity. In this scheme the charged particles that escape from the confinement region would be slowed down by electrical fields and would be collected on high-voltage electrodes. This scheme is most suitable to reactor concepts which operate at high ion temperatures. Oliphant and Ribe (6) have proposed an electromagnetic scheme for converting the kinetic energy of charged particles into electricity. This direct energy-conversion scheme is based on plasma expansion against a magnetic field and is applicable only in the case of pulsed mode operation.

Because only about 20% of the energy release in D-T fusion appears as charged-particle energy, the impact of direct energy-conversion schemes is marginal. For example, consider a case in which a thermal energy-conversion system of 40% efficiency is used for recovery of the neutron energy and a direct energy-conversion system of 70% efficiency is used for recovery of the charged-particle energy. The overall recovery efficiency for such a system would be 46%, about 15% higher than that for the thermal conversion system itself. However, energy conversion is not the total energy cycle in fusion reactors. Fusion reactors will inherently require input power to establish the fuel conditions necessary for fusion power production. Therefore, a fraction of the gross electrical output of the plant must be recirculated to sustain the fusion process. The amount of recirculating power required in fusion power plants can be appreciable. Thus, even when direct energy-conversion is assumed for a portion of the fusion energy release, the overall plant efficiency (that is, the ratio of the net electrical power output to nuclear power release) of D-T fusion power plants would be comparable to the overall plant efficiencies of fossil and fission power plants.

III. REACTOR CONCEPTS

As a point of departure the four fusion reactor concepts that have been most thoroughly considered in reactor design studies are described. They are the mirror, the theta-pinch, the tokamak, and the laser pellet concepts.

A. The Mirror Reactor Concept

The principles of the mirror confinement scheme are illustrated in Figure 1. In this open configuration, the magnetic field seen by a charged particle is weaker in the central region than in the end or "mirror" regions. A particle moving from the central region into the mirror region tends to have its direction of motion reversed as a result of the increase in field strength. Thus, particles are trapped by the mirror effect and the inherent tendency of the plasma to escape through the ends along field lines is inhibited. The mirror reactor (7) would operate in a driven steady-state mode, that is, continuous energy input would be required to sustain the plasma burn. The fusion power density would be maintained at a steady state by continuously feeding fuel into the reacting plasma and by continuously removing spent fuel. Note that in the mirror concept plasma end-losses provide an inherent spent fuel removal mechanism. The energy required to startup and sustain the plasma burn would be provided by injecting beams of energetic neutral fuel particles into the plasma. The neutral beam injectors would also serve as the fuel source in reactors.

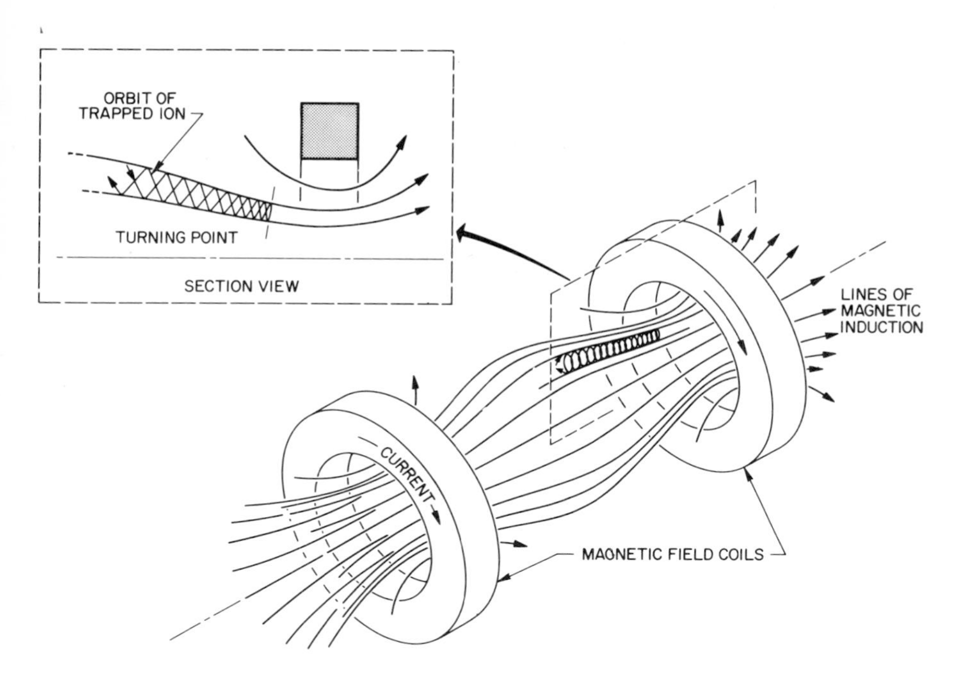

FIG. 1. Principles of the mirror confinement scheme.

Once startup has been achieved fusion power would be produced in a continuous fashion. Thus, the confining field coils would also operate in a continuous fashion. These coils must be superconducting to minimize the joule heating losses, which would be unacceptably high if normal conductors were employed for this purpose. As in all D-T fusion reactor concepts, the plasma would be surrounded by a blanket to utilize the fusion neutrons. The power associated with the fusion neutrons appears as heat in the blanket and would be converted to electricity via a thermal energy-conversion system. The power leaving the plasma in the form of energetic charged-particles (this includes essentially all the power associated with the injected neutral beams) is fed to a direct energy-conversion system based on electrostatic concepts. Unreacted fuel would be recovered from the direct converter and recycled to the plasma region.

B. The Theta-Pinch Reactor Concept

The theta-pinch concept employs a pulsed magnetic field for plasma heating and confinement. The principles of the theta-pinch confinement scheme are illustrated in Fig. 2. Both open and closed geometry have been considered for this concept. Note that a current (provided by a capacitor discharge) flows in the poloidal or "θ" direction and results in a constriction or "pinch" effect that both heats and confines the plasma. The theta-pinch reactor would operate in a pulsed mode.

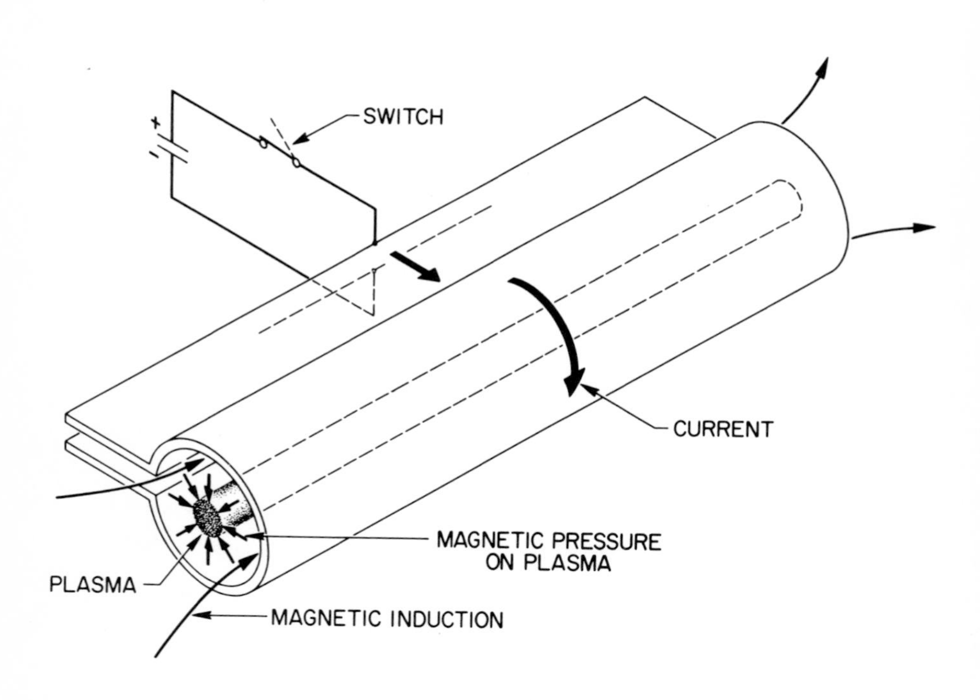

FIG. 2. Principles of the theta-pinch confinement scheme.

The reference theta-pinch reactor (8) will be adopted as a model for this concept. The reference theta-pinch reactor (RTPR) incorporates magnetic confinement in toroidal geometry. Heating is accomplished in two stages: an implosion-heating stage followed by a compression-heating stage. The implosion heating and compression coils employ normal conductors and operate near room temperature. The associated joule losses are tolerable because of the high fusion power density associated with this concept. Implosion heating occurs on a time scale of around 100 nanoseconds and compression-heating occurs on the time scale of about 20 milliseconds. The plasma burn in this concept continues for about 80 milliseconds. After cessation of the burn, plasma cooling takes place followed by flushing and refueling. It is assumed that a layer of neutral gas injected between the hot central plasma and the first wall of the blanket cools the plasma in around 2 seconds. This cooling stage is required in order to minimize the instantaneous heat flux at the first wall. The cycle time in the RTPR is taken as 10 seconds on the basis of estimated radiation damage effects to the blanket first wall.

Nearly all of the pulsed energy initially supplied to the compression coil and about 60% of the alpha-particle energy are directly recovered by plasma expansion against the magnetic field. This direct-conversion energy is returned to the compression-heating energy storage and transfer system. The remainder of the fusion energy appears as heat in the blanket and is converted to electricity through a thermal converter.

C. The Tokamak Reactor Concept

The principles of the tokamak confinement scheme are illustrated in Fig. 3. Note that an axial current is induced in the plasma by a changing magnetizing flux to provide:

1) a pulsed poloidal magnetic field that works together with a steady-state toroidal field to confine the plasma; and 2) initial plasma heating that arises from the associated ohmic heating within the plasma ring. In addition to the poloidal and toroidal fields, a tokamak plasma also requires a pulsed transverse field to provide control on the position of the plasma column. To limit joule losses to acceptable levels, the toroidal coil system must be superconducting in a tokamak reactor. The pulsed coil systems in tokamak reactor studies are usually taken to be superconducting, however, it has not yet been ascertained that these coils systems must be superconducting.

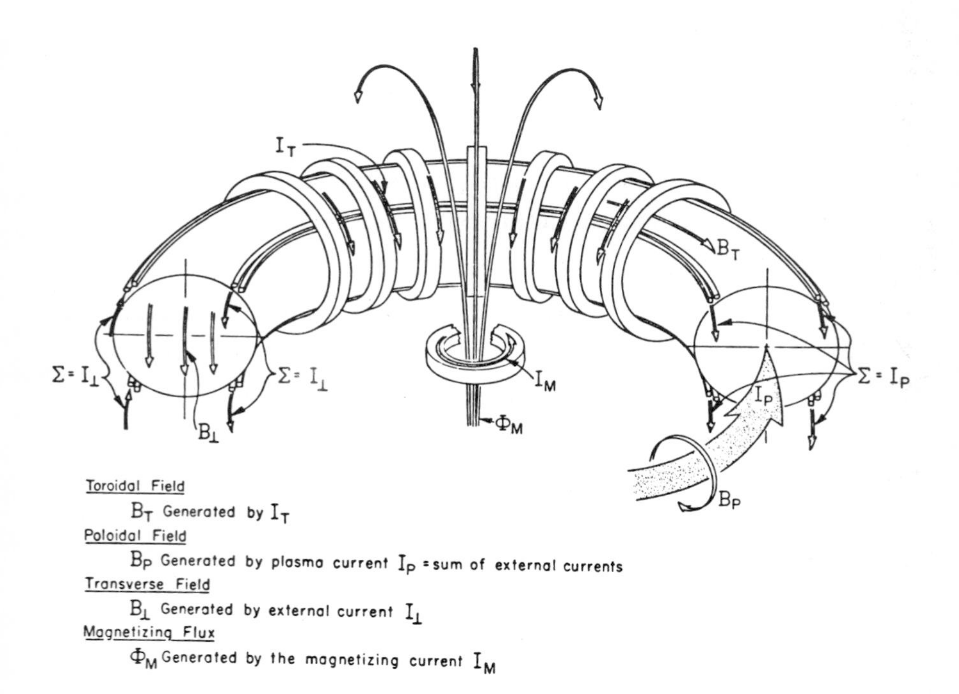

FIG. 3. Principles of the Tokamak confinement scheme.

It is generally assumed that the intrinstic ohmic heating process in tokamak plasmas will not provide sufficient heating to bring the plasma up to the ignition temperature and, therefore, auxiliary heating will be required to achieve ignition. A number of auxiliary heating techniques are currently being investigated. However, it appears that neutral beam injection is the most promising method for reactor applications. Once ignited the tokamak reactor would operate in a quasi-steady-state mode, that is, plasma fueling and spent-fuel removal would be required during cyclic burning phases. In current design studies (9-14) it is generally assumed that: 1) fueling would be accomplished by injecting solid fuel pellets into the plasma; and 2) spent fuel removal would be accomplished by guiding charged particles out of the plasma chamber along diverted magnetic field lines generated by coils called "divertor" coils.

In tokamak reactor studies, the burning phase is in the range, hundreds to thousands of seconds, with a duty factor of $\sim$90%. The length of the burning phase will be limited by either: 1) plasma quenching resulting from enhanced plasma radiation associated with impurity buildup within the plasma; or 2) the available magnetic flux through the center of the torus which determines the duration of the axial current. In this context, impurity refers to relatively high-Z material that may enter the plasma as a result of erosion processes at surrounding surfaces. It appears that the fusion energy in tokamak reactors can be practically recovered only through a thermal converter.

D. The Laser-Pellet Reactor Concept

In this concept plasma heating in the form of laser light is delivered to a fuel pellet in around 10^{-9} seconds within a cavity surrounded by a blanket. Theoretical analyses indicate that symmetric illumination of a hydrogen fuel pellet will lead to high-pellet compression ratios ($\sim 10^4$ times liquid density, that is $\sim 10^{26}$ cm^3) with an associated implosion heating process that results in ignition of the central region of the pellet. Ablation of the pellet surface generates pressure by momentum transfer and implodes the pellet. Once ignition of the central region of the pellet occurs, a thermonuclear burn front propogates radially outward and ignites additional fuel.

For a given fusion energy release per pellet microexplosion, the system output power will depend on the achievable cavity pulse rate and on the number of reactor cavities. Cavity pulse rates in the range one shot per 10 seconds per cavity to 100 shots per second per cavity have been considered in reactor studies (15,16). The allowable pulse rate will be determined, to a large extent, by the time required to reestablish the necessary cavity environment permitting subsequent pellet injection and efficient laser-beam penetration following each microexplosion. The number of reactor cavities that a single laser system can serve will be determined by laser pulse-rate capabilities and optical considerations. It is usually assumed that only thermal conversion of the fusion energy is practical for laser-pellet concepts.

A number of laser pellet concepts have been investigated (17-20). These concepts differ primarily in the proposed scheme by which energy deposition from the pellet microexplosion is accommodated at the cavity inner wall. Two promising schemes appear to be the "wetted-wall" concept (18) and the swirling-lithium vortex concept (17). In the wetted-wall concept, illustrated in Fig. 4, the laser cavity is defined by a porous metal wall that is wetted with lithium. The lithium serves as an ablative layer that protects the metal wall from the direct effects of energy deposition from x rays and pellet debris. The ablative layer, about 1 mm thick, is restored

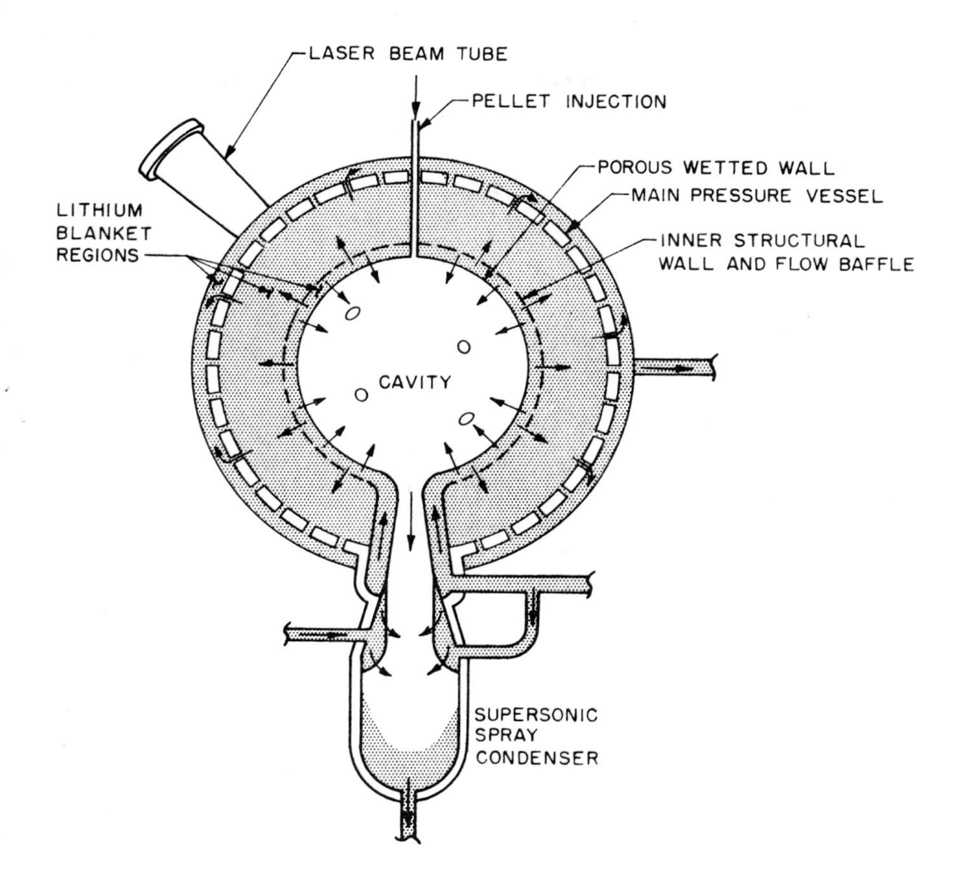

FIG. 4. The wetted-wall laser-pellet concept.

between pulses by radial inflow of lithium from the blanket region. In the swirling-lithium vortex concept, illustrated in Fig. 5, there is no structural cavity wall. The laser cavity is provided through the formation of a vortex in the swirling lithium. Note that the vortex cavity extends the full length of the pressure vessel and, thereby, permits nearly symmetric laser irradiation of the pellet. Inert gas bubbles are introduced into the lithium to attentuate the stress transmitted to the containment vessel.

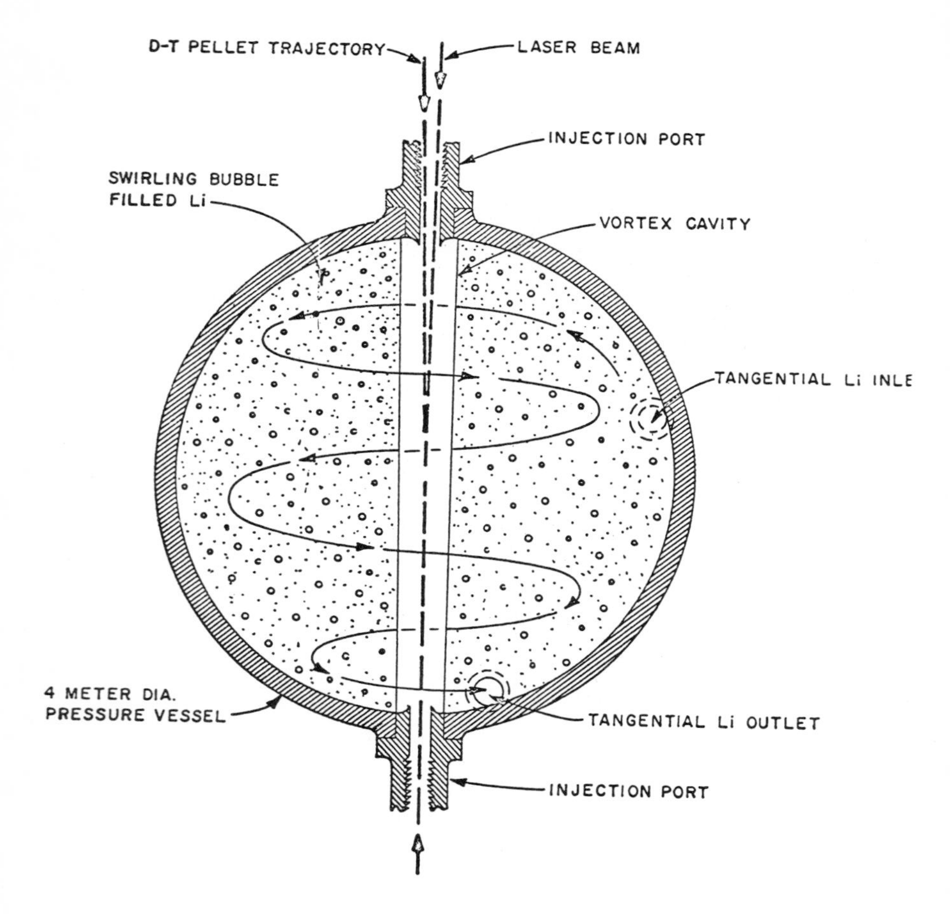

FIG. 5. The swirling-lithium vortex laser-pellet concept.

E. Materials Functional Requirements

In order to illustrate the functional requirements which materials must satisfy in a fusion power plant, consider the tokamak reactor design developed by Fraas (10). Fig. 6 indicates the physical dimensions of this toroidal reactor. A segment of the reactor blanket and shield regions is shown in detail in Fig. 7, and Fig. 8 is a schematic of the

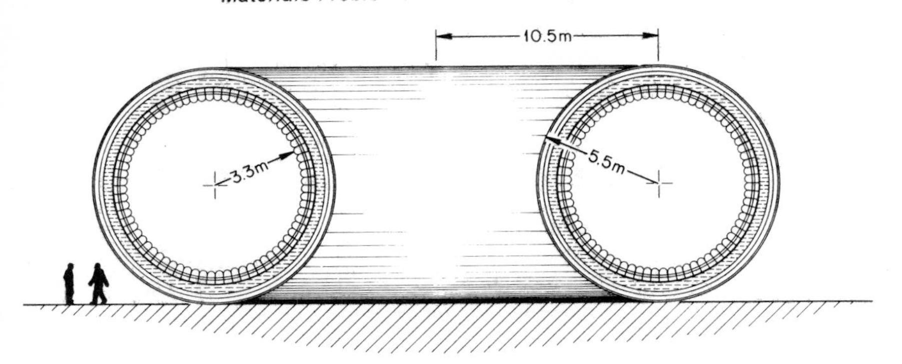

FIG. 6. Cross section through the blanket and shield region of a Tokamak reactor.

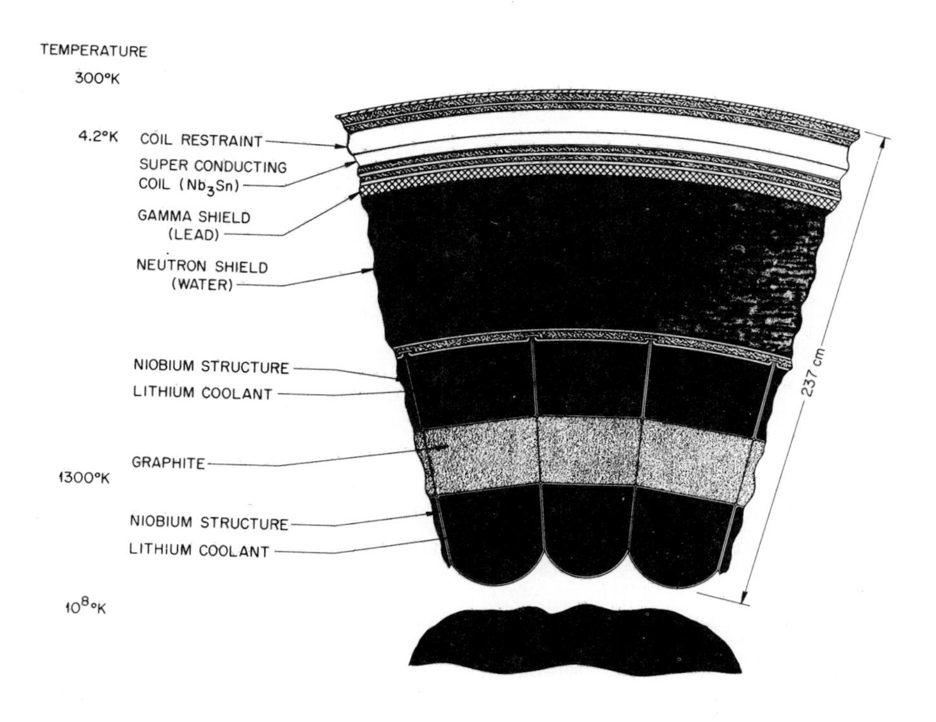

FIG. 7. Segment of the blanket and shield region of a Tokamak reactor.

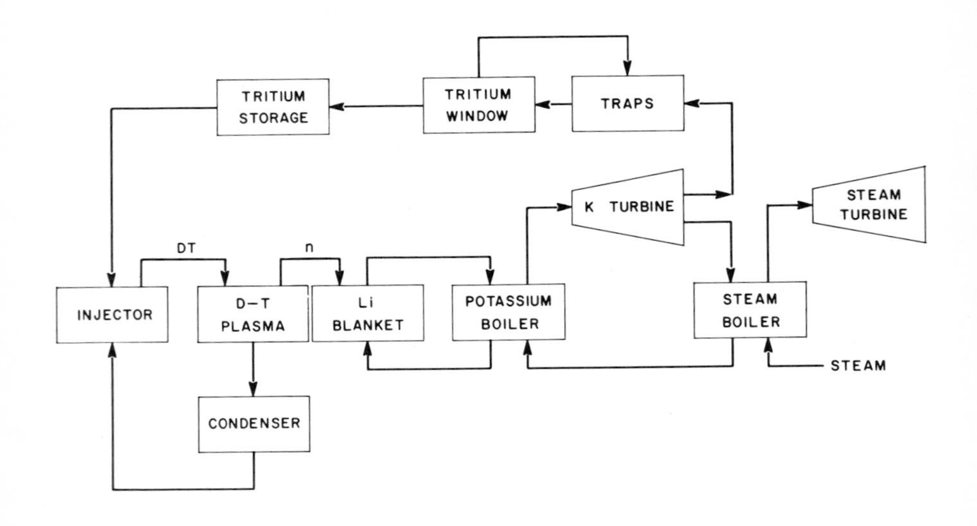

FIG. 8. Schematic of the power handling and fuel handling systems in a fusion power plant.

power handling and fuel handling systems. In this design it is assumed that lithium serves both as the primary blanket coolant and as the source of tritium regeneration via the reactions $^{6}Li(n,\alpha)t$ and $^{7}Li(n,n'\alpha)t$. The lithium in the blanket cooling system heats banks of potassium boiler tubes installed within the blanket. Niobium serves as the structural material and is chosen because of its refractory properties and corrosion resistance to liquid lithium. Graphite serves both as an neutron moderator and as a reflector. It is emphasized that the design illustrated in Fig. 6, 7, and 8 is by no means optimized and a number of alternative configurations, coolants and structural materials are currently being considered. Table 1 illustrates the range of materials currently being examined for fusion reactor applications.

TABLE 1

SUMMARY OF MATERIALS BEING CONSIDERED FOR FUSION REACTOR APPLICATIONS

Application	Materials
Breeding	Liquid Lithium
	Molten Salts (Li_2BeF_4, LiF)
	Ceramic Compounds (Li_2O, Li_2C_2)
	Aluminum Compounds (LiAl, $Li_2Al_2O_4$)
Structural	Refractory-Based Alloys (Nb, V, Mo)
	Iron-Based and Nickel-Based Alloys
	Aluminum-Based Materials
	Silicon Carbide
Cooling	Liquid Lithium and Potassium
	Molten Salts
	Helium
Moderation	The Breeding Materials
	Graphite
Neutron Multiplication	Beryllium
Normal Conductors	Copper and Aluminum
Superconductors	NbTi, Nb_3Sn, V_3Ga

IV. SURFACE EROSION

Surface erosion associated with plasma-particle bombardment can significantly limit the useful lifetime of several components in fusion reactors including:

1. the structural first wall in all reactor concepts
2. the first-wall insulator in the theta-pinch reactor
3. the collector electrodes of the direct converter in the mirror reactor
4. the divertor chamber of the Tokamak reactor
5. the optical components in the laser-pellet reactor.

In this context it appears that erosion rates in the range of ∿0.1 to 1 mm per year may give reason for concern. Moreover, in Tokamak reactors impurity buildup arising from first-wall erosion can severly limit the achievable burn time. For example, it has been suggested (21) that without means for controlling impurity buildup, the burn time in Tokamaks could be limited to one particle confinement time. In this context first-wall erosion rates as low as several microns per year might be of concern in Tokamak reactors. Note that impurities are not expected (22) to accumulate in mirror plasmas and, therefore, surface erosion does not represent an impurity problem in mirror reactors. The presence of impurities in the theta-pinch plasma requires either an increased compression-field strength or a shorter compression-field risetime to achieve ignition (23). Thus, impurities arising from surface erosion are a potential problem in the theta-pinch concept. In general, low-Z impurities are less deleterious to Tokamak and theta-pinch performance than are high-Z impurities.

Surface erosion can result from a number of processes; however, it appears that the processes of potential concern in fusion reactors will be plasma-particle sputtering and exfoliation resulting from the rupture of gas bubbles beneath the surface, that is, the bursting of radiation-induced blisters. Since the magnitude of the instantaneous particle fluxes during the plasma burn can differ by many orders of magnitude among the various concepts, it will be necessary to examine the dose-rate dependence of the erosion phenomena over a wide range of dose rates.

The plasma-particle fluxes anticipated in the various concepts and their implications relative to surface erosion are considered below.

A. Mirror Reactors

Table 2 gives estimated (24) plasma-particle fluxes at the blanket first wall and at the collector electrodes of the direct converter of a mirror reactor. Because of the relatively high particle energies associated with the plasma-particle fluxes at surfaces in the mirror reactor, it appears that wall erosion associated with light-ion sputtering will not represent a serious design limitation. On the other hand, wall erosion due to blistering may significantly limit the structural integrity of the first wall and electrical integrity of the collector electrodes.

B. Theta-Pinch Reactors

It is expected that plasma particle-fluxes to the first wall of the theta-pinch reactor will be negligible during the burn. During the plasma cooling period ($\sim$2 sec), the first wall of the RTPR will be bombarded by neutrals and radiation with a total energy flux of $\sim$130J/cm^2 or 13 W/cm^2 averaged over a 10-sec cycle time. It appears that the neutral gas blanket proposed for plasma cooling in the RTPR must reduce the mean energy of the neutrals well below 100 eV if serious erosion of the first-wall insulator by sputtering is to be avoided.

TABLE 2

ESTIMATED PLASMA-PARTICLE FLUXES AT THE BLANKET FIRST WALL AND AT THE COLLECTOR ELECTRODES OF A MIRROR REACTOR (24)

Particle Species	Mean Energy (keV)	Particle Flux (cm^{-2} sec^{-1})
	Blanket First Wall	
D^{+a}	550	5.4×10^{14}
T^{+a}	580	3.2×10^{14}
He^{++a}	1200	3.6×10^{13}
$D^{o}+T^{ob}$	550	2.0×10^{14}
	Collector Electrodes	
D^{+}	100	4.3×10^{13}
T^{+}	100	2.5×10^{13}
He^{++}	200	2.7×10^{12}

[a]These fluxes are associated with particles that do not enter the direct converter but, rather, deposit their energy on a portion (∿10%) of the first wall.

[b]These neutral fluxes result from charge exchange processes.

C. Tokamak Reactors

The magnitude and energy spectrum of the plasma particle fluxes at the first wall in Tokamaks will depend on the confinement characteristics and the feasibility of proposed schemes for protecting the first wall, e.g., divertors and neutral gas blankets (25,26). Divertors would protect the first wall by reducing the magnitude of the particle flux bombarding the wall. Neutral gas blankets would reduce the mean energy of the particles striking the first wall and, in this sense, would serve the same purpose as the neutral gas blanket proposed for plasma cooling in the RTPR.

Table 3 gives estimated (27) plasma-particle fluxes at the blanket first wall of a Tokamak reactor. The fluxes reported in Table 3 are based on an assumed divertor efficiency of 90%; that is, it is assumed that the particle flux at the first wall is 10% of the flux at the plasma edge. The flux of iron ions arises from first-wall erosion processes that introduce iron impurities into the plasma. Calculations (27) indicate that the first-wall erosion rate in the Tokamak reactor model would be $\sim$0.06 mm/yr based on sputtering and blistering phenomena associated with the plasma-particle fluxes summarized in Table 3. Such an erosion rate should not seriously limit the lifetime of the first wall. Note that in the absence of first-wall protection via the divertor, the first-wall erosion rate would be $\sim$0.6 mm/yr and might severely limit the first-wall lifetime. It is emphasized that, presently, estimates of plasma-particle fluxes at the first wall of Tokamaks are quite crude and, therefore, the associated erosion rates must also be viewed as crude estimates.

Recently, two novel schemes have been proposed for reducing the effects of plasma contamination from wall erosion in Tokamaks. One scheme is based on employing a honeycomb structure at the first wall. Calculations (28) indicate that, relative to a smooth first wall, a factor of 3 to 4 reduction in sputtering yield might be achieved with a honeycomb first wall. The second scheme is based on the introduction of a fibrous carbon curtain between the plasma and the metal first wall (29). The proposed advantages of the carbon curtain are (a) only low-Z, carbon impurities would enter the plasma and (b) the curtain would protect the metal first wall from plasma-particle fluxes. Both schemes are of potential interest and should be pursued. Moreover, these schemes might also be applicable in theta-pinch reactors.

TABLE 3

ESTIMATED PLASMA-PARTICLE FLUXES AT THE BLANKET FIRST WALL OF A TOKAMAK REACTOR (27)

Particle Species	Mean Energy (keV)	Particle Flux (cm^{-2} sec^{-1})
D^+	23	6.4×10^{13}
T^+	23	6.4×10^{13}
He^{++}[a]	23	4.7×10^{12}
He^{++}[b]	100	1.7×10^{11}
Fe^+	23	2.5×10^{12}

[a]These alpha particles are thermalized.

[b]These alpha particles are unthermalized and are assumed to have a mean energy of 100 keV.

D. Laser-Pellet Reactors

Table 4 summarizes the plasma-particle fluxes anticipated (30) at the cavity first wall of a laser-pellet reactor. Optical components in the vicinity of the first wall will be subjected to relatively large fluxes of high-energy particles and, therefore, the effects of blistering on these optical surfaces must be investigated. Currently, there is no information on blistering of such materials.

TABLE 4

ESTIMATED PLASMA-PARTICLE FLUXES AT THE CAVITY FIRST WALL OF A LASER-PELLET REACTOR (30)

Particle Species	Mean Energy (MeV)	Particle Flux[a] (cm^{-2} sec^{-1})
D^+	0.3	4.0×10^{14}
T^+	0.4	4.0×10^{14}
He^{++}[b]	0.6	4.3×10^{13}
He^{++}[c]	2.0	7.3×10^{13}

[a]Based on a cavity pulse rate of 1.2 shots per second per cavity.

[b]These alpha particles are part of the expanding plasma.

[c]These alpha particles have escaped from the expanding plasma.

V. *RADIATION EFFECTS*

Experience with fission reactors indicates that neutron-induced atomic displacements and transmutations, especially gas production via (n,α) and (n,p) reactions, can lead to deleterious changes in the engineering properties of materials and can, thereby, limit the service life of structural components. The correlation between neutron-induced phenomena and radiation effects is at present highly empirical and by no means definitive. Nevertheless, estimates of the atomic displacement and transmutation rates expected in fusion reactor materials are useful for anticipating the relevant radiation effects and for planning materials testing experiments.

Neutronic effects in fusion reactors are conveniently related to a quantity designated the "neutron wall-loading," which is defined as the energy flux of the fusion neutrons incident on the first wall of the blanket. Thus, the neutron wall-loading is a measure of the fusion neutron source strength normalized to the area of the first wall. For example, with the ∿14.1-MeV neutrons produced by the fusion of deuterium and tritium, a neutron wall-loading of 1 MW/m^2 is equivalent to a fusion neutron source strength of 4.43×10^{13} $n/(cm^2$ sec) at the first wall. The neutron source strength given in units of $n/(cm^2$ sec) should not be confused with the neutron flux at the first wall, which includes both source and reflected neutrons. Neutron wall-loadings in current reactor design studies are in the range 1 to 4 MW/m^2.

The following discussion examines the neutron-induced phenomena and the associated radiation effects expected in

1. the metal first wall of all reactor concepts
2. the first-wall insulator of the theta-pinch concept
3. graphite when employed as a neutron moderator in blankets
4. the copper compression coil of the theta-pinch concept
5. the superconducting magnets of the mirror and Tokamak concepts.

A. Metal First Walls

Neutron-induced displacements and gas production can result in dimensional changes and reductions in strength and ductility that severely limit the useful lifetime of the first wall. Table 5 summarizes displacement and gas production rates for a number of alternate first-wall materials (note that these calculations are normalized to a neutron wall-loading of 1 MW/m^2). Note also that the displacement rates reported in Table 5 are not all based on the same displacement model. Therefore, the differences in the displacement rates for the various materials should not be assigned great significance.

TABLE 5

CALCULATED ATOMIC DISPLACEMENT AND GAS PRODUCTION RATES FOR ALTERNATE FIRST-WALL MATERIALS AT A NEUTRON WALL-LOADING OF 1 MW/m^2

	Atomic Displacement Rate (dpa/yr)[a]	Helium Production Rate (appm/yr)[b]	Hydrogen Production Rate (appm/yr)
SAP (31)	14	410	780
Silicon carbide (32)	not reported	1800	580
Vanadium (31)	12	56	105
Stainless steel (31)	10	200	540
PE-16 (33)	15	156	not reported
Niobium (31)	7	24	80
Molybdenum (31)	8	49	98

[a]Displacements per atom per year.

[b]Atomic parts per million per year.

The following points are noted with regard to these displacement and gas production rates:

1. For a given neutron wall-loading, the displacement rate is relatively insensitive to the choice of first-wall material. Moreover, at the neutron wall-loadings of interest, ∿1 to 4 MW/m^2, the annual displacement rates anticipated in the first wall of a fusion reactor can be achieved in current high-flux fission reactors (34,35).

2. The gas production rates are a strong function of the first-wall material and, generally, are well beyond the capabilities of fission reactors (35). However, Wiffen and Bloom (34) note that the high helium production rates characteristic of nickel-bearing alloys in fusion reactors can be achieved in fission reactors with suitably high thermal fluxes. The high helium production rates with thermal neutrons result

from a two-step thermal-neutron capture sequence starting with ^{58}Ni. The helium production rates expected in fusion reactors can also be simulated by alpha-particle implantation and by doping with tritium, which then decays to ^{3}He (36).

3. The instantaneous displacement and gas production rates during the plasma burn of the theta-pinch and laser-pellet concepts, are, respectively, 2 and 6 orders of magnitude higher than those associated with the mirror and Tokamak concepts. The instantaneous displacement rates characteristic of the theta-pinch concept can be simulated using heavy ion bombardment. The instantaneous displacement rates characteristics of the laser-pellet concept could, in principle, be approached using neutral beam devices delivering $\sim$20 mA of deuteron beam at an energy of $\sim$200 keV (37); however, the applicability of such a technique requires much further investigation.

An experimental program to develop potential first-wall materials must assess the combined effects of displacements and gas production on swelling and ductility. Moreover, in pulsed applications the effects of fatigue must be evaluated. Finally, the effects of radiation creep must be examined. This phenomenon will control creep rates during reactor operation at temperatures for which thermal creep is unimportant and may result in elongations to failure that differ from those predicted by postirradiation testing.

Until intense sources of fusion-energy neutrons [$\sim 10^{14}$ n/(cm^2 sec)] become available, first-wall materials testing will rely heavily on fission neutron irradiations and ion bombardments. Note that the radiation effects produced by ion bombardments are limited to a small region of material (a few microns) near the target surface. Study of these regions can yield important information on the effects of irradiation on microstructures and swelling due to void formation. However, ion bombardments are generally not suitable for evaluating the effects of irradiation on the mechanical properties of materials.

B. First-Wall Insulator

The combined effects of neutron-induced atomic displacement and transmutations can result in dimensional changes and degradation of electrical and thermal properties that severely limit the useful lifetime of the first-wall insulator in the RTPR. Dudziak (38) has calculated atomic displacement and transmutation rates for Al_2O_3 as the first-wall insulator in the RTPR, and these calculations are summarized in Table 6. The specification of Al_2O_3 as the first-wall insulator in the RTPR was made for illustrative purposes and other materials are being considered; however, the magnitude of the displacement and transmutation rates calculated for Al_2O_3 can be viewed as representative of ceramic insulators in the RTPR radiation environment. In developing experimental programs to assess the performance of potential first-wall insulators for the theta-pinch concept, it will be necessary to take cognizance of the following points:

1. there is no existing neutron-generating facility that can produce transmutations in potential insulators at the rates specified in Table 6
2. the instantaneous displacement and transmutation rates during the burn are $\sim$2 orders of magnitude greater than the cycle-averaged rates
3. synergistic effects between surface phenomena and bulk phenomena may be extremely important.

For the above reasons it appears that ion bombardment is currently the most useful technique for simulating radiation effects at the first-wall insulator.

TABLE 6

DISPLACEMENT AND TRANSMUTATION RATES IN Al_2O_3 AS THE FIRST-WALL INSULATOR IN THE RTPR (38)

Neutron wall-loading (MW/m^2)	2.0
Atomic displacement rate (dpa/yr)[a]	30[b]
Hydrogen production rate (appm/yr)[c]	970
Helium production rate (appm/yr)	1680
Carbon production rate (appm/yr)	1340
Magnesium production rate (appm/yr)	830

[a]Displacements per atom per year.

[b]Estimated from results on other materials.

[c]Atomic parts per million per year.

In addition to the cumulative radiation effects mentioned above, the intense neutron and photon fluxes during the burning stage can temporarily alter the dielectric properties of the first-wall insulator. Since the design dielectric properties (dielectric strength $\geq$100 kV/cm) are only required during the implosion heating stage, a recovery time of $\sim$10 sec is available in the RTPR design. A major materials requirement for the theta-pinch concept is the development of a first-wall insulator, which exhibits rapid recovery of dielectric properties following the plasma burn.

Finally, the RTPR will be pulsed $\sim 10^6$ times per year. Therefore, synergistic effects between radiation phenomena and fatigue must be evaluated (39).

C. Graphite in Blankets

It appears that the principal radiation effects arising in the graphite will be dimensional changes and creep. Gray and Morgan (40) have used fission reactor data to predict the behavior of graphite in several fusion reactor designs. They project graphite lifetimes in the range of ∿1 to 30 yr depending on fluence and operating temperature. However, they caution that such extrapolations can be misleading and suggest several relevant experiments. Table 7 summarizes the maximum displacement and transmutation rates in the inner graphite region of the RTPR blanket design (38). (These rates fall off by a factor of ∿2 in traversing the graphite region.) Approximately 77% of the helium generation comes from the $^{12}C(n,n')3\alpha$ reaction. It is precisely the high helium generation rates and the accumulation of beryllium as an impurity that cast doubt on the validity of extrapolating graphite behavior in fission reactors to predict graphite behavior in fusion reactors.

TABLE 7

MAXIMUM DISPLACEMENT AND TRANSMUTATION RATES IN THE INNER GRAPHITE REGION OF THE RTPR BLANKET DESIGN (38)

Neutron wall-loading (MW/m^2)	2.0
Distance from first wall to graphite (cm)	10.5
Thickness of graphite region (cm)	4.3
Displacement rate (dpa/yr)[a]	12
Helium production rate (appm/yr)[b]	1240
Beryllium production rate (appm/yr)	280

[a]Displacements per atom per year.

[b]Atomic parts per million per year.

D. Copper Compression Coil

The principal radiation effect anticipated in the compression coil of the RTPR is increased electrical resistivity arising from neutron-induced displacements and transmutations. Table 8 summarizes the displacement and transmutation rates calculated at the inner radius of the compression coil (38). Note that these are the maximum rates and fall off by nearly 2 orders of magnitude in traversing the coil. Bunch et al. (41) have assessed the radiation-induced resistivity changes in the RTPR compression coil and have concluded that (a) the maximum resistivity increase associated with the nickel and zinc transmutation products would be ∿1%/yr (at a neutron wall-loading of 2 MW/m^2) and would decrease nearly exponentially with coil radius, and (b) the resistivity increase associated with displacement damage would saturate at ∿19% in a relatively short time (perhaps several months) and would be uniform throughout the compression

TABLE 8

MAXIMUM DISPLACEMENT AND TRANSMUTATION RATES IN THE COMPRESSION COIL OF THE RTPR (38)

Neutron wall-loading (MW/m^2)	2.0
Distance from first-wall to compression coil (cm)	95
Thickness of coil region (cm)	40
Displacement rate (dpa/yr)[a]	1
Nickel production rate (appm/yr)[b]	130
Zinc production rate (appm/yr)	86

[a]Displacements per atom per year.

[b]Atomic parts per million per year.

coil. A 20% increase in the resistive losses of the compression coil would decrease the net electrical output of the RTPR by ∿3.5%. This is a significant penalty and, therefore, definitive experiments on the magnitude of the displacement-induced resistivity increase are warranted. The feasibility of employing thermal annealing to recover the design resistivity value of the compression coil should also be considered.

E. Superconducting Coils

Neutron and photon interactions within the superconducting coils of the mirror and Tokamak concepts can (a) increase the resistivity of the superconducting and matrix materials with an associated decrease of the current density in the coil, (b) alter the dielectric and mechanical properties of the electrical insulators, and (c) alter the mechanical properties of the structural components. It appears (42) that the limiting radiation effect will be displacement-induced resistivity increases in the matrix material (copper or aluminum); however, there is a need for more experimental information on the performance of potential insulators and structural materials in a radiation environment and at cryogenic temperature.

Because of the shielding required to limit nuclear heating in the superconducting coils, the neutron and gamma-ray fluxes in the inner regions of the coil are ∿6 orders of magnitude lower than those in the first wall (43) and the resulting displacement and transmutation rates are also ∿6 orders of magnitude lower than those in the first wall. Moreover, the neutron spectrum is moderated to the extent that it is well simulated by fission neutron sources (42). It is concluded that radiation effects in the superconducting coils are design considerations but should not present serious design limitations. Additional experimental data on radiation effects will be required to optimize the coil and coil-shield designs. Such experiments can be performed with existing neutron sources (that is, fission reactors and D-T accelerator sources such as the Rotating Target Neutron Source at Lawrence Livermore Laboratory).

VI. COMPATIBILITY

Preliminary assessments of the design requirements and limitations arising from corrosion considerations in fusion reactor blankets have been performed by Homeyer (44), Draley et al. (45), DeVan (39), and Grimes and Cantor (46,47). The following discussion is to a large extent based on these studies.

A. Liquid Lithium Corrosion

A number of blanket designs employ liquid lithium in combination with stainless steel (7,12) or Nb-1% Zr (8,10) as the structural materials. Also it has been suggested that vanadium-base alloys be considered as structural materials in combination with liquid lithium (48). The following points are noted with regard to corrosion in these systems.

1. Iron-based alloys will be limited to relatively low operating temperatures in flowing lithium because of mass transfer phenomena (49-52). Austenitic stainless steels will probably be limited to temperatures of $\sim 500^{o}C$, based on a 1 mil/yr corrosion allowance. However, it is not known to what extent the decreased turbulence associated with MHD effects in flowing lithium will mitigate the corrosion rates relative to those observed in the absence of magnetic fields.

2. Refractory metals in general appear to be highly resistant to corrosion by lithium even at temperatures above $800^{o}C$. Forced-convection loop tests of niobium-base alloys have exhibited negligible mass transfer rates in lithium at temperatures in the range of 1100 to $1300^{o}C$ and for operating periods of 300 to 10 000 h (53,54). There is relatively little

operating experience with vanadium-base alloys in lithium; however, the limited data show no attack or mass transfer of vanadium in lithium at temperatures approaching 900°C and for a testing period approaching 1200 h (51).

3. Attempts to develop vanadium-base alloys as a fuel cladding for the LMFBR have been hampered by corrosion associated with small amounts of oxygen in the sodium coolant (55). Enhanced corrosion of niobium by sodium has also been observed when oxygen impurities are added to the sodium (56). However, it appears that oxygen impurities in lithium do not affect the dissolution or mass transfer of these refractory metals in loop systems (57,58).

4. Contamination of niobium by oxygen during reactor operation can induce significant lithium attack. For example, it has been observed that oxygen concentrations in niobium in excess of several hundred ppm by weight result in complete lithium penetration of the niobium (57). The penetration is extremely rapid (on the timescale of hours) and would result in gross failure of the structural component. Alloying additions such as zirconium significantly increase the oxygen threshold concentration for lithium penetration of niobium (59,60). Vanadium has exhibited resistance to lithium penetration at oxygen concentrations in vanadium as high as 2200 ppm by weight (58).

5. Both niobium and vanadium alloys are deoxidized in contact with liquid lithium at 600 to 1000°C (57,58). DeVan and Klueh (58) have noted that deoxidation of vanadium by lithium could seriously degrade the mechanical properties of vanadium alloys that derive strength from matrix strengthening by oxygen.

6. The theoretical understanding of lithium corrosion phenomena and kinetics is inadequately developed.

Therefore, an extensive experimental program will be required to define the corrosion behavior of potential structural materials in lithium under the operating conditions and radiation environment expected in fusion reactors.

Graphite is not compatible with lithium and, therefore, must be clad in the presence of lithium. The use of resistive coatings to reduce MHD pressure drops in flowing lithium (24) requires that the coatings be compatible with lithium. Such coatings are not currently available.

B. Molten Salt Mixtures

Mixtures of the molten salts LiF and BeF_2 have been considered as potential breeding and cooling fluids in fusion reactors (44,46). For example, the eutectic mixture, Li_2BeF_4, is employed as the breeding material in the reactor design described by Mills (13). The following points are noted with regard to molten salt corrosion.

1. Corrosion due to chemical reactions between the molten salts and nickel-base, iron-base, and refractory alloys should be minimal (46). However, the oxide films that normally coat metals and afford corrosion protection are effectively removed by LiF-BeF_2 salt mixtures. Therefore, oxidants carried in the salt, either in the form of impurities or as the result of nuclear transmutations, can lead to corrosion of the structural metals.

2. In the context of corrosion, the transmutation reactions of primary concern are those that result in the formation of TF and fluorine:

$$^{6}LiF + n \rightarrow {}^{4}He + TF \; , \qquad (1)$$

$$^{7}LiF + n \rightarrow {}^{4}He + TF + n \; , \qquad (2)$$

and

$$BeF_2 + n \rightarrow 2n + 2\,{}^{4}He + 2F \; (\text{or } F_2) \; . \qquad (3)$$

It has been suggested (47) that the corrosive effects of the oxidants TF and fluorine can be ameliorated by adding small amounts of a redox couple (such as CeF_3/CeF_4) to the salt mixture. A similar approach (use of the redox couple, UF_3/UF_4) was found to be effective in dealing with fluorine released by the fission of UF_4 in the Molten Salt Reactor Experiment (61).

3. If molten salts are pumped across magnetic fields at high velocity, chemical destabilization of the salt can result from the action of the induced electric field. The consequence of this destabilization is to render the compounds LiF and BeF_2 quite corrosive to structural metals. Several schemes have been proposed (44,46) for reducing the magnitude of the induced electric field, including (a) subdividing the coolant channel flow area in the region of high transverse magnetic field and (b) surrounding the coolant channel with a ferromagnetic material in the region of high transverse magnetic field.

4. Although Li_2BeF_4 is chemically inert towards graphite, Grimes and Cantor (46) suggest that cladding of the graphite may be required in systems employing refractory metals since the salt can transfer graphite and thereby carburize the refractories.

5. As in the case with lithium corrosion, extensive testing will be required to define molten salt corrosion under the operating conditions and radiation environment of fusion reactors.

C. *Additional Considerations*

Two additional points require emphasis with regard to debilitating chemical reactions that may occur in systems employing the refractory metals niobium and vanadium.

1. As already discussed, oxygen pickup during operation can result in lithium penetration of niobium and

vanadium; moreover, such oxygen pickup can also embrittle these metals and, thus, is a potentially serious problem even in the absence of direct lithium contact. In this context it appears that the use of helium as a blanket coolant may provide a pathway for oxygen pickup from the outside environment. DeVan (39) has noted that several attempts to use helium as a protective environment for Nb-1% Zr corrosion loops containing lithium were unsuccessful. Not only were the loop components embrittled during service, but oxygen concentrations in the loop walls increased to the point that the walls were penetrated by lithium. DeVan estimates that the oxygen concentration in a helium heat transfer loop circuit (pressure $\sim$100 psi) would have to be maintained at a level of $\sim$0.0002 ppm to avoid difficulties with niobium and vanadium structural components. There are at present no techniques that could provide such purification of helium under recirculating conditions. However, Werner (24) has suggested that small additions of lithium to the helium might provide the required degree of oxygen gettering. In the absence of purification techniques coatings might be developed to protect the metals from oxygen contamination.

2. Scott (39) has noted that niobium and vanadium are severely embrittled by hydrogen near room temperature. There is no definitive information concerning possible hydrogen embrittlement of these materials at potential operating temperatures (800°C and above). Clearly great care must be taken to prevent excessive hydrogen concentrations in the structure during startup and cool down situations. In addition, experiments must be performed to determine hydrogen isotope behavior in niobium and vanadium under operating conditions.

VII. ENVIRONMENTAL CONSIDERATIONS

A. Tritium Containment

During normal operation a major path for tritium release to the environment is through the blanket cooling system into the steam cycle via the coolant-steam heat-exchanger tube walls. The rate of tritium diffusion into the steam cycle, R, is given by the relation

$$R = \frac{AP}{t} (\sqrt{p_c} - \sqrt{p_s}) \quad , \qquad (4)$$

where

$A \equiv$ area of coolant-steam heat exchanger

$P \equiv$ heat exchanger wall permeability

$t \equiv$ heat exchanger wall thickness

$p_c \equiv$ tritium partial pressure on the coolant side of heat exchanger

$p_s \equiv$ tritium partial pressure on the steam side of heat exchanger.

Since tritium will exchange rapidly with hydrogen on the steam side, the steam-side pressure of tritum, p_s, is essentially zero. Thus the tritium release rate to the steam cycle is given by

$$R = \frac{AP}{t} \sqrt{p_c} \quad . \qquad (5)$$

The parameters t and A in Eq. (5) will be determined primarily by thermal-hydraulic design considerations. The achievable values of the coolant-side tritium pressure will depend on the characteristics of the tritium recovery scheme. Therefore, the permeability, P, will be the controlling parameter in limiting the tritium release rate, R. The permeability of the coolant-steam heat-exchanger tube walls can be reduced by (a) the presence of coatings that act as permeation barriers and (b) low-temperature operation. Thus, tritium containment involves important trade-offs between materials development and thermodynamic efficiency.

B. Reactor Safety and Waste Disposal

The magnitude and characteristics of the radioactive inventories induced by neutron interactions in the structural material of the blanket are major considerations in assessing fusion reactor safety and radioactive waste disposal. For the wide range of materials and blanket designs being considered (62-69), the level of induced activity at equilibrium is generally in the range of about 10^9 to 10^{10} Ci for a 1000-MW(e) plant, clearly a significant level. However, the level of activity by itself is not a meaningful measure of the technological problems posed by radioactive inventories; evaluation of the associated nuclear afterheat and examination of the time-dependent behavior of the activation products are also necessary.

At shutdown the afterheat power density in the fuel of advanced fission reactors is anticipated to be at least one to two orders of magnitude greater than that expected in the structural components of fusion reactor blankets. The conclusion is that afterheat removal will be less of a problem in fusion reactors than in fission reactors. Moreover, it appears that the engineered safety features necessary to limit biological impact in the event of an accident may have to satisfy less stringent requirements in fusion reactor design than in fission reactor design. This observation does not mean that fusion reactors will necessarily be safer than fission reactors, but rather that the technology and engineering necessary to achieve a given level of safety may prove less difficult and costly for fusion reactors.

Calculations suggest that, if niobium-base structures are employed in fusion reactors, long-term solutions to waste disposal similar to those sought for radioactive wastes from fission reactors may be required. On the other hand, the use of materials such as vanadium-base alloys, SAP or SiC, might allow recycle of the blanket structure following a relatively short (less than 10 years) cooling period. Recycle of iron- and nickel-base structures might also be feasible, but the required cooling period would be at least 50 years. Within this context, it is possible that a fusion power economy might eliminate the need for long-term solutions to the radioactive waste disposal problem.

C. Resource Requirements

As a final materials consideration it is appropriate to examine the resource requirements of a fusion power economy. On the basis of the current fusion reactor design studies, it appears that the unique resource requirements of fusion power will arise from the use of:

1. beryllium as a neutron multiplier for tritium breeding purposes
2. copper (or aluminum) as a conductor in normal and superconducting magnet coils
3. helium as a cryogenic refrigerant
4. lithium as a fuel (that is, for breeding tritium) and coolant
5. titanium, vanadium, niobium, and molybdenum for structural components
6. titanium, vanadium, niobium, and tin for superconducting materials
7. lead as a coil shield material.

Table 9 gives estimates of the material inventories that might be required for a 10^6-MW(e) fusion power economy. These estimates

TABLE 9

ESTIMATED QUANTITIES OF THE UNIQUE RESOURCE REQUIREMENTS ASSOCIATED WITH FUSION POWER

Material	Application	Inventory for 10^6 MW(e) Capacity (metric megatons)	Forecast U.S.[a] Demand for Year 2000 (metric megatons)	World Reserves[a] (metric megatons)	Quantity Contained in[b] Upper 10 m of Earth's Crust (metric megatons)	Ratio of Quantity in Earth's Crust to World Reserves
Beryllium[c]	neutron multiplication	0.046	0.002	0.09	1.5(4)[d]	1.6(5)
Aluminum[e]	coil conductor	1.0-2.6	33	1000	5.1(8)	5.1(5)
Copper	coil conductor	3.2[f]-8.6[c]	7	300	2.3(5)	7.6(2)
Helium	refrigerant	0.04[f]-1.1[c]	0.02	1	4(3)[g]	4(3)
Lithium	fuel, coolant	0.95[f]-1.5[c]	0.014	0.8	2(5)[h]	2.5(5)
Titanium[i]	structure, S.C.	0.5	2	150	2.8(7)	1.9(5)
Vanadium[j]	structure, S.C.	2.4	0.03	10	6.6(5)	6.6(4)
Niobium[c]	structure, S.C.	3.3	0.01	10	9.4(4)	9.4(3)
Molybdenum[j]	structure	2.8	0.09	5.4	6.2(3)	1.1(3)
Tin[c]	S.C.	0.3	0.1	7.2	1.4(4)	2(3)
Lead[f]	shielding	11	2.5	95	9.4(4)	1(3)

[a] Ref. 70

[b] Ref. 71

[c] Based on the RTPR inventories, Ref. 8, and an overall plant efficiency of 30%.

[d] To be read as 1.5×10^4.

[e] Based on the copper requirements and adjusted on the basis of densities.

[f] Based on the UWMAK-1 inventories, Ref. 12, and an overall plant efficiency of 30%.

[g] In the atmosphere.

[h] There is approximately the same quantity of lithium in sea water.

[i] Ref. 72

[j] Based on the niobium requirement and adjusted on the basis of densities.

are based on the material inventories associated with the RTPR (8) and the UWMAK-I Tokamak reactor design (12). Note that the table includes alternate materials for the same application and, therefore, not each material listed would be required in the quantity indicated. When a given material is to be used for the same application in both the RTPR and the UWMAK-I designs, both values of the associated inventory are given. In this regard, it is noted that the significant differences in helium requirements reflect the very large helium inventories associated with the current design of the magnetic energy storage and transfer system proposed for the RTPR. It is anticipated that future designs of this system would seek to reduce the required helium inventory.

Table 9 also gives for each material (a) the forecast U.S. demand for the year 2000 (70), (b) the measured and indicated world reserves (70), (c) the quantity contained in the upper 10 m of the earth's crust (71), and (d) the ratio of this quantity to the world reserve. On the basis of Table 9, the following observations are made with regard to resource requirements in a fusion power economy.

1. Extensive use of beryllium would require a substantial expansion in the beryllium production capacity and additional exploration.

2. The aluminum and titanium demand does not appear to present any serious technological requirements.

3. The copper demand would require expansion of the copper production capacity. Exploration appears to be relatively extensive.

4. The helium demand would require expansion in the helium production capacity. Recovery from the atmosphere might be required. This appears feasible but needs technological development and more accurate cost estimates.

5. The lithium demand would require substantial expansion in the lithium production capacity and additional exploration. Lithium is abundant in sea water and could be recovered with technological development; however, more accurate cost estimates are needed for proposed recovery schemes. The use of lithium for breeding purposes only could reduce the required lithium demand substantially.

6. The demand for vanadium, niobium, or molybdenum would require substantial expansion in production capacity and additional exploration. Vanadium and niobium are relatively abundant in the earth's crust, but molybdenum is substantially less abundant than either of these.

7. The demand for tin and lead would require expansion in production capacity. Both materials are relatively highly explored.

The economic and environmental implications of these observations must be examined in future assessments of fusion power; however, it appears that there are no resource limitations that would prevent the extensive deployment of fusion power.

VIII. CONCLUSIONS

The materials requirements for fusion power have been assessed on the basis of current design studies of fusion power plants operating on the D-T fuel cycle. The major conclusions of this assessment are summarized below.

1. Surface erosion associated with plasma particle bombardment can significantly limit the useful lifetime of components in fusion reactors. Moreover, in Tokamak and theta-pinch plasmas the presence of impurities arising from first-wall erosion can severely affect the plasma performance. It appears that the erosion processes of potential concern in

fusion reactors will be plasma-particle sputtering and exfoliation resulting from the bursting of radiation-induced blisters. Currently, these phenomena are poorly understood in the context of a fusion reactor environment and, therefore, it is not possible to calculate accurate surface erosion rates in fusion reactors at present. A number of schemes are being pursued to protect both the plasma and the first wall from the consequences of surface erosion.

2. Neutron-induced atomic displacements and gas production can lead to deleterious changes in the engineering properties of materials and, thereby, limit the service life of structural components in fusion reactors. These radiation effects are expected to be most severe in the vicinity of the blanket first wall. Calculations indicate that at the first wall (a) annual atomic displacement rates would be comparable to those achieved in high-flux fission reactors, and (b) annual gas production rates would, with the exception of nickel-bearing materials, be substantially higher than those achieved in high-flux fission reactors. Currently, intense sources of fusion-energy neutrons [$\sim 10^{14}$ n/(cm^2 sec)] are not available and, therefore, materials testing for fusion reactors must rely heavily on fission neutron irradiations and ion bombardments to simulate the fusion reactor radiation environment.

3. The instantaneous displacement and gas production rates during the plasma burn of the theta-pinch and laser-pellet reactor concepts are, respectively, 2 and 6 orders of magnitude greater than those associated with the mirror and Tokamak reactor concepts. The rate dependence of the radiation effects must be investigated. At present, it appears that this rate dependence can be examined only by ion bombardment techniques.

4. In general, the theoretical understanding of corrosion phenomena and kinetics is inadequately developed. Therefore, an extensive experimental program will be required to

define the corrosion behavior of potential structural materials in lithium, lithium salts, and helium under the operating conditions and radiation environment expected in fusion reactors.

5. During normal operation a major path for tritium release to the environment is through the blanket cooling system into the steam cycle via the coolant-steam heat-exchanger tube walls. The permeability of the heat-exchanger tube walls will be the controlling parameters in limiting the tritium release rate to the environment. The permeability of the tube walls can be reduced by the presence of coatings that act as permeation barriers and by low temperature operation. Thus, tritium containment involves important trade-offs between materials development and thermodynamic efficiency.

6. The magnitude and characteristics of the radioactive inventories induced in the structural material of the blanket are major considerations in assessing the technological requirements associated with fusion reactor safety and radioactive waste disposal. At shutdown, the afterheat power density anticipated in the fuels of advanced fission reactors will be 1 to 2 orders of magnitude greater than that expected in the first wall of fusion reactor blankets. Thus, it appears that afterheat removal will be quantitatively less of a problem in fusion reactors than in fission reactors.

7. Calculations suggest that niobium-based structures in fusion reactors may require long-term waste disposal solutions similar to those required for spent-fuel wastes in fission reactors. On the other hand, the use of a SAP, SiC, or vanadium structure might allow recycle of the blanket structure following a relatively short cooling period (<10yr). Recycle of iron- and nickel-base structures might also be feasible but the required cooling period would be relatively long (at least 50 yr).

8. On the basis of the current fusion reactor design studies, the unique resource requirements for fusion power will arise from the use of such materials as beryllium, copper, helium, lead, lithium, molybdenum, niobium, tin, titanium, and vanadium. While in many cases the use of these materials would require a substantial expansion in production capacity and additional exploration, it appears that there are no resource limitations that would prevent the extensive deployment of fusion power. However, the economic and environmental implications of using these materials must be examined in further detail.

ACKNOWLEDGEMENTS

This chapter contains text, tables, and figures which appeared in the paper, "The Technological Requirements for Power by Fusion," by Don Steiner, Nuclear Science and Engineering, October 1975. Permission for reproducing these items was granted by the American Nuclear Society to whom the author wishes to express his grateful appreciation.

This research was sponsored by ERDA under contract with the Union Carbide Corporation.

REFERENCES

1. J. D. Lawson, "On the Economics of Thermonuclear Reactors," AERE-GP/M-173, Atomic Energy Research Establishment, Harwell, Berks, England (1955).

2. J. D. Lawson, "Some Criteria for a Useful Thermonuclear Reactor," AERE-GP/R-1807, Atomic Energy Research Establishment, Harwell, Berks, England (1955).

3. N. G. Basov, P. G. Kriukov, S. D. Zakharov, Yu. V. Senatsky, and S. V. Tchekalin, IEEE J. Quantum Electronics, OE4, 864 (1968).

4. F. Winterberg, Nucl. Fusion, 12, 353 (1972).

5. R. F. Post, "Mirror Systems: Fuel Cycles, Loss Reduction and Energy Recovery in Nuclear Fusion Reactors," Proc. Intern. Conf. Nuclear Fusion Reactors, British Nuclear Energy Society, Culham Laboratory, Abingdon, England, CONF-690901-7, U.S. Atomic Energy Commission (1969).

6. T. A. Oliphant and F. L. Ribe, Nucl. Fusion, 13, 529 (1973).

7. W. Werner, G. A. Carlson, J. Hovingh, J. D. Lee, and M. A. Peterson, "Progress Report No. 2 on the Design Considerations for a Low Power Experimental Mirror Fusion Reactor," UCRL-74054-2, Lawrence Livermore Laboratory (1974).

8. "An Engineering Design Study of a Reference Theta-Pinch Reactor (RTPR)," LA-5336 or ANL-8019, A Joint Report by Argonne National Laboratory and Los Alamos Scientific Laboratory (1974).

9. J. T. D. Mitchell and R. Hancox, "A Lithium Cooled Toroidal Fusion Reactor," CLM-P 319, Culham Laboratory, Abingdon, England (1972).

10. A. P. Fraas, "Conceptual Design of the Blanket and Shield Region and Related Systems for a Full Scale Toroidal Fusion Reactor," ORNL-TM-3096, Oak Ridge National Laboratory (1973).

11. K. Sako et al., "Conceptual Design of a Gas Cooled Tokamak Reactor," JAERI-M 5502, Japan Atomic Energy Research Institute (1972).

12. G. L. Kulcinski and R. W. Conn, "The Conceptual Design of a Tokamak Fusion Power Reactor, UMWAK-I," Proc. 1st Topl. Mtg. Technology of Controlled Nuclear Fusion, San Diego, CONF-740402, Vol. I, p. 38, U.S. Atomic Energy Commission and the American Nuclear Society (1974).

13. R. G. Mills, "A Fusion Power Plant," MATT 1050, Princeton University (1974).

14. T. H. Jensen, C. C. Baker, and G. R. Hopkins, "Scaling of Circular and Non-Circular Cross Section Tokamak Power Reactors," Proc. 1st Topl. Mtg. Technology of Controlled Nuclear Fusion, San Diego, CONF-740402, Vol, I, p. 165, U.S. Atomic Energy Commission and American Nuclear Society (1974).

15. R. Hancox and I. J. Spalding, "Reactor Implications of Laser Ignited Fusion," CLM-P310, Culham Laboratory, Abingdon, Berkshire, England (1972).

16. F. T. Finch, E. A. Kern, and J. M. Williams, "Laser Fusion Power Plant Systems Analysis," Proc. 1st Topl. Mtg. Technology of Controlled Nuclear Fusion, San Diego, CONF-740402, Vol. I, p. 142, U.S. Atomic Energy Commission and American Nuclear Society (1974).

17. A. P. Fraas, "The Blascon—An Exploding Pellet Fusion Reactor," ORNL-TM-3231, Oak Ridge National Laboratory (1971).

18. L. A. Booth, "Central Power Generation by Laser-Driven Fusion," LA-4858-MS, 1, Los Alamos Scientific Laboratory (1972).

19. J. Nuckolls et al., Nature, 239, 139 (1972).

20. T. Frank, D. Freiwald, T. Merson, and J. Devaney, "A Laser Fusion Reactor Concept Utilizing Magnetic Fields for Cavity Wall Protection," Proc. 1st Topl. Mtg. Technology of Controlled Nuclear Fusion, San Diego, CONF-740402, Vol. I, p. 83, U.S. Atomic Energy Commission and American Nuclear Society (1974).

21. D. Duchs et al., Surface Effects in Controlled Fusion, p. 102, H. Wiedersich, M. S. Kaminsky, and K. M. Zwilsky, Eds., North-Holland Publishing Company, Amsterdam (1974).

22. R. W. Moir, "Physics of Mirror Reactors and Devices," UCRL-75324, Lawrence Livermore Laboratory (1974).

23. R. A. Krakowski, T. A. Oliphant, and K. I. Thomassen, "Ergonic Optimization and Parameter Study of the RTPR Burn Cycle," Proc. 1st Topl. Mtg. Technology of Controlled Nuclear Fusion, San Diego, CONF-740402, Vol. I, p. 112, U.S. Atomic Energy Commission and American Nuclear Society (1974).

24. R. W. Werner, "Materials Problems in Steady State Fusion Reactors with Direct Conversion," UCRL-75336, Lawrence Livermore Laboratory (1974).

25. B. Lehnert, Nucl. Fusion, 13, 781 (1973).

26. G. K. Verboom and J. Rem, Nucl. Fusion, 13, 69 (1973).

27. G. L. Kulcinski and G. A. Emmert, Surface Effects in Controlled Fusion, p. 31, H. Wiedersich, M. S. Kaminsky, and K. M. Zwilsky, Eds., North-Holland Publishing Company, Amsterdam (1974).

28. S. N. Cramer and E. M. Oblow, "Feasibility Study of a Honeycomb Vacuum Wall for Fusion Reactors," ORNL-TM-4708, Oak Ridge National Laboratory (1974).

29. G. L. Kulcinski et al., "A Method to Reduce the Effects of Plasma Contamination and First Wall Erosion in Fusion Reactors," UWFDM-108, University of Wisconsin (1974).

30. J. M. Williams, "Laser CTR Systems Studies," LA-5145-MS, Los Alamos Scientific Laboratory (1973).

31. M. A. Abdou and C. W. Maynard, "Neutronics and Photonics Study of Fusion Reactor Blankets," Proc. 1st Topl. Mtg. Technology of Controlled Nuclear Fusion, San Diego, CONF-740402, Vol II, p. 87, U.S. Atomic Energy Commission and American Nuclear Society (1974).

32. G. R. Hopkins, "Fusion Reactor Applications of Silicon Carbide and Carbon," Proc. 1st Topl. Mtg. Technology of Controlled Nuclear Fusion, San Diego, CONF-740402, Vol. II, p. 437, U.S. Atomic Energy Commission and American Nuclear Society (1974).

33. W. G. Price, Jr., "Blanket Neutronic Studies for a Fusion Power Reactor," Proc. 5th Symp. Engineering Problems of Fusion Research, Princeton University (1973).

34. F. W. Wiffen and E. E. Bloom, "Effects of High Helium Content on Stainless Steel Swelling," ORNL-TM-4541, Oak Ridge National Laboratory (1974).

35. G. L. Kulcinski et al., "Comparison of Displacement and Gas Production Rates in Current Fusion and Future Fusion Reactors," to be published in Proc. ASTM Conf., 7th ASTM Intern. Symp. Radiation Effects on Structural Materials, Gatlinburg, Tennessee (1974).

36. W. V. Green, E. G. Zukas, and D. T. Eash, "The Tritium Trick," LA-DC-12996, Los Alamos Scientific Laboratory (1971); see also Proc. Intern. Working Sessions on Fusion Reactor Technology, Oak Ridge National Laboratory, CONF-710624, p. 199, U.S. Atomic Energy Commission (1971).

37. W. Bauer, E. P. Eernisse, P. L. Mattern, R. G. Musket, R. M. Schmieder, and G. J. Thomas, "Neutron Beam Sources: Applications in Science and Technology," SLL-74-8204, p. 94, Sandia Laboratories (1974).

38. D. J. Dudziak, "Transmutation and Atom Displacement Rates in a Reference Theta-Pinch Reactor," Proc. 1st Topl. Mtg. Technology of Controlled Nuclear Fusion, San Diego, CONF-740402, Vol, II, p. 114, U.S. Atomic Energy Commission and American Nuclear Society (1974).

39. L. C. Ianniello, Ed., Fusion Reactor First Wall Materials, WASH-1206, U.S. Atomic Energy Commission (1972).

40. W. J. Gray and W. C. Morgan, "Projection of Graphite Behavior: Comparison of Result on Four Conceptual Fusion Reactor Designs," Proc. 5th Symp. Engineering Problems of Fusion Research, p. 59, Princeton University (1973).

41. J. M. Bunch et al., "An Evaluation of Major Material Problems Anticipated for the Reference Theta-Pinch Reactor (RTPR)," Proc. 5th Symp. Engineering Problems of Fusion Research, Princeton University (1973).

42. G. M. McCracken and S. Blow, "The Shielding of Superconducting Magnets in a Fusion Reactor," CLM-R 120, Culham Laboratory, Abingdon, Berkshire, England (1972).

43. J. T. Kriese and D. Steiner, "Magnet Shield Design for Fusion Reactors," ORNL-TM-4256, Oak Ridge National Laboratory (1973).

44. W. G. Homeyer, "Thermal and Chemical Aspects of the Thermonuclear Blanket Problem," Technical Report No. 435, Massachusetts Institute of Technology (1965).

45. J. E. Draley, B. R. T. Frost, D. M. Gruen, M. Kaminsky, and V. A. Maron, "An Assessment of Some Materials Problems for Fusion Reactors," Proc. 1971 Intersociety Energy Conversion Engineering Conf., p. 1065, Boston (1971).

46. W. R. Grimes and S. Cantor, "Molten Salts as Blanket Fluids in Controlled Fusion Reactors," ORNL-TM-4047, Oak Ridge National Laboratory (1972).

47. S. Cantor and W. R. Grimes, Nucl. Technol., 22, 120 (1974).

48. D. Steiner, Nucl. Fusion, 11, 3 (1971).

49. W. N. Gill, R. P. Vanek, R. J. Jelinek, and C. S. Grove, Jr., AIChE J., 6, 139 (1960).

50. E. E. Hoffman, "Corrosion of Materials by Lithium at Elevated Temperatures," ORNL-2676, Oak Ridge National Laboratory (1959).

51. M. S. Freed and K. J. Kelly, "Corrosion of Columbian Base and Other Structural Alloys in High Temperature Lithium," PWAC-355, Pratt & Whitney Aircraft Corp. Middletown, Connecticut (1961).

52. J. O. Cowles and A. D. Pasternak, "Lithium Properties Related to Use as a Nuclear Reactor Coolant," UCRL-50647, Lawrence Livermore Laboratory (1969).

53. J. H. DeVan, A. P. Litman, J. R. DiStefano, and C. E. Sessions, "Lithium and Potassium Corrosion Studies with Refractory Metals," Proc. IAEA Symp. Alkali Metals Coolants, Nov. 28-Dec. 2, 1966, Vienna, International Atomic Energy Agency (1967).

54. C. W. Cunningham and B. Fleischer, "Fuels and Materials Development Program Quarterly Progress Report," ORNL-4600, Oak Ridge National Laboratory (1970).

55. R. L. Klueh and J. H. DeVan, J. Less-Common Metals, 30, 9 (1973); see also J. Less-Common Metals, 80, 25 (1973).

56. R. L. Klueh, Corrosion, 27, 342 (1971).

57. R. L. Klueh, "The Effects of Oxygen on the Corrosion of Niobium and Tantalum by Liquid Lithium," ORNL-TM-4069, Oak Ridge National Laboratory (1973).

58. J. H. DeVan and R. L. Klueh, Nucl. Technol., 24, 64 (1974);

59. J. R. DiStefano, "Corrosion of Refractory Metals by Lithium," ORNL-TM-3551, Oak Ridge National Laboratory (1969).

60. C. E. Sessions and J. H. DeVan, "Effect of Oxygen, Heat Treatment, and Test Temperature on the Compatibility of Several Advanced Refractory Alloys with Lithium," ORNL-TM-4430, Oak Ridge National Laboratory (1971).

61. C. F. Baes, Jr., Nucl. Metall, 15, 625 (1969); see also Proc. Symp. Reprocession of Nuclear Fuels, CONF-690801, U.S. Atomic Energy Commission (1969).

62. D. Steiner, "The Neutron-Induced Activity and Decay Power of the Niobium Structure of a D-T Fusion Reactor Blanket," ORNL-TM-3094, Oak Ridge National Laboratory (1970).

63. J. D. Lee, "Some Observations on the Radiological Aspects of Fusion," Fusion Technology Monographs/Energy 70, Proc. Symp. Intern. Soc. of Energy Conversion Engineers, Las Vegas, Nevada, September 1970, CONF-709012, U.S. Atomic Energy Commission (1970); see also UCRL-72309, Lawrence Livermore Laboratory.

64. S. Blow, "Transmutation, Activity and After-Heat in a Fusion Reactor Blanket," AERE-R6581, Atomic Energy Research Establishment, Harwell, Berkshire, England (1971).

65. D. J. Dudziak, Nucl. Technol., 10, 391 (1971).

66. D. Steiner and A. P. Fraas, Nucl. Safety, 13, 5 (1972).

67. W. F. Vogelsang, G. L. Kulcinski, R. G. Lott, and T. Y. Sung, Trans. Am. Nucl. Soc., 17, 138 (1973).

68. D. J. Dudziak and R. A. Krakowski, "A Comparative Analysis of D-T Fusion Reactor Radioactivity and Afterheat," Proc. 1st Topl. Mtg. Technology of Controlled Nuclear Fusion, San Diego, CONF-740402, Vol. I, p. 548, U.S. Atomic Energy Commission and American Nuclear Society (1974).

69. D. W. Nigg and J. N. Davidson, "The Induced Activity and Decay Power of the Structure of a Stainless Steel Fusion Reactor Blanket," Proc. 1st Topl. Mtg. Technology of Controlled Nuclear Fusion, San Diego, California, CONF-740402, Vol. I, p. 578, U.S. Atomic Energy Commission and American Nuclear Society (1974).

70. Mineral Facts and Problems, Bureau of Mines Bulletin 650, U.S. Department of the Interior (1970).

71. H. E. Goeller, Oak Ridge National Laboratory, Private Communication (1974).

72. "Fusion Power: An Assessment of Ultimate Potential," WASH-1239, U.S. Atomic Energy Commission (1973).

CHAPTER 3

FUSION - FIRST WALL PROBLEMS

R. Behrisch
Max-Planck-Institut für Plasmaphysik
EURATOM Association
D-8046 Garching/München, Germany

I. INTRODUCTION

One of the most promising, but also most uncertain energy sources for the future is the fusion reactor. In this reactor the energy contained in the heavy hydrogen isotopes D and T together with the element Li will be set free in a controlled manner for peaceful uses. This will take place if a 50% D and a 50% T gas is heated to become a thermonuclear plasma, i.e. at a temperature $T \gtrsim 10$ keV and will be confined for a time τ at a density n, where $n\tau > 10^{14} cm^{-3}$sec (Lawson criterion 157),(1,2). The most promising configuration for magnetically confining such a hot thermonuclear plasma on our planet in respect to building a fusion reactor with positive energy balance is, as it looks today, the tokamak scheme. This has been developed in the early 60's in the USSR by L.Artsimovich and his colleagues (3,4). By adapting this tokamak scheme in the USA, Europe and Japan, large progress has been achieved in the early 70's toward getting a hydrogen plasma under the conditions which are needed for a fusion reactor (5).

However, more progress is necessary in several important areas of plasma physics and materials physics before a fusion reactor can be realistically designed. In the following, I will deal only with one of these problems, i.e. the first wall problem, which actually lies between plasma physics and materials physics.

The first talk in the series will deal with the plasma physics view. I will outline the principle of a tokamak and its extrapolation to a fusion reactor and thus introduce the first wall problem. This restriction to a tokamak is because of its simplicity and as it is most relevant at present (6). The other confinement schemes have similar first wall problems (7). In the second talk I will deal with basic surface and materials investigations and will outline the implications of the measured data for a future fusion reactor.

II. THE FIRST WALL PROBLEM IN A TOKAMAK (first talk)

A. Principle of a Tokamak

The operation principle of a tokamak (4,5) is shown schematically in Fig. 1. The plasma, i.e. the hot hydrogen gas is contained in a large toroidal magnetic field B_T of 3 to 5

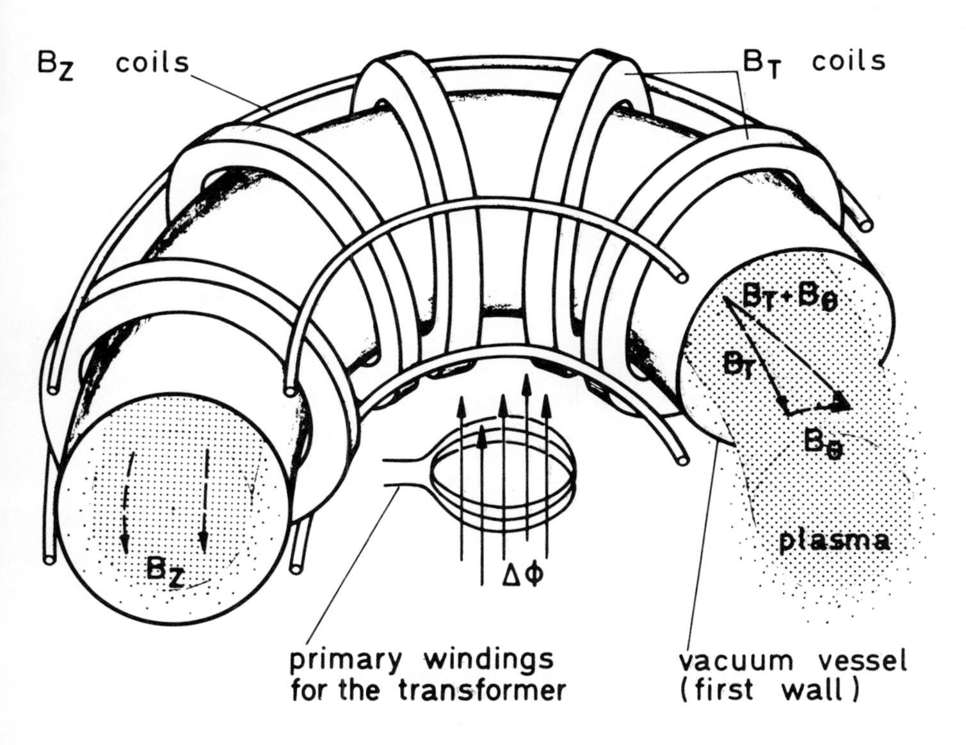

FIG. 1. Schematic of a tokamak. B_T: main toroidal magnetic field, B_Θ: poloidal magnetic field of the discharge current B_z: vertical magnetic field.

tesla. After preionization it is heated by an ohmic current induced in it by a slowly changing magnetic flux $\Delta\Phi$ through the center of the torus. The tokamak can thus be regarded as a single turn winding of a transformer. The poloidal magnetic field B_Θ of the plasma current superimposed on the toroidal field is essential for MHD-stable magnetic confinement. Further a small vertical magnetic field B_z is necessary to prevent the expansion of the plasma ring. In addition to the ohmic heating, the tokamak plasma may be heated by injection of energetic neutral beams (40 to 150 keV) and/or by electromagnetic radiation and/or by adiabatic compression (5).

The surfaces of the toroidal vessel, which directly see the plasma, are called the first wall. As the magnetic confinement is only poor, plasma is continuously leaking toward the first wall (4,5,8). In order to keep the plasma from directly touching the first wall a diaphragm generally of a refractory metal - the limiter - is introduced into the discharge vessel of today's tokamaks.

If a DT-plasma can be obtained, which fulfills the Lawson criterion, thermonuclear reactions will take over and will keep the plasma hot. These reactions are shown in Fig.2. The energy produced in the reactions is contained to 80% in the 14.1 MeV neutrons and to 20% in the 3.5 MeV α-particles. The neutrons will leave the plasma and hit directly the first wall. The α-particles will be confined and slowed down in the plasma. Their energy will be given to the plasma but finally it will also hit the first wall in the form of the plasma losses. Actually, the plasma parameters will adjust so that the losses are just equal to the α-energy. Besides the effects of the neutrons, these plasma losses to the first wall are the inherent causes of the first wall problems.

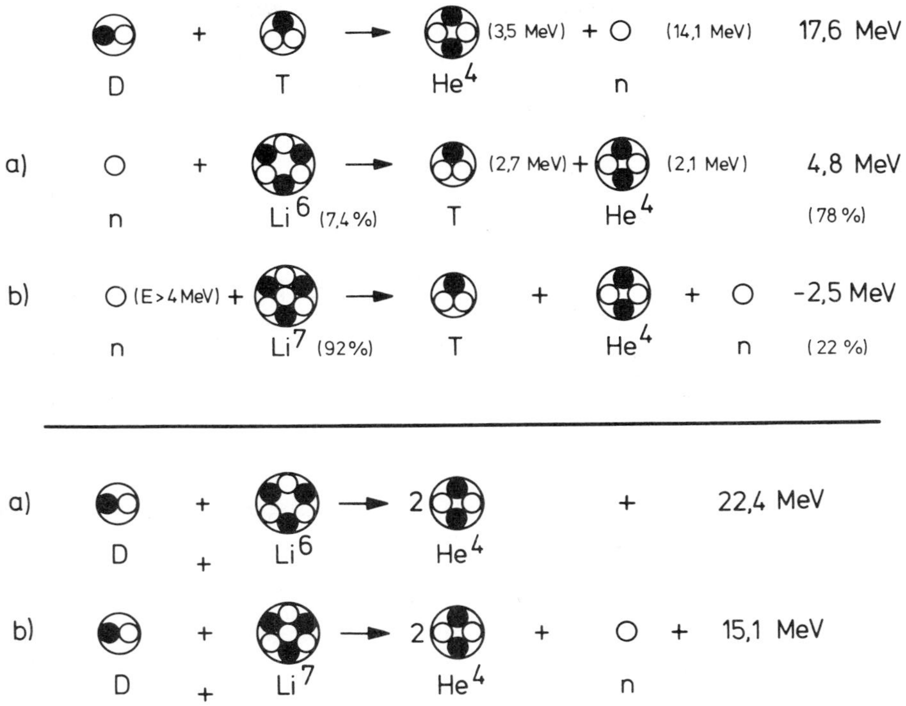

D + T → He^4 (3,5 MeV) + n (14,1 MeV) 17,6 MeV

a) n + Li^6 (7,4 %) → T (2,7 MeV) + He^4 (2,1 MeV) 4,8 MeV (78 %)

b) n (E>4 MeV) + Li^7 (92 %) → T + He^4 + n −2,5 MeV (22 %)

---

a) D + Li^6 → 2 He^4 + 22,4 MeV

b) D + Li^7 → 2 He^4 + n + 15,1 MeV

FIG. 2. Fusion reactions in a D,T-plasma of a reactor, and the reactions of the 14.1 MeV neutrons in the blanket to breed the tritium. A fusion reactor will actually burn D and Li to He.

B. The Plasma Wall Interaction

The details of the plasma wall interactions are still not yet known; however, the following general picture can be drawn (9 to 15):

i) the neutrons, the electro-magnetic radiation, and neutral atoms are not confined by the magnetic field. They leave the plasma and hit the first wall.

ii) The confinement for the plasma-ions and -electrons is only weak. They continuously leak out of the confining field and hit the limiter and/or the divertor plates as well as partly the first wall.

There will be a temperature difference between the plasma at $\sim 10^8$K and the first wall at $\sim 10^3$K which can only be maintained due to the confining magnetic field and due to the density difference of $\sim 10^{14}$cm^{-3} in the plasma and $\sim 5 \cdot 10^{22}$cm^{-3} in the solid. Most of todays tokamak experiments indicate that the temperature gradient between the plasma and the first wall will occur predominantly right at the first wall. This means that thermal equilibrium between the plasma and the first wall will not occur and the plasma-wall interaction has to be understood from the different atomic processes at the first wall. However, theoretical estimates indicate that the temperature gradient may also be shifted into a layer of high density ($\sim 10^{16}$cm^{-3}) cold plasma with a transition to neutral gas between the plasma and the first wall (cold gas blanket) (16). This would largely reduce the plasma wall problem by shifting the interactions to very low energies ($\sim$ 1 eV).

The understanding of the plasma-wall interaction is necessary in respect to the two aspects:
- the plasma is modified by the solid wall via the recycling plasma atoms and the introduction of wall atoms (impurities).
- The first wall is modified by the wall bombardment from the plasma due to the material loss and loss of integrity.

As it looks today, the modification of the plasma is the more severe problem for achieving ignition in the plasma, while the modification of the first wall will be a limiting factor for the lifetime of a fusion reactor.

C. *Particle and Energy Fluxes between the Plasma and the First Wall*

1) Neutrons

The current of 14.1 MeV neutrons from the plasma to the first wall will depend on the design and operation parameters of a future fusion reactor. Todays design studies use a total energy load from the plasma to the first wall of about 100 W/cm^2,which corresponds to a neutron current of $3.5x10^{13}$ neutrons/cm^2sec (5,7,10,12,17 to 19). Most of them penetrate the first wall and are slowed down in an about 1 m thick structure, the blanket which will surround the plasma (7,2). The slowing down takes place in collisions with the nuclei of the blanket. The energy thus transferred to the nuclei causes lattice defects in the bulk and sputtering at the surfaces (14,19). As the neutrons are scattered back and forth several times the total neutron flux in the first wall is composed of the 14.1 MeV neutron source current from the plasma and the scattered neutron flux which has a broad energy spectrum and is about a factor 5 to 10 higher in intensity (7,10,17 to 20).

Additionally, nuclear reactions induced by the neutrons in the blanket as (n,γ); (n,p,γ); (n,α,γ),... are the sources of an intense γ-radiations also toward the first wall - called backshine (20).

2) Electromagnetic Radiation

The electromagnetic radiation from the thermonuclear plasma consists of line radiation, electron capture radiation, and bremsstrahlung from the plasma atoms and of synchrotron radiation from the gyrating plasma electrons (21,22). In a pure hydrogen plasma the ions are fully stripped at the temperatures necessary for fusion reactions to occur. Thus only bremsstrahlung and synchrotron radiation will be present. Estimates show (2,21,22) that at the temperature of 10 to 20 keV this radiation is about an order of magnitude smaller than the α -energy and gets larger than the α energy at temperatures in the 100 keV range. However, every hydrogen plasma contains impurities, i.e. some ions of atomic numer $Z \geq 2$. As the bremsstrahlung increases proportional to Z^2, small amounts of impurities are sufficient to radiate more energy than dissipated by the α -particles. Additionally, high Z atoms in the plasma push away the hydrogen thus leading to a decrease in fusion reactions.In Fig.3 the tolerable impurity concentrations calculated for these conditions by different authors are drawn up (2,14,23). They show that the maximum tolerable impurity concentration becomes $< 10^{-4}$ at% for high Z impurities, while it may be a few at% for low Z materials. As the impurities are introduced from the first wall this leads to the request of a low Z material.

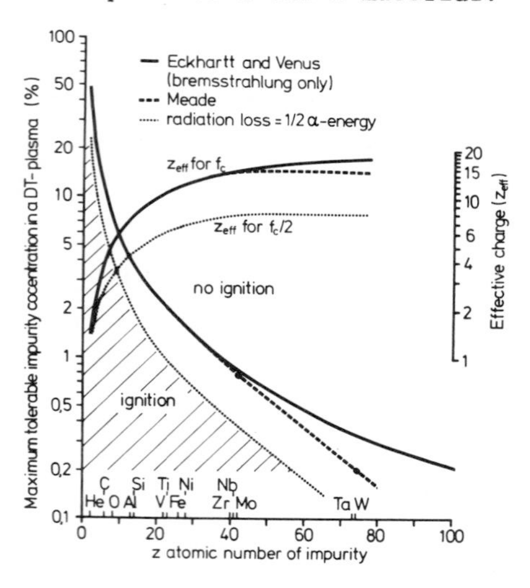

FIG. 3. Maximum tolerable impurity concentrations for a DT fusion plasma if all the α-energy or one half the α-energy is lost by radiation.

In todays plasma experiments the radiation losses from the plasma, mainly from the impurities, constitute a major energy loss (30 to 50%) of the energy introduced by ohmic heating. (24,25).

3) Electrons

The number of electrons leaving the plasma will be the same as the number of ions and they will presumably have about the same energy as the ions. Due to potential differences between the plasma and the first wall they may be accelerated or decellerated. In todays tokamak discharges additionally some very energetic electrons in the MeV-range (runaway electrons) are observed. The runaway electrons preferentially hit the limiter locally, leading to evaporation. They may also be trapped in small mirror fields between the coils of the toroidal magnet and then hit the first wall and cause damage (26). Their formation depends largely on the starting conditions of the discharge and the plasma density and they can hopefully be avoided in a fusion reactor.

The total contribution of the electrons to the energy loss in a fusion reactor is difficult to estimate. It is expected to be only a small fraction of the α-energy. In todays experiments the electrons, especially the runaway electrons, contribute an energy loss of about 10-30% of the ohmic heating.

If the first wall is made of a metal the radiation and the electrons will cause at the surface only desorption and a temperature increase. This may be tolerated if their distribution on the first wall is uniform in time and space. For non-uniform distributions local overheating, evaporation and especially pulse evaporation becomes a further serious first wall problem (10,27). For first walls of insulating materials with localized bindings as well as for sorbed layers on metal surfaces the radiation and the electrons may additionally cause a decomposition and desorption.

4) Ions and Neutrals

Neutral atoms are not confined in the plasma. For the ions the confinement in the magnetic field is not perfect and they continuously leak out of the plasma. In today's plasma experiments the particle confinement time is only $\frac{1}{5}$ to $\frac{1}{10}$ of the discharge time which means that on the average all plasma particles hit the first wall and are recycled i.e. reemitted from the first wall into the plasma 5 - 10 times during one discharge of 0.1 to 1 sec (4,5,7,8).

The processes which determine first wall bombardment depend on the question, whether a cold gas blanket can be built up or not. The model to estimate the ion - and neutral fluxes, which will be presented in the following assumes the one extreme case of no cold gas blanket,i.e. a very low particle density and no collisions in the volume between the plasma and the first wall. This gives the lowest fluxes at the highest energy compared to the other extreme of a cold gas blanket with nearly thermal equilibrium at the first wall. The plasma wall interaction processes for the case of no gas blanket are shown schematically in Fig.4 (14,15). The ions diffusing from the plasma to the first wall may be backscattered or implanted and trapped and released later. As the backscattered and

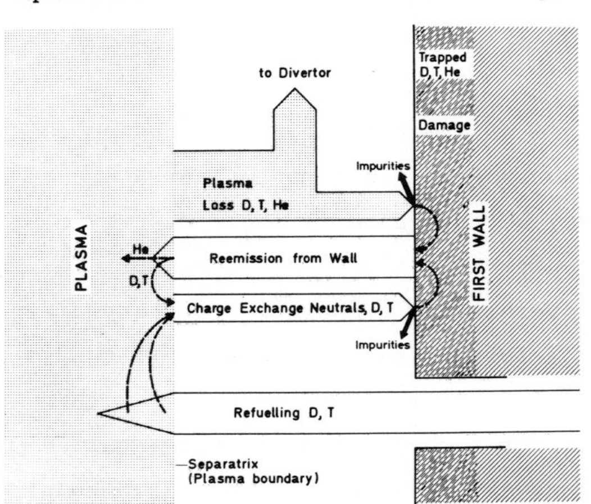

FIG. 4. Wall bombardment processes including recycling and refuelling.

released hydrogen atoms are mostly neutral (28) neutral atoms are recycled into the plasma. Additionally, neutral gas may be introduced through the first wall in order to increase the plasma density. Fast neutrals may also be injected for additional heating during start up of a tokamak. All these neutral hydrogen atoms coming to the plasma undergo predominantly resonance charge exchange with the plasma ions and we then get an additional flux of fast neutrals to the first wall. Here the fast neutrals will undergo the same process as the ions. It turns out that this current of fast charge exchange neutrals will be of the same magnitude or even higher than the loss of ions from the plasma to the first wall. Helium ions leaving the plasma and reemitted from the first wall as neutrals will not contribute to this resonance charge exchange neutral current. The energy of the neutrals may lie in the 1 to 10 keV region corresponding to the temperature near the plasma boundary. It will be smaller ($\sim$ 1eV) if a neutral gas blanket can be achieved.

In Fig.4 it is also shown that this flux of ions and fast charge exchange neutrals to the first wall causes additionally the release of wall atoms by sputtering and/or desorption. Further the first wall becomes implanted with the gas of the plasma ions leading to a modification of the surface properties and the surface structure topography.

D. *Divertor*

If a cold gas blanket cannot be achieved it has been suggested very early in fusion research to decrease the particle fluxes between the plasma and the first wall and thus also impurity influx into the plasma by scraping off continuously the outer layer of the plasma and guide it into a separate volume, the divertor (29). Several different geometries for such divertors have been proposed (toroidal (29,30), poloidal

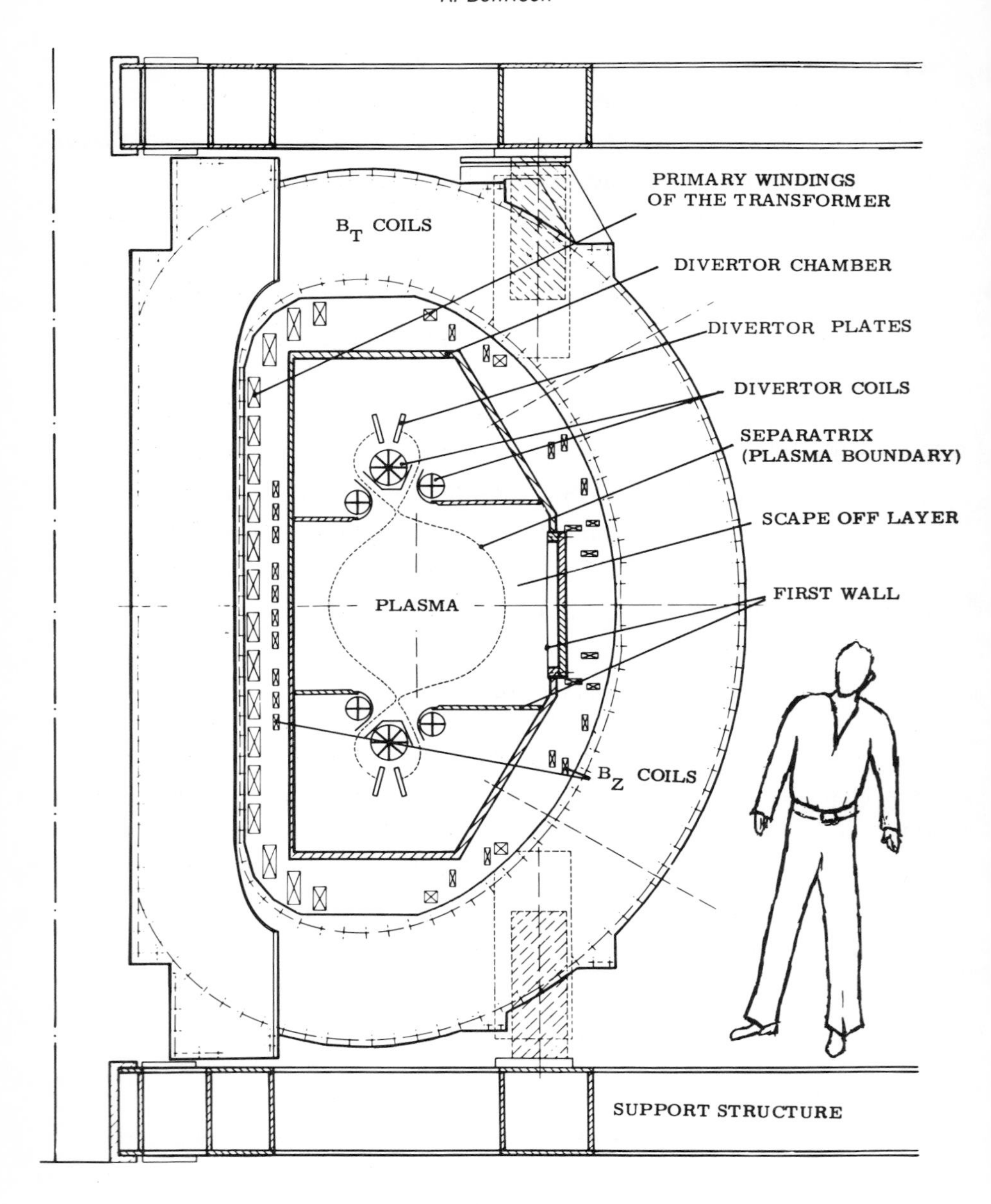

FIG.5. Cross section of the divertor of the 'ASDEX'.

(31,32) and bundle (33) as well as different operation modes have been suggested (screening and unload) (30). The operation of the divertor favoured for tokamaks, i.e. the axisymmetric poloidal divertor is shown schematically in Fig.5 as the example of the planned divertor experiment ASDEX in Garching (34). A similar experiment the PDX is being built up in Princeton (35). Most of the ions diffusing out of the plasma boundary or separatrix into the scrape off layer will be guided during the movement around the torus into the upper or the lower divertor chamber. Thus the divertor acts also in the same way as the limiter in todays tokamaks, it will define the plasma boundary. The necessary magnetic divertor field configuration is produced by the electrical currents in the divertor coils shown.

If the divertor will operate, plasma ions as well as impurities will be lost into the divertor chamber. The plasma atoms have to be refuelled which will be possible only by neutral atoms or clusters. However these neutrals will cause new wall bombardment by recycling neutrals and this will partly compensate the decrease of the bombardment by the plasma ions lost into the divertor. Therefore a divertor may not appreciably decrease the wall bombardment by plasma ions and neutrals. It may hopefully be useful in extracting together with the plasma ions:

- the impurities which are released from the first wall
- He-ions produced in the plasma by the D-T reaction
- and acts as a remote limiter.

At the entrance slits of the divertor and especially inside the divertor on the collector plates the ion bombardment will be much higher than at the first wall.

E. *Particle Balance Equations for the Fluxes of Ions and Neutrals*

If a gas blanket cannot be achieved, i.e if the plasma is surrounded with a very low density gas the ion- and neutral fluxes may be estimated from particle balance equations derived from the flux diagram of Fig.4 and some further assumptions on the atomic processes at the first wall as well as at the plasma boundary. With the additional information on the energy distribution and angular distribution of the wall fluxes in todays tokamak experiments and their extrapolation to a fusion reactor (13,36) we can learn which processes are the dominant ones at the first wall.

Particle balance equations as derived at several places (13,14,15,35,36,37) give during the steady state operation of a tokamak the following total currents to the first wall.

The deuterium and tritium current:

$$F_{DT} = \frac{n_{DT}}{\tau_{DT}} \; \frac{\eta_{DT}(1-f) + (1-\eta_{DT})\,A_1}{(1-r_{DT}A)(1-f)+(1-r_{DT})A_1} \tag{1}$$

The helium current:

$$F_{He} = \frac{\eta_{He}}{1-r_{He}\,\eta_{He}} \; \frac{f}{2} \cdot F_1 \tag{2}$$

The neutron current:

$$F_n = \frac{f}{2}\,\alpha\,F_1 \tag{3}$$

where n_{DT} is the density in the plasma and τ_{DT} the mean particle lifetime of D and T ions in the plasma. η_{DT} is the fraction of ions passing the separatrix which will not reach the divertor and f is the fractional burnup with respect to F_1, the D,T throughput$^{+)}$. 'A_1' is the probability that a charge exchange fast

(+) In a previous publication of the author and B.B.Kadomtsev (14),f has been defined in respect to the D,T density n_{DT} in the plasma. This gave slightly different formulae.

neutral is released from the plasma for a neutral particle fuelled into the plasma boundary, while A is the same probability for neutrals reflected or reemitted from the first wall. r'_{DT} is the reemission plus reflection probability. In steady state we will have finally $r_{DT} \simeq 1$. The subscript He means same quantities for He. 'α' is the neutron flux enhancement at the first wall due to scattering of the neutrons in the blanket. Neutron flux calculations give values for α between 5 and 10 (17,18,19). F_1 is the total fuelling flux of D,T gas, fF_1 will be burnt to He and neutrons while the rest ends up in the divertor: For constant particle density in the plasma the refuelling flux is given by:

$$F_1 = \frac{n_{DT}}{\tau_{DT}} \frac{1-Ar_{DT} - \eta_{DT}\, r_{DT}(1-A)}{(1-Ar_{DT})(1-f)-(1-r_{DT})A_1} \qquad (5)$$

For $r_{DT} = 1$ this gives

$$F_1 = \frac{n_{DT}}{\tau_{DT}} \frac{1-\eta_{DT}}{1 - f} \qquad (5a)$$

The D,T current to the first wall consists of two parts. The first part constitutes of those ions diffusing out of the plasma and not reaching the divertor and the neutral current induced by them by the recycling process. The second part are the recycling neutral currents due to the refuelling flux F_1. If the divertor is not very effective the first term will be large, and the second term will be small(screening divertor)while it is just opposite if the divertor is very effective (unload divertor). This means that F_{DT} will always be of the order of n_{DT}/τ_{DT} except that fuelling can be made in a way that A_1 becomes very small as hopefully by the injection of very fast neutrals or large fast clusters on pellets. The currents per unit first wall area if uniformly distributed will then be given by:

$$\phi_{wall} = (F_{DT}+F_{He}+F_n) \frac{V}{O} \qquad (6)$$

where V is the plasma volume and O the first wall surface.

For todays plasma experiments as well as for the parameters used in fusion reactor design studies this gives for the D,T current values of $\phi_{wall} \simeq 10^{15}$ to $10^{16}/cm^2 sec$, i.e.of the order of 100 μA to 1 mA/cm^2. If a sputtering yield $S \simeq 10^{-2}$ atoms per ion is taken this results in an impurity introduction of 10^{13} to 10^{14} $atoms/cm^2 sec$. The He flux to the first wall will be smaller by one to two orders of magnitude as probably $f \lesssim 10\%$ and $\eta \sim 10\%$.

The energy of the particle fluxes will range from a few eV up to several keV with a maximum close to the plasma temperature near the separatrix. There will be a component of higher energy particles near the energy of the injected neutral beams as well as a tail in the distribution of the α-particles ranging up to 3.5 MeV. All particles will have a broad distribution in angles of incidence.

The total contribution of the ion and neutral fluxes to the first wall to the energy loss of the plasma is about 1/2 of the ohmic heating energy in todays plasma experiments. It will presumably decrease for larger experiments due to the smaller volume to surface ratio.

F. *Neutral Gas Blanket*

The particle balance equations and the fluxes derived hold only for one extreme case i.e. free particle flow in the space between the plasma boundary formed by a limiter or divertor and the first wall. The neutral gas blanket is the other extreme case. It means that neutral gas at a density of $\sim 10^{16} cm^{-3}$, and at a temperature of $\lesssim 1$ eV can be built up and maintained in the region near the first wall. The boundary between the plasma and the neutral gas will be established by ionisation and recombination from both sides (16). A limiter may only be necessary for safety reasons but the hot plasma would generally not touch the first wall.

Collisions in the gas blanket will dominate and ions will not be able to hit the first wall. The final fluxes on the first wall will be given by 2nkT where n is the particle density and T the temperature of the neutral gas close to the first wall. They will be lower in energy but accordingly higher in intensity compared to the fluxes obtained from the particle balance equations.

G. *Effects Caused by the Particle Bombardment at the First Wall*

The particle fluxes to the first wall have several consequences (see also fig.5):

- They will cause severe radiation damage and transmutations in the first wall material.
- Gas ions will be backscattered or will be stopped and accumulated in the near surface region of the first wall. They may be released again at saturation.
- The accumulated gas will lead to the formation of gas bubbles in the surface layers, to surface blistering and finally the disappearance of the blisters leading to a sponge like surface structure.
- Surface atoms will be removed by sputtering leading to a contamination of the plasma and to a thinning of the first wall.

At present the plasma contamination is regarded as the most serious problem. Absolute numbers about the build up of impurities can be obtained from the particle balance equations. The results depend largely on the plasma-particle confinement time τ_{DT}, the plasma temperature near the first wall as well as the impurity diffusion coeffcients and their confinement time τ_i. A rough estimate shows that for $\tau_i \sim \tau_{DT}$ a steady state impurity concentration due to DT sputtering alone is given by the sputtering yields S_{DT}(36,13,14,15). As these yields may reach values of 10^{-2} to 10^{-3} impurity concentrations may easily extend beyond the tolerable limits.

The thinning of the first wall by all different erosion processes in a time 't' is given by

$$d = \frac{1}{N} \frac{V}{O} t \sum F_j \overline{S}_j \tag{7}$$

where d is the wall thickness removed, N is the atomic density in the wall material, F_j is the wall flux and $\overline{S}_j$ the mean sputtering yield of the component j averaged over the angular and energy distribution of the bombardment. Putting numbers in equation (7) and assuming no redeposition of material once sputtered, we get erosion yields of 0.1 to 1 mm/year. They will be decreased if the wall bombardment takes place at very low energies (< 10 eV) which would be the case if a neutral gas blanket can be achieved.

H. *Demands for the First Wall Material*

The energy load as well as the bombardment by the different particles lead to several demands for the first wall material. These are summarized in the following:

1) Due to the intense neutron bombardment with an energy of 14.1 MeV (several $10^{13}/cm^2sec$) and with lower energies (several $10^{14}/cm^2sec$) we demand:

a) Low activation material with a low after heat and only short half lifes of transmutation products.

b) Low transmutation cross sections, i.e. low H and He production and small change of the alloying during the lifetime of a reactor.

c) Low neutron radiation damage in the temperature range where it will be used, i.e. low swelling, low embrittlement and only little loss of ductility.

d) The absorption cross section for X-rays produced in the blanket(backshine) should be low to prevent overheating.

2) At the first wall surface the α-energy given to the plasma will be deposited in the form of radiation, electrons and particle bombardment, presumably not uniform in space and time. This leads to the demands:

a) The first wall material should withstand high thermal loads continuously (good thermal conductivity) as well as heat pulses (high melting point), without loss of mechanical, thermal, electrical and other properties.

b) Low yields by physical and chemical sputtering by ions, neutrals and neutrons are necessary. No major composition change should occur.

c) Trapping of incident H and He should be small and the gas should not diffuse in the material (otherwise the materials may fill up with gas).

d) Blistering should not occur or it should disappear already at low doses.

e) It should be a low Z material in order that higher concentrations can be tolerated in the plasma.

3) As the fusion reactor will be large in size and operate at a high temperature in order to get good thermal efficiency we demand:

   a) A material with good mechanical strength up to $\sim 1000^{o}C$

   b) Vacuum tight materials with no major degassing even if being kept for longer times at high temperatures.

   c) The material should withstand the pulsed thermal load in pulsed reactors (time scales of usec to h).

   d) It should be capable of easy machining, coldworking and welding.

   e) Considerable experience should be available with this material.

Up to now the materials which have been considered for the first wall are:

- Ion and Nickel base stainless steels
- Carbides such as SiC and B_4C
- Refractory metals and alloys: Nb + 1% Zr, Mo,V.

All these may not fulfill the demands expected for a CTR first wall material. We may however have to use them and take into account their limited properties in the reactor design. However, if it turns out that a fusion reactor will be the major energy source in the future, it may be of benefit to start a large materials development program.

II. INVESTIGATIONS OF DIFFERENT ATOMIC PROCESSES AT THE FIRST WALL

A. First Wall Processes

The elementary processes which will contribute at the first wall to the plasma wall interaction are firstly those which determine the plasma particle balance, i.e. backscattering, trapping and reemission; secondly, those which determine the impurity release, i.e. evaporation, physical and chemical sputtering, desorption, blistering, and finally those which influence the energy balance, i.e. energy accommodation for the particles, absorption and reflection for electrons and electromagnetic radiation. For all of these processes reliable data are needed mostly for low energy light ion bombardment and soft x-ray and synchrotron radiation at a total energy load of 10 - 100 W/cm^2. Plasma wall interaction is however not yet understood and new results in plasma physics experiments may alter the demands on measured results performed surface physicists. Synergistic effects due to the simultaneous wall bombardment by different ions at different energies and by radiation may also become important to investigate.

In the following, some of the recent experimental and theoretical work in this respect is reviewed.

B. Ion and Neutral Backscattering

Light ions with an energy of a few keV impinging on a solid mostly penetrate the surface and are slowed down in collision with the atoms of the solid. The trajectories of the ions are determined by the collisions with the nuclei while the energy is mostly lost to the electrons(40).Depending on the type of ions and their energy as well as the target material some trajectories will be bent back and thus part of the ions will leave the surface, i.e. become backscattered.

Total backscattering yields and backscattered energy as well as energy- and angular distributions of the backscattered atoms have been calculated analytically(41,42)and by computer simulation programs which are able to follow the individual ion trajectories (43 to 46). Experimental determination of backscattering yields and distributions are difficult, as in the energy range of interest most of the backscattered particles are neutral. The most successful way of ionising the neutrals is a gas filled stripping cell, as used also in neutral particle measurements at plasma experiments (47). Such a cell could be calibrated to energies as low as 130 eV (48).

The results for total backscattering yields and backscattered energy of hydrogen ions on different materials are shown in fig.6 (41 to 44, 49 to 51). The values are plotted as a function of a dimensionless universal energy ε given by (52)

$$\varepsilon = \frac{M_2}{M_1+M_2} \frac{a}{Z_1Z_2e_o^2} E$$

where M_1, Z_1 and M_2, Z_2 are the masses and charge numbers of the incident ions and the target atoms, e_o is the elementary charge (e_o^2 = 14.39 eV Å);'a' is the Thomas Fermi screening length given by: $a = 0.486\ (Z_1^{2/3} + Z_2^{2/3})^{-1/2}$ and E is the incident energy in eV. In the two lower scales of Fig.6 also absolute energies for stainless steel (Fe) and Mo are introduced. The measured numbers show only for some materials reasonable agreement with the calculated values.

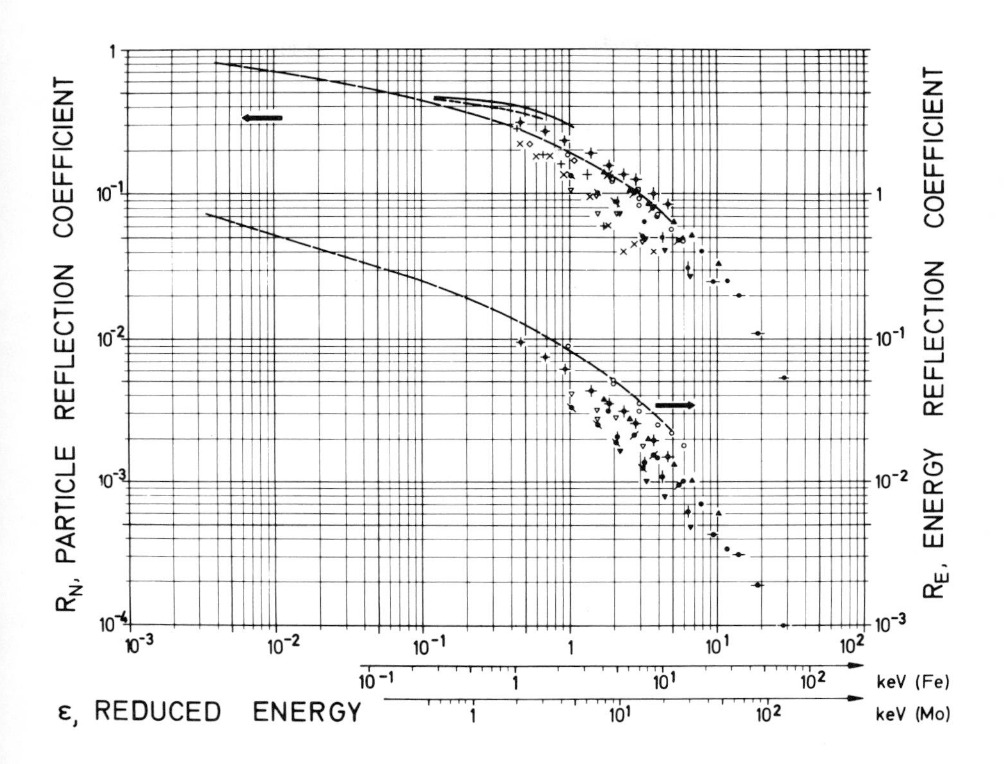

FIG.6. Particle and energy reflection coefficients as determined by different authors: ○ *H→SS,* □ *D→SS,* ▽ *H→Nb (annealed)* ▽ *H→Nb (bef.ann.),* ◇ *D→Nb (bef.ann.)(ref.49),* ● *H→SS,* ▼ *H→Nb,* ▲ *H→Cu,* -●- *H→Al,* ⧫ *H→Mo,* ● *H→Ag,* ● *H→Ta,* ⧫ *H→Au (ref.50),* + *H→Zr,* × *H→Ti (ref.51),* ——— *ref 42,* — — — *ref 43,44,* - - - - - *ref.41.*

Fig.7 shows an energy distribution of the backscattered positive, negative and neutral atoms (49). We see that more than 90% of the backscattered atoms are neutral in this low energy range, while negative and positive ions have nearly equal intensities. All spectra show that the backscattered particles may have an energy up to nearly the incident energy, while a maximum is seen at around 1 to 2 keV. The position of this maximum does not depend strongly on the incident energy. This means for lower bombarding energies the maximum is shifted more close to the incident energy (46).

An example of an angular distribution of the atoms backscattered into different energy ranges for normal incidence is shown in Fig.8 (53). While those atoms backscattered with low energy show nearly a cosine distribution, the high energy particles show predominant backscattering at glancing angles (80° to the normal).

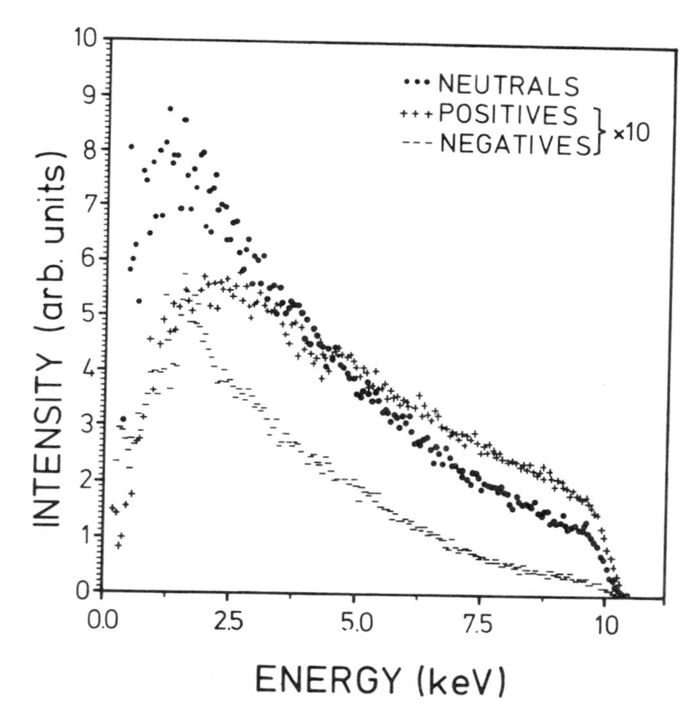

FIG. 7. Energy distribution of the positive and negative and neutral atoms backscattered from a stainless steel (304) surface for 10 keV D bombardment. The ion beam was at normal incidence and the backscattered particles had been observed at an angle of 45 deg. to the normal (ref. 49).

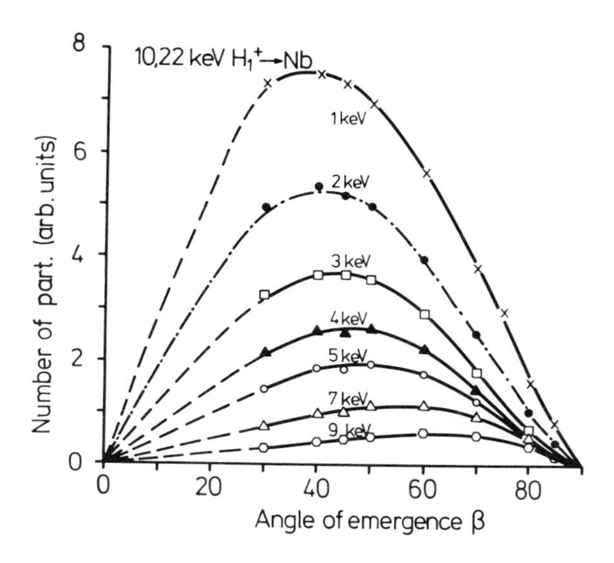

FIG.8. Angular distribution of protons backscattered from a polycrystalline Nb target into different energy ranges (ref.53).

The implication of these results to a fusion reactor are: Part of the ions leaving the plasma are directly backscattered from the first wall into the plasma as neutrals with energies up to close the incident energy. They may penetrate deep into the plasma and create more energetic charge exchange particles from the plasma than neutrals released with thermal energy from the first wall.

C. *Trapping and Reemission*

Those ions, not directly backscattered come to rest in the solid. After slowing down they generally occupy interstitial positions and may diffuse further. Depending on the solubility, the diffusibility, and the barrier at the surface they may either diffuse into the bulk of the solid, they may partly leave the surface, or they may be trapped in the implanted layer, generally at damage sides (54,55).

1. High Solubility

In the case of a high solubility as for hydrogen in titanium and zirconium all ions coming to rest in the solid are trapped if the temperature is high enough for good diffusion, but low enough so that thermal desorption is negligible (51,55, 56). This is shown in Fig.9 for different bombarding energies at normal incidence on room temperature targets (51). The measurements have been performed by weight gain at ion doses of $3\ 10^{19}$ to $5\cdot 10^{20}\ cm^{-2}$. In order to correct for the sputtered material this was also measured by collection on Al probes around the targets, which were subsequently analysed by Rutherford backscattering.

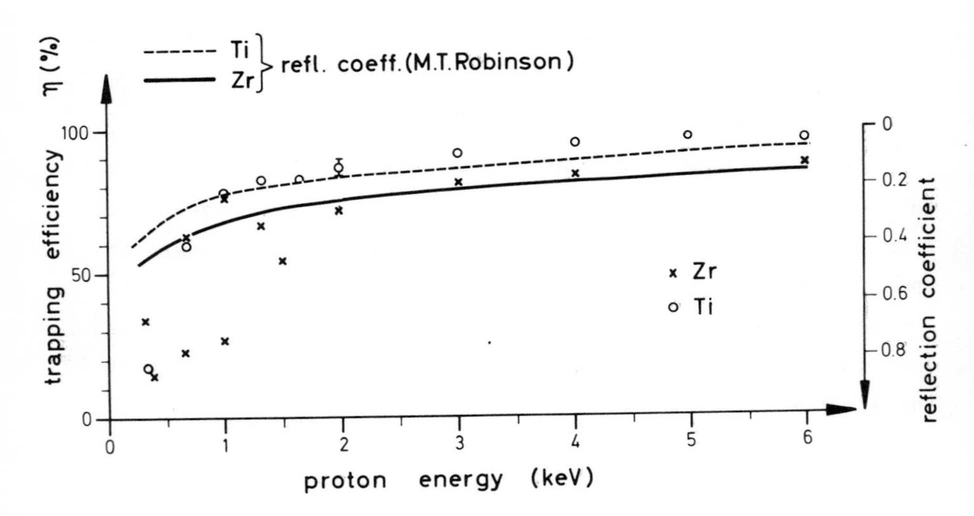

FIG. 9. Trapping efficiency for 0.3 to 6 keV protons bombarding Zr and Ti surfaces (ref.51).

As some hydrogen atoms are directly backscattered, trapping is always smaller than one. Backscattering yields determined from these measurements are in good agreement with calculated values (see Fig.6). The scatter in the points below 1.5 keV is attributed to an oxide layer on the surface which may have a different thickness for different targets used. As H does not diffuse well in the oxide trapping is decreased if a major part of the incident gas ions is stopped in the oxide layer.

2) Low Solubility, Low Dose

If the solubility of the incident gas ions in the solids is low, as for He bombardment of most metals and for hydrogen bombardment of several metals as for example stainless steel or molybdenum, the ions come to rest at interstitial positions in the solid and they will diffuse until they become trapped relatively stable at damage sites, predominantly vacancies (54). An example of depth profiles of such a nonsoluble gas in a metal is shown in Fig.10 for ^{3}He implanted into polycristalline Nb (57). The profiles have been obtained using the $^{3}He(d,p)^{4}He$ nuclear reaction. However, also several other techniques are possible for such measurements (58, 59, 60, 61).

The implantation profile follows within the statistics of the measurements mostly the calculated distributions.However, it cannot be concluded from these measurements whether the He is trapped at the damage distribution or at the range distribution as the calculated distributions for both are relatively close together.

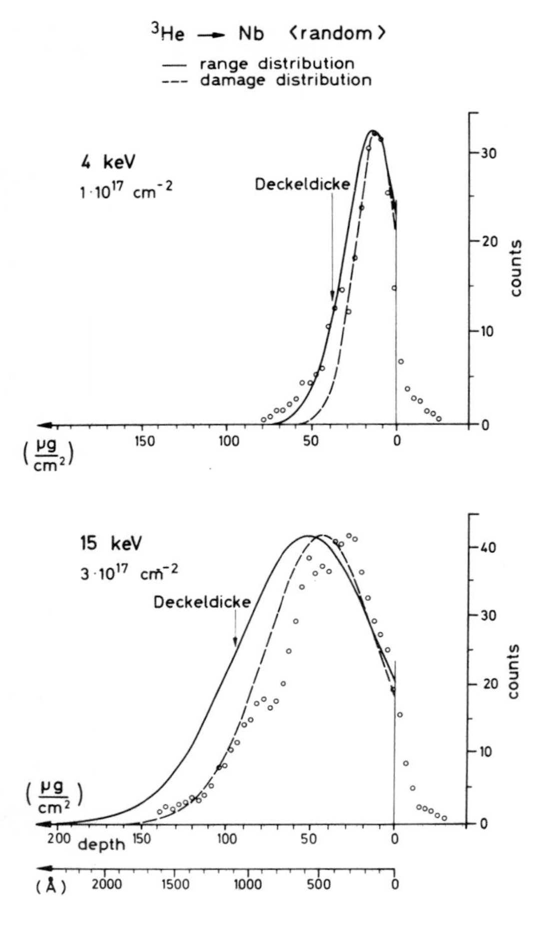

3. Low Solubility, High Dose

If the bombardment dose is increased the implanted layer becomes finally saturated, in the case of He on room temperature Nb at about 50 atomic percent. After saturation, additionally to the reflection, reemission starts and trapping becomes finally zero (55, 62). This is shown again for the case of ^{3}He on Nb in Fig.11 and Fig.12 (57, 62,63). As the range of the ions increases with the incident energy, the total amount of trapped gas also increases with energy.

FIG.10. Depth profiles of ^{3}He implanted into niobium at room temperature. The points are measured, while the lines represent the results of analytical calculations (ref.57).

The maximum concentration reached within the range of the ions is independent on the implantation energy, but depends on the target temperatures. The gas emitted after saturation is presumably released with an energy corresponding to the temperature of the first wall.

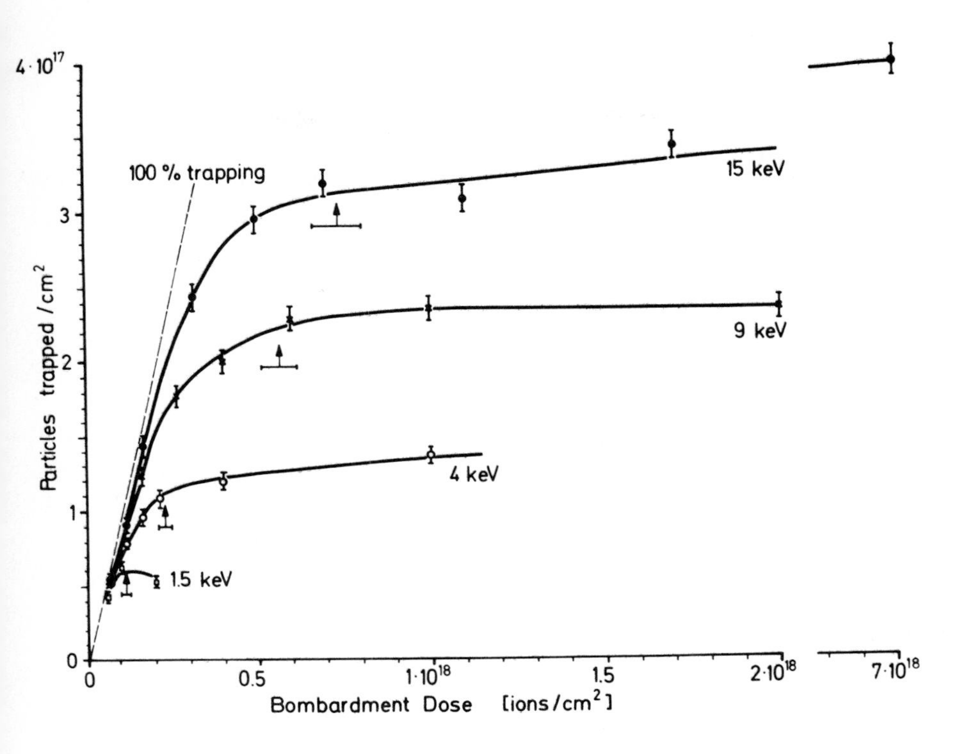

FIG.11. Trapping of ^{3}He in niobium at room temperature as a function of the bombarding dose at different energies (ref.62).

The consequences for fusion reactors are:

a) high concentrations of gas can be built up in the surface layers of the first wall. The total amount trapped in the first wall may be comparable to the amount of gas in the plasma.

b) Reemission will change with bombardment dose and target temperature. However, in any steady state operation of a fusion reactor for reemission and reflection a factor one may finally be assumed (except at the holes in the CTR vessel).

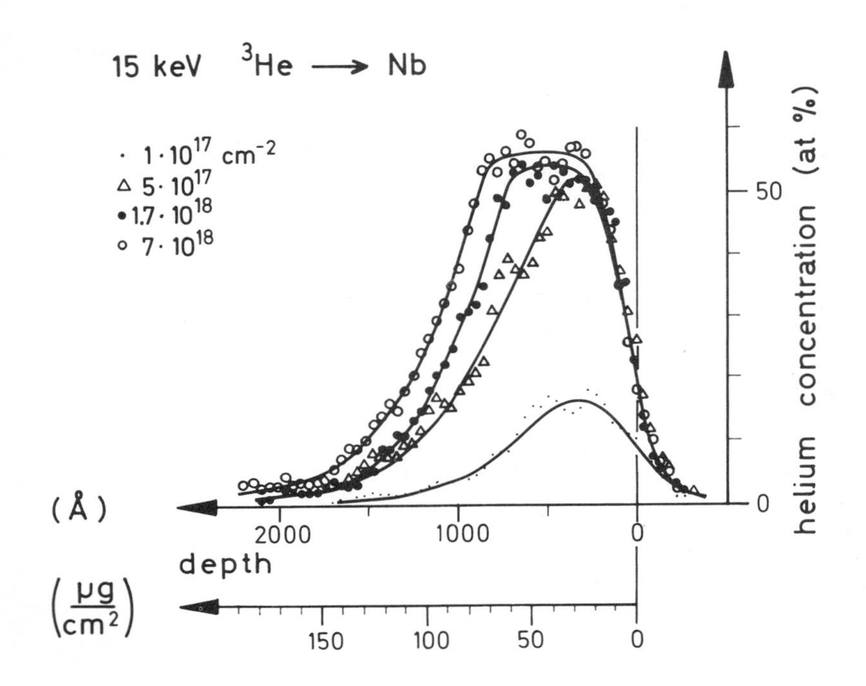

FIG.12. Implantation profile of ^{3}He in niobium for different bombardment doses (ref.57, 63).

D. Bubbles and Blistering

(very High Implantation Doses, Saturation)

If the incident gas ions are not soluble in the material, transmission electron microscope observations have shown that at bombarding doses of $\gtrsim 10^{16}/\text{cm}^2$ corresponding to 5 - 10 atomic percent gas injected into the solid, the gas starts to coalesce to small bubbles of 10 to 30 Å diameter inside the implanted layer (63,64).

If the ion bombardment is increased further, blisters, i.e. bending up of a surface layer can be observed (38,39,55,63-69). The critical doses for the appearance of blisters lie between

10^{17} to 10^{18} ions/cm^2. They depend on the bombarding energy, the current density and angular and energy spread of the bombardment, as well as the materials and its temperature (67 to 70).

Blistering and the blistering mechanism has been widely investigated within the last years, because it had been regarded as the most dangerous source for first wall erosion in fusion reactors (38,39,63). It was found that the bending-up of surface layers is basically caused by the release of lateral stress produced in the implanted surface layer (57). The pressure in the gas bubbles formed by the injected gas in the solid also contributes, but is not the major driving mechanism. Thus, blisters break up at a depth corresponding to the total range of the injected gas and not at the mean projected range where the maximum gas concentration occurs.

This blistering mechanism has lead to the conclusion that blistering may be a transient phenomenon which may disappear if the surface adapts to the ion bombardment by forming a structure in which stress can no more build up (14,71,72). In fact, such a new surface structure develops if more than one deckeldikke (= thickness of the covers of the blisters) is sputtered away. Further, blistering is considerably reduced if the bombardment is performed at a broad energy distribution and/or a broad angular distribution of the incident ions (71,72). The doses needed to observe the different surface structures for normal incidence are seen in Fig.13 again for helium bombardment of niobium (72). The figures 14-16 show some surface structures of stainless steel and niobium with blisters and after disappearance of the blisters as observed in the scanning electron microscope. A picture for high temperature blistering is also included. The temperature can largely modify the blister appearance (63,67,68,69,70,72,74). At medium temperatures

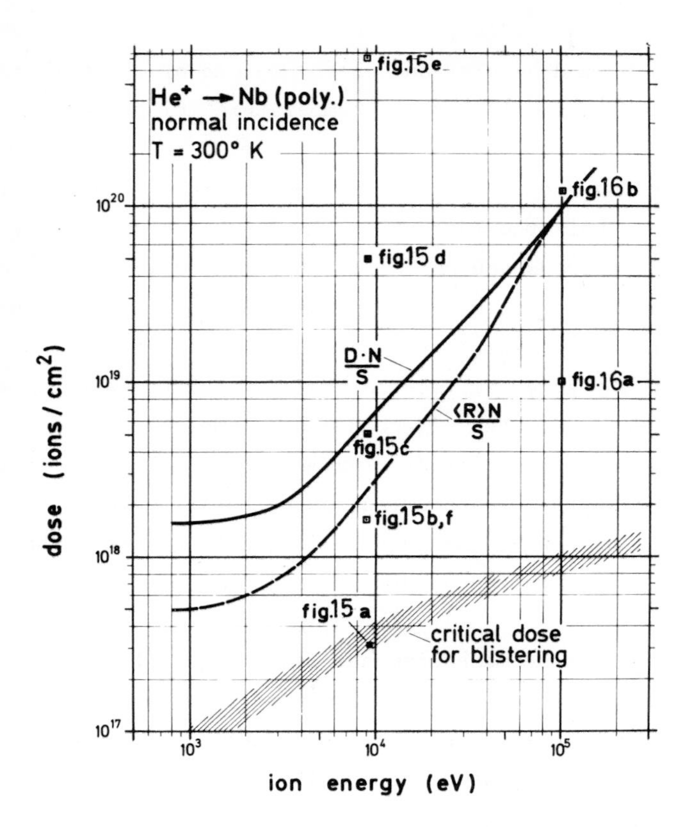

FIG. 13. Critical doses for blistering and sputtering away one deckeldicke (ref. 72).

(500 to 800°C) blistering generally is increased and large exfoliation of the blister covers is observed. At a further increase in temperature blistering is decreased and an equilibrium surface structure comparable to Fig.13.b is observed at lower bombarding doses.

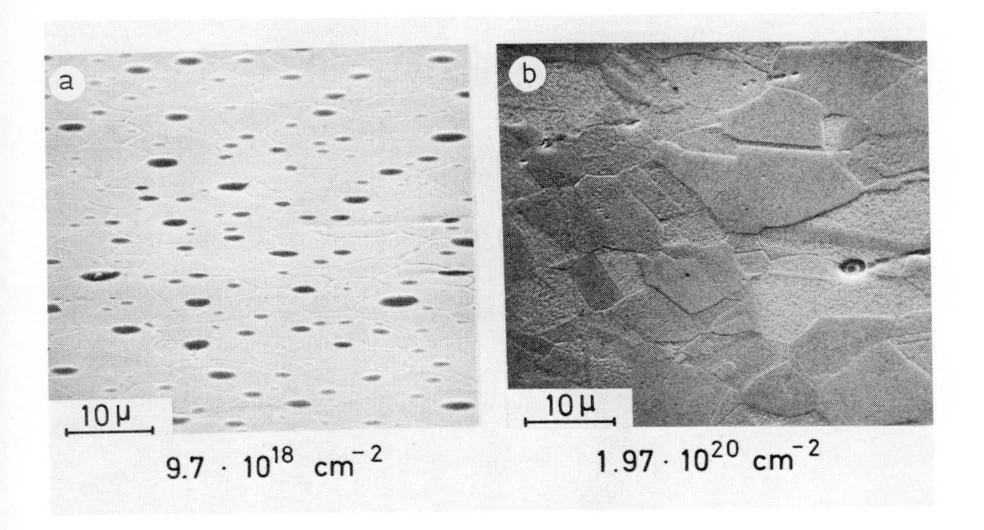

FIG.14. Surfaces of 304 stainless steel bombarded with different doses of 7.5 keV hydrogen ions at room temperature and normal incidence. Blistering is observed only at low bombarding doses (a). At higher doses the surface becomes rough and the differently oriented grains are eroded at different rates (ref.14 and 73).

The consequences for a fusion reactor are:

Blistering may possibly contribute to wall erosion in a fusion reactor only at the beginning of the operation, as it is a transient phenomenon which disappears at sufficiently high doses.

The angular and energy distribution of the ions bombarding the first wall may prevent blistering from the beginning.

Due to the increased wall temperature an equilibrium structure at the surface may form already after the first few discharges.

Finally, the sputtering yields measured by weight loss generally have been obtained with doses well above the dose for blisters and thus include blistering.

The spongelike structure found after high dose ion bombardment

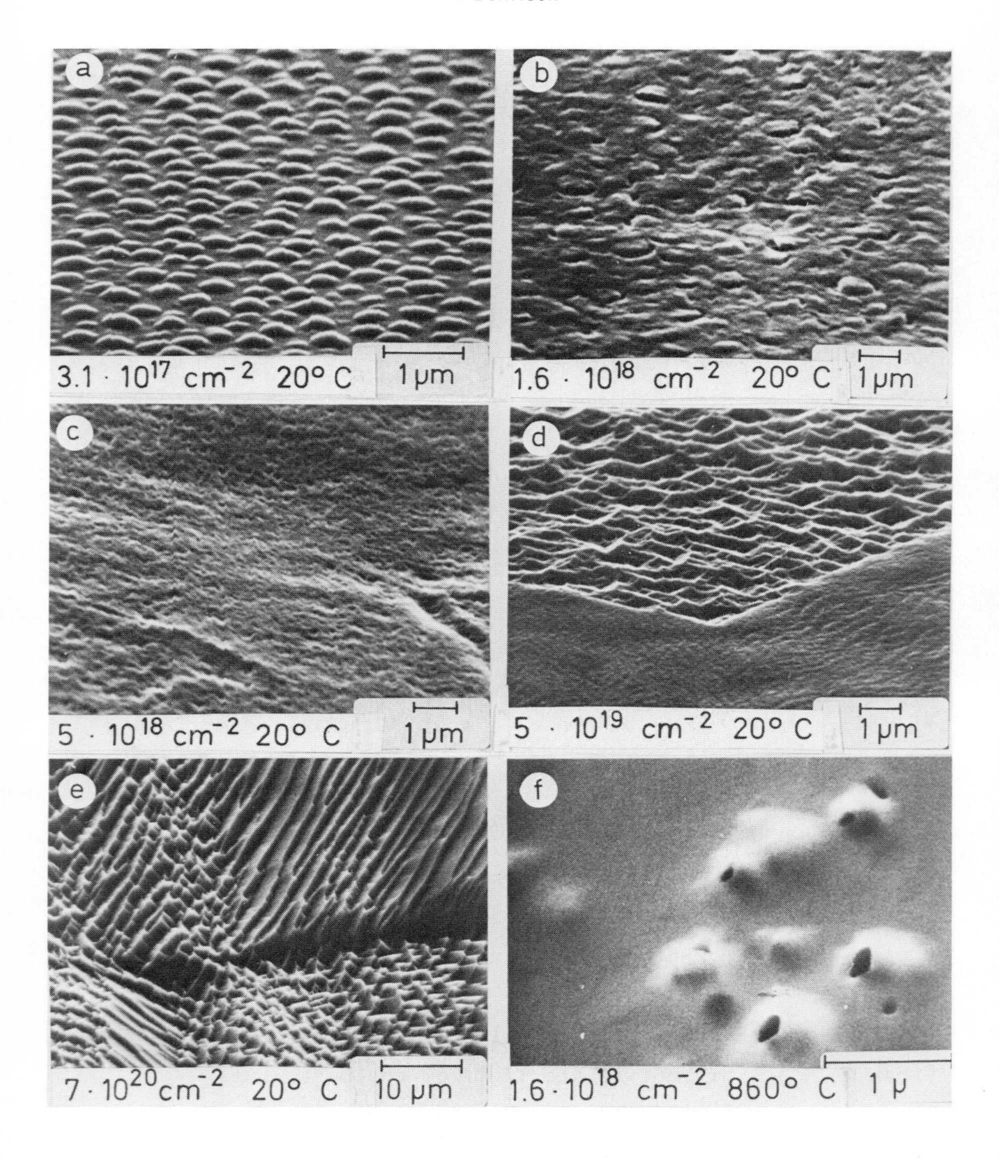

FIG.15. Surface of polycrystalline Nb bombarded with increasing doses of 9 keV He-ions at normal incidence (see Fig. 10). It can be seen that blistering is a transient phenomenon disappearing at high doses (ref.72). Fig.12 e shows a blister kind surface structure at 860° bombardment (74).

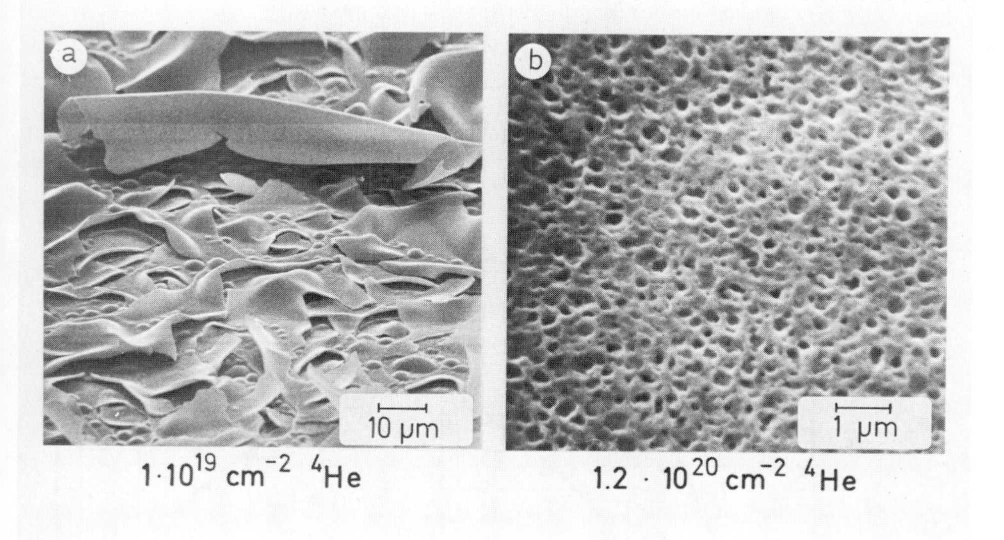

FIG. 16. Surfaces of polycrystalline Nb at room temperature bombarded with different doses of 100 keV He at normal incidence. Blisters disappear after about one deckeldicke has been removed by sputtering (ref. 72).

will have different mechanical, thermal and electrical properties compared to the unbombarded material, which has to be taken into account in designing a fusion reactor.

E. Sputtering

1) Physical Sputtering

Physical sputtering is the removal of surface atoms from a solid via a collision cascade initiated in the near surface region by incident energetic particles as ions, neutrons or electrons. Thus sputtering can be regarded as radiation damage in the surface. During the spread of the cascade the surface stays cold contrary to surface erosion by evaporation. Only in very energetic and dense (high Z target materials) cascades a thermal spike may be found (75). This may be occur in 14 MeV neutron bombardment of high Z materials.

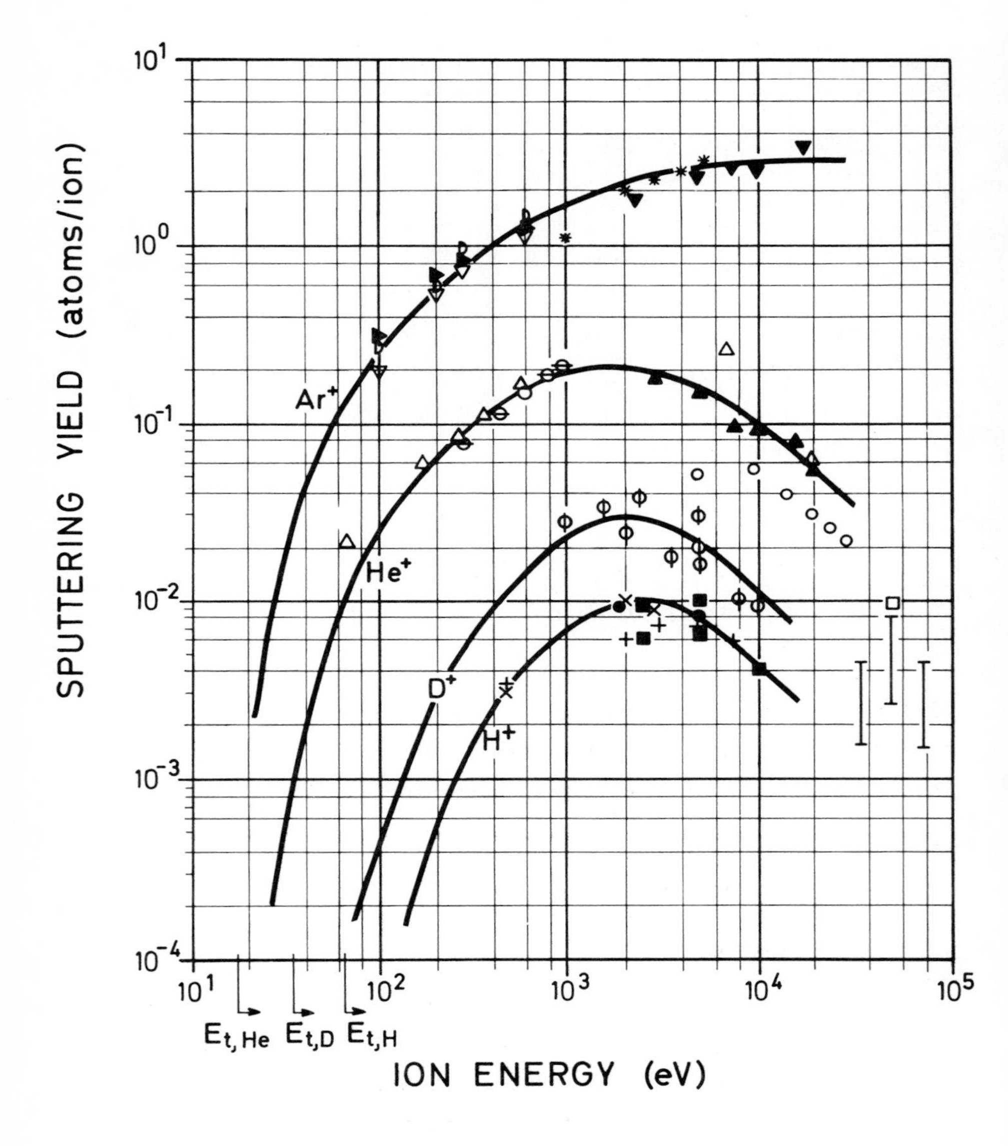

FIG. 17. Sputtering yields of stainless steel, iron and nickel, measured by different methods (see ref. 82).

The sputtering yields (atoms per incident particles) are proportional to the energy deposited by the incident particles in the surface layer in nuclear motion and inversely proportional to the surface binding energy (76). These yields have been investigated since more than 100 years (76 to 81), however, nearly no data is available for the parameters of interest in fusion research (10,14). This has basically two reasons: The yields for low energy light ion bombardments are low and intense ion beams are difficult to obtain. In Fig.17 the sputtering yields measured for normal incidence of different ions on stainless steel, nickel and iron are summarized (82). They have been obtained partly by weight loss and partly by measuring the decrease of a thin film by Rutherford backscattering. Below a threshold energy, E_t of 10 to 100 eV no sputtering occurs (83). Yields then increase with energy to a maximum which occurs for hydrogen at an energy of 1 to 10 keV. The decrease at a further increase bombarding energy is due to the decrease of the deposited energy in the surface region. Fig.18 shows some very recent sputtering yields of low Z materials which have been discussed as first wall materials in the last two years (84). For these materials the maximum in the sputtering is considerably shifted to lower energies.

The dependence of the sputtering yields on the angle of incidence has hardly been investigated. The few measurements performed have confirmed the theoretically expected increase of the yields with $(\cos\vartheta)^{-f}$ where $1 < f < 2$ and ϑ the angle of incidence with respect to the normal (10).

The atoms removed from the surface are predominantly neutral. Their energy and angular distribution has not yet been measured for the parameters of interest. It is expected that the mean energy will be low(1 eV to 10 eV) and that the atoms will be emitted in a nearly cosine distribution (77 to 81).

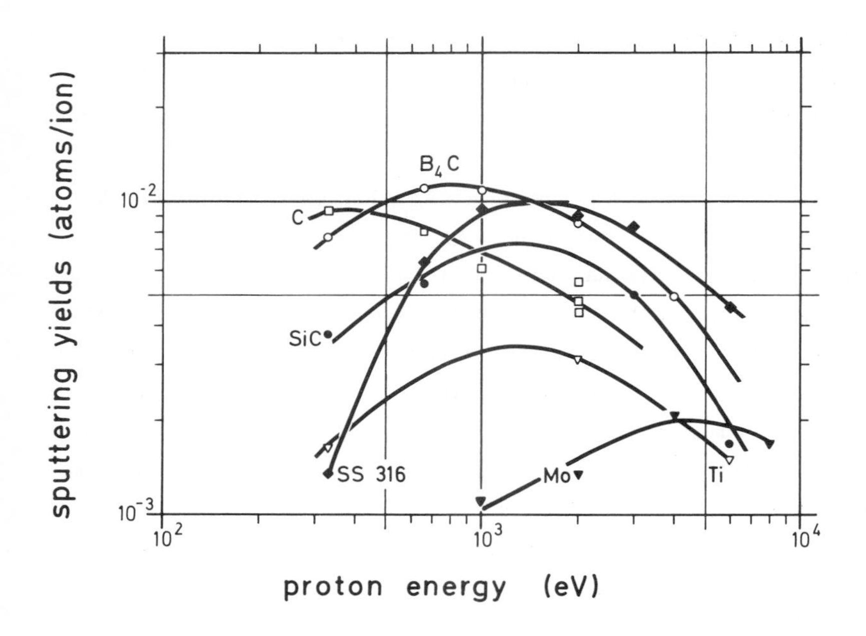

FIG. 18. Sputtering yields of different low Z materials (ref. 84).

Results on neutron sputtering yields have achieved great attention due to extremely high yields of 0.3 atoms/neutron found in some experiments (85). However, the latest more careful investigation at several places have led to the conclusion that neutron sputtering yields are below 10^{-4} atoms/neutron in agreement with theoretical prediction (10, 14, 86).

2) Chemical Sputtering

If the bombarding ions can form a volatile compound with the target material as for hydrogen bombardment of carbon, much larger yields than expected from the collisional theory may occur. At room temperature bombardment of carbon with hydrogen only small yields and no CH_4 formation could be measured. At increased target temperature however, CH_4 is found and simultaneously a large increase in sputtering.This is shown in Fig.19

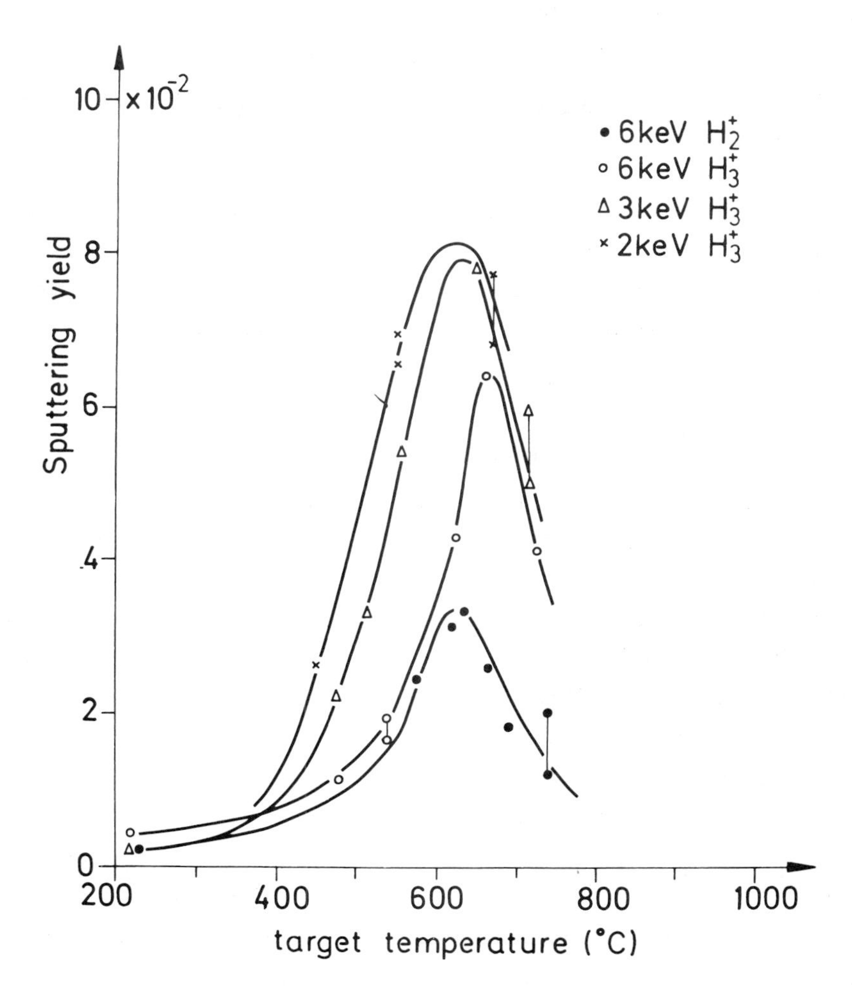

FIG. 19. Temperature dependence of the sputtering yield for pyrolytic graphite for hydrogen bombardment at different energies (from ref. 88).

for different bombarding energies at normal incidence (88). The steep increase in the yields at 400 to 500°C has been explained by increased mobility of H on the surface, so that CH_4 can be formed. The decrease of the yields at temperatures above 700°C can be understood by increased out diffusion of hydrogen so that the probability for the formation of CH_4 decreases.

Similar effects have been found for silicon carbide and boron carbide. However, in these cases the surfaces seem to become depleted of carbon after some dose, and chemical sputtering disappears (89).

Finally, in all compounds and alloys a composition change due to preferential sputtering by ion bombardment is expected and has partly been observed. However, systematic investigations for the parameters of interest in respect to CTR have not yet been performed.

The consequences for a fusion reactor are:

Sputtering at the first wall of a fusion reactor cannot be avoided except if the wall bombardment takes place at very low energies ($E < 10$ eV).
During the start of a fusion reactor also sputtering of a sorbed layer (ion desorption) may be a large source for impurity introduction in the plasma (90).
Low Z materials show exceptionally high yields at low energies which may compensate the advantage of a higher concentration of low Z materials in the plasma.
The yields for light ions are generally below 3×10^{-2} atoms/ion, while they are of the order of one for heavy ions. Thus heavy ions once introduced and heated in the plasma, which will come back to the first wall with energies above 100 eV, will cause considerable sputtering. Even if the impurity concentration is low, this sputtering may be higher than sputtering by hydrogen atoms and will cause a further fast increase of the impurities in the plasma.

Neutron sputtering yields are low. As neutron fluxes to the first wall will be much lower than the ion-and neutral fluxes the contribution of neutron sputtering to first wall erosion can be neglected (10,91).

F. *Evaporation*

1) Steady State of Evaporation

The evaporation rate $\dot{n}$(atoms/cm^{2}sec) of atoms from a surface heated in equilibrium to a temperature T is given by

$$\dot{n}(T) = 3.5 \cdot 10^{20} \frac{p[\text{torr}]}{\sqrt{M \cdot T[K]}} \left[\frac{\text{atoms}}{\text{cm}^2\text{sec}}\right]$$

where p is the vapour pressure of the material and M its atomic mass. Vapour pressures and evaporation rates for different materials of interest are given in Fig.20 (92,93). If the impurity introduction by evaporation should be smaller than by sputtering this gives a maximum first wall temperature of about 850°C for stainless steel while for the refractory metals the maximum operation temperature ranges up to 2000°C. However, small contents of impurities in the material as oxygen can largely increase the evaporation yields.

2) Pulse Evaporation

It may be difficult in a fusion reactor to provide that the energy deposited on the first wall will be uniform in space and time. A local increased energy deposition will cause an increased surface temperature and thus increased evaporation. If the amount of energy E is coming to a first wall area A in the form of a heat pulse of duration τ, the maximum increase in wall temperature ΔT will depend on the density ρ, the thermal conductivity k and the specific heat c_p of the solid and is given by (10):

$$\Delta T = \frac{E}{A} \frac{1}{\sqrt{\tau}} \sqrt{\frac{4}{\pi c_p k \rho}}$$

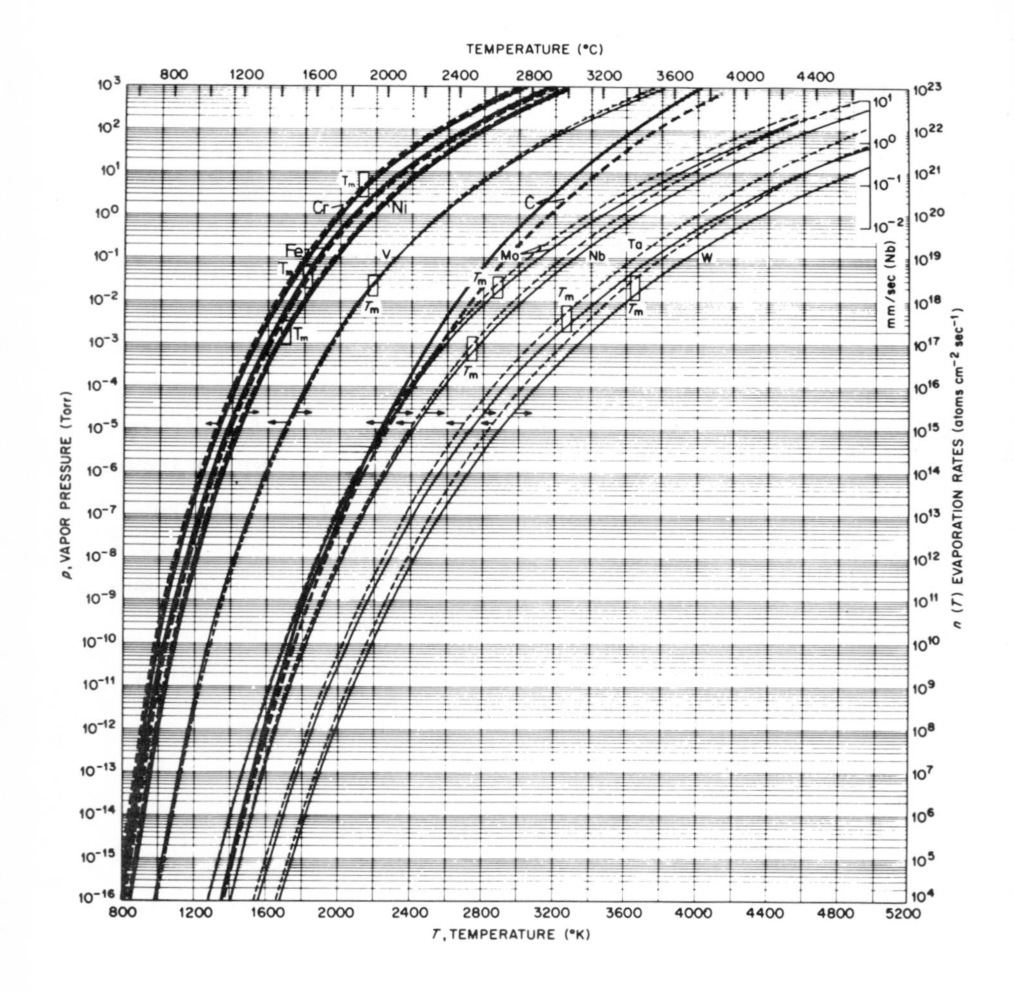

FIG. 20. Vapour pressure and evaporation rates for different first wall materials (ref. 10).

Though the temperature increases only with the square root of the deposition time, small values of τ as they may occur in a disruptive instability can lead locally to extremely high wall temperatures (10). These temperature pulses firstly cause a pulse evaporation, but may also very efficiently crack brittle materials.

Pulse evaporation yields have been estimated for Niobium (10) to be about

$$n \sim 0.2 \, \tau \, \dot{n}(T_{max})$$

which are generally small if T_{max} is not too high. No measured values seem to be available.

The consequences for a fusion reactor are:

As the surface temperature of the first wall is given by the power density from the plasma, the first wall thickness and the maximum possible cooling at the rear side, this imposes a very stringent limit on reactor design (94).
Pulse evaporation can be made small only if $c_p \cdot k \cdot \rho$ are large as for refractory materials and τ is not too short.

IV. CONCLUSIONS

An attempt has been made to review some of the relevant elementary atomic processes which are expected to be of significance to the first wall of a fusion reactor. Up to the present, most investigations have been performed at relatively high ion energies, typically E >5 keV and even in this range the available data are very poor.
If the plasma wall interaction takes place at energies of $E \gtrsim 1$ keV the impurity introduction and first wall erosion which will take place predominantly by sputtering, will be large and may severely limit the burning time of the plasma. The wall bombardment and surface erosion will presumably not decrease substantially by introducing a divertor.

The erosion can only be kept low, if the energy of the bombarding ions and neutrals can be kept below the threshold for sputtering of 1 to 10 eV.

It has not yet been shown that such a cold gas blanket can be achieved experimentally. In the cold gas blanket, energy accommodation factors for the DT gas at the first wall are very essential. In addition more studies are needed to examine the probability of obtaining a cold gas blanket in plasma devices.

ACKNOWLEDGEMENTS

Most of the material presented in this review had been extensively discussed with my colleagues, especially J.Bohdansky, W.Eckstein, J.Roth, B.M.U.Scherzer, H.Verbeek and H.Vernickel. Mrs.Polster has kindly provided the free hand drawing of the schematic of the tokamak and Mrs.Daube has typed several versions of the manuscript with great care.
They are all kindly acknowledged.

REFERENCES

In the following only the latest publications and some key references are listed. A complete list of all related publications would make up several hundred references.

First References

1. J.D.Lawson, Proc.Phys.Soc. (London) B70, 6 (1957).
2. D.M.Meade, Nucl.Fusion, 14, 289 (1974).
3. L.A.Artsimovich, Plasma physics and Controlled Nucl. Fus. Research, Proc.3rd Int.Conf.Novosibirsk, 1968, IAEA Vienna, 157 (1969).
4. L.A.Artsimovich, Nucl.Fusion,12,215 (1972).
5. H.P.Furth, Nucl.Fusion, 15, 487 (1975).
6. D.Steiner, Nucl.Sci. and Eng., 58, 107 (1975).
7. Fusion Reactor Design Problems, Proc.of an IAEA Workshop, Culham 1974, Nucl.Fusion special supplement 1974.
8. E.Hinnov, J.Nucl.Mat., 53, 16 (1974).
9. M.Kaminsky, IEEE Trans.Nucl.Sci., NS 18, 208 (1971).
10. R.Behrisch, Nucl.Fusion, 12,691 (1972).
11. H.Vernickel, Course on the Stationary and Quasistationary Toroidal Reactors, Erice 1972, EUR 4999e, p 303 (1973).
12. G.M.McCracken, Fusion Reactor Design Problems, Nucl.Fusion Special Suppl. p 471, (1974).
13. H.Vernickel, Proc. 1st Top.Meeting on the Techn.of Contr. Nucl.Fusion, San Diego, CONF 740402-P2, Vol.II, 347 (1974).
14. R.Behrisch and B.B.Kadomtsev, Plasma Physics and Contr. Nucl.Fusion Research, Proc. 5th Int.Conf. Tokyo 1974, IAEA, Vienna II, 229, (1975).
15. B.M.U.Scherzer, J.Vac.Sci.Techn. 13, 420 (1976).

16. B.Lehnert, Nucl.Instr.Meth., 129, 31 (1971) and references in this articles.

17. D.Steiner, Nucl.Appl.Techn., 9, 83, 1970.

18. D.Steiner, Nucl.Fusion, 14, 33, 1974.

19. G.L.Kulcinski, Plasma physics and Contr.Nucl.Fusion Research Proc. 5th Int.Conf.Tokyo, 1974, IAEA Vienna, 251,(1975).

20. D.Steiner, B.N.E.S. Nuclear Fusion Reactor Conference, p 483 (1969).

21. D.Rose, Nucl.Fusion, 9, 183 (1969) (older refrences are found in this article).

22. E.Hinnov, PPPL, Matt-77 7 (1970).

23. D.Eckhartt, E.Venus, JET Techn.Note, 9 (1974).

24. H.Husan, K.Bol, R.A.Ellis, PPPL, Matt.-1098 (1975).

25. P.Ginot (TFR Group), Proc.Int.Conf.on Surface Effects in Contr.Fusion Devices, San Francisco 1976, J.Nucl.Mat.(1976).

26. P.H.Rebut, R.Dei-Cas, P.Ginot, J.P.Girard, M.Huguet, P.Lecoustey, P.Moriette, Z.Sledziewski, J.Tachon and A.Torossian, J.Nucl.Mat., 53, 16 (1974).

27. J.F.Crastron, R.Hancox, A.F.Hobsen, S.Kaufmann, H.T.Miles, A.A.Wave, J.A.Wesson, 2nd Int.Conf.Peaceful Uses Atomic Energy (Proc.Conf.Geneva 1958) 32 U,N, 414 (1958).

28. R.Behrisch, W.Eckstein, P.Meischner, B.M.U.Scherzer, H.Verbeek, Proc.5th Int.Conf.Atomic Coll. in Solids, Gatlinburg, Vol.I, 315 (1973).

29. L.Spitzer, Phys.Fluids,1, 253 (1958).

30. F.H.Tenney, A short course in Fusion Power, Princeton University 1972.

31. R.G.Mills, F.H.Tenney et al., Princeton Reference Design Reactor, Princeton 1974.

32. B.Badger et al., UWMAK, Wisconsin Tokamak Reactor Design, UWFDM-68, Nov.1973.

33. C.Calven, A.Gibson, P.E.Stott, Proc.5th Europ.Conf.Contr. Fusion and Plasma Physics, Grenoble p.6 (1972).

34. R.Allgeyer et al.Proc.6th Symp. on Eng.Probl.of Fusion Research, San Diego 1975.

35. D.M.Meade, Joint European-US Workshop on large Tokamak Designs, Culham (1974).

36. D.Düchs, G.Haas, D.Pfirsch, H.Vernickel, J.Nucl.Mat.53,102, (1974).

37. S.E.Lysenko, G.N.Popkov, see ref.7.

Second Talk

The elementary processes at the first wall have been discussed in some detail at two recent international conferences (38 and 39).

38. Proc. of the Conference on Surface Effects in Controlled Thermonuclear Fusion Devices and Reactors, Argonne, Ill. Jan.1974, J.Nucl.Mat., Vol.53 (1974).

39. Proc.of the Conference on Surface Effects in Controlled Thermonuclear Fusion Devices and Reactors, San Francisco, Cal., Febr.1976, J.Nucl.Mat., (1976).

40. N.Bohr, Mat.Fys.Medd.18, No 8 (1948).

41. R.Weißmann and P.Sigmund, Rad.Eff., 20, 65 (1973).

42. J.Bøttiger and K.B.Winterbon, Rad.Eff.20, 65, (1973).

43. M.T.Robinson, Proc.3rd Nat.Conf.Atomic Coll.with Solids, Kiew 1954.

44. O.S.Oen, M.T.Robinson, Nucl.Instr.and Meth.132.

45. J.E.Robinson, Rad.Effects,23, 29, 1974, 677 (1976) (see also ref.39).

46. T.Ishitani, R.Shimizu, K.Murata, Jap.J.Appl.Phys.11, 125 (1972).

47. C.F.Barnett and J.A.Ray, Nucl.Fusion 12, 65 (1972).

48. F.E.P.Matschke, W.Eckstein, H.Verbeek,Verh.DPG IV, 10,47 (1975) and to be publ.

49. W.Eckstein, F.E.P.Matschke, H.Verbeek, (see ref.39).

50. G.Sidenius, Phys.Lett.49 A, 409 (1974), and Nucl.Instr. Meth., 132, 673 (1976).

51. J.Bohdansky, J.Roth, M.K.Sinha, W.Ottenberger, see ref.39 and Proc.9th Symp.Fus.Techn.,Garmisch, June (1976).

52. J.Lindhard, M.Scharff, H.Schiøtt, Mat.Fys.Medd.33, No14, (1963).

53. H.Verbeek, J.Appl.Phys.46, 2981 (1975).

54. E.Kornelsen, Rad.Eff.,13, 227 (1972).

55. G.McCracken, Rep.Progr.Phys.38, 241, (1975).

56. O.C.Yonts, R.A.Strehlow, J.Appl.Phys.33, 2902 (1962)

57. R.Behrisch, J.Bøttiger, W.Eckstein, U.Littmark, J.Roth and B.M.U.Scherzer, Appl.Phys.Lett.27, 199 (1975).

58. R.S.Blewer, Appl.Phys.Lett.53, 593, 1973.

59. D.A.Leich, T.A.Tombrello, Nucl.Instr.Meth.108, 67 (1973).

60. J.Bøttiger, S.T.Picraux, N.Rud.Ion Beam Surface Layer Analysis, p.811, (1975).

61. J.Roth, R.Behrisch, B.M.U.Scherzer, Appl.Phys.Lett.25,643, (1974).

62. R.Behrisch, J.Bøttiger, W.Eckstein, J.Roth and B.M.U.Scherzer, J.Nucl.Mat.56, 365 (1975).

63. J.Roth,Application of Ion beams to Materials, Inst.of Physics, Conf.Series Number 28, p.280 (1975).

64. G.McCracken, see ref.7 and private comm.

65. W.Primack, J.Appl.Phys.34, 3630 1963.

66. M.S.Kaminsky, Adv.Mass.Spectrom.3, 69 (1964).

67. S.K.Erents, G.M.McCracken, Rad.Eff.18, 191 (1973).

68. M.S.Kaminsky, S.K.Das, Rad.Eff.18, 245, 1973.

69. W.Bauer, G.J.Thomas, Proc.Int.Conf.on Defects and Defect Clusters in bcc Metals and their Alloys, p 255 (1973).

70. J.Roth, R.Behrisch, B.M.U.Scherzer, M.Risch, J.Roth, J.Nucl.Mat.53, 147 (1974).

71. J.Roth, R.Behrisch, B.M.U.Scherzer, J.Nucl.Mat.57, 365, (1975).

72. R.Behrisch, B.M.U.Scherzer, M.Risch, J.Roth, Proc. 9th Symp.Fusion Techn., Garmisch, June (1976).

73. R.Behrisch, J.Bohdansky, G.H.Oetjen, J.Roth,G.Schilling, H.Verbeek, J.Nucl.Mat. 60,321 (1976).

74. J.Roth, T.Picraux, W.Eckstein, J.Bøttiger, R.Behrisch, Ref.39.

75. P.Sigmund, Appl.Phys.Lett.25,169 (1974).

76. P.Sigmund, Phys.Rev. 184, 383 (1969).

77. G.K.Wehner, Ad.Electr.Ectro.Phys.7, 239 (1955).

78. R.Behrisch, Erg.exakt Nature 35, 295, 1964.

79. M.Kaminsky, Atomic and Ionic Impact Phenomena on Metal Surfaces, Academic Press, Springer 1965.

80. G.Carter, J.S. Colligon, Ion Bombardment of Solids, Elvere Publ.Comp. 1968.

81. N.W.Pleshivtsev, Cathode Sputtering, Atomisdat., Moskau, 1968.

82. H.v.Seefeld, H.Schmidl, R.Behrisch, B.M.U.Scherzer, see ref.39.

83. E.Hotstron, Nucl.Fusion, 15, 544 (1975).

84. J.Bohdansky, J.Roth, M.K.Sinha, Proc.9th Symp. Fusion Techn., Garmisch, 1976.

85. M.S.Kaminsky, J.Peavey, S.Das, Phys.Rev.Lett.32, 599,(1974).

86. R.Behrisch, Nucl.Instr.Meth.132, 293, 1976.

87. J.Roth, J.Bohdansky, W.Poschenrieder, M.K.Sinha, see ref.39.

88. J.Roth, J.Bohdansky, M.K.Sinha, private comm.

89. S.A.Cohen, see ref.39.

90. R.Behrisch, H.Vernickel, Proc.7th Symp.Fusion Techn., Grenoble 1972.

91. R.Hultgren, R.L.Orr, P.D.Andersen, K.K.Kelley, Selected Values of Thermodynamic Properties of Metals and Alloys, John Wiley (1963).

92. R.E.Honig, D.A.Kramer, R.C.A. Review, 30, 285 (1969).

93. A.Fraas, ORNL-TM-4999 (1975).

Chapter 4

ELECTRICAL INSULATORS FOR MAGNETICALLY CONFINED FUSION REACTORS*

Frank W. Clinard, Jr.

University of California
Los Alamos Scientific Laboratory
Los Alamos, N. M. 87545

* Work performed under the auspices of the USERDA.

I. INTRODUCTION

Electrical insulators are essential materials in all magnetically-confined fusion reactor concepts. The implosion-heated theta-pinch reactor requires an insulating liner on the first wall to stand off voltages generated during the heating stage of the D-T burn. The mirror machine needs electrical insulators in the neutral beam injector and direct conversion power generating systems. The Tokamak reactor also requires injector insulators, and may need an insulating ring in the torus to facilitate ohmic heating. Plasma contamination problems in Tokamaks may necessitate the use of a low-Z liner material such as graphite, carbide, ceramic, or glass on the first wall; in this case insulating properties are not needed. All of the above reactors will require magnetic coil insulators, and may use insulating material in the neutron blanket to reduce eddy current and liquid lithium pumping losses.

The environment for fusion reactor ceramics may include high temperatures, pulsed or steady mechanical and electrical stresses, hot D-T gases, liquid lithium, and intense irradiation (neutrons, ions, or photons). Because some of the environmental conditions in fusion reactors have never before been encountered (e.g., intense 14 MeV neutron irradiation), it is not always possible to predict with assurance the magnitude of problems to be expected. Indeed, simulation of not-yet-achieved operating conditions is in itself a challenging requirement.

Most CTR insulator problems can be conveniently divided into electrical, structural, and chemical effects; such a division is utilized in this chapter. An exception is the topic of insulators for magnet coils, which is treated separately. Research to date has been principally directed toward the theta-pinch first wall application, where insulator requirements and environmental conditions have been analyzed in some detail, and this is reflected in the references cited. However, the results of such studies are in many cases applicable to other CTR insulator uses as well.

II. *ELECTRICAL EFFECTS*

A. *High Temperature*

For several CTR applications, insulators must exhibit adequate electrical resistivity and dielectric strength at high temperatures. In the Reference Theta-Pinch Reactor (RTPR), first-wall temperature at the time of electrical stress is $\sim$1000 K, and voltage is applied in an $\sim$0.1 μsec pulse every 10 sec (1). Resistivity required is $\geq 10^3$ Ω-cm (2), and design dielectric strength for the 0.3 mm thick insulating liner is 100 kV/cm. Resistivity of a refractory insulator such as Al_2O_3 (the reference insulator for the RTPR) decreases rapidly with rising temperature as electrons and holes are thermally activated into conducting states. However, the magnitude of the decrease (from $\sim 10^{15}$ Ω-cm at room temperature to $\sim 10^9$ Ω-cm at 1000 K, depending on the form and purity of the material (3)) is insufficient to approach the minimum acceptable value. Thus in the theta-pinch first-wall application, thermally-induced degradation of resistivity should not be a problem.

The temperature-dependence of dielectric strength is a function of duration of applied voltage. Current theories (4) ascribe d.c. or "thermal" breakdown to a runaway condition in which conduction causes local Joule heating, which in turn results in a further decrease in resistivity. Thus d.c. dielectric strength, like resistivity, should decrease as temperature rises. However, if voltage is applied for times short compared with that required to cause heating, breakdown will occur only at higher voltages where a field-induced electron multiplication mechanism ("electronic" breakdown) can operate (2). Thermal effects are not strongly related to this breakdown mechanism, and pulsed-voltage dielectric strength should be approximately temperature-independent. This has been shown in studies of Al_2O_3 (Fig. 1), where breakdown strength is reduced significantly for d.c. voltages between room temperature and 1000 K, but is essentially independent of temperature for short-time voltage pulses.

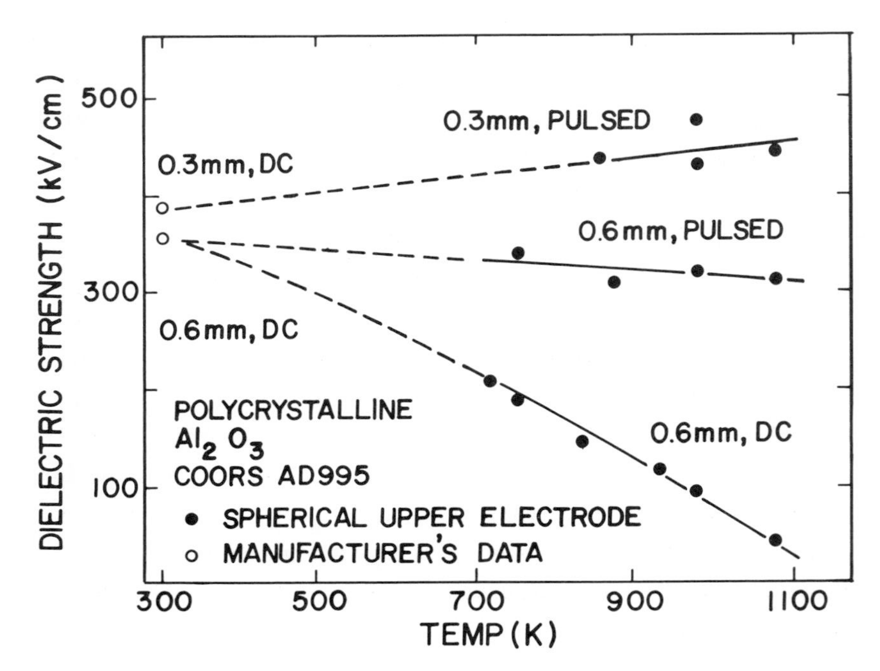

FIG. 1. Dielectric strength as a function of temperature under pulsed and d.c. voltage conditions, for a polycrystalline alumina in two sample thicknesses. [*From Bunch and Clinard (5)*].

The characteristic pulse length beyond which the d.c. breakdown mechanism is controlling depends on the insulator in question. For Al_2O_3 at 973 K, electronic breakdown extends to at least 10^{-2} sec, while for an enamel glass at 873 K, such breakdown extends only to $\sim 10^{-4}$ sec (5). It appears that for most insulators, the RTPR voltage pulse length of ~ 0.1 μsec is short enough to avoid thermal breakdown, so that room-temperature dielectric strength values should be at least roughly applicable for this elevated-temperature insulator usage. Therefore the dielectric strength of Al_2O_3 seems adequate to meet the theta-pinch first-wall requirement. However, such will not be true for applications involving long-pulse or d.c. fields such as are visualized for neutral beam injectors and direct converters. For these uses, auxiliary cooling may be necessary to keep the insulators near room temperature.

Ceramics subject to d.c. voltages at elevated temperatures may suffer electrical and structural degradation as a result of differential migration of cations and anions (6). It is not anticipated that this will be a problem for theta-pinch first-wall usage, since voltage is applied for only a small fraction of the time. Again, however, it may be desirable to keep insulators for d.c. applications near room temperature to avoid this phenomenon.

B. *Ionizing Radiation*

The effect of ionizing radiation on insulators is to excite electrons and holes into conducting states and thereby increase electrical conductivity. Radiation-induced conductivity can be expressed (2) by the relation $\Delta\sigma = K\dot{\gamma}$, where $\Delta\sigma$ is in units of $(\Omega\text{-cm})^{-1}$ and $\dot{\gamma}$ is the rate of absorption of ionizing energy in rad/sec. K is a constant characteristic of the material, and incorporates such parameters as scattering and trapping mean free paths. For the RTPR, $\dot{\gamma}$ due to background gamma radiation between burn stages is $\sim 2 \times 10^4$ rad/sec. Since first-wall insulating properties are needed only at the time of implosion heating (not during burn), this flux is applicable to considerations of prompt radiation effects.

Measured values of K for some common refractory insulators are between 7×10^{-14} and 6×10^{-17} (2). Therefore radiation-induced conductivity is expected to be roughly in the range from 10^{-9} to 10^{-12} $(\Omega\text{-cm})^{-1}$. This is negligible in comparison with the conductivity criterion described earlier. The effect of enhanced conductivity on pulsed-voltage dielectric strength should also be small, since this parameter is not normally dependent on conductivity, and at any rate thermally-induced conductivity, which is known to be tolerable, will likely exceed that from irradiation.

Delayed ionization effects (those 10 sec after intense irradiation during burn) are more difficult to calculate. Here annealing rates are important, and these involve thermal detrapping and recombination kinetics, which can be quite complex. However, even during intense irradiation, when $\dot{\gamma}$ from bremsstrahlung, neutron, and gamma radiation is approximately 7×10^8 rad/sec, conductivity will not be raised to the maximum acceptable level of 10^{-3} $(\Omega\text{-cm})^{-1}$. Moreover, calculations (2) indicate that recovery will probably be essentially complete 10 sec later ($\sigma < 5 \times 10^{-14}$ $[\Omega\text{-cm}]^{-1}$). Thus for the theta-pinch first wall, neither the relatively low level of ionizing radiation during electric stress nor the higher level ten seconds earlier appears likely to cause significant degradation of dielectric properties. For d.c. applications, the consequences of a degradation in properties by ionizing radiation must be viewed with caution, as is the case with thermal effects.

C. *Structural Degradation*

In the preceding discussion it was implicitly assumed that the structure of the ceramic in question was unaltered by the reactor environment. This will generally not be true, since intense neutron irradiation will likely cause significant structural damage. At the RTPR first wall, the average neutron flux will be $\sim 10^{15}$ n/cm^2 sec, the instantaneous flux $\sim 10^{17}$ n/cm^2 sec, and the fluence in a year 3×10^{22} n/cm^2. The neutron energy spectrum will include ~20% at 14 MeV, 25% between 1 and 14 MeV, and 55% below 1 MeV (1). In Al_2O_3, this irradiation will result in ~30 displacements per atom per yr; although the vast majority disappear via recombination, a high level of stable damage is nevertheless to be expected. Also, neutron-induced transmutations will create in Al_2O_3 roughly 1000 appm/yr of gaseous and metallic transmutation products, and stoichiometry will be shifted slightly by differential burnup rates for the two sublattices.

In assessing the electrical effects of structural damage in insulators, the nature of the damage must be specified, and in ceramics this is not easy. By analogy with metals, one might expect that the predominant defect would be voids. However, the process of making voids in ceramics can be complex; for example, in Al_2O_3 cation and anion vacancies must condense in the stoichiometric ratio 2:3. Other defect species may be favored over voids. Hobbs and Hughes (7) have found in γ-irradiated alkali halides* evidence of cation metal colloids, interstitial loops, anion molecules, and high-pressure anion gas bubbles. Pore-like defects have been found in fission neutron-irradiated refractory ceramics such as ZrO_2 (9), Y_2O_3 (10), and Al_2O_3 (10) (Fig. 2). The presence of metallic colloids in these materials should be detectable by evaluation of the Knight shift in nuclear magnetic resonance. Such studies are currently under way (11). Gas-release studies have shown that the defects are probably not full-density oxygen bubbles (12).

Electrical effects of general radiation-induced disorder (point defects, dislocation loops, clusters, etc.) may not be great, and in fact resistivity could be increased by a consequent shortening of electron mean free path. On the other hand, the consequences of the presence of voids, colloids, or gas bubbles could be serious. All of these defects represent sites for electric field discontinuities which could enhance local breakdown. Micron-sized bubbles could sustain an electron avalanche sufficient to produce ionization events and electron multiplication at the bubble wall; this could also enhance breakdown (2). Measurements of dielectric strength as a function of

*These compounds undergo displacive damage on the anion sublattice during photon irradiation, by the Pooley-Hersh mechanism (8).

FIG. 2. Pore-like defects in Al_2O_3 after irradiation to 4.1×10^{21} n/cm^2 ($E_n > 0.1$ MeV) at 1025 K. [From Clinard, Bunch, and Ranken (10).]

sintering pore content in ceramics (13) show the deleterious effects of such flaws.

If major structural changes such as transformation to an amorphous state occur, drastic changes in electrical properties might result. However, results of irradiations of a number of ceramics to high damage levels (14) suggest that this is not likely, especially at elevated temperatures. Experimental measurements are needed on materials containing defects similar to those expected to result from exposure to intense fusion neutron irradiation. The importance of accurate simulation of such damage (taking into consideration total damage energy, rate effects, nature of displacement damage, transmutation products, and ionizing/displacive energy ratio) cannot be overestimated.

III. STRUCTURAL EFFECTS

A. Fatigue and Fracture

Electrical insulators in fusion reactors may be subjected to significant thermal and mechanical stresses. Strength of refractory ceramics is often high, but their brittleness necessitates the use of special design techniques where possible, such as putting the material under compression rather than tension. Although tensile loads can be minimized, they are often unavoidable in practice, and the usual fracture mode is tensile crack propagation. Finite element stress analysis is useful in detecting areas of stress concentration; these can then be reduced or avoided by design changes. Since the distribution of fracture strengths among a number of identical ceramic components is statistical, a testing regimen should be utilized to reject those which might fail prematurely.

Ceramics under stress at both low and high temperatures can exhibit delayed fracture, or static fatigue. This results from slow propagation of microcracks to a critical size beyond which the stress in the vicinity of the crack tip becomes catastrophically large. At low temperatures, propagation usually but not always (15,16) proceeds by corrosive interaction of the crack tip with moisture in the atmosphere. At high temperatures, thermally-activated internal processes such as grain boundary sliding may control crack propagation, although reaction with the atmosphere can be important here also (17). To a first approximation, cyclic fatigue effects such as those observed in metals should be unimportant in ceramics. This is because intergranular plastic deformation is usually small, so that extrusions and intrusions should not be formed by cyclic stressing. However, if grain-boundary sliding or other mechanisms involving significant movement of material are operating, cyclic stresses might cause a different redistribution of materials near the crack tip from that which would result from a steady stress. In that case the two modes of stress application could be characterized by different fatigue behavior.

The strength of some ceramics is impressive. Gazza (18) found short-time modulus of rupture values of 7 to 9×10^8 Pa at 1590 K for a developmental Si_3N_4-10 wt.% Y_2O_3 alloy. The high strength was attributed to the formation of a refractory grain boundary phase. Si_3N_4 demonstrates a high resistance to static fatigue, even in lower-strength forms. Krakowski et al. (19) have carried out thermal, stress, and static fatigue calculations for a theta-pinch reactor first wall made of a commercial grade of Si_3N_4. A thermally-induced tensile stress of up to $\sim 3 \times 10^8$ Pa was shown to be imposed on the first wall every 10 sec. The stress-time relationship was then combined with a theory of slow crack growth (20,21) and experimental crack propagation data (22), and it was found that the insulator would have to contain an initial flaw size of ~ 200 μm in order that service life at a reference temperature of 1000 K be reduced to a year. Flaws of this size should easily be detected by either quality assurance measures or proof testing. It thus appears that a high-strength ceramic can meet the structural requirements of the RTPR first wall.

The result of neutron irradiation will be to alter the microstructure of the insulator, as discussed earlier. If damage consists primarily of point defects, clusters, and loops, the effect on strength may not be large. However, if voids, bubbles, or colloids are formed, strength may be degraded by resulting internal strains, stress concentration effects, and modulus changes. The weakening that results from the presence of porosity (23) gives an idea of what might be expected with aggregated irradiation defects. On the other hand, a fine dispersion of flaws may have the effect of arresting microcracks before they can grow to dangerous dimensions, and thus conferring added strength and toughness by mechanisms similar to those which strengthen brittle composites (24).

B. *Swelling*

The phenomenon of neutron irradiation-induced swelling is rather well understood in metals, where vacancies created during displacement events can condense into voids, the volume of which constitutes swelling volume. In Al_2O_3 and other ceramics, pore-like defects are observed after irradiation (Fig. 2), and are accompanied by macroscopic swelling (10). However, as described earlier, these defects may not be voids, but rather cation colloids or anion gas bubbles. Other flaws such as interstitial dislocation loops or anion molecules may also be present, and swelling may in some cases be primarily attributable to these rather than to the larger defects (25).

Void swelling in metals occurs primarily between 0.3 and 0.5 T_m,* as a result of the energetic and kinetic properties of vacancies (26). If swelling in ceramics is attributable to other defects, it would not be surprising to find a different swelling temperature range for these materials. Such is the case for Y_2O_3-stabilized ZrO_2, where swelling is at a maximum at 0.30 T_m and is near zero at 0.22 T_m and 0.36 T_m (27). This observation has special significance in that the upper limit to swelling temperature (1025 K) is at or below the minimum operating temperature for some fusion reactor applications (e.g., the RTPR first wall insulator). It is possible that at still higher temperatures void formation may occur (7); however, the effect on swelling behavior is not clear. For metals, it is anticipated that swelling in a fusion environment may be more severe than that in a fission environment; this is because transmutation-induced gas generation rates will be higher with the more energetic neutron spectrum of the former, and gas atoms are known to stabilize voids (28). Since other defects may be responsible for swelling in ceramics, the role of gases here is difficult to predict.

*T_m = absolute melting temperature.

Ceramics can vary greatly in their swelling responses to irradiation. Fission reactor fluences which cause 3 vol.% swelling in Al_2O_3 induce no swelling in Y_2O_3 (Fig. 3).

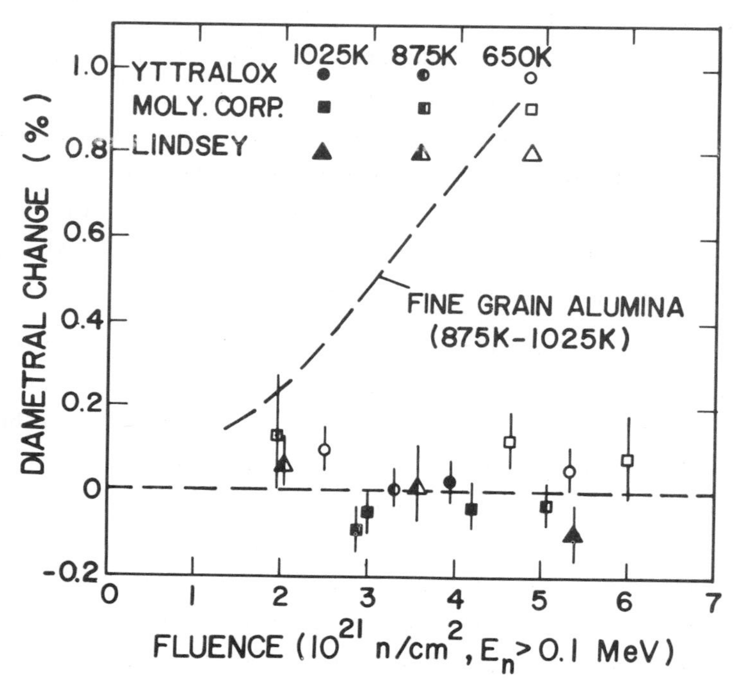

FIG. 3. Swelling of polycrystalline Al_2O_3 and three forms of Y_2O_3 as a function of neutron fluence, after irradiation in EBR-II fission reactor at three temperatures. [From Clinard, Bunch, and Ranken (10).]

The implication is either that the defects which cause swelling in Al_2O_3 are not stable in Y_2O_3 under these conditions, or that dimensional changes accompanying their formation are near zero. Since the type of aggregated defect found in Al_2O_3 was rare in Y_2O_3 (10), the former postulate may be most nearly correct. Once swelling in ceramics is understood, it may be possible to adjust parameters such as composition and microstructure to minimize this effect, as has been done with metals (29).

C. *Reduction in Thermal Conductivity*

An irradiation-induced degradation of thermal conductivity can result in higher ceramic temperatures and thermal stresses. For the theta-pinch first-wall application, a reduction of this parameter by a factor of two would result in an ∿40% increase in stress (30). Indeed, in studies with Al_2O_3, Reichelt et al. (31) found roughly a factor of two reduction in Al_2O_3 after irradiation at ∿1000 K and measurement of thermal conductivity at 973 K. Some evidence of saturation at a dose of $\sim 10^{22}$ n/cm^2 was found.

Since heat is carried primarily by phonons in insulators, the radiation-induced degradation must be due to increased scattering of these lattice waves. Calculations by Klemens et al. (32) have shown that at elevated temperatures point defects, rather than defect aggregates, are primarily responsible for this scattering. Thus in the study by Reichelt et al. the conductivity-degrading species are probably not those seen in Fig. 2, but rather accompanying defects too small to resolve in the electron microscope.

Al_2O_3 may be abnormally sensitive to reduction in thermal conductivity, judging by its swelling performance. The extent of the problem with other ceramics can only be evaluated after measurements are made on heavily-irradiated samples.

D. *Physical Sputtering*

Fusion reactor ceramics may be bombarded by D, T, and He ions, heavy impurity ions, and neutrons. Impact of these particles can cause sputtering of matrix atoms, with resultant thinning, surface roughening, or other deleterious consequences. In Tokamak reactors, loss of constituents of the first wall to the plasma could result in increased radiative losses such that burn time per pulse is unacceptably reduced (33).

The material parameter primarily responsible for sputtering response of solids is surface binding energy (34). Since this factor does not vary greatly among solids, one might expect that ceramics and metals would have similar sputtering yields.* A survey of experimental results for metals and their oxides by Kelly and Lam (35) has shown that this is usually the case. The only high-performance refractory oxides evaluated, Al_2O_3 and MgO, exhibit slightly lower sputtering yields than do the parent metals, and generally behave according to theory (34). Some other oxides show higher yields than do their parent metals, and may lose oxygen preferentially or show evidence of thermal sputtering. It appears that candidate insulators for CTR applications may be expected to resist sputtering roughly as well as do metals, but these insulators should be checked for evidence of the special behavior just mentioned.

Sputtering yield is strongly dependent on energy and mass of bombarding particle. For light ions (e.g., D^+, T^+, He^+), maximum yield (10^{-3} to 10^{-1}) is at about 1-10 keV, or roughly the energy of plasma particles during burn. Heavy ions show a much higher yield, and no maximum in the energy range of interest for fusion reactors (36). It is highly desirable to use a gas blanket (1) or divertor (37) to reduce the energy or number of ions reaching the first wall. Such devices cannot reduce 14 MeV neutron sputtering; however, recent results give yields on the order of 10^{-4} for these particles (38), which is tolerable.

The magnitude of plasma contamination problems is dependent on the atomic number of impurities. For Tokamak reactors, which are particularly sensitive to impurities, this has lead to the proposal that such materials as SiC or Si_3N_4 be considered for first walls (39), or that a graphite blanket be used to protect a metal wall (40). A thin layer of low-Z ceramic or glass bonded to a metal wall might also be useful, if thermal stresses at end of burn can be tolerated.

*Number of lattice atoms sputtered per bombarding particle.

E. *Blistering*

When energetic gas ions penetrate the surface of a solid and permeability is insufficient to allow the gas at the penetration depth to be accommodated, blistering and exfoliation can result. As with physical sputtering, the result can be thinning and surface roughening. The magnitude of blistering effects depends on such parameters as target temperature (primarily through its effect on gas diffusivity and solubility) and energy and angular spread of bombarding ions. When these parameters are unfavorable the sputtering yield can be as high as 0.8 (41), thus presenting a severe materials problem where gas ion fluxes are large, such as at the first wall. The blistering behavior of a number of ceramics and glasses has been studied (42, 43). Mattern et al. (43) found in 150 keV He^+ irradiation experiments on a series of glasses with varying helium diffusivity that for $D > 3 \times 10^{-8}$ cm^2/sec, no surface deformation was observed, whereas below this value blisters were formed. Such findings suggest that if insulators which exhibit high gas diffusivity are used, blistering problems can be reduced or eliminated. Materials which have intrinsically low gas diffusivity can perhaps be altered so as to increase the value of this parameter. Fowler et al. (44) found that tritium diffusion in Al_2O_3 can be increased roughly five orders of magnitude by the addition of 0.2 wt.% MgO. In the unalloyed ceramic, diffusivity at 1000 K was 10^{-12} cm^2/sec; when alloyed, $D \simeq 10^{-7}$ cm^2/sec (extrapolated).

For the RTPR, the gas blanket is expected to minimize the flux of energetic gas particles on the first wall, so that blistering should not be a major problem. With other fusion reactor applications such as a ceramic first wall or liner for Tokamaks, proper material choice with respect to blistering behavior will be important.

IV. CHEMICAL EFFECTS

A. Chemical Erosion

Insulators exposed to hydrogen isotopes can suffer near-surface chemical changes and loss of volatile species thus formed. At higher particle energies such as those characteristic of plasma burn, penetration is significant and chemical reactions can occur in the bulk. Gruen et al. (45) found that 15 keV H^+ and D^+ ions injected into Al_2O_3 form OH complexes within the material. This could degrade near-surface electrical, structural, or chemical properties of the insulator. Hoffman (12) examined an Al_2O_3 probe which had been repeatedly exposed to hot plasma in an experimental confinement device, and found that the surface layer was partially reduced; however the intrinsic surface resistivity was as high as that for undamaged material. The probe showed evidence of surface melting (Fig. 4), but no blistering was observed.

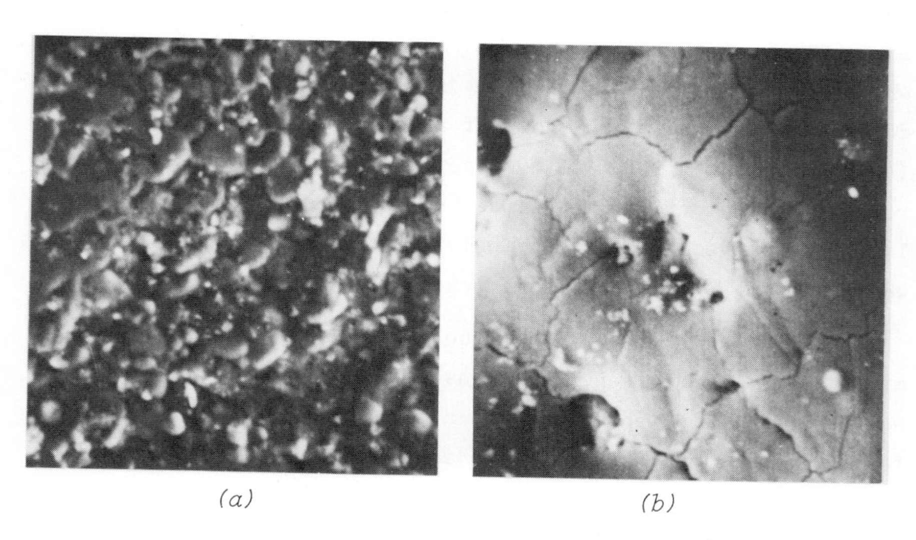

(a) (b)

FIG. 4. Surface of a polycrystalline alumina probe (a) before and (b) after exposure at room temperature to $\sim 10^{20}$ H^+ *and* D^+ *ions/cm²* *of typical energy ∿300 eV. 850X. [From Hoffman (12).]*

Energetic hydrogen isotopes which penetrate the material may subsequently diffuse to the surface and react with the insulator as thermalized atomic gas. Such chemical attack is also of concern where a ceramic first wall is protected by a gas blanket; in this case gas particles at the wall should be neutral and at a fairly low temperature, but a significant proportion will be in atomic rather than molecular form. The latter is the less-corrosive state. Recombination into molecules may occur at the surface in competition with chemical reactions with the substrate.

No quantitative results are available on chemical erosion of ceramics by atomic hydrogen. For each material the following must be considered: predominant chemical reactions, reaction probabilities, and properties of the reaction products (e.g., specific volume, volatility, electrical resistivity). Studies by Balooch and Olander (46) give an idea of what to expect for materials which react chemically with atomic hydrogen. These investigators exposed pyrolytic graphite to 2500 K atomic hydrogen at substrate temperatures between 400 and 2200 K. It was found that below 800 K CH_4 was formed, and above 1000 K, C_2H_2. Between 800 and 1000 K the predominant reaction was recombination of H into H_2. Reaction probabilities as high as 4×10^{-3} were measured. If all D-T fuel particles in the RTPR impinged on a first wall as low energy neutral atomic gas and reacted to form volatile species* with a probability of 4×10^{-3}, the wall thinning rate would be 10^{-2} mm/yr; this is tolerable. Of course, accurate calculations of erosion rate must await experimental results for ceramics.

In the Balooch and Olander study, sticking probability was $\sim 4 \times 10^{-2}$. For an application where high-energy protons impinge on a surface, penetrate, and subsequently diffuse back to the surface, sticking probability can be near unity and high erosion rates may result (47). Thus chemical erosion could

*The reaction assumed is $Al_2O_3 + 12\ H \rightarrow 2\ AlH_3 + 3\ H_2O$.

be much higher for Tokamak first walls than for theta-pinch first walls, for the same material.

B. *Corrosion by Liquid Lithium*

It would be helpful for some fusion reactor applications if ceramics could withstand direct exposure to liquid lithium at elevated temperature. However, most of these materials suffer significant attack, at least at 1366 K (49, 50, 51), and must be protected. Exceptions are Y_2O_3, ThO_2, and alloys of these compounds, which show only slight attack. Thermodynamically, some nitrides might be expected to resist corrosion. Singh and Touhig (51) found that two such materials, Si_3N_4 and sialon, endured exposure at 673 K with no appreciable weight loss; however, significant grain boundary attack was observed. It was postulated that segregated impurities could have been responsible for the preferential attack. This suggests that ceramics might be developed which could withstand direct exposure to liquid lithium. At present, however, all reactor designs call for protective cladding.

C. *Indirect Reduction by Liquid Lithium*

A metal layer between ceramic and liquid may not always protect the ceramic. Selle and DeVan (52) have shown that for at least one ceramic-metal system, partial reduction of the insulator can result. These investigators held Al_2O_3/Nb/Li samples at 1273 K for 3000 hr, and observed such an effect. They postulated the following reactions enabling oxygen dissolved in niobium to be transported through the metal and reacted with lithium:

$$2\ \mathrm{Nb} + 3\ \mathrm{Al_2O_3} \rightarrow 2\ \mathrm{NbAl_3} + 9\ \mathrm{O}\ (\text{in Nb})$$
$$9\ \mathrm{O}\ (\text{in Nb}) \rightarrow 9\ \mathrm{O}\ (\text{in Li}).$$

In order for such a reduction to occur, the metal must be permeable to the anion, and the anion-lithium compound must be stable. Niobium is particularly permeable to oxygen and nitrogen; most other metals should not so easily support the reduction process.

Also, some other ceramics (e.g., Si_3N_4, Y_2O_3) should prove less reactive.

It appears that this problem can be avoided by proper choice of metal and ceramic; however, each candidate system should be evaluated with the possibility of indirect insulator reduction in mind. If both ceramic and metal prove susceptible to reduction, a diffusion barrier which exhibits low permeability to anions and high resistance to lithium (e.g., tungsten) could be plated onto the exposed surface of the metal.

V. INSULATORS FOR MAGNETIC COILS

Mirror and Tokamak reactor designs call for superconducting magnets, while those in the theta-pinch machine utilize normal conductors operating near room temperature. Insulators used in superconducting coils need exhibit only a moderate dielectric strength, but normal coil designs call for rather high breakdown strengths. Good mechanical strength is needed for both, to withstand the large forces generated in high-field magnets.

Both organic and inorganic insulators are being considered for these applications. Superconducting coils are shielded from intense irradiation, so that rather low fluences (roughly 10^{18} n/cm^2 and 10^9 rad) are expected during magnet lifetime. These levels should present no difficulties with most inorganic insulators; however, such doses are high enough to cause structural and electrical degradation of some organics used for magnet insulation (53). This results from fracture of long-chain polymer molecules, an increase in cross-linking, and molecular decomposition. It will be important to use relatively radiation-resistant organics for these applications, and to minimize local areas of higher-than-average radiation loading due to the presence of low-absorbance regions between plasma and magnet.

Inorganic insulators can be used in superconducting magnets, although coil fabrication and assembly may be more difficult than with organics. Since no radiation shielding is used in the RTPR, coils for that machine must be ceramic-insulated.

The chemical environment of magnet insulators is not severe. However, ozone attack of organics and the effect of moisture on long-term strength of ceramics must be taken into account.

VI. SUMMARY

The severe environment of a fusion reactor exposes electrical insulators to a host of potential problems. If these ceramics are chosen with care, short-term behavior should be adequate to allow reactor operation as planned. However, accurate prediction of long-term performance is not now possible, primarily because of uncertainties with respect to structural radiation damage and chemical erosion, and their effect on electrical and structural properties. It is in these areas that in-depth studies are most needed.

ACKNOWLEDGMENTS

The author wishes to thank J. M. Bunch, J. G. Hoffman, G. F. Hurley, and R. A. Krakowski for helpful discussions on many of the topics addressed in this manuscript. Additionally, appreciation is expressed to J. M. Bunch and J. G. Hoffman for permission to cite their results before publication.

REFERENCES

1. An Engineering Design Study of a Reference Theta-Pinch Reactor (RTPR), Report LA-5336/ANL-8019, 1974.

2. V. A. J. van Lint, J. M. Bunch, and T. M. Flanagan, presented at the International Conference on Radiation Effects and Tritium Technology for Fusion Reactors, Gatlinburg, Tenn., 1975. Proceedings to be published.

3. O. T. Ozkan and A. J. Moulson, J. Phys. D: Appl. Phys. 3, 983 (1970).

4. J. J. O'Dwyer, The Theory of Electrical Conduction and Breakdown in Solid Dielectrics, Clarendon Press, Oxford (1973).

5. J. M. Bunch and F. W. Clinard, Jr., Proceedings of the First Topical Meeting on the Technology of Controlled Nuclear Fusion, Report CONF-740402-P2, (1974), p. 448.

6. Evidence of electrical degradation is reported in Quarterly Report on the Space Electric Power R and D Program, Report LA-5021-PR (1972) and in ref. 27.

7. L. W. Hobbs and A. E. Hughes, Radiation Damage in Diatomic Materials at High Doses, Report AERE-R 8092, 1975.

8. D. Pooley, Proc. Phys. Soc. 87, 245 (1966); H. N. Hersh, Phys. Rev. 148, 928 (1966).

9. F. W. Clinard, Jr., D. L. Rohr, and W. A. Ranken, presented at the Annual Meeting of the American Ceramic Society, Cincinnati, Ohio, May 1976.

10. F. W. Clinard, Jr., J. M. Bunch, and W. A. Ranken, op. cit. ref. 2.

11. J. M. Bunch, private communication (1976).

12. J. G. Hoffman, private communication (1976).

13. R. Gerson and T. C. Marshall, J. Appl. Phys. 30, 1650 (1959).

14. H. Matzke and J. L. Whitton, Can. J. Phys. 44, 995 (1966).

15. D. P. H. Hasselman, Ultrafine Ceramics (J. J. Burke, N. L. Reed, and V. Weiss, eds.) Syracuse Univ. Press (1970) p. 297.

16. J. T. A. Pollock and G. F. Hurley, J. Mat. Sci. 8, 1595 (1973).

17. R. Kossowski, J. Am. Cer. Soc. 56, 531 (1973).

18. G. E. Gazza, Bull. Am. Cer. Soc. 54, 778 (1975).

19. R. A. Krakowski and F. W. Clinard, Jr., presented at the Annual Meeting of the American Nuclear Society, Toronto, Canada, June 1976.

20. S. M. Wiederhorn, Fracture Mechanics of Ceramics Vol. I (R. C. Bradt, D. P. H. Hasselman, and F. F. Lange, eds.) Plenum Press, New York (1974) p. 613.

21. D. P. H. Hasselman, E. P. Chen, C. L. Amman, J. E. Doherty, and C. G. Nessler, J. Am. Cer. Soc. 58, 513 (1975).

22. A. G. Evans, L. R. Russell, and D. W. Richerson, Met. Trans. 6A, 707 (1975).

23. R. L. Coble and W. D. Kingery, J. Am. Cer. Soc. 39, 377 (1956).

24. K. Kendall, Proc. Roy. Soc. A341, 409 (1975).

25. J. M. Bunch, private communication (1976).

26. P. G. Shewmon, Science 173, 987 (1971).

27. Quarterly Report on the Space Electric Power R and D Program, Report LA-5113-PR (1972).

28. W. N. McElroy and H. Farrar IV, Proceedings of the International Conference on Radiation-Induced Voids in Metals, Report CONF-710601, (1972) p. 187.

29. J. J. Laidler, Ibid., p. 174.

30. R. A. Krakowski, R. L. Hagenson, and G. E. Cort, private communication.

31. W. H. Reichelt, W. A.Ranken, C. V. Weaver, A. W. Blackstock, A. J. Patrick, and M. C. Chaney, Proceedings of the Thermionic Conversion Specialists Conference, (1970) p. 39.

32. P. G. Klemens, G. F. Hurley, and F. W. Clinard, Jr., (unpublished results).

33. D. Steiner, Nucl. Sci. and Eng. 58, 107 (1975).

34. P. Sigmund, Phys. Rev. 184, 383 (1969).

35. R. Kelley and N. Q. Lam, Rad. Effects 19, 39 (1973).

36. R. Behrisch, Nucl. Fusion 12, 695 (1972).

37. G. A. Emmert, J. M. Donhowe, and A. T. Mense, J. Nucl. Mat. 53, 39 (1974).

38. Results presented by: O. K. Harling, M. T. Thomas, R. L. Brodzinski, and L. A. Rancitelli; M. Kaminsky and S. K. Das; R. G. Meisenheimer; and L. H. Jenkins, G. J. Smith, J. F. Wendelken, and M. J. Saltmarsh at the International Conference on Surface Effects in Controlled Fusion Devices, San Francisco, CA, 1976. Proceedings to be published.

39. Fusion Reactor Design Studies--12-Month Progress Report, Report GA-A13430 (1975).

40. G. L. Kulcinski et al., A Method to Reduce the Effects of Plasma Contamination and First Wall Erosion in Fusion Reactors, Report UWFDM-108 (1974).

41. S. K. Das and M. Kaminsky, Proceedings of the 5th Symposium on Engineering Problems of Fusion Research, Princeton Univ. (1973).

42. W. Primak, Facies of Ion-Bombarded Surfaces of Brittle Materials, Report ANL-75-66 (1975).

43. P. L. Mattern, J. E. Shelby, G. J. Thomas, and W. Bauer, The Effects of Gas Transport Properties on Blister Formation in He^+ Implanted Glass, Report SAND 75-8758 (1976) and op. cit. ref. 38.

44. J. D. Fowler, R. A. Causey, D. Chandra, and T. S. Elleman, op. cit. ref. 2.

45. D. M. Gruen, R. B. Wright, P. Finn, and B. Siskind, presented at the Annual Meeting of the American Ceramic Society, Washington, D. C., May 1975.

46. M. Balooch and D. R. Olander, J. Chem. Phys. 63, 4772 (1975).

47. M. Balooch and D. R. Olander, private communication.

48. D. Elliott, D. Cerini, and L. Hays, Liquid MHD Power Conversion, Space Programs Summary No. 37-41, Vol. IV., Jet Propulsion Laboratory, Pasadena, CA (1966).

49. L. G. Hays and D. O'Conner, A 2000°F Lithium Erosion and Component Performance Experiment, NASA Report No. 32-1150 (1967).

50. D. S. Jesseman, G. D. Roben, A. L. Grunewald, W. L. Fleshman, K. Anderson, and V. P. Calkins, Preliminary Investigations of Metallic Elements in Molten Lithium, Report NEPA-1465 (1950).

51. R. N. Singh and W. D. Touhig, J. Am. Cer. Soc. 58, 70 (1975).

52. J. E. Selle and J. H. DeVan, op. cit. ref. 45.

53. G. Pluym and M. H. Van der Voorde, Proceedings of the International Conference on Magnet Technology, Oxford (1967) p. 341.

Chapter 5

RADIATION EFFECTS IN STRUCTURAL MATERIALS FOR FUSION REACTORS

F. W. Wiffen

Metals and Ceramics Division
Oak Ridge National Laboratory
Oak Ridge, Tennessee 37830

I. INTRODUCTION

The radiation effects to be expected in the metal first wall and other structural components in the high-flux regions of future Controlled Thermonuclear Reactors (CTRs) have been discussed by several authors and workshop sessions in the past few years [1—3]. These treatments are in general agreement that the radiation effects of greatest concern are swelling and and mechanical property changes, especially reduced ductility.

These problems are relatively insensitive to confinement system or details of reactor design and are potential problems for all possible material choices. Their importance to the prospects of power from fusion stems from the requirement of absolute integrity of the first wall. Any breach in the wall would release breeding or coolant fluids into the plasma cavity and quench the plasma. A further restriction is that the affected segment of the reactor cannot simply be plugged to stop coolant flow (the solution used for leaks in steam generators). The high heat flux on the first wall makes this impossible and requires repair or replacement, with the attendant economic penalties of long reactor shutdown periods. The required structural integrity thus sets a very high value on a knowledge of expected radiation effects. Consideration in this chapter is restricted to bulk radiation effects, and surface radiation effects are excluded. We are thus concerned only with the neutron irradiation, as other radiation species (i.e., ions, photons) will be surface-limited and will not produce effects that penetrate more than a few microns below the surface nearest the plasma.

Current CTR conceptual designs have used the available, sparse radiation effects data to minimize radiation damage by choice of operating temperature [4] and by choice of material [5]. One attempt has also been made to assess acceptable property changes for an austenitic stainless steel, and then to use available fission reactor irradiation effects data to predict the useful lifetime of key CTR components [6,7]. The common result of these studies has been to emphasize the importance of the radiation effects on material properties and the inadequacy of the available data.

Although there is an extensive body of data available on the effects of neutron irradiation on the properties of metals, this data must be used with some care in predicting the effects of the fusion reactor environment. The difficulty is with the different neutron spectra in fission and fusion reactors; the average energy of fission source neutrons is near 2 MeV while the energy of source neutrons from the D-T reaction is 14.06 MeV. This difference is especially important for (n,p) and (n,α) transmutation reactions as these reactions in many metals have thresholds near 10 MeV. The resulting gases can have a very important influence on the metal properties. Differences in displacement damage resulting from the differences in the two neutron spectra are believed to be relatively unimportant.

If differences in the neutron spectra are kept in mind, the data from the fission reactor development program can be usefully applied to estimate the radiation effects expected in components of fusion reactors. It is generally believed that the differences in the neutron spectra will render radiation effects in fusion reactors more deleterious to reactor operation than irradiation in a fission reactor. Thus, a "window" of temperature and fluence conditions that result in acceptable properties, determined by fission reactor experiments, might be narrowed when the experiments are repeated in a fusion reactor. The fission reactor experiments, however, will eliminate some materials and conditions and will have defined the limits and identified the critical experiments to be performed on promising materials under conditions that better simulate fusion reactors. Fission reactor radiation effects programs will also provide guidance in the approaches to be taken in tailoring alloys to minimize the deleterious effects of neutron irradiation.

Two of the major omissions of radiation effects programs in fission reactor development were not predicting (1) high-temperature embrittlement due to helium and (2) volume instability (swelling) due to void formation. Both of these failures resulted from incomplete experimentation and extrapolation of results in temperature and neutron fluence. In the CTR bulk radiation effects program we are fortunate in having facilities that will closely simulate all or part of the fusion reactor environment. By careful use of all of these techniques and by the appropriate correlation with experiments that can be performed in neutron facilities with a D-T spectrum, materials can be evaluated at the correct temperature, fluence, stress state, and environment. The uncertainty of neutron spectra will be resolved as the higher-flux, high-energy neutron sources are developed, and these will provide the continued correlation to verify simulation experiments and will ultimately provide the proof testing to verify material selection and qualification.

It is the purpose of this paper to examine the components of the fusion reactor environment that affect the behavior of structural metals, to describe briefly the atomistics of neutron interactions with metals, and to use this information and the experience of past irradiation effects studies to predict the important property changes that will result from neutron irradiation during the service life of metal components in CTRs. Discussion of predicted radiation effects are divided, for convenience, into swelling, postirradiation measurement of mechanical properties, and *in situ* mechanical properties measurements. Finally, a brief discussion is included of the place of the major classes of irradiation facilities in the CTR materials irradiation program.

II. THE FUSION REACTOR ENVIRONMENT

The potential radiation effects on CTR structures can be judged only after a structural material and set of operating conditions has been specified. However, the long time necessary to generate the material performance data requires that experiments be initiated now, and the conceptual designs are useful in defining the classes of materials of interest to CTR and the potential operating conditions of the system. As these designs are continually being refined and as new designs are still evolving, the materials evaluation programs must remain as flexible as possible to satisfy the requirements of ultimate, not current, CTR designs.

Illustrative parameters that describe the environment in the CTR high-flux region are given in Table 1. Although several classes of materials are listed in this table and have been suggested for possible CTR application, the list is not complete.

TABLE 1

THE CTR RADIATION ENVIRONMENT

1. MATERIALS IN THE HIGH-FLUX REGION

   Structural Wall — Conventional alloys of Fe-Ni-Cr
   Refractory metals V, Nb, Mo
   Ceramics SiC, C

   Shield or Sacrificial Wall — Graphite
   Oxide, Carbide, Nitrides

   Insulators (Theta Pinch) — Al_2O_3 or better insulator

   Coolant Channels — Same as structural material

2. TYPICAL SUGGESTED TEMPERATURES

   Type 316 stainless steel 280–600°C

   Nb-1 Zr 420–1050°C

3. NEUTRON FLUX AND WALL RESPONSE

   Typical wall loading 1 MW/m^2

   Flux 3.7×10^{14} n/cm^2 sec (20% have $E > 10$ MeV)

   Wall response in one year at 100% duty factor

	316 SS	Nb-1 Zr
dpa	10	7.2
He, ppm	200	24
H, ppm	520	80

Limiting this discussion of radiation effects to materials in the high-flux reactor regions does not mean that these are the only problem areas. However, the most serious radiation effects problems probably relate to the first-wall structure and only these are treated in this paper.

Type 316 stainless steel and Nb—1 Zr have each been considered in more than one conceptual reactor design. These are used as examples of structural materials for near term, conservative designs where minimum extrapolation of materials technology is required (316) and for longer-term designs with more ambitious thermal efficiency goals (Nb—1 Zr). The range of temperatures specified for these two materials in the different designs is shown. This immediately defines a goal of any radiation effects program — to qualify these materials within the temperature range shown, or to restrict further the range based on radiation effects on the alloys.

Current reactor conceptual designs show neutronic wall loadings in the range 0.5 to 3.0 MW/m^2. (EPR power levels may be lower.) The production rate of the radiation parameters displacements per atom (dpa), helium, and hydrogen are given in Table 1, taken from Kulcinski, Doran, and Abdou [8].

Other components of the environment at the first wall can also affect the irradiation response. The stress level will undoubtedly influence the radiation response, as will the duty cycle and pulse repetition rate in reactors that do not operated in a steady-state mode. The reactor coolant may play a secondary role in the radiation response; for example, by affecting material chemistry through the mass transfer of interstitial elements into or out of the metal. These secondary considerations are neglected in this paper. Also neglected are possible synergistic interactions between bulk radiation effects and surface effects, compatibility considerations, and other possible CTR materials effects.

III. THE ATOMISTICS OF NEUTRON BOMBARDMENT

Two processes affecting the mechanical and physical properties occur in materials under neutron irradiation.

One process is nuclear reaction, resulting from neutron capture by a target atom and then decay of the excited nucleus. The transmutation reactions that yield helium, (n,α) reactions mainly, are of the greatest importance. The cross sections for the reactions are fairly well known and typical yields in CTR conditions are given in Table 1. These nuclear reactions are dependent on neutron energy only, and are unaffected by other components of the reactor environment such as temperature and stress state.

The second process is atomic displacement. Displacement damage in a metal is created when a neutron strikes a lattice atom and displaces it from its equilibrium postion. The unoccupied lattice position is a vacancy; and the displaced atom, when it comes to rest at a non-lattice position, is called an interstitial. The damage is further multiplied when the neutron transfers a large amount of energy to the struck atom and it in turn displaces many other atoms before itself coming to rest. Atomic distributions produced by energetic neutron bombardment are illustrated schematically in Fig. 1, showing Seeger's [9] description of the damage process. This damage production process is not temperature dependent. However, the rearrangement of the damaged zone after the damage production event is strongly dependent on temperature, stress, and other variables.

The high mobility of individual vacancies and interstitials at CTR operating temperatures results in most of these point defects being annihilated soon after they are created. A small fraction of the defects created, however, is retained by forming vacancy or interstitial clusters. It is these relatively stable defect configurations that will influence the properties of the CTR components. Mobile interstitial atoms precipitate in a two-dimensional morphology, in the form of disks or partial extra atom planes bounded by dislocations, illustrated

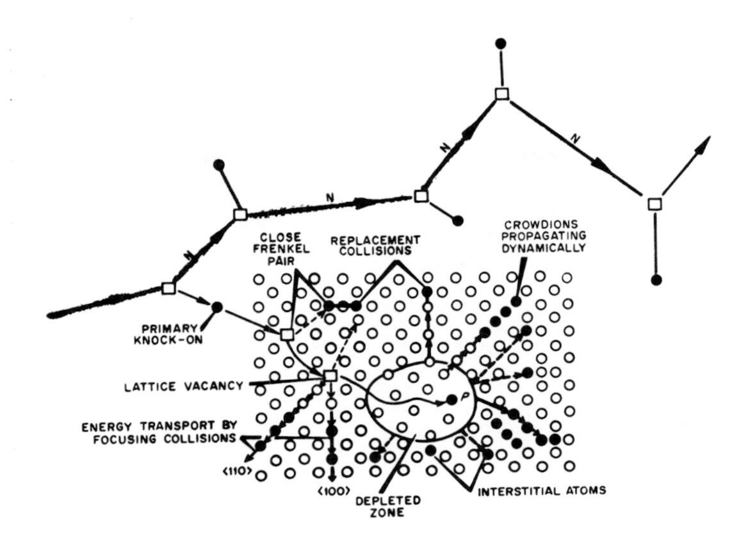

FIG. 1. Schematic representation of atom displacement processes occurring during neutron irradiation of a solid metal. Interstitials are transported long distances by the collision sequences shown and leave a vacancy-rich depleted zone. Figure adapted from Seeger [9].

schematically in Fig. 2(a). Vacancies can precipitate in two morphologies, shown in Figs. 2(b) and (c). Under some conditions, the vacancies form dislocation loops, as illustrated in Fig. 2(b). At somewhat higher temperatures vacancies can precipitate into a three-dimensional morphology, the cavity shown in Fig. 2(c). Two special classes of cavities must be considered in evaluating CTR damage. Cavities that are essentially empty, formed by the precipitation of vacancies alone, are referred to as voids. Cavities that form by precipitation of both vacancies and insoluble gases (e.g., helium) are called bubbles and can exist in equilibrium with an internal pressure, P, given by

$$P = \frac{2\gamma}{r} \tag{1}$$

where γ = surface tension of the metal and r = bubble radius. As a result of the internal pressure, voids and bubbles can be

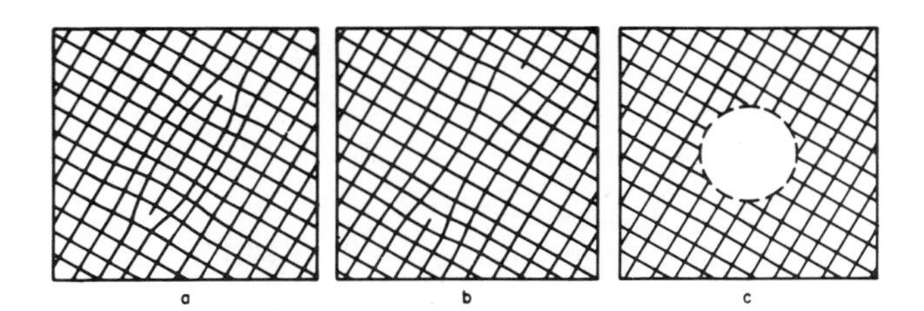

FIG. 2. Defect clusters that are stable in metals irradiated at elevated temperatures fall into three classes. A dislocation loop of the interstitial type is shown in (a), a vacancy dislocation loop in (b), and a three-dimensional cavity in (c).

distinguished by their postirradiation annealing behavior, with bubbles growing and voids shrinking on high temperature annealing.

IV. PREDICTED CTR RADIATION EFFECTS

First-wall radiation effects of interest here are those that will set limitations on the reactor operation. These limitations may restrict operating temperatures, allowable stresses, and useful lifetime of the first wall, perhaps through total allowable neutron fluence or total lifetime strain cycles. Major properties of concern, and the way in which they could limit CTR behavior, are listed in Table 2. These potential radiation effects can be divided, for the purpose of discussion, into three categories: (1) swelling, (2) mechanical property changes as deduced from postirradiation tests, and (3) in-reactor changes in mechanical properties. The swelling is important as it relates to changes in dimensional tolerances in the reactor and as a source of stress (as swelling strains are accommodated) during reactor operation. Postirradiation mechanical property testing will give some measure of the ability to withstand stresses imposed by cyclic changes in operating conditions (such as by temperature changes between "burn" and "off" phases of the duty

TABLE 2

BULK RADIATION EFFECTS ON PROPERTIES THAT MAY LIMIT CTR FIRST-WALL LIFETIME

Property	Limitation
Swelling	Dimensional Tolerances Strain Imposed on System
Tensile Properties	Stress Limitation Strain to Failure
Thermal Creep	Deformation Rate Component Time to Failure
Radiation Creep	Deformation Rate
Fatigue and Creep Fatigue	Strain per Cycle Cycles to Failure

cycle) and imposed during shutdown and start-up between periods of regular operation. In-reactor tests are required to determine deformation rates during loading under constant or cyclic stresses.

A. Irradiation-Produced Swelling

Swelling of CTR structural materials can result from the condensation of vacancies, to form voids, the condensation of insoluble gases [helium from (n,α) reactions] to form bubbles, or, more likely for materials under CTR irradiation conditions, cavities that contain some gas at less than the equilibrium pressure. General considerations that relate to swelling in CTR structures are given in point form in Table 3. The generally accepted description of the void swelling processes in metals is as follows. While self-interstitials and vacancies are produced in equal number by neutron collision with the lattice, the preferential attraction of dislocations for

TABLE 3

SWELLING

Cause:

Vacancies and gas atoms precipitate
Net volume increase results

Critical Parameters:

Fluence — dpa and gas production (He)
Temperature of irradiation
Composition
Metallurgical state

Modifying Parameters:

Damage Rate
Stress state
dpa/He ratio
Duty Cycle ?

Measure:

Overall dimensional changes
Dimensional changes between ion-bombarded and unbombarded regions of a sample
Density by immersion
Microstructure by TEM

self-interstitials provides a biased sink that results in an excess of free vacancies in the lattice. These vacancies then provide the driving force for nucleation and growth of cavities and the resultant swelling. The role of transmutation-produced helium in this process is not clearly understood, but it is likely that the helium can enhance swelling by its effect on both the nucleation and growth of cavities. An example of the microstructure seen in a highly irradiated sample, where significant swelling has occurred, is shown in Fig. 3. The dependence of swelling on temperature, microstructure, and amount of helium produced neutronically during the irradiation is shown in Fig. 4.

Fission reactor studies and heavy-ion bombardment experiments have shown that swelling due to cavity formation

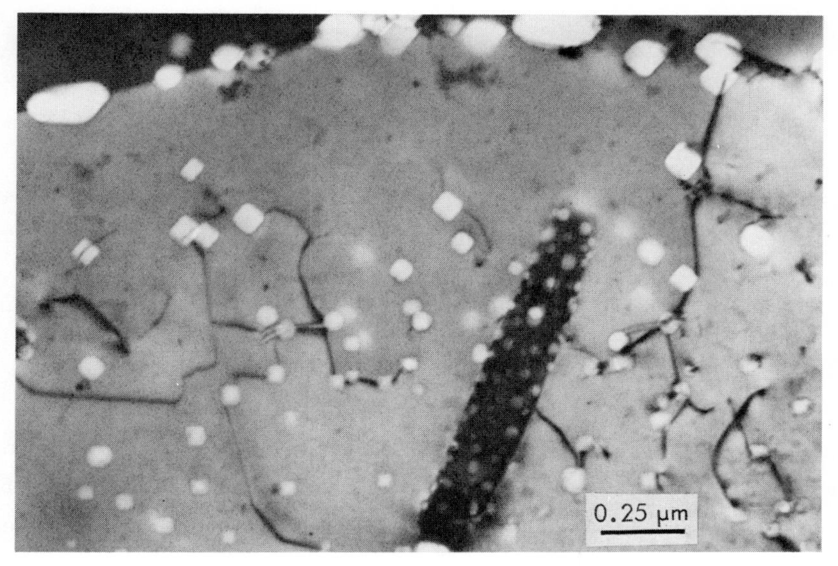

FIG. 3. Microstructure of solution-annealed type 316 stainless steel irradiated at 575°C in the HFIR to a fluence of 4.2×10^{26} n/m^2 (>0.1 MeV) producing 29 dpa and 1790 appm He. The white features are helium-filled cavities and account for about 3% volume increase. From [10].

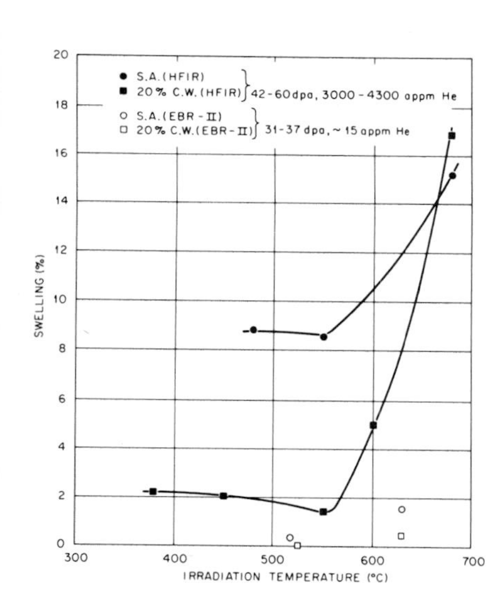

FIG. 4. Swelling in type 316 stainless steel as a function of irradiation temperature. Sample conditions are S.A. — solution annealed and C.W. — cold worked. These results show that cold work is effective in suppressing swelling at temperatures below 650°C for irradiations in either reactor. Swelling in HFIR, where the helium production rate is high, is much greater than in EBR-II with its low helium production. The helium is produced in a two-step thermal neutron capture sequence leading from ^{58}Ni to ^{56}Fe. From [11].

is a general phenomenon in solids. Values of swelling range up to 20% for reactor irradiation and up to ~100% for simulated reactor conditions. For a given alloy, the amount of swelling depends on irradiation temperature, gas content of the sample (usually helium is important), total accumulated damage, and rate at which the damage is introduced. Although probably important, there is only limited experimental evidence on the effect of stress or the effect of pulsed vs steady-state irradiation on the total amount of swelling.

Swelling is also highly dependent on the chemistry, both the major chemistry variation between alloys and the minor variations possible within the specification range of a single alloy. Although numerous examples exist, the possible variations are typified by the results of two recent ion-bombardment simulation experiments. Johnston et al. [12] have shown that for a given experimental condition, swelling within commercial alloys based on the Fe-Ni-Cr system can vary by at least a factor of 100, with some correlation between swelling and alloy nickel content. An example of the effect of minor chemistry variation is provided by the work of Bloom et al. [13], showing that even within the type 316 stainless steel specification, swelling can vary by at least a factor of 8, and that minor alloying modification near the type 316 composition can lead to a very low-swelling alloy.

The above results clearly indicate that much can be done to develop low-swelling alloys. In addition to control of composition, cold working (to produce a high overall sink concentration) is effective in swelling control of type 316 stainless steel under LMFBR irradiation, and is also effective at high levels of irradiation-produced helium [10,11]. Other possibilities include swelling control through control of grain size or distribution of a second phase within the matrix.

Swelling can be measured on very small samples. The techniques used include measurement of dimensional changes, measurement of density using the Archimedes' technique, and determination of the cavity volume fraction by measuring the size and concentration of cavities as revealed by transmission electron microscopy (TEM). Current practice is to use the density determination for a rapid and accurate method of comparison of neutron-irradiated samples. For ion-bombardment simulation experiments, rapid comparison can be made by step-height measurements [12], a technique based on dimensional changes. Transmission electron microscopy results are necessary to understand the swelling mechanisms and are vital in a search for methods of improving alloy swelling resistance. Supplemental investigative tools that are being developed to aid in cavity and swelling characterization include positron annihilation, small-angle x-ray scattering and small-angle neutron scattering.

Irradiation experiments to produce samples for swelling studies are relatively simple. Small samples are used, and only irradiation parameters need be controlled and known during the experiment. These parameters generally only include irradiation temperature, fluence, and the chemical and stress environment at the sample.

B. Postirradiation Mechanical Properties

Changes in the microstructure of a metal are usually accompanied by changes in the mechanical properties. In the case of neutron irradiation, the microstructural changes are the formation of cavities, dislocation loops and dislocation networks. In most materials, this produces a stronger matrix and results in reduced ductility. Complications to this simple guideline abound. Among the most important for CTR structures is the debilitating effect of transmutation-produced helium.

The helium usually has little or no effect on strengthening processes, but causes severe ductility reductions by promoting intergranular failures. In extreme cases, failures occur before the yield strength of the matrix is reached. Other complications to the lattice hardening — ductility reduction guidelines include softening of cold-worked structures through irradiation-enhanced recovery, and irradiation effects on the precipitate structure and stability. Frequently changes in the precipitate structure also result in softening. Some of the general considerations that relate to the postirradiation evaluation of mechanical properties of CTR materials are given in Table 4.

TABLE 4

MECHANICAL PROPERTIES AFTER IRRADIATION

Cause:

- Microstructural changes control matrix strength
- Insoluble gases affect fracture mode
- Ductility reduced by several processes
  - plastic instability, DBTT shift,
  - reduced fracture strength

Critical Parameters:

- Fluence — dpa and gas production
- Temperature of irradiation
- Composition
- Preirradiation microstructure

Measure:

- Tensile, creep, or fatigue test
  - Strength values
  - Ductility values
  - Microstructure and fractography

While the swelling and mechanical properties depend on the same irradiation parameters, their dependence on these parameters can be quite different, and is not generally the same between different alloy systems. As with the swelling, irradiation temperature and neutron fluence are the most important determinants of postirradiation mechanical properties. However, postirradiation mechanical properties are also a strong function of the test parameters. The neutron spectrum effects are greater for mechanical properties than for swelling because of the effect of helium, produced by high-energy neutrons through (n,α) reactions, on the high-temperature fracture mode.

Postirradiation mechanical property measurements can be performed in a variety of tests on various specimen geometries. Although further miniaturization is possible, specimens that have been used successfully have usually had one dimension of 20 mm or greater. In irradiation experiments to produce samples for postirradiation mechanical properties measurements, only the irradiation parameters (temperature, fluence, chemical environment, stress state) need be known. (This is in contrast to the more complex measurements required for *in situ* property measurements to be described in the next section.) Possible test techniques used to evaluate the mechanical properties after irradiation are given in Table 4.

The major changes that irradiation can produce in the tensile properties of a metal are shown by the data in Fig. 5. In this experiment samples of 20% cold-worked type 316 stainless steel were irradiated at temperatures between 350 and 700°C to neutron fluences producing about 50 dpa and 4000 appm helium. The results in Fig. 5 show that this irradiation has significantly reduced both the yield and ultimate tensile stress over the range of temperatures investigated. This weakening results from recovery of the cold-worked structure and precipitation that occurs during irradiation. While this decrease in tensile strength is important, and must of course be taken account of in reactor design, of even greater importance is the loss of ductility produced by irradiation. Figure 5 shows that for the lowest irradiation and test temperature, 350°C, the ductility of

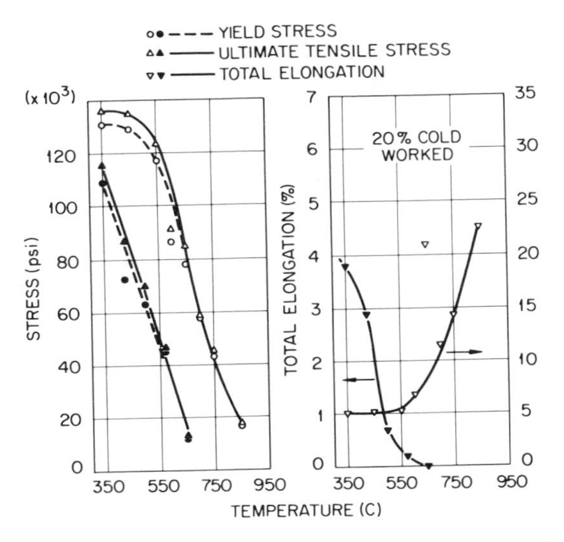

FIG. 5. The tensile properties of 20% cold-worked type 316 stainless steel as a function of test temperature. Open symbols are for unirradiated material, and closed symbols for samples irradiated at a temperature near the test temperature. The strain rate was 0.0028 min^{-1}. The HFIR irradiation produced about 50 dpa and 4000 appm helium in the samples. From [14].

the irradiated sample was only slightly less than that of the control. At increasing temperatures the tensile elongation continually decreases, dropping to zero at 650°C. There data clearly set an upper temperature limit on the use of this alloy under similar service conditions, although the minimum allowable ductility has not yet been defined by reactor designs. The dependence of tensile ductility on reactor operating time must be known to predict reactor component lifetimes. The type of data used for this purpose is shown in Fig. 6, where 575°C tensile ductility is plotted vs neutron fluence for irradiation near 600°C. These data show that the ductility is rapidly reduced under these conditions and that helium accumulations of less than 100 appm (less than 0.5 MW-years/m^2 of CTR operation) can severely impair the material properties.

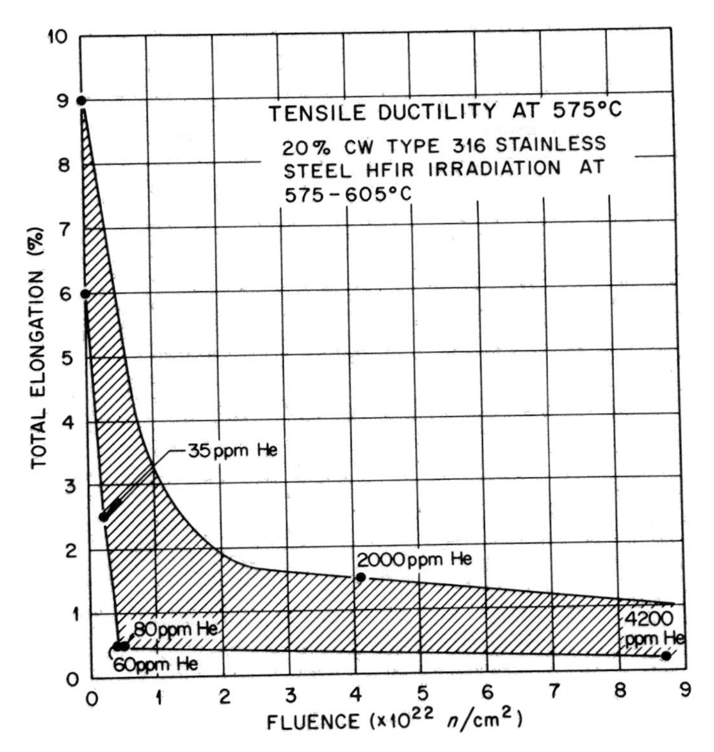

FIG. 6. The tensile ductility of 20% cold-worked type 316 stainless steel irradiated in the HFIR. Helium levels produced by the irradiation are shown on the data points, and compare to ~200 appm He produced per year of CTR operation at a wall loading of 1 MW/m^2.

C. *In Situ Property Measurements*

Some properties of interest to CTR materials application require measurement while the material is under irradiation. Of the properties requiring *in situ* measurement, the best known example is irradiation creep. Other properties that may require *in situ* measurement include low-cyle fatigue of metals and the electrical properties of insulators. A point form treatment of irradiation creep is given in Table 5.

TABLE 5

IRRADIATION CREEP

Cause:

Biased flow of point defects control deformation
Deformation rates exceed out-of-reactor rates

Critical Parameters:

Flux — dpa generation rate

Modifying Parameters:

Fluence
Stress
Temperature
Composition and metallographic state

Measure:

Stress, temperature
Strain with high precision

In situ measurements are required when changes in the property of interest is a result of the flux of point defects produced by the irradiation, rather than of the microstructure that has resulted from the precipitation of point defects. In the case of irradiation creep, the deformation rate of a stressed material is controlled by the biased flow of point defects to dislocations. Measurements of irradiation creep need to be done in a fully-instrumented test assembly where irradiation temperature, flux, stress and strain can all be measured and/or controlled with a high degree of accuracy during the irradiation. Experience has shown that these experiments are exceedingly difficult. Since the creep rate will be a function of flux, temperature, stress, and accumulated fluence, variation of all of these must be included in any program to define fully the deformation rate for a single material. There are possibilities of obtaining part of the data necessary to understand irradiation creep in simpler experiments. Pressurized tubes, loaded springs, and stress relaxation experiments have all been used successfully to obtain creep behavior without *in situ* measurements. These techniques, however, will not provide as complete a definition of the creep laws as will uniaxial creep experiments that are continuously controlled and measured.

The requirements of *in situ* fatigue or electrical properties measurements have not been defined, but they will likely be as complex as the radiation creep experiments.

Irradiation creep data generated by irradiation of pressurized tubes [15] are shown in Fig. 7. These data show that, for 20% cold-worked type 316 stainless steel, irradiation creep is the dominant creep mechanism for temperatures up to about 600°C at the displacement rates found in EBR-II. The irradiation creep rate is must less dependent on temperature than is thermal creep. At the lower temperatures the strain is linear with atom displacements (irradiation time) and with stress [15].

V. CTR RADIATION EFFECTS FACILITIES

Now that the predicted bulk radiation effects that will be of importance in metal components of CTRs have been reviewed, it is useful to examine the existing and proposed facilities that can be used in the experimental investigation of these effects. In considering these facilities, three points are important:

(1) None of the irradiation sources available or proposed for the CTR program will meet all of the program needs. All facilities fall short on one or more of the requirements of neutron flux, neutron spectrum, availability, test volume, and duty cycle.

(2) Although neutron sources that match CTR conditions would be useful, it is unlikely that the CTR program will succeed

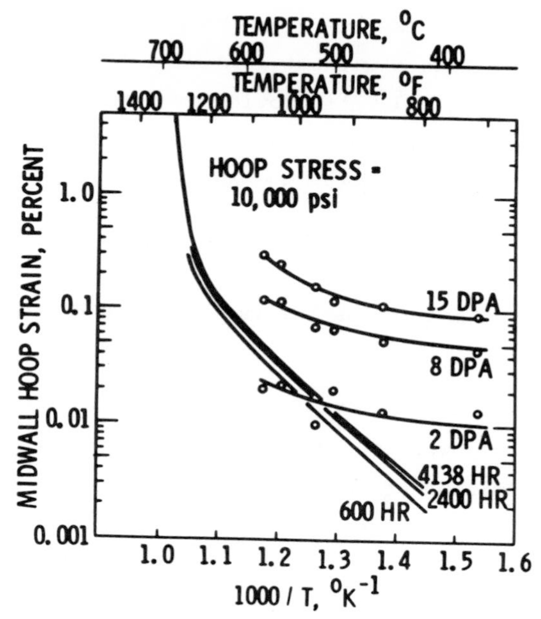

FIG. 7. Creep of pressurized tubes of 20% cold-worked type 316 stainless steel. Lines labeled with test times are for unirradiated tests, data points and lines labeled with displacement level are for pressurized tubes irradiated in the EBR-II. The times of control tests correspond approximately to the time at temperature of the irradiated specimens. (Ref. 15. Figure supplied by E. R. Gilbert, Westinghouse Hanford Company.)

or fail based on the availability of any single irradiation source.

(3) The CTR bulk radiation effects program will require the use of all facilities.

The characteristics and suggested uses of the three main classes of irradiation facilities are given in Table 6. The uses outlined in the table suggest that the main value of the ion bombardment facilities will be in the rapid screening of the microstructural response to irradiation of a large number of samples. The technique is restricted mainly in that it produces zones only a few microns deep. The advantages and restrictions of the ion-bombardment methods are discussed elsewhere by Taylor [16].

TABLE 6

CTR RADIATION EFFECTS FACILITIES

ION BOMBARDMENT	
(a) Characteristics:	Available now, temperature control good Very high damage rates Transmutation products can be included Restricted to thin samples
(b) Uses:	Swelling and microstructural response Radiation creep studies
FISSION REACTORS	
(a) Characteristics:	Available now, temperature control difficult Large volumes Damage rates ≈ CTR Transmutation products require special simulation Adaptable
(b) Uses:	Generate base data Match CTR conditions in Ni-containing alloys Screening studies
HIGHER ENERGY NEUTRON SOURCES	
(a) Characteristics:	Low flux, small volume Spectrum approximates CTR
(b) Uses:	Correlation studies to verify simulation Mechanism studies

Fission reactors are the only source in which to irradiate large numbers of samples big enough for the evaluation of all properties of interest, and to achieve high damage levels in these specimens. These reactors must be used to build the data base on radiation effects in all alloy systems of interest. The characteristics of these reactors are discussed by Horak [17]. One special case of the use of fission reactors must be elaborated on, because of its unique importance to the CTR program. This case is the high helium production rates achieved in alloys that contain nickel during irradiation in a spectrum containing thermal energy neutrons. Helium is produced by the reaction sequence of thermal neutron captures:

$$^{58}Ni + n \longrightarrow {}^{59}Ni + \gamma$$

$$^{59}Ni + n \longrightarrow {}^{56}Fe + {}^{4}He \quad . \qquad (2)$$

With this reaction sequence, nickel-containing alloys can be irradiated in existing reactors to the close simulation of CTR service conditions. Future higher-energy neutron sources, of which many varieties have been described [18], may ultimately match the neutron flux expected in CTRs. Until these conditions are achieved, however, the few available high-energy neutron sources must be used primarily to perform the correlation experiments that can be used to verify the simulation of CTR irradiation in the ion bombardment and fission reactor irradiations.

VI. SUMMARY

Neutron irradiation of structural materials in the high-flux regions of a CTR, especially the first wall of the reactor, will result in swelling due to cavity formation and loss of ductility through both lattice hardening and the effect of transmutation-produced helium on the fracture mode. These phenomena are generally expected, although the intensity and relative importance will be a strong function of the material and reactor operating conditions. A consideration of the effects of the 14 MeV D-T fusion neutrons on metals suggests that although the 14 MeV neutrons will produce damage at a higher rate than lower-energy neutrons, there is no basis for anticipating different forms of damage, except for the differences that are due to the higher rate of transmutation reactions.

Irradiation experiments designed to evaluate materials for CTR service must be conducted in a number of facilities, since none of the ion bombardment, fission reactors, or high-energy neutron sources adequately simulates the CTR environment. Used together in a well-coordinated evaluation program, the combined results from all of these facilities can be expected to lead to an understanding of the radiation effects that will occur in CTRs and to result in the development and qualification of materials for CTR service.

ACKNOWLEDGMENT

This research was sponsored by the Energy Research and Development Administration under contract with Union Carbide Corporation.

REFERENCES

1. F. W. Wiffen, "Radiation Damage in CTRs," p. 140 in *Proceedings of the Internation Working Sessions on Fusion Reactor Technology*, USAEC Report CONF-710624 (1972).

2. L. C. Ianniello, ed., *Fusion Reactor First-Wall Materials*, USAEC Report WASH-1206 (1972).

3. G. L. Kulcinski, compiler, "Bulk Radiation Damage to First-Wall Materials," p. 479 in *Fusion Reactor Design Problems*, IAEA (1974).

4. A. P. Fraas, "Conceptual Design of the Blanket and Shield Region and Related Systems for a Full-Scale Toroidal Fusion Reactor," ORNL-TM-3096 (May 1973).

5. R. G. Mills, ed., "A Fusion Power Plant," Plasma Physics Laboratory, Princeton University Report MATT-1050 (August 1974).

6. G. L. Kulcinski, R. G. Brown, R. G. Lott, and P. A. Sanger, "Radiation Damage Limitations in the Design of the Wisconsin Tokamak Fusion Reactor," *Nucl. Technol.* 22, 20 (1974).

7. B. Badger et al., "UWMAK-I, A Wisconsin Toroidal Fusion Reactor Design," UWFDM-68, University of Wisconsin, November 1973.

8. G. L. Kulcinski, D. G. Doran, and M. A. Abdou, "Comparison of Displacement and Gas Production Rates in Current Fission and Future Fusion Reactors," p. 329 in *Properties of Reactor Structural Alloys After Neutron or Particle Irradiation*, ASTM-STP 570, ASTM, 1975.

9. A. Seeger, "The Nature of Radiation Damage in Metals," p. 101 in *Radiation Damage in Solids*, IAEA, 1962.

10. F. W. Wiffen and E. E. Bloom, "Effect of High Helium Content on Stainless Steel Swelling," *Nucl. Technol.* 25, 113 (1975).

11. P. J. Maziasz, F. W. Wiffen, and E. E. Bloom, "Swelling and Microstructural Changes in Type 316 Stainless Steel Irradiated Under Simulated CTR Conditions," to be published in *Radiation Effects and Tritium Technology for Fusion Reactors*, ERDA Report CONF-750989 (1976).

12. W. G. Johnston, J. H. Rosolowski, A. M. Turkalo, and T. Lauritzen, "An Experimental Survey of Swelling in Commercial Fe-Cr-Ni Alloys Bombarded with 5 MeV Ni Ions," *J. Nucl. Mater.* 54, 24 (1974).

13. E. E. Bloom, J. O. Stiegler, A. F. Rowcliffe, and J. M. Leitnaker, "Austenitic Stainless Steels with Improved Resistance to Radiation-Induced Swelling," *Scripta Met.* 10, 303 (1976).

14. E. E. Bloom and F. W. Wiffen, "The Effects of Large Concentrations of Helium on the Mechanical Properties of Neutron-Irradiated Stainless Steel," *J. Nucl. Mater.* 58, 171 (1975).

15. E. R. Gilbert and J. F. Bates, "In-Reactor Creep of 20% Cold-Worked 316 Stainless Steel Pressurized Tubes," *Trans. Amer. Nucl. Soc.* 22, 183 (1975).

16. A. Taylor, "Scope of the Use of Energetic Ions to Study Radiation Effects," p. 842 in *Proceedings of the International Conference on Radiation Test Facilities for the CTR Surface and Materials Program*, ANL/CTR-75-4 (1975).

17. J. A. Horak, "Use of Fission Reactors for CTR Bulk Radiation Effects Studies," p. 830, *ibid.*

18. *Proceedings of the International Conference on Radiation Test Facilities for the CTR Surface and Materials Program*, ANL/CTR-75-4 (1975).

Chapter 6

TRITIUM-RELATED MATERIALS PROBLEMS IN FUSION REACTORS*

R.G. Hickman

Lawrence Livermore Laboratory, University of California
Livermore, CA. 94550

I. INTRODUCTION

Our goal here is to describe the pressing materials problems that must be solved before tritium can be used to produce energy economically. Unlike most of the other materials problems that have been discussed in the previous lectures of this series, the tritium problem is intimately connected with on-site chemical processing. Therefore, to fully appreciate it, one must have some understanding both of tritium and its peculiarities and

*This work was performed under the auspices of the U.S. Energy Research & Development Administration, under contract No. W-7405-Eng-48.

of the chemical processing requirements of a thermonuclear fusion power plant.

Tritium and deuterium are heavy isotopes of hydrogen that can undergo nuclear fusion with relative ease. This process is expressed by the reaction:

$$T + D \rightarrow {}^{4}He + n + 17.6\ MeV. \tag{1}$$

Tritium is a necessary fuel in thermonuclear reactors and is therefore at the heart of all fuel-cycle problems.

Because tritium has a half-life of only 12.3 years (1), it is not found naturally in useful concentrations and must, therefore, be bred, usually in nuclear reactors. This is expensive (about $10,000/g), but the cost is not firm because new production reactors have not been built for decades.

Tritium is a hazard to man and the environment. Although the biological effectiveness of its radiation is generally taken as unity (the same as gamma rays), tritium is believed to be three times as toxic as gamma rays in chronic, low-level doses (2). Tritium is absorbed easily through the lungs or skin. The maximum permissible concentration (mpc) in air is about one part per trillion. A lethal dose may be received from a single breath of tritiated water vapor or from a single drop of tritiated liquid on the skin. The biochemical effects of tritium have been studied extensively (3) and are fairly well understood.

The chemical behavior of tritium is nearly identical to that of normal hydrogen: it burns, explodes, and can be incorporated into biological molecules; it permeates and embrittles metals as hydrogen does. Embrittlement may also be caused by helium produced in the radioactive decay of tritium (4).

II. *BREEDING TRITIUM*

Tritium is usually bred in nuclear reactors using lithium as the breeding material. Lithium can absorb a neutron and produce tritium and helium:

$$^{6}Li + n \rightarrow T + He. \quad (2)$$

Fusion reaction (1) produces a neutron that can be used to create more tritium via reaction (2). Other important breeder reactions include that of ^{7}Li,

$$n + {}^{7}Li \rightarrow He + T + n' \quad (3)$$

and the neutron-multiplying reaction of beryllium,

$$n + {}^{9}Be \rightarrow 2He + 2n'. \quad (4)$$

The product neutrons in reactions (3) and (4) can also be used to breed tritium via reaction (2).

Several classes of breeding materials have been considered:

- pure lithium metal with isotopic enrichments
- lithium alloys, including intermetallic compounds (LiAl, Li_4Sn, and Li_7Pb_2)
- molten salts (Li_2BeF_4 and $LiNO_3/LiNO_2$ eutectic)
- solid ceramics ($LiAlO_2$, Li_2O, Li_2C_2, Li_3N, LiD, and $Li_2Be_2O_3$)

These materials may be used singly or in combination, with or without neutron multipliers and reflectors. Their effectiveness depends on the design of the reactor. In general, as atoms other than neutron multipliers are added to lithium to form compounds or alloys, the breeding ability of the combination decreases. Thus, LiAl is a poorer breeder than pure lithium, which in turn is poorer than Li_2Pb_7, where lead is a neutron multiplier.

The breeding material is contained in a blanket layer just outside the innermost wall of the reactor. Most of the neutrons produced are absorbed in the blanket; it is here that tritium is produced and that neutron energy is deposited. This general scheme is shown in Fig. 1.

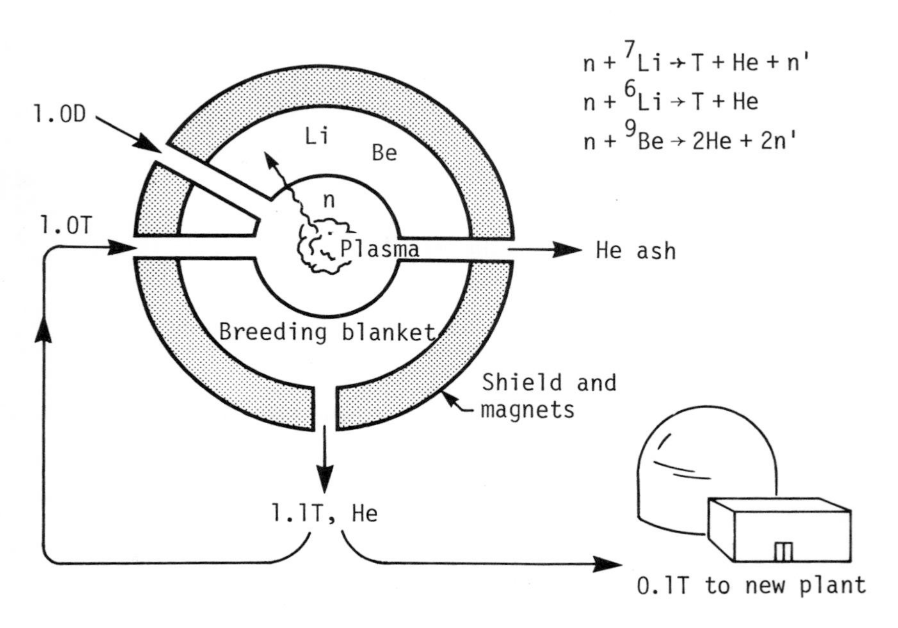

FIG. 1. Schematic diagram of net tritium breeding in a fusion power plant.

III. RECOVERING BRED TRITIUM

The various breeding materials show vastly different behaviors with respect to the release of tritium. The rate of tritium recovery from the breeding material usually depends on the concentration of tritium in the blanket. The concentration must climb to its steady-state value before recovery can proceed at a steady rate. This behavior is shown graphically in Fig. 2, where the shaded area gives the required start-up inventory. This is the

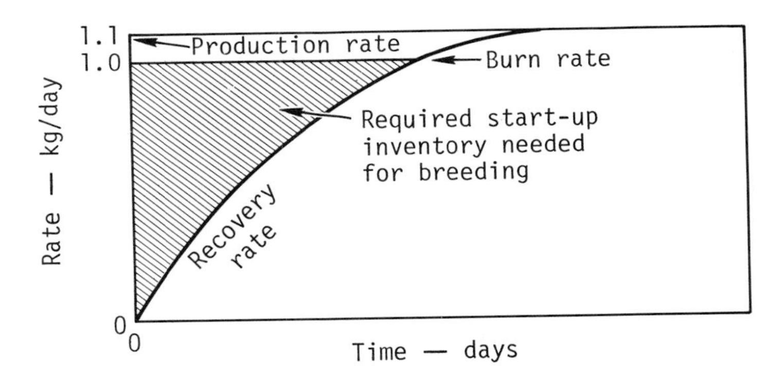

FIG. 2. Graphic representation of tritium start-up-inventory requirements as determined by recovery processing.

inventory required to keep the plant going until the recovery rate equals the burn rate, i.e., the break-even rate.

Assuming that the rate of recovery of tritium from the blanket is directly proportional to the tritium concentration in the breeding material, it can be shown that the ratio of break-even inventory I_{be} to steady-state inventory $C_{ss}V$ is given by

$$\frac{I_{be}}{C_{ss}V} = 1 - \left\{\left(1 - \frac{1}{\alpha}\right)\left[1 - \ln\left(1 - \frac{1}{\alpha}\right)\right]\right\} , \qquad ($$

where α is the breeding ratio (5). Evaluated numerically for di ferent values of α, this expression gives the data shown in Tabl

TABLE I.

TRITIUM INVENTORIES AS A FUNCTION OF BREEDING RATIO

α	$I_{be}/C_{ss}V$
1.01	0.944
1.05	0.807
1.10	0.691
1.20	0.535
1.30	0.431
1.40	0.356

We can see that if it is important for a plant to become very quickly self-sufficient in tritium, the blanket materials chosen should have neutronic properties that allow high breeding ratios. In general, this means avoidance of non-breeder materials that capture neutrons, and inclusion of neutron multipliers such as Be or Pb. The required break-even inventory is further reduced by choosing materials that have a low steady-state tritium concentration C_{ss} and by minimizing the volume V of breeding material. This requires the use of solid materials that show low (but not zero) solubility and high diffusivity for tritium. The solubility should not be zero because in a radiation field, all solids develop internal voids that would then act as permanent sinks for tritium. In general, these relationships between solubility, diffusivity, and radiation stability in solid breeding materials are unknown. Breeding materials are expected to be in powder or liquid form rather than in massive solid form so that processing kinetics will be fast.

There are three major means of recovering tritium:

- permeation,
- trapping by adsorption, and
- trapping by absorption.

Molten salt extraction from liquid lithium followed by gas sparging is being investigated and may also be feasible.

Permeation

In the permeation recovery process, tritium, in liquid-lithium, high-pressure-helium, or potassium-vapor coolants, contacts one side of a selectively permeable membrane, such as niobium. The popular Nb-1% Zr alloy might also be used as a membrane material, but due to the strong chemical affinity of Zr for tritium, it could show premature embrittlement from helium. Other possible permeation membranes are V, Ta, and Pd. On the other side of the membrane, the tritium partial pressure is lower so the tritium permeates the membrane. The rate of this continuous process is usually $R_1 = K_1\left(P_1^{1/2} - P_2^{1/2}\right)$, where K_1 is a constant and P_1 and P_2 are the tritium partial pressures on the two sides of the membrane. Ample evidence (6) shows that when the permeation rate is surface-limited, it is better described by $R_1 = K_2(P_1 - P_2)$. In either case, as shown in Fig. 3, the recovery rate increases as P_1 (or the blanket tritium inventory) increases. As just described, it is zero at start-up.

At one time, it was suggested that adding an excess of normal hydrogen to the very dilute tritium stream might increase the recovery rate by some kind of sweeping action during the permeation process, but this has since been shown to be a hindrance rather than a help (7).

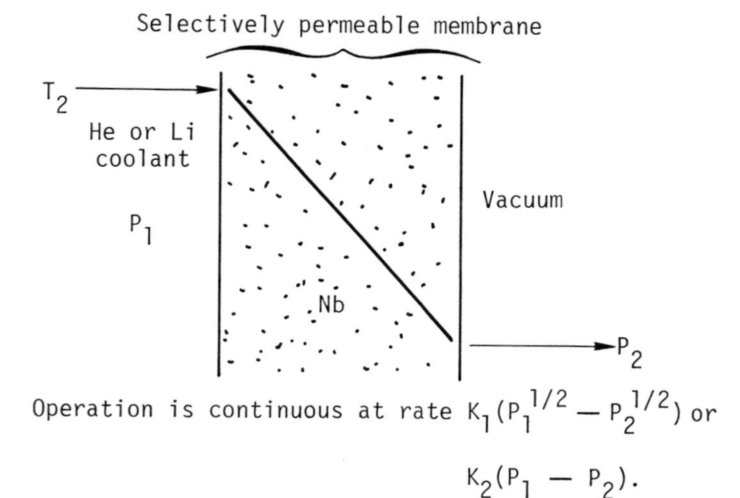

FIG. 3. Schematic diagram of a permeation-type process for recovery of tritium from a He or Li coolant stream.

Absorption

As shown in Fig. 4, the trapping processes are generally cyclical and require regeneration of the trap absorbents and adsorbents. Molecular tritium can be absorbed from a vacuum or from flowing helium or lithium streams if they come into contact with certain metals (typically cerium, scandium, or titanium) to form a stable hydride. The dissociation pressures of a few of the more stable binary hydrides are shown in Fig. 5. Uranium has been shown to be capable of removing 99.9+% of the D_2 from a flowing stream of argon containing 1% D_2, but this is not enough for T_2 recovery in a breeding cycle recovery process. On the basis of chemical reactivity, price, and dissociation pressures at both high and low temperatures, cerium has been chosen as a promising regenerable getter (tritium absorber) (8). Some ternary hydrides might be as good or better, but our knowledge of these more complex systems is very scanty in comparison. Long-term poisoning effects need to be studied.

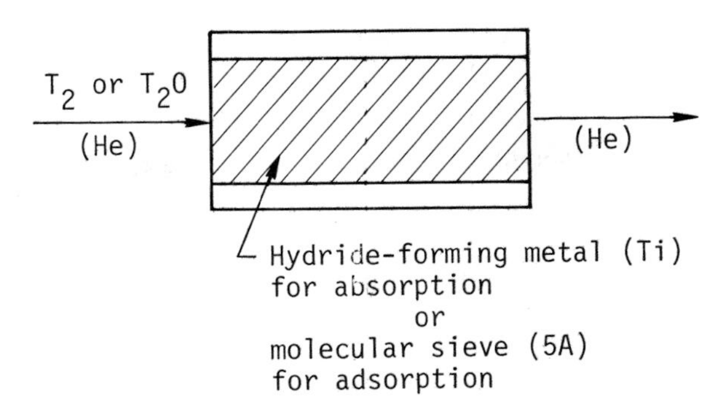

FIG. 4. Schematic diagram of a trapping-type process for recovery of tritium from a He stream.

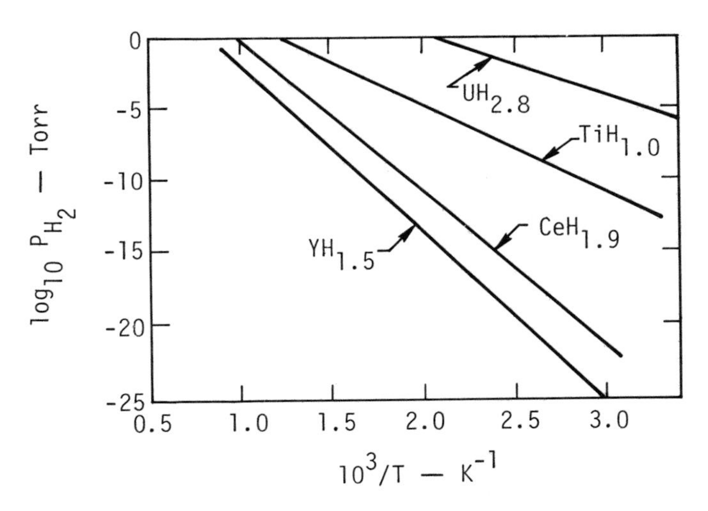

FIG. 5. Equilibrium hydrogen pressure over some substoichiometric binary metal hydrides as a function of temperature.

Adsorption

Tritium or tritiated vapors (e.g., tritiated water vapor) can be removed from flowing gases by selective adsorbents: molecular sieves, silica gel, or activated charcoal. These adsorbents are particularly effective when operated at low temperatures, but recovery processes capable of operating at high temperatures are usually preferred for reasons of thermodynamic efficiency. This is particularly true if the reactor coolant stream functions as part of the recovery cycle and if a significant fraction of that stream must be processed to supply the needed tritium.

In both trapping processes, first-order kinetics require the recovery rate R_2 to be proportional to $(P_1 - P_e)$, where P_1 is the partial pressure of the tritium-bearing species in the gas and P_e is the equilibrium partial pressure of the species being trapped at that particular temperature. Again, the recovery rate depends strongly on P_1, which is a principal factor in determining the blanket's tritium inventory.

Recovery of tritium from liquid lithium coolant is expected to be difficult when the lithium concentration is kept below 10 ppm by weight. For this concentration, the blanket inventory could be about 5 kg ($50,000,000), but as shown in Fig. 6, the tritium partial pressure available to drive a separation process would be only 10^{-6} Torr at 1000 K. Thus, large processing equipment might be required.

Figure 7 shows that adding aluminum to the lithium substantially raises the partial pressure. Tritium has been produced in this alloy for defense purposes for many years (9,10). The phase diagram for the system (Fig. 8) clearly shows that these intermetallic compounds, LiAl and Li_2Al, probably could not be used as solid powders because of sintering effects at several hundred degrees Celsius.

An encouraging recent development is the renewed activity with $LiAlO_2$ powders. The tritium-breeding reaction is

$$n + {}^6LiAlO_2 \rightarrow \frac{1}{2}\, T_2O + He + \frac{1}{2}\, Al_2O_3. \qquad (6)$$

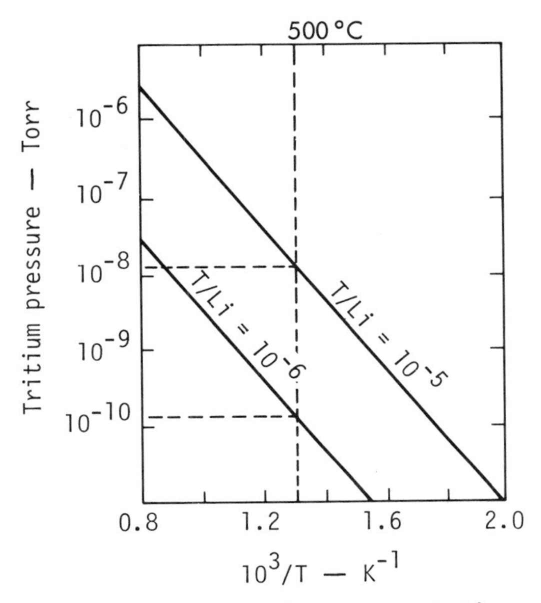

FIG. 6. Graphic representation of the pressure-composition-temperature relationship for the hydrogen-lithium system in the dilute range.

Tritium and tritiated water show a low solubility in the solid powder. Particles given the equivalent of a 1-h exposure to the neutron flux of a fusion reactor have released tritium in a mean time of 4.3 h at 600°C (10,11). It appears on the basis of recent work, that at high temperatures, tritium blanket inventories could be kept to a few hundred grams in a large power plant using the oxide reaction. There are still a few possible problems and unanswered questions, however.

The high temperatures required to release the tritiated water quickly from $LiAlO_2$ may make the use of refractory metals or carbides in structural members necessary, but these materials are chemically attacked by water vapor at high temperatures. Our current plan in response to this water vapor problem is to construct the blanket from lithium beryllate (12) and excess

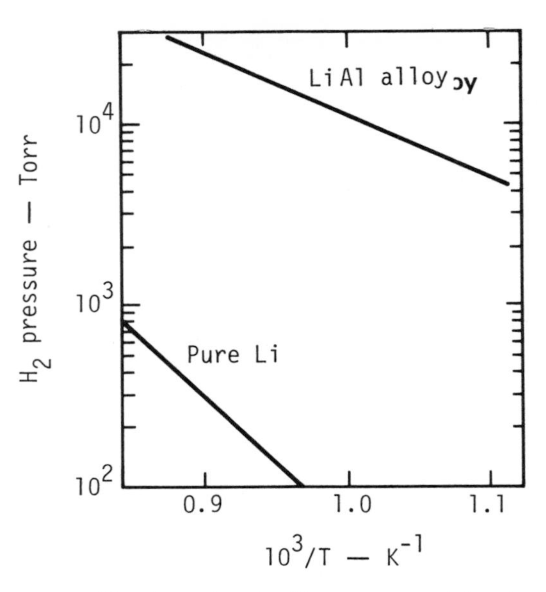

FIG. 7. The influence of alloying lithium with 50% aluminum on the equilibrium partial pressure of hydrogen in the two-phase (high hydrogen content) region.

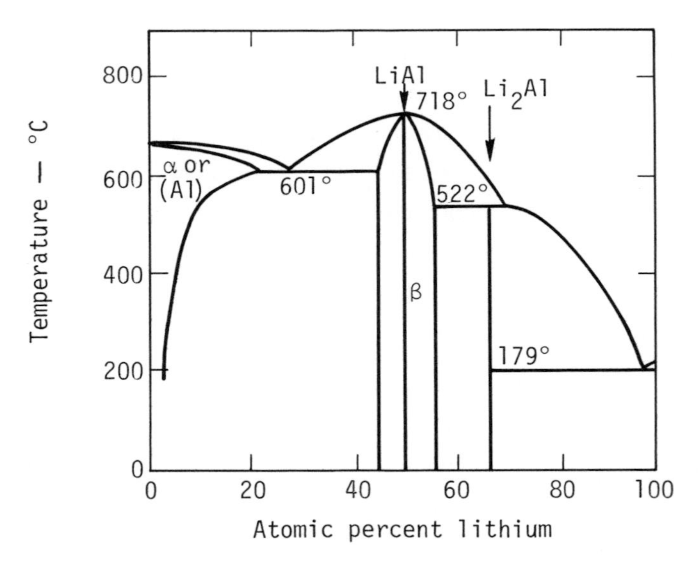

FIG. 8. The lithium-aluminum phase diagram.

beryllium as a partially sintered mix of powders. The breeding reaction would then have the following chemical result:

$$n + \frac{1}{2} Li_2Be_2O_3 + Be_{(XS)} \rightarrow \frac{3}{2} BeO + He + \frac{1}{2} T_2 \quad (7)$$

The excess beryllium would effectively scavenge the oxygen freed in the breeding reaction, would be compatible with the breeding compound, and would act as a neutron and energy multiplier in the blanket.

The availability of beryllium for this blanket design is a big question. Without further exploration and discovery, current supplies could soon be exhausted (13). Also, continuous irradiation of either material ($Li_2Be_2O_3$ or Be) may induce swelling and internal void formation, which could permanently tie up a much larger amount of tritium than presently expected. The presence of Be could aggravate this behavior due to its high (n,α) cross section. Nothing has been done experimentally on this particular problem. In general, these problems on breeding and recovery are common to all fusion reactor concepts and are independent of the plasma confinement concept.

IV. CONTAINING TRITIUM

Environmental control of tritium in fusion reactors presents a challenging set of problems. As we have noted above, tritium behaves just like hydrogen: it permeates metals and plastics, it readily becomes airborne, and once mixed with normal water, it cannot be conveniently separated. These potential escape routes are shown in Fig. 9.

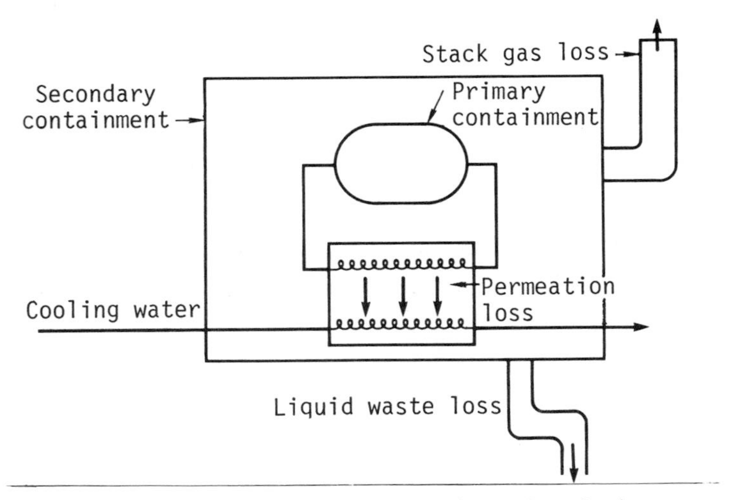

FIG. 9. Schematic drawing of a fusion power plant showing tritium escape routes.

Permeation Losses

The high tritium concentrations or partial pressures that provide the high recovery rates desired from the breeding blanket may also cause high rates of permeation loss to the environment. For example, the reference design of the theta-pinch reactor calls for a liquid-lithium breeding blanket. To keep permeation losses to less than 30 Ci/day, the partial pressure of tritium has to be no more than 10^{-10} Torr above the lithium (14). This is an incredibly low partial pressure to maintain for a product in an industrial situation, and to date, no way of doing this has been found.

An obvious way to reduce some permeation losses is to use hydrogen-impermeable heat exchangers. This can be done by dropping the coolant operating temperature (and efficiency), by adding additional heat exchange loops, or by adding new materials to the heat exchangers. Some of these new materials are barrier metals (copper, tungsten) (15), barrier oxides (oxidation films,

BeO) (7,16), and metal getters (titanium, cerium), and, as shown in Fig. 10, all might be placed into laminated (triplex) tubing to reduce or eliminate permeation losses. The virtue of oxide barriers is that, as shown in Fig. 11, they are far less permeable than the least permeable of metals (tungsten). However, they suffer from the disadvantage of being difficult to fabricate. This is especially true if pore-free thin films are required as the intermediate layer.

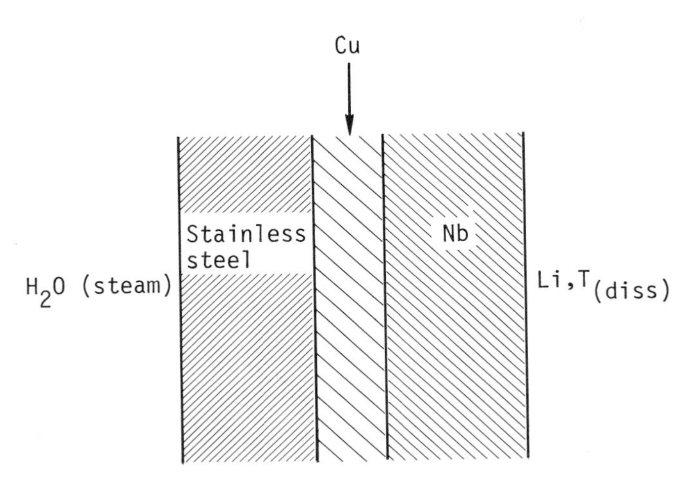

FIG. 10. Schematic wall cross section of a possible triplex heat-exchanger tube designed to reduce tritium escape via the permeation mechanism.

Liquid Loss

High-level tritium in waste water is usually immobilized by reacting it to form concrete, which can then be encapsulated and buried (17). In the future, it may be possible to remove the HTO from waste water selectively by laser isotope-

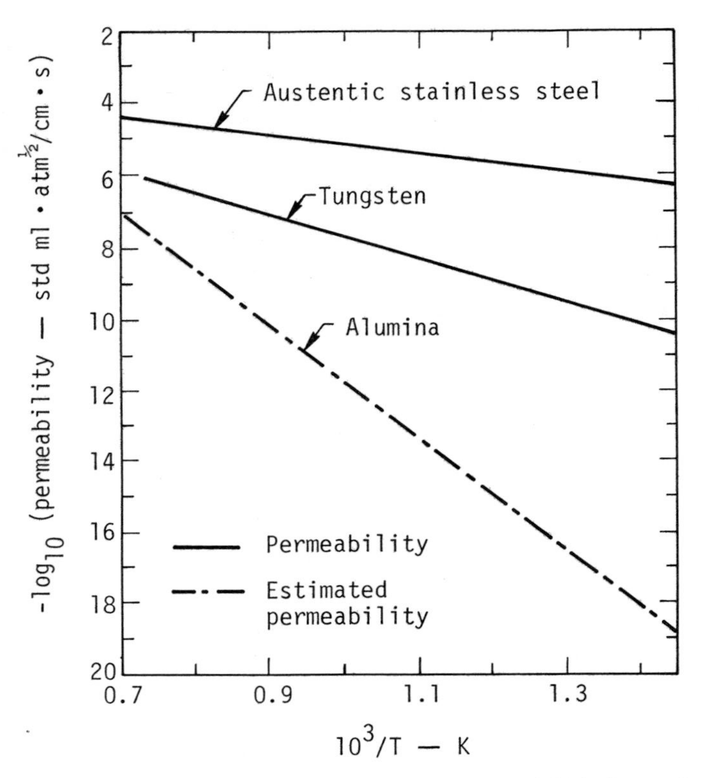

FIG. 11. The relative permeabilities of stainless steel, tungsten, and alumina.

separation techniques and thus greatly reduce the amount of solid waste to be buried. Progress to date has been slight however (18,19).

Gaseous Loss

Tritium in gaseous wastes can be reduced to low levels with trapping similar to those used for recovering tritium from the breeding blanket. Tritiated hydrocarbons may be adsorbed directly or oxidized to trap the tritium as tritiated water on molecular sieves. This oxidation would normally be done at a few hundred degrees Celsius in an oxygen-rich stream flowing over precious-metal catalysts. The gaseous species HT, DT, and T_2 can be trapped directly by cryogenic absorption or adsorption. Usually, however, they are trace contaminants in air and can therefore be catalytically converted to water or reacted with certain hot transition metal oxides (CuO, MnO_2) to form water. Tritiated water is then trapped on molecular sieves at or below room temperature. The removed fraction of tritiated water can be increased by deliberately swamping the gas flow with water vapor after the first trapping, but this adds greatly to the liquid-waste disposal problem. The general scheme is shown in Fig. 12.

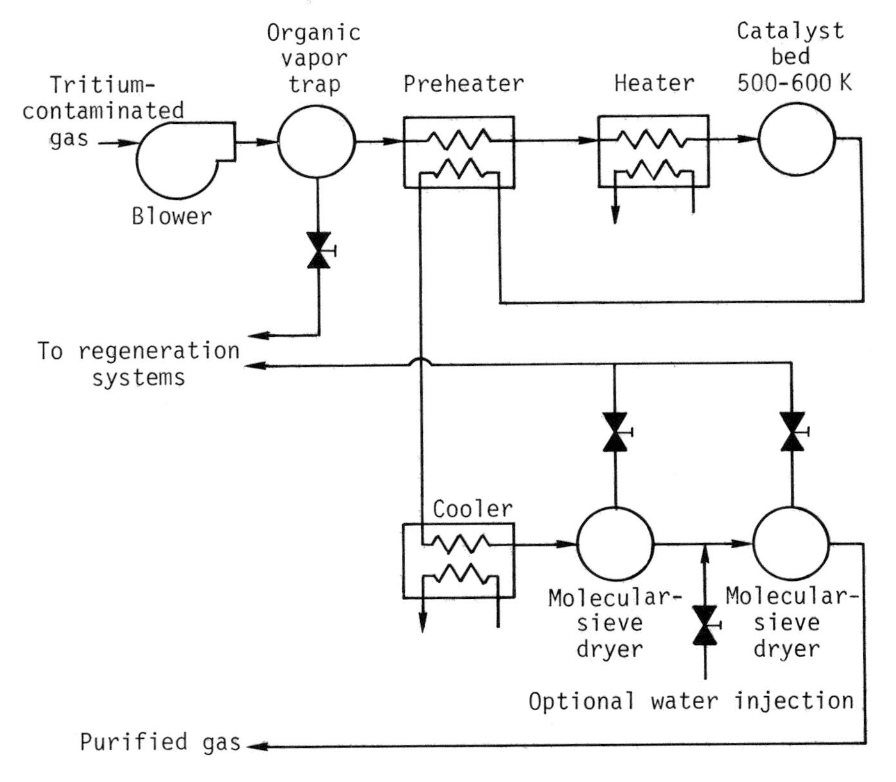

FIG. 12. Schematic diagram of a system using catalytic oxidation for removing tritium from other gases.

Two new state-of-the-art gaseous-effluent control systems have been built in the U.S. recently. They both employ catalytic oxidation, followed by adsorption of tritiated water on a molecular sieve (20,21). To date, they have not seen enough operation to demonstrate satisfactory long-term performance, but they certainly represent our present best hopes for keeping the gaseous release of tritium and its compounds to an acceptable minimum. If these systems do not perform up to expectations, much additional work on process design, catalyst development, and molecular sieve development will be required. Again, these containment problems are common to all fusion reactor concepts.

V. FUEL RECYCLING

Recycling the fuel in a fusion reactor is necessary because most of it does not react before it is exhausted from the plasma region. Fortunately, suitable materials and standards of construction exist for every part of the fuel-recycling system where large amounts of tritium must be handled at relatively high pressures. Other parts of the fuel cycle need substantial additional materials research and development before fusion power can become a reality.

Several compounds must be removed from the fuel before it can be reused:

- helium ash
- hydrogen from the D(T,H)T reaction [which is present in smaller quantities than in the normal D(T,n)He reaction]
- methanes
- traces of air

The fuel-recycling loop is fairly simple and is represented in Fig. 13.

Recycling problems depend more on the reactor concept than do other problems discussed so far. Mirror reactors are steady-state devices with continuous neutral beam injection and exhaust, and neutral injection requires separate deuterium and tritium streams. Tokamaks operate in a pulsed mode (about a 1-min cycle) with neutral beam or pellet injection and with between-cycle exhaust. The theta-pinch reactor is short-pulsed (a few seconds per pulse), and mixed isotopes in the fuel supply are desirable. Finally, laser-driven microexplosions are very short pulses (milliseconds), and several pulses per second are expected. All the reactors operate under high-vacuum conditions. In a mirror reactor, the fuel exhausts directly to an electrical direct convertor that recovers most of the kinetic energy of the escaping particles.

Neutral beam injection allows preheated (high-energy) particles to be fed into the plasma across magnetic field

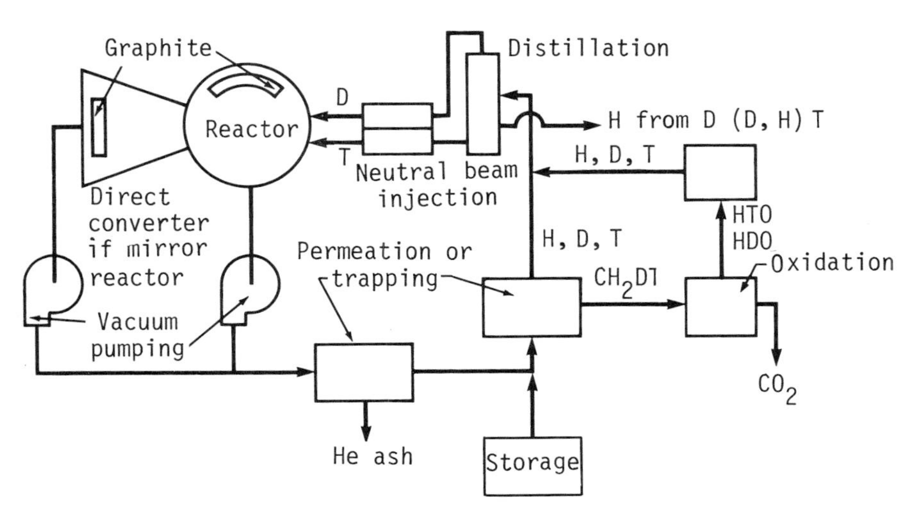

FIG. 13. Simplified schematic diagram of the fuel-recycling loop on a fusion reactor.

lines. Some necessary charge-exchange reactions are velocity-dependent rather than energy-dependent, so isotopic purity increases the efficiency of neutral beam injection. The separation of deuterium and tritium (along with the removal of hydrogen) will undoubtedly be accomplished by means of cryogenic distillation, an existing technology. Purely physical operations, such as distillation, must be supplemented by chemical methods to break down the DT, HD, and HT molecules as in Fig. 14. This is routinely done now,

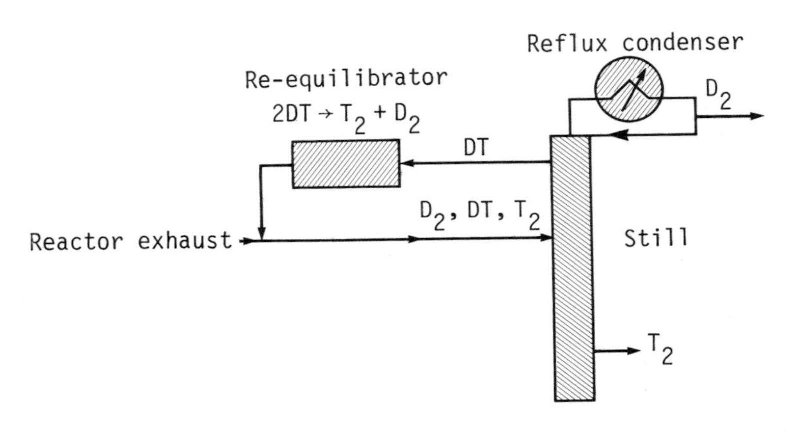

FIG. 14. Diagram of a cryogenic distillation column with a catalytic disproportionator.

but efficiency should be improved. In particular, it would be worthwhile to study the effect of the tritium radiation field on the kinetics of re-equilibration in a solution of partially separated diatomic molecules. Development of catalysts to promote the exchange reactions at liquid-hydrogen temperatures would further reduce costs and tritium inventory requirements.

Carbon First Wall

The innermost reactor wall may be lined on the plasma side with a graphite curtain to reduce neutron-radiation damage and to lower the atomic number of materials sputtered from the first wall (22). In mirror reactors, the direct-converter collection electrodes may also be made of graphite because of its high-temperature properties. Energetic particles escaping from the plasma will be implanted in the curtain materials. A certain amount of the implanted tritium will be lost from the fuel stream because of chemical interaction with the curtain materials, and this poses another unexplored inventory problem. Very crude calculations suggest the possibility of permanently trapping tens of kilograms of tritium in graphite components.

Methane Formation

In low-pressure plumbing systems, tritium reacts with hydrocarbons and other carbon compounds to form methane, by an unknown, radiolytic mechanism. This is shown in Fig. 15 on the basis of data recently taken by G. Morris at the Lawrence Livermore Laboratory (LLL). The presence of methane in the fuel supply is undesirable because it can plug the cryogenic still needed for hydrogen isotope separation and can introduce impurities into the plasma.

Some insight into the mechanisms involved may be obtained from the recent research of Balooch and Olander (23), showing that hydrogen is added to C, CH, CH_2, and CH_3 free radicals on the surface of graphite at 506 K. Some of their experimental data and theory are shown in Fig. 16 and Eq. (8). The theory is obtained from an approximate solution to the time-dependent non-

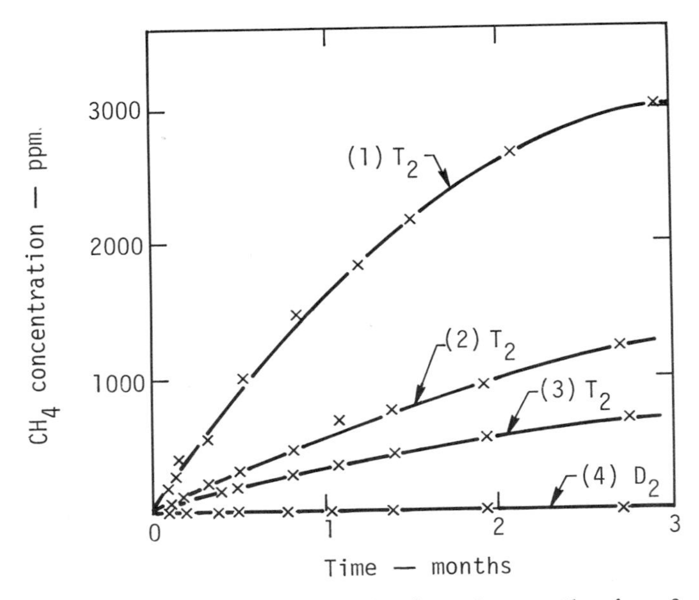

FIG. 15. Data showing the synthesis of tritiated methanes in relatively clean stainless-steel tritium-filled vessels: (1) 25°C, drawn tubing, freon washed; (2) 100°C, drawn tubing, vapor degreased; (3) 25°C, machined surface, vapor degreased, vacuum baked at 100°C for 24 h; (4) 100°C, drawn tubing, vapor degreased.

linear differential equation describing the interactions of a chopped molecular beam of hydrogen atoms with the surface.

$$\varepsilon_{CH_4} = \frac{3K_1K_2k_3\eta(C_0/H)^2\sin\phi_{CH_4}}{\omega + (H^2D/2)^{1/2}\omega^{1/2}}, \tag{8}$$

where:

ε = the apparent reaction probability = $[CH_4]/[H_2]$ at the detector,

K_1 = the equilibrium constant for H, C, and CH on the surface,

K_2 = the equilibrium constant for H, CH, and CH_2 on the surface,

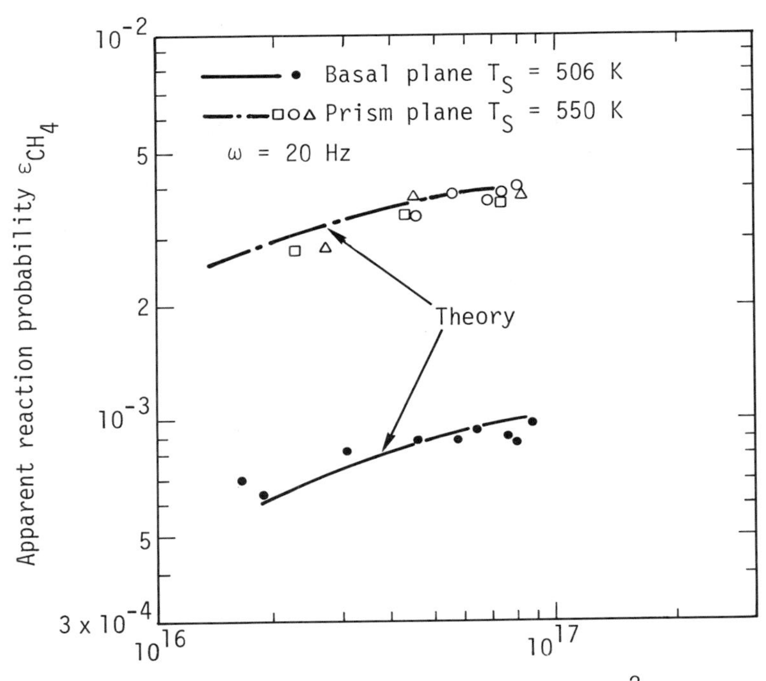

FIG. 16. Data and theory of methane formation by reaction of graphite with chopped atomic hydrogen beam, showing the variation of methane reaction probability with beam intensity. The different symbols for the prism plane represent replicating runs.

k_3 = the rate constant for $CH_{2(ads)} + H_{(ads)} \rightarrow CH_{3(ads)}$,

η = the sticking probability of H atoms in the beam to the surface,

C_0 = the bulk concentration of H atoms just below the surface,

H = the solubility coefficient relating C_0 to the surface concentration,

ϕ = the methane phase lag relative to the hydrogen at the detector,

ω = the modulation frequency of the beam, and

D = the diffusion coefficient of hydrogen atoms in the bulk graphite.

In the case of tritium in a fuel system, each decay produces a large number of dissociations in the gas phase, and also an uncombined free tritium atom that remains from the original T_2 molecule. These free atoms may react in much the same fashion as those in the molecular beam experiment. Whether additional methane will be formed by energetic implanted tritons or deuterons reacting with graphite is unknown.

VI. LASER-FUSION FUELING

The materials problems associated with safety and containment in laser fusion are identical to those for CTR. Likewise, the requirement to breed an excess of tritium persists, and since the energy-producing reactions are the same, the motivation is still strong to operate the breeding blanket at a high temperature so that it may be a part of a thermal conversion cycle. The fuel-cycle problems, however, are a bit different, and as one might expect, they involve materials problems.

The conceptual designs of laser-fusion power plants are in a very primitive stage compared to those using magnetically-confined plasmas. In general, however, the fueling scheme preferred (for reasons of simplicity) is the injection of frozen DT pellets at a high repetition rate (10 to 100 Hz). Other parts of the laser-fusion fuel cycle would be much the same as those already described for magnetic-confinement systems.

Cryogenic Targets

Heating and compressing solid pellets of DT to fusion conditions requires lasers of enormous power. The laser requirements are relaxed by a significant factor if the frozen DT is in the form of a spherical hollow shell (24). Liquid hydrogen droplets with internal cavities have recently been produced (25), but the sphericity and concentricity of the internal and external surfaces were inadequate for use as laser-fusion targets. Improvements on the existing techniques and development of other methods would be of very great interest.

Remarkably, most of the cryogenic properties of DT are unknown. Reasonable estimates have been made (26), but experimental verification has not yet been provided. Although we speak generally of DT, in any real DT system, three different molecular species (D_2, DT, and T_2) actually co-exist. Each species shows different physical properties because the isotope effect is quite large. It is this difference that forms the basis of isotopic separation by cryogenic distillation. Figure 17 contains the simplest possible ternary phase diagram for DT. It assumes regular solution behavior between species and phases and ignores complications that could arise from the varying amounts of ortho and para D_2 and T_2 that might be present. As can be seen, liquid that freezes will change composition, and depending on the kinetics of the phase change, spatial segregation of the molecular species could occur during formation of a pellet.

Current Targets

Present targets are mostly microspheres filled with high-pressure DT gas. Curiously enough, more goal-directed

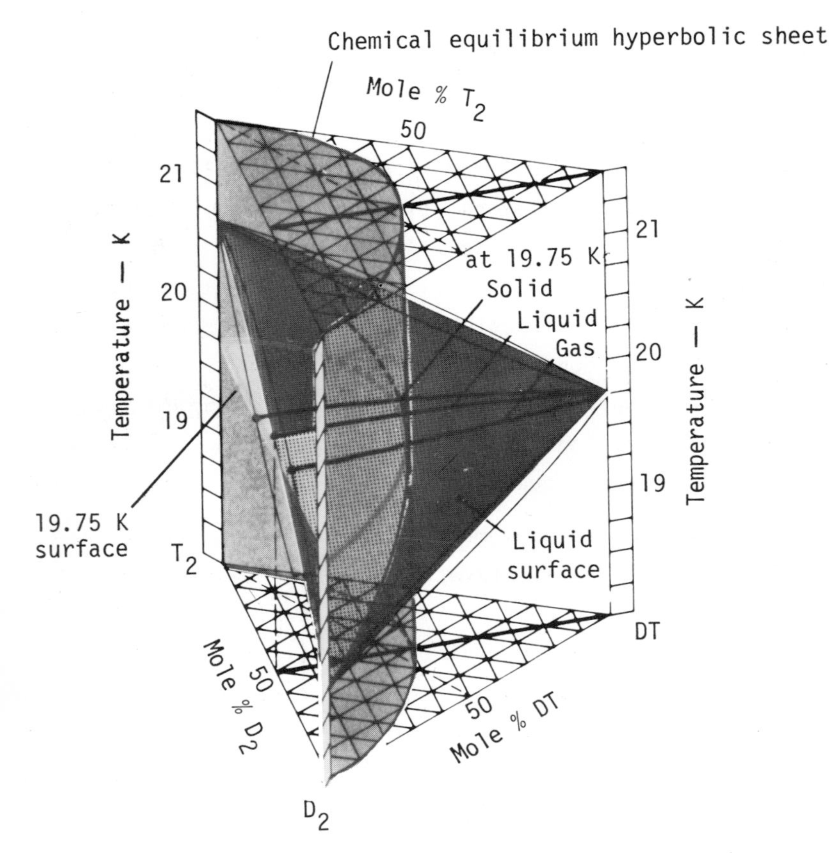

FIG. 17. Estimated ternary phase diagram for the DT system in the vicinity of the phase boundaries.

effort probably has been expended on these relatively simple targets than on all the other materials-related tritium problems confronting us in CTR with magnetic confinement. A photograph of typical DT-filled microspheres in an interference microscope is shown in Fig. 18. Those that are found to have the proper size, wall thickness, and degree of perfection are mounted for subsequent testing in the laser facility. A mounted target is shown in Fig. 19. Only about 1 microsphere out of 10,000,000 can pass the rigorous selection tests and go on to be shot in the laser (27).

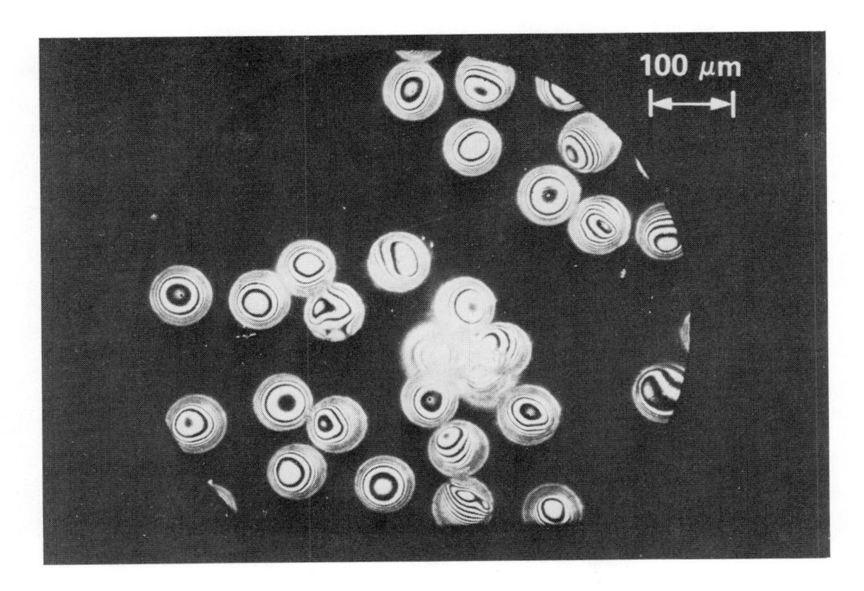

FIG. 18. Photograph of DT-filled glass microspheres under an interference microscope. Only one shows acceptable sphericity.

Development of techniques to give a higher survival rate would be of obvious benefit.

Pressurization is accomplished by heating the microspheres in a suitable pressurized vessel so that DT dissolves in and diffuses through the glass until the internal and external pressures are equal. Cooling under pressure traps the gas inside the microsphere. As soon as the external pressure is released, the trapped gas begins to escape slowly by diffusion. The rate is reduced by storage in a refrigerated location, but the exasperating aspect of this off-gassing is that it varies by as much as a factor of 100 from target to target and is predictable only on a statistical basis (28).

Figures 20 and 21 show two well-established phenomena. The first is that the outgassing transient for a batch

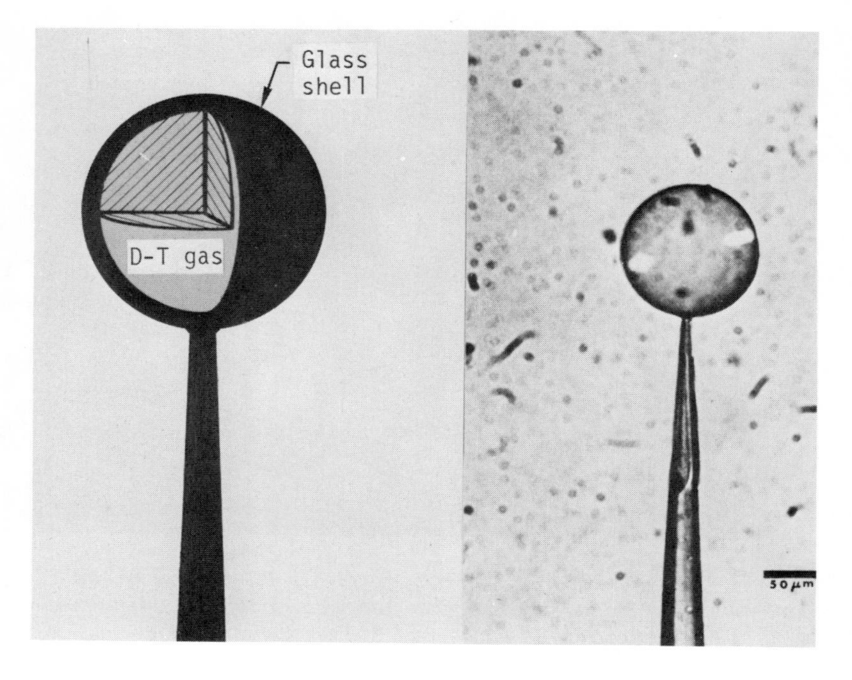

FIG. 19. A gas-filled microsphere mounted on a 5-μm stalk for further experimentation.

of spheres is non-exponential. (The reasons for this are not well understood.) The second is the variability between targets, which is believed to result from differences in glass composition. Long-term storage of surplus targets for later experiments is of questionable practicality because nondestructive absolute determination of the amount of tritium in a single target is not presently possible and interpretation of laser-fusion experiments depends strongly on this information. Work in this area should continue.

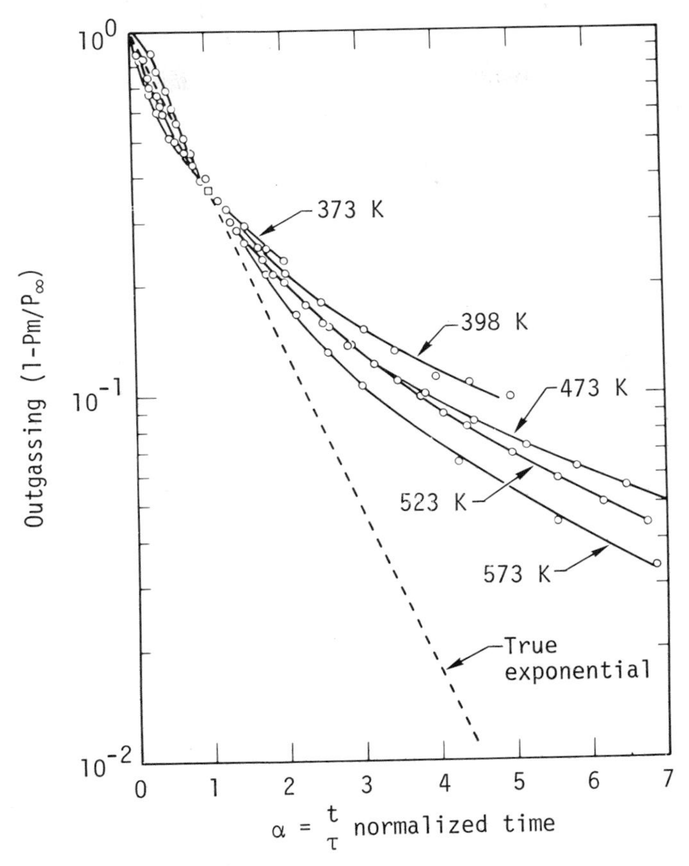

FIG. 20. Unexplained non-exponential outgassing behavior of pressurized laser targets.

VII. CONCLUSIONS

Materials problems in fusion reactors of all kinds arise in the context of breeding, recovering, containing, and recycling tritium. Technically, all these problems are probably solvable, but the economic feasibility of the solutions is still in question. The treatment of materials problems here is necessarily very general, but the references should provide a useful guide to the current literature on these topics.

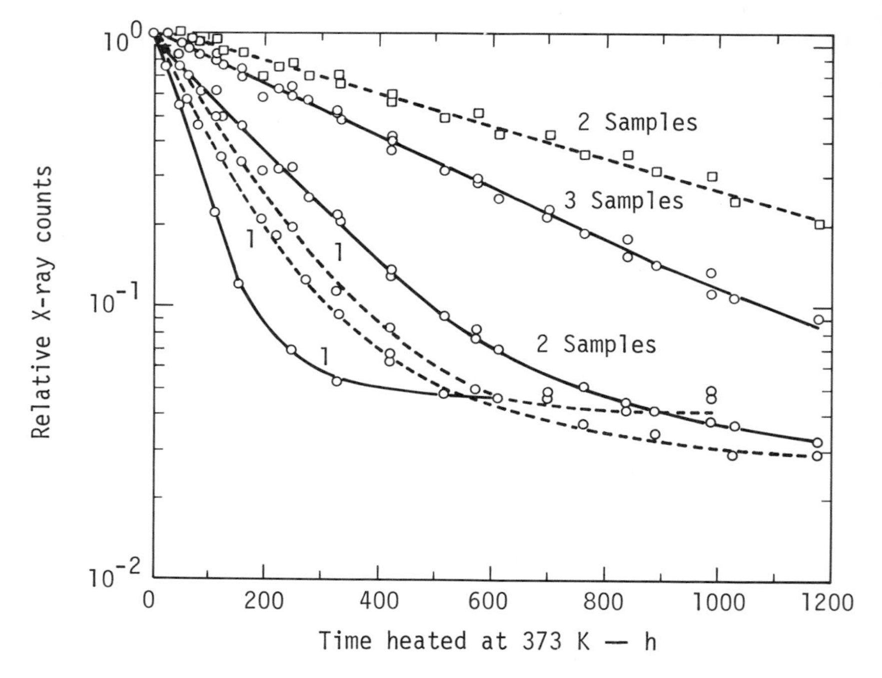

FIG. 21. Non-repeatability of outgassing behavior for tritium-filled microspheres.

REFERENCES

1. A.A. Moghissi and M.W. Carter, eds., Tritium, Messenger Graphics, Phoenix, Arizona (1971).

2. R.L. Dobson, Low-Level Chronic Exposure to Tritium: An Improved Basis for Hazard Evaluation, Lawrence Livermore Laboratory, Rept. UCRL-77372 (1975).

3. E.A. Evans, Tritium and Its Compounds, John Wiley & Sons, New York (1974).

4. D. Kramer, K.R. Garr, A.G. Pard, and C.G. Rhodes, A Survey of Helium Embrittlement of Various Alloy Types, Atomics International, Rept. AI-AEC-13047, Canoga Park, California (1972).

5. R.G. Hickman, Nuc. Tech., 21, 39 (1974).

6. P. Colombo, M. Steinburg, and B. Manowitz, Tritium Storage Development, Brookhaven National Laboratory, Rept. BNL-19981 (1975).

7. R.G. Hickman, Technology of Controlled Thermonuclear Fusion Experiments and the Engineering Aspects of Fusion Reactors, Ed. E. Linn Draper, AEC Document No. CONF 72-1111 (April 1974) p. 105.

8. J.L. Maienschein, Applicability of Chemical Getter Beds to Scavenge Tritium From Inert Gases, Lawrence Livermore Laboratory, Rept. UCRL-52034 (1976).

9. V.A. Maroni, et al., Liquid Metals Chemistry and Tritium Control Technology Annual Report, July 1974-June 1975, Argonne National Laboratory, Rept. ANL-75-50 (1975).

10. J.R. Powell, R.H. Wiswall, and E. Wirsing, Tritium Recovery from Fusion Blankets Using Solid Lithium Compounds, Brookhaven National Laboratory, Rept. BNL-20563 (1975).

11. J.H. Owen and D. Randall, in Proceedings of the International Conference on Radiation Effects and Tritium Technology for Fusion Reactors, American Nuclear Society, Oct. 1-3, 1975, Gatlinburg, Tennessee (in press).

12. T.R. Galloway, Tritium Containment and Blanket Design Challenges for a 1 GWe Mirror Fusion Central Power Station, Lawrence Livermore Laboratory, Rept. UCRL-77968 Part 2 (to be published).

13. J.R. Powell, Beryllium and Lithium Resource Requirements for Solid Blanket Designs for Fusion Reactors, Brookhaven National Laboratory, Rept BNL-20299 (1975).

14. J.E. Draley, et al., An Engineering Design Study of a Reference Theta-Pinch Reactor (RTPR): Environmental Impact Study, Los Alamos Scientific Laboratory and Argonne National Laboratory, Repts. LA 5336, Vol. II, and ANL-8019 (1975).

15. E.H. Van Deventer and V.A. Maroni, in Proceedings of the International Conference on Radiation Effects and Tritium Technology for Fusion Reactors, American Nuclear Society, Oct. 1-3, 1975, Gatlinburg, Tennessee (in press).

16. R.A. Strehlow and H.C. Savage, Nuc. Tech., 22, 127 (1974).

17. J. Blumensaat and H.D. Roehrig, Trans. Am. Nuc. Soc., 20, 728 (1975).

18. C.J. Kershner, Tritium Effluent Control Project Progress Report, Mound Laboratories, Rept. MLM-2217, Miamisburg, Ohio (1975).

19. T.J. Whitaker, J.R. Morrey, L.L. Burger, and R.D. Scheele, Deuterium and Tritium Separation in Water by Selective Molecular Excitation, Battelle Pacific Northwest Laboratories, Rept. BNWL-B-432, Richland, Washington (1975).

20. C.J. Kershner, in Proceedings of the Symposium on Tritium Technology Related to Fusion Reactor Systems, Eds. W.H. Smith, W.R. Wilkes, and L.J. Wittenberg, U.S. Energy Research and Development Administration, Rept. ERDA-50 (1975).

21. P.D. Gildea, in Proceedings of the International Conference on Radiation Effects and Tritium Technology for Fusion Reactors, American Nuclear Society, Oct. 1-3, 1975, Gatlinburg, Tennessee (in press).

22. G.R. Hopkins, in Proceedings of the First Topical Meeting on the Technology of Controlled Nuclear Fusion, Vol. 2, Eds. G.R. Hopkins and B. Yalof, USAEC Rept. No. CONF-740402-P2 (1974) p. 437.

23. M. Balooch and D.R. Olander, J. Chem. Phys., 63, 4779 (1975).

24. J. Nuckolls, L. Ward, A. Thiessen, and G. Zimmerman, Nature, 239, 139 (1972).

25. C.A. Foster, C.D. Hendricks, and R.J. Turnbull, App. Phys. Lett., 26, 580 (1975).

26. C.K. Briggs, R.G. Hickman, R.T. Tsugawa, and P.C. Souers, Estimated Viscosity, Surface Tension, and Density of Liquid DT from the Triple Point to 25 K, Lawrence Livermore Laboratory, Rept. UCRL-51827 (1975).

27. P.C. Souers, R.T. Tsugawa, and R.R. Stone, Rev. Sci. Instrum., 46, 682 (1975).

28. R.T. Tsugawa, I. Moen, P.E. Roberts, and P.C. Souers, Permeation of Helium and Hydrogen from Glass Microsphere Laser Targets, Lawrence Livermore Laboratory, Rept. UCRL-76832, Rev. 1 (1975) (submitted for publication in J. Appl. Phys.).

Chapter 7

MATERIAL LIMITATIONS IN FUSION LASERS

Alexander J. Glass
Head, Theoretical Studies, Y Division
Lawrence Livermore Laboratory
Livermore, California 94550

and

Arthur H. Guenther
Chief Scientist
Air Force Weapons Laboratory
Kirtland AFB, New Mexico 87117

I. INTRODUCTION

It is certainly true that the problems of laser fusion today are in large part problems of the laser, and the problems of the laser are in large part materials problems. This is a fact which is well appreciated by the people who are concerned with the building of high power lasers and with the development of materials for those lasers. Unfortunately, it is not always appreciated by other people working in the more general field of laser fusion. Everyone who becomes involved with a large laser soon develops an appreciation for the problems associated with the use of optical materials and components at high power densities and the constraints that are placed on the design and ultimate performance of high power laser systems.

The authors of this paper have been involved in the problems of optical materials for high energy and high power lasers for a number of years. A series of symposium proceedings addressing this specific subject have been published by the National Bureau of Standards, and will be referred to below. Where reference is made to the damage literature, unless otherwise stated, it is to papers published in the proceedings of these American Society of Testing and Materials Laser Damage Symposia.

This paper is of necessity a superficial overview, because each topic touched on lightly is an adequate subject for a lengthy presentation in itself. In the spirit of this lecture series, we have tried to eliminate most of the mathematics, but hopefully not eliminate the physics and material considerations appropriate in describing the response of materials to intense coherent optical radiation.

II. LASER FUSION APPROACHES

Let us start by giving a brief introduction to the role of the laser in fusion research. There are essentially three approaches to controlled fusion that involve lasers. They are as follows: One application is in the injection of a plasma into a confining geometry, particularly into an appropriate magnetic field geometry. As we shall see in a moment, this primarily involves the in situ formation of a plasma within the confinement geometry. This is principally with reference to low density plasmas as in a Tokamak. At an intermediate density, such as one might achieve in a theta pinch, it is possible to use the laser as an energy source to heat the plasma within the magnetic field. But the application to which we shall give most attention is plasma compression, specifically inertial confinement schemes, and that is what people generally mean when they speak of laser fusion.

Plasma injection within a confinement geometry is based on the fact that an optical wave will pass through the magnetic field undeflected. As an example, it is possible to inject a neutral pellet of solid material, and vaporize it within the field, thereby creating a clean isolated plasma. This requires an energetic pulse of moderate duration, from microseconds to milliseconds. The wavelength generally used for this application is a relatively long wavelength, 10.6 μm being typical. For heating a magnetically confined plasma, one must ask what plasma density is desired. The plasma density that can be contained in a magnetic field is limited by the magnetic field, i.e., the plasma pressure is roughly equal to the magnetic pressure, which is the square of the magnetic field in appropriate units, and for the types of magnetic fields available. Consider a field on the order of 100 kilogauss present in a typical theta pinch configuration, the associated plasma density is about 10^{17} particles per cubic centimeter, a relatively low density plasma. This again dictates the use of a laser with the wavelength on the order of 10 μm or longer, in order to couple effectively with the plasma. Here, because the plasma is confined for a relatively long period, we can use a long pulse or even a CW laser.

When we turn to the principal approach in laser fusion, the compression of an inertially confined plasma, we find the requirements are significantly different. It is ultimately the requirements of the experimental approach which dictate the parameters of the laser. As we review these laser performance parameters, we shall see that we have a particular class of materials problems to contend with.

III. INERTIAL CONFINEMENT

In the inertial confinement approach, the target, consisting of a deuterium-tritium mixture, is compressed from an initially high density, typically of the order of solid density. It must be compressed significantly, on the order of 100 to 1000 times the initial density. The time in which the experiment can be carried out, that is, the time which one has to compress the plasma is limited by about the time it takes for sound waves to propagate across the dimension of the pellet. In order to operate over a longer period, and thereby reduce some of the operating constraints on the laser, it is required to increase the energy of the laser pulse. The energy required increases roughly proportionately to the mass of the pellet, since we need to approach a temperature on the order of 7 to 10 kV in the plasma in order for the deuterium-tritium fusion reaction to efficiently proceed. It is easy to see that the required energy therefore increases as the cube of the radius, while the pulse width only increases linearly with the radius. Thus moving toward larger pellets still yields an ever increasing demand on the peak power of the laser. Accordingly, lasers for laser fusion by this approach must be high power, short pulse lasers. High power here means the laser's power is measured in terawatts (10^{12} watt). At present, pellets are on the order of 100 μm in diameter, so the time of the pulse is on the order of 100 picoseconds. The energies to be absorbed are therefore on the order of hundreds of joules.

Some years ago, in one of the first discussions of this subject, John Nuckolls and Lowell Wood (2) discussed what one could expect by way of thermonuclear gain, that is, thermonuclear energy release divided by the laser energy absorbed in a laser fusion target, as a function of how well the pellet was compressed. The material must be compressed sufficiently so that the alpha particles that are released in the deuterium-tritium reaction reinvest their energy in the compressed material, in order to realize an efficient burn. In Figure 1, we see the computed values of thermonuclear gain vs. pellet compression for various values of incident laser energy. It is seen that break-even (Gain = 1) requires 1000 J if a compression of 10^4 can be achieved, and 10^4 J, if the compression is 10^3. Let us examine two approaches to the problem, of focusing ten thousand joules on a target in a reasonable time.

Laser fusion research in the United States is being carried on in a number of places, as indicated in Table 1. There are also several significant programs outside the United States, including the USSR, the UK, France, Japan and West Germany. Let us view some of the technology that is involved (3). In Figure 2, we see one module of a carbon dioxide laser at the Los Alamos Scientific Laboratory capable of radiating 2500 joules in one nanosecond. In this module there are two lasing volumes powered from a single, cold-cathode electron gun, in an electron beam-stabilized discharge. Plans call for the combination of four such modules into a single system capable of

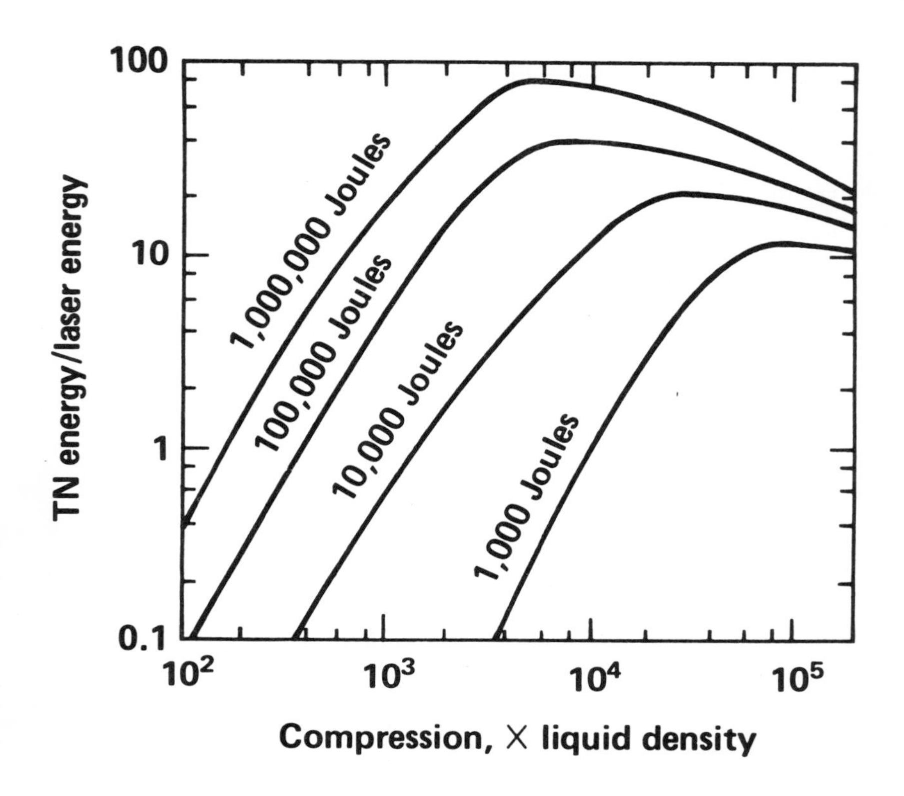

FIG. 1. Thermonuclear Gain vs. Target Compression for Various Values of Incident Laser Energy

emitting 10 kj in about one nanosecond, at 10.6 μm. This system is shown in Figure 3. We begin to see a feature characteristic of large fusion lasers, namely, one laser per building, or one building per laser. This is a far cry from helium neon laser technology, but the scale of the equipment is not inconsistent with modern power-plant technology. We also see that it becomes increasingly complex; each of these large laser sources radiates two beams which must be combined, transported into the target chamber and focused onto the target. This involves the use of windows, focusing optics, and many other diverse optical components. A more advanced system under consideration at

TABLE 1

MAJOR FACILITIES FOR LASER FUSION RESEARCH IN THE US

Facility	Location
<u>1.06 μm</u>	
KMS Fusion	Ann Arbor, Mich
Lawrence Livermore Laboratory	Livermore, Calif
Naval Research Laboratory	Washington, DC
University of Rochester	Rochester, NY
<u>10.6 μm</u>	
Los Alamos Scientific Laboratory	Los Alamos, NM
*Math Sciences Northwest	Seattle, Wash

*Devoted to long-wavelength heating of confined solenoidal pinches.

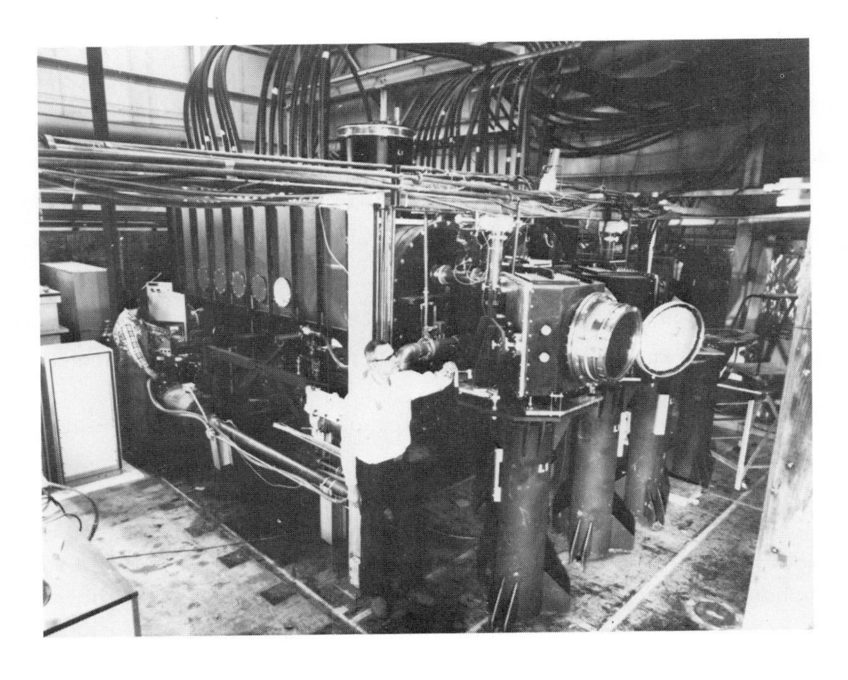

FIG. 2. LASL's 2.5 kJ Dual-Beam Amplifier for E-Beam-Controlled CO_2 Laser Fusion Feasibility Experiments. This Building Is Currently Being Outfitted for the Initial Experiments in Late 1977.

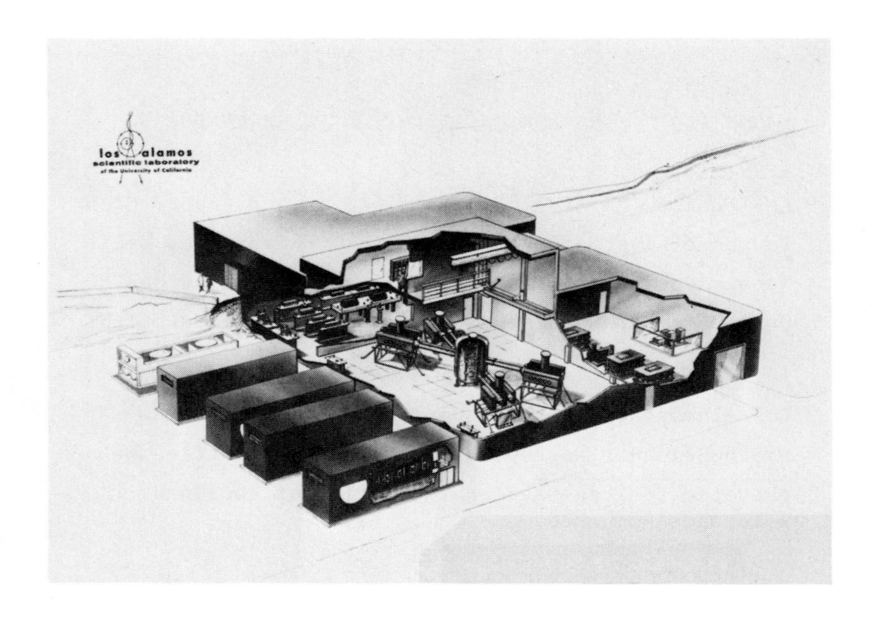

FIG. 3. Cutaway View of LASL's Laser Fusion Laboratory Showing Assembled Array of 4 Dual-Beam 2.5 kJ CO_2 Laser Amplifiers Directing 8 Beams for Focusing Into the Target Chamber for Feasibility Experiments.

Los Alamos is illustrated in Figure 4. Here we see one module for a high energy gas laser facility. Operated together, six modules are designed to emit 10^5 joules in one nanosecond. The complete High Energy Gas Laser Facility is shown in Figure 5.

IV. GLASS LASERS

As noted in Table 1, most of the present work in laser fusion emphasizes the use of neodymium glass. This has raised many questions, since neodymium glass lasers are not very efficient. They have essentially zero capability for high repetition rate, using current technology, and therefore no one proposes that the neodymium glass laser will ever be the working heart of a genuine laser fusion power plant. On the other hand, the carbon dioxide laser has a much higher efficiency. It has

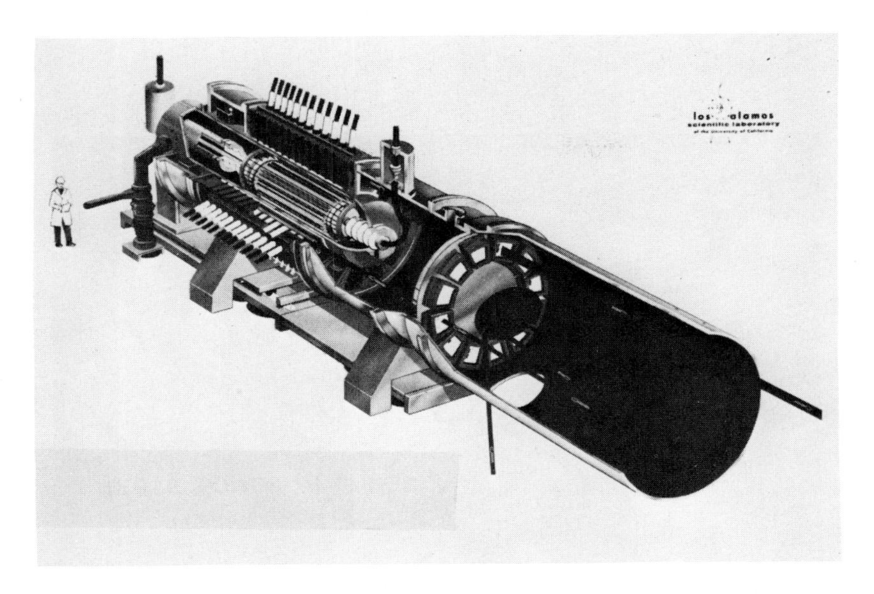

FIG. 4. High Energy Gas Laser Facility (HEGLF) Amplifier Designed by LASL for 16kJ Output at One Nanosecond From CO_2, E-Beam-Controlled Laser for "Break-Even" Experiments at 100kJ Total Design Output Energy.

the capability of high repetition rate, since the active lasing material can be replenished. However, there are serious questions relating to the coupling of 10.6 μm light to dense plasmas, which make the carbon dioxide laser seem less attractive from the standpoint of compression heating (4). Neodymium glass, of all solid materials, seems the most attractive for single-shot, proof-of-principle experiments. This is true because glass is an amorphous solid; thus, it is possible to store a great deal of energy in the form of an inverted population in the material per unit volume. Glass has been with us for a long time and is available in rather excellent optical quality. In fact, aside from the ultra-pure materials that are used in the semiconductor industry, optical laser glass is among the purest and most homogeneous materials available in large quantities. Because it is an amorphous material, glass is a

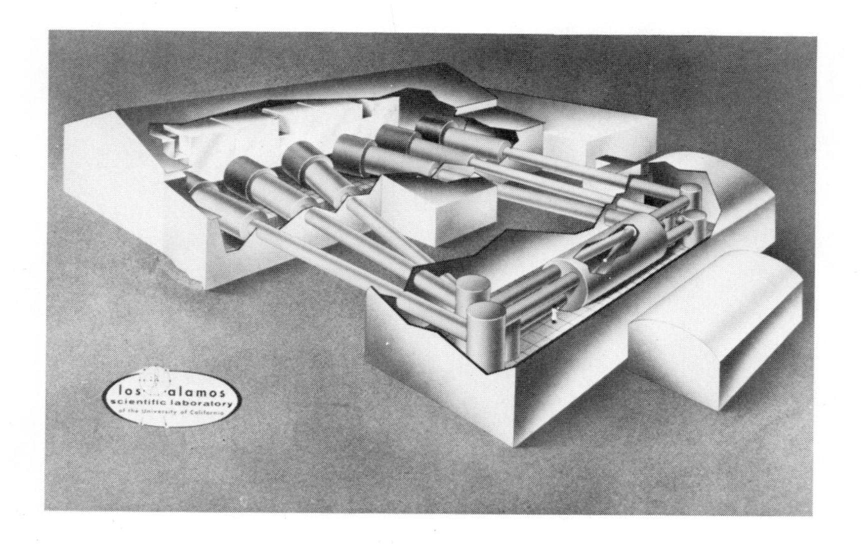

FIG. 5. Cutaway View of LASL's High Energy Gas Laser Facility (HEGLF) Now in Conceptional Design Stage, Showing Assembly of Amplifiers and Beam Paths to the Target Chamber for Focusing on DT Targets for "Break-Even" Experiments.

relatively forgiving or flexible environment for the ion, and as Dr. Weber will discuss in the next paper, this allows one to design laser materials on an engineering basis. As a laser host, glass has an added advantage, again associated with the fact that it is an amorphous material, and that is that the neodymium ion, in fact, any ion in a laser glass, exhibits a broad fluorescence bandwidth. As a result, one has a great deal of flexibility with regard to the pulse duration that one can achieve. Laser pulses in neodymium glass have been made as short as 10^{-13} seconds, so the achievement of 10^{-10} second pulse is not at first a significant problem.

Let us next examine some of the large glass laser systems used in the laser fusion program at the Lawrence

Livermore Laboratory. The Cyclops system, a one-beam, one-terawatt system, is shown in Figure 6. The pulse originates in

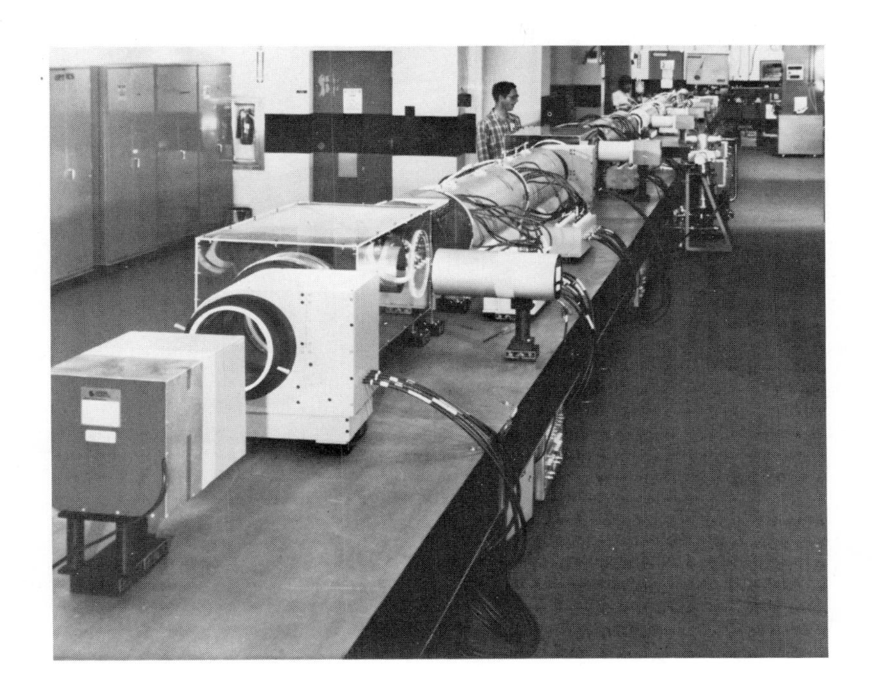

FIG. 6. Cyclops, a One-Terawatt Laser at Lawrence Livermore Laboratory.

a Nd-YAG oscillator, then amplified through YAG and glass rod pre-amplifiers, and finally propagates down the single beam line, being successively expanded and amplified in a series of disk lasers. This technology is based on the disk laser amplifier, an example of which is shown in Figure 7. A set of glass disks is placed in the flashlamp cavity so that the incident beam falls on the disk surface at Brewster's angle. At this angle there is no reflection, a very important feature. The reason for going into a disk geometry results from the

FIG. 7. Disk Laser Amplifiers.

limitation imposed on laser rod amplifier size by the requirement of uniform pumping.

There are many additional components in such a laser, including Faraday rotators, thin film polarizers, beam expanding telescopes, spatial filters, and turning mirrors.

The Cyclops laser is now being supplanted by the Argus system. Argus, shown in Figure 8, is a two-beam system generating about one and one-half terawatts in each beam. The pulse originates in a master oscillator of Nd-YAG, and is again amplified through a succession of rod and disk amplifiers, having been initially split into two beams. The details of this technology are recorded in the Lawrence Livermore Laboratory fusion annual reports (5). The two beams are combined in a complex optical system to provide uniform illumination over the entire surface of a spherical target. The entire room in which Argus

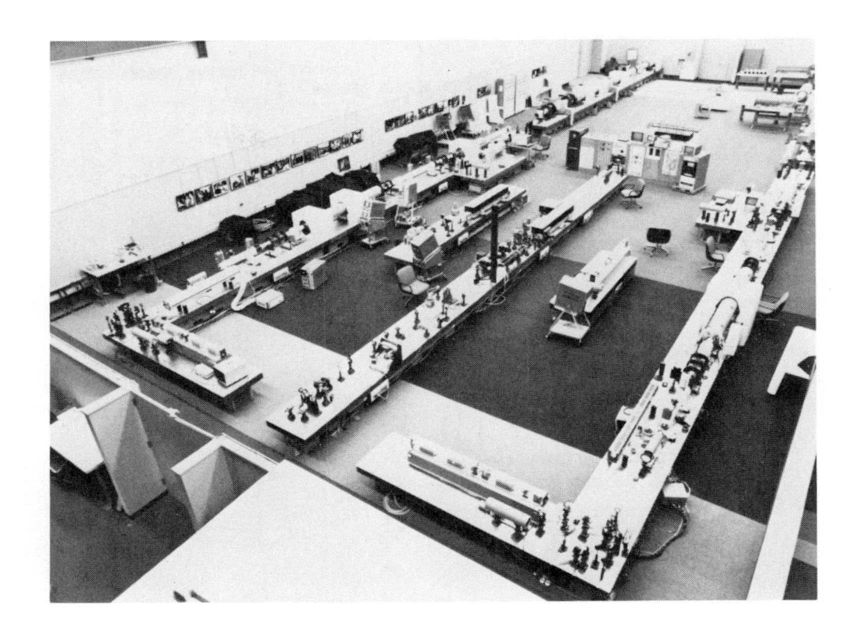

FIG. 8. Argus, a Two-Beam System Designed to Radiate 1.5 Terawatts from Each of Two Apertures, in 100 psec.

operates is an optical table, which is mounted independently from the main building to provide vibrational isolation. It is a laminar flow facility as well. Air is introduced at the top, flows to the floor and is evacuated under the walls. The Argus system only provides three terawatts in 10^{-10} seconds, i.e., a few hundred joules. What does a ten kj neodymium glass laser look like? A schematic of the SHIVA system, under construction at this time, is shown in Figure 9. The system consists of 20 arms (beams), each arm of which is comparable to the Cyclops system, or one arm of the two-beam Argus system. The 20 beams are then combined with folding optics to arrive at the target chamber. The SHIVA laser satisfies the rule of one building per laser. It is a very large laser, indeed.

Whether it is CO_2, iodine (another candidate system), or neodymium glass, all fusion lasers have certain

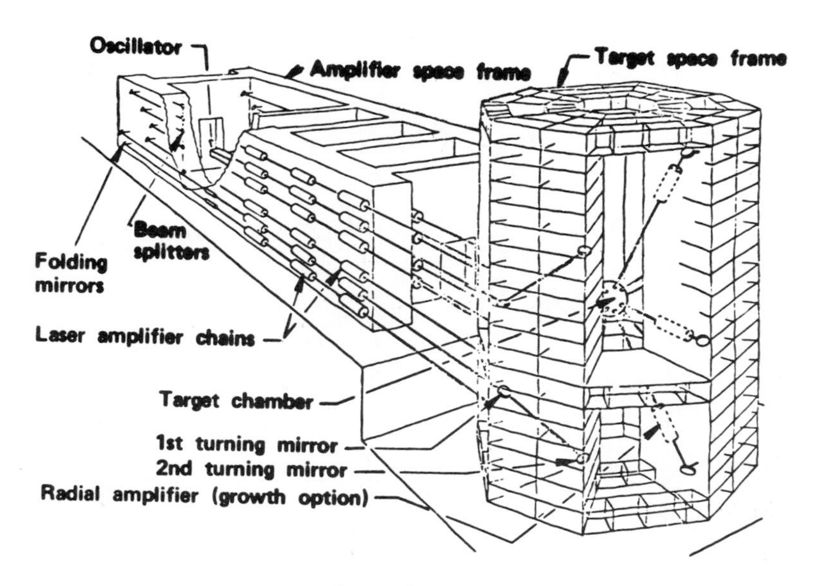

FIG. 9. Schematic Drawing of the SHIVA Laser System, Under Construction at Lawrence Livermore Laboratory.

common features. By virtue of the plasma physics of the target, we're dealing with short, intense pulses of light. Because of materials constraints which determine how much energy one can actually put through a single aperture, we're talking about systems with many apertures. In the CO_2 systems shown previously, there are four or eight apertures, depending on the particular system design. In the SHIVA system, there are 20 apertures. Planning calls for a possible extension of that system to 42 apertures. In each arm there are many, many components, and many exposed optical surfaces. For the system at KMS, which is comparable to the Cyclops system, there are roughly 100 coated surfaces per arm in the system. Thin film coatings, as we shall see, are the most vulnerable elements in the system. In SHIVA there are on the order of 200 individual optical surfaces in each arm. Accordingly these systems are very vulnerable to laser damage, which brings us to the main topic of this paper.

V. LASER INDUCED DAMAGE OF OPTICAL MATERIALS

What constitutes laser damage? When we talk about laser damage, we mean something more than just the physical destruction of the laser material. Laser damage is construed as any effect which degrades the performance of a high power laser system due to the interaction of the intense light with the optical materials in the system. We subsume all phenomena of this kind under the heading of laser damage. There are, however, several kinds of damage effects. Damage can be classified as catastrophic or non-catastrophic, i.e., irreversible or reversible, depending on the state of the material after the interaction has occurred. Furthermore, there are two fundamental mechanisms for the interaction of intense light with matter. One is thermal, due to some weak absorption or possibly even strong and localized absorption (the materials we are dealing with are nominally highly transparent materials or else we would not be using them in the laser system). The other is electrical, since, at the power densities we are considering, the electric field associated with the light wave is on the order of megavolts per centimeter. The electrical nature of light is manifested, generally by the formation of a plasma within the material. These are the two fundamental forms of interaction of the light with the matter. It is also useful from a material standpoint to characterize a damage mechanism as being either intrinsic or arising from some extrinsic factor. Intrinsic damage results from an innate property of the material, and therefore there is nothing one can do about it if one chooses to use that material. In general, one cannot actually operate at the intrinsic limits of the material. Most laser damage phenomena are dominated by extrinsic properties. Extrinsic damage arises from dirt, impurities, or imperfections. These are aspects of the material which are amenable to improvement, primarily by improving the process whereby the material is made, or the optical component is finished, or the finished component is handled and utilized. Thus, it is very important to identify whether a damage phenomenon is intrinsic or extrinsic in nature. If it is extrinsic, then the best approach may be to improve the fabrication process. If the effect is intrinsic, any improvement must result from the choice of an alternative material. In Table 2, we see some examples of laser damage effects, classified as catastrophic vs non-catastrophic, and thermal vs electrical.

TABLE 2

EXAMPLES OF DAMAGE EFFECTS

	Catastrophic	Non-Catastrophic
Thermal	Pt Inclusions in Glass (Extrinsic)	Thermal Lensing of Windows (Extrinsic/Intrinsic)
Electrical	Avalanche Ionization (Intrinsic)	Small-Scale Self-Focusing (Intrinsic)

In the early days of neodymium glass lasers, that is, in the 1960's, when pulse durations of interest were from 10 to 100 nsec, the damage limiting factor in the glass itself arose from the presence of platinum inclusions. Platinum is transported into a glass melt in the form of platinum oxide from the platinum crucible in which the glass was made, clearly an extrinsic effect. Damage is catastrophic, since when these little platinum particles absorb the light, they explode, or melt the glass in the vicinity of the particle, and that, in turn, creates a stress which causes a crack in the glass -- a thermal effect. Currently working with shorter pulses in neodymium glass lasers, people have learned how to control the manufacture of the glass in order to keep the platinum inclusions to a minimum, and to keep them to a smaller size. In some cases the glass is made in platinum-free crucibles. Even when it is made in platinum, inclusion formation is prevented by a combination of controlling the atmosphere and the thermal cycle under which the glass is melted and finally by selecting the final glass product to find those regions which are most free of damaging inclusions. Thus, the platinum inclusion problem can be circumvented. Accordingly, the limiting form of catastrophic damage in glass is associated with the avalanche ionization process. This is an intrinsic electrical process, although it may be mediated by impurities.

For high average power lasers, thermal lensing is the limiting damage effect. In such lasers, optical components such as windows will absorb some of the laser light, will heat up, and there will be induced a thermal distortion of the beam. When the laser is turned off, the window returns to its original form so that the damage is non-catastrophic or reversible. There are both extrinsic and intrinsic contributions to the responsible residual absorption, and it is clearly a thermal effect.

The most pernicious effect for high power, short pulse neodymium glass lasers is small-scale self-focusing. This manifests itself due to the instantaneous interaction of the intense light with the optical material. The refractive index of the glass is increased in regions of high optical intensity. This usually leads to a breakup of the light beam into a myriad of tiny filaments. Small-scale self-focusing arises from an intrinsic property of the glass. It is described by a single property of the glass, the non-linear refractive index, which is characteristic of a given composition of glass. In order to eliminate or reduce that effect, one has to search for a new glass composition. Dr. Weber will discuss this further in the next paper.

For this discussion we shall concentrate on short pulse lasers (less than 10 nanoseconds) operating at high peak power density, on the order of 10^9 watts per square centimeter. We shall exclude mechanical failure. This is not to imply that mechanical failures don't occur. In Figure 10, we see a laser damage phenomenon that is clearly catastrophic, arising from flashlamp explosion. This is a laser damage phenomenon one clearly wants to avoid, but it does not fall within the scope of this paper.

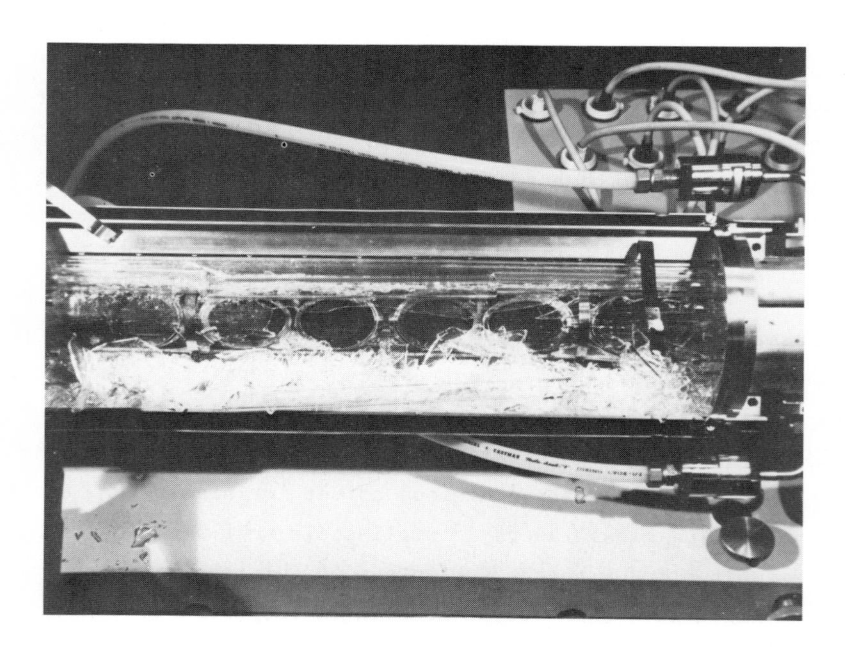

FIG. 10. Flashlamp Explosion in a Disk Laser.

Let us first consider damage arising in the bulk material. In general, the bulk damage phenomena set the limits that it is even conceivable to achieve. We have previously discussed platinum in laser glass and thermal lensing as bulk phenomena. Avalanche ionization is the ultimate intrinsic bulk damage mechanism for short pulse effects, i.e., the formation of plasmas within the material itself leading to irreversible damage. In a well designed, high performance laser, that effect does not occur, primarily because small scale self-focusing occurs first.

In looking at this effect, we shall follow a pattern which is typical of the analysis of any damage phenomenon. We ask first, what is the physical effect under investigation? In this case, it is the dependence of the refractive index on the local intensity of light. What are its consequences? This effect causes initially uniform beams - beams which are uniform in phase and intensity - to break up into little filaments. What are the relevant parameters? There is one parameter characteristic of a material which describes the magnitude of the effect, the non-linear index coefficient, n_2. What are typical values? In the next paper, Dr. Weber will give typical values of that parameter, and some ways of modifying or finding new materials for which performance can be improved. This is the typical cycle of a laser damage study. It is highly pragmatic, with the objective being to identify the damage phenomenon, find its causes, and from the causes determine the remedy.

Let us now consider non-linear light propagation in more detail. The refractive index is given by

$$n = n_o + n_2 < E^2 > \qquad (1)$$

where n_2 is the index non-linearity. The index non-linearity is typically 10^{-13} esu, which means that if we propagate 10^{10} watts per square centimeter, which is about the intensity at which avalanche ionization occurs, the local index change is 5 parts per million. However, good optical glasses display local fluctuations of refractive index on the order of parts per million anyway. Why then are we concerned about such a small effect? In order to focus the light out of a laser with good quality, one demands interferometric quality in the wave front. We demand uniformity of phase over the whole aperture. If we have a meter of glass in the system, operating at 1.06 μm, at half the operative power density 10^{10} w/cm^2, we accumulate two and one-half waves of optical distortion due to this non-linear index. As a consequence of this non-linearity, any little ripple or spatial variation in intensity or phase across the beam will grow exponentially. At the focus of a lens, the high spatial frequencies are mapped away from the center of the focal plane, so we have a loss of focusable energy, and it is this last effect that is the most serious consequences of the non-linear index.

Erlan Bliss of the Lawrence Livermore Laboratory has carried out some elegant experiments on ripple growth in neodymium glass. Using a shearing plate interferometer, he imposed a two line per millimeter ripple across a laser beam. In Figure 11, we see the intensification of the ripple pattern. The beam was propagated through 24 cm of glass at 5 gigawatts per square centimeter. One sees that the modulation depth is significantly increased. This is an example of the instability of the beam against the growth of small ripples or high spatial frequencies. Let us consider the growth of an intensity fluctuation of spatial frequency κ. It grows exponentially, according to

$$I_\kappa = I_\kappa(o) \exp g(\kappa) \ell, \qquad (2)$$

where ℓ is the optical path length, and $g(\kappa)$ is the growth rate. The most unstable frequency grows with a growth rate given by

$$g\ell = B, \qquad (3)$$

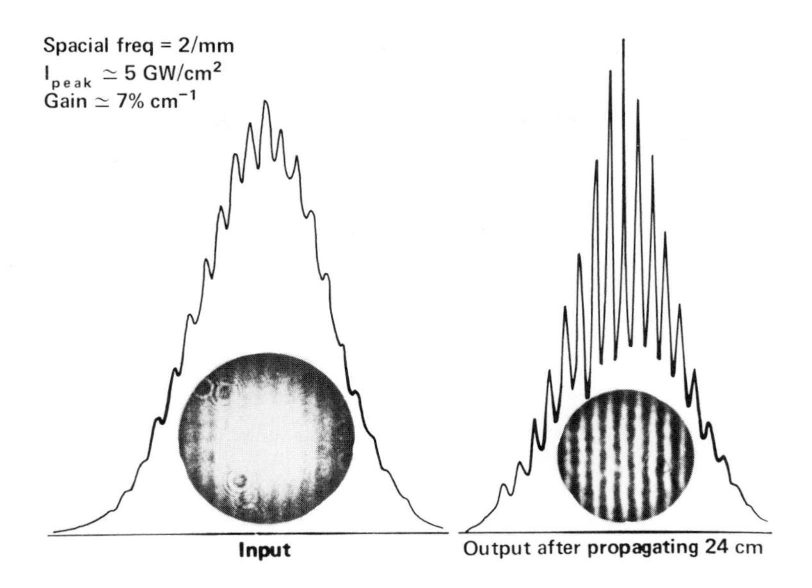

FIG. 11. Ripple Growth in Laser Glass

$$\text{where } B = \frac{2\pi}{\lambda} \left(\frac{n_2}{n_o}\right) \int < E^2 > d\ell, \tag{4}$$

simply called the B integral, and is used to quantify the exponential growth rate. We see that B is the relative phase shift along the center of the beam induced by the non-linear index change. There are three means by which one can suppress the growth of this effect. One is to make $I_K(o)$ as small as possible. This requires cleanliness, which implies the total absence of dust, imperfections, scratches, surface inhomogeneities, bulk inhomogeneities, and fluctuations. It is very hard to achieve, but nevertheless, one strives as much as possible to achieve it by enclosing, for example, the entire system is a controlled atmosphere to prevent the precipitation of dust onto exposed optical surfaces. The problem is exacerbated since there are so many surfaces in the system. Since it is not possible to make $I_K(o) \equiv o$, the next remedy is to keep the effect from exponentiating. We do this by periodically filtering the beam. The spatial filter is a lens with a pinhole at the focal plane, followed by another lens at the corresponding focus to recollimate the beam. High spatial frequencies are clipped off by the pinhole, thereby setting I_κ back to zero for sufficiently large values of κ. Low frequency perturbations, those that are of significant sizes compared to the structure of the beam, are not attenuated and therefore continue to grow. The ultimate solution to this problem is to reduce the non-linear index, which is an intrinsic limit. This requires new materials, and is one of the principal subjects of the next paper.

The ultimate limit for bulk material is avalanche ionization, which is a process in which a nearly free electron gains energy in the electric field because of its intermittent collision with the atoms in the material (7). This can be analyzed in a simple way. A free electron which experiences intermittent collisions gains energy from the electric field at the rate

$$\frac{dW}{dt} = \frac{e^2E^2}{m} \frac{\tau_k^2}{1+\omega^2\tau_k^2} \tag{5}$$

where τ_k is for large angle scattering, or the average time for the interruption of the electron motion. If this energy gain process is equated to some loss rate, R_o, we will obtain a threshold for electron avalanche. If $R_o < dW/dt$, the electron keeps gaining energy until it is sufficiently energetic to cause a secondary ionization. The secondary electron then subsequently gains energy and an avalanche ensues. This simple analysis yields some relevant predictions. One is the frequency dependence for the electric field at which the avalanche occurs. Balancing gain and loss, we find that the breakdown field should have the frequency dependence,

$$E_B \sim (1 + \omega^2\tau_k^2)^{1/2} \tag{6}$$

This is a clear cut prediction for the frequency dependence of the breakdown field. This prediction, however, is not in complete agreement with experiment. One can also estimate what the loss rate is. Let us take typical collision times as being 10^{-15} seconds, and the typical energy loss per collision on the order of 0.1 eV. Equating gain to loss, we discover that breakdown fields should be on the order of 10^6 to 10^7 volts per centimeter at optical frequencies. This prediction is confirmed. In Table 3, we see optical breakdown fields for various dielectrics measured by L. Smith (8) and reported at the 1975 Damage Symposium.

TABLE 3

MEASURED VALUES OF OPTICAL BREAKDOWN FIELDS IN MV/cm

Material	E_B	Material	E_B
NaF	10.8	LiF	12.2
NaCl	7.3	RbI	3.4
NaBr	5.7	ED-4	9.9
KF	8.3	YAG	9.8
KCl	5.9	Fused Silica	11.7
KBr	5.3	CaF_2	14.4
KI	5.9	KDP	22.3

Although these figures may be uncertain to 50 percent, nevertheless they are all on the order of 1 to 10 megavolts per centimeter. And, indeed, the argument is so simple that with that conclusion you know you can't be off by more than an order of magnitude. Bloombergen (loc. cit.) has noted that it is quite remarkable that all transparent dielectrics break down optically at about the same field strength, and, from this simple argument, one sees why. It is beyond the scope of this discussion to go into the details of the electron avalanche process and the current controversies in theory (9). As one notes, something as basic as the frequency dependence fails to satisfy the simplest argument. That is not a very satisfying state of theory, and that is only part of the dissatisfaction with the existing theory of electron avalanche process.

In practice, catastrophic failure generally appears at exposed surfaces, rather than in the bulk. Surfaces are dirty, because they are exposed to the environment, and there is also a possibility of adsorption of water and atmospheric gases, and other contaminants. In addition, surfaces are subject to the tender ministrations of the optical technician, who uses the process called polishing, by which he hopes to improve the surface. One of the events that has transpired due to the study of damage phenomena is that optical technicians have become aware that many of their time-honored processes for polishing surfaces are extremely deleterious as far as high power optical components are concerned. Jewelers' rouge, for example, which is an iron-rich compound, was widely used for optical polishing until it became clear that it created an absorbing layer on the surface of the polished component.

Surfaces also carry scratches, at which there is field intensification. Even in the absence of scratches, there are Fresnel reflections at the surface, which can lead to standing waves. This is particularly true in multilayer dielectric films. Because of the presence of very strongly absorbing layers of extremely small dimension, due to surface contaminants, one can have both electrical avalanche at the surface, and absorption induced thermal damage. Recently, N. Boling presented results on surface damage for a variety of optical glass surfaces (10). His results are shown in Table 4. In the table, we see that for damage accompanied by the formation of a luminous plasma, the lower index materials damage at a greater

TABLE 4

SURFACE DAMAGE THRESHOLDS IN OPTICAL MATERIALS AT 1.06 μm. PULSE DURATION IS 30 NSEC.

Material	Index	Damage Threshold (J/cm^2) No Plasma	Plasma
FK-6	1.43		123
Fused Silica	1.46		210
AO Laser Glass	1.51		105
Schott Glass	1.51		110
OI Glass	1.55		100
EY-1	1.61		60
Sapphire	1.75	14	30
SF-6	1.78	25	34
E-4	2.10	7	36
E-11	2.10	8	37

threshold than the higher index materials. In addition, in the higher index materials, one sees the development of surface damage without plasma formation. This is undoubtedly some type of thermal damage, due to surface absorption. Since higher index glasses tend to be softer, they may not be finished as smoothly, and thus may have more scratches on the surface or more embedded polishing material.

There is experimental evidence to demonstrate that the surfaces of optical materials can be much more absorbing than the bulk values (11). For example, in alkali halides used for windows at 10.6 μm, the bulk absorption coefficient will be typically as low at 10^{-4} to 10^{-5} per centimeter. The surface absorption coefficient, in a region of 1 to 10 microns depth near the surface, may be from 1 to 10 per centimeter. There may be as much absorption in the surface layer as there is in the entire window, if the window is typically of the order of centimeters thick. Surface absorption can be reduced by treating the surface. One can etch the surface, or remove the surface layer with an ion beam, or clean the surface chemically, and obtain much lower surface absorption, and a higher damage threshold. In general, the improved surface will then degrade if exposed to the atmosphere, or exposed to any environment other than vacuum, due to the absorption of materials from the surroundings. Surface absorption is not as severe a problem in optical glasses as it is in the crystalline materials used as window materials at 10.6 μm. In optical glass the principal surface damage mechanism is avalanche ionization at the surface.

Surface absorption is just one parameter of importance in determining the threshold for surface damage. If we consider damage with accompanying plasma formation, an interesting empirical relationship is found to describe damage thresholds for a large class of transparent dielectrics. This relationship was put forth by Bettis, Guenther and Glass (12) at the 1974 Damage Symposium. They found that for the previously published data on alkali halide bulk damage, the breakdown field could be expressed as

$$E_B = A\ N/(n^2-1), \tag{7}$$

where E_B is the observed breakdown threshold electric field, N is the number density of ions, and n is the optical refractive index. They found that the same value of A fit both Na- and Rb-halides, while for K-halides, a slightly lower value of A fit the data. Their results are shown in Figure 12. The postulated dependence of breakdown field on index arises from the fact that $(n^2-1)/N$ is equal to the polarizability of the atom. Equation 7 thus merely states that breakdown occurs more easily for higher polarizability. The connection of this postulated dependence with avalanche theory has as yet not been quantified.

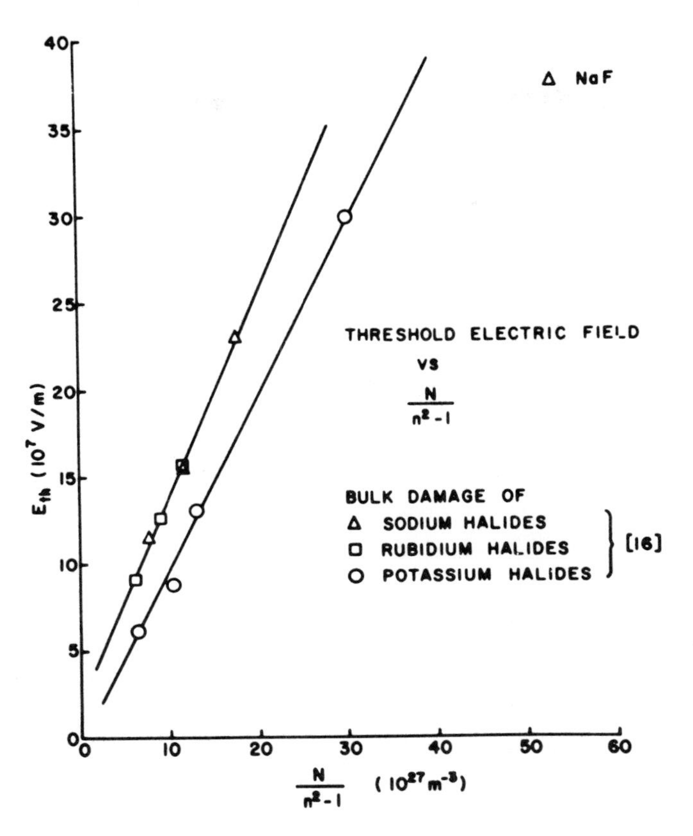

FIG. 12. Bulk Breakdown Threshold vs. $(N/(n^2-1)$ *for Alkali Halides.*

A much more exhaustive investigation of this relationship has been done more recently by House, Bettis and Guenther of the Air Force Weapons Laboratory, along with R. R. Austin of Perkin-Elmer, and is reported in the 1975 Damage Symposium (13). They found that the data available on surface damage also fit a relationship of the form given in Equation 7, if a correction factor was added to take account of surface roughness. Equation 7 was again obtained for the bulk damage of numerous optical materials. In going to the surface, we know that surface roughness has an influence on the damage threshold. The empirical relation which was obtained takes the form,

$$\sigma^{0.5} E_B \sim 1/[\sigma_s^{0.5} (n^2-1)] \quad (8)$$

where σ is the rms surface roughness, in Angstroms, and σ_s is the average lattice spacing. Equation 8 expresses an intriguing relationship. There are some very clear cut statements in that relationship, but it is not clear why it has the form that it does. A summary of experimental results obtained on a variety of surfaces with a 1.06 μm laser operating in a 40 nsec. pulse is shown in Figure 13. The data shown were obtained on SiO_2 surfaces, except for the indicated points for MgF_2 overcoating on SiO_2 substrates. It is seen that each type of surface preparation yields a different slope for the relation, but the power law dependence is generally observed. It is also noted that the bulk value obtained by Fradin (14) represents an extrapolation of Equation 8 to a value of σ given by the lattice spacing. The multiplicity of parameters involved in characterizing a surface begins to be evident in Figure 13. In Figure 14, we see further evidence of the proposed scaling relation. This is for surfaces of pure materials with bowl-feed optically finished surfaces. The indicated materials include crystals and optical glasses. Here the indicated relationship holds to a remarkable degree for a very large class of materials, and we note that in LiF we are talking about a breakdown field which is ten times that of ZnSe and ZnS. Thus the proposed empirical relation holds over an order of magnitude in breakdown field values. That is, the optical breakdown threshold in watts per square centimer varies over a range of 100 for the data presented in the figure. As we have seen, that is about the available range in transparent dielectrics - two orders of magnitude in the intensity. Thus, this is a very compelling relationship and seems to hold quite well for a large class of materials.

Let us turn now to an examination of thin film damage. In Figure 15, we see data for a variety of materials.

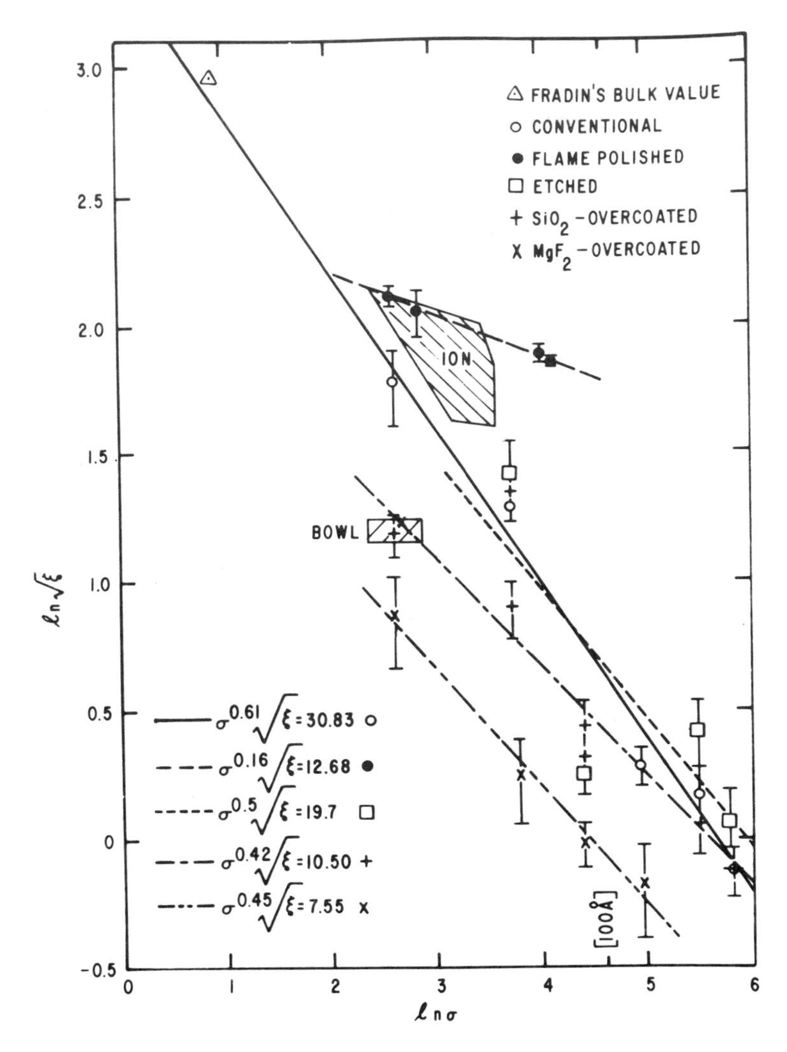

FIG. 13. Dependence of Observed Surface Breakdown Field on Surface Roughness. ξ is incident power density at threshold, in relative units, σ is rms surface roughness in Angstroms. Substrates are all fused silica.

Here are many of the same materials but now deposited as half-wave films on SiO_2 substrates. We see that again the empirical relationship given by Equation 8 holds over something like an

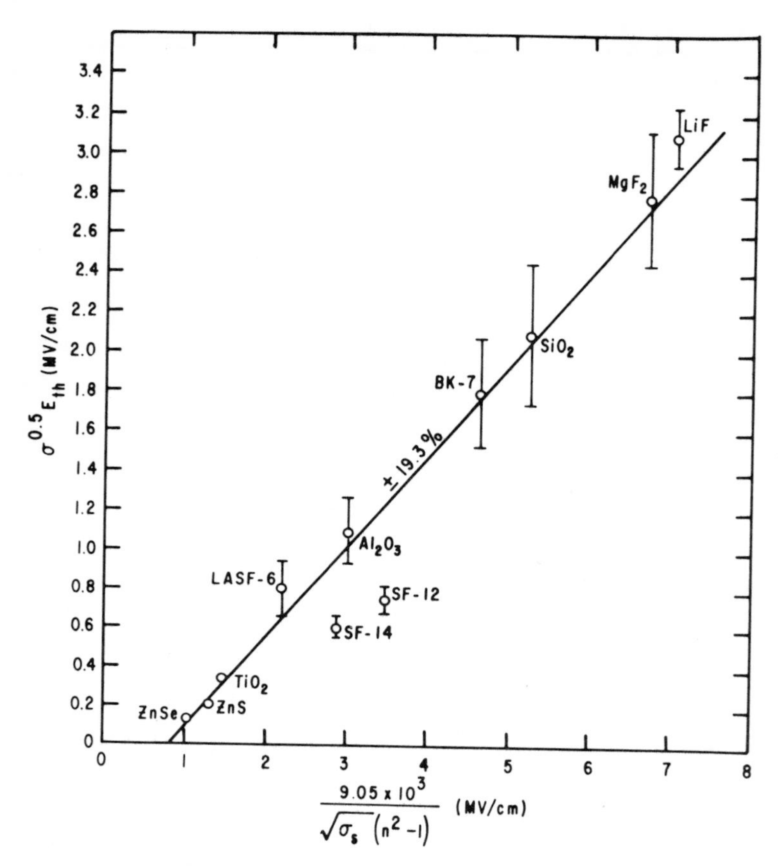

FIG. 14. Dependence of Breakdown Field on Surface Roughness (σ) and Material Parameters for Bowl-Feed Optically Furnished Surfaces.

order of magnitude in the electric field. We also note that the thin film values are typically significantly lower than the corresponding surface values. Those films that are considerably off the curve are identified as inhomogeneous films (MgF_2, LiF, BaF_2), that is, films that are not perfectly uniform in their physical character. There are two points for silicon dioxide films -- one on a silicon dioxide substrate (a) and another on a BK-7 (b) substrate. Once again, the dependence of the breakdown field value on surface roughness is clear. In Figure 13, we saw

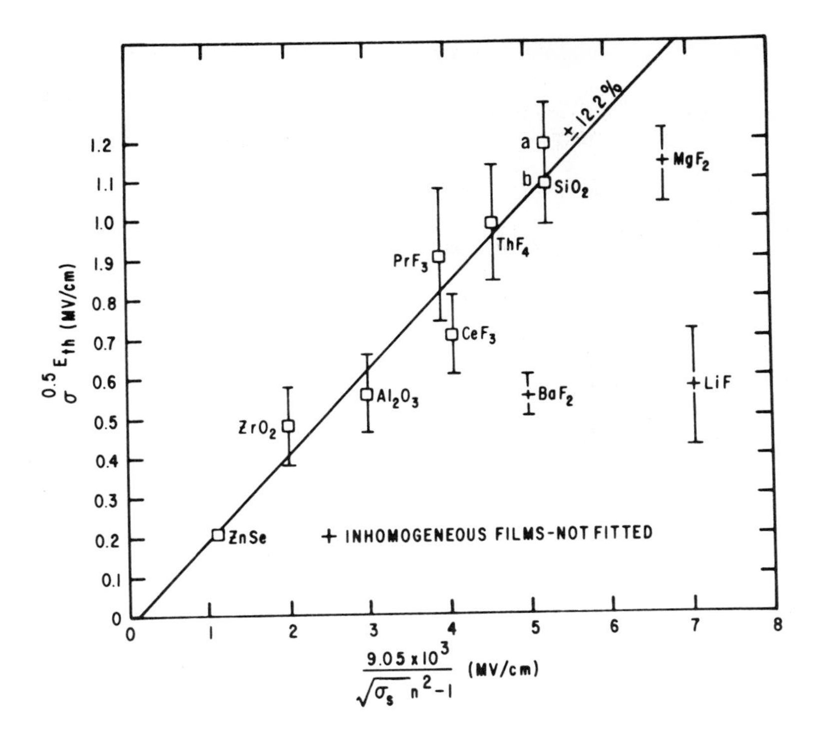

FIG. 15. Same as Figure 14, for Half-Wave Films of the Indicated Materials, Deposited on SiO_2 Substrates. (See text for further explanation.)

the variation of the breakdown field with the surface roughness from several hundred angstroms down to a value of something like 2 1/2 angstroms, which is the lattice spacing. This latter point represents the bulk damage threshold. It is quite remarkable that the surface damage, when extrapolated back to a roughness figure corresponding to the lattice spacing, fits the relation given by Equation 8. The damage experiments represented by Figures 13-15 were carried out at 1.06 μm. We ask, how does the surface damage threshold come to depend on the surface roughness over a range of rms roughness from 10 angstroms to > 100 angstroms, when all the distances are much, much smaller than the wavelength of the incident light. One hypothesis is that the dependence arises from

FIG. 16. Scanning Electron Micrographs and Diffraction Patterns of ZnS Coating on KCl Substrate Deposited by RF Sputtering and by Evaporation.

variation in the way electrons just beneath the surface are scattered by the rough surface, i.e., the diffuse scattering of subsurface electrons. This is hypothesized even for dielectrics, since just before the plasma is formed at the surface, there are many conduction electrons near the surface. This may be the reason why the surface roughness is significant in the surface avalanche process down to dimensions of the order of the lattice spacing.

Thin film optics represent the Achilles heel of large laser systems. As previously noted, thin film optics are used in fusion lasers for polarizers, beam splitters, anti-reflection coatings, apodizers, turning mirrors, etc. In order to achieve either anti-reflection or high-reflection properties, we are usually talking about multi-layer stacks, 30 to 40 layers in some cases, of various dielectric materials. Because they are multi-layer stacks, there are very profound standing wave effects, and the electrical field within the stack is very sensitive to its design. Thin film coatings yield high or low reflection because of the creation of standing wave patterns, and the cancellation of the reflected waves from successive surfaces. Thus the coating design as well as the coating composition must be specified in characterizing the damage properties of a particular coating. We saw before that the characterization of a surface required many parameters. In thin film coatings, the number of parameters increases almost without bound. There are a large number of variables--the material the coating is made of, the substrate on which it is deposited, the technique by which the substrate was prepared, and most important the conditions under which the coating was deposited (substrate temperature, pressure, rate of deposition, angle of deposition, background gas, etc., etc.). The design and function of the coating is important, because that is what determines the standing wave pattern. The individual vendor is a very significant factor; in fact, there is some evidence that the day of the week, the week of the month, and the month of the year in which that particular vendor made that particular coating is also a significant variable. The morphology of the coating is important, whether the film was deposited as a single crystal, in polycrystalline form, or in amorphous form. The porosity of the film is also important. If the film is porous, it tends to absorb material from the environment. The resultant defects, impurities and other imperfections in the coating are also of great significance.

Let us look at a few results to obtain some feeling for the numbers involved, because in large glass laser systems, at present, the limiting optical element that determines the performance and durability of the system is thin film optics. Gill and Newman from Los Alamos reported results recently on titanium dioxide films (15). Their results are shown in Table 5. The spot size to which the laser beam was focused was

TABLE 5

DAMAGE RESULTS FOR QUARTER WAVE TiO_2 FILMS, AT 1.06 μm IN 30 PSEC PULSES

Deposition Method	Damage Threshold (J/cm^2)
RF-Sputtered	10.0 - 11.2
	8.0 - 11.5
	6.2 - 9.6
	5.1 - 9.8
	5.5 - 9.2
e-gun Deposited	3.3 - 6.8
	4.1 - 4.5
	2.2 - 3.6

500 μm. About 40 shots were taken on each sample. Although there is significant variation from sample to sample, we see that the RF sputtered films were significantly more resistant to damage than the electron gun deposited films. A possible explanation for this difference is seen in Figure 16, taken from the work of Golubovic, et al (16), presented at the 1975 Damage Symposium. Here we see ZnS films deposited on KCl substrates, both by sputtering and e-gun evaporation. The sputtered film is polycrystalline, as one can see in the electron micrograph and accompanying diffraction pattern. The evaporated film is seen to be amorphous, and evidence of crystal structure cannot be resolved on the scale of the electron micrograph or detected in the diffraction pattern. There is some evidence of porosity, and there may be other factors here as well, but it is clear, at least in this experiment, that there is a significant difference in the crystal morphology of the film, depending on how the film

was prepared. So it is not surprising that observed damage thresholds were different.

Recent results obtained by D. Milam (17) at the Lawrence Livermore Laboratory using a 1.06 μm, 125 picosecond laser, are shown in Table 6. Here multilayer dielectric films of various designs from several vendors were damage tested. The variables involved include the identity of the vendor, the application of the coated element, and the polarization and angle of incidence of the optical radiation.

TABLE 6

OBSERVED DAMAGE THRESHOLDS FOR VARIOUS APPLICATIONS
125 PSEC, 1.06 μm

Vendor	Device	Polarization		$\frac{J}{cm^2}$
A	Polarizer	S	Brewster	> 9
B				5.4
C				1.2
C				1.3
A	99% REFL	P	45°	> 8
B	At 45°			3.8
C				3.5
A	99% REFL	1	0°	3.0
	At 0°			3.8
				3.8
				4.0

We see that the damage threshold is on the order of a few joules per square centimeter at this pulse length. The variation from vendor to vendor for what are essentially identical coatings is pronounced. This gives one some feeling for the state of the art in the design, reproducibility, and fabrication of thin film optics.

VI. THE FUTURE

In conclusion, we should discuss a bit about the future. Thin film optics must be improved, since they are presently the weakest link in the large glass laser system. In the next paper, Dr. Weber will be talking about some really significant improvements in the performance of the laser materials and of the optical glasses used in the energy delivery systems associated with the laser. We won't be able to capitalize on those improvements, however, unless we can strengthen the weakest link in the system, and that is the thin film optics. It is fairly clear that at present, damage in thin film optics is dominated by extrinsic factors, mainly those introduced in the fabrication and handling process. This is evidenced by the variation from vendor to vendor, and the variation resulting in morphology. Clearly the major improvement in thin film optics is going to come about from improving the process whereby the film is made. Before that improvement can be made, a substantial amount of research is required in order to identify the way in which process variables govern structural variables, and in turn how structural variables influence the damage threshold. That is the present state of affairs for both neodymium glass laser systems and carbon dioxide laser systems.

We have said nothing at all about metal substrate mirrors, but these are of great interest and importance for both CO_2 and Nd-glass lasers. Looking into the future, we want to see what the problems of the next generation of lasers will be. Let us just mention one of them. From the standpoint of plasma physics, it is fairly clear that one wants to go to shorter wavelength lasers for laser fusion. As we go to shorter wavelength lasers, we run into the problem that we don't have a large number of good materials that transmit well in the UV, and it is essential, with a high-power laser, that the material transmit not only at the laser frequency but also be highly transparent at twice the laser frequency. The reason is that, assuming there are states of appropriate symmetry available, two-photon absorption is allowed. That is, due to the intense photon field the material responds nonlinearly, and absorption can occur at twice the optical frequency. Two-photon absorption is described by the relation

$$dI/dz = -\alpha_2 I^2 \quad (9)$$

where α_2 is the coefficient of two-photon absorption. This coefficient has been measured for several solid dielectrics, and although values are uncertain, its magnitude is on the order of 10^{-8} cm per watt. If the photon energy is more than half the band gap in the material, then that is the order of magnitude of the two-photon absorption coefficient. Thus, if we want to keep the bulk absorption below a value of 10^{-4} per cm, we have to keep the intensity below 10^4 watts per sq cm. In this discussion, we have been considering flux levels of 10^{10} watts per sq cm, so it is clear that if two-photon absorption is energetically allowed in a material, that material is not acceptable for use as a high-power laser component. We are concerned primarily with windows, since future short wavelength lasers will probably be gas lasers, so we are not as worried about this particular process in the active laser material itself.

In Table 7, taken from the work of Duthler (18), we see the band gap energy for a variety of short wavelength window materials. We see that they are mostly fluorides. All the band gaps are on the order of 9 to 13 eV. Also shown are some of the existing rare gas eximer lasers wavelengths, as well as the N_2 molecular laser and frequency doubled neodymium glass. We see that even if we had a 1,000 to 2,000 Angstrom laser, and it really looked attractive from a standpoint of laser fusion, in order to avoid the two-photon limit we would have to have a

TABLE 7

SHORT WAVELENGTH WINDOWS AND LASER CANDIDATES

	Window Material	Band Gap	Window Material	Band Gap	
W	LiF	13 eV	CsF	10 eV	
I	MgF_2	11 eV	LiCl	10 eV	
N	KF	10.9 eV	CaF_2	10 eV	
D	NaF	10.5 eV	SrF_2	9 eV	
O	RbF	10.4 eV	BaF_2	9 eV	
W					
S					
L	Laser	λ (nm)	hν (eV)	2 hν (eV)	
A	Ar	126	9.8	19.6	NO WINDOW ↑
S	Kr	146	8.5	17.0	
E	Xe	174	7.2	14.4	
R	N_2	337	3.7	7.4	
S	2 x Nd	532	2.3	4.6	

window with a band gap greater than 14 electron volts. Looking at our table, we find there are no readily available materials with band gaps greater than 14 electron volts. This says something about the future of 1,000 Angstrom lasers for laser fusion, or for any high power application. In fact, it tells us that we are generally restricted, at high power, to wavelengths longer than ∿2500 Angstroms.

We have only given the reader a little bit of the flavor of the problems of laser materials for high power lasers. The discussion has been restricted to one particular application of high power lasers, namely, laser fusion. Great advances have been made in this field over the past few years, but further advances are clearly needed if laser fusion is ever to become a practical reality.

REFERENCES

1. "Laser Induced Damage in Optical Materials, 1975" NBS Special Publication 435, U.S. Government Printing Office, Washington, D.C. (1975), A. J. Glass and A. H. Guenther, Editors.

   See also "Laser Induced Damage in Optical Materials, 7th ASTM Symposium," to be published in Applied Optics 15 (1976), A. J. Glass and A. H. Guenther, for an extensive bibliography.

2. J. Nuckolls, L. Wood, A. Thiessen and G. Zimmerman, Nature 239, 139 (1972).

3. The authors are indebted to Dr. Keith Boyer of the Los Alamos Scientific Laboratory for the use of these illustrations.

4. R. Kidder and J. Ziule, Nuclear Fusion 12 (1972).

5. Annual Report of the Laser Program, UCRL-50021-75, A. J. Glass, Editor (1976) and UCRL-50021-74, J. I. Davis, Editor (1975), available from NTIS, U.S. Department of Commerce, Springfield, VA 22151.

6. E. S. Bliss, D. R. Speck, J. F. Holzrichter, J. H. Erkilla, and A. J. Glass, Appl. Phys. Letters 25, 448 (1974).

7. N. Bloembergen, IEEE J. Qu. Elec. QE-10, 375 (1974).

8. W. Lee Smith, in Ref. 1.

9. Marshall Sparks, in Ref. 1.

10. N. L. Boling, G. Dube, and M. D. Crisp, "Laser Induced Damage in Optical Materials; 1973," A. J. Glass and A. H. Guenther, Editors, NBS STP 387, Washington, D.C. (1974).

11. J. S. Loomis and E. Bernal, in Ref. 1.

12. J. R. Bettis, A. H. Guenther and A. J. Glass, "Laser Induced Damage in Optical Materials; 1974," A. J. Glass and A. H. Guenther, Editors, NBS STP 414, Washington, D.C. (1975).

13. J. R. Bettis, R. A. House, A. H. Guenther and R. R. Austin, in Ref. 1.

14. D. W. Fradin and M. Bass, in Ref. 10.

15. B. E. Newnam and D. H. Gilkey, Topical Meeting on Optical Interference Coatings, Asilomar (1976).

16. A. Golubovic, W. Ewing, R. Bradbury, I. Berman, J. Bruce, in Ref. 1.

17. D. Milam, private communication.

18. C. J. Duthler, in Ref. 1.

Chapter 8

OPTICAL MATERIALS FOR NEODYMIUM FUSION LASERS

M. J. Weber

Lawrence Livermore Laboratory
University of California
Livermore, California 94550

I. *INTRODUCTION*

Problems associated with optical materials impose severe limitations on the performance of high-power lasers. Damage to optical components and thin-film coatings, and self-focusing and beam breakup arising from the propagation of intense laser beams have been discussed by Drs. Glass and Guenther in this volume. In this chapter we will be concerned with intrinsic limitations of transmitting dielectric materials. In particular, we will examine ways in which, by compositional changes, these limitations can be reduced and laser performance increased.

The focus will be on materials for large, optically-pumped neodymium glass lasers (1). These lasers are presently being employed in several major laser-fusion laboratories and hold the most promise for providing the power, pulse shapes, and frequencies most useful for physics experiments essential to understanding laser-plasma interactions and fusion by inertial confinement (2). Since these lasers suffer from low overall efficiency and low repetition rates, they are not presently viewed as sources for a laser fusion power plant. The knowledge gained in the development of improved optical materials for Nd lasers should, however, also be of value for other high-power lasers.

Several different optical materials are used in large solid-state lasers. These include passive materials for lenses, windows, and polarizer substrates and active materials for the lasing medium and Faraday rotators. Glasses have been used for these applications because of the requirements of high optical quality and large sizes (beam apertures of 30 cm). Crystalline materials are used for components such as the oscillator rod, Pockels cell shutters and switchouts, and for frequency conversion to higher harmonics by nonlinear processes.

Of the many properties which enter into the selection of optical materials, the nonlinear refractive index is of prime importance. Intensity-dependent changes in the refractive index cause small-scale and whole-beam self-focusing and loss of focusable energy on target. For the lasing medium, one is concerned with several spectroscopic parameters including wavelengths and intensities of absorption and emission bands, excited-state lifetimes, and quantum efficiencies. Requirements on optical quality of materials include refractive index homogeneity ($\sim 10^{-6}$), low birefringence ($\lesssim 5$ nm/cm), no absorption at the laser wavelength, and no bubbles, striae, or metallic inclusions. Other physical and chemical properties of interest are the hardness and capability for optical finishing to $\sim 1/20$ of a wave and the stability of the material and environmental durability. The laser material is exposed to and heated by intense xenon flashlamp radiation. Therefore thermal properties such as expansion, conductivity, change of refractive index with temperature, stress-optic coefficient, and solarization are also factors. Since all of the above properties are affected in varying degrees by the composition of the material, they must be considered when tailoring a material for a specific application.

Finally, the selection of optical materials is also governed by their costs, since alternate approaches to increasing output power may be possible. Optical materials are a large, but not dominant, cost of large laser systems. Improvements in their performance can, however, be translated into significant increases in power to fusion targets, hence materials research has great leverage. Recent work which illustrates this potential is reviewed below. This includes studies of (1) the refractive index nonlinearity, which is of interest for all transmitting components, (2) the lasing media, which has the longest optical pathlength, and (3) Faraday rotator materials.

II. *NONLINEAR REFRACTIVE INDEX OF MATERIALS*

The instantaneous index of refraction of a material is given by

$$n = n_o + \Delta n \quad , \tag{1}$$

where n_o is the refractive index normally measured with low-intensity light of given polarization and propagation directions and Δn is the change in index induced by an intense optical beam. Δn is expressed in terms of the time-averaged optical electric-field amplitude E as $n_2\langle E^2\rangle$, where the nonlinear refractive index coefficient n_2 is in electrostatic units. The intensity-dependent change in index is small, being in the order of parts per million for most optical materials at intensities of 10^9 - 10^{10} W/cm^2. But the optical pathlength through materials in an amplifier chain of a large fusion laser may be long,

$\gtrsim$ 1 m, and therefore the total index change can cause a phase shift of the optical wavefront of several waves.

The spatial profile of the laser beam is more intense in the center. In addition, material imperfections or "dirt" can cause local beam irregularities. These spatial intensity variations grow as exp B, where B is an integral over the optical pathlength L given by

$$B = \frac{2\pi}{\lambda} \int_0^L \Delta n \, d\ell \quad . \tag{2}$$

This breakup integral is a measure of the beam degradation and and has contributions from all transmitting optical components. To minimize B, one seeks to minimize the following quantities:

n_2/n -- for rods, windows, Faraday rotators ($\perp$ beam)

n_2/n^2 -- for laser disks (at Brewster's angle)

$n_2/n(n-1)$ -- lenses

In all cases, low-n_2 materials are advantageous.

Self-focusing of laser beams in solids arises from three mechanisms: optical Kerr effect, electrostriction, and thermal effects, when absorption is present (3). For short ($\lesssim$ 1 ns) optical pulses, only the Kerr process is significant. Furthermore, it has been shown that for silicate glasses, the dominant contribution to n_2 is electronic in origin and that for glasses have different linear refractive indices the ratios of the nonlinear polarizability contributions from electronic and nuclear motions are approximately the same (4).

Predicted versus Measured Refractive Index Nonlinearities. The electronic contribution to n_2 originates from the third-order electronic susceptibility $\chi^{(3)}$. Wang (5) has developed an empirical relationship between $\chi^{(3)}$ and the linear susceptibility. Following this approach, expressions can be derived which relate n_2 and the linear refractive indices in the long-wavelength limit. By using dispersion data and a single adjustable parameter, measured n_2 values for silicate glasses spanning at range of ~ 10 can be fitted satisfactorily (6).

Recently, n_2 measurements have been made for a wider class of materials including several low-index, low-dispersion glasses and crystals. Measurements were made using time-resolved interferometry (7,8). The results, obtained using 100 ps, linearly-polarized 1064 nm laser pulses, are shown in Fig. 1 together with predicted values. The materials

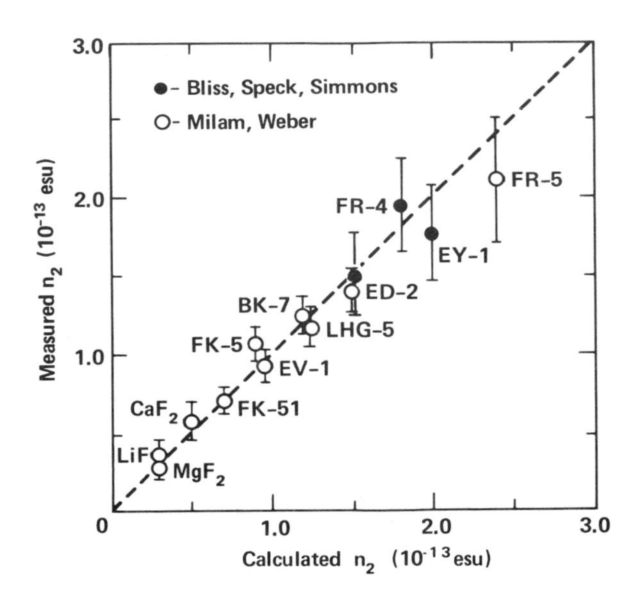

FIG. 1. Calculated versus measured nonlinear indices.

examined included silicate (ED-2) and phosphate (EV-1, LHG-5) Nd laser glasses, borosilicate glass (BK-7), fluorosilicate (FK-5), fluorophosphate (FK-51), Faraday rotator glasses (EY-1, FR-4,

FR-5), and three fluoride crystals. Within experimental error, the data is accounted for with a single adjustable parameter. Since low n_2 values are characteristic of materials having low index and low dispersion, these properties provide guidance in the search for improved optical materials.

Materials with Small Refractive Index Nonlinearities. Optical glasses are conveniently grouped using their refractive index n_d and Abbe number ν as coordinates (the latter being a reciprocal dispersion defined by n_d-1/n_F-n_c). Regions of known optical glasses are shown on a $n_d - \nu$ plot in Fig. 2.

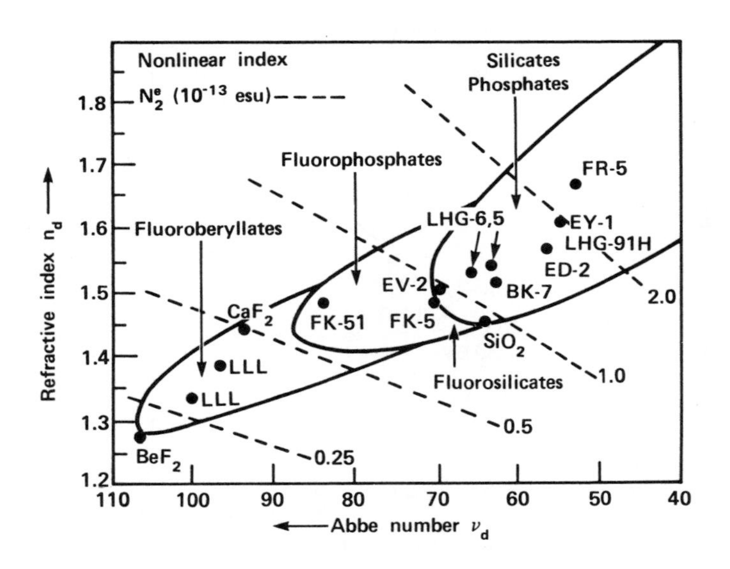

FIG. 2. Refractive indices of optical glasses

Superimposed are dashed lines of constant n_2 predicted from the n_d and ν values (9). The most promising glasses for fusion lasers are those in the lower left-hand corner of Fig. 2.

Most commercial optical glasses are oxide glasses,

principally silicates and phosphates, with values of 1.45 - 2.0 for n_d and 20 - 84 for ν. Fluorosilicates (e.g. FK-5) and fluorophosphates (FK-51) are near the extremes of low n_2 for present optical glasses. Since the most polarizable ions in glasses are the oxygen and high-atomic-number cations, the replacement of some of the oxygen by fluorine and the use of low-atomic-number cations result in lower index glasses. The lowest index glasses reported are pure fluoride glasses with beryllium fluoride as the network former (10). The ion refractivities of Be^{2+} and F^- are among the smallest known for cations and anions. BeF_2 forms a glass analogous to SiO_2, the bonding being weaker (lower melting point) and more ionic for the former. Pure BeF_2 and SiO_2 glasses are included in Fig. 2 and represent the extremes of each glass type. The addition of network modifiers such as alkalis, alkaline-earths, aluminium and other ions changes the optical properties and yields the $n - \nu$ regions shown.

Because of their low predicted n_2 values, fluoride glasses are the most promising candidates for both passive components such as lenses, windows, and substrates and, as discussed later, as hosts for laser and Faraday rotator ions. Fluoride crystals such as CaF_2, LiF, and MgF_2 are also of interest for passive components since they can be prepared in large sizes. The figure of merits for several fluoride materials, based upon reduction of the B integral and relative to the presently used BK-7, are given in Table 1.

Fluoride materials, in addition to their low n_2, generally have large bandgaps and hence transmit over a large spectral range. Thus they are potentially useful for high-power lasers operating at other wavelengths.

TABLE 1

REFRACTIVE INDICES AND RELATIVE FIGURE OF MERIT FOR LENS MATERIALS AT 1064 nm

Glasses	n_D	$n_2(10^{-13}$esu)	$n(n-1)/n_2$
Borosilicate (BK 7)	1.517	1.24	1.0
Fused silica (SiO_2)	1.458	0.95	1.1
Fluorosilicate (FK 5)	1.487	1.0	1.1
Fluorophosphate (FK 51)	1.487	0.69	1.7
Fluoroberyllate -	∿ 1.35-1.40	∿ 0.4	2.0
Crystals			
CaF_2(cubic)	1.434	0.57	1.8
LiF (cubic)	1.392	0.35	2.5

III. *NEODYMIUM LASER MATERIALS*

The performance of a laser amplifier is determined by maximizing the ratio of the output power to the beam breakup. For an amplifier rod, this ratio is

$$\frac{P_{out}}{B} \propto \frac{\lambda n}{n_2} \frac{\alpha D^2}{1-1/G} \tag{3}$$

where α is the gain coefficient of the lasing medium, D is the effective beam diameter, and G is the small-signal gain. The gain coefficient is equal to $\sigma(N_u-N_\ell)$, where σ is the stimulated emission cross section and (N_u-N_ℓ) is the population inversion. For an amplifier of a given diameter D and required gain G, the performance in Eq. (3) is improved by increasing α and decreasing n_2 of the laser material. This is for the pumped-limited case where the inversion achievable is limited by the maximum energy deliverable to the flashlamps without seriously shortening their lifetime. For large disk amplifiers, parasitic losses due to amplified spontaneous emission become a greater limitation than pumping. This constraint is approximated by $\alpha\hat{D}$ = constant, where $\hat{D}$ is the major axis of the disk. As the parasitic-limited case is approached, increasing α is not as important as decreasing n_2.

The rate of energy extraction from an optically-pumped laser material is governed by σ, $N_u - N_\ell$, and the beam intensity I. For short pulse ($\sim$ 100 ps) operation, because of the limitation imposed on I by beam breakup and damage, one is unable to extract all the energy stored in present laser glasses. An increase in σ is therefore desirable to increase the output power. At the other extreme of long ($\gtrsim$ 1 ns) pulses, the amplifier gain may become depleted. The saturation flux is given by $h\nu/g_{eff}\sigma$, where g_{eff} is the effective level degeneracies, and is raised by decreasing σ. In this limit, pumping efficiency and energy storage are important criteria.

Glasses, because of their varied nature, provide flexibility in tailoring the refractive index and laser gain characteristics for the specific modes of operation discussed above. The Nd^{3+} lasing properties are affected by the absorption and emission spectra, quantum efficiency, and fluorescence decay rates which, in turn, are all affected by the host material.

Most Nd laser glasses in the past have been silicates because of their excellent physical and optical properties. Borates have also been considered, however if a large amount of borate is present, it causes nonradiative quenching of the Nd^{3+} fluorescence which reduces the quantum efficiency. Recently several phosphate glasses have been developed which have acceptable physical properties and higher gain coefficients and lower nonlinear indices than the best silicate glasses. These glasses are currently being tested in fusion lasers. Because of their lower n_2, fluorophosphate glasses are even more promising for parasitic-limited applications. Fluoroberyllate glasses are still in an early stage of research and development and have not yet been melted and cast in large sizes. While there are other inorganic glass formers, the above glasses appear to offer the greatest potential for providing the combination of high gain coefficients and low nonlinear indices for fusion lasers.

Studies have been and are being made of the spectroscopic properties of Nd in many different glass forming networks. With the advent of lasers, thousands of glasses were melted in the 1960's to investigate the effects of changes in glass network and network modifier ions on the spectroscopic and laser parameters of Nd. This work to a large degree was empirical and was hampered by two difficulties: first, the spectroscopic properties and interactions of rare-earth ions were not completely understood in a crystalline environment without the complications of an amorphous host; and second, the specific uses and applications of lasers were less well defined and therefore it was not always certain which properties should be optimized. Today, both of these situations are clearer.

Neodymium enters a glass as a network modifier ion and has oxygen or fluorine ligands. The local field at the Nd site is determined by these and more distant network and network modifier ions. Because of the differences in the symmetry and strength of the bonding, there are large site-to-site variations in the effective crystal field as seen by the Nd^{3+} ions. This results in inhomogeneous line broadening and site-dependent transition probabilities. Studies in which the glass network and network modifier ions are systematically altered have demonstrated that the spectroscopic properties of Nd glasses can, within limits, be tailored to specific applications by varying composition (11). Examples of the variations of the cross section, fluorescence linewidth, and radiative lifetime associated with the $^4F_{3/2} \rightarrow {}^4I_{11/2}$ emission of Nd^{3+} are given in Table 2 for silicate (Si), phosphate (P), fluorophosphate (FP), and fluoroberyllate (FB) glasses. The range of variation of these properties for the latter two glasses is smaller because investigation of the full compositional space is still in progress.

TABLE 2

MEASURED VARIATIONS IN Nd LASER GLASS PROPERTIES

Glass	Cross section ($10^{-20} cm^2$)	Linewidth $\Delta\lambda$(nm)	Lifetime τ(μs)	Relative total absorption
Si	1.0-3.1	28-35	330-950	0.75-1.0
P	1.8-4.7	19-28	320-560	0.8-0.95
FP	1.7-3.9	22-29	350-470	0.8-0.9
FB	2.4-3.1	18-24	590-670	0.77-0.82

A large number of new glasses have been evaluated further for amplifier applications. Predictions of gain coefficient are based upon measurements of absorption and fluorescence spectra and decay properties combined with computer calculations of flashlamp pumping. The results for several silicate and phosphate glasses are in good agreement with the actual performance in a laser cavity (12). Since the information required for this evaluation can be obtained from small samples of the laser material, this considerably simplifies the comparison and selection of Nd glass for use in large laser amplifiers.

In practice, the absorption spectra of candidate glasses is recorded from 1000 to 250 nm. Using the measured line strengths, a set of optical intensity parameters is derived from a least-squares fit of calculated and measured values. These parameters are used to calculate the fluorescence line strengths and branching ratios for the $^4F_{3/2} \rightarrow {^4I_J}$ transitions of Nd^{3+}. The stimulated emission cross section is determined from the calculated line strength and the linewidth measured from the fluorescence spectrum. The radiative lifetime is also calculated and compared with the observed fluorescence decay rate to determine the radiative quantum efficiency. The latter is dependent upon the Nd concentration and the host glass.

Using the per-ion absorption spectrum, a computer program calculates the fractional absorption of xenon flashlamp power by the material as a function of optical depth (thickness times doping) and current density in the flashlamp. Relative values for the spectral range 400-950 nm and for a flashlamp current density of 1000 A/cm^2 are shown in the final column in Table 2. The program contains a model of the flashlamp output spectrum that is modified by rerunning some of the light back through a layer of plasma to approximate the multiple passes that actually occur inside the laser pumping cavity. The output curves of fractional absorption as a function of optical depth are then used in conjunction with decay-rate information and cross section to predict the relative behavior of different materials in disk laser amplifiers.

A comparison of spectroscopic properties, predicted gain coefficients, and relative figures of merit for several Nd laser glasses is given in Table 3. The glasses are listed with decreasing refractive indices from left to right. The spectroscopic parameters are representative of each glass

TABLE 3

COMPARISON OF PROPERTIES AND FIGURES OF MERIT FOR Nd LASER GLASSES

	Si	P	FP	FB
Refractive index n_D	1.57	1.54	1.49	1.35
Nonlinear index $n_2(10^{-13}esu)$	1.4	1.1	0.7	$\sim$ 0.4
Cross section $\sigma(10^{-20}cm^2)$	2.9	4.1	2.5	2.9
Fluorescence peak (nm)	1062	1054	1054	1047
Fluorescence linewidth (nm)	34	25	31	24
Radiative lifetime (μs)	350	340	460	610
Relative absorption	1.0	0.94	0.88	0.77
Relative gain coefficient Example: small disk	1.0	1.4	0.8	0.9
Relative figure of merit				
Pump-limited rod: $n\alpha/n_2$	1.0	1.8	1.5	2.7
Parasitic-limited disk: n^2/n_2	1.0	1.2	1.8	2.6

type but are not necessarily the extreme values possible. Phosphate glasses, because of their large oscillator strengths and narrow linewidths, have the largest cross sections. (The linewidths in Table 3 are the effective linewidths which take into account the asymmetry in the emission band; they therefore differ from the full-width, half-maximum linewidths in Table 2). The fluoroberyllate glasses also have narrow linewidths but the oscillator strengths of radiative transitions are less. This is also reflected in the longer lifetimes and smaller relative absorption (for equal Nd/cm^3) of the FB glasses. For most operating conditions, the phosphate glasses provide the greatest gain coefficient and therefore are attractive for short-pulse, pumped-limited use. When beam breakup is considered, however, the fluoroberyllate glasses have the best figure-of-merit because of their low predicted n_2 value. This is also the case for parasitic-limited operation. Note the wavelength of peak gain is dependent on the glass type. In general one cannot mix glasses in different stages of a long amplifier chain. An exception is phosphate and fluorophosphate glasses. Here the former would be used in the initial pump-limited rod and small disk amplifier stages and the latter used for the large parasitic-limited disk amplifiers. Today these two Nd glasses constitute the best combination available for short-pulse, high-peak-power fusion lasers.

IV. *FARADAY ROTATOR MATERIALS*

Lasers for fusion experiments must be protected from back-scattered light and, in the case of multibeam experiments, from scattered and transmitted light. Optical isolation can be obtained using the Faraday effect. A Faraday rotation material is placed between two polarizers oriented at 45° and a magnetic field applied to rotate the plane of polarization of light transversing the material by 45°. Forward-going light passed through the second polarizer and onto the target. Back-reflected light, however, undergoes an additional 45° rotation and is rejected by the second polarizer.

The requirements for a good Faraday rotator material include low absorption at the laser wavelength, low birefringence, high homogeneity, and high damage threshold. The angle of rotation of the polarization is given by $\theta = V\ell H$, where V is the Verdet constant, ℓ is the optical pathlength, and H is the component of the magnetic field in the direction of light propagation. For a 45° rotation in a given H field, the Verdet constant of the material should be as large as possible to reduce ℓ and the beam breakup integral B. A figure of merit for a Faraday rotator material is

$$\frac{P_{out}}{B} \propto \frac{\lambda n}{n_2}\frac{D^2}{\ell} \rightarrow \frac{nV}{n_2} \quad , \tag{4}$$

where λ is the laser wavelength.

Faraday isolators have been built using diamagnetic and paramagnetic materials (13). Classically, the Verdet constant for a diamagnetic material is proportional to the dispersion $dn/d\nu$. But as noted in Section II, materials having large dispersion (low Abbe number) also have large n_2 values. This, combined with the fact that larger V's are possible with paramagnetic materials, favors the latter for use in fusion lasers. The total Verdet constant for a material is composed of a positive diamagnetic contribution and a negative paramagnetic contribution. Since to minimize n_2 one selects low-dispersion host materials, the diamagnetic contribution is also reduced.

The paramagnetic Verdet constant is approximated by

$$V(\lambda) = -\frac{A}{T}\frac{N}{\lambda^2-\lambda_t^2} \quad , \tag{5}$$

where T is the temperature, N is the number of active paramagnetic ions per unit volume, and A and λ_t (the effective transition wavelength) are constants characteristic of the ion and host material. Rare earths have been used extensively as the paramagnetic ion. Of these, Tb^{3+}, Ce^{3+}, and Eu^{2+} are particularly useful for the visible-near infrared spectral region since they have no absorption bands in this region and have large

Verdet constants per ion. The rotation occurs via virtual $4f^n - 4f^{n-1}5d$ transitions. Since the location and transition matrix elements of the 5d bands exhibit only small variations with host, the specific Verdet constant of a given rare earth is approximately the same for all materials in the long wavelength limit ($\lambda >> \lambda_t$). Therefore to increase V, the concentration N should be made as large as possible. Increasing the rare earth content in a glass will concurrently increase the linear and nonlinear refractive indices, however this is generally a smaller effect. As evident from Eq. (5), the paramagnetic V can also be increased by reducing the temperature.

Glasses having small n_2 values are obvious choices as hosts for paramagnetic rotator ions. Relative figures of merit of some commercial and prospective Faraday rotator glasses are tabulated in Table 4.

TABLE 4

PROPERTIES AND RELATIVE FIGURE OF MERIT OF FARADAY ROTATOR MATERIALS AT 1064 nm

Glasses	-V(min/Oe-cm)	n	$n_2(10^{-13}$esu)	Vn/n_2
FR-4(Ce-phosphate)	0.031	1.556	1.9	0.45
FR-5(Tb-silicate)	0.071	1.678	2.1	1.0
Tb-fluorophosphate	$\gtrsim$0.05	1.51	~0.9	~1.5
Tb-fluoroberyllate	$\gtrsim$0.05	~1.35	~0.5	~2.4
Crystals				
$LiTbF_4$	0.13*	1.49	~0.7	4.9
CeF_3	0.11**	1.61	~1.2	2.6

*Estimated from V value at 633 nm provided by A. Linz (private communication).

**Measured by C. Layne, R. Morgret, and W. Moss.

The specific V for Ce^{3+} is approximately 80% of that for Tb^{3+} at 1.06 μm. The larger V for FR-5 is due to the larger amount of terbium which can be incorporated into this glass while still preserving good glass-forming properties. The maximum Tb concentrations possible for the fluorophosphate and fluoroberyllate glasses are not known and are currently under investigation. The entries for these glasses in Table 4 are based upon an assumption that a Tb content ≈ 70% of that in FR-5 may be obtained. The actual figure of merit achievable may be higher or lower than this estimate.

The rare earth density in crystals can be significantly higher than that in most glasses. By combining this feature with a fluoride host, superior rotator materials are possible. Data for two examples, $LiTbF_4$ and CeF_3, are included in Table 4. Although n_2 for these material has not been measured, based upon estimated values, the projected figures of merit are higher than for glasses. These crystals have been grown with good optical quality and only in diameters of a few cm.

V. *CONCLUSIONS*

The most severe limitation to the performance of present-day high-power fusion lasers is self-focusing and beam breakup arising from intensity-dependent changes in refractive index. Tables 1,3,4 illustrate that factor of two or greater improvements in focusable laser output are possible by the development of low-index fluoride materials. While the existence and properties of fluoride glasses have been known for many years, these glasses have been of only limited interest to the optical designer. However, because of their small refractive index nonlinearities, they are of prime interest for fusion laser applications. Currently, research is being devoted to exploring the range of compositions possible for fluorophosphate and fluoroberyllate glasses, to determining the effects of compositional changes on the refractive indices and the spectroscopic properties of neodymium, and to identifying problems associated with the melting, casting, and finishing of large optical components. These results will dictate the final cost and, coupled with actual performance evaluation, will establish the ultimate merits of these new materials.

The materials research presented here is oriented toward optical materials for neodymium fusion lasers. The general approach has wider applicability and should be useful in identifying and optimizing optical materials for lasers operating under other conditions and requirements. For example, more powerful, shorter-wavelength lasers are desired for achieving laser-induced fusion. The beam breakup integral in Eq. (2), however, is inversely proportional to wavelength and thus will become increasingly important for operation at shorter wavelengths. In addition, little study has been made of the frequency dispersion of the nonlinear index n_2. Finally, it should be noted that the ability to propagate more intense beams through optical components will in turn require higher damage thresholds for thin-film coatings used on various components. Hence concominant developments of optical coating materials and techniques may be necessary to fully utilize new optical materials.

ACKNOWLEDGMENTS

I am indebted to many colleagues associated with the Laser Fusion Program at the Lawrence Livermore Laboratory for their stimulation and contributions to the work reported here. In particular, I wish to thank Drs. R. Jacobs, C. Layne, D. Milam and R. Saroyan for collaborating on experiments, Drs. J. Emmett, A. Glass, W. Hagen, and J. Trenholme for helpful discussions, and Ms. B. Flow for assistance in preparing this manuscript.

REFERENCES

1. For a description of current high energy glass lasers, see, J. A. Glaze, Proc. S.P.I.E. 69, 45 (1975) and J. A. Glaze, W. W. Simmons, and W. F. Hagen, Proc. S.P.I.E. (1976).

2. J. L. Emmett, J. Nuckolls, L. Wood, Scientific American 230, 24 (1974).

3. A. Feldman, D. Horowitz, R. M. Waxler, IEEE J. Quant. Electron QE-9, 1054 (1973) and references therein.

4. R. Hellwarth, J. Cherlow, T. T. Yang, Phys. Rev. B11, 964 (1975).

5. C. C. Wang, Phys. Rev. B2, 2045 (1970).

6. N. Boling, A. Glass, and D. Owyoung, (to be published).

7. E. S. Bliss, D. R. Speck, W. W. Simmons, Appl. Phys. Lett. 25, 728 (1974).

8. D. Milam, M. J. Weber, J. Appl. Phys. (June 1976).

9. A. J. Glass (private communication).

10. For a survey of fluoride glasses, see W. Jahn, Glastechn. Ber. 34, 107 (1961).

11. R. R. Jacobs, M. J. Weber, IEEE J. Quant. Electron. QE-12, 102 (1976).

12. J. B. Trenholme, Laser Fusion Annual Report, Lawrence Livermore Laboratory, UCRL-50021 (1974) and private communication.

13. C. F. Padula and C. G. Young, IEEE J. Quant. Electron. QE-3, 493 (1967).

14. N. F. Borrelli, J. Chem. Phys. 41, 3289 (1964).

Chapter 9

CRITICAL MATERIALS PROBLEMS IN THE EXPLOITATION OF FISSION ENERGY

S. F. Pugh
Head of Metallurgy Division
A.E.R.E. Harwell, Oxfordshire
England OX11 ORA

I. *INTRODUCTION*

A. *Constraints On Selection Of Materials*

Successful exploitation of fission energy demands the evolution of new materials and the reoptimisation of conventional materials as essential and major parts of the development program.

For nuclear power plants, additional constraints are placed on the choice of materials and their composition while there are also additional stringent requirements for performance in service. For example, for services in the core it is necessary to choose new materials or modify the composition of existing alloys according to the nuclear properties of the constituent atoms. Certain constraints are also placed on the quantities of particular materials which can be used in the core in the interest of neutron economy. Such restrictions often preclude the use of fuel cladding sufficiently thick and strong to resist collapse on to the fuel under the coolant pressure.

For materials in and around the core of the reactor it is necessary to consider radiation effects particularly those arising from transmutation and from displacement of atoms by fast neutron collisions and the subsequent displacement cascades. Particular problems involving atomic displacements have included the distortion of graphite moderator blocks in gas cooled reactors, reduction in fracture toughness of the beltline region of PWR pressure vessels, and void formation in fast reactor cladding and in-core structural materials. Important transmutation effects include the generation of helium causing a reduction in creep ductility in the AGR cladding, and the generation of fission products in nuclear fuel giving rise to a volume increase which is accommodated by strain of the cladding. Fission fragments dissipate their energy entirely within the fuel and inner layer of the cladding. The main significant effects are densification of slightly porous UO_2 and an augmentation of fission product diffusion.

In and around the core and for the whole of the primary circuit the required standards of integrity of the containment are higher than those optimised for conventional pressure vessel systems. Farmer's criterion (1) quantifies the acceptable probabilities of various types of reactor accident. There has also been a growing awareness of the high levels of gamma radiation field which build up round the primary coolant circuit of water cooled reactors and the consequent radiation exposures to personnel during inspection and repair operations (2,3). The economies of scale which appear to be very rewarding in reactor systems introduce the new development problems of the fabrication and quality assurance of heavy section steel vessels and pipework.

Although all the constraints enumerated above apply in the core region, in many respects the problems arising there are perhaps not as critical as those elsewhere. The need to remove and re-load fuel elements periodically implies access to the core and hence considerable sections of the in-core structures could also be made removable. This procedure has already been used for several thermal reactors and it might be necessary routinely in fast reactors, although it is likely to be a costly operation also giving rise to extra radioactive waste.

Outside the core region the properties of materials that are important include corrosion resistance, resistance to plastic yielding and resistance to crack growth and catastrophic fracture, fabricability from the original ingot stage to the final welded structure and availability in a proven quality of known consistency at an acceptable cost. Clearly these latter requirements also apply to the in-core region also. For each reactor type it has been necessary to pass through stages where new materials were introduced into reactor construction. While laboratory testing was essential to eliminate unsuitable materials the final developments have required controlled operation of several prototypes for large scale statistical testing and to achieve a realistic environment. For thermal systems such stages have been successfully passed for a range of new materials including UO_2 fuel, zircaloy cladding and pressure vessel steels. Further improvement may be difficult to achieve economically.

In a concise description of materials problems limiting the exploitation of fission energy, it is possible to treat only general major areas with some indication of their relevance. Problems with individual reactor components or individual reactors are not critical since they can be avoided by changes in design.

The comments in this note describe a personal view written in order to indicate the main areas in which research and development of materials might be profitable.

B. *Are Materials Problems Critical?*

So far in the exploitation of fission energy no requirement for a material has appeared which could be said to be critical in the sense that failure to provide a suitable material has held up the exploitation of fission energy. It is not possible either to foresee any critical requirements for future systems. In many cases what appears to be a materials problem can find an engineering solution. Whether the cost of exploitation of fission energy will remain competitive with other sources of energy is more likely to be the critical question. Competitiveness depends on the future cost of alternative energy sources, the value to society of a particular source of energy, the future cost of uranium and the possible increased plant and operational costs needed to meet tighter safety requirements. A balance also has to be achieved between installation cost and cost of maintenance in service. The economics can also be dominated by interest rates on capital and inflation rates.

Materials properties do however set limits to the engineering parameters of a system since it is the engineers' task to design a plant to use materials to the limits of their endurance. Any improvement in the properties of materials can be exploited by increasing the efficiency of an installation, reducing the cost or making it more reliable. The fact that nuclear power can be produced in commercial thermal reactors at a competitive cost is the best evidence that there are no critical problems at present in the exploitation of fission energy.

II. REACTOR SAFETY

A. Technical Aspects

All technical assessments have shown consistently that the probability of major releases of radioactive materials from thermal reactor systems is extremely low. Death by natural causes is still by far the most common while the automobile accident remains the most likely cause of accidental death in industrialised communities even neglecting secondary effects caused by the distribution of lead and other forms of atmospheric pollution, (4). The reactor industry has had a very good health record. Statistics for one group showed that between 1962 and 1974 while 66 had been killed in road accidents on public roads none had died as a result of the particular hazards of atomic energy (5).

There is a considerable materials in-put to questions of safety of nuclear plant since many of the accident chains at some stage involve the ability of material to withstand imposed stresses. With engineered safeguards such as the emergency core cooling system, however, even large ruptures occurring in the plant can occur without major fission product release.

B. Non Technical Aspects Of Reactor Safety

The main difficulties in assessing non technical aspects of safety are that it is not possible to predict and quantify activities determined by individuals and social groups. The non technical hazards include negligence during design, fabrication or operation of nuclear plant, sabotage of reactors, acts of war, theft of the plutonium used in fast reactor fuel cycles, and damage to nuclear plant caused by criminals or terrorist groups (6). In addition, the wider availability of enriched fissile materials makes possible the proliferation of nuclear weapons. Ultimately the acceptance by society of these hazards will depend on the need for this particular source of energy and the extent to which the social factors can be brought under control.

III. THERMAL REACTOR FUEL ELEMENTS

A. Gas Cooled Reactors

The four metals aluminium, magnesium, zirconium and beryllium have very low neutron capture cross sections. Each has therefore been investigated as a cladding and structural material for thermal reactors. A magnesium alloy is very successfully used as natural uranium metal fuel cladding in UK power reactors. Failure rates have been only about 10^{-4} and, with on-load refuelling, cladding perforation has not caused significant loss of fuel life or unscheduled shut downs, (7,8). Power density is however limited by the need to avoid overheating of the fuel elements during upset or accident conditions. One channel fire has in fact occurred, (9).

Of the low capture cross section metals, only beryllium had sufficient strength and oxidation resistance at the proposed clad operating temperature of 600°C originally set for the UK advanced gas cooled reactors. The fabrication of pure beryllium into tubes was established at the production plant scale. A sudden drop in the world price for enriched uranium caused stainless steel cladding to become competitive in view of its lower cost and the possibility of increasing cladding temperature (10). Beryllium is an inherently brittle material (11, 12) which becomes even more brittle under irradiation; caused by both displacement damage and transmutation processes producing helium. It would therefore be essential to design beryllium alloy cladding to be free standing in the reactor to avoid fuel-clad mechanical interactions. Because of the relatively high neutron absorption of the elements in austenitic steel the "strong can" concept had to be abandoned for economic fuel designs, and mechanical interaction was accepted. A considerable effort over the past twelve years has therefore been directed towards improving the high temperature creep ductility of the AGR cladding steel, and it was possible to demonstrate that materials with high initial creep ductility retained a major part of that ductility during and after irradiation.

Gas cooled reactors have achieved very high coolant temperatures suitable for direct application for process heat in the chemical industry and for driving gas turbines. Fuel temperatures are well above those for which metal cladding could survive, hence graphite is chosen for the moderator and in-core structural material. Pyrocarbon layers are used for the fuel cladding with an interlayer of silicon carbide to prevent the escape of certain fission products. Prismatic and pebble forms of fuel element have operated successfully at temperatures up to 1300°C in prototype reactors, (13). The main materials problems in high temperature reactors will probably arise in the heat exchangers.

B. *Water Cooled Reactors*

Of the metals with low neutron capture cross section aluminium has been extensively used as cladding for materials testing reactors and high flux reactors which are water cooled, but only zirconium alloys have sufficient resistance to corrosion-erosion in rapidly flowing water at temperatures up to 350°C. Attack of the inner wall of zircaloy cladding by water introduced with the fuel has been an important cause of failure. Fuel fabrication routes have therefore been improved to eliminate water. The main remaining cause of systematic failure of the cladding is from mechanical pellet-clad interaction (PCI) formerly aggravated in some instances by densification of the fuel during the early stages of irradiation to leave axial gaps between pellets (14).

LWRs are designed for operation with an annual shutdown for refuelling, and to operate with some failed fuel elements with consequent escape of radioactive fission products into the primary coolant circuit. There is therefore an increase in exposure of plant personnel and the public to increased radiation doses, though still within the permitted levels. When fission product escape levels become too high either a reduction in power or an early shut down of the reactor for refuelling is necessary. There is not enough information gathered to permit estimation of the total economic penalties arising from fuel clad perforation. Elimination of fuel failure would, however, be consistent with the requirement to reduce radiation to the population to "as low as practicable levels", would reduce the amount of active effluent from the plant, and increase the average fuel burn-up per reprocessing cycle, (14).

C. *Nuclear Fuel*

The U-magnox fuel element is unusual in that the uranium metal bar is the main structural element providing all the strength of the fuel element and supporting the cladding which at the higher operating temperatures of up to 435°C is very soft. All other fuel elements currently exploited use uranium dioxide pelleted fuel which cracks in service and must be held in place by the can. The main nuclear penalty with UO_2 arises from the lower density and not from neutron capture in oxygen which is negligible. Combined with heavy water as a moderator and coolant UO_2 can form the basis of a natural uranium fuelled system. Its resistance to chemical attack by water makes UO_2 the best fuel for all water cooled reactors.

IV. FAST REACTOR FUELS AND IN-CORE STRUCTURES

A. *Void Formation*

Special problems in the use of materials in the core of a fast reactor arise from radiation effects caused by the high fast neutron flux. In 1966 voids were observed for the first time in cladding material from the core of the Dounreay fast reactor, (15). This phenomenon has been subjected to intensive study throughout the world in the past 10 years and its main characteristics are now fairly well known and understood, (16). Because void swelling is both temperature and dose dependent and because these vary throughout the core of the reactor changes in volume are not uniform and give rise to significant distortion. Enough work has already been done, however, to show that in principle there are samples of material which have a very high resistance to void formation, (17). Variation in the composition and structure of materials within a given conventional specification has been shown to give rise to marked differences in rates of void formation. Some of this variation is certainly caused by variations in the levels of minor alloying elements and impurities.

No existing fast reactor has a neutron flux higher than any of those proposed for commercial fast reactors. To find the effects of full life dose on components intended for service in the core therefore requires very long irradiation times. Much progress has been made using heavy ion beams from accelerators to irradiate specimens to full life level of displaced atoms. For cladding material irradiation for less than one day is required and displacement levels for the full life dose of core components have also been achieved. The heavily damaged layer is at the end of the tracks of the incident ions and can be examined by transmission electron microscopy which is ideally suited to the study of void formation. All the main features of void formation can be reproduced in this way (17,18). For example the preferred fuel cladding material is a cold worked austenitic steel. Ion beam irradiation has shown experimentally that cold worked 316 steels swell at much lower rates than those typical of annealed material, (19). Fig. 1a, is a transmission electron micrograph of a partly recrystallised 316 steel after irradiation to 30 displacements per atom at 525°C using 20 MeV C^{++} ions. The recrystallised grain contains voids while the unrecrystallised grain and the region near the grain boundary are free of voids.

Under some circumstances voids form in arrays with the same structure as the host material. They therefore constitute superlattices with "lattice parameters" about one hundred times greater than the host. The voids are also fairly uniform in diameter and are more resistant to annealing than are random arrays, (20, 21). The range for void formation is therefore extended to higher temperatures. The super lattice forms probably to reduce the amount of additional free energy of the system arising from the

Recrystallised grain showing voids (white) — *Grain with dislocation network and free from voids*

(a) 316 steel, 30 displacements per atom at 525°C produced by 20 MeV C2+ beam. After J. A. Hudson

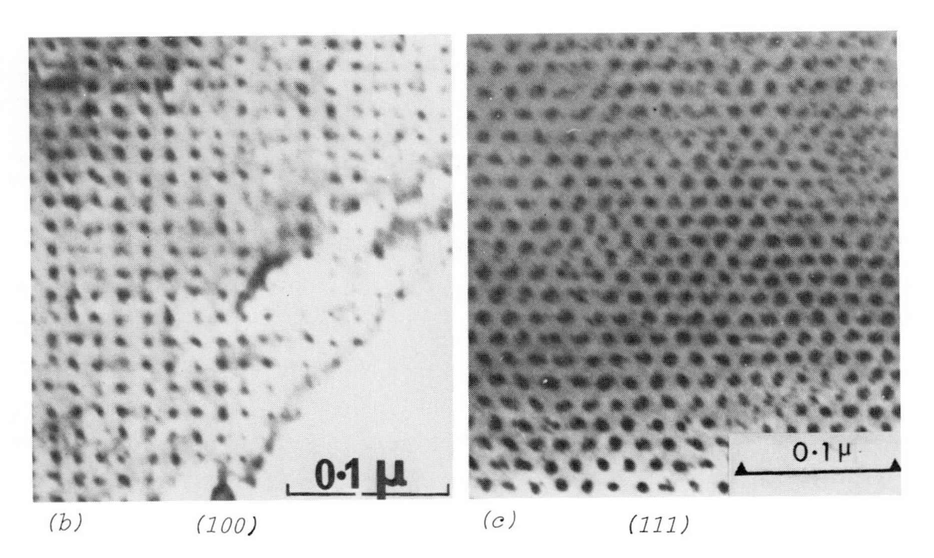

(b) (100) — *(c) (111)*

Projections of a bcc void lattice in molybdenum formed by bombardment at 870°C with 2 MeV nitrogen ions to 100 dpa
After J. H. Evans

FIG. 1 TRANSMISSION ELECTRON MICROGRAPHS SHOWING VOID FORMATION BY HEAVY ION IRRADIATION

elastic interaction between voids, (22,23). Void lattices were first discovered in molybdenum specimens after irradiation with nitrogen ions. Figs. 1b and 1c show two projections of a bcc void lattice in molybdenum formed by irradiation with 2 MeV nitrogen ions to 10^{16} particles/mm^2 i.e. about 100 dpa, at 870^oC.

Correlation of the results of ion beam irradiations with those obtained from neutron irradiation has been excellent, provided allowance is made for the effects of high dose rates. In view of the structure and composition sensitivity of void formation there remains a need to further establish the value of simulation experiments. The susceptibility of batches of materials to void formation could then be determined in advance of their use for manufacture of in-core components.

B. *Radiation Creep*

The chief concern arising from the void swelling process is the possible distortion of the core causing changes in reactivity and difficulty in loading and unloading fuel elements. One possible solution is that of mechanical restraint of the core to prevent distortion. High stresses might then be expected to build up in the core. A second radiation effect is here highly relevant, namely radiation creep. It has now been demonstrated both under fast neutron bombardment (24) and by using proton irradiation (25) that significant rates of radiation creep occur in materials under stress in a high energy radiation field at temperatures below those at which thermal creep is significant. This mechanism prevents the build up of high stresses in the core. Radiation creep is different in many ways from conventional thermal creep in that there is a linear stress dependency, only a very small temperature co-efficient and the creep rate does not appear to be very sensitive to the metallurgical structure. Thus radiation creep can prevent build up of high stress even in materials which have a very high resistance to thermal creep (25).

It is relevant to remark here that the satisfactory performance of thermal reactors in many cases depends on the existence of significant levels of radiation creep. For example coated particle fuels with layers of silicon carbide and pyrocarbon would fail under the high stresses developed by different rates of radiation growth if it were not for the reduction of the stresses by radiation creep, (26,27). Blocks of graphite moderator must accommodate large internal strains arising from different growth rates in temperature and flux gradients and between grains because of the different crystal orientations. Macroscopic cracking and even break up of the material into individual crystals would occur if radiation creep were not effective in reducing the internal stress level, (28).

There are several possible mechanisms for radiation creep. Calculations have been undertaken to indicate quantitatively their relative importance, (29). Elucidation of the mechanism by direct observation of dislocation movement in specimens under irradiation is not particularly effective because the preferred movements are obscured by the much larger random movements of the dislocations.

Study of radiation creep using charged particle irradiation is difficult. It is necessary to damage the specimen uniformly and to control its temperature very accurately say to $0.1^{o}C$. To obtain reasonably uniform damage even in a thin specimen requires the use of light ions such as protons which unfortunately give atomic displacement rates which are no greater than those obtained in existing fast reactors. The advantage of the simulation technique here is the ability to change stress, temperature, and flux with continuous simultaneous measurement of the creep strain. Success has recently been achieved using 4 MeV protons from an accelerator. Because of the fluctuations in power of the proton beam it is necessary to use special techniques for controlling the temperature of the specimen. With this equipment it has been possible to measure changes of creep rate of nickel at $500^{o}C$ on switching the 4 MeV proton beam on and off (Fig. 2a), and also to measure the

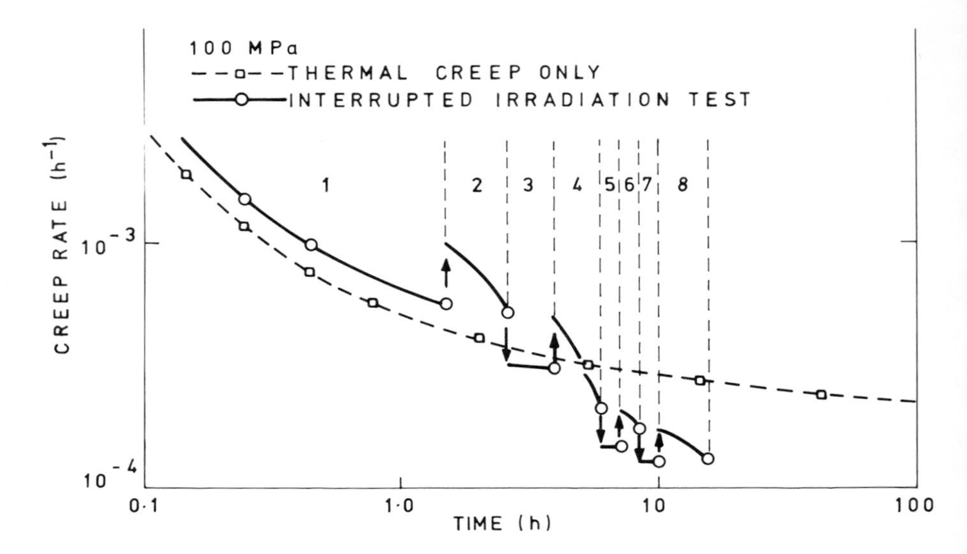

(a) Nickel specimen showing increased creep rate in periods 2,4,6 and 8 when the beam is switched on

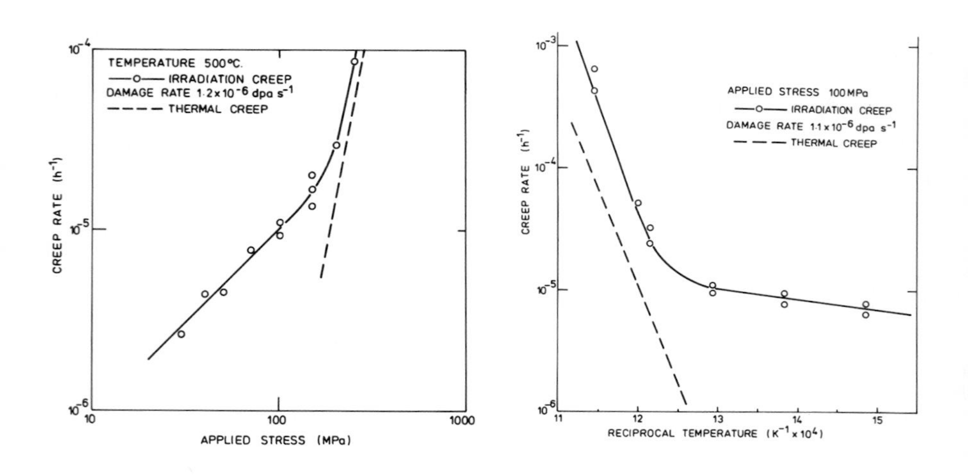

(b) Stress Dependence in 321 steel

(c) Temperature dependence

FIG. 2 IRRADIATION CREEP AT 500°C IN A 4 MeV PROTON BEAM. after R. J. McElroy

stress dependence (Fig. 2b) and temperature dependence of irradiation creep in 321 steel (Fig. 2c), (25), after prior cold work.

Study of radiation creep is at a relatively early stage and experimental data is somewhat scarce. Creep experiments in-pile are tedious and costly if complete control and recording of all the relevant experimental parameters are attempted. There is a particular shortage of data at high neutron dose levels which in fact is likely to continue for many years since there is at present no technique for doing tests at accelerated displacement rates. In view of the importance of radiation creep in fast reactor technology intensive further study will be required.

C. *Radiation Embrittlement*

A third radiation effect of considerable relevance is that of radiation embrittlement at very high displacement doses. Much of the plastic distortion of the core will occur by a process of irradiation creep and there is reason to suppose that under such conditions the materials may behave in a superplastic manner. It is however necessary to know the behaviour of the core towards sudden application of high transient stresses arising from pressure and temperature transients. This study should include measurement of crack growth rates under creep and fatigue conditions and fracture toughness, and the effect of the sodium environment.

D. *Fast Reactor Fuel Elements and Fuels*

Both swelling and irradiation creep of the cladding tends to reduce the stresses built up in the cladding as a result of swelling of the fuel. Thus both processes could be beneficial in prolonging the life of fuel elements.

For sodium cooled fast reactors uranium plutonium carbide fuel would be preferred to uranium-plutonium oxide for its higher heavy metal density and hence improved neutron economy. The development of carbide fuels will probably be delayed until oxide fuel has been well established. Because of the intensive work on UO_2 fuel over the past 25 years it is unlikely that more than minor further improvements could be made in that material, but much remains to be done to develop carbide fuels.

E. *The Fast Reactor Fuel Cycle*

An essential part of the exploitation of fast reactors is the re-cycling of plutonium in the total fuel cycle. These operations are largely concerned with process technology, instrumentation and radiation protection. The fabrication of plutonium bearing fuel must be done remotely using a process producing a low level of dust and waste. Fabrication routes suitable for uranium fuel may therefore not be the best for plutonium bearing fuel. If a different fabrication route is chosen then it is necessary to prove the resultant fuel by statistical radiation tests to find whether the behaviour of the fuel under irradiation is significantly changed by the change in fabrication route.

In the re-processing of spent fuel a number of materials problems arise, one of which is to find by examination of the highly irradiated fuel the distribution of fission products and transuranic elements in the fuel and cladding. The objective here is to optimise the head end processes in the re-processing plant to make a clean separation. The currently favoured technique for long term storage of high level waste is to incorporate it in a borosilicate glass, (30). The glass must be stored for many centuries and in the initial stages will be both at a high temperature and be subjected internally to a high gamma-ray field. It is necessary to investigate the effects of these conditions on the structure of the glass. Recently an interesting process has been proposed for the storage of radio krypton. The inert gases are insoluble in solids but they can be introduced by firing them as energetic charged particle beams into solids. Once implanted below the surface of a refractory solid they cannot escape and therefore provide a means of long term storage which does not require refrigeration.

V. *STEEL REACTOR PRESSURE VESSELS*

A. *Modes of Failure*

The properties of the ferritic steels chosen for primary containment in light water and other reactor systems is such that a flaw free structure could not fail mechanically, (31). The size and number of flaws in the structure in the as fabricated condition and the rate of crack growth are therefore very important failure parameters. Among the other types of failure that might arise on an extremely low probability level are complete melting of the core with subsequent melting through the bottom of the main reactor pressure vessel, or a major pressure rise in the core to an order of magnitude above the working pressure. Reactor systems are designed to tolerate a complete breaking off of one of the main coolant or outlet pipes without melt out of the core by introducing engineered safeguards such as the emergency core cooling system in the light water reactors.

B. *Cracking in Welds*

The presence of flaws in primary containment structures arise mainly in the welding process when cracks may appear both in the weld itself and in the heat affected zone adjacent to the weld. Part of the present procedure for constructing vessels is to search for such flaws and when found they are machined out and re-welded. Neither ultrasonic inspection nor other NDT techniques provide absolute guarantee of finding all flaws nor can the nature of the feature giving rise to signals be unambiguously determined, (32). The quality of the vessel would clearly be improved if fabrication could proceed without the introduction of any flaws. There would also be a considerable reduction in fabrication cost. Because the fabrication of crack free welds is such an important feature in the whole of reactor technology systematic further attempts to improve the weldability on pressure vessel steels and weld materials would be justified. Recent work has shown a connection between hydrogen embrittlement and temper embrittlement in ferritic steels (33) which give rise to fracture along prior austenite grain boundaries particularly in martensitic steels (Fig. 3a) but also to a smaller extent in bainitic pressure vessel steels. The role of trace element segregation in both temper embrittlement and in stress relief cracking of welds in the heat affected zone (34) is now also becoming clearer.

Much of the difficulty in weld development arises from the large scatter in incidence of cracks in materials which are nominally the same and welded under similar conditions. UK work has indicated that the variation in trace element levels in commercial batches of steel cover a range in which there is significant variation in resistance to stress relief cracking. The rate at which the steel tempers during the stress relief anneal also affects stress relief cracking. (34).

C. *Fracture Toughness*

If the steel has a very low yield strength then failure can occur by plastic yielding or by fracture in the region of the flaw. Since yield strength can be traded for fracture toughness it is difficult to decide where the optimum balance lies to maximise the strength of the structure. Unfortunately the optimum values are not the same for flaws of different sizes, and the fracture toughness of ferritic steels is a function of temperature, and may even be accompanied by a total change in fracture mode from cleavage at low temperatures (Fig. 3b) to ductile dimpled fracture at high temperatures (Fig. 3c).

For light water reactor vessels the high stresses are applied to the vessel when the reactor is at operating temperatures and the steel is in the upper regime of fracture. The dominating role of sulphur present in the steel as manganese sulphide is here well established. The sulphide inclusions are plastic at the hot working temperature and so become elongated and confer

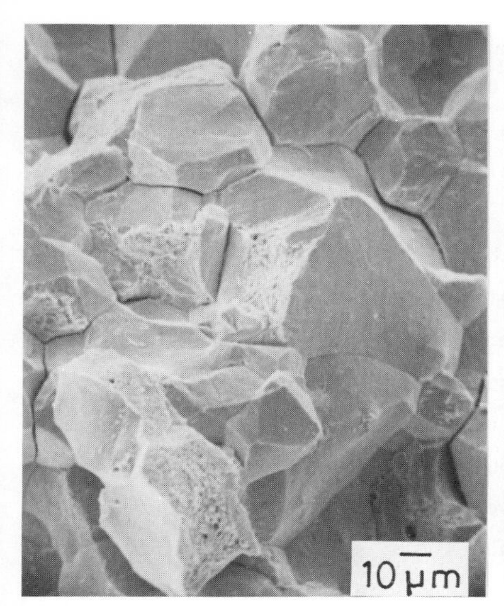

Fig. 3a Fracture in temper embrittled EN30A steel

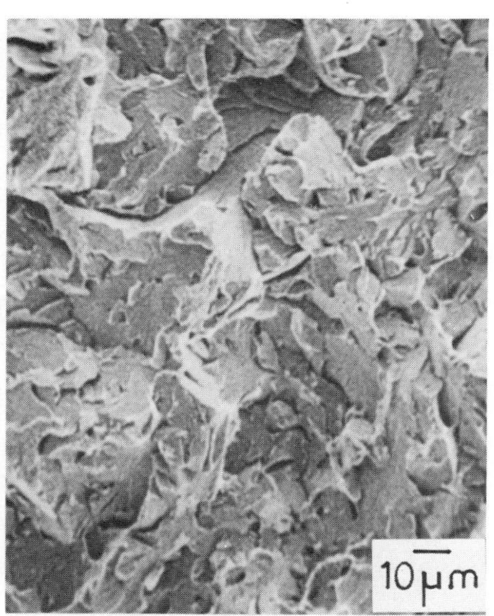

Fig. 3b Cleavage in A533B steel impact tested at - 196°C After B. C. Edwards

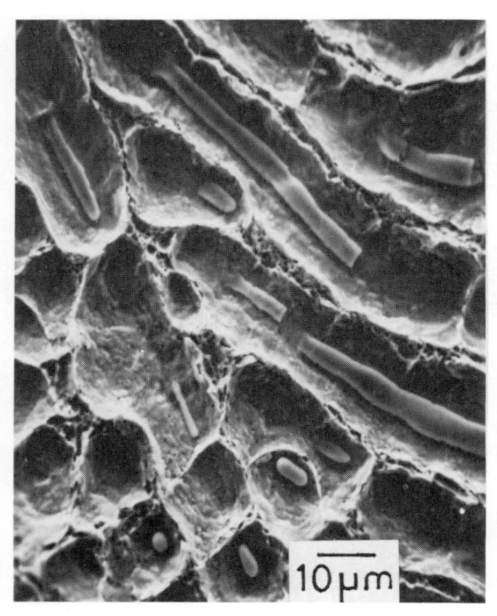

Fig. 3c Ductile dimpled fracture in A533B impact tested at 255°C

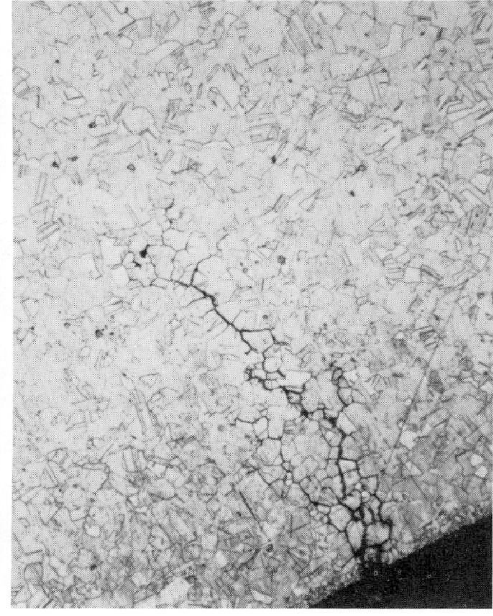

Fig. 3d Stress Corrosion Alloy 800, 350°C, 150 MPa in 4% aqueous caustic soda

directionality on the fracture toughness. The upper shelf energy for fracture in a plane parallel to the long axis of the sulphide stringers (Fig. 3b) is lower than that for fracture transverse to the stringers. Additions to steel which will combine with the sulphur to form a compound which is harder or more brittle at the hot working temperature so that the formation of stringers does not take place is being investigated. Reduction in the sulphur levels has been shown by experience to increase the danger of hydrogen embrittlement and explosive fracture of large ingots. It has empirically been argued that the sulphides provide a sink for hydrogen so that high instantaneous concentrations of free hydrogen cannot occur. Further work on hydrogen embrittlement perhaps exploiting the findings about the synergistic effects of trace impurities might ultimately allow the production of a steel with both a very low sulphur level and freedom from susceptibility to hydrogen embrittlement.

The production of heavy section steel plate with satisfactory mechanical properties and the successful fabrication of a welded structure from those plates is a major technological achievement in the metallurgical and process technology fields. There is no doubt that the steel makers have considerable knowledge concerning the relevant factors involved in obtaining the required structure and mechanical properties. Much progress has been made over the past decade particularly in achieving relatively fine austenite grain sizes in heavy section steel. Any further progress is likely to be difficult to achieve and will require a wide systematic study of the factors controlling strength, fracture toughness, and weldability. It seems likely that a much tighter control of composition of the steel would be required especially with respect to those elements which segregate on prior austenite grain boundaries and affect the fracture toughness and weldability.

D. *"Dry" And "Wet" Fatigue*

Cracks in a structure which are not in themselves sufficiently large to cause catastrophic failure under the stresses imposed during service may nevertheless be potential sources of failure because of the possibility of crack growth under alternating stress levels. Fatigue strength of steels in a dry environment is not a highly structure sensitive property and design data are well established. More recently however it has been shown that in the presence of water crack growth rates in pressure vessel steel for a given stress regime are up to 50 times greater than those for the dry condition. These increased rates are found only if long periods are allowed for the individual stress cycles more nearly approaching the timescale for stress transients under service conditions, (35). Also if the mean tensile stress is high it has been shown that significant crack growth occurs for smaller values of the alternating stress amplitude, (36). Published information on wet fatigue is almost entirely confined to examination of one plate namely HSST plate No.2. Some laboratory work has indicated that crack propagation is along grain boundaries and therefore segregation of impurity elements on to those grain boundaries might have an effect on crack growth rates. Experience with conventional steam plant has not indicated that there is any new problem arising from wet fatigue but clearly more work is required in particular to show the effect of variation in metallurgical composition and structure on crack growth rates.

VI. *THE PRIMARY COOLANT CIRCUIT*

A. *The Use Of Austenitic Steels And Nickel Alloys*

In conventional steam plant ferritic steels are used extensively if not exclusively in many installations. The choice of austenitic steels and nickel alloys for superheater sections, of conventional plant and gas cooled reactors is based on their higher creep strength and improved general oxidation resistance above 500^oC.

In the light water reactors austenitic steels and nickel base alloys are used at much lower temperature levels of about 300^oC. They are chosen mainly for their lower rate of general corrosion in water and also their resistance to pitting corrosion. Experience has shown that even with great care to obtain good quality materials and to make good welds the incidence of leaks caused by stress corrosion cracking in water environments is unduly high. A recent investigation (37) has indicated that residual stresses in welds play a large part in cracking although the choice of non stabilised steels is also a contributing factor. Austenitic materials in circuits in which the free oxygen level in the water can be kept to a low concentration are less susceptible to stress corrosion cracking. In the primary circuit of a PWR radiolytic oxygen levels in the water are controlled by an over pressure of hydrogen. In direct cycle systems such control cannot be used so effectively. The welding of heavy section austenitic steels, particularly those which are fully austenitic, and nickel alloys is difficult to achieve without the incidence of cracking in the welds. Here again there is an as yet unexplained batch to batch variation in weldability (38) which again might arise from variation in the levels of certain trace elements which segregate on to grain boundaries.

B. *Steam Generators And Condensers*

Austenitic steels and nickel alloys are used for the tubing of steam generators in the indirect cycle water cooled reactors, and HTRs. Very careful metallurgical and operational control is required to avoid the incidence of stress corrosion cracking (39), (Fig. 3d). Failure to control the water conditions will cause enhanced cracking. Here again much of the incidence of cracking in the indirect cycle systems is on the secondary coolant side where control of water chemistry is more difficult. Both general corrosion rate and grain boundary attack can also be important since thin walled tubes are used.

In direct cycle reactors there is an expected incidence of contamination of the working fluid with cooling water leaking from the condensers. Such contamination may be particularly severe where seawater cooling is used. Even in direct cycle gas cooled reactors using gas turbines for power generation extremely large areas of pipe work carrying cooling water are exposed to the main coolant. Here the leakage of water into the primary circuit causes oxidation of the graphite in the core. For these applications therefore, titanium alloy tubing is attractive in providing higher reliability, though at increased cost.

C. *Circuit Activity*

An unusually high integrity is required in the primary cooling circuit because of the difficulties and high cost of inspection and repair in the presence, after a few years of operation, of high γ radiation fields. These fields arise principally from the circulation of corrosion products in the water, their activation in the core and deposition round the circuit. Quite large amounts of radioactive corrosion product may be deposited in steam generators in indirect cycle systems. For example, in the Stade reactor about one inch of deposit was discovered after about $2\frac{1}{2}$ years of operation, (39). In conventional plant it is usual to block off or repair a large fraction of the steam tubes during the life of the plant. Several campaigns to repair reactor plant in the past have been achieved only by exposing large numbers of personnel to the permitted radiation dose levels, (3). Any improvement in the performance of materials in this context so that the need for inspection and repair becomes less frequent would improve the economics of energy production from fission. Improved techniques for remote inspection and remote repair would also be of great value and may be necessary if the ICRP regulations become more stringent.

Transfer of radioactive corrosion products round the coolant circuit potentially represents one of the major problems in water cooled reactor technology. Any particular elements giving rise to hard gamma rays after neutron adsorption should be avoided. Although ferritic steels have higher general corrosion rates in water circuits than austenitic steels, the latter usually give rise to a higher gamma activity per unit weight of corrosion product.

D. *Fast Reactor Steam Plant*

In fast reactor steam raising plant the current concept utilises two sodium cooled circuits and a final steam raising circuit. Although the final sodium to water and sodium to steam heat exchangers are not in a high intensity γ-radiation field, these units still present problems similar to those found in primary coolant circuits. There is still a need to achieve an extremely high integrity to avoid the accidental mixing of sodium and water. The consequences of a single leak could be very great since the interaction of sodium and water produces chemically aggressive products at high temperatures. To guarantee absolute freedom from leaks in a steam generator of conventional design is difficult. Designs aimed at reducing the consequences of a perforation include highly subdivided modules (40) perhaps with each unit being capable of rapid isolation, and double walled heat exchangers.

Hitherto austenitic steel and nickel alloys have been chosen for fast reactor steam plant because sodium temperatures of about 600° were popular. There has, however, been a trend to lower temperatures and ferritic steels have now become a feasible alternative. In the UK for example ferritic steel is the current choice for the CFR steam plant. The arguments for and against choosing ferritic or austenitic steels and nickel alloys are however quite complex and much can be said on either side.

VII. SECONDARY CONTAINMENT

Because of the very low probabilities required of accidents involving major release of fission products from the core of nuclear reactors, it is necessary to consider protecting the plant from events such as the impact damage from a crashing aircraft, and acts of sabotage or war. The main material for protecting the plant against external missiles and internal missiles arising from plant failure is reinforced concrete. Several investigations into the ability of reinforced structures to withstand impact damage have been reported, (41). An improvement in the fracture toughness of reinforced concrete would be useful. Work in this direction involves choosing the optimum amount and orientation of the reinforced steel wires and in the use of fine ceramic fibres to toughen the cement matrix.

VIII. SUMMARY

The development of nuclear power systems has required many new materials and modifications in the technology of conventional materials. The fact that thermal reactors have been built and are economically attractive as sources of energy is an immediate indication that there are no materials problems of sufficient magnitude to prevent the economic exploitation of fission energy at the present time. There may remain some concern that extraneous changes could cause such systems to cease to be attractive. Among these changes are possibilities of alternative energy sources, tightening of international standards for radiation exposure to the population, the reduction in supply of the more concentrated uranium ores and increased difficulties in disposing of radioactive effluents.

Although in the current technology there are no critical materials problems, improvements in the service performance of materials could be exploited with economic benefit. Desired improvements include reduction in failure rates of fuel elements and improvements of the integrity of containment structures to reduce the need for inspection and repair.

The need to choose materials for fuel and core structures according to their nuclear properties and resistance to radiation damage has stimulated intensive research and development in many countries. While the development of materials for thermal systems is at an advanced stage, much remains to be done in the development of materials for service in the cores of fast reactors. Further study is required of the effects of high levels of atomic displacement in causing void formation, radiation creep and perhaps embrittlement. In addition the development of carbide fuels for improved neutron economy is at only an early stage.

The development of conventional materials for the primary pressure containment circuits and steam raising plant has not perhaps attracted the same level of research interest. Perhaps the difficulty of the task of producing heavy section ferritic and austenitic steels, and steam raising plant to a much higher level of reliability than is achieved in conventional structures by established technology has been somewhat underestimated. Ferritic and austenitic steels and nickel alloys have very different water corrosion characteristics and fracture modes. Problems arise in the weldings of these materials, and all give rise to radioactive corrosion products when used in the primary coolant circuit. The elimination of stress corrosion cracking is an important task but luckily the austenitic steels have high fracture toughness in the fast fracture mode so that leaks rarely develop into breaks. Further improvement in the quality of well established materials and fabrication processes is desirable but unlikely to be achieved easily. Fracture surface examination by Auger and related methods and the use of high voltage microscopes (42,43,44) and accelerators in studying radiation damage will here play a significant part.

ACKNOWLEDGEMENTS

I wish to acknowledge the assistance of my colleagues at A.E.R.E. Harwell in providing illustrations for this chapter and for comments on the text.

REFERENCES

1. F. R. Farmer J. Inst. Nucl. Eng., 16, (2) (1975)

2. Office of Operations Evaluation rep. "Nuclear Power Plant Operating Experience During 1973", DRO, USAEC OOE-ES-004 (Dec 1974) also designated WASH-1362.

3. A. S. Chiesetz, Trans. Amer. Nuclear Soc., 18, 355 (1974)

4. N. Rassmusen, WASH 1400, Oct. (1975)

5. UKAEA Annual Report 1974-75. HMSO, 569, London SE1 9NH.

6. Lord Avebury. Debate in the House of Lords, London 29·Jan. 1976 quoted in Atom, UKAEA London (233) 71 March (1976)

7. K. P. Gibbs and D. R. Fair, Nucleonics, 24 (9) 48 (1966)

8. S. F. Pugh et al. J. Brit. Nucl. Energy Soc., 11 (4) 313 (1972)

9. G. E. Betteridge and W. A. J. Wall, AHSB, R82 (1965)

10. H. K. Hardy, L. M. Wyatt and S. F. Pugh, J. Brit. Nucl. Energy Soc., 2 (2) 236 (1963)

11. S. F. Pugh, Phil. Mag., 7, 45 823 (1954)

12. S. F. Pugh, AERE report M/R 1290 (1953), also in Rev. de Met., L1 (10) 683 (1954)

13. N. Piccinini Advances in Nucl. Sci. and Tech. 8 Academic Press 256 (1975)

14. P. E. Bobe USAEC report NUREG - 0032 Jan (1976) Fuel Performance of Licensed Nuclear Power Plants through 1974)

15. C. Cawthorne and E. J. Fulton, Nature 216 (5115) 575 (1966)

16. A. D. Brailsford and R. Bullough T.P. 482 1972 and J. Nucl. Mat. 44 121 (1972)

17. J. A. Hudson, D. J. Mazey and R. S. Nelson, "Voids formed by Irradiation of Reactor Materials" p 215 Proc. BNES Conf. 24 and 25 March 1971

18. R. Bullough, B. L. Eyre and K. Krishan AERE R7952 and Proc. Roy. Soc. London A 346 81 (1975)

19. T. M. Williams, ibid ref 17. p 205

20. J. H. Evans, Nature 229 403 (1971)

21. J. H. Evans, Radiation Effects 10 55 (1971)

22. V. K. Tewary and R. Bullough J. Physics F. 2 269 (1972)

23. A. M. Stoneham, J. Phys. F, 1 (6) 1971

24. D. Mosedale and G. W. Lewthwaite, Proc. Iron and Steel Inst. Conf., "Creep Strength in Steel and High Temperature Alloys", Sheffield, 1972, The Metals Society, 169 (1974)

25. R. J. McElroy, J. A. Hudson and R. S. Nelson, Proc. Int. Colloq. Measurement of irradiation enhanced creep in nuclear materials, JRC Petten Netherlands 1976.

26. J. W. Prados and J. L. Scott, Nucl. Applications, 3, 488 (1967)

27. H. Walther, Nuclear Eng. and Design 18, 11 (1972)

28. B. T. Kelly and J. E. Brocklehurst, Proc. 3rd Conf. on Industrial Carbon and Graphite, Soc. Chem. Ind., London p 363 (1971)

29. R. Bullough and M. R. Hayns J. Nucl. Mat. 57 348 (1975)

30. J. P. Olivier, ibid ref 13 p 141

31. G. C. Robinson et al., Paper 74-Mat-5 ASME Pressure Vessel and Piping Conf., Florida, June (1974)

32. Report to Amer. Phys. Soc., Rev. Mod. Phys., 47 suppl 1 (1975)

33. B. C. Edwards et al., AERE R8298 (1976) to be published in Acta Met.

34. J. M. Brear and B. L. King, Proc. Inst. Met. Conf. "Properties of grain boundaries," Jersey April (1976)

35. T. Kondo et al., Corrosion Fatigue, NACE-2, Houston, Tx 77001, 539 (1972)

36. P. C. Paris et al., ASTM STP 513 141 (1972).

37. U.S. report NUREG 75-076 Oct (1975)

38. M. Ward and P. L. Norman, Proc. BNES Conf. 25 and 26 Sept (1974) p 319. A Status Review of Alloy 800

39. L. Stieding, ibid ref 38, p 345.

40. N. G. Robin et al., Fast Reactor Power Stations, BNES London 205 (1974).

41. Papers in Session J5, Vol 4 Trans. 3rd SMiRT conf on Structural Mechan. in Reactor Tech., London 1-5 Sept (1975)

42. M. J. Makin Phil. Mag 18 637 (1968)

43. M. J. Makin JERNKONT. ANN Vol 155 p 509 (1971)

44. K. Urban 4th Int. Congr. of Electron Microscopy p 159 Toulouse Sept 1975

B
V
2
J

II. MATERIALS FOR HIGH TEMPERATURE APPLICATIONS

Materials problems envolved in the structural parts of open cycle magnetohydrodynamics channels and high temperature gas turbines and jet engines are so closely related that their inclusion into a single section seems warranted.

Magnetohydrodynamic generators offer attractive possibilities for this Country's energy needs because they can directly utilize our vast supplies of coal and because they are such efficient converters of this form of energy into electricity. Unfortunately, the heart of the MHD system, the generator channel, has a limit of only several hundred hours of operation due to the degradation of the electrodes and insulators in the severe service environment imposed. These materials must withstand high current densities, arcing, temperatures in the range of 2800°K, corrosion from potassium and sulfur compounds, erosion from combustion products at nearly Mach one velocities, thermal fatigue, thermal stresses and creep.

While the magnetohydrodynamic generator became a laboratory reality relatively recently (1960), jet aircraft engines have been used since 1942. However, since that time research by materials scientists has only been able to increase the operating temperature of the blades by 10°C per year. Now a series of new approaches in high temperature materials development has been introduced which promises the next lèvel of performance improvement over the superalloys, the refractory metals and oxide dispersion alloys. These include fiber reinforced alloy matrices, directionally solidified composites and rapidly solidified, splat cooled supersaturated fine grained alloys and glassy metals.

Our aim, in the next three chapters, is to examine the progress made in the development of high temperature materials which would be suitable for use in MHD converts and advanced turbines and jet engines.

Chapter 10

MATERIALS PROBLEMS IN OPEN CYCLE MAGNETOHYDRODYNAMICS

H. K. Bowen

Department of Materials Science and Engineering
Massachusetts Institute of Technology
Cambridge, Massachusetts 02139

and

B. R. Rossing

Materials Science Division
Westinghouse Research Laboratories
Pittsburgh, Pennsylvania 15235

I. INTRODUCTION

In the late thirties and early forties a concentrated effort was devoted to the development of two types of gas turbines, the mechanical turbine using moving blades, and the electrical turbine making use of the interaction between a conducting gas (plasma) and a magnetic field. These new turbines showed promise of substantial fuel savings per unit of power produced. In our age of increasing environmental awareness fuel savings also mean a reduction in thermal pollution and conservation of natural resources. The mechanical turbine found rapid application in jet engines which dominate the aircraft propulsion field. More recently, the gas turbine has found wide spread use in the electrical power industry (1).

Through better understanding of the physics of high temperature gases and progress in high temperature technology, the electrical turbine or magnetohydrodynamic (MHD) generator became a laboratory reality around 1960. At that time, the economic and technical potential of MHD power generation was recognized by electrical manufacturers, some utilities, and military agencies. These different groups provided the early support for research in the field of MHD power generation. However, as a result of the construction of the first economical nuclear power plants, and of the slowing of the development of MHD power generation, the commercial interest of MHD began to wane (1). Two commercial concerns, Westinghouse and AVCO-Everett, have continued their development work from the laboratory successes in the early sixties to the present time.

In 1968 the Office of Science and Technology formed an MHD panel to evaluate the potential of MHD power generation. In its recommendations, the panel confirmed the soundness of the MHD power generation principles and underlined the valuable possibility of increased fuel utilization through higher efficiency and noted especially that if coal were used, fossil reserves could be extended. There have been several subsequent evaluations (1-3) which have recommended a national program including an accelerated plan for the development and testing of materials. The comprehensive volume describing a national effort is a book by Heywood and Womack which describes the British program (4). The specific materials requirements are outlined in the proceedings of a recent workshop (5).

The principle of the MHD channel is shown in Fig. 1, where the high velocity hot gas at velocity V interacts with the perpendicular magnetic field B to create a DC-current flux J in the third orthogonal direction. The active elements in the system are the electrode walls, anode and cathode, which transfer the electrons from the working gas to the external load (R_L).

MHD is the only advanced energy conversion technique that can directly use coal as a fuel. A schematic of the total system is given in Fig. 2. Pulverized coal is combusted with preheated air in the burner. To increase the electrical conductivity of the combustion products potassium compounds (e.g., K_2CO_3) are added. The "seeded" gas at ∿2800°K passes through the magnetic field where 10-20% conversion of thermal to electrical power occurs. On the low temperature phase of the combined cycle, a steam plant with about 40% conversion makes the overall process from 50-60% efficient. Because of the high temperatures of the working fluids and because of the corrosive conditions, ceramic materials are important throughout the system.

The cooled gases pass through regenerative heat exchangers, necessary to preheat air to 1200°C in order to maintain high combustion temperatures; and finally the gases are cleaned of the seed, sulfur compounds and oxides of nitrogen. The potassium seed provides an efficient "getter" for sulfur in the fuel and condenses principally as K_2SO_4.

The extremely high plasma temperatures exiting from the combustor at nearly Mach one velocities are necessary to provide high electrical conductivity of the gas; but this temperature is also above the fusion and volatility temperature of most of the components in coal ash. Most U. S. coals have ash contents of ∿10%. It is assumed that ash removal efficiencies in the combustor will not exceed 90%; thus thousands of pounds per hour of an

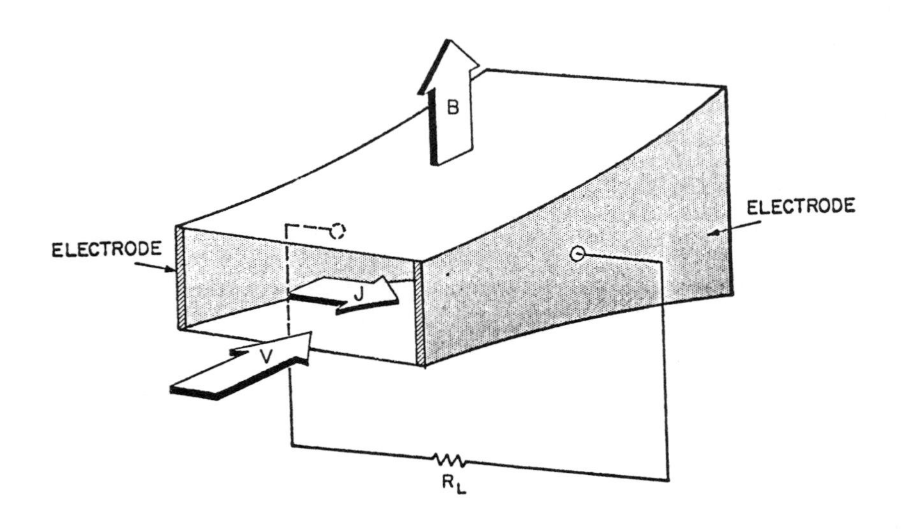

FIGURE 1. MHD Generator. V = plasma flow, J = current flow, and B = magnetic flow.

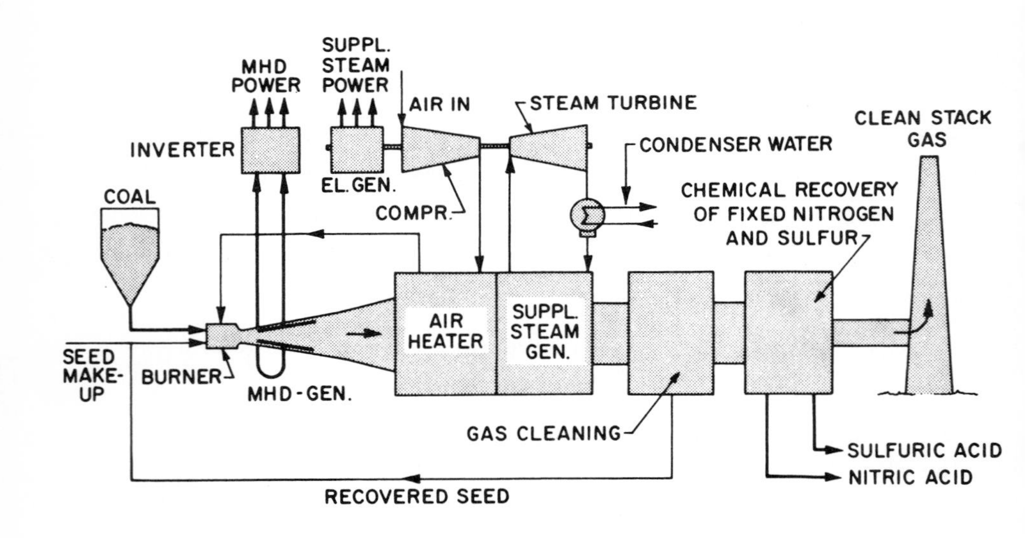

FIGURE 2. Schematic of MHD power plant.

iron-aluminum-silicate ash (slag) will be rushing down the channel of a 1000 megawatt plant. But even more critical is the approximately equal mix of slag and seed which must pass through the air preheater. This lower temperature unit must be designed such that this corrosive mix does not erode or plug heat transfer surfaces.

II. MHD SYSTEMS

The MHD generator uses the hot combustion gases as the working fluid which in this case is electrically conducting. Reasonable plasma conductivities, 0.01 - 0.3 mho/cm, can be obtained at temperatures of 2000 - 3000°K by adding 1% potassium. For base-load power extraction near sonic gas velocities, strong magnetic fields, > 3 Tesla, and current densities of 1 amp/cm^2 are required. For these conditions and for the low electron carrier density, the Hall field becomes important and may be comparable in magnitude to the induced field. The presence of this field places constraints on the channel construction and configuration, e.g., segmentation of the electrode walls as alternating electrode insulator pairs. In addition, there are three primary modes of loading the generator, i.e., making the electrical connections for the external circuit. They are called the Faraday, Hall and Diagonal generators and are shown schematically in Fig. 3 (6).

In the Faraday generator, the MHD duct is segmented along the flow axis and each pair of transverse electrodes is separately connected to a load. The induced voltage ($\bar{J} \times \bar{B}$) causes current transverse to the flow which works through its Lorentz force against the expanding gas to transfer power to the electrodes.

FIGURE 3.

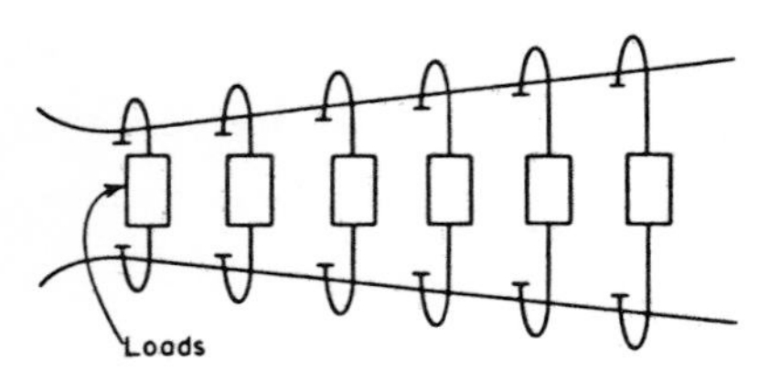

A linear Faraday Generator with segmented electrodes.

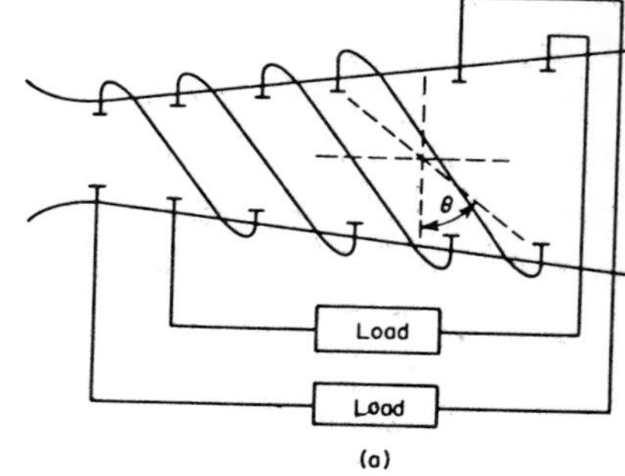

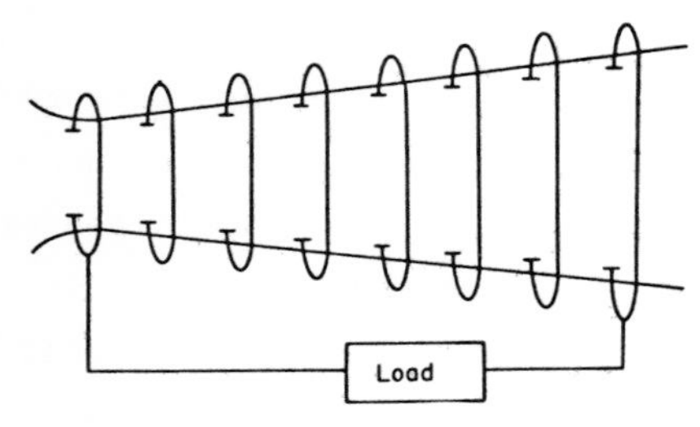

A linear Hall generator.

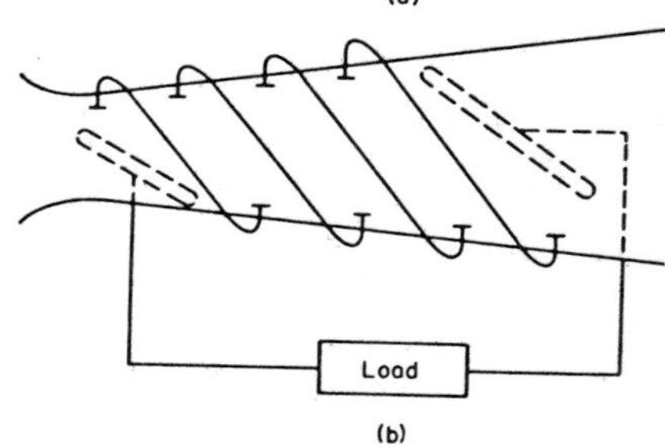

(a) A diagonally connected generator: (b) diagonal-electrodes.

A base load generator would require thousands of electrodes, each connected to a separate inverter (DC- to AC-power) (6).

In the Hall generator, the opposing electrodes are short circuited so that in this case the Hall field is used to induce current between a pair of electrodes located at the inlet and at the outlet of the channel. Because of the current path through the length of the generator, the resulting Joule dissipation is larger than in Faraday generators operating under similar conditions. The disadvantage is that lower power extraction efficiencies are possible for conditions expected in commercial plants; however, the simple electrical connections are a strong plus for this design. Each transverse section should be located on equipotential lines allowing for construction of the channel from a series of window frames which are stacked in the plasma flow direction.

When the Faraday generator is designed such that anodes and cathodes at the same equipotential are connected, the generator is operated in the diagonal mode. Single or multiple loads can be made. When operating at the design point, the diagonal generator functions like the Faraday but with no axial or Hall current.

Generators with a radial flow expansion of the hot gas through a magnetic field, called disk generators, operate as Hall generators. By using vanes to cause swirl to the radial flow, the disk generator operates like the linear diagonal. Recent experiments on the disk generator with $\sim$45° swirl have given 17% enthalpy extraction - this conversion efficiency indicates the great promise for MHD (7). The disk generator has an added advantage in the construction of anode and cathode walls because it does not require the same complex segmentation as linear generators do.

The size of MHD generators and the duration of continuous operation is shown graphically in Fig. 4. Note that a base load system would be designed to deliver > 300 MW of electrical power and be able to operate for up to one year. Because the funds and some aspects of the technology have been unavailable, the size and the duration of operation of MHD systems has been far short of the goal. In addition to domestic programs on coal-fired facilities at AVCO-Everett, Westinghouse, and UTSI, there are two important programs indicated in Fig. 4 which are designed to test performance without the complications of coal combustion. The AEDC test, designed to give maximum power in a short duration, is geared to show efficiency of scale and prove high enthalphy extraction when losses due to the high surface to volume ratio in

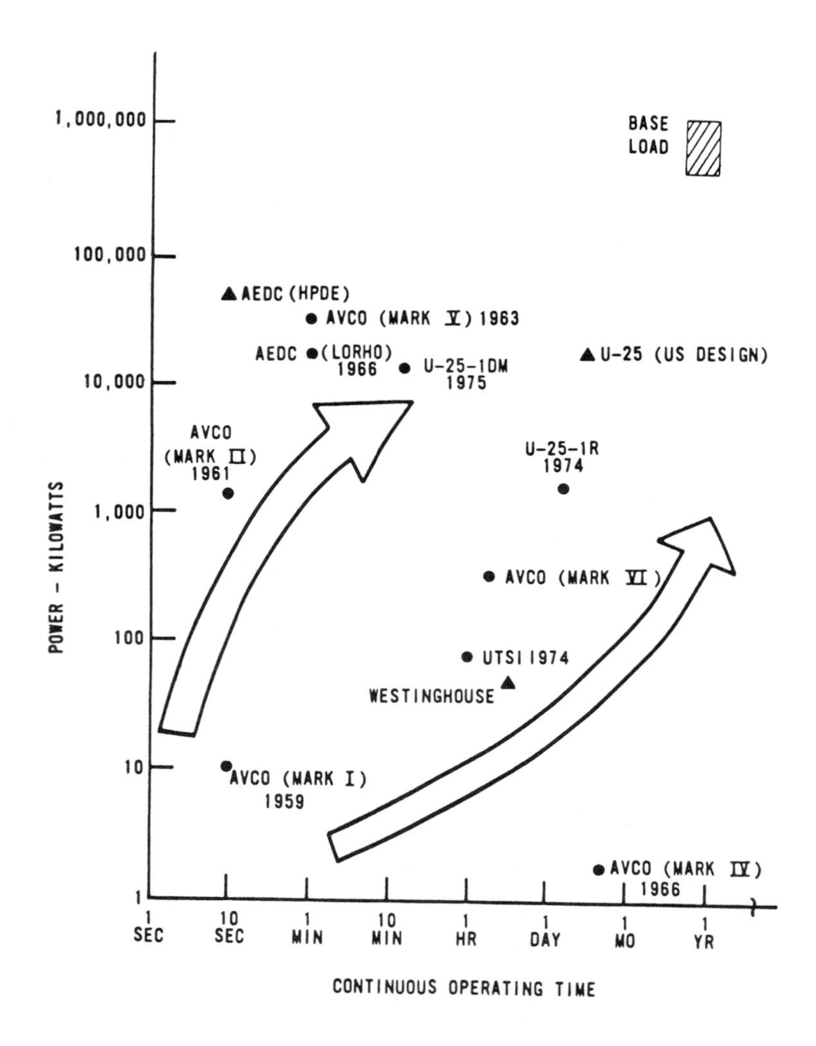

FIGURE 4. Electrical power generated by MHD channels and duration of experiment (after Jackson et al (2))

smaller ducts are not a factor. The other experiment indicated as U-25 (U.S. Design) is a window frame design to be tested at the Soviet MHD plant in Moscow.

The criticality of materials which serve as electrodes and insulators in these ducts cannot be overstated if these systems are to reach their potential. Another area of critical materials needs is in the air preheater. Fig. 5 shows the range of power generating plant efficiencies for the combined MHD-steam turbine cycle as a function of the air preheat temperature. For example, air preheat temperatures possible in heat exchangers made from metals would yield efficiencies less than 45%; while temperatures over 1200°C (2200°F) are required for ∿50% plant efficiencies.

In addition to the basic differences in the MHD channel design (Hall, Faraday, etc.), there are differences in the integrated system for various fuels and preheating schemes. Fig. 6 includes the schematics for the three systems of most interest. The clean-fired system (Fig. 6a) is of interest to test efficiency, DC- to AC-inverters, integration of both MHD and steam cycles, etc. and represents the Soviet U-25 plant with the exception of a separately fired preheater. The advantage of this system is that there is no coal slag/ash. Some MHD proponents have suggested that this system could be used with a coal gasification process. However the most realistic systems are those involving direct combustion of coal. The most simple and the system which perhaps should be constructed first, is the one with a separately fired preheater (Fig. 6b). This would simplify enormously the problems of large quantities of seed and slag passing through the heat transfer channels in the regenerators. The regenerators for the indirectly fired system represent a technology which is currently available for the steel/glass industry. For efficiency reasons the directly fired preheater, i.e., using the combustion products from the channel, is more desireable, but is also complex and

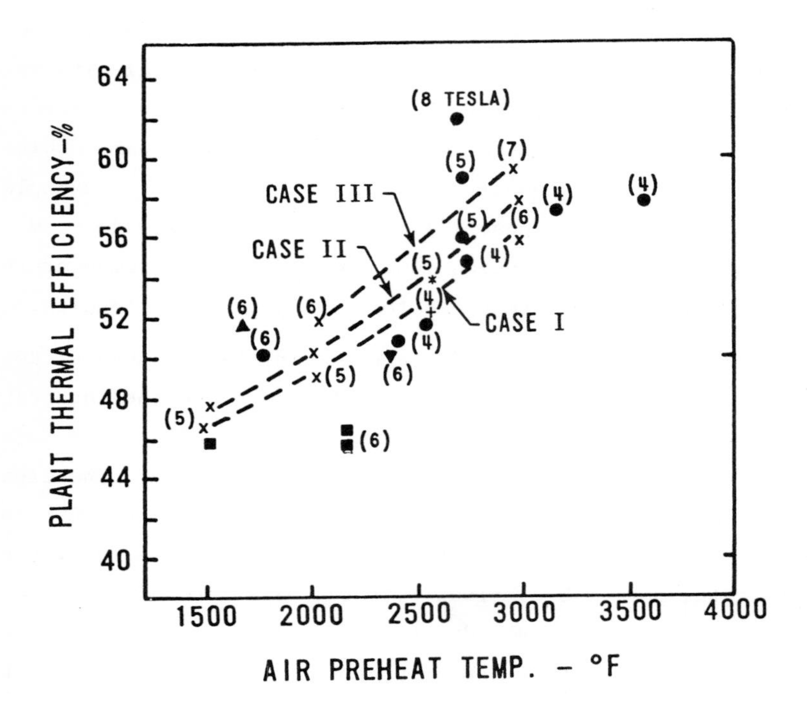

FIGURE 5. Variation of overall plant efficiency with air preheat temperature (after Hals and Jackson reproduced in ref. 2).

FIGURE 6.

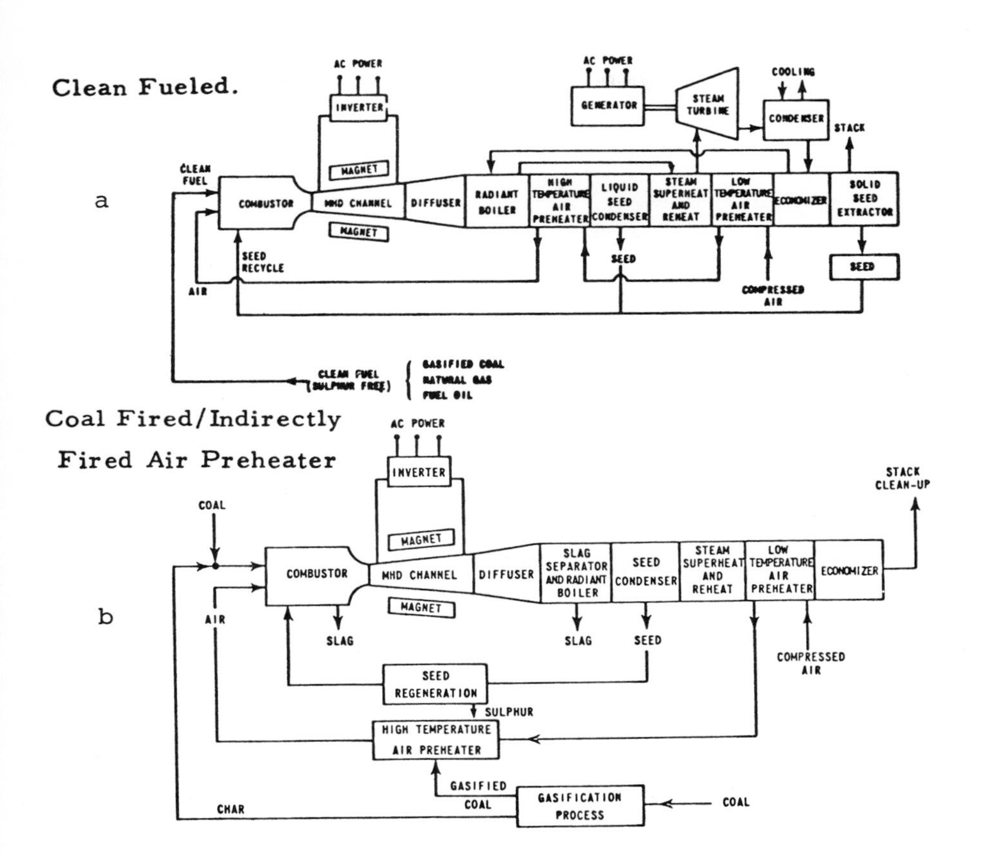
Clean Fueled.
a
AC POWER
INVERTER
MAGNET
CLEAN FUEL
COMBUSTOR
MHD CHANNEL
DIFFUSER
MAGNET
RADIANT BOILER
HIGH TEMPERATURE AIR PREHEATER
LIQUID SEED CONDENSER
STEAM SUPERHEAT AND REHEAT
LOW TEMPERATURE AIR PREHEATER
ECONOMIZER
SOLID SEED EXTRACTOR
AC POWER
GENERATOR
STEAM TURBINE
COOLING
CONDENSER
STACK
SEED RECYCLE
AIR
SEED
SEED
COMPRESSED AIR
CLEAN FUEL (SULPHUR FREE)
GASIFIED COAL
NATURAL GAS
FUEL OIL
Coal Fired/Indirectly Fired Air Preheater
b
AC POWER
INVERTER
MAGNET
COAL
COMBUSTOR
MHD CHANNEL
DIFFUSER
MAGNET
SLAG SEPARATOR AND RADIANT BOILER
SEED CONDENSER
STEAM SUPERHEAT AND REHEAT
LOW TEMPERATURE AIR PREHEATER
ECONOMIZER
STACK CLEAN-UP
AIR
SLAG
SLAG
SEED
COMPRESSED AIR
SEED REGENERATION
SULPHUR
HIGH TEMPERATURE AIR PREHEATER
GASIFIED COAL
GASIFICATION PROCESS
COAL
CHAR

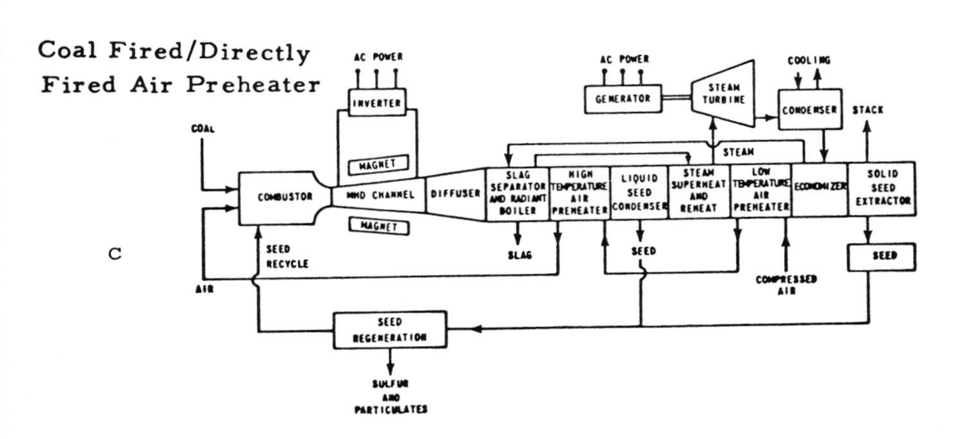
Coal Fired/Directly Fired Air Preheater
c
AC POWER
INVERTER
MAGNET
COAL
COMBUSTOR
MHD CHANNEL
DIFFUSER
MAGNET
SLAG SEPARATOR AND RADIANT BOILER
HIGH TEMPERATURE AIR PREHEATER
LIQUID SEED CONDENSER
STEAM SUPERHEAT AND REHEAT
LOW TEMPERATURE AIR PREHEATER
ECONOMIZER
SOLID SEED EXTRACTOR
AC POWER
GENERATOR
STEAM TURBINE
COOLING
CONDENSER
STACK
STEAM
SEED RECYCLE
AIR
SLAG
SEED
SEED
COMPRESSED AIR
SEED REGENERATION
SULFUR AND PARTICULATES

difficult to make a reality. The parameters considered by ERDA as the first order conditions for a coal-fired, base load plant are given in Table I (2).

TABLE 1

KEY PARAMETERS OF BASE-LINE REFERENCE DESIGN FOR COAL FIRED MHD/STEAM COMBINED CYCLE WITH PARAMETRIC VARIATIONS (2)

	Base Case	Parametric Variation
Application	Base Load	Base Load
Thermal Input Power (MW)	2000	600, 1200
Coal Type	Illinois #6	Montana Rosebud
Coal Treatment	None	Drying
Combustor Type	Direct	Direct
Slag Carry-over	10%	5%, 20%, 100%
Oxidizer	Preheated Air	Preheated Air
Preheat Temperature (°C)	1300	1100, 1400, 1700
Preheater Type	Direct	Indirect (@1700°C)
MHD Generator Type	Faraday	Diagonal
Inlet Static Pressure (Atm)	7.5	6, 10, 15
Seed/%	1.0	1.5
Max. Mag. Field (T)	6	5, 7
Bottom Cycle	Steam 3500/1000/1000	Steam 3500/1000/1000

III. MATERIALS REQUIREMENTS FOR THE MHD CHANNEL

The materials requirements for the MHD channel have been described in the past (1,4,8,9); however, much of the present thinking for coal-fired systems is due to interactions of several scientists on an ERDA sponsored working group (10). These are described in detail in the next chapter by Rossing and Bowen. In addition to the basic electrical requirements, the most important considerstions are related to the interactions of the electrodes and insulators with the coal combustion products (mineral ash) and the potassium seed. Here we wish to emphasize the constraints on the materials selection because of transverse current transfer and because of interelectrode fields. The first is related to problems of electron emission from cathodes; the second to maintaining the electrical strength of insulators.

The requirement that the core of the plasma carry an average current density of ~ 1 amp/cm^2 narrows the list of potential electrode materials not only from the standpoint of electronic conduction in the solid, but because electrons must be transported across the thermal boundary layers to and from the solid. The most significant of these are the cathode problems, electron emission. Under similar operating conditions a ceramic anode can conduct about 10 times the current of a cathode before arcing occurs.

The surface potential barrier or work function at the cathode-plasma interface limits efficient electron transfer and results in a buildup of charged species in the plasma adjacent to the electrode surface producing large voltage drops across the boundary layer and subsequent current instabilities and discharges (arcing) through the otherwise insurmountable surface potential barrier. That is, if enough current cannot flow, the electrons are "evaporated" from the surface by the high temperatures at the base of the arcs.

Much of the available data on the emission characteristics of high temperature materials are shown in Table 2. In fact, very little is known, and it is obvious from the available data that electron emission is a serious problem. Assuming that the Richardson equation is valid, $J = -AT^2\exp[-\phi/kT]$, the current density is given by a pre-exponential A (in the simple theory this is a constant equal to 120 amps/cm^2), the absolute temperature T and the work function ϕ. The experimental determined value of A is strongly dependent on the surface chemistry, i.e., contamination; however, if the ideal expression is assumed (A = 120 amps/cm^2) current densities of 1 amp/cm^2 are possible at T >1800°K for ϕ = 3eV and T >2000°K for ϕ = 3.5eV. The measured current densities are all well below these theoretical values.

This problem is further exacerbated by the fact that the interaction between the magnetic field and the transverse current into the electrode causes focussing of the current. While the average may be 1 amp/cm^2, the local current at the edges of anodes or cathodes may be three times that value. The presence of arc spots (10-50 amps/cm^2) on metal electrodes leads to a material loss rate greater than 10^{-4}-10^{-5} gm/coulomb (11). For a 10 hour test and an average current density of 1 amp/cm^2 this leads to the unacceptable average metal loss rate of 0.36-3.6 gm/cm^2.

There are only two practical solutions to this dilema: higher surface temperatures to accomodate electron emission and *insitu* replenishment of the electrode material to replace that which is eroded away (12). The only materials which appear to meet the first solution are zirconia based ceramics which have been developed in this country by Rossing. Experience in the U.S.S.R. shows loss rates for ZrO_2-CeO_2, ZrO_2-In_2O_3 or ZrO_2-CaO electrodes can be in the range of 10^{-6} gm/coulomb for cathodes at T <1800°C and current densities 1 amp/cm^2 (13). For $J \simeq 1$ amp/cm^2 this gives a recession rate of 10^{-6} cm/sec or 0.036 cm in 10 hours. If

TABLE 2. THERMIONIC EMISSION OF MHD MATERIALS

Electrode Material	Work Function (eV)	Temperature (°K)	Atmosphere	Emission Density (A/cm^2)	Reference
SiC	3.95	1800	Vacuum	2×10^{-3}	26
		1700	Vacuum	4×10^{-4}	
		1800	Cesium Vapors	0.2-0.3	
$MoSi_2$	3.7	1700	Vacuum	4.5×10^{-3}	27
		1650	Cesium Vapors	~ 0.1	
ZrO_Z-Y_2O_3(15%)	3.1	1920	Vacuum	$\sim 10^{-2}$	27
		1920	Cesium Vapors	~ 0.1	
ZrO_Z(0.9)-CaO(0.1)	3.1	1850	P_{O_2} 10^{-8} atm	5×10^{-4}	27
		1850	P_{O_2} 10^{-2} atm	$\sim 10^{-6}$	
ZrB_2 (doped with LaB_6)	<3	1700	Vacuum	~ 2	27
Seeded Coal Slag	1.8-2.4	1600-1900	K-Seeded Combustion Gases	10^{-2}-10^{-1}	28
UO_2	~ 3	-	-	-	29
TiB_2	3.95	-	-	(A = 35)	30

operated in a truly diffuse mode (no arcing) the loss rate should only be time dependent and not current dependent. Thus higher surface temperatures are necessary and ZrO_2 with Y_2O_3 additions have the best properties for the material nearest the plasma.

Due to the probable erosion rates in the channel and especially due to the cathode emission problems, replenishment of the duct walls during long term operation will be necessary (12). The experience to date in MHD channels is shown in Table 3 (12).

TABLE 3

ELECTRODE REPLENISHMENT EXPERIMENTS (12)

	Replenishment Mat'l	"Electrode"	T_{surf} (°C)
AVCO			
Louis	$ZrSiO_4$ + CaO/MgO + Fe_2O_3	ZrO_2	∿1700
Brogan	$ZrSiO_4$ + MgO + Fe_2O_3	inconel	600
Rosa	ZrO_2 + CaO, MgO	ZrO_2	>1400
Petty	slag	inconel	∿1000
Stickler	slag	ZrO_2/inconel	<1400
UTSI	slag	copper	∿300
Westinghouse	slag	ZrO_2-CeO_2	>1500
U.S.S.R.			
U-02	slag	copper	∿300
U-02/U-25	ZrO_2 (combustor)	ZrO_2	∿1700
Poland	slag	stainless steel	600

It does appear to be a feasible operation with coal slags and a specific example is discussed in the companion chapter by Rossing and Bowen. Experiments to observe current transfer to the coal slags have shown the arcing mode operative (14,15), because the shear forces maintain a slag layer thickness with a surface temperature in the range of 1400-1600°C (16,17). Because this layer is continuously replenished, arc erosion losses can be tolerated. The experience using zirconia replenishment does not appear to be as successful as the experience with coal ash.

The final critical materials problem is in maintaining the interelectrode insulator. Besides the potential for contamination by potassium or by iron in coal slags, electrical discharge in the plasma may initiate failure in the insulator. The theory for this has been developed by Oliver (18). In Fig. 7 the current into the

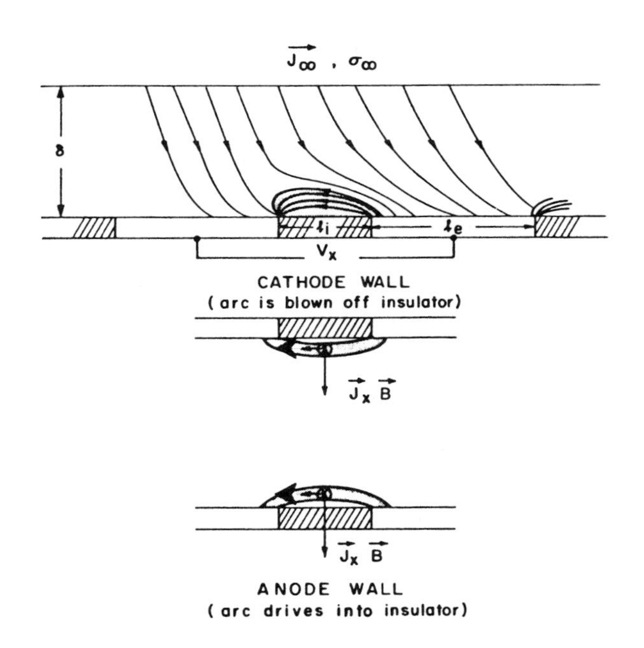

FIGURE 7. Interelectrode arcing - J_∞, σ_∞ are core current density and electrical conductivity; δ the boundary layer thickness, ℓ_i and ℓ_e the axial length of the insulator and electrode. (After Oliver (18))

adjacent electrodes is shown from the plasma core with current density J_{∞} and electrical conductivity σ_{∞} and through the boundary layer δ. Because of the potential difference between adjacent electrodes, arcing between electrodes can occur. Two destructive effects can result. First, the arc can be driven by the Lorentz force into the wall yielding immediate destruction of the insulator. Second, smaller arcs or arcs on the cathode wall can cause local heating of the insulator which leads to loss of the electrical strength.

Thus, in order to design the electrode module for long term use, data on the principle failure modes of the insulator must be generated, (e.g., loss of resistance due to Fe and K penetration and electrical strength loss--dielectric breakdown). For example, comparison of single crystal (sapphire) and polycrystalline alumina samples has shown that the electrical breakdown strength decreases from $>10^6$ volt/cm at room temperature to $\sim 10^4$ volt/cm at 1400°C. Above 1000°C the breakdown mechanism appears to be due to thermal run away. There are at least two important consequences of these observations. First, the absolute resistivity will be the important factor in determining the allowable electric field strengths for interelectrode insulators; and second, although the values of the allowable field strengths at high temperatures are larger than required for MHD generators, the breakdown is very sensitive to the local thermal fluctuations described above; that is, loss of insulation can be due to Joule heating or due to localized heating from arcing above the insulator surface.

The electrical resistivity of iron doped $A\ell_2O_3$ has been measured to determine the allowable levels of iron contamination. The resistivity decreases at 1600°C and $P_{O_2} = 10^{-2}$ atm from 10^5 ohm·cm for pure to 8×10^3, 2.5×10^3 and 2×10^2 ohm·cm for 0.08, 0.5 and 4.4% additions of Fe_2O_3 respectively. Since the dielectric breakdown appears to be thermally triggered at temperatures in excess of 1000°C, electrical strengths will therefore scale according to the inverse of conductivity and thus the inverse of the iron concentration as determined from the isothermal section at each temperature of interest.

IV. MATERIALS REQUIREMENTS FOR THE AIR PREHEATER

Preheating the combustion air is required to obtain the high flame temperatures and high electrical conductivity of the gas. For thermal efficiencies of 50-52% the air must be preheated to 1200°C while a thermal efficiency of 60% would necessitate preheat temperatures of about 1650°C. These high preheat temperatures would not be required if the combustion gas could be enriched with oxygen, but oxygen enrichment would be too costly for a full scale plant. Fig. 8 shows a schematic of a high temperature regenerative air preheater. Along the length of the heater there is a temperature gradient during each half of the cycle, for example, when the MHD exhaust gases are passing through the chamber. The air to be heated would then flow in the reverse direction. Initial heating of the air occurs in a lower temperature metallic preheater so that it enters the ceramic regenerator at a temperature of several hundred degrees and exits the preheater at 1000 to 1200°C. This schematic unit itself would represent one of probably six units that would be cycled during the continuous operation of the MHD system and thus will require large ceramic valves and ducts to direct the flow of the hot gases.

In selecting materials for regenerative air preheaters, consideration should be given to: a large air flow for 1000 megawatt station ∿700 kg/sec; thermal cycling of ∿200°C in 3-30 minute cycles; a duty cycle of several years; a non-plugging configuration

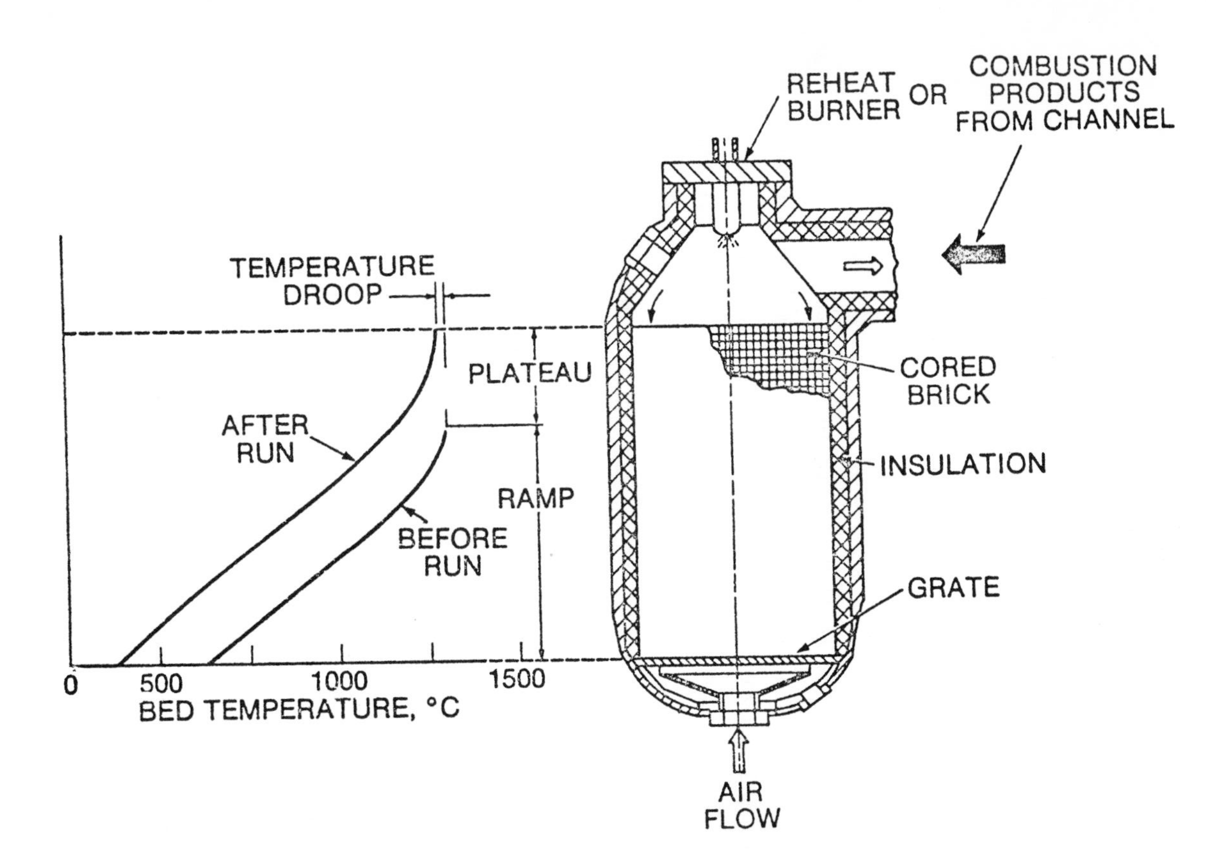

FIGURE 8. Schematic of indirectly fired regenerative air preheater (after Decoursin, FluiDyne Engineering).

for large volume coal slag and seed (K_2SO_4) carry over; the ceramics must be resistant to K_2SO_4 corrosion, to compressive creep and to thermal shock and thermal fatigue; and finally the materials must be low cost. In tailoring materials to meet these requirements, trade-offs will frequently be necessary. For example, high mechanical strength and resistance to corrosion require high density bodies, but the resistance to thermal shock generally increases with increasing porosity and microcracks. For mechanical strength and resistance to erosion, cored brick or stacked-plate, checker configurations are preferred, while more efficient heat transfer is accomplished by the use of pebble beds. The most important consideration is the cost of the preheater because of the large volume of refractories which will be required.

The preheater refractories may present the most difficult ceramic engineering problem because of the large volume of material involved and because the seed-slag mixtures are very corrosive in the temperature range where the oxides are fluid enough not to plug gas flow channels. Studies by Cutler (20-22) and Rossing (23) have provided the laboratory data for understanding these problems, while tests at FluiDyne Engineering (24) have been in a large simulation system. The most complete analysis of the preheater for coal-fired systems and the auxillary equipment and hot-gas ducting is given by Heywood and Womack (4). Recent reviews have also been done by Bates and Ault in reference 5 and by Bates (25) on the ceramic refractory requirements.

V. *SUMMARY*

The success of MHD power generation with its high conversion efficiency is intimately connected with the science and engineering of ceramic materials. While the system requirements seem overwhelming, recent development programs which have involved ceramics research groups have demonstrated creative solutions and alternative design schemes. The success of this technology is dependent on the link between the MHD system engineer and the materials specialist.

REFERENCES

1. J. F. Louis, et al, "Open Cycle Coal Burning MHD Power Generation, An Assessment and a Plan for Action," published by the Office of Coal Research, Dept. of the Interior, June, 1971.

2. W. D. Jackson et al, "Considerations in the Development of Open Cycle MHD Systems for Commercial Service," 6th International Conference of MHD Electrical Power Generation, June, 1975.

3. "An Overall Program for the Development of Open Cycle MHD Power Generation," Electrical Power Research Institute, Report No. EPRI-SR-12, June, 1975.

4. J. B. Heywood and G. J. Womack, Ed., Open Cycle MHD Power Generation, Pergamon Press, Oxford, England, 1969.

5. A. L. Bement, Ed., Engineering Workshop on MHD Materials, sponsored by NSF-ERDA, published by M.I.T., Nov., 1974.

6. J. F. Louis, et al, "MHD Generators: A Status Report," M.I.T. Report, Mar., 1976.

7. J. F. Louis, et al, M.I.T. reports on "Critical Experiments in MHD Power Generation," Nov., 1975.

8. B. R. Rossing and T. P. Gupta, "The Role of Processing in the Development of Generator Components," in reference 5.

9. H. K. Bowen, "MHD Channel Materials Development Goals," in reference 5.

10. J. L. Bates, A. L. Bement, H. K. Bowen, H. P. R. Fredrickse, B. R. Rossing and S. J. Schneider.

11. J. D. Cobine, Gaseous Conductors: Theory and Engineering Applications, Dover Press.

12. H. K. Bowen, "Stabilization of MHD Electrodes," Proceedings of 2nd U.S.A.-U.S.S.R. Symposium on MHD, Washington, D.C., June, 1975.

H. K. Bowen, et al, "High Temperature Electrodes," Proceedings of the 1st U.S.S.R.-U.S.A. Colloquium, Moscow, February 25-27, 1974.

13. N. M. Zykova, et al, "Investigation of Electrode Regions of a Discharge on an Oxide Ceramic in a Plasma Composed of Combustion Products," Sov. J. of High Temperature Physics, 13, 569 (1975).

14. D. B. Stickler and R. DeSaro, "Replenishment Processes and Flow Train Interaction," 15th Symposium on Engineering Aspects of MHD, May, 1976.

15. J. K. Koester, et al, "In-Channel Observations on Coal Slag," ibid.

16. D. B. Sticker and R. DeSaro, "Replenishment Analysis and Technology Development," 6th International Conference on MHD Electrical Power Generation, June, 1975.

17. M. Martinez-Sanchez, M.I.T. Annual Report, June, 1976.

18. D. A. Oliver, 6th International Conference on MHD Electrical Power Generation, Vol. I, pp 329-344.

19. T. Pollak, Electrical Conduction in Iron-Doped Alumina, M.S. Thesis, M. I. T., May, 1976.

20. W. D. Callister, et al, "Corrosion of MHD Preheater Materials in Coal Slag-Seed Mixtures," 14th Symposium on the Engineering Aspects of MHD, April, 1974.

21. T. I. Oh, I. B. Cutler, and R. W. Ure, Jr., "Corrosion of MHD Preheater Materials in Seeded Coal Slag Liquid," 6th International Conference on MHD Power Generation, June, 1975.

22. L. N. Shen, et al, "Corrosion of MHD Preheater Materials in Liquid Seed Slag," 15th Symposium on Engineering Aspects of MHD, May, 1975.

23. B. R. Rossing, et al, "Slag-Refractory Behavior Under Hot Wall Generator Conditions," 15th Symposium on the Engineering Aspects of MHD, May, 1976.

24. J. E. Fenstermacher, et al, "Progress on the Testing of Refractories for Directly-Fired MHD Air Heater Service," 15th Symposium on the Engineering Aspects of MHD, May, 1976.

25. J. L. Bates, "Behavior/Requirements for MHD Support Systems," Workshop on Ceramics for Energy Applications, J. R. Schorr, Ed., Battelle Columbus Labs, Ohio, Nov., 1975.

26. B. S. Kulivarskaya, et al, "Thermionic Emission of Certain Refractory Materials," Mat. Kanala MGD-Generators, 205 (1969) in "Materials for MHD Generator Channels," Joint Publications Research Service, No. 53439, Aug., 1971.

27. A. M. Anthony, et al, "Properties Thermigues, Electrigues et Thermoelectrongues de Diveis Materiaux D'Oxydes Refractaires," in Elect. MHD Proc. Symposium, 5, 3019 (1968).

28. C. K. Petersen and R. W. Ure, Jr., "Thermionic Emission Characteristics of Seeded Coal Slags," 15th Conference on Engineering Aspects of MHD, Philadelphia, Pa., May, 1976.

29. V. S. Fomenko, Handbook of Thermionic Properties, G. V. Samsonov, Ed., Plenum Press, Data Division, New York, 1966.

30. G. V. Samsonov, et al, Radiotekhn i Elektron, 2, 631 (1957).

Chapter 11

MATERIALS FOR OPEN CYCLE MAGNETOHYDRODYNAMIC (MHD) CHANNELS

B. R. Rossing

Materials Science Division
Westinghouse Research Laboratories
Pittsburgh, Pennsylvania 15235

and

H. K. Bowen
Department of Materials Science and Engineering
Massachusetts Institute of Technology
Cambridge, Massachusetts 02139

I. INTRODUCTION

Open cycle magnetohydrodynamics (MHD) is at a stage of development where the scientific principles have been proven but there has been no demonstration of the long term, high performance necessary for commercial acceptance. As efforts proceed toward operation of pilot plant scale (and larger) systems, attention becomes focused on those technological areas that limit the rapid development of this advanced energy conversion system. The development of an MHD generator (also termed the channel or duct) that can efficiently convert the thermal and kinetic energy of coal combustion gases into electrical energy for thousands of hours must be designated as the critical pacing item in the MHD system. The heart of the generator is an electrode system which allows for current transfer between the hot plasma core and the external load. The combination of high temperatures, high velocity gases containing reactive chemical species and electrochemical effects impose a very severe environment upon the electrode and insulator elements that comprise the electrode system. Therefore, the major obstacle in realizing the desired generator lifetimes is the degradation* of the electrode and insulator materials.

There are many materials that have the required properties to function as either electrodes or insulators; however, the combination of high temperatures, slag and seed, high velocity gases, high thermal fluxes, high voltages, etc., will degrade the performance of most materials to unacceptable levels in only a few hours. The longest operation of electrodes using clean fuels is slightly more than 300 hours. While there is a general lack of long-duration testing under coal-fired conditions, it may be estimated that presently available materials may be operable for 100-200 hours. Therefore, the lifetime of electrodes and insulators must be increased by more than an order of magnitude to make coal-fired open-cycle MHD power generation a commercial reality. This must be accomplished by taking a holistic approach to the entire electrode/plasma system and to those processes that would degrade channel materials during operation. Materials selection, electrode design and generator operating conditions should be contingent upon considerations to avoid, or at least, minimize the

*Degradation will be used here in a broad sense to cover those processes leading to eventual inoperability of either electrodes or insulators.

effects of these processes. This article will review the present state-of-the art of MHD channel materials with specific attention to the degradation processes that may limit electrode-insulator (and generator) lifetimes.

II. ELECTRODE/PLASMA SYSTEM

The MHD channel can be viewed as an array of different materials arranged to allow for the transfer of current from (and to) the plasma core to (and from) an external load in some desired manner. The channel wall from this perspective consists of an electrode system consisting of two major subelements, i.e., the electrodes which carry current from the plasma to the external load and the insulators which block the passage of current in certain directions and in certain portions of the channel. In viewing both electrodes and insulators from a functional standpoint, any condensed slag layer must be viewed as part of both the electrode and insulator, while any current leadout material should be regarded as part of the electrode. The current transfer process should be regarded from the perspective of the entire electrode/plasma system in order to properly comprehend the functioning of electrodes and insulator elements. The electrode/plasma system, illustrated schematically in Fig. 1 is composed of a number of dissimilar phases (solid, liquid, and gas) which must be viewed as a single entity having nonhomogeneous properties. A system must, therefore, be designed to accommodate the effects of these nonhomogeneities as current is transferred through a series of layers and interfaces. At the same time, current transfer must be blocked or at least limited in the axial direction of the channel (or in the insulating sidewalls of a Faraday generator). Furthermore, the current transfer process

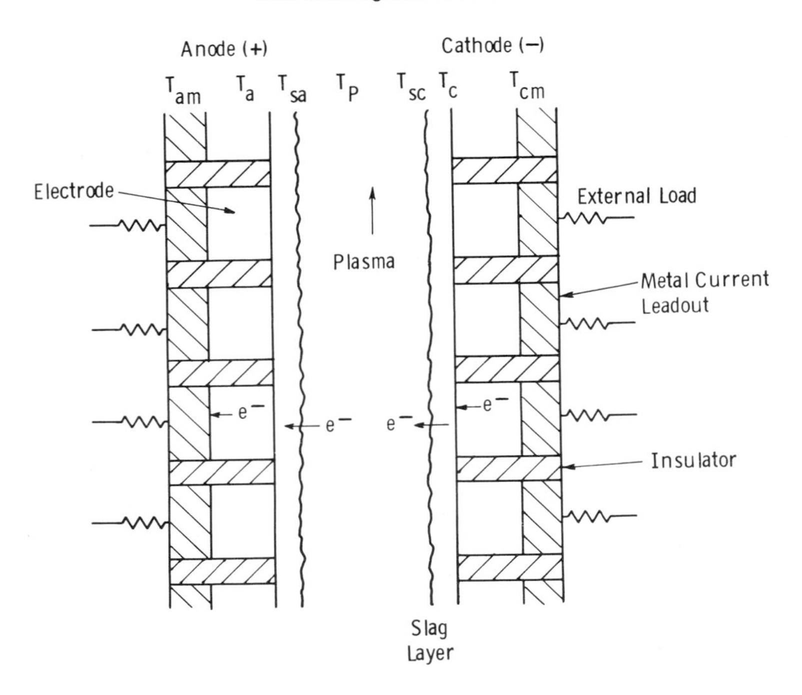

FIG. 1. Schematic of Electrode/Plasma System

itself can lead to further nonhomogeneities through the electrochemical segregation of charged chemical species. For example, positive potassium ions are driven toward the cathode wall (away from the anode wall) by the electric field and negative oxygen ions are driven toward the anode wall (and away from the cathode wall) by the electric field. With slag-coated walls the segregation will be of greater complexity; however, it is clear that the channel environments seen by cathodes and anodes can be quite different, requiring different criteria for materials selection and operating conditions. In addition, there will be variations to a lesser extent in channel conditions in the axial direction which may require further variations in both materials and designs.

III. DEGRADATION OF THE ELECTRODE/PLASMA SYSTEM

In order that the current transfer process functions satisfactorily for long-term electrode stability (and generator operation) several requirements must be imposed upon channel materials. First, current must be transferred at densities of 1 to 2 amps per cm^2 of electrode area in future commercial generators. Current constrictions caused by plasma electromagnetic fields will produce localized current densities that can be much higher (1). At these high current densities, several sources of instability are possible in the electrode/plasma system. First, any resistive regions will give rise to significant voltage drops and joule heating. In the plasma, the gas boundary layer can be poorly conducting, particularly in contact with heavily cooled metal electrodes. In this extreme case, the large voltage drops plus the poor thermoelectron emission capabilities of the cold metal will give rise to large electric arcs which can erode material by localized melting and/or vaporization. These arcs can be avoided by operating electrode surfaces at very high (>2000 K) temperatures, i.e., to reduce gas boundary layer voltage gradients and to increase thermoelectron emission. Current transfer at these elevated temperatures is through a 'diffuse' mode. With increasing electrode temperatures there is a gradual transition from 'arc' to 'diffuse' modes of current transfer. Hot electrode operation with diffuse mode current transfer predominating is preferable to minimize arc erosion and electrode degradation. The Soviet MHD program, after a decade of electrode testing, has proceeded to the use of very hot walls. Under clean fuel conditions the Soviets have concluded that hot wall channel operation is preferable for maximizing electrode stabilization.

Secondly, very resistive regions can occur both in refractory oxide electrodes and in condensed slag layers. Both classes of materials are based upon oxides having resistivities that vary exponentially with temperature. As can be seen in Fig. 2, these materials may have adequate conductivity (>0.1 ohm^{-1} cm^{-1}) at temperatures above 1500 K, but the large increase in resistivity with decreasing temperature (due to the large temperature gradients in actual electrodes) will produce large Joule losses and thermal instabilities in the solid. These instabilities

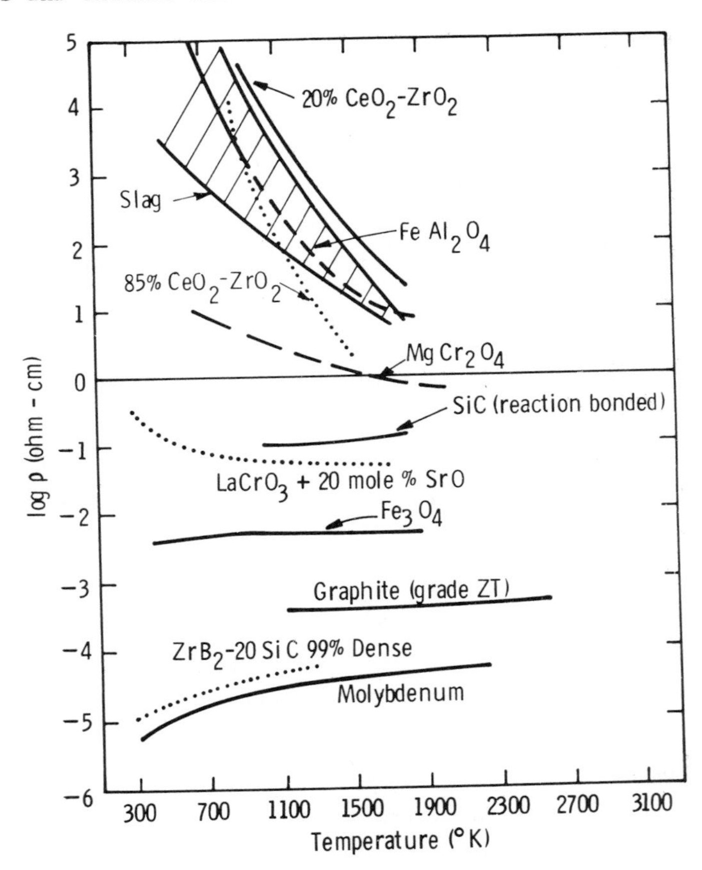

FIG. 2. Resistivity of Potential Electrode Materials

include current channeling, electrolysis and arcing. A solid slag layer on a cold metal surface can be particularly resistive (slag conductivities at 1000 K are of the order of 10^{-5} ohm^{-1} cm^{-1}).

Thus in selecting electrode materials (or in tailoring the electrical properties of slag layers) the important parameters of electrical conductivity, ($\sigma = \sigma_o \exp(-Q/RT)$, are a low activation energy Q and a high value of σ_o. Bowen (2) has indicated that this is as important a factor as the electronic transference number (fraction of current carried by electrons), in facilitating stable current transfer from the plasma to the current leadout.

Another source of electrical instability at current densities in excess of 1 amp/cm^2 occurs in electrode materials when current transport is carried by charged ions rather than electrons. This type of conduction (ionic) is common in many refractory oxides, such as ZrO_2, that have wide band gaps and crystal structures amenable to rapid diffusion by oxygen anions. If the slag layer and anode and cathodes shown in Fig. 1 are ionic conductors, then the following prime reactions occur at the various interfaces:

$$(T_{cm})\ 1/2\ O_2 + 2e^- \rightarrow O^= \quad \text{(metal-cathode)}$$
$$(T_c)\ O^= \rightarrow 1/2\ O_2 + 2e^- \quad \text{(cathode-plasma)}$$
$$(T_a)\ 1/2\ O_2 + 2e^- \rightarrow O^= \quad \text{(plasma-anode)}$$
$$(T_{am})\ O^= \rightarrow 1/2\ O_2 + 2e^- \quad \text{(anode-metal)}$$

where O_2 is an oxygen molecule, $O^=$ represents an oxygen ion in the oxide electrode and e^- is an electron available in either the metal leadout or the plasma. Of the four reactions, the two occurring at the metal-electrode interfaces are probably more deleterious to the stability of the electrode. At the cathode side, the oxide can be reduced by an electrochemical reaction (electrolysis) to the metal (e.g., ZrO_2 to Zr metal). Even if complete reduction does not occur the structural integrity of the body could be impaired through formation of microcracks during the reduction process.

At the anode side, excess oxygen is available to combine with the current leadout to form a resistive layer at the interface. For a completely ionically conducting oxide electrode under a current density of 1 amp/cm^2 approximately 2 x 10^{-3} moles/hr/cm^2 of oxygen is released at the anode surface. The metal oxide reaction product created by the reaction with the metal (anode) current leadout would grow at the rate of approximately 1 mm/hr. Even with very electronically conducting materials, current leadout selection must be made with care to avoid the formation of reaction products that are resistive and that form a weak interfacial bond that would be subject to failure during operation or thermal cycling.

The penetration of potassium seed compounds into the pore structure of channel materials is a serious problem in itself. These compounds, primarily K_2CO_3, condense in the temperature region of 1200-1600 K. These compounds then react with moisture in the atmosphere forming hydrated compounds resulting in volume expansions of 40 to 60%. These volume changes generate stresses large enough to cause extensive cracking and even disintegration of the ceramic. Under the influence of the electric field, even higher quantities of potassium are driven into the cathode. The analysis of several U.S. electrode materials after a 100-hour test in the Soviet Union revealed that while little potassium was found in the anodes, the cathodes contained significant quantities of seed (3). The resulting degradation of the anode wall was minimal while the cathode had suffered severe damage due to reactions related to seed penetration. Ceramic channel materials can be corroded away at very rapid rates if their surfaces are operated at temperatures where liquid seed or slag-seed mixtures are condensed on their surfaces.

The use of ceramics as electrodes or insulators in the generator duct also poses serious problems with respect to the thermal stress damage of these materials. Studies on a number of materials indicate that, given the range of thermal conductivities of most oxide ceramics, the thickness of electrodes and the insulators would be relatively small, ranging from a few to 25 mm in thickness. The temperature gradients imposed on these thin sections are severe, of the order of 100°C/mm or more. These gradients will impose high thermal stresses on these materials which can lead to failure in use at temperature. In addition, there is a danger of failure due to thermal shock during startup and shutdown.

Thermal stress damage resistance can be built into a given ceramic by development of suitable microstructures. Microstructural features that are desired include 15 to 25% porosity, a wide variation in grain sizes (4), and the presence of second phases and microcracks (5). These microstructures are in conflict with those required for both corrosion and erosion resistance, where high density, in particular, is required.

Channel insulation materials, either the inter-electrode insulators or the insulating sidewalls, serve the role of blocking the flow of current in parts of the channel. Good electrical resistivity (>100 ohm-cm) and good dielectric strength (>4 kV/m) are required at operating conditions. The penetration of seed or seed-slag compounds into the insulator will degrade these properties. Again microstructure control through ceramic processing is crucial in avoiding this problem. Other degradation processes with insulators include several discussed for electrodes, i.e., corrosion, erosion, and thermal shock resistance.

IV. MATERIALS FOR CLEAN FUEL FIRED MHD CHANNELS

The development and evaluation of electrodes using 'clean' fuels has involved three classes of materials: metals, oxide ceramics, and nonoxide ceramics. This work is reviewed in some detail in the literature (6-10) and will only be summarized in this discussion.

Water-cooled metal electrodes are attractive on the basis of ease of fabrication, electrical properties, mechanical properties, thermal conductivity, and so on; the cold surfaces, however, promote high thermal and electrical losses and permit seed condensation, which is believed to cause erratic and degenerative electrical performance. The most deleterious effects, however, arise from nonuniform current transfer (arcing) from the plasma across the cool boundary layer to the cold electrode surface. This arcing is localized and produces enhanced erosion because of the extreme localized heating. The combination of arcing and seed condensation leads to electrochemical erosion of the material and subsequent power losses. Seed condensation can be avoided by use of wall temperatures in excess of 1573 K. At these temperatures, however, most metals encounter severe oxidation. The only metallic compositions that have demonstrated promise at temperatures above 1273 K are cobalt-based superalloys tested by Bone (11) in the VEGAS generator, the oxidation resistant tungsten alloys used by the Soviets (12) in their U-25 generator, and a chromium-based cermet developed by Reynolds Aluminum Company (13). Bone estimates the maximum use temperature of the cobalt alloys as 1493 K, and the Soviets indicate that the tungsten-chromium alloys have been used in the 1273 to 1773 K temperature regime. The chromium electrode patented by Reynolds Aluminum has additions of thorium dioxide dispersed through it to enhance thermolectron emission. It is claimed to have good slag resistance and that the volatility of chromium offers no difficulties.

A number of nonoxide ceramics have been developed and evaluated as electrodes in clean fuel-fired systems. These materials are attractive in that they possess excellent electrical and thermal properties; they are limited, however, by their poor oxidation resistance above 1773 K. The most promising compositions of this class (SiC, Mo_2Si, ZrB_2-SiC) rely on the formation of a silica or a silicate protection layer for oxidation resistance. The reaction of this silica-based layer with potassium forms a fluid potassium oxide-silicon dioxide glass which limits the use of these materials in the MHD channel to temperatures less than 1673 K (8). The thermelectron emission of silicon carbide is also inadequate, and additions of such materials as molybdenum, titanium, and chromium have been used to improve this property (14).

Therefore, while metals are the most promising electrode materials in cold (<1273 K) wall channels, and oxidation resistant refractory alloys and nonoxide ceramics show promise in moderately hot (1273 to 1773 K) channels, the only class of materials that show any promise in hot (>1773 K) channels are the refractory oxides. At these high temperatures both the problems of destructive arc erosion and seed condensation are avoided. Generator thermal and electrical losses are also reduced by operating the walls at these temperatures. Much effort has been directed toward developing refractory oxides with electronic conduction. The refractory oxides typically have very large band gaps, resulting in poor electrical conductivity. Two compositions have been identified that have both high-temperature stability and adequate electrical conductivities to serve as hot MHD electrodes. These compositions are based either on zirconium oxide or on the rare earth perovskite-structured chromites. The development and properties of each are reviewed in detail elsewhere (7-9, 15-17).

Compositions based on zirconia received early attention as potential MHD electrode materials. Additions of calcium oxide or yttrium oxide were found not only to stabilize zirconium oxide but also to provide for conductivities that are adequate to meet electrode requirements above 1373 K. Because this conduction is almost entirely ionic there is degradation of these compositions with time by electrolysis at even moderate ($\sim$1 A/cm^2) current densities (7, 18, 19). This deficiency can be avoided, in part, by the use of electrode designs incorporating metallic fibers, pins, or screens which will reduce the local zirconium oxide current density with the metal carrying the major portion of the current (12, 20). Although calcia- or yttria-stabilized zirconia electrodes can function satisfactorily for ten or even hundreds of hours, they are not the solution for hot electrodes for commercial MHD plants. The conduction must be predominantly electronic in nature to prevent electrolysis and electrode degradation.

Additions of rare earth oxides such as cerium oxide, praseodymium oxide, neodymium oxide, and lanthanum oxide produce varying degrees of electronic conductivity in zirconia compositions. The Soviets have spent considerable effort in developing an electronically conducting zirconia, based on ternary compositions of zirconia and rare earth oxides. Although they have revealed no information as to the composition of these electrodes, it has been claimed that they have operated successfully for hundreds of hours at temperatures close to 2273 K and current densities of more than 2 A/cm^2.

It appears that the major obstacle to the full development of zirconia-based compositions as suitable electrode materials is the identification of a suitable current lead material. Since the conductivity of these zirconia-based compositions are very temperature dependent, they become resistive and subject to large voltage drops and joule heating at temperatures below 1273 K. Proper selection of suitable materials to carry the current ultimately to copper wire becomes important. Selection of these materials must be made on the basis of compatibility with the passing of direct current at moderately high current densities. One possible solution (21) is a composite electrode consisting of an electronically conducting zirconia bonded to a ceramic (Ta_2O_5 doped CeO_2) current lead material that has high conductivity at moderate (873 to 1473 K) temperatures. The ceramic current lead material is, inturn, bonded to an oxidation resistant alloy (similar to the Inconel series).

The rare earth chromites when doped with alkaline earth oxides are unique as oxides in that they possess both high electronic conduction, even to room temperature (see Fig. 2), and very high melting points. While these materials possess other very desirable features, such as a small variation in resistivity with temperature and the possibility of achieving the current lead contact at low temperatures, they are limited in maximum use temperature by the preferential volatilization of chromium (22). The extrapolation of vaporization data for $La_{.8}Sr_{.2}CrO_3$ to higher temperatures predicts that vaporization losses would limit this material's use to approximately 2023 K (where recession rate is 1 cm/yr). These vaporization losses can be reduced by the substitution of other cations for La, Sr, or Cr (22, 23) in order to raise the use temperature of this material. Lanthanum chromite is subject to two additional degradation processes; reactions with condensed potassium compounds forming potassium chromate (24) and poor erosion resistance at high temperatures. Improved $LaCrO_3$ electrodes have been developed by control of microstructure to prevent seed reactions and by the addition of fused zirconia crystals to improve the erosion resistance (25).

In summary, in the past fifteen years much experience has been accumulated with the use of electrodes whose surface temperatures vary from a few hundred to 2273 K. Although substantial progress has been made, a workable electrode for long-duration use in clean fuel-fired generators does not exist today. Some of the most serious problems are located not at the hot surface but rather at the cold surface of the electrode. Finally, operation of electrodes using clean fuels is most favorable under hot wall conditions where seed reactions and arc erosion are minimized.

Insulation materials have not been the subject of studies as extensive as those of electrode materials. Reviews of the literature (6-8) also indicate that insulator development has been limited to the following oxides: alumina, magnesia, and the zirconates of calcium and strontium. Beryllium oxide has been used in at least one study (16), but it generally has been discounted due to its toxicity. Aluminum oxide has excellent high-temperature insulation properties but reacts with potassium at temperatures above 1573 K to form $K_2O \cdot 11Al_2O_3$ (potassium β-Al_2O_3) (27, 28). Strontium and calcium zirconates are very refractory [melting points >2573 K], possess good seed resistance, but suffer from poor thermal shock resistance due to a very low thermal conductivity. Magnesia has, thus, been generally used as the insulating material in hot-or moderately hot-wall channels, while alumina has been used in cold-wall channels.

V. ELECTRODE SYSTEMS FOR COAL FIRED MHD CHANNELS

The use of coal as fuel and the presence of significant quantities of slag in the MHD system requires a re-evaluation of the materials used in the generator channel. The ash constituents that are volatilized into the plasma in the combustor condense on the cooler channel walls as a potassium-enriched slag layer. A schematic representation of a slag-coated channel wall is shown in Fig. 3. This slag coating increases the complexity of the electrode/plasma system as shown by the variation in chemistry and properties across the channel wall in Fig. 3.

The slag coating will affect both the performance of the generator and the durability of channel materials. Since the effect of the former has been covered in the previous chapter (29), the following discussion will center on the stabilization of the electrode system, i.e., the deterrence of degradation processes, under coal fired MHD conditions.

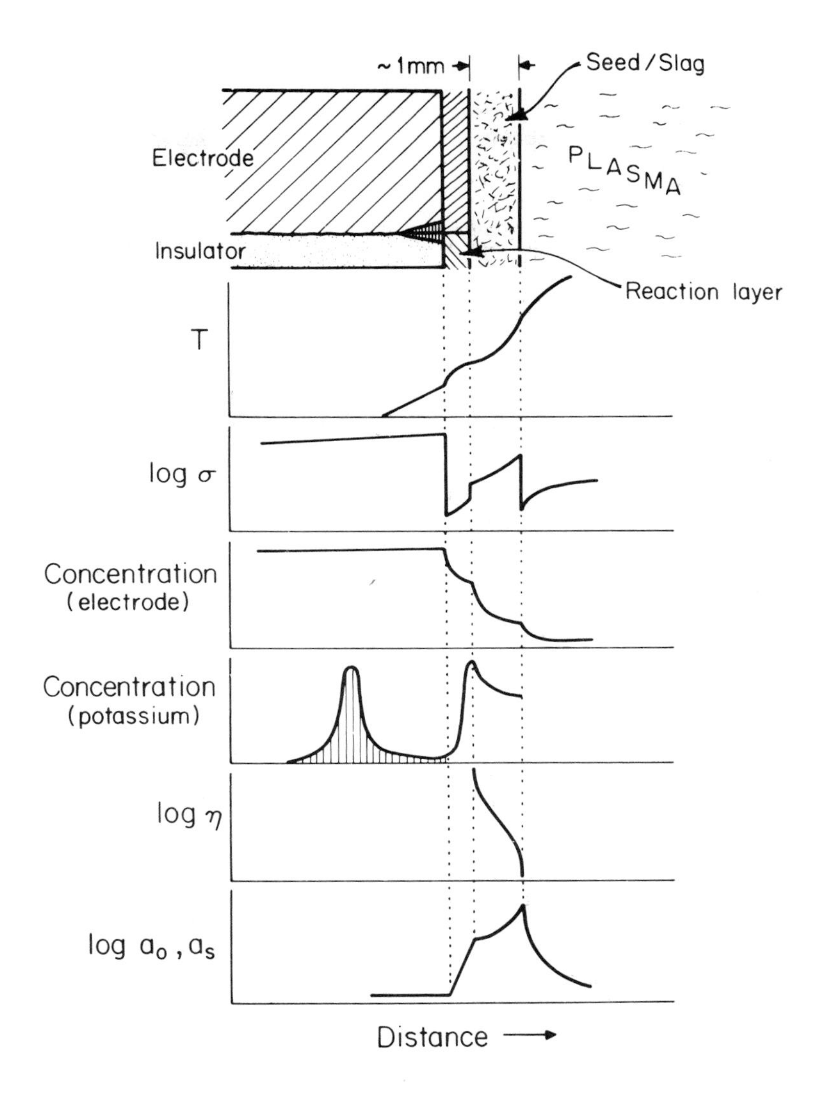

FIG. 3. Schematic of Slag-Coated Channel Wall Showing Variation in Chemistry and Properties Through the Wall (2).

The lifetime of the coal fired MHD generator will be limited by either the degradation of current transfer processes in the electrode/plasma system or by the loss of channel materials due to excessive reactions with the channel environment.

The possible degrading interactions between condensed phases can be inferred from Fig. 3 as follows:

1) dissolution, penetration or reaction of the electrode by the slag;
2) dissolution, penetration or reaction of the inter-electrode insulator by the slag;
3) interdiffusion or penetration of electrically conducting ions from the electrode into the insulator;
4) penetration of slag and potassium compounds down the electrode-insulator interface; and
5) formation of electrically insulating or debonding reaction products at the electrode-current leadout interface.

The electrical instabilities that can occur include arcing, electrolysis, joule heating and current channeling in electrodes and loss of resistivity and dielectric strength in insulators. It is apparent that there is a strong coupling between these chemical and electrical factors so that many of these degradation processes are truly electrochemical in nature.

In order to deal with possible deleterious reactions and degradation of the electrode system, three different approaches can be taken; each is based upon the operation of channel walls within a specific temperature regime: 1) cold wall, i.e., below the slag-seed solidification temperature, $T < 1473$ K; b) hot wall ($1473 > T > 1973$ K) where the slag-seed forms a liquid layer on the wall; and c) super hot wall ($T > 1973$ K) where little of the mineral ash or seed condenses on the surface.

By operating in the cold wall mode, the chemical degradation is minimized because the solid slag is slow to react with wall materials. Also metals and non-oxide ceramics can be used as electrodes. However, serious instabilities in the current transfer process occur due to arc mode current transfer and to the poor electrical conductivity of solid slag. Also, migration of oxygen through the slag layer under the influence of the electric field may cause excessive oxidation of the anodes.

On the other hand, operation in the super hot wall mode with the rejection of most (>80%) of the slag in the combustor will minimize the condensation of slag. While current transfer is through a diffuse mode at these temperatures and arcing is avoided, the selection of materials is limited to a short list of refractory oxides having low vaporization rates and corrosion resistance to slag-seed condensates. Analysis of materials tested under these conditions indicate that the slag-seed condensates are in the form of isolated droplets (30), that recession rates are dramatically reduced over those predicted from tests in molten slag-seed mixtures (30), and that several oxides show resistance to this environment up to at least 1973 K (31).

The intermediate temperature range, the hot wall channel, which allows for liquid slag-seed contact with channel materials, will maximize the chemical dissolution of these materials. Refractory corrosion is generally a diffusion-limited process; thus the rate limiting diffusion process in the slag-seed and the chemical driving forces are important considerations. Thus, the primary mechanism for chemical compatibility is to control the compositions of the gas phase, the slag and the electrode (insulator) to minimize concentration gradients to reduce the chemical driving forces.

Taking this approach, Bowen and his co-workers (32) have identified electrode/insulator materials in the Fe-Al-O system as possible channel materials which would maintain small concentration gradients with $FeO \cdot Fe_2O_3$-Al_2O_3-SiO_2 based coal slags. The electrodes (see Fig. 4) would be based on solid solutions of $FeAl_2O_4$-Fe_3O_4 spinels and the insulators would be of Al_2O_3. The electrode would be graded $FeAl_2O_4$ rich at the slag interface to more highly conducting Fe_3O_4-rich compositions at the iron current leadout interface.

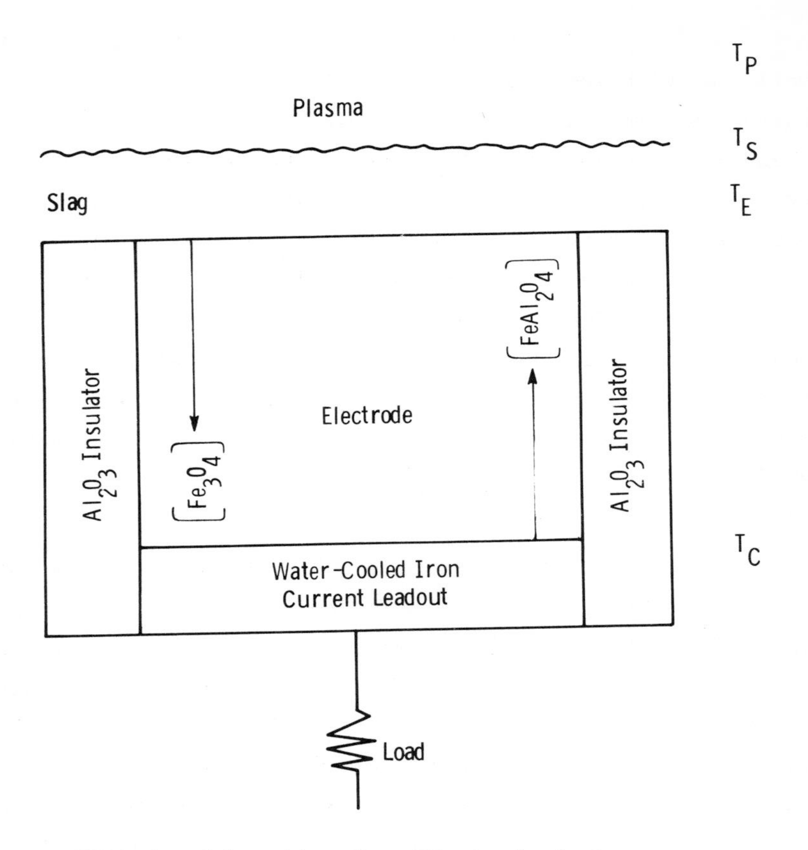

FIG. 4. Schematic of an Electrode System Based on the Fe-Al-O System

This spinel-alumina electrode module offers the following distinct advantages for current extraction from coal fired MHD systems:

1) The complete electronic conductivity and low temperature dependence of the graded spinel meets design requirements.
2) Operation temperatures will allow optimization of slag properties without aggravated dissolution of the module.
3) There will be no resistive reaction layers at the hot or cold electrode faces as in other systems where reaction and/or electrolysis are often unavoidable.

4) Dissolution of spinel and alumina can be minimized by control of the slag chemistry.
5) Diffusion of iron into the Al_2O_3 dielectric can be modified by slag composition, temperature, and the oxygen partial pressure of the combustion gases.
6) Grading of the spinel layer as indicated in Fig. 4 will minimize the driving force for interdiffusion with alumina at the hot end of the module adjacent to the slag.

Using the same approach, other electrode systems ($MgAl_2O_4$-Fe_3O_4/$MgAl_2O_4$ (33), SiC/SiAlON) have been proposed that would have electrical and chemical stability with a slag layer.

In summary, the corrosive conditions in the coal fired MHD duct are both severe and not clear at this time. The identification of both promising materials and optimum generator operating conditions is dependent upon serious attention to degradation processes in the slag containing MHD channel.

VI. SUMMARY

In the last fifteen years open cycle MHD power generation has moved from small laboratory generators generating a few kilowatts of power to pilot plants where tens of megawatts can be generated. Before this energy conversion process can be commercially realized, the lifetime of major components must be demonstrated. The MHD generator, the heart of the MHD system, is limited to several hundred hours of operation due to the degradation of electrodes and insulators. The use of coal as the fuel in this system intensifies the severity of the channel environment. In order to realize the generator lifetimes required for commercial plants, serious attention must be paid by both materials and generator specialists to prevent or, at least, minimize the degradation of the electrode system. There is much work that must be done. First, extensive long-duration testing under a variety of coal fired conditions is needed to better define the problems themselves. This work must be complemented by slag modeling, slag property and slag-refractory corrosion studies. The current transfer processes in the slag containing generator must also be better defined. Once the events occurring in the MHD channel are better understood, the direction for materials development and use will become more clearly defined.

REFERENCES

1. C. D. Maxwell, S. T. Demetriodes and S. R. Davis, Resistive Coatings Over MHD Electrodes, 14th Symposium on Engineering Aspects of MHD, UTSI, Tullahoma, Tennessee, April (1974).

2. H. K. Bowen, MHD Channel Materials Development Goals, Engineering Workshop on MHD Materials, A. L. Bemert, Ed., (sponsored by the Office of Coal Research and the National Science Foundation), Massachusetts Institute of Technology, Cambridge, November 20-22, 1974.

3. W. D. Jackson, A. E. Scheindlin, et al., Joint Test of U.S. Electrode System in the U.S.S.R. U-02 Facility, 15th Symposium on Engineering Aspects of MHD., U. of Pennsylvania, Philadelphia, PA, May (1976).

4. F. H. Norton, Refractories, (1968).

5. R. Goodof and D. R. Uhlmann, Thermal Shock Resistance of Air Preheater Materials, 14th Symposium on Engineering Aspects of MHD, April (1974).

6. J. F. Louis, (Ed.) Open Cycle Coal Burning MHD Power Generation: An Assessment and a Plan for Action, MIT Report prepared for OCR, June (1971).

7. J. B. Heywood, W. T. Norris and A. C. Warren, Electrodes and Insulators, Chapter 5 in Open Cycle MHD Power Generation, J. B. Heywood and G. J. Womack, Eds., (1969).

8. A. I. Rekov, Materials for MHD Generator Channels (Translation available through NTIS as report #JPRS 53939) (1960).

9. G. Rudins, U.S. and U.S.S.R. MHD Electrode Materials Development, R-1656-ARPA (ARPA Order No. 189-1, 6L10 Technology Assessment Office) (1974).

10. Energy Conversion Alternatives Study (ECAS), Westinghouse Phase I Final Report Vol. 2 - Materials Considerations, NASA-CR-134941, Vol. 2, (1976).

11. T. Bone et al., Experiences with the Open Cycle MHD Facility VEGAS, 12th Symposium on Engineering Aspects of MHD, Argonne National Lab, Chicago, April (1972).

12. A. E. Sheindlin, The U-25 MHD Pilot Plant Part II MHD Generator Studies, (English Translation available from NTIS as Report ERDA-TR-102 pl).

13. P. Goolsby, Reynolds Aluminum Company, U.S. Patent 3,710,152.

14. V. G. Gordon et al., Oxide and Carbide Materials as MHD Electrodes, 4th Int'l. Conf. on MHD Electrical Power Generation, Warsaw (1968).

15. D. B. Meadowcroft, Electronically-Conducting Refractory Ceramic Electrodes for Open Cycle MHD Power Generation, Energy Conversion 8, p. 185 (1968).

16. D. B. Meadowcroft, Some Properties of Strontium-Doped Lanthanum Chromite, Brit. J. Appl. Phys., 2, p. 1225 (1969).

17. G. P. Gokhshteyn, A. A. Safanov and V. P. Lyibimov, On the Physical and Chemical Behavior of the Electrodes of an MHD Generator Made of ZrO_2-Y_2O_3 and ZrO_2-CeO_2, in MHD Method of Producing Electrical Energy, (V. A. Kirillin and A. Y. Sheyndlin, Eds.) (1972).

18. G. P. Gokhshteyn and A. A. Safonov, Operation of a Refractory Ceramic Electrode in an MHD Generator, Inorganic Materials, 7, p. 582 (1971).

19. R. E. W. Casselton, Blackening in Yttria Stabilized Zirconia Due to Cathodic Processes at Solid Platinum Electrodes, J. Applied Electrochemistry, 4, p. 25 (1974).

20. G. Johnson, E. P. Tuffy and D. Balfour, Development and Testing of High Temperature Electrodes Based on Stabilized Zirconia, 4th Symposium on MHD Power Generation, Warsaw (1968).

21. 1972 Annual Report of the Institute of High Temperatures of the U.S.S.R. Academy of Sciences, 1973. (English Translation available from NTIS as Report JPRS 59983).

22. D. B. Meadowcroft and J. Wimmer, Volatile Oxidation of Lanthanum Chromite, Paper 87-B-73, 75 Annual Meeting of the American Ceramic Society, April (1973).

23. A. M. George et al., Improved Lanthanum Chromite Ceramics for High Temperature Electrodes in Open Cycle MHD Systems, 15th Symposium Engineering Aspects of MHD, U. of Pennsylvania, Philadelphia, PA, May (1976).

24. M. Berberian, R. W. Ure and I.B. Cutler, The Preparation, Electrical Resistivity and Chemical Stability of Doped $LaCrO_3$ Compounds, 6th Intl. Conf. on MHD Electrical Power Generation, Washington, D.C. June (1975).

25. V. G. Gordon, High Temperature Institute, Moscow, US/USSR Cooperative Program in MHD (1975).

26. K. Fushini et al, Development of a Long Duration MHD Channel, 5th Intl. Conf. on MHD Electrical Power Generation, Munich (1971).

27. A. Nagahiro et al., Preparation of Test of Refractory Materials for Heater Bed in Regeneration of MHD Power Plant, 4th Intl. Conf. on MHD Electrical Power Generation, Warsaw (1968).

28. A. Dubois and M. Hamar, Corrosion of Refractories by Potassium Seed Combustion Gases, Intl. Conf. on MHD Electrical Power Generation, Paris (1964).

29. H. K. Bowen and B. R. Rossing, Materials Problems in Open Cycle Magnetohydrodynamics, this volume.

30. B. R. Rossing et al., Corrosion Resistance of MHD Generator Materials to Seed/Slag Mixtures, 6th Intl. Conf. on MHD Electrical Power Generation, Washington, D.C. June (1975).

31. B. R. Rossing, J. A. Dilmore and H. D. Smith, Slag-Refractory Behavior Under Hot Wall Generator Conditions, 15th Sym. on the Engrg. Aspects of MHD, U. of Pennsylvania, Philadelphia, PA, May (1976).

32. T. O. Mason et al., Properties and Thermochemical Stability of Ceramics and Metals In An Open-Cycle Coal Fired MHD System, 6th Intl. Conf. on MHD Electrical Power Generation, Washington, D.C. (1975).

33. S. Schneider, National Bureau of Standards, Private Communication (1976).

Chapter 12

ALLOYS FOR HIGH TEMPERATURE SERVICE IN GAS TURBINES

Nicholas J. Grant

Department of Materials Science and Engineering
Massachusetts Institute of Technology
Cambridge, Massachusetts

I. INTRODUCTION

In the field of high temperature materials, recent prior history must record superalloys as one of the truly outstanding developments among materials generally. Such developments don't just happen; the critical need for materials to sustain the rapid growth of gas turbines, with emphasis on the jet engine, resulted in phenomenal growth in temperature - time performance, in the stresses sustained, in fatigue resistance, and even in oxidation and corrosion (abetted at later stages by coatings developments).

If one but remembers that the first Allied jet flew successfully in 1942 in England, yielding net thrust of about 800 pounds, operating at a blade temperature of about 650° C, for a plane which weighed in for the trial run at 4000 pounds, and now sees routinely jet engines developing over 60,000 pounds thrust, with blade temperatures approaching 950° C, propelling planes at the three-quarters million pound levels, for expected life times that are nudging stationary gas turbines, then the growth in performance is remarkable.

Offshoots of superalloy developments for jet engines have in turn been of benefit in many other applications, including turbines for trains, buses, trucks, for increasingly large land based power applications, in the marine world, progressively in nuclear power systems, and hopefully, one day in the automobile.

It is our purpose to examine these developments, and the developments which parallel them in other classes of materials, including some only now under way.

II. THE SUPERALLOYS

Basically, the superalloys are of three types, and are based on Ni, Co, or Fe or combinations of all three, and containing Cr in all earlier alloys at the 17 to 28 percent level for oxidation and corrosion resistance. Recent, more highly alloyed materials[1] are lower in chromium to permit higher volume content of the hardening phases. Fe-base alloys while of interest for lower strength applications, and where oxidation and corrosion resistance demands are high, have never developed the strength levels achieved by Ni and Co-base alloys above about 650° C.

The major strengthening modes include aging by γ' in nickel base alloys, with compositional variations of $Ni_3(Al,Ti)$, a coherent aging phase for correctly chosen conditions;[1] carbide dispersion as the major feature of most cobalt base alloys;[2] and Laves phases in both cobalt and nickel base materials. Combinations of these strengthening modes are of course possible and sometimes used.

For compositions of some representative alloys, see Table I.[1,2,3]

TABLE I

COMPOSITIONS OF SOME Ni AND Co-BASE SUPERALLOYS
ALL VALUES WEIGHT PERCENT

Alloy	C	Cr	Ni	Co	Mo	W	Ti	Al	Ta	Other
S-816 (w)*	.40	20	20	bal.	4	4	-	-	4Cb	-
Mar M-509(c)	.60	23.5	10	bal.	-	7	0.2	-	3.5	.01B, .5Zr
Nim 80A (w)	.10M	20	bal.	2M	-	-	2.3	1.2	-	.008B, Tr.Zr
Nim 90 (w)	.13M	20	bal	18	-	-	2.4	1.4	-	Tr.B, Zr
Nim 115 (w)	.2M	15	bal.	15	4	-	4	5	-	.02B, .04Zr
U-700(w)	.06	15	bal.	18.5	5	-	3.5	4.5	-	.03B
Astroloy (w)	.06	15	bal.	15	-	4	3.5	4.4	-	.03B, .04Zr
Rene 100 (c)	.18	9.5	bal.	15	3	-	4.2	5.5	1 V	.015B,.06Zr
B-1900(c)	.10	8	bal.	10	6	-	1	6	4.3	.015B,.08Zr
IN-100(c)	.18	10	bal.	15	3	-	5	5.5	1 V	.015B,.05Zr
Mar M-200(c)	.15	9	bal.	10	-	12.5	2	5	1 Cb	.015B, .05Zr

*(c) Cast; (w) Wrought; Tr (trace); M (maximum)

Complex cored castings were developed throughout the sixties, permitting blade cooling. In the meantime, directionally solidified (DS) castings were made successfully, eliminating all but longitudinal grain boundaries; this helped with thermal fatigue and avoided early intergranular cracking. Also in the picture are single crystal DS blades which also offer certain advantages, but perhaps at costs not now attractive for broad use.

The DS developments were with standard alloys; this had the disadvantage that in spite of overall performance improvements and increased safety factors, no important change in temperature capability was achieved. Overaging takes place at the same temperature for both polycrystalline and DS alloys. In gas turbine developments the key feature is increasing operating temperatures for reasons of performance and efficiency; thus while DS materials are very attractive, they are not a substitute for higher temperature capability.

In a sense then, ingot produced alloys reached developmental plateau by the mid-sixties; cast superalloys have now also reached a temperature plateau, at a point below 1000° C.

Typical stress rupture values at 732 to 983° C are given in Table II for representative, leading cobalt-base and nickel-base superalloys, both cast (c) and wrought (w).

TABLE II

Stress-Rupture Properties at Selected Temperature for Wrought and Cast Superalloys

Alloy	Temp.° C	Stress for 100 hours (psi)	Stress for 1000 hours (psi)
IN-100 (c)	732	97	83
	815	73	55
	983	25	15
Nim 115 (w)	732	85	70
	815	58	45
	983	16.5	9.5
U-700 (w)	815	58	43
	983	16	7.5
B-1900 (c)	815	73	55
	983	26	15.5
Mar M-509 (c)	815	39	33
	983	17	13
Mar M-200 (c)	815	76	60
	983	27	19

Examination of Table II and the voluminous data on superalloy properties would indicate that based on superalloy technology, gas turbine blades today meet requirements for long life (>10,000 hours) at stresses above 20,000 psi, at temperatures approaching 950° C, uncooled, with superior performance attained by cast alloys with their coarse grain size. Directionally solidified blades offer safer performance due to the elimination of transverse grain boundaries, and a preferred orientation of coarse grains. The aim through alloy and process development is to reach blade operating temperatures of 1000° C in the near future and 1100° C in the longer term.

Disk materials have gradually evolved from the early 16 Cr - 25 Ni - 6 Mo stainless steels to alloys A-286, to Rene 95, and Waspalloy, and, as will be shown later, to still higher alloy content superalloys based on powder metallurgy techniques. High yield strength at temperatures up to 700° C is desired by the nature of disk performance. High yield strength at room and sub-zero temperatures in excess of 265,000 psi are desired, coupled with excellent ductility and toughness. Maintaining yield values of 150,000 psi above 650° C, with a vanishingly small creep component, is the aim.

III. REFRACTORY ALLOYS

For a while it almost seemed as though developments with the refractory metals would solve all of the gas turbine requirements for (much) higher temperature blade performance[4].

Starting perhaps with chromium (melting point of about 1900° C) and including Nb, Mo, Ta, W (melting points from 2460 to 3400° C), strength at temperatures well above 1100° C was excellent. (See Table III for typical high temperature strength values)[5,6]. The four higher melting metals, when suitably alloyed, developed recrystallization temperatures well in excess of 1200°C, often up to 2000° C or higher for the Ta and W-based alloys. This means that use of such alloys up to 1200 to 1600° C would not result in high temperature modes of creep[7]; the alloys would behave in the low temperature mode, work hardening without recovery or recrystallization under stress and temperature. Accordingly yield strength values are the significant measures of strength. Chromium, on the low end of the melting point scale among these elements, was of interest because of a hope that oxidation resistance could be developed to a practical level for performance to at least 1200° C. The remaining four elements, and their ductile (at high temperatures) alloys never exhibited oxidation resistance potential beyond very short times at all temperatures above about 1000° C, with performance sharply worse as temperature and time increased.

Actually Nb and Mo-base alloys received the most attention because of greater availability, lower cost, and more attractive specific gravity for blade applications. Excellent strength values in creep and stress rupture from 1100° C to 1500° C were obtained for a number of alloys based on each alloy system. Problems were of several types. Molybdenum (and Cr and W) suffered from high values of TTDB fracture (transition temperature for ductile to brittle fracture). Pure Mo could be produced to remain ductile significantly below room temperature; however, alloying (necessary for improvements in both oxidation resistance and strength) invariably increased the TTDB well above room temperature (in excess of 600° C for some alloys) without, unfortunately, improving adequately the oxidation resistance. The same brittle fracture problem, except that it persisted to higher temperatures, held for Cr and W-base alloys.

TABLE III

High Temperature Tensile Strength Values for Refractory Metal Alloys[5,6]

Alloy	Temp. ° C	Tensile Strength, ksi
W- 4 Re - 0.35 Hf - 0.35 C	1270	115
	1650	90
	1925	70
Mo - 0.6 Hf - 0.5 C	980	105
	1220	78
	1650	30

Whereas Nb and Ta were blessed with much better low temperature ductility, significant alloying here, too, raised the TTDB too high for safe use. Alloying also did not solve the oxidation problem. In addition to the above, Nb and Ta, being much more reactive metals than Cr, Mo, and W, readily reacted with C, O, N, and H both in processing and in use, leading to other embrittlement problems.

While a great deal of talent, dollars and effort were devoted to a search for successful coatings for long time use at temperatures above about 1100° C, none of the many coatings which were developed and evaluated turned out to be useful over an important temperature interval. Among the more promising coatings for the Mo-base alloys was $MoSi_2$, which was extensively tested and evaluated, but finally without success.

OTHER ALLOY SYSTEMS

When it was determined in the sixties that refractory metal alloys would not provide a solution for attainment of high temperature performance, it also became evident that for a significant time period beyond about 1970, the workhorses for high temperature applications would continue to be based on nickel and cobalt-base alloys. It is interesting to re-examine how some of these developments originated, what progress they have made to date, and what they offer for the future.

IV. OXIDE DISPERSED ALLOYS (OD ALLOYS)

Starting with the unusual high temperature strength of the first OD alloys[8,9], namely SAP (Sintered Aluminum Powder), it was shown that the $Al\text{-}Al_2O_3$ alloys would maintain interesting engineering strength values to 500° C in contrast to a top limit of about 150° C for the best conventional age-hardened aluminum alloys.

It was demonstrated at that time that the important alloy and structural features which made possible OD alloys were the following[10]:

a) fine oxide particle size: from about 50 to 1000 Å

b) fine oxide interparticle spacing: 0.2 micron or finer

c) an insoluble, non-reactive oxide (for example Al_2O_3

in Al, or ThO_2 in Ni)

d) a high level of stored energy in the alloy, for example, through cold work and solid solution alloying.

The first two of these features are of course true also for all aging systems, which easily meet such requirements. Item (c) is in contrast to aged systems in that the very factors which promote age-hardening (a fine, coherent precipitate), also lead to overaging and even solution with increasing temperature, and therefore loss of strengthening. A fine insoluble, refractory oxide will not overage, and, combined with item (d), will stabilize the cold worked structure by pinning dislocations and dislocation sub-boundaries, preventing total recovery, and, of course, recrystallization. These OD alloys are perhaps best labeled "oxide-stabilized" alloys, since the oxide itself, being unbonded to the matrix, does not directly contribute importantly to the strength[11].

The Al-Al_2O_3 SAP type alloys are a special case in that the Al_2O_3 is naturally formed on the fine aluminum flake powders in increasing amounts as the surface area of the metal is increased through comminution or attrition, yielding a series of alloys with increasing oxide content from a fraction of a percent to 14 to 18 percent by volume[12]. Nevertheless, the transfer of knowledge from the Al-Al_2O_3 system progressed rapidly and smoothly to include alloys of Mg, Cu, Ag, Fe, Ti, Pt, Cb, and others[13].

Numerous methods were developed to obtain the desired oxide content in various metallic bases, both for pure metals and their alloys (see reference 10 for other methods). Briefly (to mention only a few) these methods including the following:

a) mechanical blending of an atomized or comminuted fine powder (1 - 5 microns) with the selected very fine refractory oxide (.01 to 0.1 micron)[14];

b) internal oxidation of a dissolved minor element, the element being both much more reactive with dissolved oxygen than the matrix metal, and capable of forming a stable, refractory, fine oxide, for example, 1 percent

Al in Cu[10,15];

c) selective reduction of oxides in an oxide mixture to yield an alloy base with a finely dispersed refractory oxide: for example, reduction of MoO_2+NiO+ThO_2 to give Ni(Mo)-ThO_2[16];

d) decomposition of salts and salt mixtures or solutions to yield ultimately a fine metal-metal oxide blend, or a metallic coating on very fine refractory oxides, or fine mixtures of metallic and oxide powders[16,17].

The oxide content of early alloys tended to be high (up to 10 v/o) in order to attain a fine oxide interparticle spacing (typically 0.2 to 0.3 microns)[14,16]; this led to strong alloys with fair ductility which were difficult to form into shapes due to a relatively poor oxide dispersion.

Through the sixties the oxide content of newly developed alloys was gradually reduced. This was possible because of acquired ability to produce finer and finer oxides (down to 50 to 500 Å) with more uniform dispersion. As such, the most recent OD alloys generally show less than about 3 v/o oxide[17,18]. With this much lower but better dispersed oxide, it was observed that there was a large improvement in the hot and cold workability of the alloys. One could, for example, achieve up to 90 percent cold reduction without intermediate anneals[19,20]. This results in a combination of a high level of stored energy through cold work combined with a dislocation pinning action by the fine, stable oxides. The stabilized dislocation networks result in excellent properties to temperatures as high as 1200° C.

The OD alloys are generally characterized by relatively low strength values at low and intermediate temperature, but with relatively excellent strength values at the highest temperatures. Further, the OD alloys based on Ni and Co have higher melting temperatures than their superalloy equivalents and are very much more stable structurally above 1000° C. Table IV lists some representative properties of these alloys.

TABLE IV

A. Tensile Properties of OD Alloys: Longitudinal Tests

Alloy	Temp.° C	Y.S.(ksi)	UTS(ksi)	Elong.%	Note
TD-Ni	20	80	95	7	1
Ni-Co-Mo-BeO[18]	20	160	190	6	2
Ni-Cr-Al-Y_2O_3[21]	815	40	43	8	2
Ni-Cr-Al-Y_2O_3	870	48	49	6	3
Ni-Cr-Al-Y_2O_3	1095	16	16.5	5	3

B. Stress for 100-Hour Rupture Life, psi

Alloy	815° C	982° C	1095° C	Note
TD-Ni	26,000	12,000	10,000	1
Ni-CO-Mo-BeO[18]	25,000	10,000	5,500	2

(1) TD-Ni: Ni - 2 v/o ThO_2 - as-cold reduced, 90 percent
(2) Ni-20 Co - 8 Mo - 2 v/o BeO: as-cold worked, with intermediate anneals[18]
(3) Ni - 16 Cr - 5 Al - 1% Y_2O_3: recrystallized - mechanically alloyed material[21]

More recently an aim has been to increase strength values at low and intermediate temperatures without loss of properties at the very highest temperatures. This results unfortunately in a tendency toward lower melting temperatures as the alloy content and alloy complexity of the matrix is increased.

Recent alloy developments are of two types: one is based on highly increased alloy content in the solid solution range to minimize loss of melting temperature while striving for maximum high temperature stability and combined solid solution strengthening plus strengthening through cold work[16,18]; the second is to go to complex two and three phase alloys, including aging type superalloys, to increase in particular the low and intermediate temperature strength values but with an unavoidable decrease in melting temperature and perhaps structural stability at the highest temperatures. In the latter case, the alloys are prepared by a process described as mechanical alloying[21,22], since a mixture of elemental and master alloy powders is blended with the desired oxide and then heavily ball-milled in a protective atmosphere at room temperature. Alloying takes place during progressive attrition of the mixture, combined with build-up and break-down of the alloying ingredients on the balls. An intimate mixture is attained; the oxides are generally well dispersed; and reactive elements such as Cr, Al, Si, etc., can be part of the alloy mix without being converted to undesired non-refractory oxides.

What is the status of OD alloy developments? TD-Ni, TD-NiCr, and DS-Ni have all seen service at rewardingly high temperatures and at stresses of interest to gas turbine designers. Strength properties up to about 800 or 900° C are only fair compared to those of the superalloys, but improve sharply relative to superalloys as temperatures exceed 1000° C and up to 1200° C. At the moment, the OD alloys are best suited for vane applications where temperatures are high (>2000° F) and operating stresses are relatively low (less than 5000 psi). Thermal fatigue behavior is satisfactory. Some of the OD alloys containing Cr and/or Al in significant amounts show excellent oxidation resistance above 1100° C; such alloys are TD-NiCr, Ni-Cr-Al-Y_2O_3, Co-Cr-Al-Y_2O_3, and others. Not unlike superalloys, coatings are also a way of life with these alloys.

Figure 1 demonstrates the unique strengthening enjoyed by OD alloys in contrast to conventionally alloyed materials which suffer coarsening and solution of the strengthening phases at high temperatures. In Figure 1 we examine the relative strengthening of OD alloys versus conventional alloys, both referred to the base pure metal (Ni,Cu,Al,etc.). At a ratio of test temperature (T_T) to the melting temperature (T_m) of about 0.5 to 0.7 the conventional alloys weaken rapidly, approaching an "f" (strengthening) ratio of one at $T_T/T_M = 1$. The OD alloys continue to gain strength, relatively, up to at least 0.9 T_T/T_M. This gain is of course associated with the unusual stability of the OD alloys at these

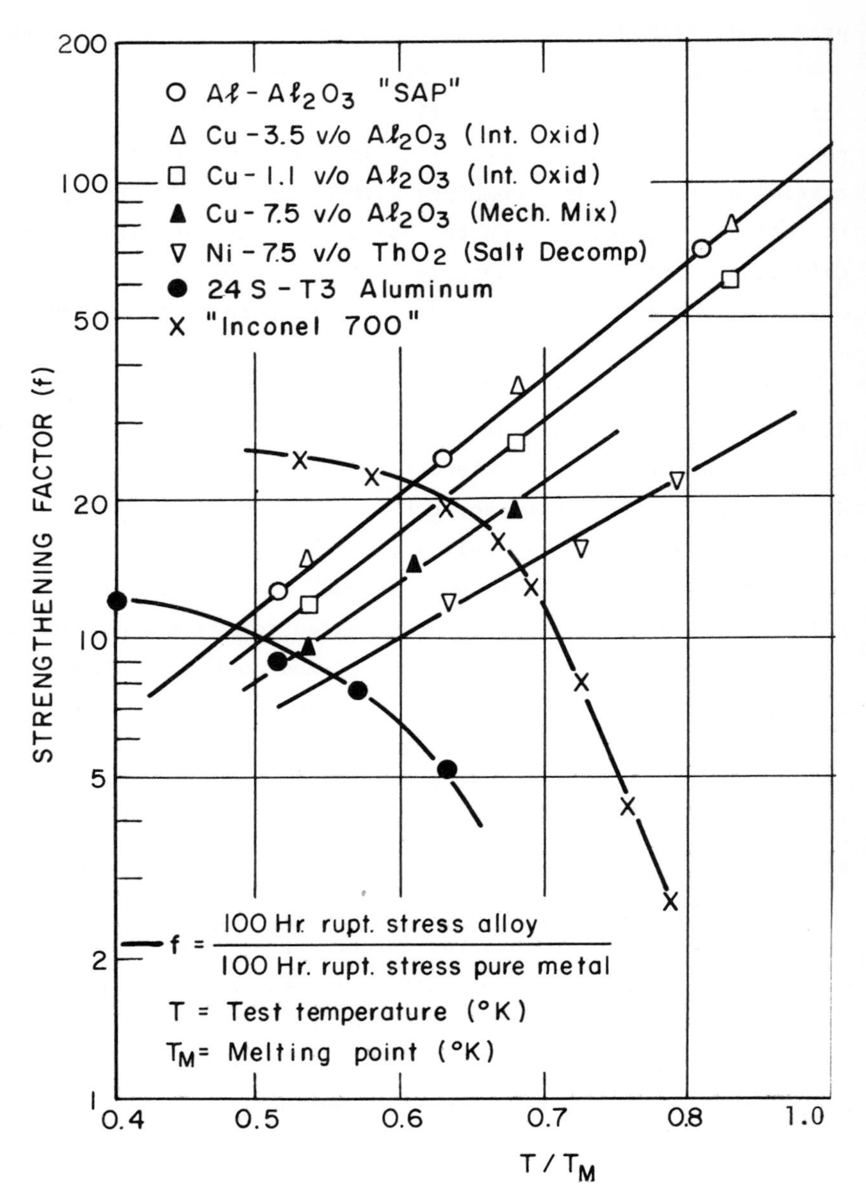

FIG. 1. Relative strengthening as a function of the homolgous temperature for conventional alloys and OD alloys.

very high temperatures. That stability is based on the retention of prior cold work by the structure stabilizing effect of the finely dispersed fine oxides.

Mechanical alloying has introduced more highly alloyed base compositions (Nimonic 80A, for example); while such alloying has increased the low and intermediate temperature strength values significantly, the melting temperatures are lower. The combination of γ' strengthening, coupled with the OD strength contributing modes, could result in specific structural instability during temperature excursions into the solution temperature range for γ', which could coarsen through a series of high temperature cycles.

Perhaps the single most severe criticism of the OD alloys is that the processing methods are slow and time consuming; many of the starting materials are costly, and the price per pound of OD material has been too high to warrant extensive applications of the alloys. Further, if the OD alloys are used in the as-cold worked condition, they suffer from severe texture problems which result in differences of 2 to 1 or greater between longitudinal and transverse tension strength values and stress rupture values. Ductilities are even more severely affected. The more recent recourse to the use of directional recrystallization of the highly cold worked OD alloys produces a coarse grain size which maintains a preferred texture. This grain coarsened structure is quite strong (typically at the highest temperatures) but will have lost the benefits of the fine grain size and stored energy of cold work at low and intermediate temperatures (see Table IV).

While there are still some on-going disagreements regarding the specific mechanism(s) of strengthening of OD alloys, the more critical studies should be focused on processing variables to accomplish the following three goals.

1. A much cheaper alloy, attainable by eliminating one or more processing steps now used for OD alloys.

2. A highly alloyed material whose structure will remain stable for long periods of time at temperatures well above 1000°C.

Obviously, single phase, solid solution strengthened alloys would be of interest. Achievement of oxidation resistance through such alloying would also be of merit.

3. An alloy capable of being processed without the recrystallization step to coarsen the grain size. For sheet materials this is a difficult practice; for all shapes and forms production of the desired texture and grain structure is difficult. The low strength values at the lower temperatures, which are a result of recrystallization, limit some of the applications of OD alloys.

V. *FIBER REINFORCED ALLOYS*

Success at the lower temperature levels with fiber reinforced structures has been outstanding. Fiberglas- epoxy structures are seeing extensive and highly successful use in military and civil aviation, in sporting goods, in household items, marine applications, etc. Temperatures are limited for now to several hundred degrees, primarily because of temperature limitations of the polymer or epoxy base material. The fiberglas is an excellent inexpensive material ideally suited for fiber reinforced structures.[23]

Boron and graphite (carbon) fibers are now common fiber reinforcing materials, with a significant preference for the graphite material because of a significant price advantage per pound. In other respects, the two fibers are reasonably similar in performance. As the price has decreased, usage has gone up proportionately. These fibers also are enjoying successful applications in resin-based systems[24], but with the use limitation again associated with the low temperature stability of the polymer.

Slowly improving technology and price considerations have made boron and carbon fiber reinforcement of aluminum and magnesium alloys practical and useful applications with a net advance in operating temperatures above 400° C. Military aircraft have been testing these materials advantageously in increasing amounts as pertinent technology permits. With fiber loading volumes of 30% tensile strengths for 6061 Al-Thornel 50 bars were 560 M Pa at 20° C. At temperatures above about 150° C to 500° C, these materials are superior to Ti - 6Al - 4V and Rene 41[25,26].

Efforts to move to still higher temperatures with fiber reinforced alloys have been relatively unsuccessful to date. Both boron and carbon react far too readily with iron, nickel, cobalt, and titanium; the reactions between matrix and fiber can start as low as 600° C in longer time tests, and are rapid at 800° C and above[27]. Coatings and barriers have not been particularly successful.

There are, of course, other strong, high stiffness fiber materials which are being examined. Fine cold drawn molybdenum and tungsten alloy wires look attractive to temperatures beyond those permissible with boron and carbon, but again there are problems of fiber-matrix reactivity above about 800 or 900° C.

One potential application is that of reinforcing titanium alloys with molybdenum (alloy TZM) wire. Since titanium alloys are generally not used in applications above about 550° C due to the high reaction rate of titanium with the atmosphere, TZM wires remain inert in Ti - 6Al - 4V alloys[28]. For 0.2 percent plastic creep in 100 hours at 540° C a 26 v/o TZM reinforcement of Ti - 6 - 4 increased the creep stress to 45,000 psi from about 10,000 psi for the unreinforced product.

Negative features with respect to the use of Mo and other refractory metal wires are cost, high specific gravity, poor oxidation resistance, and reactivity with many of the metal base matrices at high temperatures.

For a significant number of years, major efforts have been made to develop non-reactive or negligibly reactive fibers which can be used to reinforce Ni and Co-base alloys to achieve high strength levels at 1000° C and higher. The two most promising materials are silicon carbide and aluminum oxide (α Al_2O_3, the high temperature structure).

Figure 2 shows the tensile strength values of representative fiber materials with increasing temperature. The four types of

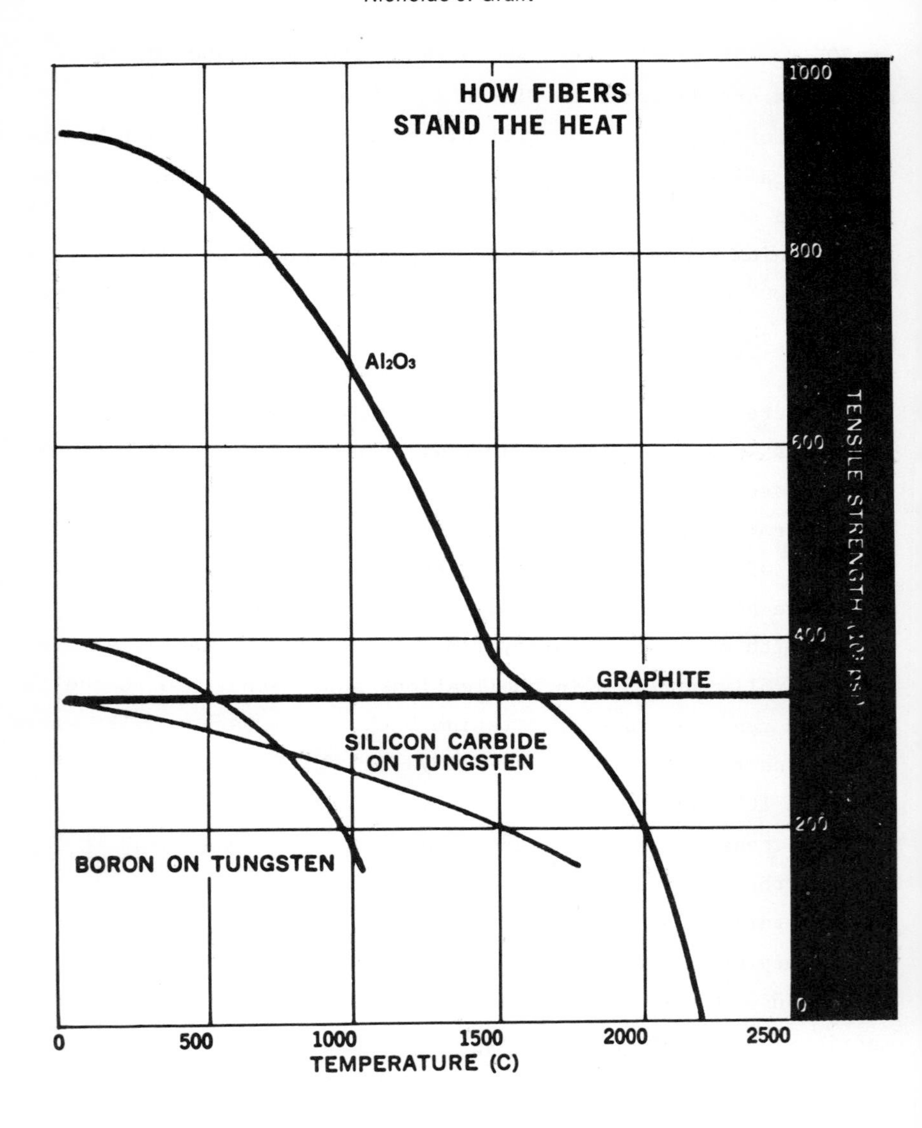

FIG. 2. Tension strength versus temperature for four refractory fibers.

fibers are our strongest and best known materials[29]. Whereas silicon carbide (bare fiber) shows excellent strength values to 1500C those of Al_2O_3 are quite remarkable. In spite of a rapid decrease of strength from 20 to 1500° C, the strength at 1500° C is still an amazing 380,000 psi. At those high temperatures Al_2O_3 begins to show measurable plasticity in slow creep. Graphite of course would be ideal from considerations of strength over an amazingly large temperature range were it not for poor oxidation resistance and high reactivity with metals above about 600° C.

The silicon carbide is much less reactive than all materials previously studied except Al_2O_3; however, producing long lengths of silicon carbide is expensive by currently used techniques. The least reactive of all the promising fibers is Al_2O_3. Growth of Al_2O_3 short fibers (less than about 1 inch) has been reasonably successful but costs are prohibitively high. Production of long filaments and fibers has been singularly unsuccessful if one takes into account the cost of the produced fibers.

Fiber reinforced materials have their own set of problems. Properties are highly directional. For boron reinforced aluminum, tested in stress rupture at 300° C, correct fiber alignment gave a stress of 70,000 psi for 100 hours life; 5% misalignment dropped that value to 36,000 psi; and 10% misalignment dropped that value further to 16,000 psi.

Whereas the fibers are strong in pure tension, they are quite weak in shear. Various fiber weaves and patterns are possible and are used, but for a particular volume content of fibers in the reinforced body, the high strengths of correctly aligned fiber reinforced bodies is not approached by cross woven structures.

Non-destructive examination of fiber reinforced products is difficult, slow, expensive, and hardly fool-proof. Delamination between fibers and matrix is a repetitive problem.

For the time being, it is perhaps fair to state that this class of material is not yet a contender for temperature applications above about 800° C. At lower temperatures near 500° C, the probability of producing strong, light weight, rigid bodies is excellent; cost is a factor, however, which calls for ever cheaper fibers. For applications below about 200° C, the fiber reinforced non-metallic systems are an outstanding success.

VI. DIRECTIONALLY SOLIDIFIED COMPOSITES (EUTECTIC COMPOSITES)

Selection of simple eutectic or near eutectic compositions, or quasi-eutectic systems, provides, in a directionally solidifying system, at selected narrow growth rates, the frontal solidification of two phases in parallel to produce a variety of interesting composite structural phase arrangements[30]. The two phases may be fine rods embedded in the second phase, or composed of parallel lamallae of the two phases (see Figure 3),or complex combinations of structural styles of either or both phases. Long rods or unfaulted lamallae are desired if stress distribution in

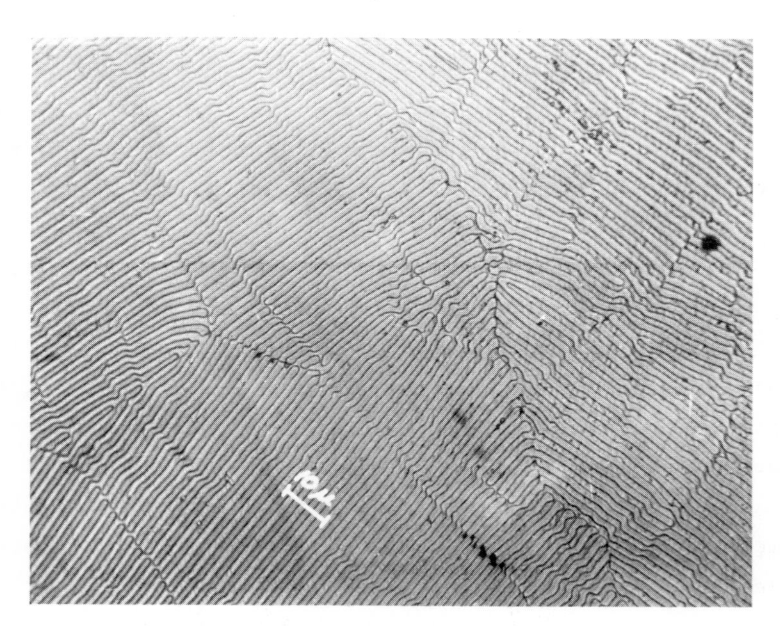

FIG. 3. Transverse section of unidirectionally solidifed eutectic alloy Al-$CuAl_2$. X750.

the structure is to be uniform[31] and continuous; cellular growth is avoided. The combination of the two phases may be ductile - ductile, ductile-brittle, or brittle-brittle. Combinations may be metallic, intermetallic, ceramic, etc.

What makes the DSC alloys of interest is an indicated potential for about an 100° C increase in operating temperatures at high stress levels[32]. Figure 4 demonstrates that potential, utilizing the Ni_3Al - Ni_3Cb DSC alloy as a comparison against the strongest wrought and cast nickel and cobalt-base superalloys[33]. As of this date, only a relatively small number of alloy systems has been studied in any detail[34]. Among the more interesting combinations are the ductile metallic systems based on Ni, Ni-Cr, and Co-Cr, reinforced by, for example, Ni_3Cb or TaC in rod form; the oxidation resistance is fair to poor. Among the lamallaer structures the $\gamma/\gamma'-\ \delta$ structure of Ni_3Al-Ni_3Cb shows particularly high strength values at high temperatures. Figure 5 shows tensile data for representative Ni-base superalloys, for a simple DSC ductile matrix alloy (Ni - Ni_3Nb) and for the more brittle Ni_3Al-Ni_3Nb[35]. The temperature advantage of the latter alloy is readily seen. Problems here in a sense are not too different than those associated with the fiber reinforced materials. Properties are highly directional, oxidation resistance will require much attention and improvement, joining will be very difficult, NDT procedures are not now available. Growth of the aligned phases through tapered sections, through blade roots and platforms, etc., must be worked out for each particular blade design.

An advantage as of now is the ability to utilize precision casting technology for growth of the DSC structures with real success for many of the compositions developed to date.

It is clear that this is a new, developing materials technology. Inadequate alloy studies have been made, testing for broad classes of properties (impact, fatigue, shear properties, etc.) will be necessary, joining and NDT developments must be undertaken. The potential however for an increase of 100° C in

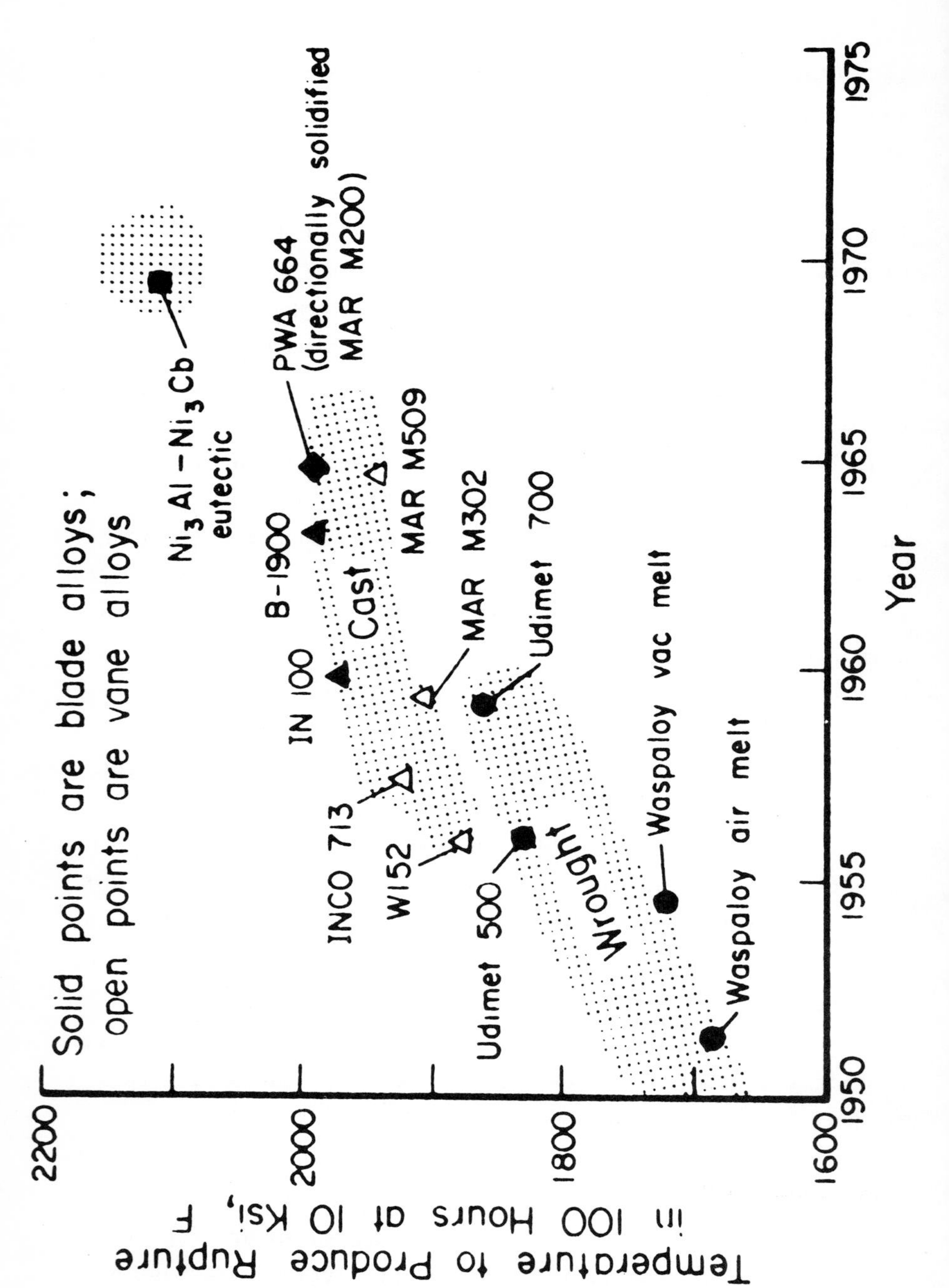

FIG. 4. Improvements in temperature capability with time for various superalloys versus the Ni_3Al-Ni_3Cb DSC alloy.

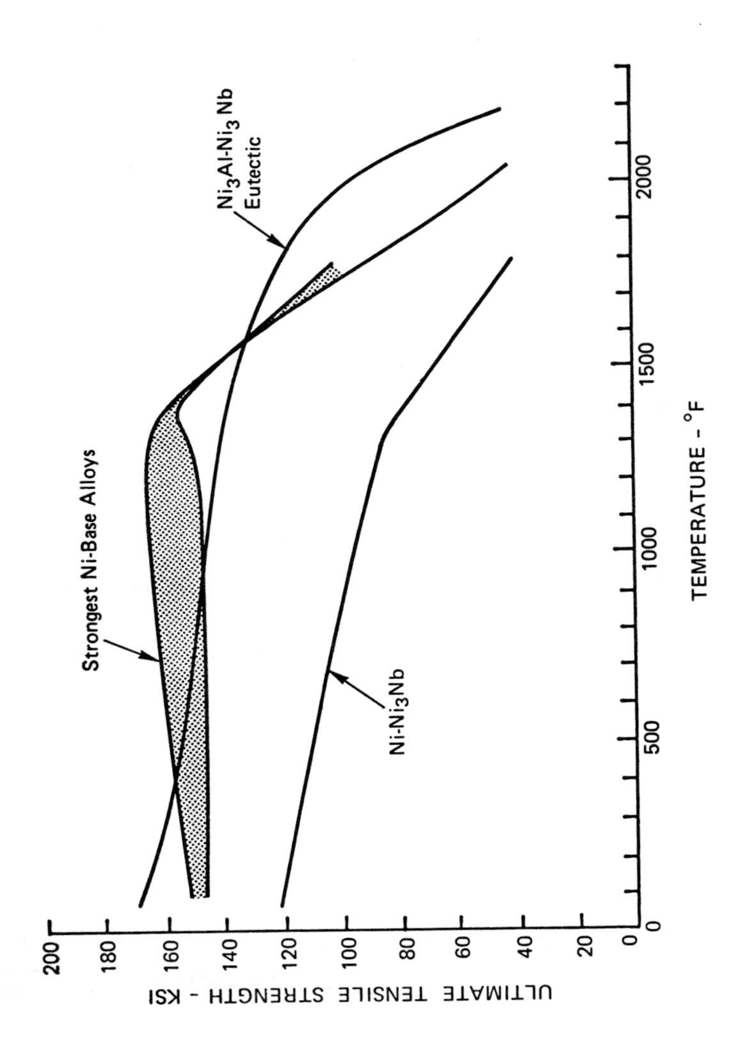

FIG. 5. Temperature dependence of the longitudinal tensile strength of two DSC alloys versus a representative Ni-base superalloy.

operating temperature is a tremendous incentive to stimulate the additional studies.

VII. RAPID QUENCHING FROM THE MELT - POWDER METALLURGY AND SPLAT COOLING

The latest of alloy development programs and the last to be covered in this chapter has to do with the potential offered through controlled rapid solidification of metals from the melt.

Many, if not most, of our problems with metallic materials arise from coarse, heavily segregated structures (grain size, excess phase particles, inclusions, etc.) due to slow solidification of the homogeneous melt. Many of the undesired phases which exist in ingot products or castings are due purely to segregation which produces chance compositions in the last remaining melt at dendrite and grain boundaries[36]. As was indicated in the superalloy section, in wrought high alloy compositions the limitation on further alloy development was the inability to hot work the coarse grained, segregated ingot product. With precision castings, due to the unusually slow cooling during solidification, the formation of very coarse structures in high alloy materials led to strong but brittle structures with very poor toughness characteristics. The problem was to overcome these limitations.

Figure 6 shows an experimentally determined plot of dendrite arm spacing (DAS), the primary measure of micro and macroscopic segregation, versus the cooling rate through the solidification range for aluminum and its alloys[37,38]. The cross-hatched areas merely show that ingot size, powder size, foil thickness, metal pouring temperature, etc., all have an effect on the final DAS, thus the spreads and overlaps which are shown.

Typically, ingots and precision castings solidify at rates of 10^{-3} to 10° C sec^{-1}. Conventional gas atomization to produce standard -100 mesh powders (-149 microns) results in average quench rates of about 10^{2} ^{o}C sec^{-1} with significant refinement of DAS (or ultimate grain size). Water or steam atomization

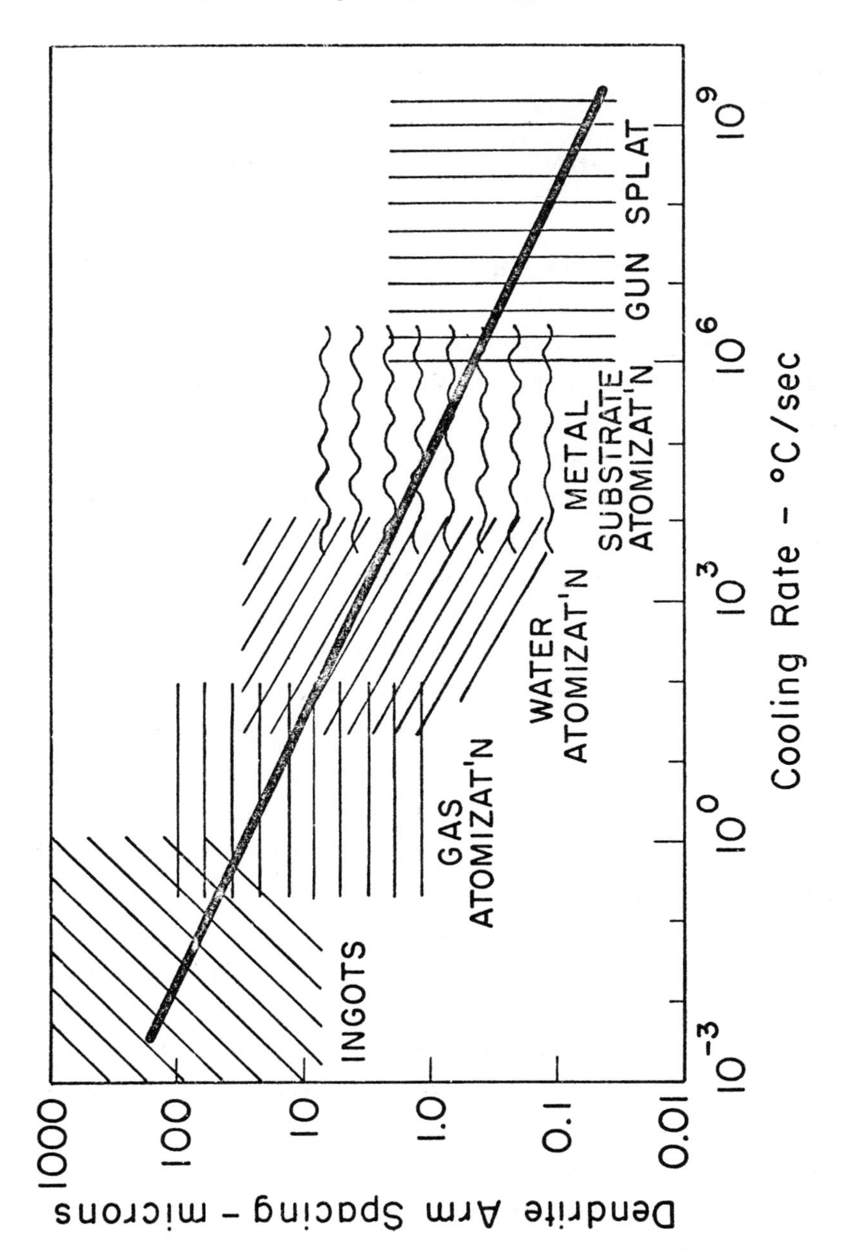

FIG. 6. Dendrite arm spacing as a function of cooling rate for aluminum and aluminum alloys. Cross hatched areas indicate variations due to size and shape of particulates and other processing variables.

yields 10^3 °C sec^{-1}; splatting against a high conductivity metallic substrate (average foil thickness of 50 to 100 microns)produces quench rates of 10^4 to 10^6 °C sec^{-1}; splatting similarly but to produce very thin flakes between 0.1 and 10 microns is shown to produce quench rates of 10^7 to 10^9 °C sec^{-1}.

Throughout this quench rate range, dendrite arm spacing (and subsequent grain size) decreases for a given metal from perhaps 100 to 1000 microns to 0.01 micron; for selected compositions, quench rates greater than about 10^5 or 10^6 °C sec^{-1} can and often do, in fact, produce glassy metals. Particle sizes of excess phases decrease into fractional micron sizes, or the involved alloying elements are retained in highly supersaturated solid solutions[39]. Segregation phases, almost always undesired, usually do not form. In a mixture of carbides normally found in high speed tool steels, or in cobalt base superalloys, high quench rates and a balanced stoichiometry will result in a totally different carbide mix, with the thermodynamically favored carbides forming in preference to those carbides which formed from mass action considerations associated with segregation.

Perhaps for the first time the materials engineer or scientist has almost total control over the structure and phase types in his alloys.

Some notable findings are perhaps worth listing merely to indicate some of the exciting and intriguing developments which have been observed and reported.

1. Glassy alloys based on Fe and Fe-Ni, with high concentrations (20-25 atom percent) of P, C, Si, B, etc., 80Pd-20Si, 60 Cu-40 Zr, and other compositions show fracture strength values between about 300,000 and 550,000 psi[40,41,42]. Ductilities in tension are nil to extremely small.

2. In 2024-T6 aluminum, quenching from the melt at rates of about 10^4 - 10^5 °C sec^{-1} results in a fine grained, wrought (extruded) product with 20% increases in yield strength and ultimate tensile strength, with no loss of ductility; a 20% increase

in stress for a fatigue life of 10^6 cycles compared to the ingot bar product is achieved; improved hot creep properties are observed; and elimination of the brittle segregation phase (AlCuFeMn) takes place[43].

3. In modified 7075 aluminum alloys, similar high quench rates of a composition containing about 1% each of Fe and Ni resulted in very rapid aging at 20° C, (versus 120° C for ingot material). Yield strength values of about 95,000 psi (versus 76,000), ultimate tensile strength of 104,000 psi (versus 86,000) with ductilities in the 10% range were attained[44]. Hot and cold workability of the extruded powder product were excellent whereas an ingot product containing such high levels of Fe and Ni would have been impossible to roll or forge.

4. IN-100, one of the strongest Ni-base superalloys, a casting alloy, when hot extruded from atomized powders produced at quench rates of only 10^{2} ^{o}C sec^{-1}, resulted in grain sizes from 4 to 15 microns. The alloy is superplastic when strained at rates of 10^{-3} to 10^{-5} sec^{-1} at temperatures of 1100 to 1150° C. Ductility values (in air) ranged up to 1200 percent[45].

This fine grained alloy developed yield strength values in the as-extruded condition of 175,000 psi, ultimate tensile strength of 242,000 psi, and room temperature ductility of 20% elongation (see Table V). Note the improvements over cast IN-100 in Table V.

5. Cast Mar M-509, one of the best cobalt base superalloys, containing about 0.6% carbon, almost doubled its yield and ultimate strenth values compared to the cast material and increased its elongation 5X to 14% when prepared by hot extrusion from atomized powders quenched at 10^{2} ^{o} C sec^{-1}. At 1000° C, this alloy showed 180% elongation in contrast to 7% for the cast alloy. In fact, increasing the carbon content to 1.0% still resulted in a ductile alloy at all temperatures from 20 to 1100° C, with improved strength levels at 20° C.

TABLE V

Room Temperature Tensile Properties of IN-100 and Mar-509: As Cast and Extruded From Atomized Powders

Alloy	Condition	0.2% Y.S. ksi	UTS ksi	Elong. %	R. A. %
IN-100	Cast (1500 μm g.s.)	136	143	4	8
IN-100	As hot extr'd (8 μm g.s.)	175	242	20	16
Mar M-509	Cast (4000 μm g.s.)	80	115	2.5	3
Mar M-509	As hot extr'd (4 μm g.s.)	130	192	14	12

Cold working of both P/M IN-100 and P/M Mar M-509 is readily accomplished and is effective in further increasing yield strength values by an additional 50 percent.

These fine grained superalloys are of course quite weak at temperatures above about 800° C and are far inferior to the same composition in the cast, coarse grained state. From cryogenic temperatures to about 700° C, however, the fine grained P/M rapidly quenched Co and Ni base superalloys are very much stronger, making ideal, strong, tough gas turbine disc materials.

The superplastic capability of these alloys in the very fine grained condition permits relatively easy pressing (slow strain rates) of discs to near final shape. Combining this feature with the high levels of strength available up to about 700° C is permitting production of the finest disc materials ever available for gas turbines.

Beyond this current state of affairs, one foresees almost unlimited potential for alloy development studies, and for combinations of as yet untried strengthening modes and phases. Higher and more complex alloying is clearly possible. The highly improved plasticity of these fine grained alloys will encourage new forming and shaping technologies. The resultant very fine grained alloys should have superb low and intermediate temperature properties. There still exists the potential to grain coarsen or directionally recrystallize the fine grained, clean structures in order to recover strength at the high temperatures.

VIII. SUMMARY

In spite of periodic roadblocks to further development of high temperature alloys, new approaches and technologies continue to emerge, which make possible renewed advances in strength, temperatures capability, and operating efficiency. Our problem is to avoid the pessimism that frequently seems to emerge as we wring out the last of useful properties from on-going alloy systems by conventional processing techniques. It is critical to keep a number of developments going at all times. One of those discussed in this chapter (or one only just now developing), will probably emerge a winner. We are not today at a point where through scientific reasoning or computer logic we can select or predict the best developments ten years in advance of their ultimate success. This means, for now, a need for continued research and development.

REFERENCES

1. The Superalloys, Eds. C. T. Sims and W. C. Hazel, J. Wiley and Sons, N. Y. (1972).

2. Cobalt-Base Superalloys - 1970: Cobalt Monograph Series, Eds. C. P. Sullivan, M. J. Donachie, Jr., and F. R. Morral, Centre d'Information du Cobalt, Brussels (1970).

3. R. F. Decker and C. T. Sims, The Superalloys, Eds. C. T. Sims and W. C. Hazel, J. Wiley and Sons, New York (1972).

4. R. I. Jaffee, C. T. Sims, and J. J. Harwood, Third Plansee Proceedings , Pergamon Press, 380 (1959).

5. W. D. Klopp and W. R. Witzke, NASA TN D-5348 (1969).

6. P. L. Raffo, NASA TN D-5025 (1969).

7. N. J. Grant and A. G. Bucklin, Trans. ASM 42, 720 (1950).

8. R. Irmann, Metallurgia, 46, 125 (1952).

9. E. Gregory and N. J. Grant, Trans. AIME, 200,

10. N. J. Grant and O. Preston, Trans. AIME, J. Mets. 9,349 (1957).

11. L. L. J. Chin and N. J. Grant, Powder Metallurgy, 10,344(1967).

12. J. P. Lyle, Metal Progress, 62, 109 (1952).

13. N. J. Grant, The Strengthening of Metals, Ed. D. Peckner, Reinhold Publishing Corp., New York, (1964).

14. K. Zwilsky and N. J. Grant, Ultrafine Particles, Ed. W. E. Kuhn, J. Wiley and Sons, New York, (1963).

15. L. Bonis and N. J. Grant, Trans. AIME Met. Soc., 224, 308 (1962).

16. J. G. Rasmussen and N. J. Grant, Powder Metallurgy, 8, 92 (1965).

17. F. J. Anders, G. B. Alexander, and W. S. Wartel, Metal Progress, 82, 88 (1962).

18. M. S. Grewal, S. A. Sastri and N. J. Grant, Met. Trans. A, 6A, 1393 (1975).

19. V. A. Tracey and D. K. Worn, Powder Metallurgy, 10,34 (1962).

20. B. A. Wilcox and A. H. Clauer, Trans. Met. Soc. AIME, 236, 570 (1966).

21. J. Benjamin and T. Volin, Met. Trans. 5, 1929 (1974).

22. J. Benjamin, Met. Trans. 1, 2943 (1970).

23. Composite Materials: Testing and Design, ASTM:STP-460, ASTM, Philadelphia, Pennsylvania (1969).

24. New High Strength Graphites, Stackpole Carbon Company.

25. H. Shimizu and J. F. Dolowy, Jr., Composite Materials, ASTM:STP 460, ASTM, Philadelphia, Pennsylvania (1969).

26. J. F. Dolowy, Jr., Refractory Composites Working Group Comm.

Anaheim, California, February (1969).

27. D. W. Petrasek and R. A. Signorelli, Composite Materials, ASTM: STP 460, ASTM, Philadelphia, Pennsylvania (1969).

28. E. A. Snajdr and J. F. Williford, Tech. Report AFML-TR-68-252, (1968).

29. M. Nicholas, Report AERE-R5681, Harwell, U. K., (1968).

30. G. A. Chadwick, Proc. Conf. on In-Situ Composites, National Materials Advisory Board Report NMAB-308-I, Jan. (1973).

31. H. Bibring, Proc. Conf. on In-Situ Composites, National Materials Advisory Board Report NMAB-3080II, Jan. (1973).

32. R. L. Ashbrook, NASA Tech. Memo, NASA TMS-71514, April (1974).

33. E. R. Thompson and F. D. Lemkey, Met. Trans. 1, 2799 (1970).

34. E. R. Thompson, D. A. Koss, and J. C. Chestnut, Met. Trans. 1, 2807 (1970).

35. E. R. Thompson, J. Comp. Materials, 5, 235 (1971).

36. N. J. Grant, High Temperature Materials Phenomena (Proc. Third Nordic High Temperature Symposium), Ed. J. Rasmussen, Polyteknisk Forlag, Denmark (1973).

37. H. Matyja, B. C. Giessen, and N. J. Grant, Jour. Inst. Metals. 96, 30 (1968).

38. N. J. Grant, Fizika, 2, Suppl. 2, 16.1 (1970).

39. M. Itagaki, B. C. Giessen, and N. J. Grant, Trans. ASM, 61, 330 (1968).

40. C. A. Pampillo and H. S. Chen, Mat. Sci. Eng..13, 181 (1974).

41. R. Maddin and T. Masumot, Acta Met, 22, 493 (1972).

42. L. A. Davis, Scripta Met, 9, 431 (1975).

43. M. Lebo and N. J. Grant, Met. Trans. 5, 1547 (1974).

44. J. P. Durand, R. M. Pelloux, and N. J. Grant, Proc. 2nd Int'l Con. on Rapidly Quenched Metals, Sec. II: Mat. Sci. and Eng. 23, No. 2 - 3, May, June (1976).

45. L. Moskowitz, R. Pelloux, N. Grant, AIME Met Soc. -IMD Hi Temp. Alloy Comm. Proc. 2nd Int'l Conf. Superalloys, (1973).

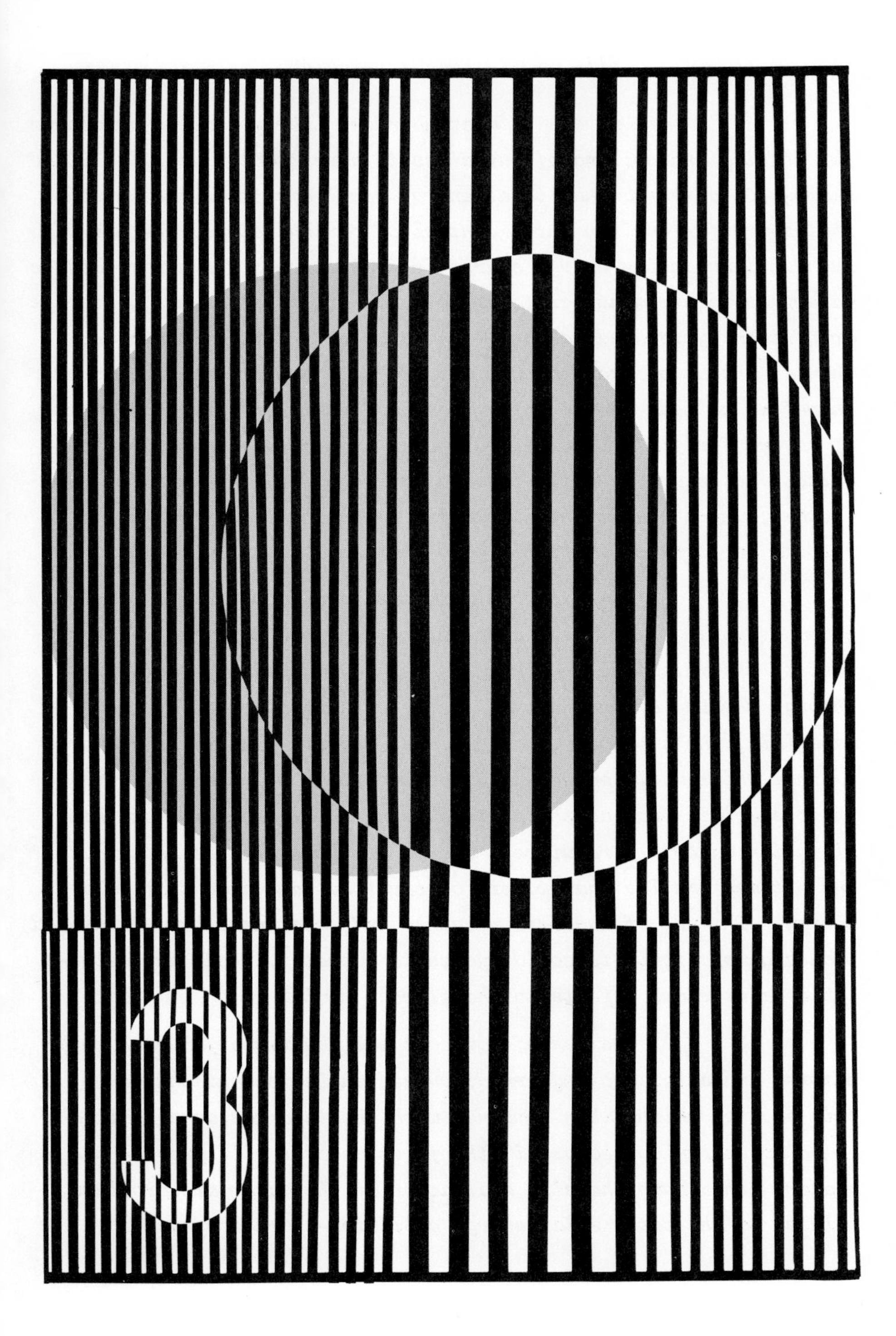
3

III. SOLAR ENERGY MATERIALS

One of the most abundant energy sources available for heating and cooling of buildings is direct solar energy. Since nearly 30% of our annual total energy requirements in the United States are for environmental control of our buildings, the use of such an inexpensive energy source is imperative. The nature of the materials requirement for solar collectors and solar energy storage are primarily related to their cost and their durability rather than to the development of new materials. Compromises must therefore be made between, for example, optimizing the transparancy of covers for flat plate collectors, the encapsulants for photovoltaic cells or the reflectance of mirrors for concentrating collectors, which are important factors effecting the efficiency of systems performance and their long term durability and cost.

Similar considerations are imposed on the solar storage system where the economics are dictated in terms of the cost of the materials and the floor space requisite for storage. Through the use of the heat of fusion of several phase change materials, the storage volume has been greatly diminished compared to that which would be required for materials which only provided their specific heat as the storage means. For example, most heat of fusion materials requires a volume 17 times smaller than rocks and 8 times smaller than water to store the same amount of solar energy. The use of nucleation agents and thickeners have permitted thousands of liquid-to-solid cycles to operate without any detectable deleterious effects to the storage media.

The use of low cost salt-hydrates for both solar heating and solar cooling are discussed in the second chapter of this section which, when combined with the information contained on solar energy collectors in the first chapter will provide the reader with a comprehensive evaluation of the materials problems related to the total solar energy collector-storage system.

Chapter 13

FUNDAMENTAL MATERIALS CONSIDERATIONS FOR SOLAR COLLECTORS

Gene A. Zerlaut

Desert Sunshine Exposure Tests, Inc.
Phoenix, Arizona 85020

I. INTRODUCTION

The inherent operational requirements of solar devices and facilities dictate that they be deployed, largely, in the very environments most degrading and hostile to their components. Therefore, the intelligent selection of functional materials for solar utilization devices must take into account their reliability and long-term durability as well as their relevant optical and thermal properties. Lifetime predictions of such devices are critical because of the trade-offs that must be made to establish optimum operating cycles. Losses in optically- and thermally-functional properties -- such as transparancy of covers for flat plate collectors and encapsulants for photovoltaic arrays, and reflectance of mirrors for concentrating collectors and heliostats for central receiver thermal power stations -- can be translated directly into decreases in efficiency and ultimately into increases in system performance costs.

Because of these considerations, this paper will examine the thermal/optical requirements for solar utilization devices and will relate these requirements to the relevant properties of candidate materials, to the durability of these properties under extended use conditions, and to the synergistic interaction of materials components during long-term environmental aging as a system.

II. GENERALIZED SYSTEMS DESCRIPTION

A. *Flat Plate Collectors*

Perhaps the best means to describe a flat plate solar collector is to first define its governing equations. The thermodynamic equations describing flat plate collectors have been discussed thoroughly by Hottel and Whillier (1), Hottel and Woertz (2), Bliss (3), Whillier (4) and Gupta and Garg (5), among others. Their analyses may be summarized by the following relationships:

$$\eta = \frac{\dot{m}C_p(T_o - T_i)}{q_i A} = \overline{\alpha\tau} - U_L \frac{(\overline{T}_p - T_a)}{q_i} \tag{1}$$

where η, the efficiency, is determined from the mass flow $\dot{m}$ of heat transfer fluid (usually water or air), the heat capacity C_p of the fluid, the inlet and outlet temperature differential $(T_o - T_i)$, the product of global (direct plus diffuse) insolation q_i and the aperture area A. The collector can then be represented by a governing plot of efficiency η versus the plate parameter $(\overline{T}_p - T_a)/q_i$, where $\overline{T}_p$ is the average plate temperature and T_a is the ambient temperature for the same data interval. This plot fits the general slope-intercept equation of a straight line y = mx + b where b, the ordinate intercept, is $\overline{\alpha\tau}$ (the time- and goniometrically-averaged product of the solar absorptance α of the plate and the transmittance τ of each of the existing covers), and where m, the slope, is U_L, the overall collector heat loss.

Because of the difficulty in determining an accurate average plate temperature, and because of the additional diagnostic information furnished, efficiency is usually plotted also as a function of two other ambient temperature differentials -- the average fluid temperature $\overline{T}_f = (T_i + T_o)/2$ and, more importantly, the inlet temperature T_i:

$$\eta = F' \left[\overline{\alpha\tau} - U_L \frac{(\overline{T}_f - T_a)}{q_i} \right], \text{ and} \tag{2}$$

$$\eta = F_R \left[\overline{\alpha\tau} - U_L \frac{(T_i - T_a)}{q_i} \right] . \tag{3}$$

The term F' in Equation (2) is the *plate efficiency factor* and is the ratio of the actual useful energy collected to the useful energy that would have been collected if the entire receiver plate were at the average fluid temperature. Similarly, F_R is the *heat removal efficiency factor* and is the ratio of the useful energy collected to the useful energy that would have been collected if the entire plate were at the entering fluid (air or water) temperature. Both F' and F_R are independent of the solar intensity, the collector operating temperature $\overline{T}_f$ and $\overline{T}_p$, and the ambient temperature T_a. They depend to a first order extent upon the plate design and, to the extent that the heat losses U_L affect the heat collected by the transfer fluid, these efficiency factors are sensitive to the package design as well, that is, to U_L.

The extent to which the performance of a flat plate collector is governed by thermal/optical properties of its components is obvious from analysis of Equations (2) and (3) -- especially when we consider the role played by the thermal emittance characteristics of the solar receiver (plate) and covers (glazings). Solar absorptance may be defined by

$$\alpha = \frac{\int_{300\text{ nm}}^{3000\text{ nm}} (1 - R_\lambda)_\lambda \, W_{s,\lambda} \, d\lambda}{\int_{300\text{ nm}}^{3000\text{ nm}} W_{s,\lambda} \, d\lambda} \qquad (4)$$

where $W_{s,\lambda}$ is the air mass 2 (or air mass 1) terrestrial solar radiation intensity at wavelength λ between λ and $\lambda + d\lambda$. The value $(1 - R_\lambda)_\lambda \, W_{s,\lambda} \, d\lambda$ is the total energy absorbed by the surface in the wavelength interval $d\lambda$ about λ, where R_λ is the directional-hemispherical spectral reflectance of the surface in the wavelength region 300 to 3000 nm. (The denominator integrates to the solar constant when air mass zero solar radiation data are employed.) The solar transmittance of cover plates is computed in a similar manner. The denominator is the same as in Equation (4) and the numerator becomes $\int \tau_\lambda W_{s,\lambda}\, d\lambda$, where τ_λ is the directional-hemispherical spectral transmittance in the solar wavelength region 300 to 3000 nm.

The total collector heat loss coefficient U_L is composed of the following plate losses -- U_{up}, the upward heat loss through the transparent covers; U_r, the downward (rearward) losses through the insulation and enclosure; and U_e, the edge, or sideward, heat losses through the insulation and enclosure sides. Generally, the upward heat losses comprise approximately 85% of the total losses U_L and the thermal emittance (infrared) characteristics of the plate and the individual transparent covers are the determining factors in U_L -- especially at higher plate temperatures. Indeed, for a single transparent cover, it can be shown from an extension of Kirchhoff's law (that treats of the relation between the emittance and absorptance of surfaces) that

$$U_{up} = \frac{1}{\frac{1}{\varepsilon_p} + \frac{1}{\varepsilon_c} - 1}\, \sigma (T_p^4 - T_c^4) \tag{5}$$

where σ is the Stefan-Boltzmann constant and ε_p and ε_c are the infrared emittance of the plate and cover, respectively, at their corresponding absolute temperatures T_p and T_c. The total emittance, measured either radiometrically, calorimetrically or spectrally, is defined by the relation

$$\varepsilon_T = \frac{1}{W_{b,T}} \int_0^\infty \varepsilon_{T,\lambda}\, W_{b,T,\lambda}\, d\lambda \tag{6}$$

where $W_{b,T,\lambda}d\lambda$ is the energy emitted by a perfect backbody per unit of area at an absolute temperature T in the wavelength interval $d\lambda$, and $\varepsilon_{T,\lambda}$ is the spectral emittance of the surface in question.

The optical and thermal properties of the plate (α, ε_p) and the transparent covers (τ, ε_c), all of which are sensitive to long duration exposure to solar ultraviolet radiation, that is, to the weathering process, are thus seen to play major roles in the ultimate performance of flat plate solar collectors. It is these properties and the durability of these properties that are emphasized here. Although insulation is not generally exposed to solar irradiation, or the weather, changes in insulation efficiency as a result of the various aging processes that can occur will be discussed in terms of collector diagnostics. The remaining factor in collector design, that of the receiver plate design (a problem in fluid heat transfer), will not be covered.

B. *Concentrating Solar Collectors*

It is widely recognized that the Hottel-Whillier relationships represented by Equations (1), (2), and (3) are essentially valid for defining the governing equations of concentrating solar collectors. The expression for a concentrating collector is

$$\frac{\dot{m}C_p(T_o - T_i)}{q_{ni}A_a} = \eta = F'\left[\alpha\tau\beta - \frac{U_L}{k}\frac{(\overline{T}_f - T_a)}{q_{ni}}\right] \tag{7}$$

where A_a is the collector aperture area, q_{ni} is the direct component of solar radiation (at normal incidence to the plane of the aperture), β is a cosine factor that is the weighted average of the angles of incidence of the concentrated rays (also called the filter factor), and k is the concentration factor which is the ratio of the aperture area A_a to the area of the irradiated portion of the solar receiver. All of the reflectance properties of the concentrating mirrors are embodied to a first order extent in the concentration factor k and to a lesser extent in the angular distribution function β.

III. ASSESSMENT OF THE DURABILITY OF MATERIALS FOR SOLAR DEVICES

A. *General Considerations*

Outdoor weathering is a degradation-promoting process that is both qualitatively and quantitatively dependent upon the constituent factors that are peculiar to a particular environment. It is a process that is exceedingly costly to industry and the consumer alike. The objective of outdoor exposure *testing* is the determination of the weathering performance of materials for any one or combination of reasons: to provide statistical data regarding the prediction of the influence of weathering on materials' properties; as a quality-control technique; or as a research tool in the development of new and/or improved materials. The essential characteristics of accurate outdoor testing are:

(1) That the test environment matches as nearly as possible that of the anticipated end use, or, *more importantly,*

that it create in the material the same effects as the anticipated end use; and

(2) That the diagnostic tests, usually of relevant functional properties, be selected to accurately assess the degradative effects that most importantly affect the choice of materials for utilization in specific environments.

B. *The Need For Accelerated Weathering Tests*

Because of the need to ensure 5- to 20-year lifetimes of solar devices now being constructed, acceptable accelerated durability testing techniques, such as DSET's EMMA(QUA) accelerated weathering machines, must be employed to predict long-term exposure tests in as short a time as possible. The essential requirement of accelerated weathering tests is that the response to accelerated testing be independent of the solar flux; i.e., that it depend only on the total dose (fluence) of sunlight (UV) deposited. When this requirement is met for solar ultraviolet irradiation, reciprocity is said to hold for that material.

C. *Response to Accelerated Weathering*

Correlation factors need to be determined for most materials when exposed in accelerated weathering devices -- whether laboratory weathering cabinets or DSET's EMMA(QUA) machines that utilize natural sunlight as source. EMMA(QUA) outdoor accelerated weathering machines are essentially Fresnel concentrators; EMMA(QUA) testing is the subject of a recent paper by Zerlaut (6). It is generally recommended that control specimens be tested by conventional, direct-exposure testing both in Florida and Arizona, or in similar climates, whenever possible. The important question in accelerated testing, which is equally germane in EMMA(QUA) testing, is "Does the test induce the same defects both qualitatively and quantitatively that would occur under direct exposure after essentially an equal deposition of sunlight (e.g., total watts, or langleys)?" In this respect, the validity of EMMA(QUA) testing is demonstrated by the curves presented in Figure 1. These graphs represent the optical property-retention correlations between EMMA(QUA) exposure for 16, 34, and 55 *days* and Florida exposure for 6, 12, and 18 *months,* respectively. These exposures correspond to increments of 80,000 langleys. Data represent the averages of specific results obtained on 47 different coating systems. It will be observed that the EMMA(QUA) results fall exactly between the 5° South Florida *washed* and *unwashed* results for each of the three exposure periods.

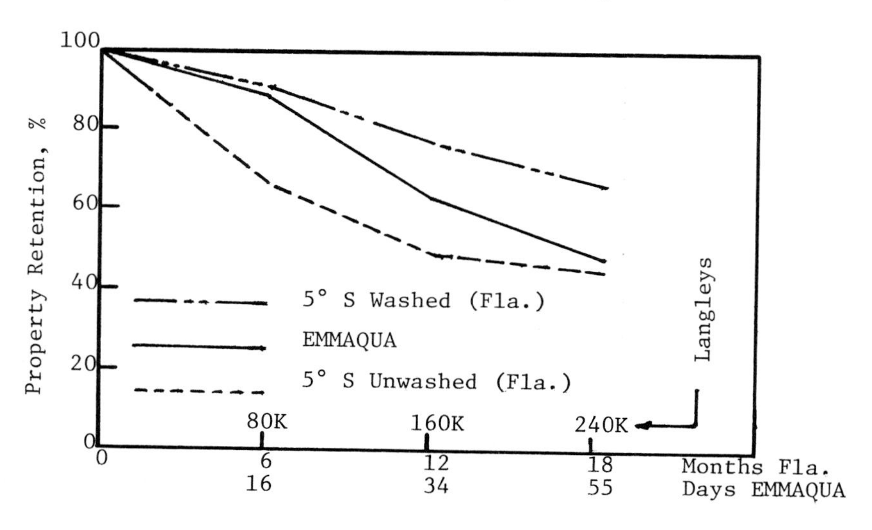

FIG. 1. Correlation of "Reflectance" Between EMMAQUA and Florida.

The EMMAQUA results during the first 80,000-langley exposure, corresponding to about six months in Florida at 5° South (spring-summer exposure), are more representative of 5° South *washed* results than an average between *washed* and *unwashed* results. This is attributed to initial dirt build-up on both washed and unwashed samples (which provides a screening action), and to photochemical induction. The latter, which, in reality, is the time required for optical defects to become statistically observable, depends on the initial production of defects and the secondary photophysical reactions that tend either to enhance or interfere with the primary processes. Induction is usually observed in direct real-time exposure testing as well as in accelerated exposure testing -- whether performed in the laboratory or out-of-doors.

The significance of these data presented in Figure 1 is that exceptionally relevant correlation can be achieved between accelerated and real-time (Florida) exposure and the fact that only 1/10th the time is required to achieve the same degree of weathering.

IV. TRANSPARANT COVERS

A. General Considerations

The purpose of utilizing one or more transparent covers is to minimize convective and radiative heat losses from the plate to the environment, the object being to perform this function with minimum attenuation of the incident solar radiation.

The external spectral transmittance of transparent materials is defined by

$$\tau_{\lambda} = [1 - \alpha - \rho_i - \rho_s] \tag{8}$$

where α is the spectral absorptance, ρ_i is the Fresnel (first surface) reflection and ρ_s is the back reflection, or scattering, by internal and surface disparities, including dirt.

The spectral absorptance of transparent media depends upon the fundamental chemical structure of the materials employed. Fresnel reflection losses, which occur at both surfaces, are a function of the complex refractive index of the material. In addition to surface disparities, back reflection (scattering) is a result of material inclusions, dispersed insoluble phases, microcrystallinity, etc.

Also of considerable importance is the thermal emittance of the covers since, from Kirchhoff's Law (see Equation (5)),

$$[\varepsilon_{\lambda} = \alpha_{\lambda}]_T \tag{9}$$

and the infrared absorptance, α_{λ} in Equation (9), with λ defined by the absolute temperature of the plate, determines the extent to which heat is lost through the covers (in an absence of antireflection coatings, etc.). The infrared absorptance of materials in the 5 to 25-micron wavelength region is a function of the chemical structure of the material and is due to fundamental vibrational, stretching and rotational frequencies of the chemical bonds present. In general, glass has a very low transmittance to long wavelength infrared (high α_λ) whereas plastics have transmittances in the 0.4 range.

B. *Monolithic Glass*

The advantages of window glass compared to plastic film are: (1) glass is self-supporting, (2) glass generally has high long-wavelength infrared absorptance (high emittance), making it opaque to thermal radiation from the plate, and (3) monolithic glass exhibits exceptional resistance to long duration weathering processes. Glass exhibits two major disadvantages -- it possesses poor impact resistance and is easily broken, and glass has a high refractive index and, therefore, exhibits high Fresnel reflection losses.

Soda-lime-silica glass (70% SlO_2, 15% Na_2O, 10% CaO) is the most appropriate for solar collectors from the standpoint of cost. Soda/lime glass, which is conventional window glass, is manufactured from one of three different processes: plate, sheet or float. Float glass will soon be the only type of glass available as U.S. manufacturers convert to this process, which produces a superior glass in terms of finish and clarity. Float is supported in the semimolten state on molten tin and produces a finish that does not require polishing. However, its solar transmittance will still depend upon the iron content and Fresnel reflection losses.

Glass compositions are available that are optically superior to soda-line glass:

- Silica glass and quartz, 99.5% SiO_2
- Borosilicate glass, 75% SiO_2, 22% B_2O_3, 3% Al_2O_3, and
- Aluminosilicate glass, SiO_2, Al_2O_3, CaO and B_2O_3.

These three types of glasses are generally more difficult to work with than soda-lime glass and are considerably more expensive.

Though more expensive, tempered soda-lime glass permits utilization of comparatively thinner covers to achieve the same strength.

Figure 2 is an equal-energy plot of the transmittance of three soda-lime glasses (7). Superimposed on this plot is the transmittance of 1/4" common plate glass. The water-white Crystal #76 glass is reported to have an iron content of 0.01%, whereas the sheet-lime glass has 0.055% iron and the float glass had a high content of between 10 and 13%. The iron content of the 1/4" common plate glass is unknown, but presumed to be quite high. As will be observed from Figure 2, the "common" presence of iron as Fe_2O_3 in glass, with an absorption band at 1.1 microns wavelength, reduces solar transmittance in proportion to its concentration. While the properties of glass vary from producer to producer, float glass generally contains more iron than plate. Some relevant properties of a number of commercial glasses are presented in Table 1.

The effect of angle of incidence on the transmission of Crown window glass, with an index of refraction of 1.538 and an absorptance of 0.02, is adapted from Edlin (8) and along with curves for two fluoropolymers, is presented in Figure 3. The data are in general agreement with the data published by Hottel and Whillier (1) for the effect of angle of incidence on the product $\alpha\tau$, with a solar absorptance of 0.98 at normal incidence. The data from Edlin are experimental (ASTM E-424) and do not show the fall-off in transmittance at 45° incidence indicated

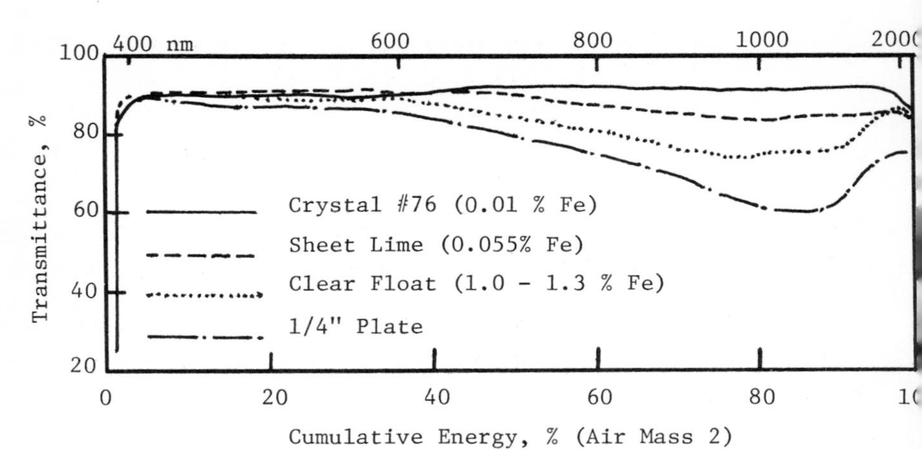

FIG. 2. Equal Energy Plot of Glass Cover Plates (Adapted from Reference 7).

TABLE 1

Relevent Properties of Glasses for Solar Collectors

Mechanical	LOF Float	LOF Sheet	ASG Float	ASG Sheet	PPG Float	PPG Sheet
Density (SpG)	2.48	2.49	-	2.49	2.50	2.49
Tensile Modulus ($PSI \cdot 10^6$)	-	10	-	10	10	10
Impact Strength	-	8–16	-	-	6	6.5
Thermal						
Conductivity ($BTU/hr \cdot ft^2$ in °F)	-	6.9	-	-	6.5	6.5
Expansion Coef. ($In/in \cdot °F \cdot 10^{-6}$)	-	5.0	-	-	4.8	4.7
U Factor ($BTU/hr \cdot ft^2$ °F)	1.1	1.1	1.13	1.09	1.0	1.0
Optical						
Refractive Index, n_o	1.515	1.520	-	1.520	1.518	1.516
Solar Transmittance, τ	0.86	0.87	0.85	0.90	0.85	0.85
Solar Reflectance, ρ	0.08	0.08	-	-	0.08	0.08

by theoretical calculation of the Fresnel losses -- in this respect, the data in Figure 3 is in agreement with the general experience reported informally by others.

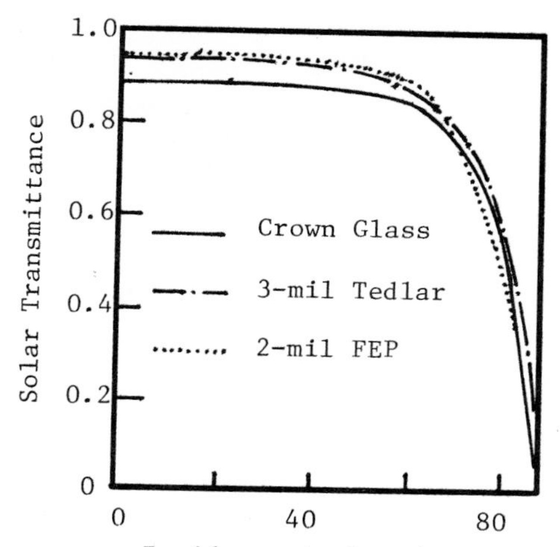

FIG. 3. Solar Transmittance of Glazings as Function of Angle (Adapted from Reference 8).

Although it is acknowledged that monolithic glass does not undergo solar ultraviolet exposure-related deterioration, the importance of dirt accumulation on the thermal performance of solar collectors cannot be overemphasized -- in contrast to the minimal effect reported by Hottel and Woertz (2) and by Whillier (4). Figure 4 is adapted from Garg (9) and shows the effect of up to 30 days dirt accumulation on the transmittance of glass plates inclined at various angles from the horizontal.

C. *Etched Glass*

The etching and surface roughening of glass is utilized to increase the external transmittance of solar energy by decreasing the Fresnel losses at incident angles less than 90°. This is done by either glass/glass abrasion or, preferrably, by fluoride etching. Glass that has been etched or finely abraded can have solar tranmittance τ values of 0.96 to 0.97, but at a higher cost.

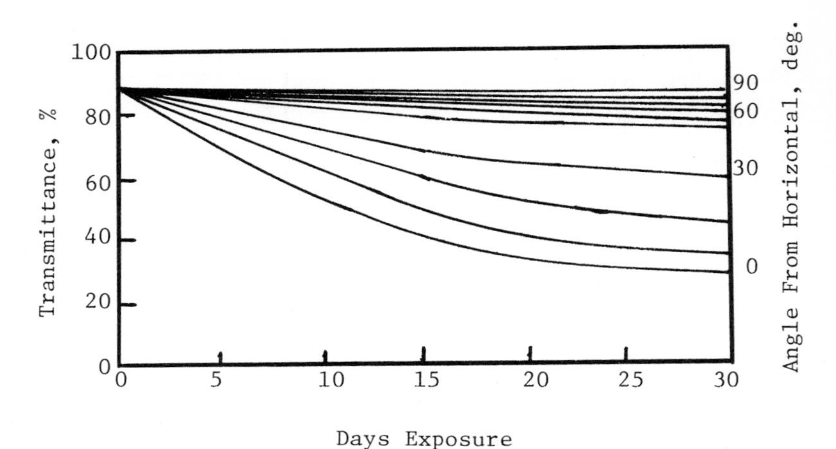

FIG. 4. Effect of Dirt on Transmittance of Glass as Function of Tilt and Exposure (Adapted from Reference 9).

D. Coated Glass

"Solar" antireflection, and long wavelength infrared-reflecting (heat mirror) coatings, are applied to collectors. Antireflection coatings effectively change the refractive index of the outermost surface in contact with air, thereby reducing the Fresnel losses. Long wave infrared-reflecting coatings possess low thermal emittance and are generally multilayer vapor- or chemically-deposited metal oxide films designed to back reflect the plate-emitted infrared radiation.

Antireflection Coatings. Although it is not a thin-film coating in the sense of those described in the following paragraphs, the glass-etch process described by Ramsey, Borzoni and Holland (10) is reported by them to reduce the two-surface reflection from 8% to 2% in the solar spectrum. The 6% increase in the energy delivered to the plate is significant. The process involves immersion of the glass in a hydrofluosilicic acid bath saturated with SiO_2, which solubilizes all glass components except the silica. The skeleton of porous silica remaining has an effective refractive index between that of glass and air; this porous layer of SiO_2 is the antireflection coating.

The objective of depositing antireflecting films on glass and other materials is to provide an optical coupling between air and the glass, thereby reducing the reflection losses. The coupling can be achieved by both single and multiple layers (stacks) of thin films. Although MgF_2 antireflecting films are widely utilized in focusing optics, and have been for many years, they have only recently been utilized on flat plate solar collector covers.

Wolter (11) and Schneider (12) derive the equations for predicting the optimum index of refraction n_{AR} and the optimum thickness t_{AR} of antireflecting coatings based on the refractive index n_o of the glass, or the medium through which the light transmission is to be increased:

$$n_{AR} = \left[n_o \frac{V^2}{(u - n_o)} + U \right]^{1/2} \tag{10}$$

$$t_{AR} = \frac{\lambda}{2\pi n_{AR}} \tan^{-1} \left[\frac{n_{AR}(U - n_o)}{n_o V} \right] \tag{11}$$

where U and V are the real and imaginary parts of the complex index of refraction of the substrate, respectively, and λ is the wavelength at which minimum reflection is to occur. Hsieh and Coldeway (13) have evaluated the use of MgF_2-coated soda-lime sheet glass specifically for flat plate collector covers. They conclude that although theoretical calculations, as above, suggest that the thin film coating should have a refractive index of 1.523 to optimumly reflect 500-nm wavelength sunlight (the peak solar wavelength), no durable coatings of this refractive index could be found. Figure 5 is a plot of surface reflectance at 70° angle of incidence for uncoated and MgF_2-coated glass. Winegarner (14) claims that because of the breadth of the solar spectrum, a single layer of MgF_2 is superior to multilayer, stacked, antireflection films in suppression of surface reflectance of solar collector covers.

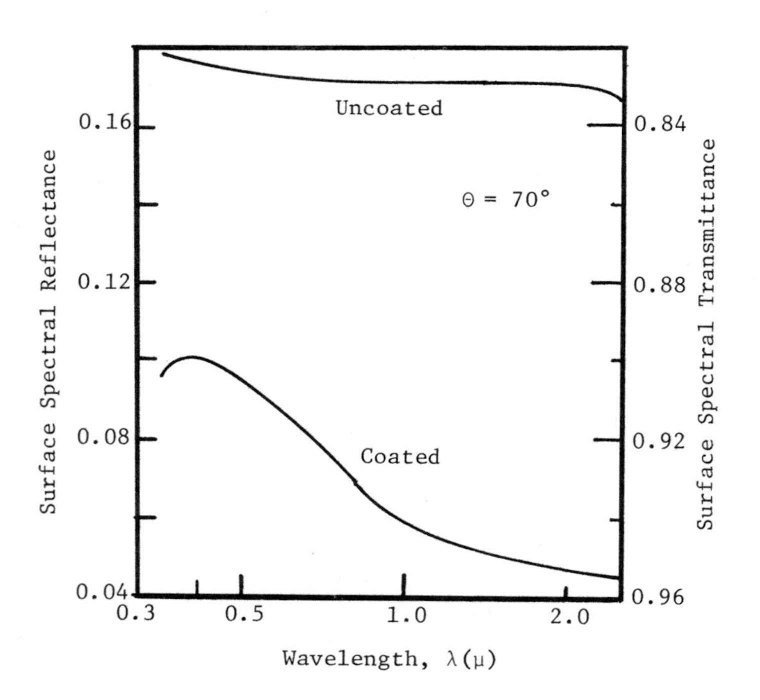

FIG. 5. Optical Properties of Uncoated and MgF_2-coated Glass at 70° Incidence (Adapted from Reference 13).

Solar Transparent, Low Emittance Coatings. Infrared-reflecting heat mirrors offer the advantage of providing cooler glazings, with the result that their durability to long-term operating conditions is significantly increased. While heat mirrors can be effective in increasing collector efficiency, less than optimum coatings (composition- and application-wise) may seriously impair the solar transmittance by exhibiting short wave reflection edges in the visible or near infrared region. Effective heat mirror will have the effect, however, of making convective losses the dominant mode as a result of creating a greater temperature differential between cover and plate.

Fan, Reed and Goodenough (15) give an excellent discussion of heat mirrors for solar energy collection. They discuss the coatings in terms of the concept of Drude mirrors, which are conductive films whose infrared reflectance is a result solely of free (conduction) electrons. They are useful for collector covers insofar as they exhibit a sharp low-wavelength edge that is a function of the mobility and concentration of the charge carriers. They claim that ReO_3 and Na_xWO_3 are the most infrared reflective but exhibit reflection down into the solar region. They also report success with n-type semiconductors such as tin-doped indium oxide (In_2O_3:Sn) and antimony-doped tin oxide (SnO_2:Sb).

Goodman and Menke (16) also discuss the efficiency improvement in flat plate collectors utilizing low-emittance, heat-reflecting cover plates. Their results are presented in Figure 6, which shows the 0.3 to 15 μ wavelength reflectance of SnO_2-coated 0.1" thick glass. They show optimum infrared reflectance with tin oxide coatings of 32 and 41 ohm/sq SnO_2.

Taylor and Viskanta (17) have studied both indium and tin oxides and conclude that doped SnO_2 can be applied to glass with a reduction of solar transmittance to only 75% while affecting an infrared reflectance of 80%. They also present an analysis of Sheklien's studies in the Soviet Union on fluorine-doped SnO_2. These authors show that the use of a selective receiver (absorber) coating with a heat mirror coating is

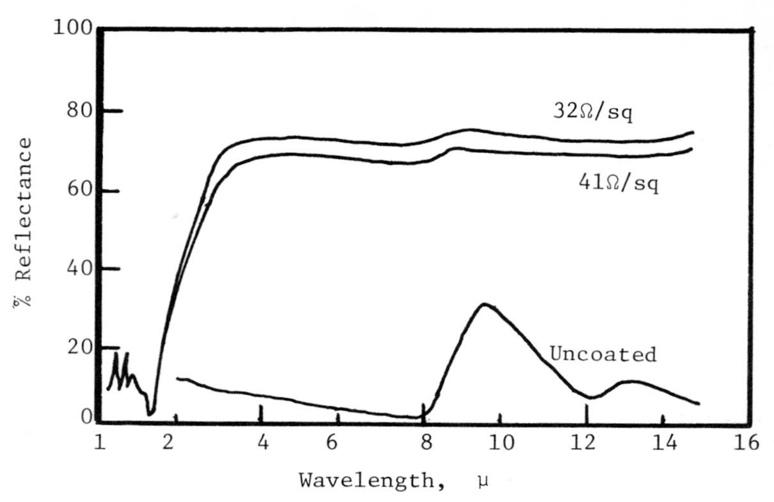

FIG. 6. Effect of Tin Oxide on Reflectance of Glass (Adapted from Reference 16).

ineffective when employed together: Not only is the combination economically unfeasible, but, thermodynamically, the coupling results in higher plate temperatures creating the potential for a spectral mismatch between plate and receiver. Perhaps more important is the fact that a selective absorber often results in a significantly lower solar absorptance and a less-significant decrease in radiative losses.

Weingarner (14) indicates that an indium-type heat mirror provides a 10% increase in collector efficiency when employed on the 3rd surface of a two-cover collector, but that only a 5% improvement is realized when the heat mirror is on the 4th surface.

Most of the candidates for low emittance heat mirror coating have high infrared dispersion; that is, the imaginary part of their complex refractive index increases significantly with wavelength and their infrared reflectance concomittanly increases with increasing wavelength.

Durability of Thin-Film Coated Glass. A number of antireflection- and heat mirror-coated glass specimens have been and are being outdoor exposure tested at the author's company. Unfortunately, the results of these tests (being performed on conventional south-facing racks and on the EMMAQUA accelerated weathering machines) are proprietary to the requestors. However, the experience with a thin-film coated aluminum foil mirror may be instructive. The author (18) irradiated an Al_2O_3-sputtered aluminized plastic material for about 400 and 1000 equivalent sun-hours of extraterrestrial ultraviolet in vacuum employing a 5000-watt mercury-xenon burner. The detailed spectra are presented in Figure 7 and the loss in transmittance is believed to be due to ultraviolet-induced bound-state defects in the thin aluminum oxide film. Irradiation produced an ever increasing displacement of the interference-peak maxima to shorter and shorter wavelengths. Admission of oxygen to the chamber not only bleached the damage in terms of increasing the average reflectance (which prompted the thesis that the damage is due to bound-state defects), but resulted in a displacement of the peak heights back toward longer wavelengths.

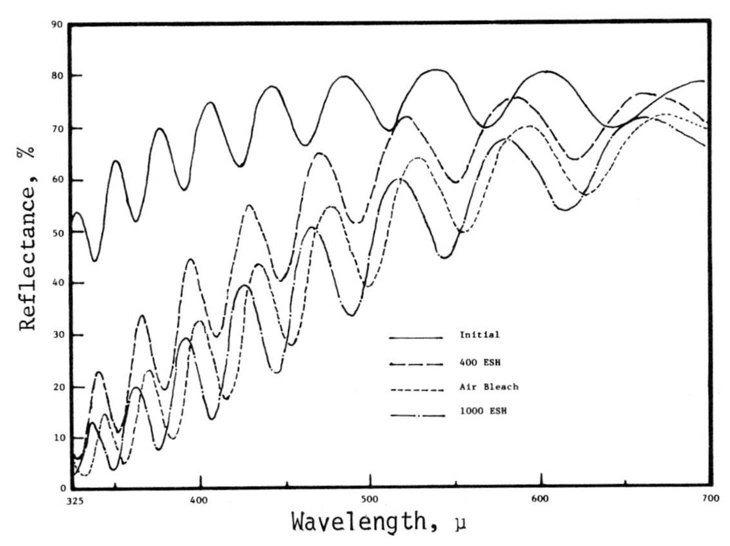

FIG. 7. Effect of Solar Radiation on Thin Film Coatings of Al_2O_3 on Aluminum (Adapted from Reference 18).

The displacement in interference peaks to shorter wavelengths is due to refractive index changes wherein the imaginary (extinction) portion of the complex refractive index is increased due to the ultraviolet-induced color centers. Admission of oxygen simply bleaches out the centers more closely associated with the surface.

It must be noted that these studies were made in vacuum where the annealing effect of oxygen is precluded. Ultraviolet irradiation in the presence of oxygen diminishes the extent of optical damage by either replacement of oxygen-depleted lattice surface sites, or by diffusion and reaction with excited species in the bulk material. However, the fundamental hole-electron pair reactions will still occur in all candidate materials having low, but finite, extinction coefficients for ultraviolet. The question is -- will the photoreactions exhibit sufficiently greater rates than the oxygen-replacement and back reactions to be statistically important (that is, observable) in time frames important to the desired durability of solar device cover materials. Although it is generally believed that MgF_2-coated glass exhibits excellent durability and resistance to weathering, the literature on exterior durability of other thin-film candidates is essentially non-existent.

E. *Polymetric Covers*

Polymeric covers may be divided into two types -- self-supporting plastic sheet and flexible polymeric films (that are essentially not self-supporting). However, from the standpoint of their photothermal and photochemical characteristics, they may be treated alike; it is only when considering their mechanical properties, and the stability of these properties that they should be treated separately.

Polymeric materials that have been employed, or as a minimum considered, for collector covers include the following:

Sheet, self-supporting

- Polycarbonate (PCO)
- Polymethylmethacrylate (PMMA)

Flexible films

- Polyethylene (PE)
- Polypropylene (PP)
- Polyvinyl chloride (PVC)
- Polyvinyl fluoride (PVF), "Tedlar"
- Polyethylene terephthalate (PET), "Mylar"
- Polytetrafluoroethylene (TFE), "Teflon"
- Polyfluoroethylenepropylene (FEP), "Teflon"

Supported/Reinforced film

- Polymethylsiloxane (silicone)
- Glass-reinforced polyester, "Kalwall" and "Filon"
- Plastic/plastic laminates, "Tedlar/Filon"

Solar Transmittance Properties. Polymeric/plastic materials can be optically distinguished from soda-lime glass in three respects:

(1) Plastics possess lower refractive index and therefore exhibit lower reflection losses -- that is, they have a higher resulting solar transmittance (see Table 2).

(2) Plastic films may be employed at thicknesses that significantly reduce fundamental extinction (absorption) at solar wavelengths, thereby also increasing solar transmittance.

TABLE 2

Relevent Properties of Plastics for Solar Collectors

Mechanical	PMMA	Teflon FEP	PVF	Tedlar 20	Mylar PET	Polycarb PCO
Density (SpG)	1.19	2.15	1.40	1.38	1.39	1.2
Tensile Strength ($PSIX10^3$)	10.5	3	7-18	16	24	8-9.5
Elongations, %	-	300	115-250	135	100	85-130
Thermal						
Conductivity (BTU/hr·ft^2 in. °F)	1.3	1.35	-	-	1.05	-
Expansion Coef. (In/In·°$FX10^{-6}$)	41	60	-	24	15	-
Optical						
Refractive Index,	1.49	1.34	-	1.45	1.64	1.59
Solar Transmit.	0.90	0.98	0.94	0.92	0.85	0.88
Solar Reflectance	-	0.01	0.04	0.04	0.04	-

(3) Plastic films/sheet are significantly more transparent in the 7- to 11-μ wavelength region permitting greater upward heat losses from the absorber plate.

Solar-region transmittance curves, adapted from Reference 10 and 20, are presented in Figure 8 for Tedlar* polyvinyl fluoride (PVF), FEP Teflon* (polyfluoroethylenepropylene), polyethylene (PE), and Mylar* polyethyleneterephthalate (PET). The UV-opaque "Tedlar" contains ultraviolet absorber to increase the weatherability of the material. The use of ultraviolet absorber, which provides a high extinction for solar ultraviolet, is very important when "Tedlar" is utilized as the outer "laminate" coating designed to protect a less UV-stable substrate plastic. Figure 9 is an equal energy plot (7) of the solar

---

*Registered TM of E.I. duPont.

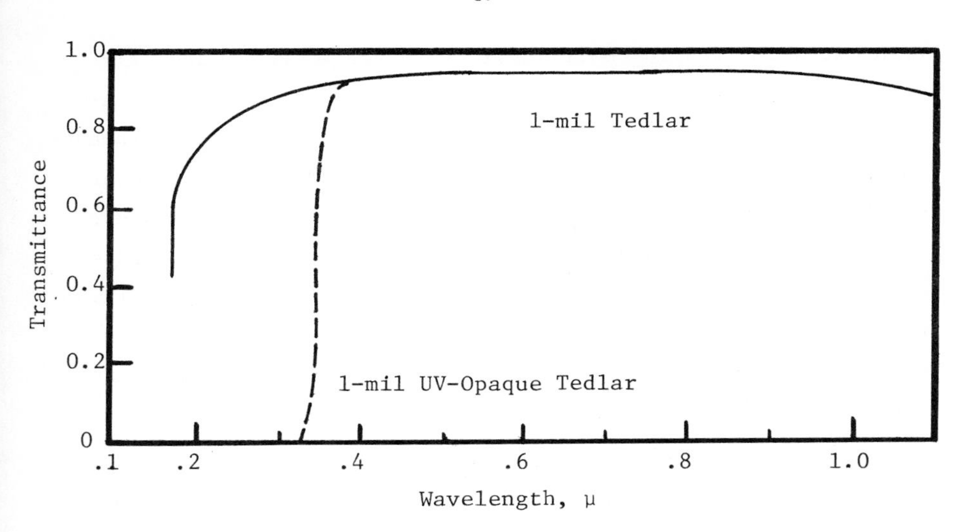

FIG. 8. Transmission Spectra of Tedlar (Adapted from Reference 10).

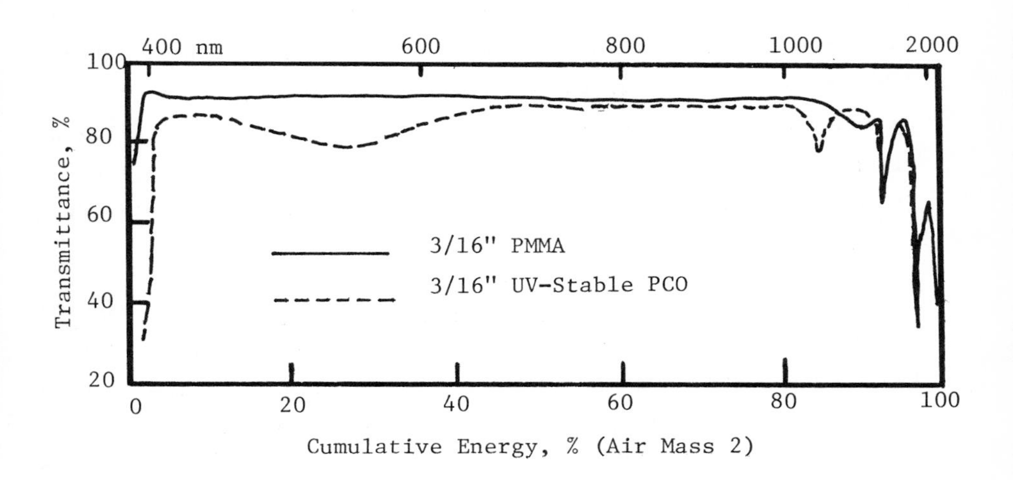

FIG. 9. Equal Energy Plot of Solar Transmittance of PMMA and PCO Sheet (Adapted from Reference 7).

transmittance of 3/16" polymethylmethacrylate (PMMA) and 3/16" polycarbonate (PCO) sheet that has been stabilized with ultraviolet absorber. The 400-nm absorption edge of the polycarbonate is due to the UV-absorber additive; however, the absorption at about 560-nm wavelength is believed by this author to be due to microcrystallinity in the relatively thick sheet that, because of the disparity between the refractive index of the crystalline sites and the amorphous environment, result in small "centers" for back scattering of visible light. This gives the polycarbonate a slightly translucent character.

The effect of angle of incidence on solar transmittance has been determined by Edlin (8) for several polymeric films, and plots for 2-mil Teflon FEP and 3-mil Tedlar polyvinyl fluoride (PVF) are presented in Figure 3.

Because of the electrostatic properties of many plastic films, compared to glass, dirt and dust have a significantly greater effect on transmittance at angles greater than normal incidence. Indeed, this problem is the subject of a G. T. Sheldahl Company project at the author's firm, one aspect of which is to study the dirt-accretion effects on second-surface metallized plastic mirrors (19). Garg (9) has presented dirt correction factors for (uncited) plastic film exposed at various angles from the horizontal.

Infrared Transmittance Properties. The binormal infrared transmittance spectra of polyethylene (PE) and "Mylar" polyethyleneterephthalate are presented in Figure 10; the spectra of "Tedlar" polyvinylfluoride and FEP "Teflon" plastics are presented in Figure 11 (20). Because an absorber plate operating in the 80 to 100°C range emits its peak energy (if it is non selective) in the 8- to 8.5-μ wavelength region (from Wein's Displacement Law), polyethylene films will permit significant upward transmittance losses from the plate to the ambient environment (Figure 10). The spectra show that although both FEP "Teflon"

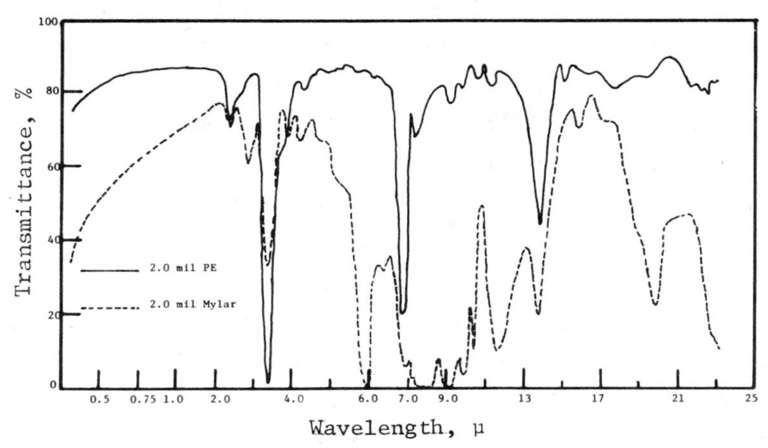

FIG. 10. IR Spectra of Polyethylene (PE) and "Mylar" (PET)

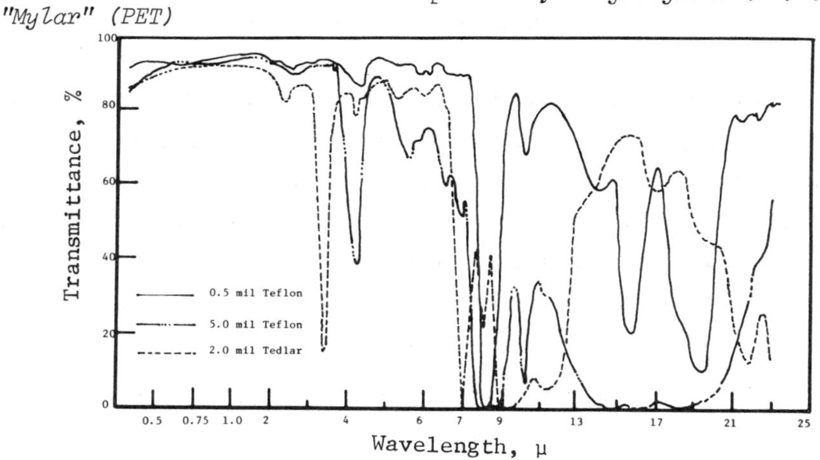

FIG. 11. IR Spectra of FEP "Teflon" and "Tedlar" PVF Film.

and "Tedlar" PVF exhibit absorption in the 8- to 10-u region, neither are particularly opaque at the critical wavelengths for high plate temperatures (7.5- to 8.5μ). Polymethylmethacrylate (PMMA), not shown, exhibits both high solar transmittance as sheet and possesses strong absorption bands in the 7- to 9-μ region.

Coated Plastics. Polymeric coatings for plastics are in general use, but commercial systems are employed primarily to impart high abrasion resistance to soft acrylic and polycarbonate sheet. DuPont's Abcite AC abrasion resistant coating is supplied only as applied to acrylic sheet. However, both Owens Illinois (Glass Resin) and Dow Corning manufacture polymonomethylsiloxane abrasion resistant coatings for acrylic, polycarbonate and glass sheet. In addition, a number of chemical firms are known to be developing competitive products. Abrasion resistant coatings are important to solar collectors in terms of the protection of the exterior, weatherable, cover plates when they are either acrylic or polycarbonate sheet.

The application of beam-sputtered, vacuum-deposited or chemically applied heat-mirror coatings to plastics would be very beneficial from a theoretical standpoint. However, the comparatively low thermal stability of plastics make the process of thin film coatings difficult. Cold sputtering, though expensive, is a possibility that is being considered. (Obviously, the application of anti-reflection coatings would not significantly increase the solar transmittance of comparatively highly transmitting plastic to offset the cost.)

Photochemical Considerations. From the viewpoint of weatherability, that is, of durability to long-term exterior "operational" exposure, plastic film and sheet may be divided into three categories:

- Those that are highly transparant to ultraviolet radiation and are thereby essentially less affected by solar irradiation.

  (Acrylics (PMMA), fluropolymers, methyl silicones)

- Those that are semitransparent to ultraviolet radiation, possess a propensity for undergoing photochemical reactions at undesirable rates, but may be protected by additives that either screen out the harmful ultraviolet in competition with the host polymer or chemically "block" photodissociation.

(UV-stabilized polyvinyl chloride (PVC), polyethylene (PE) and polypropylene (PP))

- Those that absorb ultraviolet but possess structures and main-chain bond energies that resist photodissociation. These systems are for the most part, colored and are not useable in solar systems.

  (Polyimides such as H-Film)

UV-Transparent Polymer Films. It can be generally said that the less reactive the hydrogen atoms (or chlorine, or fluorine) removed as a result of photochemical dissociation, the more reactive and the less stable, therefore, are the resulting radicals. Also, the less stable the radicals the greater their propensity for decomposition. Thus, polymers that result in reactive ultraviolet photodissociation products on irradiation tend to decompose rather than cross link. An example is polymethylmethacrylate, PMMA, in which the primary process is the photoextraction of unreactive hydrogen

$$\cdots\cdots CH_2-\underset{CH_3}{\overset{COOCH_3}{C}}-CH-\underset{CH_3}{\overset{COOCH_3}{C}}\cdots\cdots \rightarrow CH_2 + \underset{CH_3}{\overset{COOCH_3}{C}}{=}CH-\underset{CH_3}{\overset{COOCH_3}{C}}\cdots\cdots$$

and the molecular weight - and the strength - decreases. It will be recalled that PMMA is very transparent to ultraviolet and this process is, for pure polymer, slow. However, both "weathering" at high temperature (i.e., >100°F) and photochemical impurities speed the photodegradation of such polymers. Generally, photochemical reaction rates are *doubled* for each 10°C (18°F) rise in temperature; also, impurities result in greater dissociation and in "coloration" through energy transfer and photocatalytic processes. Rainhart and Schimmel (21) have recently reported 17-year life-times for "Plexiglass 55"* exposed in the New Mexico desert. Nevertheless, it is expected that the thermal environment of either "enclosed" flat plate or Fresnel-lens collectors will significantly increase the low solar transmittance change reported by Rainhart and Schimmel.

Polyvinyl flouride (PVF), with a structure

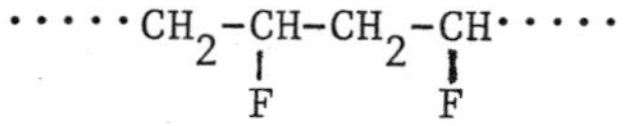

$$\cdots\cdots CH_2-\underset{\displaystyle F}{\underset{|}{CH}}-CH_2-\underset{\displaystyle F}{\underset{|}{CH}}\cdots\cdots$$

exhibits good resistance to ultraviolet irradiation because of the electronegativity of fluorine and the high C-F bond energies - which, combined with low ultraviolet absorption - makes PVF films very weather resistant. Polyvinyl fluoride films, however, are subject to thermally-induced embrittlement/cracking of stressed films on long term aging (under elevated-temperature exposure conditions). Polyfluoroethylenepropylene (FEP), like polyvinylfluoride, exhibits excellent weathering resistance - and for the same reasons. FEP films have not shown the degree of thermal-aging effects attributed to PVF, however.

Another cover-material candidate for long term weathering resistance is polydimethylsiloxane (and polymonomethylsiloxane) polymer.

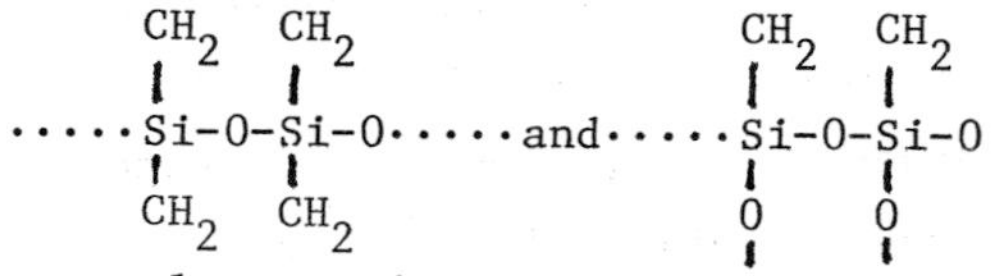

$$\underset{\text{elastomeric}}{\cdots\cdots \underset{\underset{\displaystyle CH_2}{|}}{\overset{\overset{\displaystyle CH_2}{|}}{Si}}-O-\underset{\underset{\displaystyle CH_2}{|}}{\overset{\overset{\displaystyle CH_2}{|}}{Si}}-O\cdots\cdots} \text{ and} \underset{\text{rigid thin film}}{\cdots\cdots \underset{\underset{\displaystyle \overset{|}{O}}{|}}{\overset{\overset{\displaystyle CH_2}{|}}{Si}}-O-\underset{\underset{\displaystyle \overset{|}{O}}{|}}{\overset{\overset{\displaystyle CH_2}{|}}{Si}}-O}$$

*Rohm and Haas polymethylmethacrylate

The exceptional ultraviolet transparancy, the main chain with high bond energies, and the lack of a polycarbon backbone to support the development of unsaturated structures (-C=C-C=C-), which are often "colored," all combine to give methyl silicones very high weathering/ultraviolet resistance. Indeed, free-radical reactions involving photodissociations such as

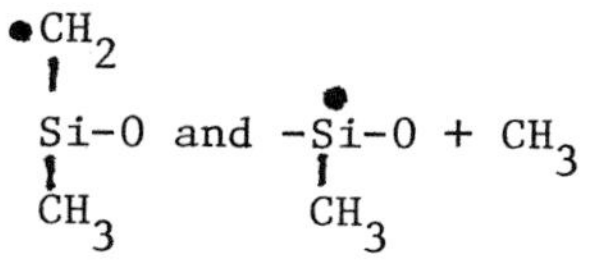

$$\begin{array}{l}\bullet CH_2 \\ \;| \\ Si\text{-}O \\ \;| \\ CH_3\end{array} \quad \text{and} \quad \begin{array}{l}\;\;\;\bullet \\ \text{-}Si\text{-}O \\ \;\;\;| \\ \;\;CH_3\end{array} + CH_3$$

which are comparatively infrequent, result in short-lived free radicals (that may be colored). These radicals disappear by quickly abstracting hydrogen, reacting to form a cross link, back-react, or react with oxygen (which is paramagnetic). Unfortunately, silicone elastomers result in comparatively weak films (poor tensile and tear strengths). On the other hand, monomethyl films are quite hard and brittle in even very thin films (although they are finding use as plastic coatings to impart abrasion resistance to otherwise "soft" materials). The elastomeric silicones may indeed find application in solar systems by reinforcing with "wire" and polymeric fiber scrim, however.

UV-Stabilized Plastics. Polyvinyl chloride undergoes chain-reaction degradation under ultraviolet irradiation, as well as discoloration,

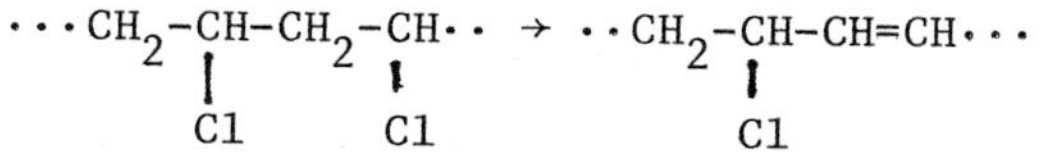

$$\cdots CH_2\text{-}\underset{\displaystyle Cl}{\underset{|}{CH}}\text{-}CH_2\text{-}\underset{\displaystyle Cl}{\underset{|}{CH}}\cdots \rightarrow \cdots CH_2\text{-}\underset{\displaystyle Cl}{\underset{|}{CH}}\text{-}CH{=}CH\cdots$$

which, on conjugation of the unsaturation, leads to a "colored" species and, by creating allylic structures, activates the decomposition of an adjacent monomer. This process is rapidly intensified with increased temperature.

Polyolefins such as PE and polypropylene (PP) undergo classical photooxidation where the mechanism proceeds through the hydroperoxide stage as

$$\text{Initiation} \quad RH \xrightarrow[\Delta]{h\nu} R\cdot + H\cdot$$

$$\text{Propagation} \quad R\cdot + O_2 \rightarrow ROO\cdot$$

$$ROO\cdot + RH \rightarrow ROOH + R\cdot$$

$$\text{Termination} \quad \left.\begin{array}{r} 2R\cdot \rightarrow RR \\ R\cdot + ROO\cdot \rightarrow ROOR \\ 2ROO\cdot \rightarrow ROOOOR \end{array}\right. \quad \text{non-propagating structures}$$

However, these polymers are finding increased utility in outdoor applications through the use of ultraviolet-light stabilizers. Although a recent review in *Plastics Engineering* (XXXII, May 1976) is completely instructive of this technology, it is noted here that UV stabilizers for polyolefins are generally of two types: (1) UV-absorbing "screening" agents and (2) excited-state quenchers. The former are represented by organic materials such as the hydroxybenzophenones and benzotriazoles which "degrade" the harmful ultraviolet to harmless thermal energy through internal hydrogen resonance. The quenchers such as the organometallic ferrocenes (and other metallocenes) not only absorb ultraviolet but quench the energy of the excited state of the carbonyl groups in polyolefin degradation.

An example of the effectiveness of ultraviolet stabilization of polyolefins is given in Figure 12. Specimens of UV-stabilized polyvinyl chloride, polyethylene and polypropylene were exposed for 320,000 langleys (equivalent to two years in

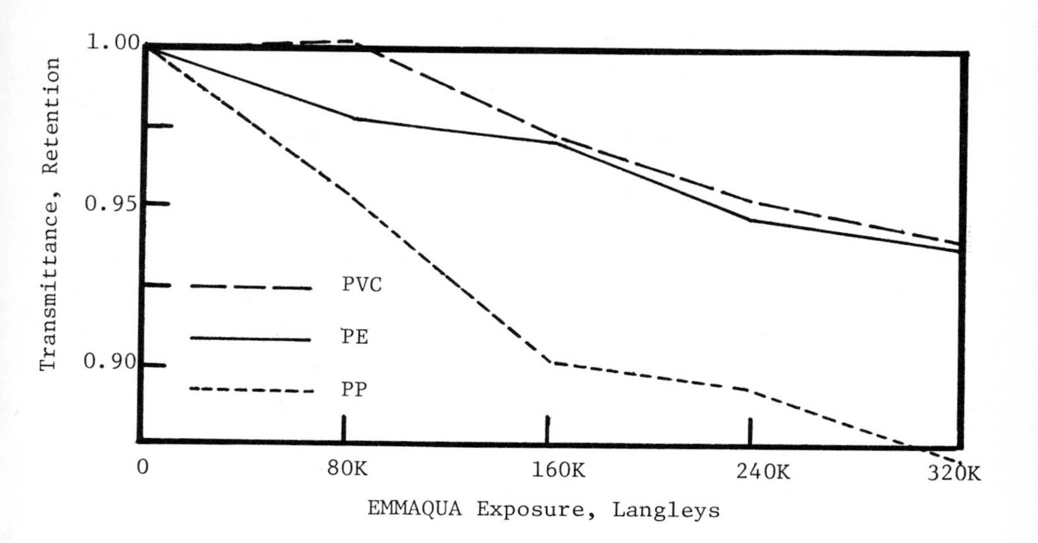

FIG. 12. Accelerated Weathering of UV-Stabilized Plastic Film.

Florida) for about two months in the EMMAQUA accelerated weathering machine. The effect of exposure is plotted as percent retention of the original transmittance. It will be seen that UV stabilization was effective in all three cases and was particularly so for the polyvinyl chloride and polyethylene films.

V. *SOLAR ABSORBER RECEIVER COATINGS*

A. *General Considerations*

The essential characteristics of an efficient solar absorber, or receiver, coating are its:

- Solar absorptance, defined by Equation (4)
- Thermal conductance, and
- Thermal emittance, defined by Equation (6)

The collection efficiency of a solar device is a first-order function of the receiver's solar absorptance, as shown from examination of Equations (1) through (3). However, the effectiveness of high absorptance is easily negated by employing "black" coatings at a thickness that permits *reradiative and convective cooling* to be too dominant -- that is, the exterior layers of the coating/surface are cooled and the energy is not transferred to the water/fluid due to self insulation. While thermal conductance is a higher order effect, it is often overlooked and can be a major problem.

Although "blackness" is primarily a function of fundamental absorption, surface characteristics play an important role in solar absorptance. Re-entrant surfaces that are characterized by rough, dendritic, particulate, or porous outer layers tend to interfere with first-surface Fresnel reflections. Absorption is increased at such surfaces due simply to multiple reflection/absorption processes. Re-entrant surfaces also increase the Lambertian nature of most surfaces.

Even though thermal emittance has a first-order effect on the upward radiative heat losses, Equation (5), it is important only at high plate temperatures. For this reason, most low and medium performance collectors utilize *black paints* with thermal emittances of 0.85 to 0.95 (e.g., swimming pool heaters, domestic hot water heaters, and well-packaged space heating collectors).

The relationship between air mass 2 solar radiation, emitted blackbody radiation and *black* coatings having both high emittance (e.g., non-selective paints) and low emittance (selective solar absorbers) is presented in Figure 13.

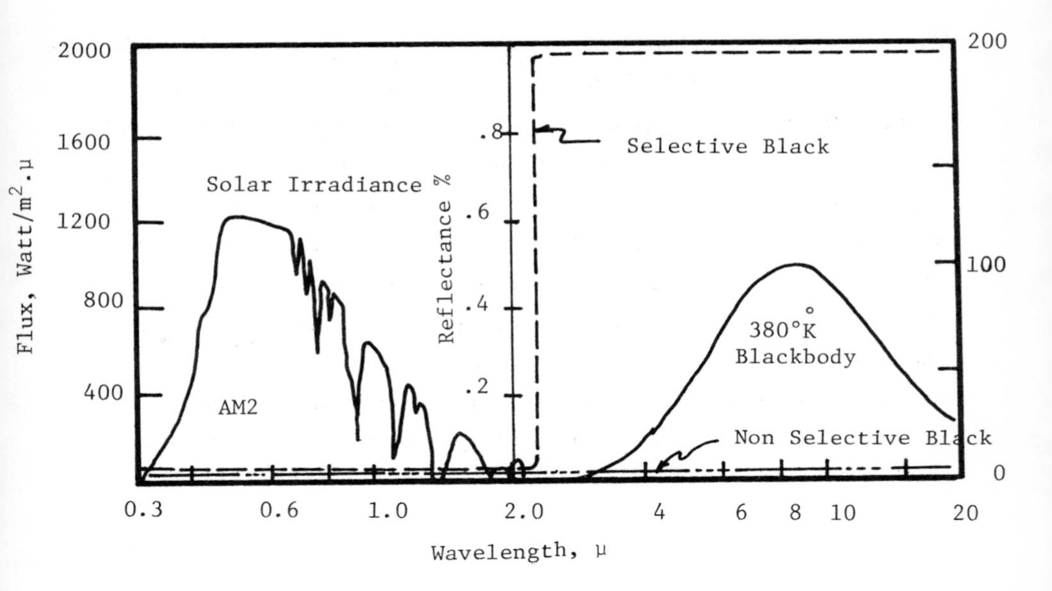

FIG. 13. Relationship Between Solar and Blackbody Radiations, and Ideal Selectivity.

The selective coating is *idealized* to absorb a maximum of solar energy at *all* solar wavelengths and to suppress a maximum of re-radiated energy at *all* blackbody wavelengths - that is, to suppress U_{up} in Equation (5) and minimize, therefore, U_L in the fundamental governing Equations (1), (2) and (3).

We observe from Equation (6) the dependence of emittance ε_T on the infrared spectral emittance, $\varepsilon_{T,\lambda}$, which, from Kirchoff's Law is, in turn, equivalent to the spectral absorptance (or unity less the reflectance).

$$[\varepsilon_{T,\lambda} = \alpha_\lambda = 1 - \rho_\lambda]_{T;\ \tau = 0} \tag{10}$$

The infrared reflectance of metals and semiconductors is high (due to conduction electrons) and hence they have a low thermal emittance. On the other hand, the absorptance characteristics, and therefore the emittance, of insulators such as organic and inorganic coatings is a function of their chemical structure (see Section IV.A). The emittance of such materials is a *bulk* process. Indeed, the author and his coworkers (22) utilized this concept to construct satellite radiator coatings having high spectral emittances in spectral regions corresponding to the maximum (peak) wavelength defined by the radiator temperature; for the most part, the absorption spectra of materials in a composite are additive.

B. Non-Selective Black Coatings

Non-selective black coatings are of three types: Paints, fused vitreous porcelain enamels, and metal conversion coatings. Paints may be either organic or inorganic; porcelain enamels and metal conversion coatings are both inorganic in nature.

The essential characteristics of black paints for solar receivers are that they possess high absorptance at angles other than normal incidence, i.e., that they are Lambertian, or optically "flat;" they should also be adherent and durable under use conditions. In the author's experience, black automotive manifold paints and instrument blacks are good candidates for studying as non-selective solar receiver coatings. The absolute spectral reflectance of 3M's Nextel Black Velvet and PPG's Duracron Super 600, two organic black coatings that are widely employed as solar receiver coatings, are presented in Figure 14 (23, 24).

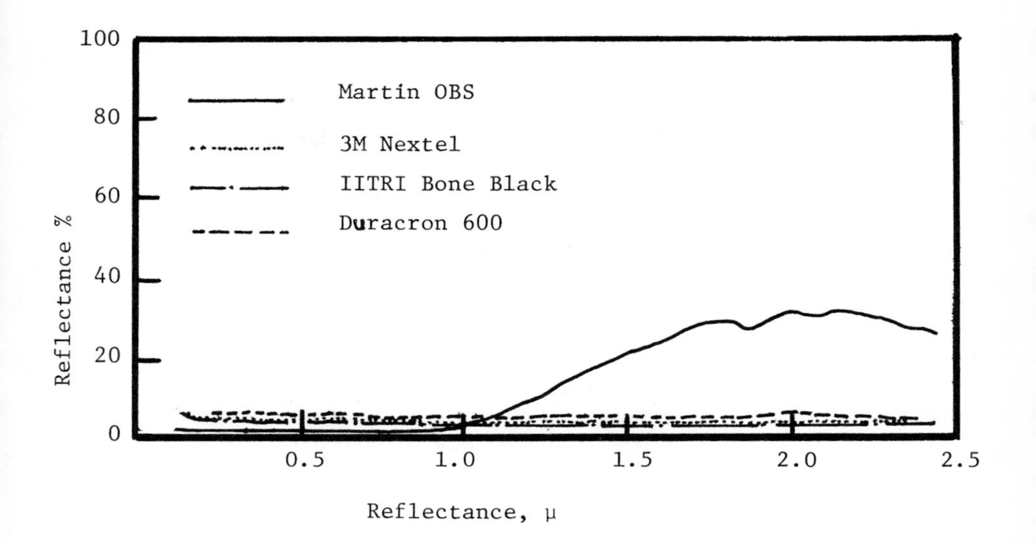

FIG. 14. Spectral Reflectance of Non-Selective Solar Absorber Paints.

The reflectance of a bone black-pigmented potassium silicate (water glass) paint (23) and of Martin Marrieta's Optical Black Surface, a black anodize aluminum coating (23, 25) are also shown in Figure 14. They possess a porous surface which largely accounts for the high absorptance and highly lambertian nature of both coatings. Indeed, the flatness of the Martin OBS coating is shown, along with other coatings, in Figure 15 (25). (It should be noted that acetylene soot and Parsons Black are included for reference and are not recommended for solar receivers.)

Fused vitreous porcelains have the advantage of high temperature and long term durability, but non-glossy, "flat" coatings are difficult to formulate and require firing at from moderate to high furnace temperatures.

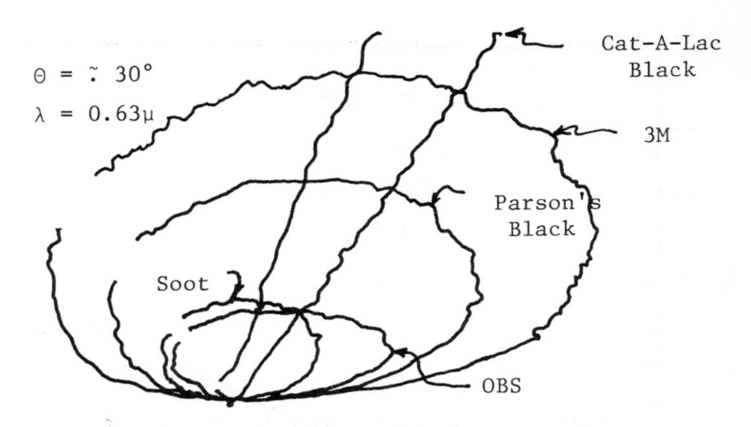

FIG. 15. Polar Distribution of Reflected Flux from Non-Selective Coatings.

C. *Selective Solar Absorber Coatings*

Selective absorber coatings may be divided into: (1) low-temperature systems for flat plate solar collectors, and (2) high-temperature coatings for focusing collectors with moderate to high concentration factors. Selective absorber coatings may be divided phenomenologically into a number of different types according to the method of preparation and the mechanisms associated with their functional performance. These types are listed in Table 3 essentially in order of their current use.

The most complete description of the phenomena associated with optical selectivity has been published in numerous communications by the Optical Sciences Center of the University of Arizona, the most instructive of which is a recent publication by Seraphin and Meinel (26). Masterson (27) has published a more concise treatment of this University of Arizona review.

TABLE 3

Selective Solar Receiver Coating Tapes (Phenomenological)

Type	Examples
Absorber-Reflector Tandems	Black Nickel (Tabor Coatings) Black Chrome (Harshaw Process) Copper Black (Ebonol* Process)
Multilayer Interference Stacks	Al_2O_3 - Mo - Al_2O_3 (Honeywell)
Powdered Semiconductor-Reflector Combinations	PbS "pigment" bound to reflector substrate such as aluminum
Wavefront Discriminators	Dendritic Tungsten and Rhenium
Resonant Scattering Systems	Transition metals dispersed in dielectric

Absorber-Reflector Tandems. These coating systems make use of a highly infrared-reflecting metal substrate (such as copper, aluminum or silver) over which a coating is deposited having high absorptance at solar wavelengths (i.e., it is black), but which is transparent to long wavelength "black-body" radiation. Thus, the tandem has the solar absorptance of the black exterior deposit and the thermal emittance of the metallic reflector substrate.

High absorption of the exterior coating may be either intrinsic in nature, or geometrically enhanced, or, as is usually the case, may be a combination of the two. This is true for a number of metal blacks (CuO being an example) where the surface geometries are of the dimensional order of the "visible" solar wavelengths, i.e., 500 nm. These black selectively absorbing/transmitting coatings are semiconductive in nature and

*Enthone Corporation, New Haven, Connecticut

their absorption is a result of interaction of photons having energies greater than the semiconductor band gap. Thus, the coating absorbs the photon as a result of raising the material's valence electrons into the conduction band, and photons of less than the band gap energy are transmitted through the material unaffected.

Black nickel (over bright nickel) was one of the first selective blacks to be employed as solar receiver coatings, and was first described in the United States by H. Tabor (27). Indeed, the term Tabor Coatings has become synonymous with these black nickel coatings, which have been used in Israel for many years. Likewise, copper oxide selective coatings (of which the Ebonol Process is an example) have been employed in Australia for some time and have been discussed in the United States by Edwards (20) and many others (28).

More recently, black chrome (29), which are $Cr-Cr_2O_3$ coatings, have been widely investigated. Although black chrome can be obtained as proprietary chemicals from Harshaw Chemical Company and E. I. DuPont Company, the non-proprietary formula has been presented in *Metal Finishing* (30). The "solar" and infrared reflectance of black chrome are presented in Figures 16 and 17, respectively (29). Also shown are the spectra of two different deposits of black nickel. The interference bands of these coatings, which are process variable, are also shown.

Lincoln et al (33) has presented the results of the development of temperature-resistant zironium oxynitride films on copper and silver by reactive sputtering.

Multilayer Interference Stacks. Multilayer interference stacks are not totally unlike the complex multiples of the tandem systems described previously. Indeed, interference stacks can be designed to be more selective than any other method

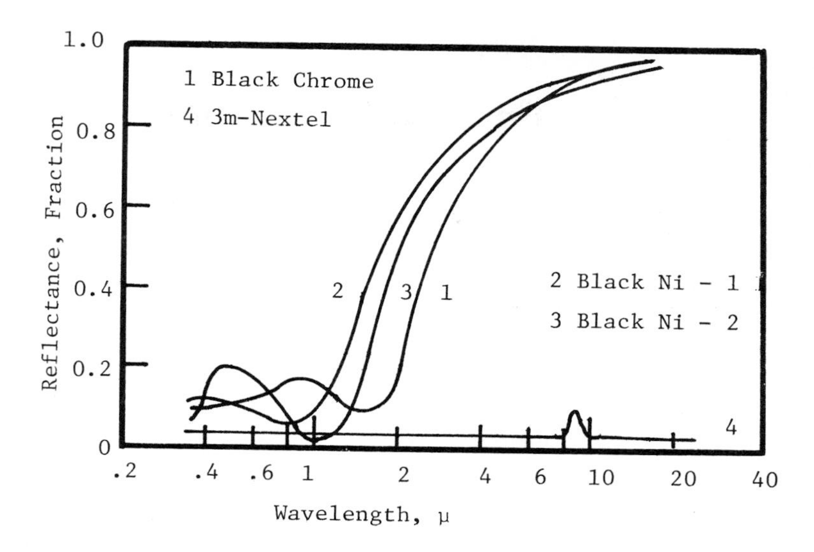

FIG. 16. Infrared Reflectance of Several Selective Coatings.

if cost and durability considerations are ignored. Seraphin (26) shows such a stack, reproduced as Figure 18. It depicts a four-layer interference stack comprised for two dielectric quarter-wave layers separated by a thin semi-transparent film. It is noted that the dielectric need not have intrinsic absorption for the stack to be an effective absorber.

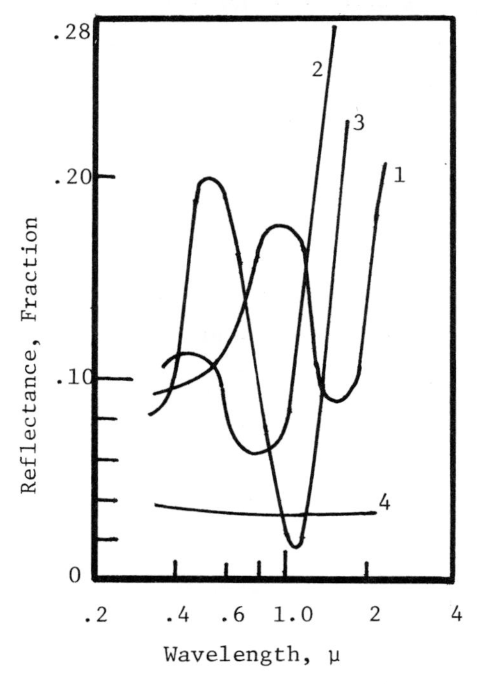

FIG. 17. Solar Reflectance of Several Selective Coatings.

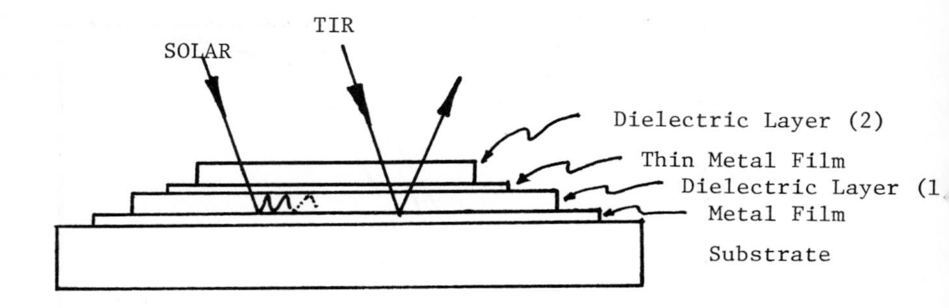

FIG. 18. Schematic of an Interference Stack.

In spite of the difficulties in preparing efficient interference stacks, Schmidt et al (32) have produced excellent high-temperature resistant multilayer coatings of Al_2O_3-Mo-Al_2O_3. Indeed, Peterson and Ramsey (28) have shown this AMA coating to maintain its selectivity at temperatures greater than 800°C. Similar high temperature characteristics have been reported by Masterson (32) for a complex tandem (with the dielectric layer having intrinsic absorption) of silicon and silver.

The reflectance spectra of these materials are not presented since they appear essentially identical in character to the spectra presented in Figure 17.

Powdered Semiconductor-Reflector Combinations. Particulate semiconductor coatings such as PbS dispersed/deposited on a reflector substrate are, in reality, poor cousins of the absorber-reflector tandems described previously. Their advantage is two fold: cost and ease of application. The author, in unreported work on satellite temperature control (1959), prepared a special lead sulfide paint from finely precipitated, purified, very-black lead sulfide dispersed in a minimum of silicone binder. The resultant paint had a solar absorptance of about 0.9 and an emittance of as low as 0.3. Williams et al (34) report emittances of 0.2 when PbS is dispersed directly on aluminum.

Other semiconductor candidates, such as indium phosphide, with its long-wave absorption edge at more than 1.5μ and high refractive index in the transparent region 2- to 20-μ wavelength, will be a candidate for such paints when the material is more available (and less costly).

The principal consideration in formulating high solar absorptance-to-emittance paint coatings is the use of a minimum of binder material. In fact, the essential condition is that the binder contain few, if any, absorption bands in the region corresponding to the peak (wavelength) blackbody radiation.

Narrow spectrum (size) microcavities enhance the absorption characteristics of the particulate systems described above.

Wavefront Discriminators. Wavefront discrimination provides wavelength discrimination by giving rise to multiple reflections in cavities, pits, dendrites, etc. This phenomenon was discussed by Tabor (35) in 1956 and requires a narrow spectrum of cavity size in order to provide a reasonably sharp spectral selectivity. Thus, they may absorb solar wavelengths but may be too small to "be seen" by long wavelength infrared, and the emittance (infrared reflectance) characteristics of the substrate prevail. Examples of such systems are dendritic rhenium whiskers reported by Seraphin (26) and dendritic tungsten reported by IBM (36).

Resonant Scattering Systems. Masterson (33) discusses the use of conductive particles such as finely divided vanadium, calcium and niobium metal dispersed in copper. These systems, which are poorly understood at this time, provide high reflectance in the infrared and broad resonance absorption in the solar spectrum.

Resonant scattering depends on the multiple reflections due to the high real portion of the complex refractive index and the ultimate (almost total) absorption due to the high absorption coefficient that characterizes the imaginary portion of the complex refractive index.

C. *Problems of Selective Solar Absorbers*

In summary, selectively absorbing/reflecting coatings can be, and have been, produced to provide moderately to reasonably good solar absorptance with satisfactory to outstanding reradiation-suppression characteristics in terms of possessing low thermal emittance. Generally speaking, the electrochemical, vapor deposition and sputtering processes employed result in systems that, except for the electrochemically deposited tandem systems, are quite expensive. Needed are highly solar absorbing paint-type selective coatings that have moderately low emittance - the expensive high temperature systems often can be justified for the smaller receiver areas required in concentrating collectors.

Two problems of selective coatings that must be overcome are their: (1) need for antireflection coatings, and (2) uncertain durability to moderate and long-term exposure under use conditions. The high absorption coefficient for solar radiation of many selective coatings gives rise to high complex refractive index and, therefore, high Fresnel reflection at solar wavelengths. This phenomenon requires the deposition of still an additional, antireflection (AR) coating to couple the system's refractive index with air (or vacuum for evacuated systems).

The durability of most selective coatings is a function of photooxidative processes at the conductive metal substrate - a process that is generally accelerated by moisture (and other contaminants) - and often can be observed in only moderate lengths of time.

VI. *SUMMARY - STAGNATION, DURABILITY AND DIAGNOSTICS*

The fundamental materials considerations that apply in solar collector technology have been related individually to the governing thermodynamic relationships that define their thermal performance. These include the optical and radiometric characteristics of both cover materials and solar receiver surfaces. In summary, however, we are left with the task of relating all of these properties to the actual performance of systems where environmentally- and operationally-synergistic phenomena are prevalent under moderate to long-term ($\geq$ one year) use conditions. Indeed, the urgent requirement for methods to assess short term reliability and long-term durability has led to the concept of stagnation testing and the diagnostic determination of collector malfunction as a result of such testing.

Stagnation testing involves, quite simply, the exterior (real-time) exposure of solar collectors in the dry condition after first filling with transfer fluid, which is permitted to evaporate in the early stages of stagnation. Depending on the length of exposure testing, the Hottel-Whillier governing plot, Equations (2) and (3), are determined initially and quarterly for the duration of the test (we have tested collectors in this manner for longer than one year).

The possible exposure-induced changes, many of which are subsystem interactive, or synergistic, are best described by the theoretical plots presented in Figure 19. The solid curve represents the initial Hottel-Whillier plot and the broken curves represent post-stagnation changes. That is, both A and B represent deleterious changes, while C and D are beneficial in nature. The specific causes for each condition may include one or more of the diagnostic interpretations given in Table 4.

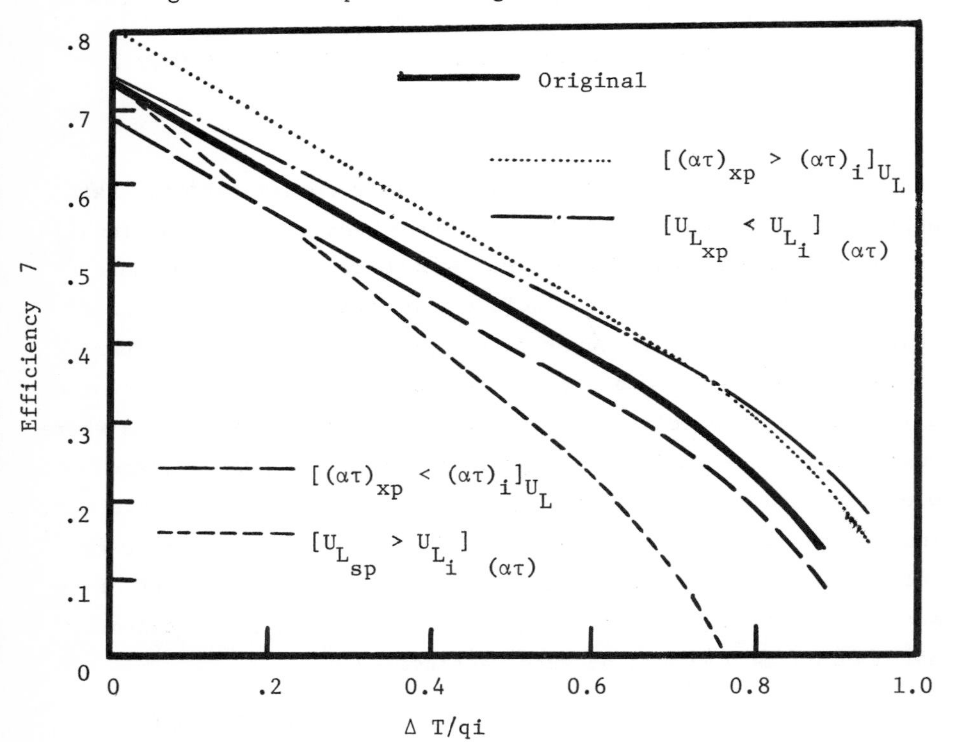

FIG. 19. Diagnostic Assessment of Collector Failure Modes.

TABLE 4

Flat Plate Solar Collector Diagnostics

Condition		Possible Causes
A. $[(\alpha\tau)_{xp} < (\alpha\tau)_i]U_L$ Deleterious	*1.	Glazing yellowed, darkened, or colored
	*2.	Glazing fogged
	*3.	Receiver bleached, i.e., faded (solar absorptance decreased)
	*4.	Glazing dirtied, contaminated (external-dust; internal-outgassing)
	*5.	Loss of bond conductance
B. $[U_{L_{xp}} > U_{L_i}]\ (\alpha\tau)$ Deleterious	*1.	Insulation compaction or destruction
	*2.	Insulation became wet
	*3.	Oxidation of receiver coating <u>increased</u> emittance (selective only)
	*4.	Contamination of selective receiver.
C. $[(\alpha\tau)_{\chi p} > (\alpha\tau)_i]$ Beneficial	*1.	Translucent glazings became more transparent
	2.	Receiver became "blacker" (unlikely)
	*3.	Dirty for initial test and cleaned for second (χp)
	*4.	Selective (only): 2nd test, less diffuse insolation
D. $[U_{L_{xp}} < U_{L_i}]\ (\alpha\tau)$	1.	Redistribution and/or increased effectiveness of insulation

*Changes in performance observed by author for these conditions.

TABLE 4 (CONTINUED)

Flat Plate Solar Collector Diagnostics

Condition	Possible Causes
	2. Spectral changes in glazing to allow less IR to radiate through
	*3. Decreased emittance (if selective) due to loss of original contaminant

REFERENCES

1. H.C. Hottel and A. Whillier, Trans. Conf. on the Use of Solar Energy, 2, 74, Univ. Arizona Press (1958).

2. H.C. Hottel and B.B. Woertz; ASME Trans., 64, 91 (1942).

3. R.W. Bliss, Solar Energy, 3, 55 (1959).

4. A. Whillier, Low Temperature Applications of Solar Energy, R.C. Jordon, ed., ASHRAE, New York (1967).

5. C.L. Gupta and H.P. Garg, Solar Energy, 11, 1 (1967).

6. G.A. Zerlaut, Proc., 21st Annual Meeting, Institute of Environmental Sciences, Vol. I, Mt. Prospect, Ill. (1975), p.155.

7. C.W. Clarkson and J.S. Herbert, Transparent Glazing Media for Solar Energy Collectors, International Solar Energy Conference (Preprint), Ft. Collins, Colo. (1974).

8. F. Edlin, Materials and Their Use in Solar Collector Systems, Proc. Solar Utilization Now, Arizona State Univ., Tempe, Az. (1975).

9. H.P. Garg, Solar Energy, 15, 299 (1974).

10. J.W. Ramsey, J.T. Borzoni, and T.H. Holland, NASA Contractor Report CR-134804, Contract NAS3-17862, (June 1975).

11. H. Wolter, Encyclopedia of Physics, 24, 461 Springer Verlag, Göttengen (1956).

12. M.V. Schneider, The Bell Systems Tech. J., 45, 1611 (1966).

13. C.K. Hsieh and R.W. Coldewey, Solar Energy, 16, 63 (1974).

14. R.M. Winegarner, Optics in Solar Energy Utilization, Proc. of Soc. of Photo-Opt. Inst. Engs, 68, 154 San Diego, Ca (1975).

15. J.C.C. Fan, T.B. Reed, and J.B. Goodenough, Proceedings of 9th Intersociety Energy Conversion Engineering Conference, San Francisco, Ca., (Soc. Auto. Eng., Inc.) New York (1974), p. 341.

16. R.D. Goodman and A.G. Menke, Solar Energy, 17, 207 (1975).

17. R.P. Taylor and R. Viskanta, Proc., First Southeastern Conference on Applications of Solar Energy, Wu et al, ed., UAH Press, p. 189, Huntsville, Alabama (1975).

18. G.A. Zerlaut, D.W. Gates and J.E. Gilligan, Proc., 20th Congress of the International Astronautical Federation, Springer Verlag, Mar del Plata, Argentina, (1969).

19. R.A. Stickley, Solar Power Array for the Concentration of Energy, Sheldahl Semi-Annual Report NSF/RANN/SE/GI-41019/PR/74/2, (July 1974).

20. D.K. Edwards and R.D. Roddick, Basic Studies in the Use and Control of Solar Energy, UCLA Reprint 62-27, NSF Grant G9505, U. Calif., (July 1962).

21. L.G. Rainhart and W.P. Schimmel, Jr., Solar Energy, 17, 259 (1975).

22. G.A. Zerlaut, W.F. Carroll, and D.W. Gates, Proc., 16th Congress of the International Astronautical Federation, Polish Tech. Pub., Warsaw (1966), p.259.

23. D.W. Gates and D.R. Wilkes, NASA - Marshall Space Flight Center, Huntsville, Alabama - Personal Communications 1975.

24. P.G. Patil, Solar Energy, 17, 111 (1975).

25. J.F. Wade and W.R. Wilson, Proc. Soc. of Photo-Optical Instrumentation Engineers; Long-Wavelength Infrared, 67 59 (1975)

26. B. O. Seraphin and A. B. Meinel, Optical Properties of Solids - New Developments, B. O. Seraphin, ed., North-Holland, Amsterdam (1975).

27. H. Tabor, Trans., Conf. on the Use of Solar Energy, 2, 1A, Tucson, Arizona (1956)

28. P. E. Peterson and J. W. Ramsey, J. Vac. Sci. Tech., 12, 471, (1975).

29. G. E. McDonald, Solar Energy, 17, 119 (1975).

30. Metal Finishing, p. 268, Metals and Plastics Publications, Inc. (1974).

31. R. L. Lincoln, D. K. Deardorff, and R. Blickensderfer, Soc, Photo-Optical Instrumentation Engineering Proc., 68, 161 (1975) Palos Verdes Estates, Calif.

32. R. N. Schmidt, K. D. Park, and J. E. Jamssen, "High Temperature Solar Absorber Coatings, Part II, ML-TDR-64-250, Wright-Patterson Air Force Base, 1964.

33. K. D. Masterson, Soc. Photo-Optical Instrumentation Engineers, Proc., 68, 147 (1975) Palos Verdes Estates, Calif.

34. D. A. Williams, T. A. Lapin, and J. A. Duffie, Trans. ASME, J. Eng., Power, 85A, 213 (1963).

35. H. Tabor, Bul. Res. Council Israel, 5A, 119 (1956).

36. J. T. Cuomo, J. F. Ziegler and J. M. Woodall, Appl. Phys. Letters, 26 (1975)

Chapter 14

SOLAR ENERGY STORAGE

Maria Telkes

Institute of Energy Conversion
University of Delaware
Newark, Delaware 19711

I. INTRODUCTION

In the United States, Heating and Cooling of Buildings requires nearly 30 percent of our annual total energy consumption. According to recent estimates, during average winter climate conditions nearly 5 times more solar energy is received on the walls and roofs of buildings, than is required to maintain comfort conditions inside. A major part of solar energy could be collected, stored and distributed to serve buildings.

This must be accomplished at an economically acceptable investment level.

Thermal energy storage is essential, to extend the warming effect of the sun to sun-less periods, at least overnight but preferably for one or more days, or possibly for longer periods. Methods of storage can involve sensible (specific) heat capacity of walls, floors and other parts of buildings. Water has been used for thermal storage in basement tanks. Heat of fusion type phase-change materials require smaller storage volumes, that can be confined within the occupied space. Heat contents, temperature-change limitations define weight, volume and cost of storage systems. Combination with solar collectors, using heat exchange fluids and circulating pumps or blowers, influence the durability and cost of thermal storage systems.

Passive or semi-passive systems have been devised, which replace major parts of walls or ceilings with solar collecting and storing components at substantial cost savings. Constant storage temperatures - using heat of fusion materials - can be combined directly with standby auxiliary systems based on the use of electric power, stored during off-peak periods, in combination with the use of Heat Pumps. Energy savings and economic values and their assessment can be used to determine future construction plans and directions.

II. HEATING AND COOLING REQUIREMENTS

Estimates have stated that if solar heating and cooling of buildings is proved to be practical, 10% of new buildings may be solar heated by 1985. (1) Ultimately up to 80% of thermal energy used for heating and cooling buildings, may be obtained from the sun. At the average yearly rate of two million housing units, 200,000 new housing units represent 10% of the yearly construction total. Programs to be attained during the next two years have been defined by ERDA. (2)

Thermal energy storage is an unavoidable neccessity to conserve daytime excess solar heat for use at night and possibly for another cloudy day and night. It is convenient to assume that storage of one million Btu would be a "yardstick" to consider for winter heating application. (3) (4)

III. THERMAL STORAGE MEDIA

Storage materials can accumulate thermal energy as specific heat ("sensible"), or as heat of fusion ("latent"), including their combined effects. Experimental work has been favoring water as the heat storage medium. Even if water costs practically nothing, a water tank and its insulation are unavoidably needed; so are space and foundation for the tank. Rocks, bricks, concrete or cinderblocks as thermal storage materials are likewise burdened with delivery charges, the need for insulation and spacing.

TABLE I

STORAGE OF ONE MILLION BTU, AT $\Delta T = 20^{o}F$

	Rocks	Water	Heat of Fusion
Specific heat, $Btu/lb^{o}F$	0.2	1.0	0.5 (avg.)
Heat of fusion, Btu/lb	--		100 (avg.)
Density lb/ft^3	140	62	100
Storage of 10^6 Btu:			
WEIGHT lb	250,000	50,000	10,000
Weight, relative	25	5	1
VOLUME (ft^3)			
+25% space	2,150	1,000	125
Volume, relative	17	8	1
Location	Basement	Basement Tank 6,000 gal	Closet 5 ft cube

Table 1 compares thermal storage data and shows the weight and volume of rock-like solids, water and heat of fusion materials for storing one million Btu heat, at $\Delta T = 20^{o}F$ temperature limit. Economic studies indicate a "cost allowance" of $2,000 to $3,000 for one million Btu thermal storage installed in place. This estimate is based on fuel costs of $5 per million Btu (requiring 10 gallons of oil at $0.50 per gallon). The initial cost of $2,000 in this way can be amortized during 400 heat-cycles, or less than 3 heating seasons, in regions of 5000 Degree Day heat loads.

IV. HEAT STORAGE MATERIALS, USING HEAT OF FUSION

The use of heat of fusion materials for heat storage has progressed considerably during the past 15 years. The writer has assembled thermophysical properties of an extensive collection of materials (5), selected from data by Rossini (6) and Kubaschewski (7). The Handbook Tables and other previous publications referenced in (5) are inadequate for critical review of materials. Summaries of space-application-needs have been issued (8), including related heat transfer calculations and heat exchanger designs, applicable to phase change thermal storage materials (9) (10) (11). Data are available giving the heat of fusion of hundreds of materials, most of them, however, cannot meet the obvious criteria of low cost, availability in large quantities, simplicity of preparation, combined with harmless applicability (non-toxic, non-flammable, non-combustible, non-corrosive) (4).

Cost limitations probably require the use of raw materials that cost not more than $500 per million Btu heat-content, allowing $1,500 for the fabrication of heat-exchangers, filled with the mixtures. For orientation we can assume that the heat content of suitable materials is 100 Btu/lb, requiring 10,000 lbs for the storage of one million Btu. This implies raw material cost limits of 5¢/lb or $100/ton. This low cost excludes all metals, most organic materials (including "waxes", that are also ruled out because they are potential fire hazards). The lowest cost materials are large volume chemicals, based on compounds of sodium, potassium, calcium, magnesium, aluminum and iron. They could be preferably in the form of salt-hydrates to take advantage of high heat of fusion of hydrate groups. Low cost compounds are probably restricted to chlorides, sulfates, nitrates, phosphates and carbonates. Costs must be realistically estimated, using technical grade materials (12). Table 2 lists selected materials.

TABLE 2

CHEMICAL PRICES
MATERIALS FOR PREPARING SALT-HYDRATES
FROM: CHEMICAL MARKETING REPORTER, MAY 1976 (12)

	\$/ton	\$/100 lb material		Solids
		anhydr.	salt-hydrate	%
Calcium chloride	90	4.50	2.25	50.5
Sodium carbonate	108	5.40	2.00	37
Disodium phosphate	350	17.50	7.00	40
Sodium sulfate	56	2.80	1.20	44
Sodium thiosulfate pentahydrate	166	--	8.30	64

V. SALT-HYDRATES

Previous reports and articles (13, 14, 15, 16), describe several low cost salt-hydrates. The "best" have high heats of fusion, are low in cost and compatible with container or heat exchanger materials.

These materials can be observed when sealed into transparent tubes or containers, a temperature sensor being used to measure temperatures. The materials can be heated until they attain a temperature 10°F above their melting point. The tubes should be positioned horizontally and can be cooled to obtain crystallization during heat exchange at ambient air temperature. All materials (Table 3) supercool and crystallization must be induced by providing crystal nuclei.

TABLE 3

SALT-HYDRATES

	+Water Mols	Melting Point, °F	Heat of Fusion Btu/lb	Density lb/ft^3
Calcium chloride	6	84-102	75	102
Sodium carbonate	10	90- 97	106	90
Disodium phosphate	12	97	114	95
Sodium sulfate	10	88- 90	108	97
Sodium thiosulfate	5	118-120	90	104

VI. PROPERTIES OF SALT-HYDRATES

Calcium chloride hexahydrate ($CaCl_2 \cdot 6H_2O$) is a hygroscopic low-cost material which forms several salt-hydrates of different crystal structure, involving transition from the hexahydrate to the tetrahydrate at 84°F and to the dihydrate at 102°F; both $CaCl_2 \cdot 6H_2O$ and $CaCl_2 \cdot 4H_2O$ have two crystal forms each (α and β). The salt, when heated above 110°F dissolves completely in its water of crystallization. When cooled, several of the four different crystals can form. If the $4H_2O$ crystals form, the heat of fusion is considerably lower than 75 Btu/lb. Calcium Chloride is rather corrosive and difficulties are encountered with four different crystal forms, when only one form is permitted.

Sodium carbonate decahydrate ($Na_2CO_3 \cdot 10H_2O$) is the "washing soda" in common use. The decahydrate contains 37% Na_2CO_3 and 63% water. It forms several salt-hydrates of different crystal habits, the decahydrate changes to the heptahydrate at 90°F and the monohydrate forms at 97°F dissolving almost completely. The material must be nucleated to form crystals of the decahydrate, eliminating the heptahydrate, a problem not solved as yet.

Disodium phosphate dodecahydrate ($Na_2HPO_4 \cdot 12H_2O$) has been subjected to 90 consecutive cycles, without any noticeable change. Fig. 1 shows the 90th cycle. (T_a is the ambient temperature, T_m the salt-hydrate, melting at 97°F) The material must be prepared, treated and used in combination with a nucleating system (17).

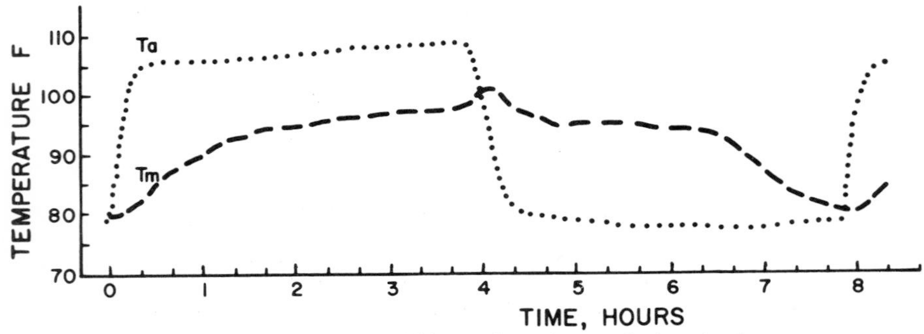

FIG. I. Disodium phosphate dodecahydrate (90th cycle)

Sodium sulfate decahydrate ($Na_2SO_4 \cdot 10H_2O$) is one of the least expensive and most available chemical materials. It is produced in very large quantities as a by-product and it can be obtained from natural resources in salt lakes and dry deposits, the supply being practically inexhaustible.

Sodium sulfate decahydrate melts at 90°F, when 34 grams of sodium sulfate dissolves in the available water, leaving some solid Na_2SO_4 residue. Being heavier, the residue (15% of total weight) settles to the bottom of the container. It is well known that if the material is stirred occasionally to mix it, total cyrstallization is obtained, without the separation of any residue, or uncombined liquid.

The writer suggested the use of sodium sulfate decahydrate for the storage of solar energy. Although its melting point, at 90°F, appeared to be somewhat on the low side, its relatively high heat of fusion of 108 Btu/lb is a definite advantage. The process involved the storage of solar energy, as the heat of fusion of sodium sulfate decahydrate, and the use of stored heat during the night and on cloudy days. (18, 19).

The difficulty of nucleation was solved by using a near-isomorphous nucleating agent: sodium tetraborate decahydrate, Borax (20, 21). When 3 to 4% Borax is added to sodium sulfate decahydrate its melting point decreases slightly to 89°F, but supercooling is eliminated as the material invariably starts crystallizing at 83 to 85°F. It is possible to obtain complete crystallization in a melt by inverting or occasionally shaking the container after crystals start to form. Tests have been made, relying on diffusion only, by limiting the vertical height of the latter to less than 1/4". In this case the material crystalized almost completely, without leaving any residue on the bottom of the container. Fig. 2 shows sodium sulfate decahydrate crystals growing in a melt containing 3% Borax, no thickener, vertical dimension one inch.

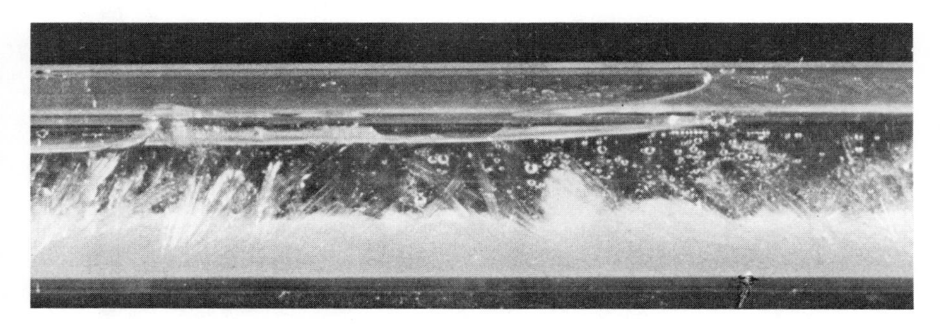

FIG. 2. The Growth of Sodium sulfate decahydrate crystals.

Experiments have been performed with various thickening agents as additives, with the aim of producing a gel in which the anhydrous sodium sulfate could not settle out by gravity. Many different thickening agents have been tested, including wood shavings, saw-dust, paper pulp, various types of cellulosic mixtures and Methocell. Additional organic materials include starch and Alginates. Inorganic materials included silical gel, diatomatious earth and other finely divided silica products. Some of these materials performed quite well for a number of cycles. Some of the organic materials were slowly hydrolyzed, or decomposed by bacterial or enzyme action. The silica gel, formed in the mixture itself, proved to be a hindrance in filling the mixture into containers, because it thickened too rapidly. Some of the silica material combined with Borax and inhibited its nucleating capability. Various phases of this research have been described (20, 21, 22, 23). The first solar-heated home, built at Dover, Mass. (Fig. 3) used this material for the storage of heat. An experimental laboratory was built in Princeton, N.J. (23) using the same heat-storage material.

FIG. 3. First Solar-Heated Home, Dover, Mass.

Eutectics are formed with many other chemical compounds, added in relatively small amounts to $Na_2SO_4 \cdot 10H_2O$. A number of salt-hydrate equilibria are reported in the literature and their crystallization characteristics have been described (24,25). Most of these eutectics are based on low cost compounds, such as sodium chloride, ammonium chloride, potassium chloride and others. All eutectics of this type require a nucleating agent, which is the same as described above for $Na_2SO_4 \cdot 10H_2O$. Practically all eutectics melt partly incongruently and require thickening agents to prevent settling of the higher density anhydrous components. (14, 16)

For the storage of cooling, a eutectic material has been developed, which melts at $55^{\circ}F$. It contains a mixture of sodium chloride and ammonium chloride, which decrease the normal melting point of Sodium sulfate decahydrate from $89^{\circ}F$ to $55^{\circ}F$. Borax is used as a nucleating agent (3-4%). An inorganic thickening agent is added to prevent settling (14, 16).

Sodium thiosulfate pentahydrate ($Na_2S_2O_3 \cdot 5H_2O$) is the photographer's "hypo", although a less pure technical grade can be used. It contains 64% $Na_2C_2O_3$ and 36% water. The pentahydrate changes to the dihydrate at $118^{\circ}F$ and to anhydrous at $120^{\circ}F$. At $126^{\circ}F$ $Na_2S_2O_3$ dissolves completely in its water of crystallization. The heat of fusion is 90 Btu/lb. In the $110^{\circ}F$ to $130^{\circ}F$ temperature range the heat content of the salt-hydrate is 5.5 times greater than that of water, based on equal weights. The advantage increases to nine-fold, based on equal volumes. (15)

VII. APPLICATION OF THERMAL STORAGE MATERIALS

The experimental laboratory building "SOLAR-ONE" (Fig. 4) at the University of Delaware used heat of fusion type salt-hydrates for thermal storage (26, 27).

FIG. 4. SOLAR-ONE, Experimental House of the University of Delaware.

Solar-heat during the winter is stored in sodium thiosulfate pentahydrate ($Na_2S_2O_35H_2O$) described above. The material is sealed into pan-shaped containers, (measuring 21"x 21"x;") made of ABS thermoformed material with 0.060" wall thickness. (Fig. 5).

One inch wide flanges support the pans on their lateral peripheries and additional supports are provided to maintain air passages for the circulation of air. The empty weight of pans is 2.25 lb; they contain 23.8 lb of sodium thiosulfate pentahydrate. The material melts congruently and dissolves completely in its water of crystallization at $126^{o}F$. The pans are completely sealed; they contain a nucleating device to induce crystallization when the stored heat is removed, by circulating room air between the pans. A total of 294 pans, weighing 661 lb, contain 7000 lb of salt-hydrate; the containers weigh only 8.6% of the total weight.

FIG. 5. Pan-shaped containers used in SOLAR-ONE

Thermal storage as the heat of fusion (90Btu/lb) is 630,000 Btu. In addition the specific heat of the solid material adds 122,000 Btu (in the temperature range of 70°F to 126°) and the thermal capacity of the melt adds 25,000 Btu (between 120° to 126°F). The specific heat of the ABS containers contributes only 15,000 Btu (between 70°F to 126°F).

Total heat exchange surface of the pan-shaped containers is 1940 ft^2. The net volume of the salt-hydrate is 67 ft^3; the surface to volume ratio is 29. The cross-sectional area of air passages is 7.6 ft^2 with the air flowing between spaced, stacked pans (28).

During the summer, cooling is stored during nights to lessen peaktime power demands from air conditioners, which can cause "brownouts" during hot summer days. The system uses a standard heat pump air conditioner, which is operated during the night, storing up "coolness" in a phase-change eutectic system, which freezes below 55°F, by circulating air at 35°F. During the day, when cooling is needed, the stored coolness is used, by blowing room air through heat-exchangers filled with the phase change material. In this way, peak demands for electric power can be diminished, resulting in "peak-shaving". The storage material is liquified after the stored coolness is extracted. As night approaches and the peak electric demand is over, the air conditioner resumes operation, cooling the storage system again, completely solidifying it by early morning. The coolness storage material is re-used every day, as this is needed in Delaware.

The thermal storage material melting at 55°F was contained in ABS tubes of 1.25" diameter, 0.030" wall thickness and 6 foot lengths (Fig. 6). A total of 620 tubes weighed 323 lbs

FIG. 6. Cooling Tubes in SOLAR-ONE

and had net salt-hydrate contents of 2500 lbs. Heat transfer surface was 1213 ft^2, net volume 27.4 ft^3 and surface to volume ratio 44. The cross sectional area of air being circulated between tubes was 6.7 ft^2 (28). The tubes were supported with an open crate-type structure, providing more than 0.5 inch spacing (as an average) between tubes.

Thermal storage was 195,000 Btu as the heat of fusion, while the specific heat of the solid (between 45° to 55°F) added nearly 9,000 Btu and of the liquid (between 55° to 65°F) added nearly 16,000 Btu, for a total of 220,000 Btu. Testing of these thermal storage systems continues (28).

VIII. ECONOMICS OF RAW MATERIALS

Five salt-hydrates described above are compared in Table 4 on the basis of raw material cost ($/100 lb salt-hydrate), heat content cost (Btu/$) and cost per one million Btu storage (14) at May 1976 prices.

TABLE 4

RAW MATERIAL COSTS (MAY 1976)

Salt-hydrate	Mp°F	$/100 lb	Btu/$	Cost of storing one million Btu,$
$CaCl_2 \cdot 6H_2O$	84*	2.25	3,300	300
$Na_2CO_3 \cdot 10H_2O$	90*	2.00	5,300	190
$Na_2HPO_4 \cdot 12H_2O$	97	7.00	1,600	620
$Na_2SO_4 \cdot 10H_2O$	89	1.20	9,000	110
$Na_2S_2O_3 \cdot 5H_2O$	120	8.30	1,100	900

* The melting point of the stable crystal form is indicated.

Table 4 shows sodium sulfate decahydrate is lowest in raw material cost (its eutectics are only slightly more expensive, but still in the "low cost" category). Sodium carbonate decahydrate is inexpensive, but its nucleation has not been established and it is corrosive. $Na_2HPO_4 \cdot 12H_2O$ with its higher melting point of 97°F may appear desirable, but at an increased cost as this is the case with $Na_2S_2O_3 \cdot 5H_2O$. These materials are "food" grade and "photo" grade, less expensive technical grade materials may be available. Costs have been based on current market prices (12) using the most advantageous method of obtaining the salt-hydrate from anhydrous ingredients.

IX. ECONOMICS OF CONTAINER/HEAT EXCHANGERS

Various types of containers have been suggested and tested including metallic heat exchangers (tubes, or "tin-cans") and plastic structures,(thermo-formed, or blow-molded containers). Container material and configuration evaluation is now in progress.

Metallic containers are relatively heavier, more expensive and subject to corrosion. Internal corrosion can be prevented by using inhibitors; both liquid and vapor phase inhibitors are needed and are relatively inexpensive. External corrosion cannot be prevented and are to be expected in humid atmospheres. The sealing or capping of metal containers is too cumbersome and any possible leakage must be totally eliminated. The use of two different metals, or alloys (solders), invariably produces corrosion and leakage of the contents.

Plastic containers are relatively less expensive, their filling and sealing can be accomplished by mass-production methods at relatively low cost. Water vapor permeation of plastic containers must be decreased to lowest levels. Closures of containers must be absolutely leak-proof and vapor-loss proof: they must be sealed because it has been demonstrated that screw-cap closures are not satisfactory. Ample experience is available on the "containerization" of chemicals in plastic containers.

It is expected that suitable plastic containers in the form of heat exchangers can be fabricated, filled and sealed at a cost of less than $1.00 per pound of container material. The containerization of 100 lb salt-hydrate may be accomplished by using about 8 lb of plastic material at a cost of less than $8.00. In this way it is expected that 10,000 lb salt-hydrate can be filled and sealed into plastic heat exchangers using less than $800 for this purpose. Added to the cost of the raw materials, shown in Table 4, the combined cost of thermal storage for one million Btu is around $900 to $1700. The average well-insulated housing unit - located in the 5000 Degree Day heat load zone - has an average winter heat load of 340,000 Btu/day or 100 kwh (thermal) per 24 hours, with average hourly consumption nearly 4 kw (thermal). For the storage of one day's average heating requirement 340,000 Btu should be available at a cost (of salt-hydrate and container) in the range of $300 to $580.

Durability of salt-hydrates has been explored with cycling tests. Alternate heating and cooling cycles have been performed in substantially identical container configurations and heat transfer rates, the same way as they will be used in houses. The rate of heat transfer during daytime "charging" periods is determined by available solar energy, usually not more than 6 to 8 hours per day during winter months. The rate of discharge may be longer, during the entire night period. Accelerating the rate of charge by rapid heating, or the rate of discharge by rapid cooling (quenching) may lead to mistaken conclusions, limited by rates of crystal growth.

Cycling tests have been conducted with sodium thiosulfate pentahydrate and the 55°F eutectic used in SOLAR-ONE since 1973. Additional cycling tests have been completed in the laboratory with sodium sulfate decahydrate (+ borax) for one thousand cycles. Samples made with thickener showed no difference between the first cycle and any subsequent cycles. Samples prepared without thickener exhibited segregation after the first few cycles, and never recovered in their performance. The production of 1000 cycles required the continuous operation of the cycling device for nearly 6 months and the tests were stopped because there was no change in the performance of the materials. Detailed description and analysis of these tests will be published.

Solar heat storage systems may be required for 200 day winter-heating periods, to perform 100 cycles per heating season, using an additional day's heat storage capacity. One thousand cycles therefore represent the cycling performance of 10 heating seasons (=years). This assumption does not imply that after 10 years the salt-hydrates must be replaced, because the salt-hydrates should last indefinitely unless the containers are damaged.

The durability of plastic containers has been well established. Plastic pipes and other fixtures have been used in buildings for decades and their use is increasing in a variety of additional applications. Water vapor permeability may be a factor, because salt-hydrates contain water (up to 60% as shown in Table 2). It is possible to diminish water loss to less than 0.1% per year, by using water impermeable barriers. In this way less than 1% water vapor loss can be expected during 10 years of operation.

X. *COMBINED SOLAR HEAT COLLECTION AND STORAGE SYSTEM*

For efficient operation of the entire solar heating system the thermal storage and solar heat collection efficiencies must be matched or balanced. The average insolation values for the winter months are well established (29) for variously tilted surfaces. The monthly Degree Days for most locations can be found in Tables. Frequency tables, for sequences of clear and cloudy winter days are available or can be established from meteorological data (30). The efficiency of solar collectors, for the winter months, can be established from published experimental data. It is expected that performance and efficiency standards will be established by the National Bureau of Standards (NBS), in cooperation with the American Society of Heating, Refrigerating and Air-Conditioning Engineers (ASHRAE). The probable amount of solar heat collection as a winter monthly average can be established by using the above mentioned data. The probable average winter heating load of the housing unit or building can be established.

Collectors may be classified according to locations and heat transfer fluids. South-facing tilted roof collectors can be used with pumped liquid heat transfer, or by using air circulated with a blower. These are usually known as forced circulation systems, involving electric motors to operate pumps or blowers. Collectors mounted on the roof require efficient insulation to prevent heat loss to an unheated attic. If the roof collectors form the ceiling of living-space, they must be even more insulated, to prevent overheating during the summer months. One of the difficulties of roof-mounted collectors is their excellent performance in collecting heat during summer months, due to their favorable tilt. A "stagnating" roof-type collector during summer months can reach internal temperatures close to $400^{o}F$. It is essential that collector materials should resist such temperatures, without any deterioration. Most roof-mounted collectors are rather expensive (present costs are around $\$10/ft^{2}$). Insulated pipes or ducts must be used to conduct heat transfer fluids to rooms to be heated, or to storage units; return pipes or ducts must also be provided. These necessary components increase collector costs to an appreciable extent.

Collectors can be mounted on the South-facing (or near-South facing) walls of buildings, actually replacing non-load bearing surfaces. The obvious advantage of such collectors is their ability to collect winter sunshine almost as well as tilted collectors. South facing vertical collectors receive much less solar energy during summer months and their stagnation temperature is considerably lower. Roof overhangs, or other shading devices, can eliminate most of the summer sun, on South-facing walls, to diminish necessary cooling. Systems of this type are known as "passive" or "autonomous" storage systems.

An interesting example of this type, Fig. 7, shows the schematic of the Solar Heated House, built in 1967 at Odeillo France (31). The South facing double glazed "window" -1- transmits sunshine to a black surface -2-, coating a concrete wall -3-, that acts as the storage unit. Room air is circulated by thermosyphon action, but a small blower was introduced to improve heat transfer. Thermal capacity of the 3 ft thick wall could be around 60 Btu/ft^2 of South facing surface, per one^{o}F temperature increase. The daily temperature "swing" could be up to 10°F resulting in a maximum storage capacity of 600 Btu/ft^2. Heat-of-fusion type storage - using sodium sulfate decahydrate (with nucleating and thickening additives) - can store nearly 10,000 Btu/ft^3. For the storage of 1,000 Btu a "Heat-storage-wall" of one ft^2 solar absorbing area would have to be only 0.1 ft (1.2 inch) thick. If desired the melting point can be decreased from 89°F to 80 or 75°F and the storage wall can form the inside wall of rooms, if this is otherwise desirable. Calculations and experimental projections are in progress with this type of installation that could initiate drastic cost reductions.

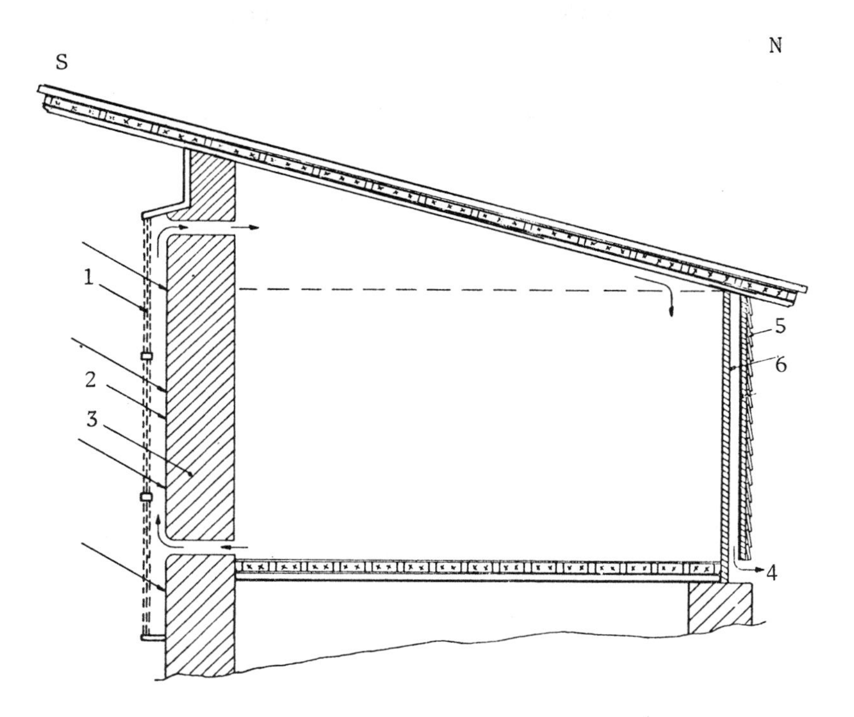

FIG. 7. Schematic of Solar Heated House built in 1967 C.N.R.S. Laboratory, Odeillo, France.

1. Two glass panes
2. Black coating
3. Concrete Wall
4. Air Outlet
5. Insulated outer wall
6. Inner Wall

XI. FUTURE APPLICATIONS

Low cost salt-hydrates may provide thermal storage systems for numerous applications to control the temperature of enclosures, greenhouses and other agricultural and commercial buildings. In addition to the storage of heat from the sun, cooling can be stored during the night when ambient temperatures are sufficiently low. Night-sky radiation cooling can be used to augment this effect.

Off-peak electric power can be stored as heat (Joule-heat), if rates are favorably low for augmenting solar heating, sharing the same storage system. Solar-assisted Heat Pumps can be used in combination with thermal storage systems, to store daytime solar heating for night operation of the heat pump. During summer heat pumps can use the type of storage system that has been installed into Solar-One, providing the storage of cooling at 55°F. It is possible that a combined system can be developed using only one thermal storage material for both heating and cooling operations to assist (or be assisted by) the heat pump.

In addition to salt-hydrates melting at moderate temperatures, thermal storage materials are available for higher temperature applications (32, 33). The development of suitable latent heat of fusion type materials and their application is progressing and could be the deciding factor in the conversion of solar energy to electrical power by thermal engines.

REFERENCES

1. NSF/NASA Solar Energy Panel, Solar Energy As A National Energy Source, Publ. University of Maryland (1972).

2. ERDA, Solar Energy Program ERDA-15 (Dec. 1975).

3. NSF/RANN, Solar Heating and Cooling for Buildings Workshop, Part I, Publ. University of Maryland, NSF/RANN 73-004, Washington, D.C., (March 21-23, 1973).

4. L. U. Lilleleht (editor), Solar Energy Storage Subsystems., NSF/RANN 75-041 (1975).

5. M. Telkes, Development of High Capacity Heat Storage Materials, R-380, Report to Instrumentation Laboratory of M.I.T. (1962, released 1963).

6. F.D. Rossini, et al, Selected Values of Chemical Thermodynamic Properties., National Bureau of Standards Circular 500 (1952, reprinted 1961).

7. O. Kubaschewski and E.L. Evans, Metallurgical Thermo-Chemistry., 3rd ed., Pergamon Press (1958).

8. D.V. Hale, M.J. Hoover and M.J. O'Neill, Phase Change Materials Handbook., NASA CR-61363 (1971).

9. S.Z. Fixler, Passive Thermal Control by Phase-Change Materials., Space/Aeronautics (1966).

10. T.C. Bannister and E.W. Bentilla, Research and Development Study on Space Thermal Control by Use of Fusible Materials., Proc. Inst. Environmental Sci. (1966).

11. P.G. Grodzka, Thermal Control of Spacecraft by Use of Solid-Liquid Phase-Change Materials., AIAA 8th Aerospace Sci Meeting, New York (Jan. 1970).

12. Chemical Marketing Reporter, Schnell Publ. Co. Weekly Magazine, New York, (Formerly called Oil, Paint and Drug Reporter) (1976).

13. M. Telkes, Energy Storage Media in Solar Heating and Cooling Workshop, NSF/RANN 73-004, Page 57-59 (Proceedings March 21-23, 1973).

14. M. Telkes, Storage of Solar Heating/Cooling, ASHRAE Journal (September 1974).

15. M. Telkes, Thermal Storage in Sodium Thiosulfate Pentahydrate, Proceedings 10th Intersociety Energy Conversion and Engineering Conference, University of Delaware, Newark, DE (August 1975).

16. M. Telkes, Thermal Storage for Solar Heating and Cooling, Proc. Workshop of Solar Energy Storage Subsystems for Heating and Cooling of Buildings, Charlottesville, VA, April 1975, Sponsored by NSF-RANN, ERDA, Univ. of Virginia and ASHRAE., NSF-RA-N-75-041 (Publ. Dec. 1975).

17. M. Telkes, Solar House Heating, A Problem of Heat Storage, Heating and Ventilating 44, 68-75 (May 1947).

18. M. Telkes, Review of Solar House Heating, Heating and Ventilating 46, 68-74 (Sept. 1949).

19. M. Telkes, Storing Solar Heat in Chemicals, Heating and Ventilating 46, 79-86 (Nov. 1949).

20. M. Telkes, Nucleation of Supersaturated Inorganic Salt Solutions, Ind. and Eng. Chem., 44, 1308-10 (1952).

21. M. Telkes, Composition of Matter for the Storage of Heat, U.S. Pat. 2,667,664 (1954).

22. F. Daniels and Duffie (editors), Solar Energy Research, Chapter on Solar House Heating by M. Telkes, Univ. of Wisconsin Press (1955).

23. A. Olgyay and M. Telkes, Solar Heating for Houses, Prog. Archit., 195-207 (March 1959).

24. A. R. Ubbelohde, Melting and Crystal Structure, Clarendon Press, Oxford (1965).

25. W. G. Blasdale, Equilibrium in Saturated Salt Solutions, Reinhold Pbl., New York (1927).

26. K. W. Böer, The Solar House and Its Portent, CHEMTECH, 394-400 (July, 1973).

27. K. W. Böer, Solar One, A Systems Approach to Solar Energy Conversion, The NATURALIST.

28. K. W. Böer, Solar One, Two Years Experience, Proceedings 10th Intersociety Energy Conversion and Engineering Conference University of Delaware, Newark, DE (August 1975).

29. ASHRAE HANDBOOK, Solar Energy Utilization for Heating and Cooling, Chapter 59 (1974).

30. ASHRAE Annual Meeting, Solar Energy Applications, Montreal, Canada (1974).

31. Ref. 3), 127-138 (1973).

32. M. Telkes, Chapter VII, Solar Thermal Energy Storage in Solar Energy for Heating and Cooling Buildings, Editors R.C. Jordan and B.Y. H. Liu, to be published by ASHRAE (1976).

33. E.G. Kovach (Editor), Thermal Energy Storage, NATO Science Committee Report, Report of Conference at Turnberry, Scotland, NATO Scientific Affairs Division, Brussels (1976).

4

IV. DIRECT SOLAR ENERGY CONVERSION

Of all of the systems which are proposed as substitutes for the utilization of fossil fuels, photovoltaic solar cells are the most attractive from the standpoint of their directness. Unfortunately, their relatively low efficiencies and high cost make them less competitive than other systems discussed in this volume. Nevertheless, recent advances which promise photovoltaic cells of high efficiency, low cost, large area, and enhanced stability have initiated an intensive research activity which has produced a technology with many different approaches and possibilities. These materials options and achievements in homojunction, heterojunction, direct and indirect gap semiconductors and Schottky barrier devices are explored in the next two chapters, with emphasis on thin film heterojunctions.

Increasing the solar efficiency of a photovoltaic heterojunction cell is a challenging endeavor in which such variables as collection efficiency; correct range of band gap energy; transport phenomena at the interface; generation and injection of holes and electrons with long diffusion lengths; the effect of impurity atoms on charge collection barriers; the growth of ultra thin oxide films; low, ohmic contacts and reducing the reflection losses demand simultaneous advances in both solid state physics and materials science.

Chapter 15

SOME MATERIALS PROBLEMS ASSOCIATED WITH PHOTOVOLTAIC SOLAR CELLS

J. J. Loferski

Division of Engineering
Brown University
Providence, R. I. 02906

I. INTRODUCTION

Semiconductor solar cells convert sunlight directly into electricity by means of the photovoltaic effect (PVE) which can be defined as the generation of an emf in a semiconductor which is exposed to ionizing radiation. For the PVE to occur three conditions must be met. 1) The radiation (light, electrons, protons, etc.) must generate current carriers in excess of the concentration present in the unilluminated semiconductor. If the radiation is light, this means that the a portion of the photons in the incident light must have energies in excess of the minimum internal ionization energy of the semiconductor, i.e. an energy in excess of the forbidden energy gap E_G of the semiconductor. Figure 1 shows how the absorption constant α of various semiconductors depends on the energy of the incident photon. As is evident from this figure, for some

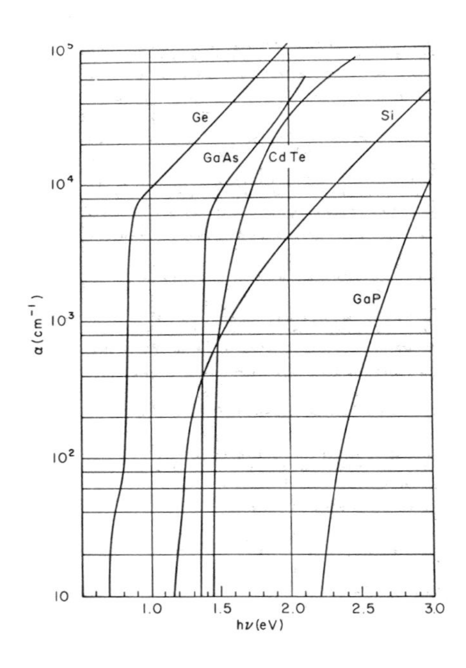

FIG. 1. Plot of ln α vs. photon energy.

semiconductors like GaAs and CdTe (direct gap semiconductors) the magnitude of α increases from a low value at the absorption edge ($h\nu = E_G$) to values in excess of 10^4 cm^{-1} when the photon energy exceeds E_G by several tenths of an electron volt. Other semiconductors like Si (an indirect gap semiconductor) have a more gradually rising curve and the value of α rises to 10^4 cm^{-1} only when the photon energy exceeds E_G by more than 1 eV. This difference in shape of α vs. $h\nu$ curves is important in solar cells since the thickness of semiconductors of the second type needed to absorb a specified fraction of solar photons having $h\nu > E_G$ is substantially larger than that of semiconductors of the first type. This point is illustrated in Fig. 2 which compares the fraction of photons with $h\nu > E_G$ absorbed in a given thickness of GaAs and Si. 2) There must exist inside the semiconductor

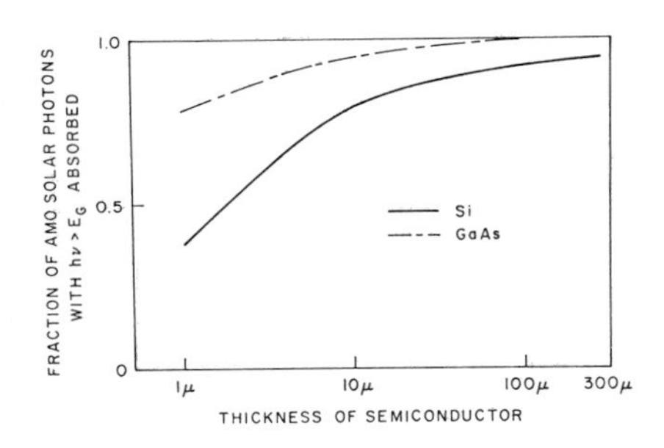

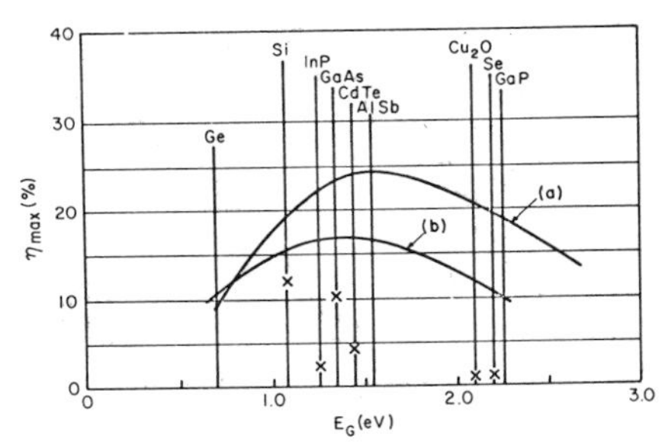

FIG. 2 Fraction of solar photons with $h\nu > E_G$ absorbed by Si and GaAs.

FIG. 3 Maximum efficiency η_{max} vs. energy gap.

an electrostatic potential difference which can separate the excess holes and electrons generated by the ionizing radiation. This potential difference can be produced by a p/n homojunction; a p/n heterojunction, or by a Schottky barrier, either purely metal-semiconductor (M-S) or metal-insulator-semiconductor (M-I-S) type. 3) Either the holes or the electrons or both must remain mobile long enough to migrate to the charge separation site, i.e. their diffusion length must be long compared to the distance between the point where the carriers are generated and the p/n, M-S or M-I-S junction.

Solar cells encounter materials related limitations in all three of these areas; a few of these limitations will be explored in this chapter.

II. SELECTION OF SEMICONDUCTORS FOR PVE SOLAR CELLS

In 1956, Loferski (1) published the results of a calculation of the dependence of the efficiency of solar cells (η_{max}) on the forbidden energy gap of semiconductors; his curves are reproduced in Fig. 3. Because he was focusing attention on the important role played by E_G in determining upper limits on η_{max} he used a number of simplifying assumptions which must be kept in mind when drawing conclusions from these curves.

a) He assumed that the short circuit current of the solar cell was equal to

$$I_{SC} = q\, n_{ph}(E_G) \qquad (1)$$

where q is the charge on the electron and $n_{ph}(E_G)$ is the number of photons in the solar spectrum with $h\nu \geq E_G$; he used the AMO (Air Mass Zero) solar spectrums. This relation assumes that the effective thickness of the cell, w, is large enough to absorb all photons with $h\nu \geq E_G$ and that for every photon absorbed, one electronic charge passes through a short circuited cell. b) He assumed that the current-voltage characteristic of the solar cell was governed by the diode equation

$$I = I_o(e^{\lambda V} - 1) \qquad (2)$$

where

$$\lambda \equiv q/AkT \tag{3}$$

$$I_o = K \exp(-E_G/BkT) \tag{4}$$

The values of A, B, and K are determined by the mechanism assumed to govern the behavior of the diode. For example, if it is an ideal diode as described by Shockley (2) then A = B = 1 and K is a function of material parameters specifically

$$K = qn_i^2 \left| \left(\frac{D_e}{\tau_p}\right)^{1/2} \frac{1}{N_A} + \left(\frac{D_h}{\tau_n}\right)^{1/2} \frac{1}{N_D} \right| \tag{5}$$

In this equation n_i is the concentration of carriers in intrinsic material; its value is given by the equation:

$$n_i^2 = (N_C N_V) \exp(-E_G/kT) \tag{6}$$

where $N_C(N_V)$ is the effective density of states in the conduction band (valence band); $D_e(D_h)$ is the minority electron (hole) diffusion constant in the p-region (n-region) of the p/n homojunction; $\tau_p(\tau_n)$ is the minority electron (hole) lifetime in the p-region (n-region); $N_A(N_D)$ is the concentration of acceptors (donors) in the p-region (n-region). Different semiconductors and indeed different samples of the same semiconductor have different values of some of these parameters. The values of some of them (N_C, N_V, E_G) are fixed once the semiconductor is specified. However, others (τ_n, τ_p, N_A, N_D, D_e, D_h) can assume a wide range of values. Consequently, to make the calculation of η_{max} vs. E_G tractable, Loferski proceeded as follows. The value of I_o was calculated for silicon using a reasonable set of values for the parameters in Eq. (5). This establishes a value for the parameter K in Eq. (5); the same value of K was then entered into Eq. (4) and I_o becomes a function of E_G and T only.

Once the values of I_{SC}, I_o and A are set, the remainder of the calculation of η_{max} vs. E_G is straightforward. Thus, the open circuit voltage is found to be

$$V_{OC} = \frac{1}{\lambda} \left(\ln \frac{I_{SC}}{I_o} + 1\right) \tag{7}$$

The voltage at maximum power V_{mp} can be calculated from the relation

$$(1 + \lambda V_{mp}) \exp \lambda V_{mp} = \frac{I_{SC}}{I_o} + 1 \tag{8}$$

and the current at maximum power I_{mp} from the relation

$$I_{mp} = \left(\frac{\lambda V_{mp}}{1+\lambda V_{mp}}\right) I_{SC} \tag{9}$$

finally

$$\eta_{max} = \frac{I_{mp} V_{mp}}{P_{in}} \tag{10}$$

Figure 3 shows the dependence of η_{max} on E_G calculated on the basis of the assumptions summarized above. Since the coefficients A and B in Eqs. (2) and (3) commonly have values lying between 1 and 2, Loferski also calculated η_{max} vs. E_G for A = B = 2; the results are also shown in Fig. 3. The value of K in Eq. (5) was set in this instance by reference to empirical observation that the value of I_o in the Si p/n junction diodes available in 1956 was approximately 10^{-8} A/cm^2 and that in such diodes A $\sim$ 2.0.

Figure 3 shows that the highest value of η_{max} one can expect from a single semiconductor p/n homojunction is of the order of 25% (for A = B = 1); that semiconductors with energy gaps in the range $1.0 \lesssim E_G \lesssim 2.2$ eV for A = B = 1 and between $1.0 \lesssim 1.7$ eV for A = B = 2 are the best choices for solar energy converters, and that η_{max} is higher for A = B = 1 than for values of these parameters greater than unity.

The curves of Fig. 3 apply mainly to p/n homojunctions. However, they also provide estimates of potentially achievable values of η_{max} for p/n heterojunctions and Schottky barrier cells. In the case of a heterojunction, the value of I_{SC} in Eq. (1) is determined by $n_{ph}(E_G)$ of that semiconductor which has the smaller bandgap, provided of course that the carriers generated in it contribute significantly to I_{SC}. The values of I_o and λ are determined by parameters of the interface, like interface state densities, electron affinites of the two semiconductors comprising the heteropair, etc. It is a reasonable assumption, however, that I_o will not have a significantly smaller value than that calculated for a p/n homojunction made from the member of the heteropair which has the smaller value of E_G. Consequently, η_{max} is limited to values less than or at best equal to those calculated for p/n homojunctions based on the smaller E_G in the heteropair.

Similar conclusions apply to Schottky barrier cells (3). In general, M-S junction cells have values of I_o larger than those calculated for a homojunction made from the semiconductor and therefore η_{max} is likely to have a smaller value than that shown in Fig. 3 for a p/n homojunction based on the same semiconductor.

In short, Fig. 3 provides reasonable estimates of upper bounds on the efficiency attainable from PVE solar cells made from semiconductors having various values of E_G, no matter what the nature of the charge collecting barrier.

More exact calculations of the limit efficiencies have been made for Si (4) and GaAs (5) in an effort to understand why the values of η_{max} were "stalled" at values substantially below those predicted by theory.

In the case of Si, Wolf pointed out that while small increases in I_{SC} were to be anticipated, the main source of increased η_{max} would be an increase in V_{OC}, i.e. a decrease in I_o. These parameters could be changed in the required directions by increasing the impurity concentration on the two sides of the Si p/n homojunction. The increased values of I_{SC} envisioned by Wolf have been realized in the cells designed and fabricated by the COMSAT Laboratories [the violet cells (b), the velvet cell (7) and the black cells (8)]. These improvements have resulted in cells having AMO η_{max} values of about 15% and AM1 (the AM1 solar spectrum has the spectral composition and distribution which would be observed at sea level with the sun at the zenith) η_{max} values of about 18%. However, attempts to increase V_{OC} have been frustrated by the experimental observation when the doping levels in Si increase above a certain limit, I_{SC} begin to decrease. The explanations advanced for this behavior will be discussed in Section IV.

In the case of GaAs, the observed I_{SC} values were substantially lower than those predicted by Eq. (1), even after reasonable allowance had been made for carrier recombination in the bulk regions of the cell. The excessive loss of carriers was traced to very high surface recombination velocities. In a direct gap semiconductor like GaAs, absorbed photons generate carriers very close to the surface; therefore, surface recombination can annihilate a larger fraction of these light generated carriers than in the case of an indirect gap semiconductor like Si. A solution to this problem (9) which has raised η_{max} of GaAs based cells to values observed for Si will be discussed in Section IV.

Solar cells have been fabricated from many semiconductors but differences in the effort expended on various materials are considerable. The most thoroughly studied material for solar cell application is of course Si, which has produced the highest efficiency cells fabricated to date. Cells of comparable efficiency have been produced on GaAs after a substantially smaller effort. Cells based on p/n homojunctions have also been fabricated in InP ($\eta_{max} \sim 8\%$) (10). Heterojunction cells of acceptable efficiencies have been made from a number of semiconductor pairs. The most thoroughly explored cell in this category is the p-Cu_2S/n-CdS thin film cell (η_{max} up to $\sim 8\%$) (11) described in more detail in Professor Bube's chapter in this book. Other cells include n-CdS/p-InP ($\eta \sim 12\%$) (12) n-CdS/p-$CuInSe_2$($\eta \sim 12\%$) (13); p-Cu_2S/n-Si($\eta \sim 5\%$) (14); p-Cu_2Te/n-CdTe ($\eta \sim 7\%$) (15); n-CdSe/p-Se ($\eta \sim 2\%$) (15), cells with M-I-S barriers have been prepared on Si single crystals; values of η_{max} up to 9.0% have been reported in such cells (17). Cells with a barrier formed between a large E_G semiconductor "window" like In_2O_3 or SnO_2 and Si have reported efficiencies up to 8% (18).

Other materials with bandgaps in the desirable range in which p/n homojunctions can be made include $CuInS_2$ ($E_G \sim 1.5$ eV, direct gap); AlSb ($E_G \sim 1.5$ eV, direct gap) and GaP ($E_G \sim 2.2$ eV, indirect gap). Many other semiconductors (ternaries, etc.) have E_G values in the required range and are potentially available for fabrication of heterojunctions and M-S or M-I-S structures. In short the photovoltaic effort is very commonly encountered in semiconductors and many of them could in principle serve as the basic material for solar cells.

III. CHARGE COLLECTION BARRIERS, ESPECIALLY MIS STRUCTURES

As we have already pointed out, the charge

collection barrier (CCB) can be a p/n homojunction, a p/n heterojunction or a metal-semiconductor barrier. Commercial silicon cells utilize a p/n homojunction barrier formed by diffusion an appropriate impurity into a single crystal wafer doped with an impurity of the opposite type. Cells in which the base wafer is n-type and the diffused skin is p-type as well as cells in which the base wafer is p-type and the diffused skin is n-type have been fabricated; the efficiency of the two types is essentially the same, even though differences could be expected because of differences in minority carrier mobilities in n-and p-Si. The mobility μ enters the picture because the minority carrier diffusion length L is related to μ through the relation

$$L^2 = \frac{\mu k T \tau}{q}$$

where τ is the minority carrier lifetime. In standard silicon cells, high efficiency requires that the L be large compared to the absorption depth of the shortest wavelength light utilized efficiently in the cell. It turns out that the values of L attainable in n-and p-Si are comparable i.e. that the value of $(\mu\tau)^{1/2}$ for minority electrons does not differ significantly from the value it has for minority holes.

During the past few years, an increasing amount of work has been devoted to the CCB formed by a metal-insulator-semiconductor (MIS) structure because it should be inherently simpler (and therefore less expensive) to fabricate than p/n homojunction cells and yet theory predicts solar energy conversion efficiencies comparable to those achieved in homojunction cells. This device is closely related to the classical metal-semiconductor (Schottky barrier) structure. Both experimental and theoretical work have shown that the efficiency which can be obtained from M-S barrier cells must necessarily be substantially lower than that which can be obtained from p/n homojunctions. This is because the work function difference between silicon and metals leads to insufficient band bending at the semiconductor-metal interface. The result is an open circuit voltage whose value is substantially smaller than that achieved in homojunction cells. By interposing a thin (several tens of Angstroms) insulator layer between the metal and semiconductor, the value of I_o can be reduced substantially and V_{OC} increased. In a Si M-I-S device, the insulator is SiO_2. Such M-I-S devices can be classified as majority or minority carrier devices depending on the work function of the metal. It is the minority carrier M-I-S structure which shows promise for photovoltaic cell application. Minority carrier M-I-S devices which exhibit a significant solar energy conversion efficiency have been fabricated on p-Si with Al as the metal and with an oxide layer having a thickness between 10 and 20 Angstroms; the i-V characteristics of such a cell are shown in Fig. 4 (17)

Theoretical i-V characteristics calculated for an Al-SiO_2-(p-type, 2 ohm cm) Si diode are shown in Fig. 5 with oxide thickness as the running parameter (T = 300°K; $\tau_{no} = \tau_{po} = 10^{-5}$ sec; other parameters entering the calculation have "standard" values). As the oxide thickness decreases, the i-V characteristic approaches the ideal Schockley diffusion diode characteristic. Consequently, the efficiency predicted by theory approaches the efficiency predicted for a p/n homojunction cell having the ideal diode characteristic A = B = 1 in Eqs. (3) and (4).

Such M-I-S cells should be possible on other semiconductors also. The mainproblem in their fabrication is of course the production of oxide layers of reproducible characteristics, having no "pin-holes" over large areas in thickness between 10 and 20 Angstroms . This is difficult with Si which has a native oxide that has been studied intensely because of the commercial importance of M-O-S transistor and

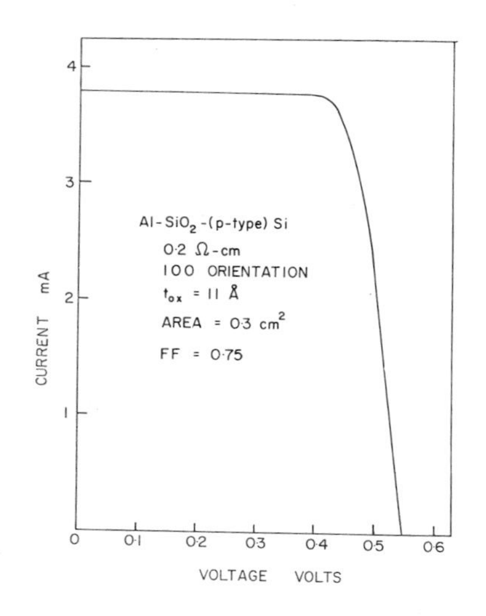

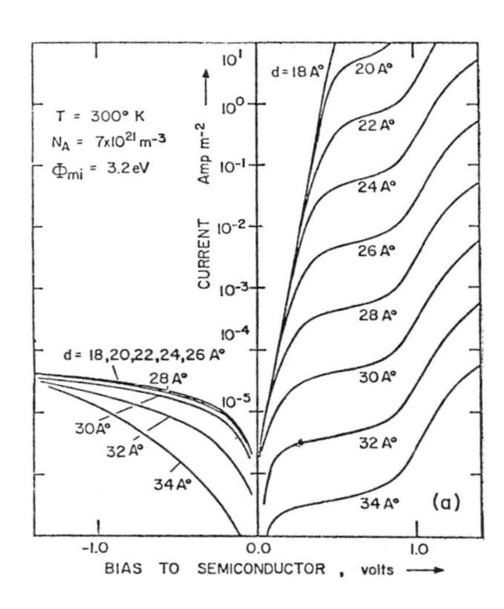

FIG. 4. i-V characteristic of an M-I-S cell made from p-Si. The metal is aluminum; the oxide layer is 11 Å thick (From Ref. 17).

FIG. 5. Theoretical i-V characteristics of MIS cells with various oxide thickness. (From Ref. 17).

other devices; it is even more difficult with other semiconductors which may not have native oxides.

The processing steps required to produce a silicon M-I-S cell like that whose i-V curve is shown in Fig. 4 are as follows. The silicon wafer surface is prepared in a conventional way, the main object of preparative techniques being to remove any residual oxide on the surface. The oxide of controlled thickness is then grown in a standard tube furnace at a temperature not exceeding 500°C. A thin (between 50 and 60 Å) layer of aluminum is evaporated in vacuo over the oxide and a a contact grid also of aluminum is deposited over the aluminum film.

Progress in M-I-S technology requires the development of information and understanding about the growth and characteristics of ultra-thin oxide films. Such M-I-S structures are especially attractive as CCB for polycrystalline Si cells since conventional diffusion methods probably cannot be used with such substrates; enhanced impurity diffusion along the grain boundaries in polycrystalline films can cause short-circuits between the CCB produced by diffusion and the substrate.

IV. FACTORS AFFECTING MINORITY CARRIER DIFFUSION LENGTHS

Excess minority carriers generated in the solar cell as a result of absorption of light must remain mobile long enough so that they can reach the CCB and contribute to the photovoltaic current as they tranverse the space charge region. These light generated carriers can be lost by recombination within the bulk of the semiconductor or at the surface.

Recombination in the bulk is characterised by the bulk minority carrier lifetime τ_B which in turn is a function of the concentration of bulk recombination centers N_r, their capture cross section σ_r and the occupancy factor f which in turn is a function of the relative positions of the Fermi level E_f and of the energy level associated with the recombination center E_r. In terms of these parameters

$$\frac{1}{\tau_B} = N_r \sigma_r f$$

Bulk lifetimes of the order of several microseconds are required for Si solar cell efficiencies in excess of 10%; such values of τ_B require that $N_r \sim 10^{11} - 10^{13}$/cc, values readily attained in single crystal Si having resistivities in excess of about 1 ohm cm.

If the resistivity of the silicon is decreased to the 0.1 - 0.01 ohm cm range [the range which is required to increase V_{OC} to values sufficient to produce 20% AMO cells (4)], the value of τ_B drops below its value in higher resistivity silicon (19, 20). This decrease in τ_B may be caused by the introduction of impurities which act as recombination centers as satellite impurities to the dopant; by an increase in the concentration of structural defects as a result of the incorporation of higher dopant concentrations or by Auger recombination. The latter sets an upper limit on τ_B which can be achieved in any semiconductor. In the case of Si, the Auger process lifetime is given by

$$\frac{1}{\tau} \simeq 1.2 \times 10^{-31} N^2$$

when N is the dopant concentration. For $N = 10^{19}$/cc ($\rho \sim 10^{-2}$ ohm cm), $\tau \approx 10^{-7}$sec. a value which would lead to a reduced value for I_{SC}. The problem posed by recombination in highly doped semiconductors is under investigation. If the observed reduction in τ_B is caused by Auger recombination, the upper limit on η_{max} may be limited to about 17% AMO. If other mechanisms are responsible for the reduced τ_B, there is reason to expect that η_{max} will approach 20% more closely.

Recombination at the free surface of the solar cell is commonly characterized by a surface recombination velocity s whose value can in principle range from zero to infinity. High values of s are encountered at the ohmic contact between a semiconductor and a metal and may also be encountered at the light receiving surface of the cell. Surface recombination losses can be reduced in a number of ways. In the case of an indirect gap semiconductor like Si, the surface of the diffused skin may have a "dead" region in its outermost. This region with an infinite recombination rate reduces the short wavelength response of the cell. Losses can be reduced substantially by making this region very thin compared to the average absorption depth of solar photons; this had led to improved efficiencies (6).

In the case of direct gap semiconductors surface losses play a more significant role because the absorption depth of solar photons is very small, typically several microns. Consequently, an extremely thin diffused skin is needed to keep this loss under control. However, such losses can be reduced significantly by incorporating potential barriers which act as minority carrier mirrors on the boundaries of the solar cell. Such potential barriers can be produced by increasing the doping level at the surfaces as shown in Fig. 6 which shows a p^+ region on the main p-region of the solar cell and an n^+ region on the main n-region of the cell. Silicon cells in which a high

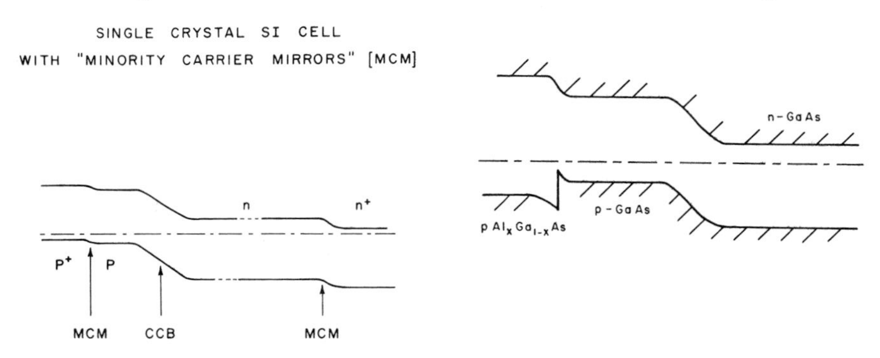

FIG. 6. Energy band diagram of cell with p^+ and n^+ end regions for reducing surface losses.

FIG. 7. Minority carrier mirror formed on GaAs cell by adding $Al_xGa_{1-x}As$. (Ref. 9).

conductivity region is interposed between the ohmic contact and the main volume in which photons are absorbed have been called "back-field cells". Such "back-fields" are especially useful in thin cells because they reflect minority carriers away from the ohmic contact toward the CCB and the short circuit current is increased significantly. Another way to form a minority carrier mirror is to incorporate a material with a larger band gap on the boundary of the cell as shown in Fig. 7 for the $Al_xGa_{1-x}As$/GaAs solar cell. Here the change in E_G results in a potential barrier at the surface which again reflects minority carriers back toward the junction (9). It is possible that a similar situation exists in the Cu_2S/CdS cell, where high short circuit currents are produced after the Cu_2S layer has been subjected to heat treatment in air. This treatment could lead to the formation of a $Cu_2S_xO_{1-x}$ layer at the light absorbing surface; the $Cu_2S_xO_{1-x}$ would have a larger E_G than Cu_2S.

Thus a number of methods for reducing surface recombination losses are available to the solar cell designer.

V. SUMMARY

Semiconductor materials for fabricating solar cells intended for large scale solar energy conversion must satisfy certain criteria. Their forbidden energy gaps must lie between about 1eV and 2.5eV; they should preferably be direct gap semiconductors though this is not absolutely necessary, it being sufficient if the minority carrier diffusion length is greater than the average absorption depth of solar photons. The solar cell must have a charge collecting barrier incorporated in it; this CCB need not be a p/n homojunction. Heterojunction solar cells and metal-insulator-semiconductor cells have led to reasonable solar energy conversion efficiencies. Recombination losses in the bulk and at the surface must be minimized if high efficiencies are to be realized.

REFERENCES

1. J. J. Loferski, J. Appl. Phys., 27, 777 (1956).

2. W. Shockley, "Electrons and Holes in Semiconductors", Wiley, N. Y., 1951.

3. R. J. Stirn and Y. C. M. Yeh, Proc. of the Eleventh IEEE Photovoltaic Specialists Conference-Scottsdale, Arizona, May 1975; p. 437.

4. M. Wolf, Energy Conversion 11, 63 (1971).

5. B. Ellis and T. S. Moss, Solid St. Electronics, 13, 1 (1970).

6. J. Lindmayer and J. F. Allison, COMSAT Tech. Rev. 3, 1 (1973).

7. J. Haynos, J. Allison, R. Arndt and A. Muelenberg, Internat. Conf. on Photovoltaic Power Generation, Hamburg, Germany, September 1974.

8. A. Muelenberg, First International Conference on Solar Electricity, Toulouse, France; March, 1976.

9. H. T. Hovel and J. M. Woodall, J. Electrochem. Soc. 120, 1246 (1973).

10. J. J. Loferski, Proc. IEEE, 51, 667 (1963).

11. T. S. TeVelde, Energy Conversion, 14, 111 (1975).

12. S. Wagner, J. L. Shay, K. J. Bachman, and E. Buehler, Appl. Physics Letters, 26, 226 (1975).

13. S. Wagner, J. L. Shay, and P. Migliorattì, Appl. Phys. Letters, 25, 434 (1974).

14. J. J. Loferski, J. Shewchun, C. C. Wu et al., Contract NSF-RANN/SE/GI 38102X, Brown University, 1976.

15. D. A. Cusano, Rev. de Phys. Applique, 1, 195 (1966).

16. See Ref. (10).

17. J. Shewchun, R. Singh, J. St. Pierre, and J. J. Loferski, Proceedings of First NSF Workshop on Polycrystalline Si Photovoltaic Cells, Dallas, Texas, May, 1976

18. W. Anderson et al., Proceedings of First NSF Workshop on Polycrystalline Si Photovoltaic Cells, Dallas, Texas, May, 1976.

19. P. A. Iles and S. I. Socloff, Proc. of the Eleventh IEEE Photovoltaic Specialists Conference, Scottsdale, Arizona, May, 1975, p. 19.

20. H. Fischer and W. Pschunder, Proc. of the Eleventh IEEE Photovoltaic Specialists Conference, Scottsdale, Arizona, May, 1975, p. 25.

Chapter 16

NON-CONVENTIONAL HETEROJUNCTIONS FOR SOLAR ENERGY CONVERSION

Richard H. Bube

Department of Materials Science and Engineering
Stanford University
Stanford, California 94305

I. *INTRODUCTION*

In this chapter we discuss several types of photovoltaic systems based on heterojunctions between different semiconductors. Most of these might be classified as "less-likely" candidates for solar energy conversion, compared to silicon or gallium arsenide, either because their efficiencies are in general not up to the minimum of 10% desired, or because the materials technology involved is not part of standard industrial practice. In the long term, on the other hand, these systems may well be described as "more-likely" candidates for terrestrial applications, since they have the potentiality of providing sufficiently high efficiencies in all-film cells, thus minimizing the quantity of material needed for large-area applications.

At the ERDA National Solar Photovoltaic Program Review Meeting held on January 20-22, 1976, a variety of systems of non-conventional nature were described, providing a summary of the state of the art as of that date. Taking a look at these systems provides us with a convenient overview of the types of materials involved. Several attempts were being made to reduce the cost of conventional single crystal homojunction cells by a variety of techniques. Efficiencies of 3.7% were reported for polycrystalline Si on graphite (1), 4% for polycrystalline Si on metallurgical grade Si (2), 10% for epitaxial Si on Si ribbon (3), 6.2% for diffused epitaxial Si on Si ribbon (4), 3% for corona discharge deposited polycrystalline Si (5), and 8.5% for a metal-oxide-semiconductor junction on epitaxial GaAs on a recrystallized Ge substrate (6).

A second group of cells described at this same meeting were all-film cells based on the heterojunction between p-Cu_2S and n-CdS. These cells have been known for over 20 years, but have never received extensive development attention. Relatively high efficiencies in the 5-8% range were reported at least 10 years ago, and there has been no significant increase in these figures in the course of recent research. New techniques and processes are now being investigated with encouraging results. The cells described at the meeting included vacuum evaporated CdS with the Cu_2S being formed by a dipping process in which the CdS film is dipped into a warm aqueous solution of a copper salt, showing efficiencies of 4.8% (7) and 6.0% (8); solution-sprayed CdS (in which the CdS film is formed by spraying a solution of $CdCl_2$ and thiourea onto a heated substrate) with the Cu_2S being formed by electrolytic deposition, showing an efficiency of 5.2% (9); and vacuum evaporated CdS with the Cu_2S being formed by reactive sputtering, showing an efficiency of 4% (10).

A third group of cells described at the meeting involved heterojunctions between n-CdS and p-type materials other than Cu_2S. These included vacuum evaporated CdS deposited on single crystal CdTe with an efficiency of 5.2% (11); vacuum evaporated CdS deposited on single crystal InP with an efficiency of 14% (12); and an all-film cell made from CdS and $CuInSe_2$ with an efficiency of 3.6% (13).

Our discussion in this chapter consists of three principal topics: (a) significant criteria in choosing and developing a photovoltaic heterojunction cell, (b) properties of the p-Cu_2S/n-CdS heterojunction cell, and (c) properties of heterojunction cells made from II-VI compounds.

II. *CHOICE OF A HETEROJUNCTION SYSTEM*

We consider general criteria that guide the choice of a potential heterojunction system for photovoltaic solar energy conversion.

A. *Bandgaps*

A heterojunction cell consists of a junction between two semiconductors with different bandgaps. The "normal" mode of operation is for illumination to be incident on the material with the larger bandgap, through which it is transmitted directly to the junction with the material with smaller bandgap; this mode is called "back wall" operation. It provides direct access to the sensitive junction region and effectively eliminates losses due to surface recombination. Heterojunctions may also be operated in a "front wall" mode in which the light is incident on the smaller bandgap material; this mode is in many ways equivalent to the normal homojunction, and absorption in the smaller bandgap material more than a diffusion length from the junction, as well as surface recombination at the smaller bandgap material's surface, must both be considered in optimizing the cell performance.

For operation in the "back wall" mode, it is desirable to have as large a difference in bandgap between the two materials as possible. The smaller bandgap material is chosen to meet the normal criteria for optimum properties of a homojunction and therefore would have a bandgap somewhere near 1.4 eV. The larger the bandgap of the larger bandgap material, the larger is the effective "window" for light to be transmitted to the active junction, and hence the greater the utilization of the solar spectrum.

Most of the photovoltaic "action" in a heterojunction cell occurs in the smaller bandgap material. It is here that the major optical absorption occurs with generation of minority carriers that then must diffuse across the junction to the larger bandgap material. Evaluation of the wavelength dependence of the optical absorption, and of the minority carrier diffusion length in the smaller bandgap material are therefore both important considerations. It is desired to have the minority carrier diffusion length exceed the penetration depth of the light. This goal is aided by using material with long minority carrier lifetimes, and with direct optical transitions yielding a rapid increase in absorption constant at the absorption edge to large values. Limitations on minority carrier collection imposed by falling short of this goal may be removed to some extent by designing the cell so that a drift field toward the junction exists in the smaller bandgap material, so that collection is aided by drift and is not dependent solely on diffusion.

Another source of concern is the loss of light by simple reflection from the surface of the cell on which it falls. This reflection loss can be minimized by coating the surface with a suitable anti-reflection film that utilizes destructive interference effects to reduce the fraction of the light reflected.

B. *Electron vs Hole Injection*

Photovoltaic effects at a heterojunction may involve the injection of either electrons or holes from the smaller bandgap to the larger bandgap material. If the smaller bandgap material is p-type, then electrons are the minority carrier produced, and it is the injection of these electrons into the large bandgap n-type material that completes the process. If the smaller bandgap material is n-type, then holes are injected. Since in general the diffusion length for holes is less than the diffusion length for electrons, the most efficient cell will probably involve electron rather than hole injection.

C. *Interface Spikes*

Since a heterojunction is composed of two different semiconductors, these two materials will in general have different electron affinities. The result is that at the junction interface (considering an abrupt junction) there will be an abrupt change in energy with either a positive or negative sign. This abrupt change in the conduction band will be given by

$$\Delta E_c = \chi_p - \chi_n \tag{1}$$

where χ_p is the electron affinity of the p-type material and χ_n is the electron affinity of the n-type material. A positive value for ΔE_c means a potential spike at the junction interface that impedes the injection of the minority carriers; a negative value for ΔE_c means a negative "spike" or simply an abrupt drop in potential that does not by itself affect the injection of carriers.

The abrupt change in the valence band is given by

$$\Delta E_v = E_{g(n)} - E_{g(p)} - \Delta E_c \tag{2}$$

where $E_{g(n)}$ is the bandgap of the n-type material and $E_{g(p)}$ is the bandgap of the p-type material. A positive value for ΔE_v implies a simple abrupt change without impedance to carriers, whereas a negative value for ΔE_v implies a spike for holes at the junction interface.

The relations of Eqs. (1) and (2) are, of course, idealized ones that may not hold for one reason or another in a real heterojunction. Values of electron affinity available in tables are normally obtained at a material surface in vacuum, and may not correspond to the actual situation at the semiconductor-semiconductor interface. Insofar as real junctions are commonly diffused, rather than abrupt, junctions, much of the detailed structure of an energy spike may be dissipated.

In terms of general criteria, however, one would choose a system in which a spike that would impede the injection of minority carriers is not present.

D. *Diffusion Voltage*

The diffusion voltage describes the total of the band bending produced at the heterojunction by the equalization of Fermi levels in the two materials. Since the diffusion voltage is a measure of the maximum open-circuit voltage realizable in photovoltaic operation, it is desirable to choose a system with as large a diffusion voltage as possible. The diffusion voltage is given by

$$V_D = [\chi + E_g - (E_F - E_v)]_p - [\chi + (E_c - E_F)]_n \tag{3}$$

where the first brackets describe quantities for the p-type

material with E_v as the energy of the edge of the valence band, and the second brackets describe quantities for the n-type material with E_c as the energy of the edge of the conduction band.

E. *Lattice Mismatch*

Although the use of a heterojunction system in the "back wall" mode eliminates the problem of surface recombination that troubles homojunction cells, the existence of a heterojunction commonly introduces interface states because of lattice mismatch between the two semiconductors. These interface states will in general provide a mode for recombination loss of photoexcited minority carriers, preventing their successful injection into the larger bandgap material.

The degree of lattice mismatch (producing dislocations at the interface) must be indeed very small ($\lesssim 0.1\%$) in order to play no significant role in carrier recombination. There are heterojunction systems with such small lattice mismatch, but this criterion, if held strictly, would greatly limit the number of systems to be considered. One of the reasons for the high performance of the n-CdS/p-InP cell (12) is probably attributable to the fact that there is excellent lattice match between cubic (zincblende) InP (a = 5.869A) and the corresponding parameter of hexagonal (wurtzite) CdS ($2^{1/2}$ a = 5.850A). Milnes and Feucht (14) list some fourteen cases of minimum lattice mismatch in III-V/IV (e.g., ZnSe / Ge), II-VI/III-V (e.g., ZnSe/GaAs), III-V/III-V (e.g., AlAs/GaAs), and II-VI/II-VI (e.g., ZnTe/CdSe) systems.

F. *Methods of Preparation*

If an all-film cell is the goal of the choice of a heterojunction for large-area terrestrial applications, then available technology for production of thin films of the material must be included in the evaluation. Many opportunities are available here, including vacuum evaporation, chemical vapor deposition, close-spaced vapor transport, sputtering, reactive sputtering, solution-spraying, electrolytic deposition, etc. Some of these methods have been described above and some will be described in the following discussion of specific systems.

G. *Electrical Contacts*

The photovoltaic cell is not complete until ohmic, low resistance contacts can be made to the cell to allow utilization of the energy generated. The choice of the specific metal or other contact material, of the surface preparation and of subsequent heat treatments, form a composite in which science and art are subtly mixed.

In an all-film cell the sheet resistance of the films commonly contributes more to the series resistance of the cell than does the electrical contact itself. This means that the electrical contact has to be distributed over the area of the film, thus decreasing the effective area for photovoltaic operation. A promising alternative is the use of one of several transparent conducting coatings (e.g., indium-tin oxide, cadmium stannate).

H. *Other Considerations*

Once the criteria for the choice of a heterojunction system are extended beyond the purely scientific and technical criteria, a number of other considerations arise. Among these are the availability of the material in sufficiently large quantities to be considered practical for large-area applications, the cost of the material and the processing required in competition with other sources of electrical energy, the toxicity of the material and dangers of public contamination, and the stability and lifetime of the cells under operating conditions.

Because it is desired that photovoltaic cells should be both financial cost and energy cost effective (i.e., provide much more recovery of investment and of energy expended in the course of its operating lifetime than was needed to produce it), the last consideration of long lifetime is essential. Typical periods considered viable are of the order of 20 years.

III. *p-Cu_2S/n-CdS HETEROJUNCTIONS*

In 1954, Reynolds et al. (15,16) reported an open-circuit voltage of 0.45 V and a short-circuit current of 15 mA/cm^2 when Cu contacts were made to single crystals of CdS, and interpreted the results as being caused by excitation from impurities in the CdS. Williams and Bube (17) investigated Cu contacts on single crystal CdS without any heating, and concluded that they were seeing photoemission from the Cu metal into the CdS. Similar results were early interpreted by others as being the consequence of a p-n junction in CdS (18-20). In 1962 Cusano (21) worked with both p-Cu_2Te/n-CdTe and p-Cu_2S/n-CdS junctions and interpreted the results in terms of a heterojunction model, an interpretation confirmed by subsequent investigations (22-24). It is evident that this system is an example of one developed by a series of historical incidents, rather than of a system to which the above criteria for choice of a heterojunction were consciously applied.

Cells with efficiency in the 5-8% range have been prepared from Cu_2S and CdS using a wide variety of preparation techniques. These include vacuum evaporation, solution spraying, sputtering, reactive sputtering, and vapor deposition. Historically, however, the fascination of the Cu_2S/CdS system was at least partially due to the ease of forming Cu_2S layers on CdS by the simple "dipping" process. A layer of Cu_2S is topotaxially formed on CdS by dipping the CdS into an aqueous solution of cuprous ions at a temperature between 75° and 100°C. This dipping process, with its advantage of convenience, carries with it considerable complexity in the nature of the ion-exchange processes by which the Cu_2S layer is formed. As Cu replaces Cd in the CdS, Cd must diffuse out; once an appreciable layer of Cu_2S is deposited, this diffusion must be through this Cu_2S layer. In this way an additional uncertainty is added to the character of the junction interface and the regions immediately adjacent to it

A. *Sources of Problems*

The development of the Cu_2S/CdS cell has been characterized by several problems that can be summarized briefly as follows.

1. What has been referred to above as Cu_2S is really Cu_xS. The Cu_xS system includes a number of different phases differing only slightly in Cu/S ratio, but with appreciable differences in desirability for photovoltaic use. Between values of $x = 1.75$ and $x = 2.0$, about eight different crystal structures have been identified, the two with the highest values of x being djurleite with $x = 1.96$ and chalcocite with $x = 2.0$. Unless the Cu_xS layer is protected from the environment, changes in composition will occur at the exposed surface through oxidation and subsequent out-diffusion of Cu to change the composition of the Cu_xS. There will also be a diffusion of Cu from the Cu_xS into the CdS, both changing the Cu_xS composition and affecting the properties of the CdS depletion layer, as discussed below.

Research at Philips in the Netherlands (25) has shown that the photovoltaic efficiency of djurleite is about 0.1 that of chalcocite, and that the efficiencies of the other phases of Cu_xS are still smaller. The efficiency is controlled by the diffusion length of electrons in the Cu_xS, which in turn is related to the conductivity of the Cu_x/S. The conclusion of this investigation is that Cu_2S is actually too insulating, and that the ideal composition is for $x = 1.995$. Of all the possible phases of Cu_xS, chalcocite has the largest electron diffusion length, the highest optical absorption, the longest wavelength cut-off, and the highest stability against oxidation.

Examples of processes developed to counteract this degradation effect of the Cu_xS are those reported by (a) AEG-Telefunken (26) in which reproducible chalcocite was obtained by a post-treatment consisting of deposition of Cu onto the Cu_xS layer followed by heating in air, and (b) Philips (27) in which CuCl was evaporated onto a single crystal CdS, was heated 2 min at 180^oC to produce Cu_2S and $CdCl_2$, and then the $CdCl_2$ was dissolved away in H_2O. A cell produced by the latter process on single crystal CdS was reported to have an efficiency in excess of 8% in spite of an approximately 30% reflection loss at the Cu_2S surface (the Cu_2S/CdS cell has been traditionally operated in the front-wall mode).

2. Electrochemical decomposition of the Cu_xS is possible if a voltage in excess of the decomposition voltage (about 0.35V) is applied across any region of the Cu_xS. Since the interface of the Cu_xS/CdS cell is not uniform, but includes "fingers" or even "islands" of Cu_xS, such electrochemical decomposition is possible. When it occurs, such a decomposition causes the growth of Cu whiskers that short out the cell. If a serious problem, this one can be minimized by operation with a voltage less than the decomposition voltage.

3. Diffusion of Cu into the CdS leads to a high-resistivity compensated layer in the CdS, the properties of which can be extremely important in determining cell parameters, as discussed in the next section.

Because degradation of the Cu_2S/CdS cells has been a serious detriment to their wider development and application, considerable attention has been paid to this aspect of their behavior, and a variety of solutions have been proposed. Recent reports on large-scale production-run cells are promising; Shirland (7), for example, reported no degradation in efficiency over 400 days of 50% duty-cycle operation at temperatures up to $80^{o}C$, and a decrease of 5% in power over 10 months of roof-top testing, possibly due to weathering. (For reference to a system not containing Cu_2S, Shay (12) reported that no change in resistivity or degradation was found for a n-CdS/p-InP cell for 24 hours at temperatures up to $550^{o}C$ in air.)

B. *Phenomena Observed for Cu_2S on Single Crystal CdS*

Investigations of the nature of the Cu_2S/CdS cell with single crystal CdS reveal a rich complexity of possible phenomena in these cells, all of which may contribute to some lack of reproducibility, various instabilities, and degradation under operation. Most of the phenomena observed in the single crystal system are also observable in thin-film cells, but frequently with quite different magnitudes. A careful categorization of real or alleged differences between thin-film and single crystal cells is needed, with subsequent deliberation on their reality and interpretation.

Figure 1 illustrates the nature of the band structure at the Cu_2S/CdS interface. Cu_2S is a p-type degenerate semiconductor with a bandgap of 1.2 eV; CdS is an n-type non-degenerate semiconductor with a bandgap of 2.4 eV. The interface is characterized by a $\Delta E_c \approx 0.35$ eV and no positive energy spike for electrons. Because of diffusion of Cu into the CdS, the depletion layer in the CdS is broadened; modulation of the depletion layer width and the local fields at the interface are responsible for many of the effects seen. Because there is a lattice mismatch of about 4% between Cu_2S and CdS, interface states at the interface are expected; recombination loss through these interface states can be a dominant loss process.

Localized states exist in the CdS in the depletion region, at least partially as the result of Cu diffusion. Photoexcitation changes the charge on these states and thereby changes the conditions determining the depletion layer. The presence of

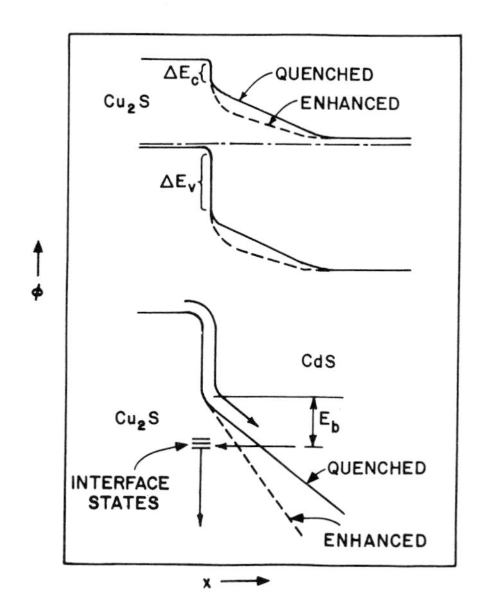

FIG. 1. Band diagram for the interface of the Cu_2S/CdS heterojunction showing the effects of charge in the depletion region modulating the interface to produce enhanced and quenched states (above), and the recombination loss path via tunneling and interface states (below).

positive charge on these states, as from captured photoexcited holes, decreases the width of the depletion layer, increases the forward current which in these cells is controlled by tunneling to interface states, and increases the short-circuit current particularly for long-wavelength excitation, for reasons to be more explicitly stated below. The presence of negative charge on these states, the normal condition of ionized acceptor states, increases the width of the depletion layer, decreases the forward current, and decreases the short-circuit current of the cell. These two states of the cell are referred to as the enhanced state and the quenched state, respectively (28-34).

The existence of effects associated with the enhanced and quenched states of the cell can be demonstrated dramatically in an experiment using two different light sources. The results are shown in Figure 2.

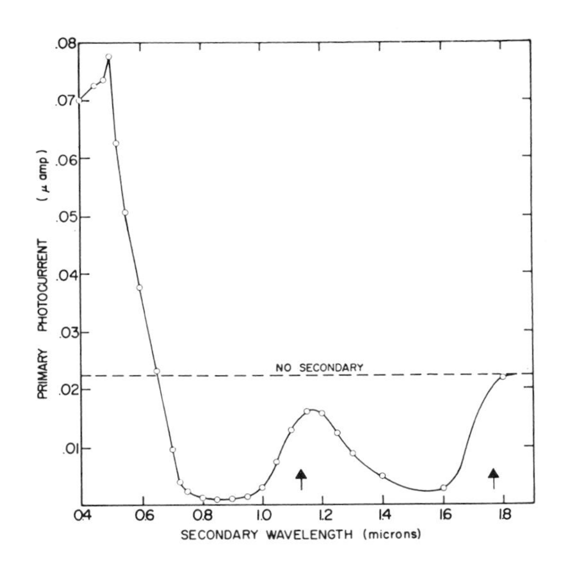

FIG. 2. Enhancement and quenching of short-circuit current for a Cu_2S/CdS heterojunction by secondary light. Arrows indicate where the onset of optical quenching of photoconductivity occurs in sensitive CdS crystals. (From Gill and Bube (30)).

A primary light source with wavelength of 0.655 μm by itself provides a short-circuit current of 0.22 μA. The total current observed when a secondary illumination source was varied in wavelength from 0.4 to 1.8 μm is plotted in Figure 2. For photon energies high enough to photoexcite directly from the localized states in the depletion layer to the conduction band, thus producing localized positive charge in these states, the secondary light contributes additional current with a maximum near the absorption edge of CdS. For smaller photon energies, however, transitions from these localized states to the conduction band are not possible, but transitions from the valence band to hole-occupied localized states are possible, thus freeing the holes for recombination elsewhere and reducing the positive charge in the depletion region. Photoexcitation with the primary light produces the enhanced condition, but additional photoexcitation with long-wavelength secondary light produces the quenched condition.

As might be expected, trapped positive charge near the interface in the CdS also causes a persistent increase in junction capacitance (33). Photoexcitation of capacitance by short-wavelength illumination (photocapacitance) is accompanied by an enhancement of the long-wavelength photovoltaic response. Similarly, a decrease in photocapacitance by exposure to long-wavelength light (optical quenching) or to heat (thermal quenching) causes a decrease in response. For illumination by white light, of course, these enhancement and quenching effects play no significant role, since the white light excitation maintains the enhanced state. They do, however, provide the explanation for a variety of effects observed upon heat treatment, particularly if probed with two-source photoexcitation.

A long heat treatment of a Cu_2S/CdS cell made with single crystal CdS causes a large decrease in response and an increase in the magnitude of the effects of enhancement and quenching. This decrease in response consists of two components: (a) a relatively small thermal component occurring on heat treatment if the cell is not subsequently exposed to light, and (b) a much larger degradation caused by exposure to light at room temperature after the heat treatment (35-38). By a short additional heat treatment above about $100^{\circ}C$, the cell can be completely restored to its condition before optical degradation with no change in depletion layer width. This thermally restorable optical degradation (TROD) effect is phenomenologically similar in several different ways to so-called "photochemical changes" produced by photoexcitation of CdS single crystal photoconductors with excess Cu, manifesting itself as a decrease in electron lifetime. Figure 3 shows the magnitude and the rapidity of this optical degradation process, both for the short-circuit current of several Cu_2S/CdS cells and for the photocurrent of a single crystal CdS:Cu photoconductor. This optical degradation effect, which involves variations by orders of magnitude for the single crystal cell, was also tested on a commercial thin-film Cu_2S/CdS cell of 1972 or earlier vintage; for this all-film cell an optical degradation under white light of 15% was observed, but the long-wavelength response was reduced by a factor of 4.

It may be concluded, therefore, that the measured short-circuit current in a single crystal Cu_2S/CdS cell is the product of two relatively independent processes: the first depends on the generation of electrons in the Cu_2S and their injection into the CdS (a process that is almost completely independent of heat treatment, degradation/restoration, or enhancement/quenching); the second depends on the properties of the CdS:Cu layer near the junction interface that controls the rate of loss of injected electrons via tunneling recombination through interface states (see Figure 1), a process that is affected by heat treatment, degradation/restoration, and enhancement/quenching.

Detailed consideration of the situation shown in the lower portion of Figure 1 leads to a model in which a decrease in tunneling width at the interface (enhancement) leads to both an increase in dark forward-bias current and an increase in short-circuit current, even though tunneling recombination is the primary mechanism for loss of short-circuit current. This result is a consequence of the fact that the short-circuit current is determined by the effect of the electric field at the interface both on the carrier density available (higher fields give smaller densities and hence smaller recombination rates) and on the

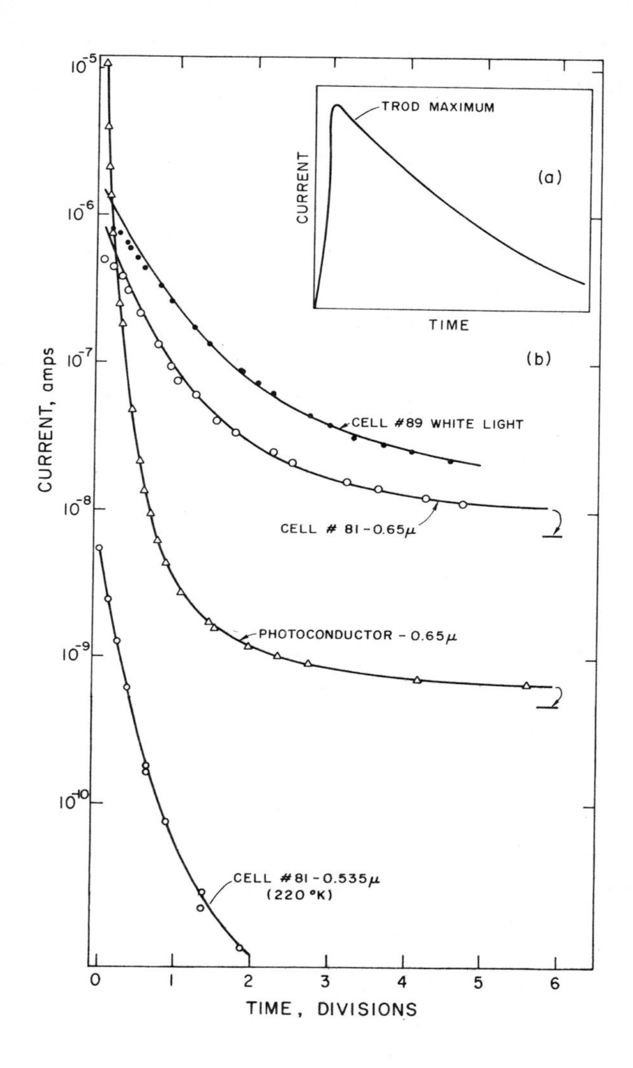

FIG. 3. (a) Schematic of thermally restorable optical degradation (TROD) caused variation of short-circuit current for Cu_2S/CdS cell with time after the beginning of photoexcitation. The rising part of the curve corresponds generally to enhancement, and the falling part corresponds to optical degradation. (b) Short-circuit current and photocurrent vs time during optical degradation. Time scales for the four curves from top to bottom are 200, 4000, 4000, and 100 sec/div, respectively. The upper three curves are for 300°K. (From Fahrenbruch and Bube (38)).

depletion layer width (higher fields give smaller depletion layer widths and hence larger recombination rates), with the former effect dominating. For the case of the forward current, however, the carrier density available for tunneling depends only on the tunneling energy; higher fields with smaller depletion layer widths increase both the effective carrier density and the tunneling probability.

IV. *II-VI HETEROJUNCTIONS*

Because of the variety of ways in which thin films of CdS can be prepared and controlled, it is reasonable to seek other p-type materials in addition to Cu_2S, which might be able to form heterojunctions with CdS and could be free of some of the complications associated with the use of Cu_2S. InP (12) and $CuInSe_2$ (13) are two such materials. It is possible, however, to also find suitable p-type materials among other II-VI compounds, the thin film preparation of which might be expected to be quite analogous to that of CdS. As a second example of actual systems under investigation, we here discuss recent work on II-VI heterojunctions, particularly the n-CdS/p-CdTe junction (39-41).

A. *General Information on II-VI Compounds*

II-VI compounds have bandgaps that cover the visible portion of the spectrum, extending from CdTe with a bandgap of 1.4 eV at the low end to ZnS with a bandgap of 3.7 eV at the high end. Of the six materials (ZnS, ZnSe, ZnTe, CdS, CdSe, and CdTe), only CdTe can be prepared in both n-type and p-type form, and of the rest only ZnTe can be prepared in p-type form. For the preparation of p-n heterojunctions, therefore, the choice of the p-type member is restricted to ZnTe or CdTe. This restriction still leaves some nine possible systems, even if the choice of heterojunction components is limited to simple binary compounds. With the inclusion of ternary compounds, the range of possibilities is extended further.

The various possibilities for II-VI heterojunctions are listed in Table I. All of these materials can be prepared both in single crystal form with various doping levels and in evaporated thin film form with conductivity controlled by co-evaporation. In terms of their photovoltaic properties, if light is incident on the larger bandgap material in the heterojunction p-ZnTe/n-CdTe will be controlled primarily by the photoexcitation of minority

TABLE I

ENERGY RELATIONS AT II-VI BINARY HETEROJUNCTION INTERFACES

Larger Bandgap Material (Bandgap,eV)	Smaller Bandgap Material (Bandgap,eV)	Diffusion Voltage, V[a]	Lattice Mismatch,%
n-CdSe (1.70)	p-CdTe (1.44)	0.57	6.3
n-CdSe (1.70)	p-ZnTe (2.26)	0.61	0.5
n-CdS (2.42)	p-CdTe (1.44)	1.02	9.7
n-CdS (2.42)	p-ZnTe (2.26)	1.06	3.9
p-ZnTe (2.26)	n-CdTe (1.44)	1.28	5.8
n-ZnSe (2.69)	p-CdTe (1.44)	1.43	12.5
n-ZnSe (2.69)	p-ZnTe (2.26)	1.47	7.1
n-ZnS (3.70)	p-CdTe (1.44)	1.62	15.3
n-ZnS (3.70)	p-ZnTe (2.26)	1.66	10.5

[a]Based on electron affinity data cited in Table 1.2 of Reference (14).

holes in the smaller bandgap material with diffusion or drift of these holes over the junction for collection; all the other systems involve photoexcitation and collection of minority electrons.

Conduction band spikes impeding electron injection of 0.19 eV and 0.38 eV are predicted for an abrupt junction for n-ZnSe/p-CdTe and n-ZnS/p-CdTe respectively; a valence band spike impeding hole injection of 0.04 eV is predicted for an abrupt junction for p-ZnTe/n-CdTe. All other systems have no carrier-impeding spikes predicted.

Table I also lists values for the diffusion potential as deduced from published values of the electron affinities of the various compounds. Knowledge of these electron affinities is important for the evaluation of the energy relationships at the junction. Research (41) indicates that adjustments in these cited values are necessary when an actual heterojunction is formed.

All of the II-VI heterojunctions except n-CdSe/p-ZnTe show appreciable lattice mismatch between the components of the hetero-junction. The existence of junction interface states as a result of this mismatch is therefore expected in all such junctions. The role that these interface states play, whether they dominate over recombination in the bulk, and the magnitude and temperature dependence of leakage currents that flow through them, all depend on the detailed energy structure of the junction and the types of processes active.

As illustrative of the type of program and problems encountered, the following sections of this paper describe research recently done on II-VI heterojunctions at Stanford University.

B. *Film Preparation and Control*

Table I indicates that the two most likely candidates for efficient photovoltaic heterojunction systems are the n-ZnCdS/p-CdTe and the n-ZnSSe/p-CdTe systems, as long as normal back-wall operation of a heterojunction is considered. Initial work has been focussed on the n-CdS/p-CdTe cell as the phototype of these systems.

Thin films of CdS and CdTe have been deposited by a close-spaced vapor transport method and by vacuum evaporation. An extensive investigation was made of films of p-CdTe deposited by the close-spaced vapor transport method (42-44) on single crystal n-CdS (45). In the CSVT method, material is transported from a source wafer to a substrate through H_2 at atmospheric pressure. The source and substrate are in contact with blocks that may be heated either by infrared lamps or by electrical heating. The growth rate, grain size, and surface morphology of such CdTe layers is a function of source and substrate temperatures, and the temperature vs time profiles of the source and substrate (39). The best heterojunction of this type prepared to date had a collection efficiency of 0.85 (i.e., 0.85 of the electrons excited in the CdTe by photons absorbed there are collected through the CdS), an open-circuit voltage of 0.61 V, and a solar efficiency of 4.0%. Thin films of other II-VI compounds have also been prepared by CSVT techniques; these include films of CdSe, ZnSe, and ZnTe.

Vacuum evaporation to date has been restricted to CdS, which has been deposited with excellent optical properties on quartz, soft glass, and indium-tin oxide coated glass substrates, as well as on p-CdTe single crystals. A Mo Knudsen cell containing CdS at 885°C was used as the source, and films of thickness 1 to 5 μm were deposited at rates of about 0.5 μm per minute. In order to obtain low-resistivity CdS films, it was necessary either to heat-treat the films in H_2 after deposition, or to co-evaporate a donor impurity. Co-evaporation of In impurity without subsequent heat treatment produces films with excellent optical properties and bulk resistivities ranging from 0.002 to 10^4 ohm-cm; the resistivity can be controlled to within a factor of two by control of the In source temperature. As the concentration of In impurity increases, the electron mobility in the plane of the film increases markedly, and the optical absorption a few hundred

Angstroms beyond the absorption edge of CdS decreases.

C. *Analysis of Heterojunction Properties*

Techniques to evaluate heterojunction performance include (a) J-V characteristics in the dark as a function of temperature in order to establish diode characteristics; (b) J-V characteristics in the light to determine short-circuit current, J_{sc}, open-circuit voltage V_{oc}, fill factor f (maximum power delivered by the cell divided by the product of J_{sc} and V_{oc}), collection efficiency g, and solar efficiency; (c) open-circuit voltage V_{oc} vs ln J_{sc} as a function of light intensity at different temperatures, to permit determination of the diode constant A at different temperatures under illumination, the temperature dependence of the reverse saturation current J_o under illumination, and the effective barrier height for forward currents E*,

$$J = J_o[\exp(qV/AkT) - 1] - g\,J_L \tag{4}$$

$$J_o = J_{oo}\exp(-E^*/kT) \tag{5}$$

where J_L is the total light-generated current; (d) junction capacitance as a function of voltage in dark and light to determine the width and variation of the depletion layer; (e) determination of minority carrier diffusion lengths using scanning electron microscopy; (f) optical transmission to determine the wavelength dependence of the absorption constant; and (g) the spectral response of J_{sc} to determine the variation of collection efficiency g with photon energy.

For heterojunctions prepared by vacuum evaporation of n-CdS films on p-type CdTe single crystals, it has been demonstrated that for 3 ohm-cm CdTe, the reverse saturation current follows Eq. (5) with J_{oo} between 10^2 and 10^3 A/cm^2 for different cells, and with $E^* = 0.60$ eV. Presumably the value of E* represents the thermal activation energy for holes in the p-CdTe at the interface in order to recombine with electrons via interface states, and is determined by the relative electron affinities of the materials making up the heterojunction and their respective doping levels. J_o follows Eq. (5) at room temperature and above, corresponding to a diode factor A in Eq. (4) between 1 and 2. At lower temperatures, transport at the junction is controlled by tunneling and not by thermal excitation.

If the open-circuit voltage for an n-CdS/p-CdTe cell is measured as a function of temperature, it is found to increase with $dV_{oc}/dT = -2 \times 10^{-3}$ V/^oK and an intercept at 0^oK of 1.20 V. The highest actually measured value of V_{oc} (at 80^oK) is slightly larger than 1.0 eV; the value of V_{oc} should saturate with increasing light intensity at a value equal to the diffusion voltage of the junction.

D. *Solar Efficiency*

The ultimately realized solar efficiency of a photovoltaic heterojunction cell is a function of many variables including the bandgap window, the collection efficiency, the reverse saturation current, the detailed transport processes at the interface, and the series and shunt resistances. Furthermore some of these may be functions of voltage; in particular there is evidence that the collection function should be expressed as g(V), with the magnitude of g decreasing with increasing forward bias (41).

An ideal representation of the dependence of solar efficiency on collection efficiency and reverse saturation current for n-CdS/p-CdTe cells is given in Figure 4. The lines drawn are equi-efficiency lines corresponding to the efficiency values indicated. For an ideal heterojunction with J_o controlled only by diffusion or recombination currents at the heterojunction, with negligible series resistance, a voltage-independent collection efficiency, full utilization of the cell area, and negligible reflection loss, the solar efficiency is predicted to be 17%

($J_o \approx 10^{-10}$A/cm^2 and g = 1.0).

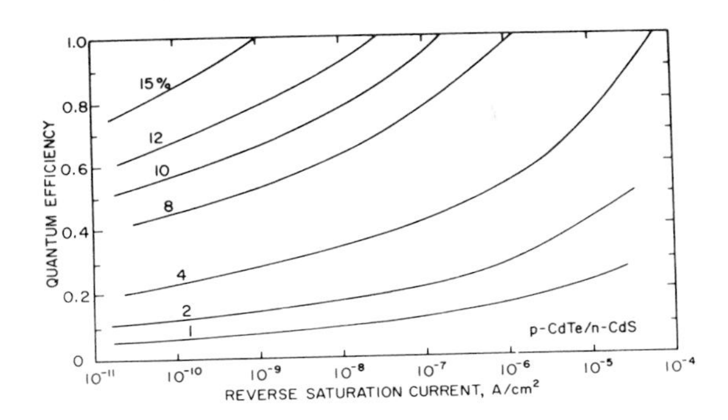

Fig. 4. Equi-efficiency curves for the n-CdS/p-CdTe heterojunction system as a function of quantum efficiency and reverse saturation current. Calculated for negligible reflection, series resistance and area utilization losses.

Table II summarizes some of our most recent results in obtaining values of photovoltaic parameters in real cells of n-CdS/p-CdTe, showing the results also of a transparent conducting coating (ITO) on the CdS film, and the results obtained from a solution-sprayed n-CdS film on single crystal p-CdTe.

If the J_o and g for the highest efficiency cell of Table II are used in Figure 4, it is predicted that an ideal cell would have a solar efficiency of about 11%. Some of the difference between the observed value of 7.3% and this predicted value lies in the fact that the observed fill factor f is less than the theoretically expected fill factor of about 0.75 for an ideal cell; this difference may be the consequence of a voltage-dependent collection efficiency g(V). Continued research has resulted in continued increases in solar cell parameters of n-CdS/p-CdTe cells, and it is expected that cells with efficiency in excess of 10% should be obtainable in the near future.

It is significant to note that the commonly accepted

TABLE II

PHOTOVOLTAIC PARAMETERS OF n-CdTe CELLS[a]

Cell	V_{oc},V	J_{sc},mA/cm^2	g[b]	f	J_o,A/cm^2	A	Solar Efficiency,%
Evap. n-CdS on single xtal p-CdTe; no ITO layer	0.59	13.9	0.85	0.55	2.5×10^{-8}	1.72	6.3[b]
Evap. n-CdS on single xtal p-CdTe; with ITO	0.62	14.4	0.83	0.62	1.3×10^{-8}	1.89	7.3[b]
Solution-sprayed n-CdS on single xtal p-CdTe; no ITO layer	0.54	14.8	0.92	0.59	6×10^{-7}	2.2	6.7[b]

[a]Measurements made using a solar simulator with 87 mW/cm^2
[b]Corrected for reflection losses

assumption that lattice mismatch and high quantum and solar efficiency are mutually exclusive is not necessarily true. In spite of a lattice mismatch of 9.7% in the n-CdS/p-CdTe system, Table II shows that collection efficiencies close to unity for J_{sc} are found in cells made both with vacuum evaporated and solution-sprayed CdS films.

V. *CONCLUSIONS*

The demand for a low-cost, large-area, high-efficiency, stable photovoltaic cell has triggered the hunt for the kind of system that can meet these criteria. Although to date Si single crystal cells have dominated the field, it is not at all clear whether adjustments can be made in these cells to meet the above requirements. Even higher efficiency can be obtained in single crystal GaAs or AlGaAs/GaAs cells, but unless one chooses to proceed with small relatively high cost cells and utilize a high degree of solar concentration, it is again not at all clear whether major utilization of these cells will meet the requirements. The consequence, at least for the long term, has been to produce a wide open field with many different possibilities, each with its own particular areas of promise and problems. As long as options are not too quickly cut off, the situation and the need is right for exciting materials research and development in the scramble to find and develop several potential candidates for different types of photovoltaic applications.

ACKNOWLEDGMENTS

I would like to take this opportunity to acknowledge my indebtedness to my students and colleagues who have worked in the photovoltaic area with me: William D. Gill, Peter F. Lindquist, Alan L. Fahrenbruch, Fredrik Buch, Kim W. Mitchell. I would also like to thank the international scholars who have been involved in this work: Professor Tadashi Takahashi of Sendai University, Dr. Stefan Kanev of the Bulgarian Academy of Sciences, Dr. Valery Vasilchenko of Tartu University, and Dr. E. H. Z. Taheri of the Atomic Energy Commission of Iran.

REFERENCES

1. T.L. Chu (Southern Methodist University), ERDA Review Meeting, Lake Buena Vista, Florida, Jan. 20-22, 1976

2. T.L. Chu, see Reference 1.

3. H. Kressel (RCA Laboratories), ERDA Review Meeting, Lake Buena Vista, Florida, Jan. 20-22, 1976

4. H. Kressel, see Reference 3.

5. R. Wichner (Lawrence Livermore Laboratory, University of California), ERDA Review Meeting, Lake Buena Vista, Florida, Jan. 20-22, 1976

6. R.J. Stirn (Jet Propulsion Laboratory), ERDA Review Meeting, Lake Buena Vista, Florida, Jan. 20-22, 1976

7. F.A. Shirland (Westinghouse), ERDA Review Meeting, Lake Buena Vista, Florida, Jan. 20-22, 1976

8. J.D. Meakin (University of Delaware), ERDA Review Meeting, Lake Buena Vista, Florida, Jan. 20-22, 1976

9. J. Jordan (Baldwin Company), ERDA Review Meeting, Lake Buena Vista, Florida, Jan. 20-22, 1976

10. E. Hsieh (Lawrence Livermore Laboratory, University of California), ERDA Review Meeting, Lake Buena Vista, Florida, Jan. 20-22, 1976

11. R.H. Bube (Stanford University), ERDA Review Meeting, Lake Buena Vista, Florida, Jan. 20-22, 1976; see also A.L. Fahrenbruch, V. Vasilchenko, F. Buch, K. Mitchell, and R.H. Bube, Appl. Phys. Lett. 25, 605 (1974)

12. J. Shay (Bell Laboratories), ERDA Review Meeting, Lake Buena Vista, Florida, Jan. 20-22, 1976; see also S. Wagner, J.L. Shay, K.J. Bachmann and E. Buehler, Appl. Phys. Lett. 26, 229 (1975); J.L. Shay, S. Wagner, K.J. Bachmann and E. Buehler, J. Appl. Phys. 47, 614 (1976)

13. L.L. Kazmerski (University of Maine), ERDA Review Meeting, Lake Buena Vista, Florida, Jan. 20-22, 1976; see also J.L. Shay, S. Wagner and H.M. Kasper, Appl Phys. Lett. 27, 89 (1975) for CdS films on single crystal $CuInSe_2$.

14. A.G. Milnes and D.L. Feucht, Heterojunctions and Metal-Semiconductor Junctions, Academic Press, N.Y. (1972), p.9

15. D.C. Reynolds, G. Leies, L.L. Antes and R.E. Marburger, Phys. Rev. 96, 533 (1954)

16. D.C. Reynolds and S.J. Czyzak, Phys. Rev. 96, 1705 (1954)

17. R. Williams and R.H. Bube, J. Appl. Phys. 31, 968 (1960)

18. J. Woods and J.A. Champion, J. Electron, Control 7, 243 (1960)

19. E.D. Fabricius, J. Appl. Phys. 33, 1597 (1962)

20. H.G. Grimmeiss and R. Memming, J. Appl. Phys. 33, 2217, 3596 (1962)

21. D.A. Cusano, IRE Transactions ED-9, 504 (1962); Solid-State Electron. 6, 217 (1963)

22. B. Selle, W. Ludwig and R. Mach, Phys. Stat. Solidi 24, K149 (1967)

23. P.N. Keating, J. Phys. Chem. Solids 24, 1101 (1963)

24. P.N. Keating, J. Appl. Phys. 36, 564 (1965)

25. Reported by J. Dieleman at the International Workshop on CdS Solar Cells and Other Abrupt Heterojunctions, April 30 - May 2, 1975, University of Delaware. See also B.J. Mulder, Phys. Stat. Solidi a13, 79, 569 (1972); a15, 409 (1973); a18, 633 (1973). Also T.S. teVelde, Solid-State Electron. 16, 1305 (1973), and T.S. teVelde and J. Dieleman, Philips Res. Repts. 28, 573 (1973)

26. Reported by K. Bogus at the International Workshop on CdS Solar Cells and Other Abrupt Heterojunctions, April 30 - May 2, 1975, University of Delaware. See also K. Bogus and S. Mattes, 9th IEEE Photovoltaic Spec. Conf. Proc., Silver Spring (1972)

27. See Reference 25.

28. W.D. Gill, P.F. Lindquist and R.H. Bube, 7th IEEE Photovoltaic Spec. Conf., p.47 (1968)

29. W.D. Gill and R.H. Bube, J. Appl. Phys. 41, 1694 (1970)

30. W.D. Gill and R.H. Bube, J. Appl. Phys. 41, 3679 (1970)

31. P.F. Lindquist and R.H. Bube, 8th IEEE Photovoltaic Spec. Conf., p. 1 (1970)

32. W.D. Gill and R.H. Bube, Proc. 3rd Internat. Conf. on Photocon., E.M. Pell, Ed., p. 395 (1971)

33. P.F. Lindquist and R.H. Bube, J. Appl. Phys. 43, 2839 (1972)

34. P.F. Lindquist and R.H. Bube, J. Electrochem. Soc. 119, 936 (1972)

35. S.K. Kanev, A.L. Fahrenbruch, and R.H. Bube, Appl. Phys. Lett. 19, 459 (1971)

36. A.L. Fahrenbruch and R.H. Bube, 9th IEEE Photovoltaic Spec. Conf., p. 118 (1972)

37. A.L. Fahrenbruch and R.H. Bube, 10th IEEE Photovoltaic Spec. Conf., p. 85 (1973)

38. A.L. Fahrenbruch and R.H. Bube, J. Appl. Phys. 45, 648 (1974)

39. A.L. Fahrenbruch, V. Vasilchenko, F. Buch, K. Mitchell and R.H. Bube, Appl. Phys. Lett. 25, 605 (1974)

40. See series of Reports NSF/RANN/SE/GI-38445X/PR/74,75 and NSF/RANN/SE/AER-75 1679/75,76

41. F. Buch, A.L. Fahrenbruch, and R.H. Bube, to be published in Appl. Phys. Lett. May (1976)

42. J. Saraie, M. Akiyama, and T. Tanaka, Japan J. Appl. Phys. 11, 1758 (1972)

43. F.H. Nicoll, J. Electrochem. Soc. 110, 1165 (1963)

44. F.H. Robinson, RCA Rev. 24, 574 (1963)

45. D. Bonnet and H. Rabinhorst, 9th IEEE Photovoltaic Spec. Conf. 127 (1972)

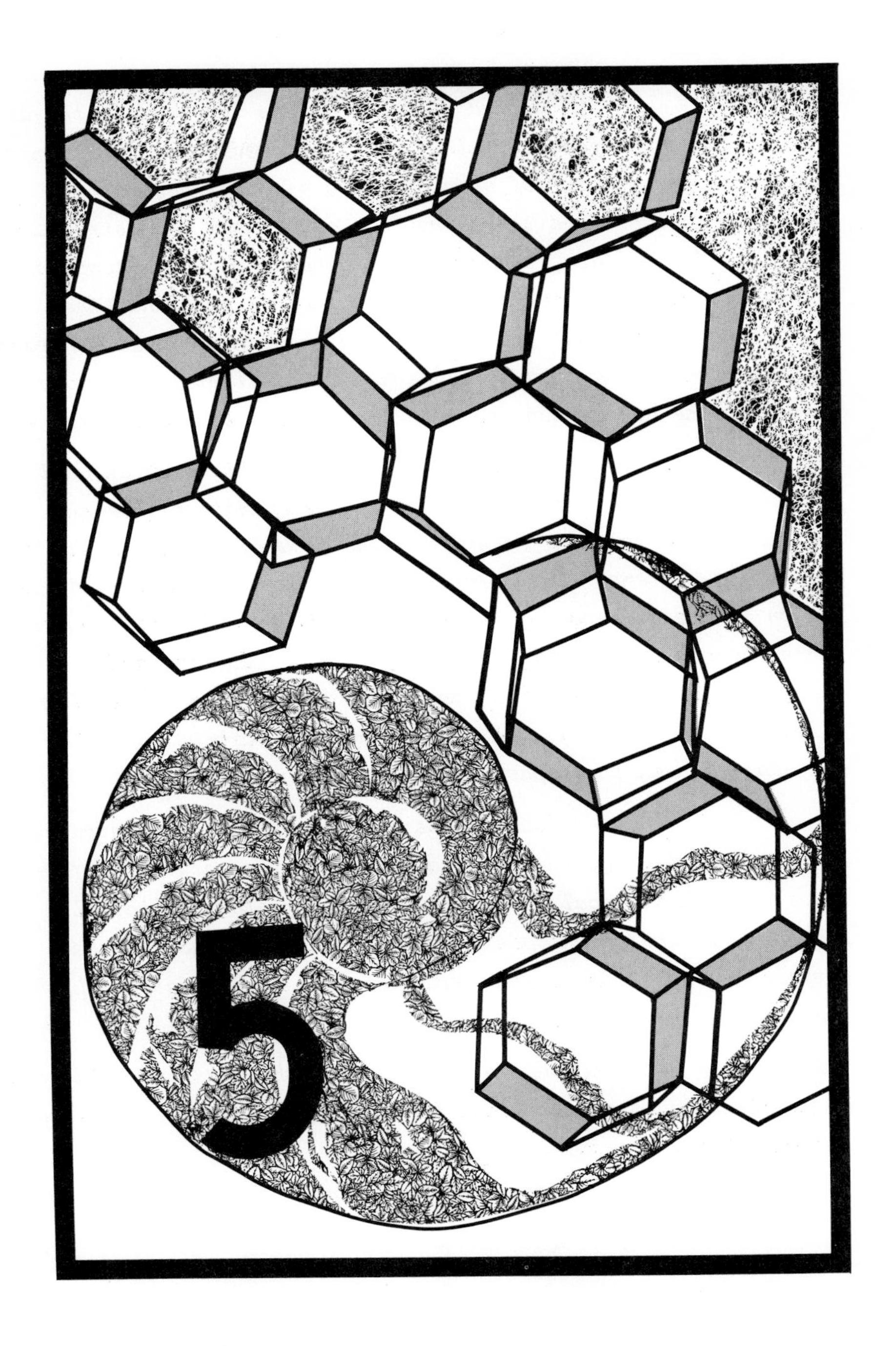
5

V. COAL AND OTHER FOSSIL FUELS

The nature of the problems involved in the use of coal and other fossil fuels for energy production are primarily related to two rather different aspects of power generation. In the first instance are those problems associated with structural materials for boilers, steam turbine generators, high temperature valves, et cetera, whose protective nature is such that they resist the detrimental features of the environment. In contrast to this, the second area is concerned with materials for catalysts for the conversion and cracking of petroleum products.

The first two chapters of this section deal primarily with the metallurgical problems emanating from the materials used for confinement and the flow of energy in steam generating systems. Such problems as: the acceleration of corrosion in the presence of sulphur both at low and high temperatures; carbon contamination leading to metal dusting and carburization; stress corrosion cracking, hydrogen embrittlement, blistering and flaking; and erosion by hard carbide particles are addressed in these papers.

The last chapter of this section illustrates, rather dramatically, the close dependence of catalytic performance on technological innovations in the field of materials science. In the recent past, each important advance in the performance of heterogeneous catalysts was made possible by the emergence of new materials.

The characterization of the many new catalysts by physical and chemical means is the subject matter of the third chapter of this section.

CHAPTER 17

CRITICAL MATERIALS PROBLEMS IN COAL CONVERSION

R. W. Staehle

Department of Metallurgical Engineering
The Ohio State University
Columbus, Ohio 43210

I. PROLOGUE

Following the presentation of this lecture a meeting sponsored by the Mechanical Failure Prevention Group was held at Battelle Memorial Institute on 21-23 April 1976. The subject was: "Prevention of Failures in Coal Conversion Systems." While I presented a paper on the first day, I was not able to stay owing to other commitments. Later I called one of the meeting organizers to ask concerning the outcome expecting that much of what was predicted at the Ohio State Conference (1) would be vindicated. He replied indignantly that "We have not even got that far. The equipment which has been supplied by industry for the prototype plants is so shoddy that major delays have prevented us from finding what the operating materials problems really are."

II. INTRODUCTION

We start this discussion with the first law of materials application: "all engineering materials are unstable." Thus, the primary efforts of design and materials engineers must be directed toward delaying the inevitable deterioration of materials for sufficiently long times that the material will perform as desired for the intended life of the equipment. A second law having as much validity as the first for large engineering systems is: "The chain is as strong as its weakest link." While the latter seems trite, arising as it does out of someone's long forgotten book of sayings, nonetheless, it is a conceptual cornerstone in the development of large engineering systems. Unfortunately much of the development effort in new systems is placed upon the novel and intriguing problems with little effort given to what is commonly referred to as off the shelf or conventional equipment.

Past history with the fossil and nuclear power industries shows that most of the down time is associated with conventional equipment and not with more exotic components. The implicit suggestion here is that whatever development work is undertaken to construct large coal conversion systems should consider reliability of all components; action should be taken to raise the integrity of such equipment in proportion to the consequences of its failure to the performance of the total system.

The full scope of potential materials problems

was considered at a meeting in April 1974 at Ohio State University on the subject: "Materials Problems and Research Opportunities in Coal Conversion" (1). This document should be carefully reviewed.

A second and related meeting was held to consider materials for MHD applications: "NSF-OCR Engineering Workshop on MHD Materials" (2). This is also an excellent volume and should be carefully reviewed.

In assessing the expected performance of structural materials in coal conversion systems the approach is little different from that which should be used in any other system. One starts first by defining the chemical species which are an implicit part of the system. The second question one asks concerns the operational influences including temperature, pressure, flow, particle velocity, heat transfer, etc. Having defined these one then moves to define acceptable materials or to develop a materials testing program. Unfortunately the common approach is to decide early on the materials of construction and then forget the question of compatibility hoping, by closing ones eyes sufficiently tightly, that any deleterious influence would somehow lie pleasantly dormant for an infinite time.

With respect to the environmental species in coal conversion equipment, these are not difficult to identify. They include primarily carbon, hydrogen, nitrogen, oxygen, sulfur, and chlorine. Impurities such as zinc, lead, tin, and others often are damaging but are unlikely.

Physical influences which are to be considered simultaneously with the chemical influences include the following: temperature, time, thermal cycling, erosion, stress, and stress cycling together with the exacerbation of creep-fatigue and stress corrosion cracking.

The above have implied damage processes which are associated with fracture or corrosion type phenomena. These phenomena may be abetted by prolonged thermal influences on the structure and properties of metals. For example, temper embrittlement operates over an intermediate range of temperatures and affects low alloy steels of the type used in pressure vessels and turbines; stainless steels sustain the sensitization phenomenon at a somewhat higher temperature range; and, finally, high alloy ferritic materials age to produce an embrittling sigma phase.

Having now defined the major considerations in materials performance, i.e., the chemical environments, physical influences, and the change of the substrate with time, now consider what tools and resources are available for describing boundary conditions within which satisfactory performance might be expected. There are four separable categories. The first resource is the large reservoir of empirical data obtained from reasonably similar industrial systems. For example, there is already a fund of information from low BTU coal gasification systems. The petroleum, gas turbine, and chemical industries provide extensive information.

Secondly, there are well developed procedures based on thermodynamic analyses which can be used to predict regimes of stability of materials. These diagrams for stability of materials in environments are utilized in various forms. One of the most common is the simple plot of free energy of formation of a compound vs. temperature.
For the

$$M + {}^{1}/_{2}\ O_2 = MO \qquad (1)$$

the free energy of formation is plotted vs. temperature and the curves for the formation of various metal oxides can be readily compared as for Figure 1. Lower values of the free energy of formation indicate the more stable oxides. Such comparative plots as in Figure 1 can be developed for the formation of other compounds such as sulfides, chlorides, carbides, nitrides, etc. all of which are of interest in coal conversion technology.

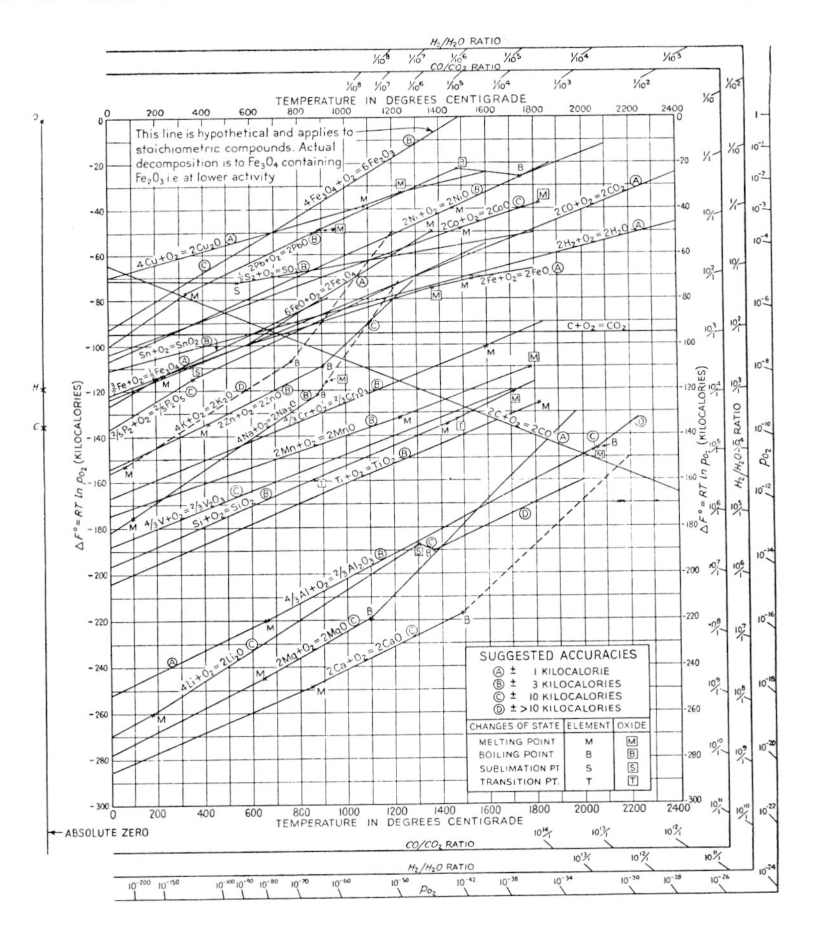

Figure 1. The standard free energy of formation of many metal oxides as a function of temperature [From F. D. Richardson and J. H. E. Jeffes, substantially as in J. Iron Steel Inst. 160, 261 (1948)].

A useful approach for assessing stability of alloys in high temperature gases is to follow the pattern of Figure 2 where the partial pressures of sulfur and oxygen are plotted with respect to the existence of principal phases of the alloy species. Here regimes of stability of chromium and nickel phases are identified from work by Quets and Dresher (3). The broad range of stability of the protective Cr_2O_3 film accounts for the stability of alloys which contain chromium as a major alloying element. The general approach epitomized in Figure 2 has been discussed by Rapp (4).

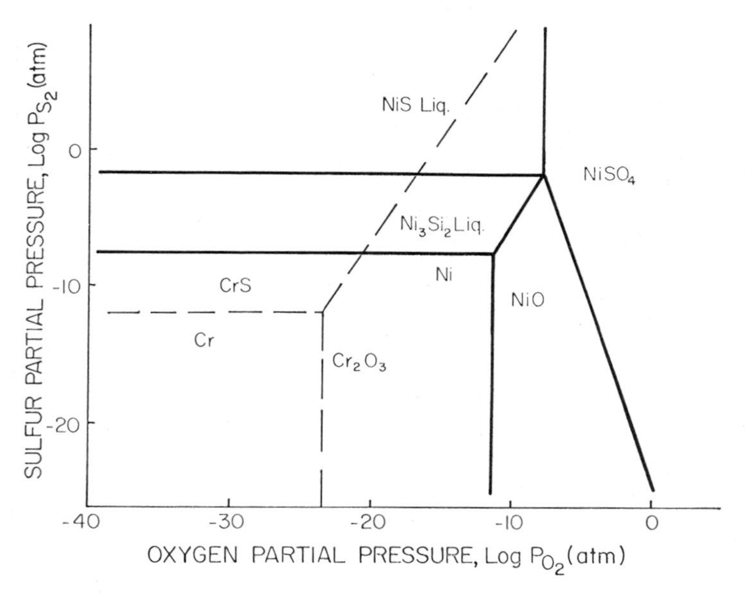

Figure 2. Superposition of Ni-S-O and Cr-S-O stability diagrams at 1200°C (3).

The comparably important diagrams for low temperature aqueous corrosion processes are the Pourbaix diagrams (5, 6). These address the question of stability of pure metals in aqueous solutions. Here important reactions are considered, e.g.

$$Fe = Fe^{++} + 2e \qquad (2)$$

$$Fe_2O_3 + H^+ = 2Fe^{+3} + 3(OH^-) \qquad (3)$$

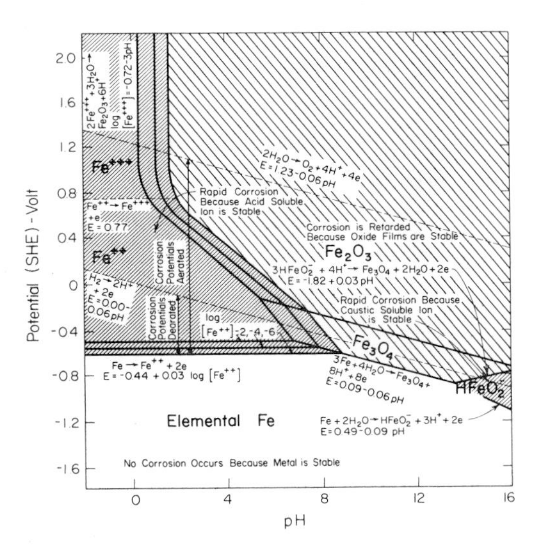

Figure 3. Annotated potential-pH diagram for the iron-water system at 25°C. The hatched regions indicate zones of various reaction types: no hatching indicates that the metal will not corrode; find hatching means that the metal tends to corrode since the ionic species are soluble; broad hatching shows regions of insoluble product layers where passivity should occur if the layers are protective. The applicable equilibria with their corresponding pH and potential dependencies are placed parallel to the respective lines. Regions are also noted where the corrosion potential can exist depending upon the state of aeration of aqueous solutions. Adapted from Pourbaix.

Equilibrium equations for these reactions are developed in terms of the electrochemical potential and pH and diagrams such as Figure 3 are developed for regimes of stability for various iron compounds. For combinations of potential and pH where ions are stable corrosion is expected; where the insoluble compounds are stable, corrosion is reduced owing to the protective nature of insoluble compounds. Figure 4 compares Pourbaix diagrams for aluminum in water and fused sulfate salts. There is substantial similarity here which, to some extent, is fortuitious but the general applicability of the diagrams is clearly demonstrated for electrolytes in general.

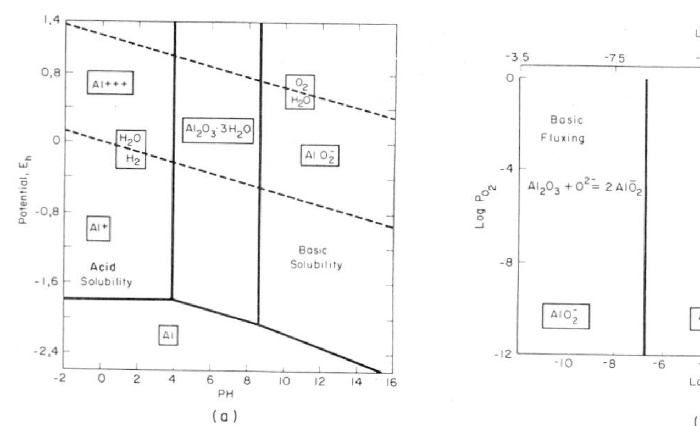

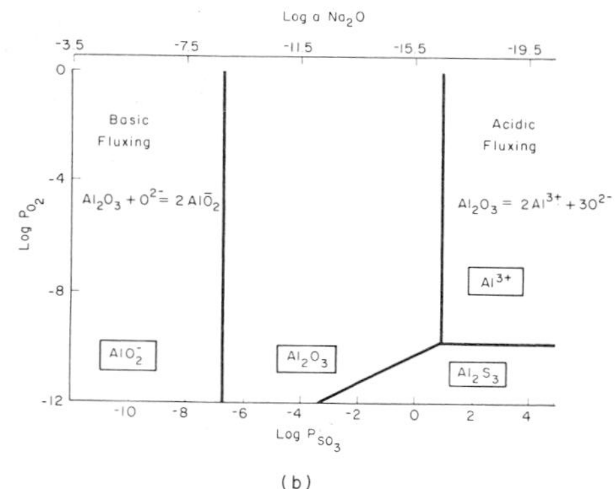

Figure 4. Comparison of thermodynaic diagrams of the stability of aluminum in water at left (a) and in molten sulfates at right (b). The water developed for 25°C and the other for 1000°C. The reversal in location of acid and base regions arises from conventions in the respective technologies.

A third approach to assessing the expected performance of materials involves laboratory testing. Advanced approaches take their leads from the diagrams of Figures 1-4 as modified by the expected engineering environments. These laboratory tests would measure general and localized attack under carefully controlled conditions.

Fourth, materials testing is conducted in pilot plant facilities and compared with the predictions from laboratory tests.

Each of these four methods for developing an understanding of materials performance produces its own feedback which suggests approaches, analyses, and experiments on the other category.

The following discussion will emphasize problems associated with metallic materials. Problems of ceramic materials in coal conversion systems have been dealt with by Wachtman et al (7). Unfortunately, the corrosion behavior of ceramic materials has never been considered in as much detail as metallic materials. The reason, of course, has to do with the use of bulk of quantities of ceramic materials where small amounts of corrosive activity is not so serious as it is for metallic materials. However, only metallic materials can serve as a pressure boundary; thus, integrity still remains as the principal concern in performance.

The following is divided into brief discussions concerning each of the chemical species; this is followed by comments on the change in substrate, physical influences, and a few commentaries on the so called weak link problem.

I have divided the consideration of the chemical species into high and low temperature circumstances. One is prone to think in the coal conversion circumstances that only the high temperature problem is at issue, and a good deal of work is directed toward the resolution of problems arising in high temperature operations. However, there are inherently low temperature environments which can be equally or perhaps even more debilitating. Such environments, for example, can occur in scrubbers.

Finally, in an overall view Table 1 summarizes the occurrence of failures of components in coal conversion units. This information was prepared by Smith at the National Bureau of Standards and was reported later in the ERDA newsletter "Materials and Components in Fossil Energy Applications" (8).

III. CARBON

In high temperature gases carbon-containing gas molecules produce two different kinds of phenomena depending upon whether the gases are oxidizing or reducing.

In oxidizing conditions mild steels oxidize in dry CO_2 environments much as they do in oxygen. However, above about 500°C and especially in the presence of small amounts of water the corrosion rate is greatly accelerated. After an initial period where parabolic kinetics is operative--characteristic of adherent films--the protective quality of the films breaks down and a rapid linear rate ensues (9, 10). Figure 5 shows this phenomenon schematically and Figure 6 shows the post breakaway rate as a function of water content and temperature. These conditions are most pertinent to coal gasification. These rapid rates can be reduced by some alloying elements including silicon and by lowering the water concentration.

In reducing environments carburizing conditions may occur and the metal is attacked by two processes. One is simple carburization. Here the surface is embrittled by the inward diffusion of carbon.

In atmospheres where reducing conditions prevail, more serious is the phenomenon of "dusting" which has been

TABLE I

TYPE OF INCIDENT BY FREQUENCY OF OCCURRENCE*

Type of Problem	No. of Items	No. of Incidences	Process A	Process B	Process C	Process D	Significance of Incident: To Particular Process	Significance of Incident: To Coal Conversion Technology	Recommendations
Sulfidation	22	20		2			(1) Thermocouple critical for control (2) Slurry grid – complete shutdown	Major problem causing short life	(1) Use coatings, better alloy, change environment
			18				(1) Severe and continuing problems with thermocouple tube and heater coil not catastrophic but severely limits life of parts	Expect to be the major problem determining life because of high sulfur coal	(2) Alonized Fe-Ni-Cr alloys are best now available
Corrosion	10	8	5	3	3		Unknown	Unknown – many critical areas, possible catastrophic	(1) Need detail diagnostic failure analysis (2) Need critical review of all processes
SCC	9	4	3	1			Great – causes shutdown and expensive repair – all Cl cracking Expansion bellows Water Lines	Great – nearly all systems have some areas where possible – identify and correct	(1) Carefully monitor environment (2) Better material selection (3) Some design change possible
Erosion/ Wear	8	6	2	1			Great – short life, shutdown Not critical – repair during maintenance	Unknown – identify areas	(1) Change design/material (2) Misalignment is cause
					3	1	Critical – causes shutdown Critical – causes shutdown	Great – urgent problem of seals on pumps	Better pump seals, filters, double seals
Fabrication Defect	3	3		3			(1) Valve, (2) Reducer – residual stress in welds – shutdown (3) Piping weld crack – total shutdown	Great – short life, catastrophic shutdown	Strict attention to welding procedures and residual stresses
Carbonization	3	1				3	Great – this is the critical materials problem for the clean coke process	Unknown – review other systems to identify critical regions	(1) Add S, water to environment (2) New alloy (3) Redesign
Design	5	5					(1) Quench pot – short life (2) Bellows – total shutdown (3) Heat exchanger – total shutdown (4) Knife shouldn't be problem (5) Sight glass – severe safety hazard	(1) Many similar quench areas – critical (2) Urgent problem – all systems (3) Many similar areas – critical (4) None (5) Critical – all systems use this	(1) Redesign (2) Redesign (3) Redesign on basis of thermal stress (4) None (5) Redesign, new material
Thermal Stress	3	3		2	1				
Other	2	2			2				
Metal Dusting	1	1	1				Generally not a problem S in environment prevents	Identify critical areas Extremely rapid, catastrophic failure when present	(1) Need better alloy (2) Add small amounts of S (3) Coatings (oxide, sulfide) may help
Refractory	1	1	1				Great significance – major shutdown, long time shutdown, expensive repair	Highest significance – identify critical areas, inspection, Q.C.	(1) Diagnostic Fail. Analysis URGENT (2) Need better refractory
Material Selection or Q.C.	1	1	1				Great – shutdown plant	Requires constant alertness	(1) Need high level of Q.C.
Unknown	2	2	1		1		Great – total shutdown Great – fire, total shutdown	Unknown Unknown	(1) Diagnostic failure analysis needed

*Most failures tabulated here occurred in 1975 and 1976, with some earlier.

reviewed by Hochman (11). The mechanism of the dusting phenomenon is not well established but seems to involve critical surface processes where CO decomposes with the next stop being the inward diffusion of carbon. Particularly characteristic of dusting is a peak in the attack in the range of 500 to 700°C

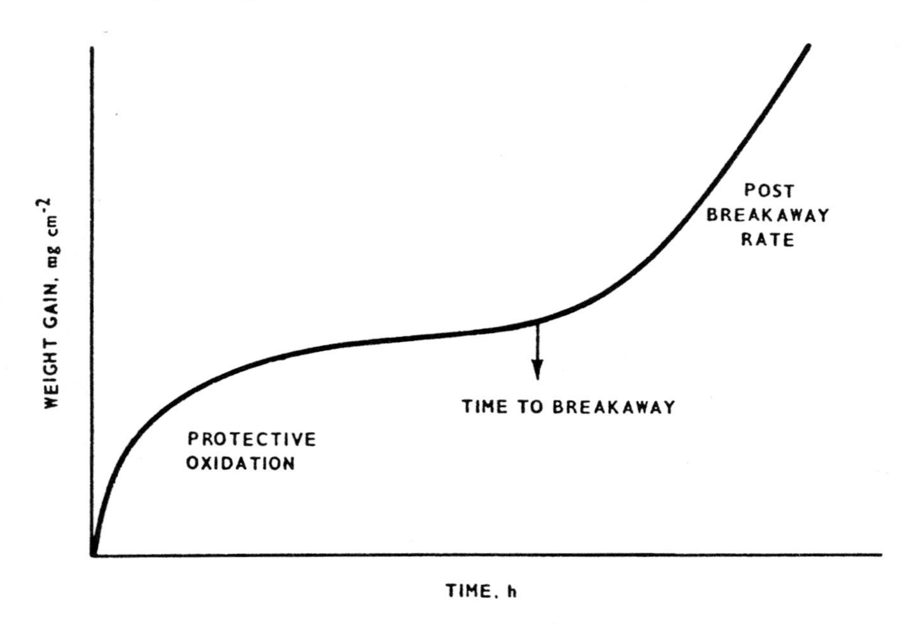

Figure 5. Schematic representation of breakaway corrosion.

depending upon the alloy and environment. Figures 7 and 8 from Hochman show the effects of environment, temperature, and alloy composition on dusting.

Dusting can be inhibited by adding sulfur species which seem to slow the decomposition of CO molecules on the surface much the same as sulfur prevents other catalytic reactions at high temperatures. Hydrocarbon gases also produce

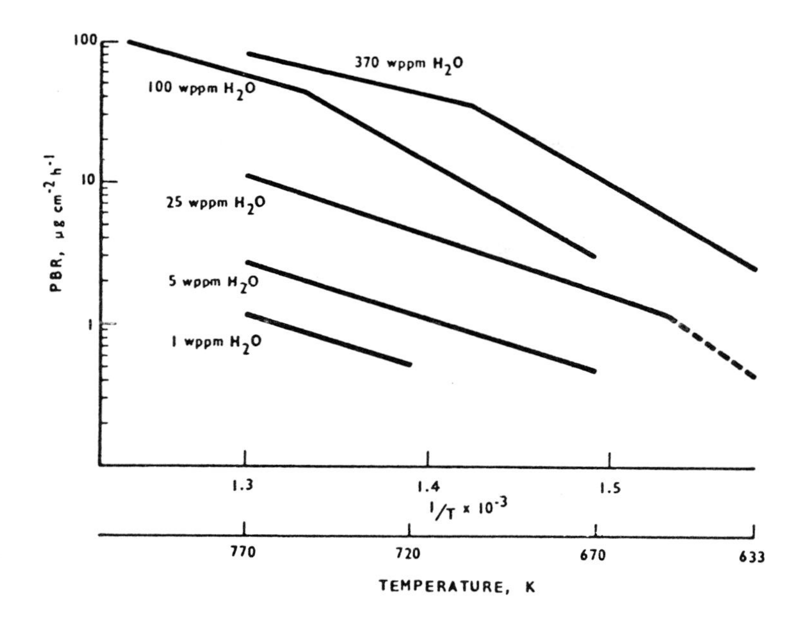

Figure 6. The effect of temperature on PBR at water levels in the range 1-370 wppm H_2O.

dusting but at higher temperatures. Water vapor and ammonia appear also to inhibit the dusting.

At low temperatures carbon species contribute to acidifying the environment when CO_2 dissolves in water to form carbonic acid. More virulent is the stress corrosion cracking of mild steels which is caused by carbonates and by dissolved CO (9, 10). Results are illustrated in Table II and Figure 9 from references 12 and 13. The general range of conditions where either of these environments cause accelerated cracking has not been established.

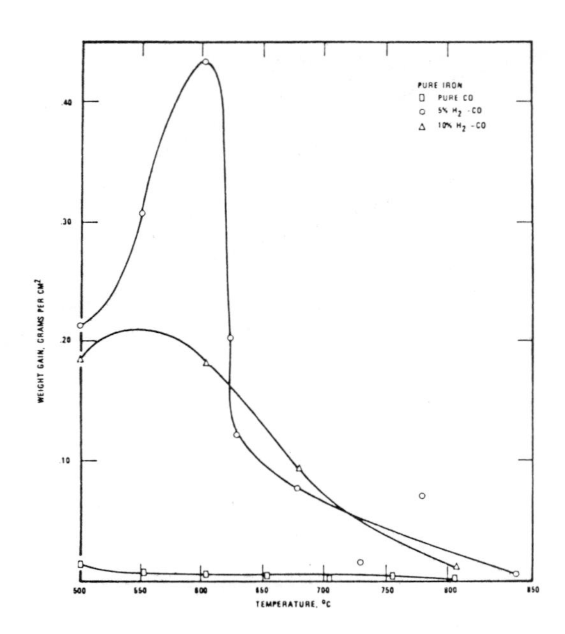

Figure 7. The reactivity of pure iron in pure CO, 5% H_2-CO and 10% H_2-CO environments.

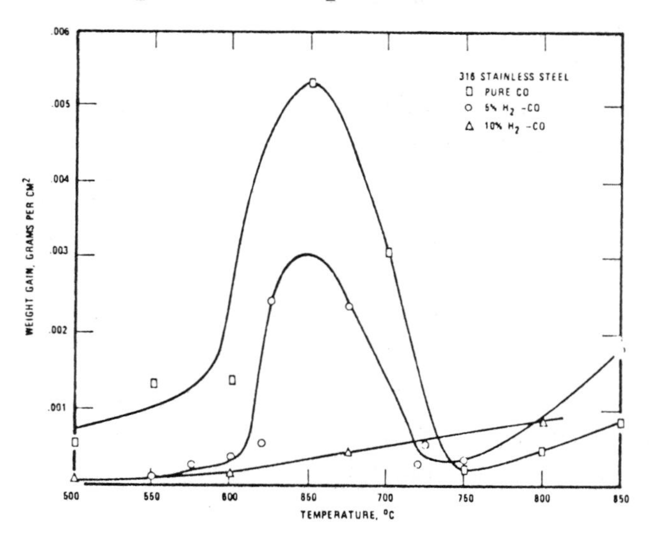

Figure 8. The reactivity of 316 stainless steel in pure CO, 5% H_2-CO, and 10% H_2-CO environments.

TABLE II.

Effect of the Amount and the Ratio of Gas Mixture on Stress Corrosion Cracking of Mn-Steel (Q.T.) at 18°C in Moist CO/CO_2

No.	Partial Pressure (kg/cm^2)			Total Pressure (kg/cm^2)	Period (Week)		
	CO	CO_2	N_2		2	4	6
1	26	-	-	26	O	-	O
2	26	14	-	40	X	X	-
3	26	14	60	100	X	X	X
4	10	16	-	26	X	X	-
5	10	16	74	100	X	X	X
6	5	18	77	100	X	X	-
7	1	17	82	100	X	X	-
8	0.1	16	84	100	-	-	O
9	0.01	16	84	100	-	-	O
10	26	1	73	100	-	-	X
11	26	0.2	74	100	-	-	X

X: Crack O: No Crack

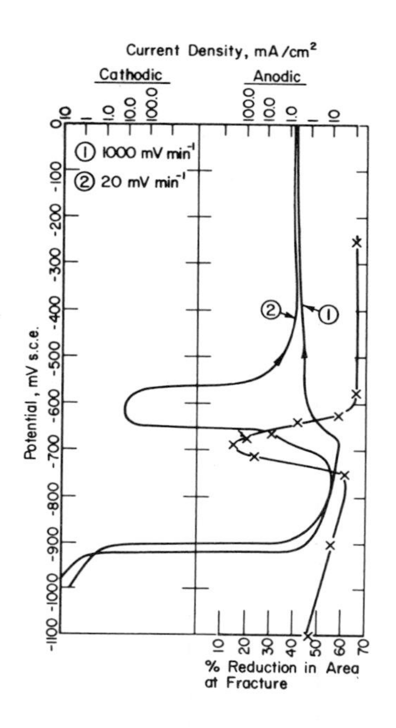

Figure 9. Polarization curves and stress corrosion test results for mild steel in 1N Na_2CO_3 + 1N $NaHCO_3$ at 90°C.

IV. SULFUR

At high temperatures sulfur-containing species seem to exert three important effects. The first involves the so-called hot corrosion phenomena which involves attack of metals in environments which contain both sulfur and oxygen. This phenomenon has been extensively investigated in connection with accelerated corrosion phenomenon in gas turbine blades where sulfur is present in the fuel. The accelerative process seems to be associated with the formation of a molten sulfate salt at the metal surface.

Figure 10 shows that the molten sulfate greatly accelerates the corrosive attack on high nickel alloys (14). This attack can be substantially mitigated adding aluminum or chromium to the alloy in order to form a protective film.

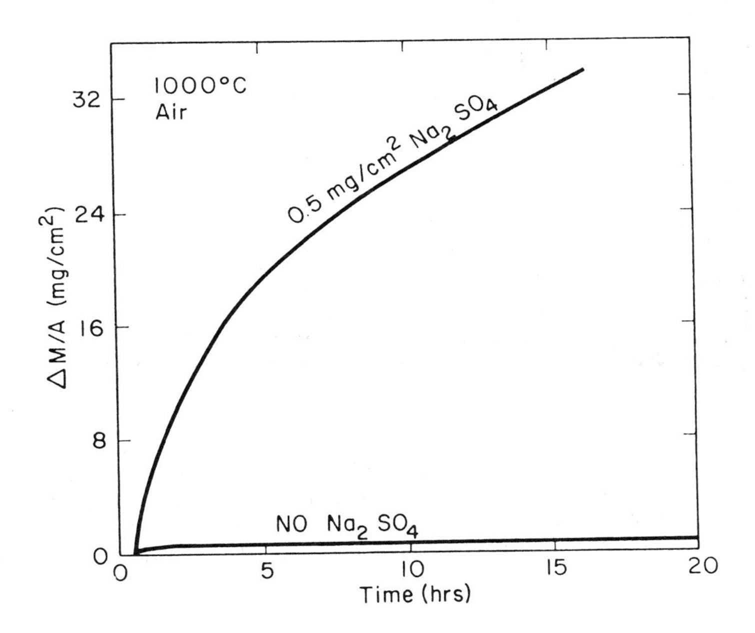

Figure 10. Comparison of the oxidation behavior of the nickel-base super alloy B-1900 (Ni + 8 Cr + Ti, 6 Al, 6 Mo, 4 Ta, 10 Co, 0.1 C, 0.07 Zi, 0.01 B) with and without Na_2SO_4 coating.

Figure 11 shows the beneficial effect of adding species which improve the quality of the protective film (15).

Sulfur gases at high temperatures seem to attack high nickel alloys rapidly along grain boundaries. For

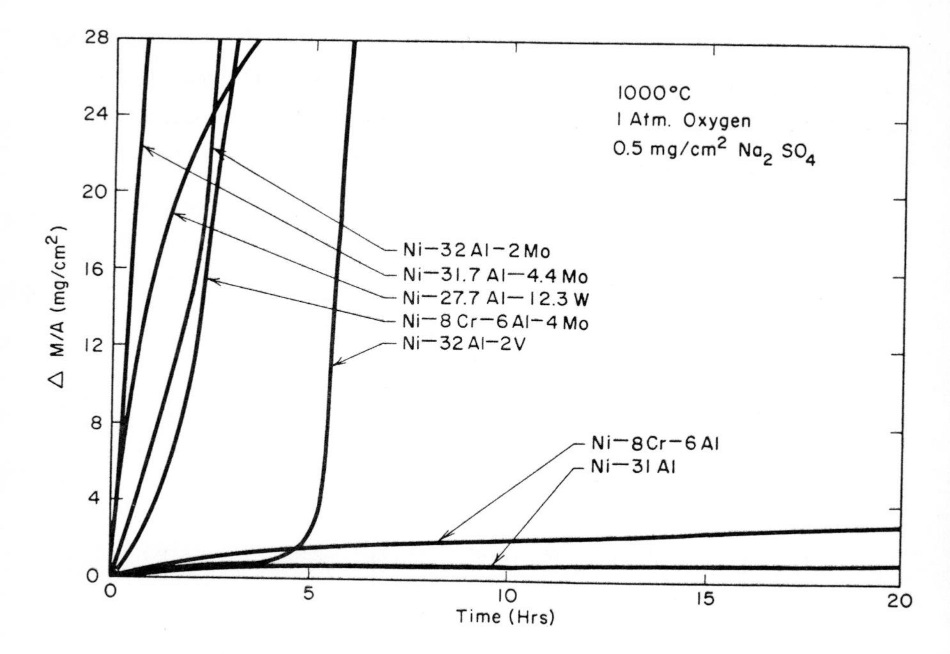

Figure 11. Weight change versus time for the oxidation of Na_2SO_4-coated (0.5 mg/an^2) specimen of Ni-Al and Ni-Cr-Al alloys as effected by the addition of V, Mo, and W.

metals exposed to H_2S environments, they are protected generally by increasing the chromium concentration as shown by Yamamoto et al in Figure 12. The beneficial effect of chromium is to be expected here according to the thermodynamic prediction from Figure 2.

At lower temperatures and in aqueous solutions sulfur exerts two important influences leading to accelerated attack--again depending upon the state of oxidation of the sulfur. In the oxidized state dissolved SO_2 acidifies the environment. Additionally, SO_2 raises the dewpoint to the range of 160°C depending upon the concentration of SO_2. The precipitated water with the SO_2 now incorporated to produce H_2SO_4 is very corrosive. Figure 13 shows clearly this correla-

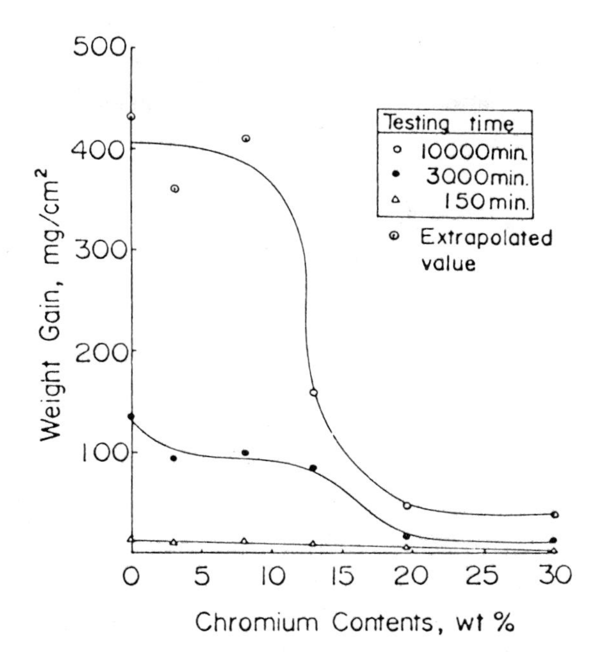

Figure 12. Amount of corrosion products (weight gain) of iron-chromium alloys at 600 C against chromium content of alloys in hydrogen sulfide.

tion between precipitation of water from a SO_2-contaminated atmosphere and the corrosion of iron (17).

If the sulfur is present as H_2S it substantially aggravates the stress corrosion cracking of iron base alloys as shown in Figure 14; and as the strength of the alloy is increased the tendency for SCC to occur is increased. Further, H_2S accelerates the entry of hydrogen in acid environments and blistering often results.

Dissolved H_2S decreases the corrosion resistance of iron. Together with HCl there is a synergistic effect even at relatively low temperatures as shown in Figure 15 (18).

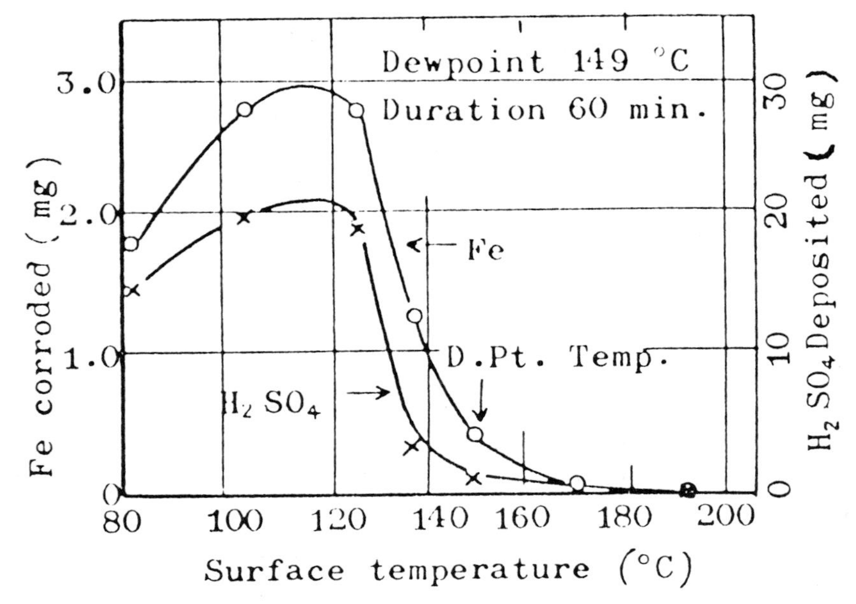

Figure 13. Effect of surface temperature on condensation and corrosion.

V. HYDROGEN

At high temperatures the most insidious effect of hydrogen is called hydrogen damage. The engineering data upon which hydrogen damage is based is the Nelson curve shown in Figure 16 (19). This figure defines the combination of temperature and pressure above which a failure will occur for an iron base alloy. This phenomenon seems essentially to result from the formation of methane molecules within the metal and the subsequent formation of blisters. Hydrogen damage can be mitigated by adding chromium and molybdenum and as a result the failure curve is raised to higher temperatures and pressures.

During thermal cycling of large vessels where hydrogen may have entered the metal there is a further form of deterioration called "flaking." This occurs when the surface temperature is rapidly reduced with the result that the hydrogen is supersaturated. This resulting supersaturation causes

Steel	Chemical composition (%)									TYS (ksi)	UTS (ksi)	H_V
	C	Si	Mn	P	S	Ni	Cr	Cu	Mo			
HT 60	0.12	0.35	1.10	0.019	0.019	0.25	0.15	0.25	0.10	82	97.5	223
HT 70	0.11	0.30	0.95	0.013	0.014	0.85	0.42	0.14	0.32	104	111	238
HT 80	0.14	0.36	1.03	0.010	0.011	1.12	0.47	0.29	0.52	113	119	277

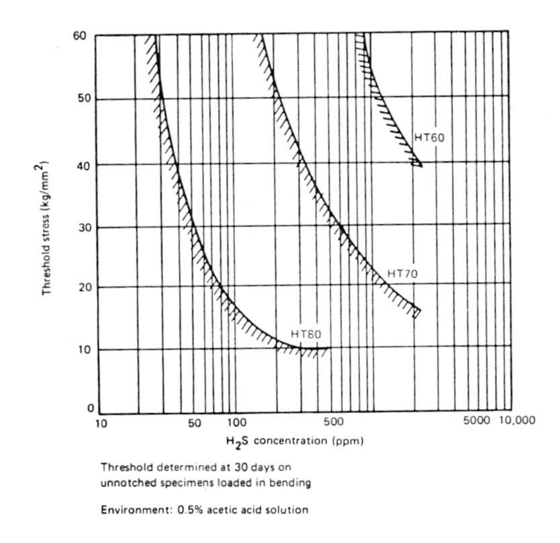

Figure 14. Relationship between applied stress and H_2S concentration for initiation of cracking in various grades of high-strength steel (after Ishizuka and Onishi).

hydrogen bubbles to precipitate and force metal flakes off the surface.

In addition, another damage process which can occur at low temperatures and high temperatures is called blistering. An example of blistering is shown in Figure 17. Blistering essentially results from relatively high hydrogen activities. These may result from the presence of hydrogen sulfide or cyanide.

At lower temperatures, hydrogen enters the metal to cause stress corrosion cracking of steels. An early manifestation of this problem in high strength steels especially

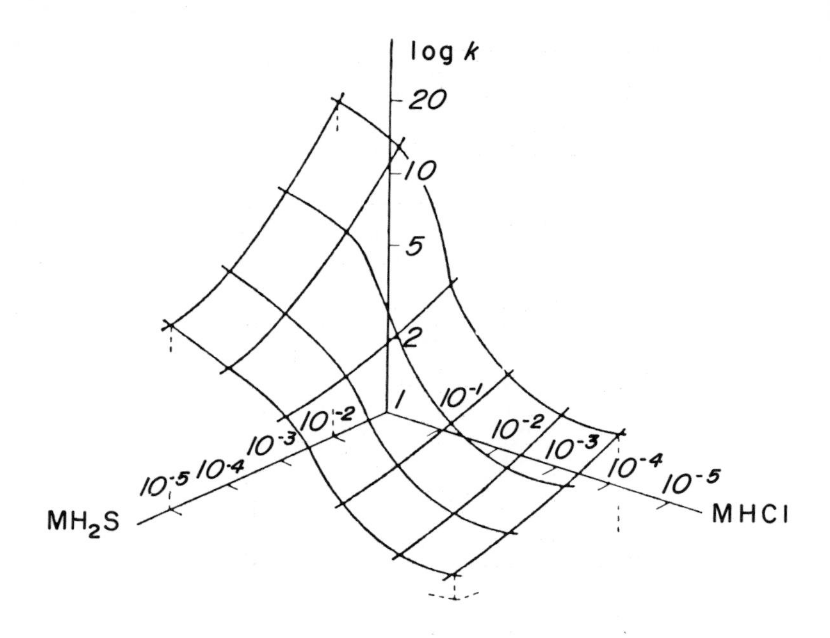

Figure 15. Dependence of steel corrosion on the H_2S and HCl content in the solution t = 25°C k-corrosion loss in g.cm^{-2}/6 hrs. 10^{-4}, M-HCl and M-H_2S. Concentration of the components in the system, gram-mole/l.

associated with landing gears was that of delayed failure. This phenomenon was extensively investigated by Troiano (20). More recent work on high strength steels has shown that the critical factor in the mechanism of crack propagation is associated with the entry of hydrogen in the material (21). This phenomenon is exacerbated as the strength of the alloy is increased. The hydrogen can be derived either from water, gaseous hydrogen or other materials such as H_2S, HCl, and HBr. A comparison of

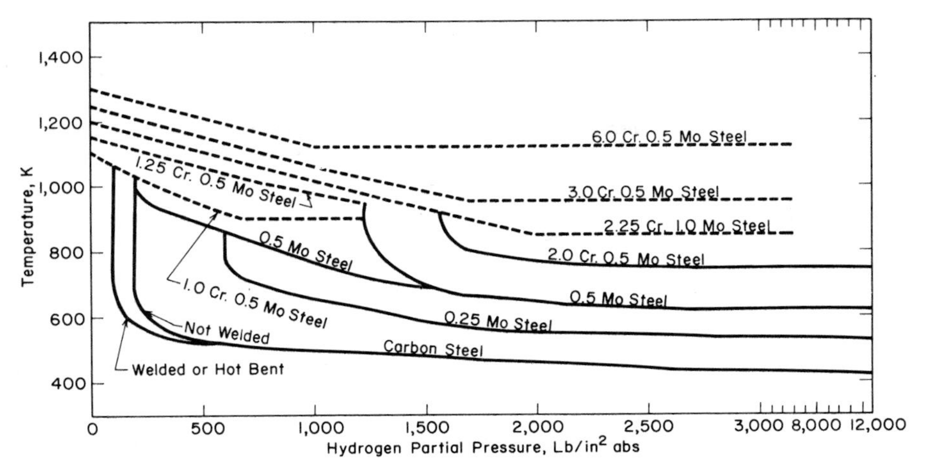

Figure 16. Effect of temperature and hydrogen pressure on the hydrogen damage failure of iron base alloys. Alloys are "safe" below the lines shown and are subject to failure above these lines. Effects of alloy additions indicated. From Nelson (19).

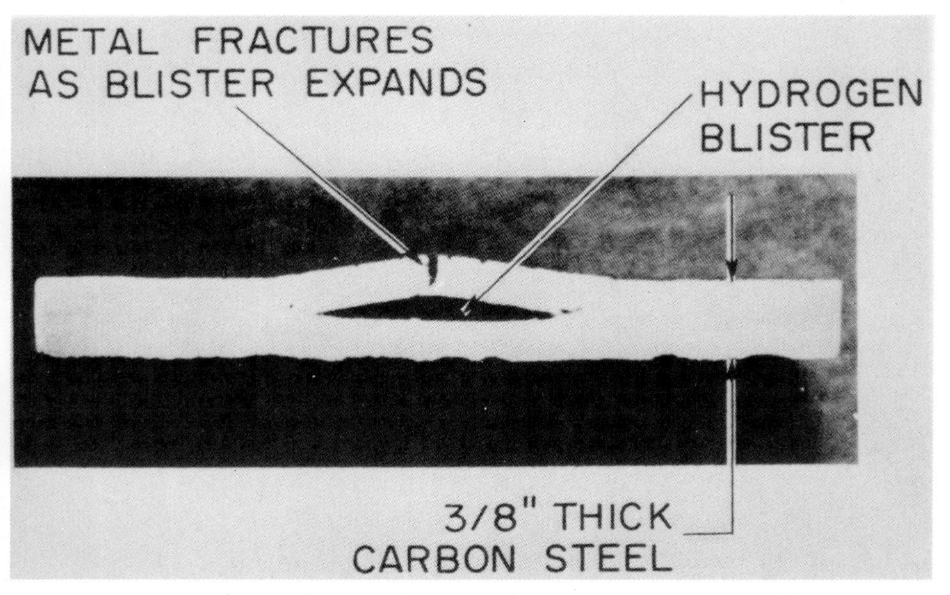

Figure 17. Hydrogen blister in a 9.5 mm thick A-283 carbon steel plate exposed to environment containing H_2S, HCN, NH_4OH, and a trace of NaCl at 130°F.

the effects of H_2O, H_2, and H_2S is shown in Figure 18 from the work of McIntyre (21). Figure 15 has already shown that H_2S accelerates SCC in lower temperature regimes for relatively low alloy strengths. Two important volumes summarize recent developments concerning effect of hydrogen on mechanical properties (22, 23).

VI. CHLORIDE

At high temperatures the halidation of metals may occur. This general subject area has been recently reviewed by Rapp (24). However, the concentration of halogen chemicals is not sufficient to be a significant factor at high temperatures in coal gasification environments.

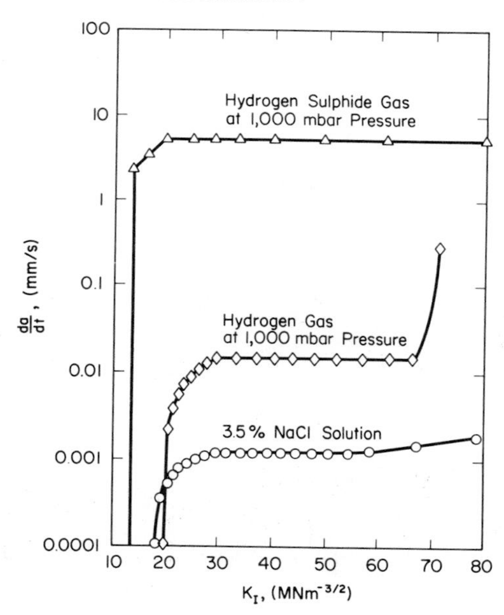

Figure 18. Crack velocity as a function of stress intensity for a 897 M39 steel at 291°K in environments of hydrogen sulfide, hydrogen, and 3.5% NaCl.

At lower temperatures the effect of chloride ions on the stress corrosion cracking of austenitic stainless steels is well known and has been extensively discussed (25). It is not necessary here to further elaborate except to point out the important interdependence of chloride cracking on the oxygen concentration which is shown in Figure 19 (26). Chloride also accelerates localized corrosion processes such as pitting and crevice corrosion owing to the debilitation this ion has on the passive films.

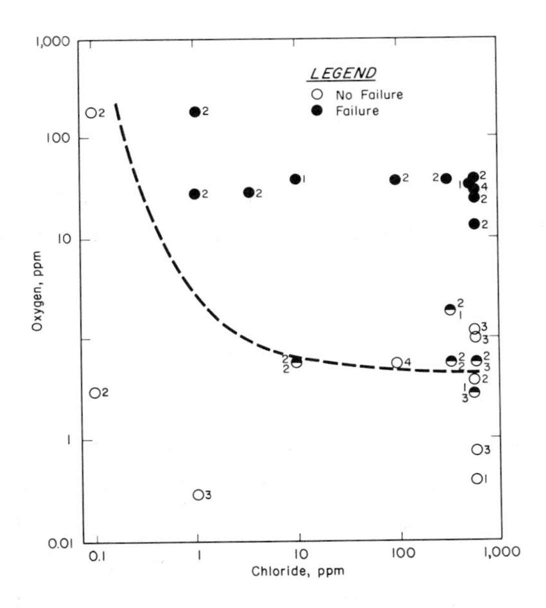

Figure 19. Relationship between chloride and oxygen content of alkaline phosphate treated boiler water for stress corrosion cracking of austenitic stainless steel.

VII. OXYGEN

At high temperatures oxygen contributes to the oxidation of metals and is also significant in combination with sulfur, carbon, and hydrogen as discussed above. The high temperature oxidation of metals has been extensively treated and will not be elaborated upon here (27, 28).

At low temperatures oxygen exerts its influence primarily when it dissolves in water. Referring to Figure 3 which shows the Pourbaix diagram of iron, the line for the H_2O/O_2 equilibrium is shown near the top and at potentials substantially higher than the H_2/H_2O equilibrium. This higher equilibrium potential for the H_2O/O_2 equilibrium is the basis for oxygen raising the potential. The influence of oxygen in exacerbating the chloride-SCC of stainless steel as shown in Figure 19 is directly related to raising the potential into a regime where SCC can occur.

VIII. OTHER SPECIES

Species such as molybdenum and vanadium have been traditionally associated with the phenomenon of catastrophic oxidation which was first studied by Fontana and Leslie (29). This phenomenon has been more recently interpreted in terms of the effect that these species exert on the equilibrium chemistry and the location of these circumstances with respect to Pourbaix diagram stability.

Other low melting species such as lead, zinc, aluminum, and gallium may accelerate damage to either grain boundary attack or a type of liquid metal embrittlement. If substantial quantities of these species can exist in metallic form such problems should be considered.

I have omitted consideration of effects of nitrogen since previous experience suggests that it does not play a significant role relative to the more aggressive reactions involving hydrogen, carbon, chlorine, oxygen, and sulfur.

IX. PHYSICAL INFLUENCES

The important physical influences in addition to the chemical ones involve primarily the effects of stress and particles as they contribute to wear or erosion.

Stress exerts its influence in a number of ways which vary depending upon the temperature. At high temperatures stress is associated with the phenomenon of creep rupture and environments may accelerate this process although environmental influences on creep have not been extensively studied.

As a variation on the subject of creep, the subject of creep fatigue is receiving significant recent attention as it applies to pressure vessels operating at elevated temperatures (30, 31). While creep fatigue has been of primary interest to nuclear vessels, it will certainly be of greater importance to pressure vessels used in coal gasification. The problem embodied in the subject "creep fatigue" involves the fact that cyclic loading superimposed upon a creep situation accelerates the creep damage; conversely, long holding times between stress cycles also exacerbate fatigue damage, i.e. increases da/dN.

At low temperatures constant stress causes stress corrosion cracking, and in the oscillating circumstances we have corrosion fatigue. Stress corrosion cracking depends upon the synergistic interaction of numerous factors; this interdependence is illustrated in Figure 20. If the product is below an arbitrary constant value, SCC will not occur in a time less than the same value and at some point this time is long enough to be effectively infinite. SCC can be eliminated if

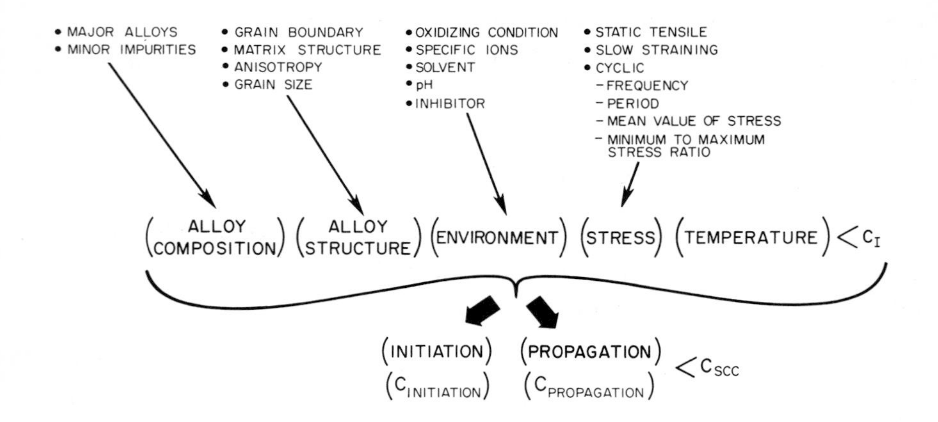

Figure 20. Five factor relationship giving critical components in stress corrosion cracking.

any of the factors in the equation is zero or if several factors can be reduced simultaneously.

Corrosion fatigue and stress corrosion cracking are part of the same continuum as illustrated in Figure 21. These two phenomena, which are ordinarily considered as separate, approach each other along the coordinates of cyclic figuring and stress ratio. In the limit of a period which is long with respect to the time required for SCC to occur, corrosion fatigue becomes stress corrosion cracking. On the other hand, as the ratio of the minimum stress to the maximum stress approaches positive unity the condition of constant load is again approached.

With respect to the question of particle induced deterioration consideration should be directed toward two cases. One is the interaction of particles on the wear process. This is particularly significant relative to valves and seals which separate zones of different pressures. There is presently very

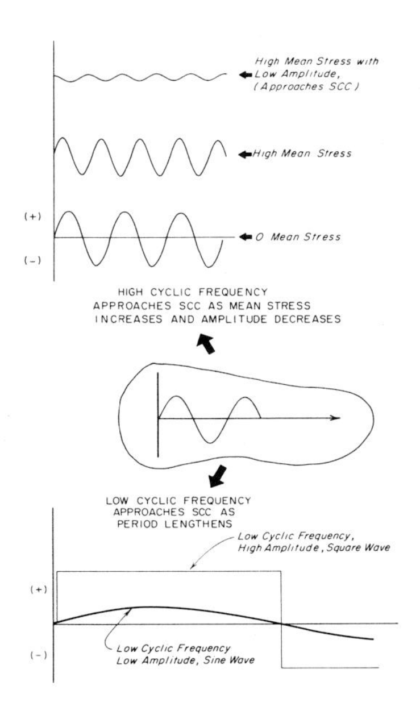

Figure 21. Schematic diagram showing how fatigue parameters approach constant load conditions, i.e. corrosion fatigue approaches stress corrosion cracking.

little information on this process but it deserves careful consideration.

The second problem of erosive wear produced by high velocity particles has been considered more extensively and is expected to be a major issue especially in components such as cycline separators and turbines. With respect to particles, parameters of size, shape, strength, and angle of impingement are critical. With respect to the substrate, hardness and simultaneous corrosion processes are critical. Here, again, this subject is just now being seriously considered by the technical community and only generalities are available at present.

The general problems of abrasive and erosive wear are discussed generally in the volume from the Ohio State Conference (1).

X. *CHANGES IN ALLOY STRUCTURE WITH TIME*

An implicit part of the performance of coal conversion equipment is its operation at relatively higher temperatures for long times. These temperatures approach those of interest to phenomena such as sensitization of stainless steels, temper embrittlement and sigma phase formation depending upon the alloys. The phenomenon of sensitization in metals has been extensively studied and relationships for the development of the sensitized condition have been determined experimentally (32). Similarly, the phenomenon of temper embrittlement has been extensively studied and the conditions of alloy chemistry, temperature and time, again have been generally developed (33, 34).

These time-dependent changes are important as they contribute to unexpected degradation at a later time in the operation of equipment. This implies a particularly important mandate in the early testing of materials: i.e. that these changed conditions should be evaluated early in laboratory testing for any enhanced susceptibilities to failure which they may imply.

XI. *THE WEAK LINK*

As engineering systems become larger and operate under progressively more punishing circumstances, the possibility of failure of any of the components increases. Further, there is an increased potential for failure owing to the fact that it becomes progressively more expensive to conduct proof testing on large scale equipment. There is a growing tendency to attempt predicting reliability without going through the prototype stage.

Such an approach is generally disastrous but it seems to be the current procedure. Such an approach in this area should be avoided.

Simply stated, the concern of this section is the following: when any one component is shut down, the entire system is shut down; and the length of time necessary to get back on stream depends on the time it takes to repair the failed component. One of the most striking examples occurs in the nuclear industry with the failure of condensers. Condenser failures have contributed substantially to the downtime of nuclear power plants. It would be thought that condensers are a part of "standard engineering;" however, they are not and are frequently subject to leaking. Failures also occur in steam turbines despite the fondest hopes of their manufacturers. Again, these are supposed to be standard items of commerce, but failures are all too frequent. When the turbine or condenser fails the entire system must be stopped until the offending equipment is repaired.

In view of such past history, it seems appropriate that more careful consideration be given to the possibility of such failures. Hopefully, a more balanced approach to design and development would be taken for these large systems.

XII. ACKNOWLEDGEMENTS

It is a pleasure to acknowledge several contributions to the development of the ideas in this paper. The suggestion to use Table 1 I obtained from Dr. John Smith of the National Bureau of Standards. I discussed the matter of contributions of ceramic materials with Dr. Jack Wachtman also of the National Bureau of Standards. Further, I am indebted to the National Science Foundation and the former Office of Coal Research now of ERDA for its support of the Ohio State Conference on Materials Problems and Research Opportunities in Coal Conversion. Their financial support and encouragement in organizing this meeting got me initially involved in the subject. Finally, I would like to acknowledge the able assistance of Dr. Arun Agrawal, Mr. Gary Welch, and Ms. Catherine Ward in helping with the manuscript.

XIII. REFERENCES

1. Materials Problems and Research Opportunities in Coal Conversion, 16-18 April 1974, editor: R. W. Staehle, Department of Metallurgical Engineering, The Ohio State University.

2. NSF-OCR Engineering Workshop on MHD Materials, 20-22 November 1974, Massachusetts Institute of Technology, editor: A. L. Bement, Energy Laboratory, Massachusetts Institute of Technology, Cambridge, Massachusetts.

3. J. Quets and W. Dresher, Journal of Materials, March 1969.

4. R. A. Rapp, paper presented at the Agard Panel on Propulsion Energetics, October 13, 1969, Dayton, Ohio.

5. M. Pourbaix, Atlas Electrochemical Equilibria in Aqueous Solutions, NACE, Houston, Texas (1966).

6. M. Pourbaix, Lectures on Electrochemical Corrosion, Plenum Press, New York (1973).

7. J. B. Wachtman, S. J. Schneider, A. D. Franklin, "Role of Ceramics in Energy Systems" presented at the ERDA/NSF Workshop on Ceramics for Energy Applications, 24-25 November 1975.

8. ERDA Newsletter, Materials and Components in Fossil Energy Applications, 1 April 1976.

9. Corrosion of Steels in CO_2, Proceedings of British Nuclear Energy Society, International Conference at Reading University, 23-24 September 1975, editors: D. R. Holmes, R. B. Hill, and R. M. Wyatt, British Nuclear Energy Society, 1974.

10. J. M. Furgeson, J. C. P. Garrett, and B. Lloyd, from reference 9.

11. R. F. Hochman, Catastrophic Deterioration of Metals at High Temperature and Carbonatious Environments, Report of Work to the National Association of Corrosion Engineers under sponsorship of the NACE Research Committee.

12. M. Kowaka and S. Nagata, "Stress Corrosion Cracking of Mild and Low Alloy Steels in CO - CO_2 - H_2O" Proceedings of the Conference on Stress Corrosion Cracking and Hydrogen Embrittlement of Iron Base Alloys, NACE, to be published in 1977.

13. J. M. Sutcliffe, R. R. Fessler, W. K. Boyd, and R. N. Parkins Corrosion 28, No. 8, pp. 313-320 (1972), NACE, Houston.

14. J. A. Goebel and F. S. Pettit, Metallurgical Transactions Vol. I, p. 1943-1970.

15. J. A. Goebel, F. S. Pettit and G. W. Goward, Deposition of Corrosion in Gas Turbines, editors: A. B. Hart and A. J. B. Cutler, Wiley and Sons, New York, 1973.

16. K. Yamamoto, M. Izumiyama, and K. Nishino, "Corrosion of Iron and Iron-Chromium Alloys By Hydrogen Sulfide," Proceedings of the 5th International Congress on Metallic Corrosion, 1974.

17. M. Kowaka and H. Nagano, "The Mechanism of Sulfur Dewpoint Corrosion," Proceedings of the 5th International Congress on Metallic Corrosion, 1974.

18. R. Baronicek, "Corrosion of Steel in Solutions of Chlorides and Sulfides" Proceedings of the 3rd International Conference on Metallic Corrosion, Vol. I, 1968.

19. G. A. Nelson, Proceedings of American Petroleum Institute, 45 190 (1965).

20. A. R. Troiano, Trans. ASM, 52, 54 (1960).

21. P. McIntyre, Stress Corrosion Cracking and Hydrogen Embrittlement of Iron Base Alloys, editors: R. W. Staehle, J. Hochmann, J. E. Slater, and R. D. McCright, to be published by NACE, Houston, to be published in 1977.

22. I. M. Bernstein and A. W. Thompson, Hydrogen of Metals, ASM, 1974.

23. I. M. Bernstein and A. W. Thompson, Effect of Hydrogen on Behavior of Materials, TSM, AIME, 1976.

24. R. A. Rapp and P. L. Daniels, "Halogen Corrosion of Metals" Advances in Corrosion Science and Technology, Vol. 5, p 55-172, (1976) Plenum Press.

25. R. W. Staehle and R. M. Latanision, Proceedings of Conference on Fundamental Aspects of Stress Corrosion, p. 214, September 9-15, 1967, Columbus, Ohio.

26. W. L. Williams, Corrosion, 13, 539t (1957).

27. K. Hauffe, Oxidation of Metals, Plenum Press, New York (1965).

28. G. R. Belton and W. L. Worrell, Editors: Heterogeneous Kinetics at Elevated Temperatures.

29. W. C. Leslie and M. G. Fontana, "Mechanism of the Rapid Oxidation of High Temperature High Strength Alloys Containing Molybdenum," Trans. ASM, 41, 1213-1247, (1949).

30. S. Manson, G. R. Halford, and M. H. Hirschber, "Symposium on Design for Elevated Temperature Environment," American Society for Mechanical Engineers, New York, 1971.

31. L. F. Coffin, Jr., Fracture, 1969 Proceedings, Second International Conference on Fracture, Brighton, April 1969.

32. H. F. Ebling and M. A. Scheil, Advances in the Technology of Stainless Steels and Related Alloys, ASTM-STP 369, ASTM, Philadelphia, Pennsylvania.

33. C. J. McMahon, "Strength of Grain Boundaries in Iron-Base Alloys," Grain Boundaries in Engineering, Proceedings: 4th Bolton Landing Conference, June 1974, Editors: J. L. Walter, J. H. Westbrook, D. A. Woodford.

34. B. J. Schulz and C. J. McMahon, Temper Embrittlement of Alloy Steels, ASTM STP 499, ASTM 1972, p. 104.

Chapter 18

MATERIALS PROBLEMS (REAL & IMAGINED) IN COAL GASIFICATION PLANTS

H. E. Frankel, G. A. Mills+, and W. T. Bakker***

TABLE OF CONTENTS

---

*H. E. Frankel, Assistant Director, Materials and Power Generation, Fossil Energy Research, ERDA.

+G. A. Mills, Director, Fossil Energy Research, ERDA.

**W. T. Bakker, Chief, Materials Branch, Materials and Power Generation, Fossil Energy Research, ERDA.

1.0 INTRODUCTION

1.1 Historical Perspective

Coal gasification technology dates to 1670, when the Reverened John Clayton, a Yorkshire clergyman, reported the generation of a luminous gas when coal was heated in a chemical retort. A century later, in 1792, William Murdoch, a Scotsman, illuminated his home with gas obtained by distilling coal in an iron retort (1). The first coal gas company to distribute its product for lighting was chartered in London in 1812, the first American company in Baltimore in 1816 (2).

Nearly every major city in the eastern United States once had its gashouse where gas was manufactured, usually from coal, for lighting and cooking. The gashouse and the bulky cylindrical storage tanks that stood nearby gradually disappeared after World War II as natural gas came to be distributed nationally by pipeline. After manufactured gas lost its markets here, the technology was further improved in Europe, where coal was then the only indigenous fuel available in any significant quantity. The processes installed most often in Europe and elsewhere were the Lurgi, Koppers-Totzek, and Winkler, which served mainly to make synthesis gas for the manufacture of ammonia and other synthetic products. Extensive coal gasification occurred in Germany during World War II to make synthetic transportation fuels via Fischer-Tropsch synthesis (3).

Current American effort in coal gasification is now centered on improving process efficiencies (4). New regulations on the emission of sulfur oxides and the difficulties that have been encountered with stack-scrubbing processes have led to renewed interest in producing a clean low-Btu gas from coal to be burned as a boiler fuel.

1.2 *Principles of Coal Gasification*

Coal consists of a multi-ring condensed aromatic structure with alkyl, carbonyl, hydroxyl, and other linkages containing sulfur and nitrogen (5). To convert coal to gaseous (or liquid) fuels, hydrogen must be added. Coal is reacted with steam (the most abundant source of hydrogen) to form gaseous products:

$$C + H_2O \rightarrow H_2 + CO \text{ (endothermic)} \tag{1}$$

Heat for this reaction is supplied by combustion of a portion of the coal:

$$C + O_2 \rightarrow 2CO \text{ (exothermic)} \tag{2}$$

In practice steam and oxygen are simultaneously injected. The gaseous mixture of CO and H_2 catalytically reacts to form methane (pipeline gas):

$$3H_2 + CO \rightarrow CH_4 + H_2O \text{ (exothermic)} \tag{3}$$

Hydrogen also reacts with coal impurities (sulfur, nitrogen, and oxygen) to form H_2S, NH_3, and H_2O. All are gases that are readily separated from methane.

The stoichiometric ratio, H_2/CO, is 3 (equation 3) but is never this high in the gas stream emerging from the steam-oxygen reaction chamber. The gas stream contains excess steam along with the reaction products (equations 1, 2). A dynamic equilibrium exists among these gases as expressed by the shift reaction:

$$H_2O + CO \rightleftarrows H_2 + CO_2 \tag{4}$$

Lowering the temperature shifts the equilibrium to the right and results in more conversion of steam to hydrogen. This, in turn, increases the H_2/CO ratio and favors more complete formation of methane (equation 3).

1.3 *Commercial and Advanced Processes*

A typical gasification process is schematically depicted in Fig. 1, where the major operations needed to produce pipeline gas are identified. Table 1 compares key features of commercially available coal gasification processes with advanced processes. No individual process is identified; the former include Lurgi (6,7), Koppers-Totzek (6,7) and Winkler (8); the latter include Hygas (6,9), Bi-gas (6,10), CO_2-Acceptor (6,11), Synthane (6,11), and Molten Salt (12).

Advanced processes expose materials to increasingly severe conditions of stress, due to higher operating pressures, and chemical attack, because of higher pressures and more erosive feedstock.

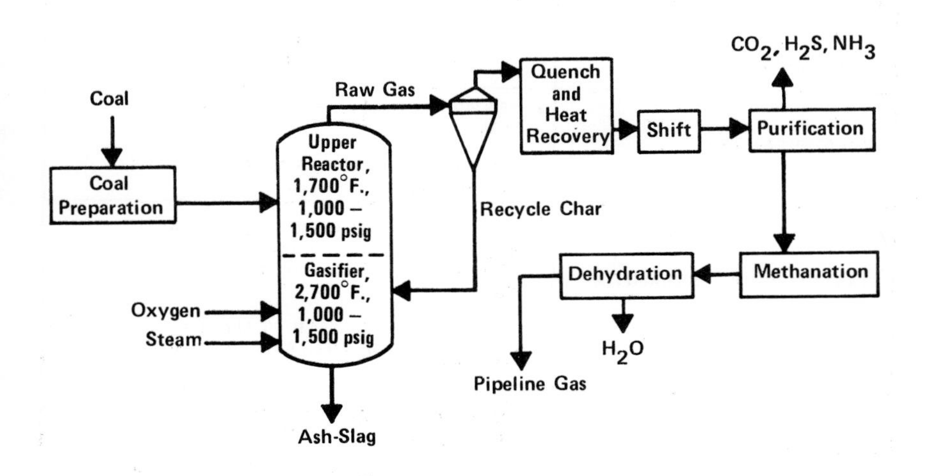

FIG. 1. Schematic flow sheet for coal gasification (Bodle and Vyas (6)). By permission of Oil and Gas Journal.

TABLE 1

COMPARISON OF COMMERCIAL AND ADVANCED
COAL GASIFICATION TECHNOLOGIES

GASIFICATION PRESSURE	• MOST ADVANCED TECHNOLOGIES ARE AT 1000 psig, THE INTERSTATE PIPELINE PRESSURE. • COMMERCIAL UNITS ARE GENERALLY ATMOSPHERIC PRESSURE; LURGI UNITS OPERATE AT 450 psig BUT ONLY ON SELECTED COALS.
COAL TYPES	• NEW PROCESSES ATTEMPT TO USE ALL DOMESTIC COALS INDEPENDENT OF SWELLING, CAKING, ASH AND WATER CONTENTS, PARTICLE SIZES, OR IMPURITY LEVELS (S, Cl CONTENTS). • COMMERCIAL PROCESSES ARE SELECTIVE, I.E., COAL MUST BE NON-CAKING, COAL MUST BE SIZED, COAL MUST BE DRIED.
METHANE FORMATION	• NEW PROCESSES ATTEMPT TO MAXIMIZE METHANE FORMATION IN FIRST GASIFICATION REACTOR. • COMMERCIAL PROCESSES USE COMPLEX CATALYTIC STEPS TO FORM METHANE FROM SYNTHESIS GAS.
THERMAL EFFICIENCY	• NEW PROCESSES ATTEMPT TO MAXIMIZE EFFICIENCY. • COMMERCIAL PROCESSES ACCEPTED PENALTIES NO LONGER CONSIDERED VIABLE.
ENVIRONMENTAL EFFECTS • WASTE PRODUCTS • EMISSIONS	• NEW PROCESSES MEET CURRENT STANDARDS. • COMMERCIAL PROCESSES REQUIRE MANY CHANGES TO MEET CURRENT STANDARDS.
OPERATOR SAFETY	• NEW PROCESSES MEET PRESENT-DAY OSHA STANDARDS. • COMMERCIAL PROCESSES REQUIRE SIGNIFICANT MODIFICATIONS.
PLANT RELIABILITY/OPERABILITY	• NEW PROCESSES ARE BEING DEVELOPED WITH RELIABILITY PROVISIONS AND FAILURE REPORTING AS PART OF THE DEVELOPMENT WORK. • COMMERCIAL PROCESSES HAVE LIMITED RELIABILITY DATA APPLICABLE TO CURRENT SITUATIONS.
PLANT PRODUCT	• NEW PROCESSES PRODUCE METHANE WITH HEATING VALUE OF AT LEAST 950 BTU/SCF. • COMMERCIAL PROCESSES CAN PRODUCE METHANE BUT GENERALLY CONTAIN MORE IMPURITIES.

The Problem

The performance of materials, either in components, in structural members, or as interior gasification reactor parts, has not been determined in the complex environments present in advanced coal gasification techniques. A failure analysis reporting system is now underway (13,14). Until the results come in, metal problems must be viewed as either real or imaginary. A materials design engineer with the assignment of specifying materials for a commercial plant does not have a data base of domestic commercial operation from which to draw. This lack of operational materials data from an integrated commercial plant is the most serious single materials problem facing the materials community today. This problem, in turn, causes significant uncertainty in estimates of capital investment and provisions for spare parts. These uncertainties, in turn, cause totally unsatisfactory uncertainty in overall process economics and therefore dim the financial prospects of virtually all coal gasification projects (14).

2.0 METALS

Metals are used throughout coal gasification plants in areas where the combined requirements of strength, ductility and corrosion/erosion resistance dictate their use. Problems will be described in the paragraphs that follow.

The approach for this chapter will be to consider the behavior of metals in increasingly complex thermal-chemical-mechanical environments. Virtually all the discussion will concern ferrous alloys and selected nickel and cobalt-base alloys.

2.1 *Thermal Exposure in Absence of Chemical Reaction*

Upon short-term* exposure to elevated temperatures, metals lose strength when compared to room temperature. Metals also experience creep. The nature of creep deformation (primary, secondary, tertiary) is strongly dependent on both time and temperature.

The problem in creep is that very long tests, extending to perhaps 100,000 hours (11 years) are needed. Accelerated tests at higher stress levels (stress-rupture tests), using various parameters, such as the Larson-Miller, Dorn, and Manson-Haferd parameters enable the prediction of long-term behavior from short-term tests (18). Given the present lack of theoretical understanding of the fundamental relationships in metals, the parameters must be used with care. The parameters require that the basic metallurgical stability of an alloy be known. Many aspects are not understood at this time. However, as an example, one case that is understood involves the inherent metallurgical instability of Type 316 stainless steel. This alloy is inherently unstable near 800°C due to precipitation of carbide phase, $M_{23}C_6$, in the grain boundaries (Fig. 2).

This behavior is common. Virtually every heat-resisting alloy is in a metastable state and transforms to a thermodynamically stable state while at the service temperature. Thus, a significant problem is that of determining the kinetics of transformations and determining the influence of these transformations on the important engineering properties (strength, ductility, creep strength, fatigue, etc.).

---

*Short means less than one hour.

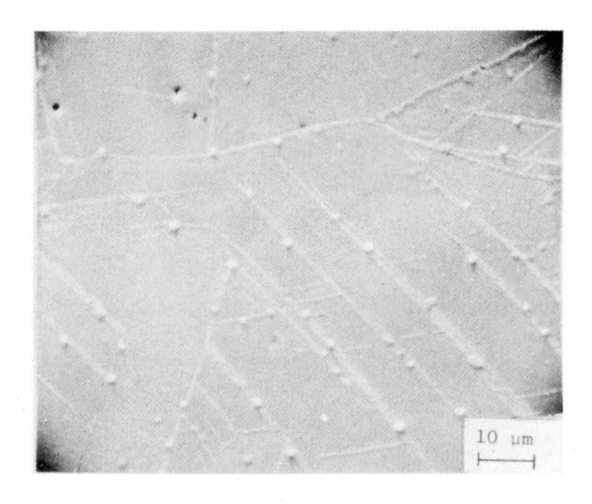

FIG. 2. Scanning electron microscope (SEM) image of Type 316 stainless steel (vacuum melted) after 15 hours at 810°C in capsules sealed in vacuum of better than 10^{-6} Torr. (J. A. Spitznagel and R. Stickler, Trans. AIME, 5, 1363 (1974)). Note precipitation of $M_{23}C_6$ carbides in grain boundaries. Copyright American Society for Metals, 1974.

2.2 *Thermal Exposure in the Presence of a Chemically Aggressive Environment*

In many parts of a coal gasification plant, metals are subjected to both chemical attack and some degree of elevated temperature. Maximum temperatures can reach ~3300°F (~1800°C) in reactors.

In other areas, as in the front end coal preparation area of the plant (Fig. 1), major corrosion problems are not expected. In this area corrosion is less of a problem than physical abrasion of metal from the cutting and grinding surfaces. This area could benefit from new materials but does not appear to be seriously constrained at this time.

The coal gasification vessel (Fig. 1) represents a major corrosion problem area requiring significant attention. Safety, legal, and insurance regulations dictate that gasifiers be made to ASME Boiler and Pressure Vessel Code specifications.

The code specifies that an allowance be made for chemical attack of the pressure vessel walls. This presents a significant problem because that allowance has several contributions: (i) a portion of the original metal is removed (uniform attack), (ii) a portion of the remaining metal is altered due to hydrogen or carburization embrittlement, and (iii) another portion of the remaining metal is degraded due to pitting and grain boundary attack.

All contributions must be determined in a manner that allows extrapolation of the overall corrosion allowance to 30 years (~240,000 hours).

The Energy Research and Development Administration (ERDA) has programs directed toward developing new alloys to satisfy the present code and developing a data base that will permit currently specified allowable stress levels to be increased.

Materials used internally in gasifier components (but not subject to the code) also are subjected to erosion, corrosion, and wear in highly complex environments. These materials and components are now under investigation under ERDA-sponsored efforts (19). One set of initial results can now be presented.

In a complex gaseous environment simulating coal conversion*, stainless steel (Type 304) experiences total degradation within 1000 hours at 1800°F (982°C). In contrast, a specialty metal (Alloy 671**) appears satisfactory (Fig. 3). It has a scale (0.0012 inch) and a substrate precipitate zone (0.0012 inch) (Fig. 4). X-ray diffraction analyses of the scale indicate Cr_2O_3. Electron microprobe analyses of the scale (after mounting) established the presence of chromium and oxygen but no nickel or sulfur (Fig. 5). After one year, the scale would be expected to

---

*Gas environment: 12% CO_2, 18% CO, 24% H_2, 5% CH_4, 1% NH_3, 1% H_2S, 39% H_2O; 1000 psi; 1800°F.

**Nominal composition: 50% Cr, 48.4% Ni; 0.35% Ti, 0.30% Fe, 0.03% each Mn, Si, C, and 0.0% Co.

be about 0.010 inch in thickness (linear extrapolation). The combined scale and precipitation zone could be about 0.020 inch. It appears that the 50% Cr alloy may well form a dense, thicker oxide than the 18% Cr alloy (Type 504).

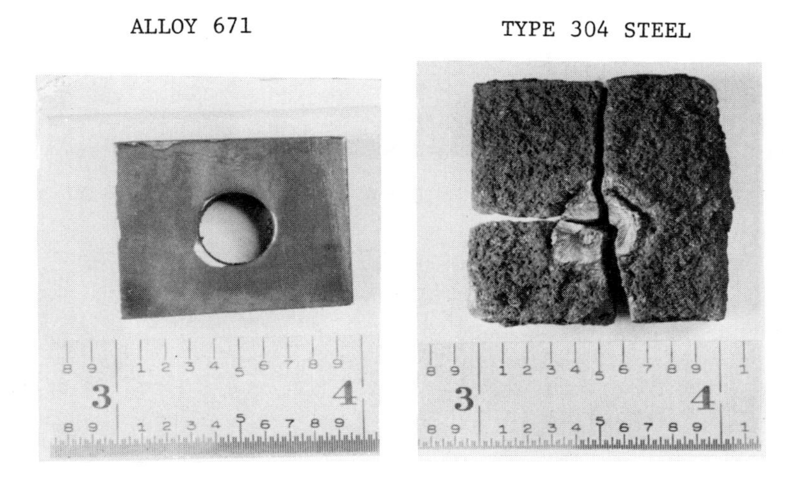

FIG. 3. Alloy 671 and Type 304 stainless steel samples after 1000 hours at 1800°F (982°C) at 1000 psi. The sample of Type 304 stainless steel was completely a matrix of reaction products.

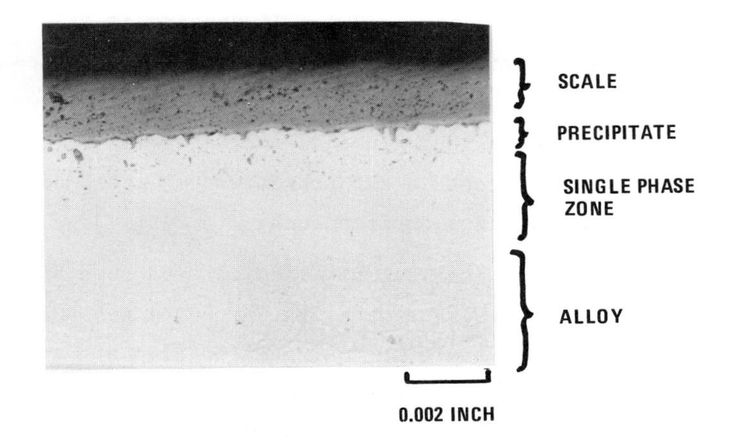

FIG. 4. Microstructure of Alloy 671 from FIG. 3.

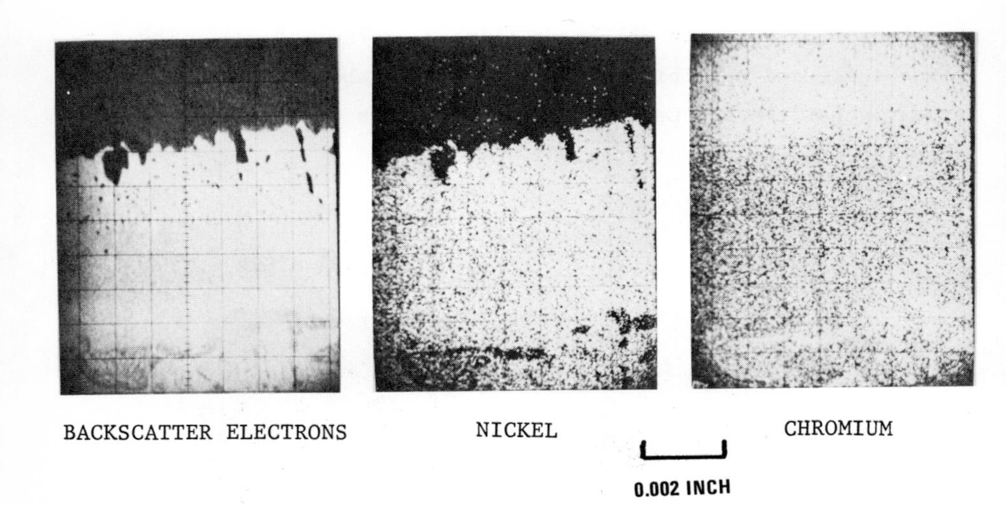

FIGURE 5. Electron microprobe images of Alloy 671 from Figure 4. Oxygen was detected only in the scale but was not photographed. Sulfur could not be detected in either the scale or in the metal.

The 671 alloy also displays a rather featureless (single-phase) zone under the scale (Fig. 3). The hardness of the single-phase is KHN-206, somewhat lower than that of the bulk alloy, KHN-286. The zone appears to be nickel-rich solid solution from the chromium-nickel phase diagram (20). It appears as if chromium migrates to the surface and, in the presence of a complex coal gasification environment, reacts with the oxygen-bearing constituents to form an adherent oxide. In addition, preliminary data (19) suggest alloys must contain ~25% chromium to withstand long-term exposure to gasification environments.

Detailed examinations of Alloy 671 and some 50 other metals are in progress (19) and data are now being generated on the behavior of commercial metals in coal gasification atmospheres. Other efforts are experimentally determining the effects of metal dusting (21), carburization, hydrogen embrittlement (a concern during cool-down), and sulfur attack.

Combined attack from H_2 and H_2S is a serious problem in petroleum refineries as well as the down-stream portions of a coal gasification plant. Corrosion in the petroleum industry is combated via use of coatings, and initial data (19) from an ERDA program indicate they may be effective in coal gasification. Corrosion is also combated by use of chromium steels. Corrosion rates of steels in many reacting media decrease in proportion to the chromium content (Fig. 6). It would seem that chromium is the answer for the corrosion problem, but this raises another problem as almost all chromium is imported from either Soviet Russia or Rhodesia (23). Estimates of domestic chromium resources are placed at three years, and this includes California and Oregon deposits which are very low grade.

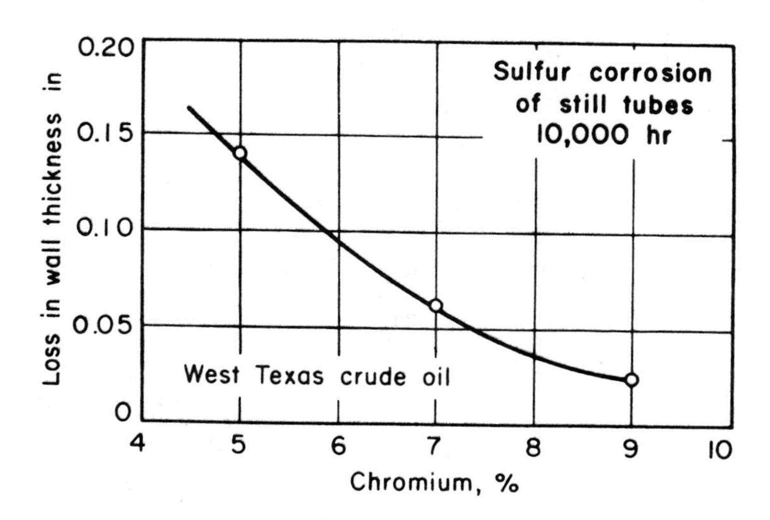

FIG. 6. Reaction rates of chromium-containing solids at 780°F in sulfurous atmospheres (American Society for Metals, (22)).

2.3 Thermal Exposure in the Presence of a Chemically Aggressive Environment and Mechanical Stress

Steady State Stress

A common problem is stress-corrosion cracking, characterized by the star pattern of transgranular cracks (Fig. 7). Solutions containing chloride ions are the worst offenders for ferrous alloys, although recent evidence suggests sulfur may play a similar role (14). The situation is not understood for the case of gasification.

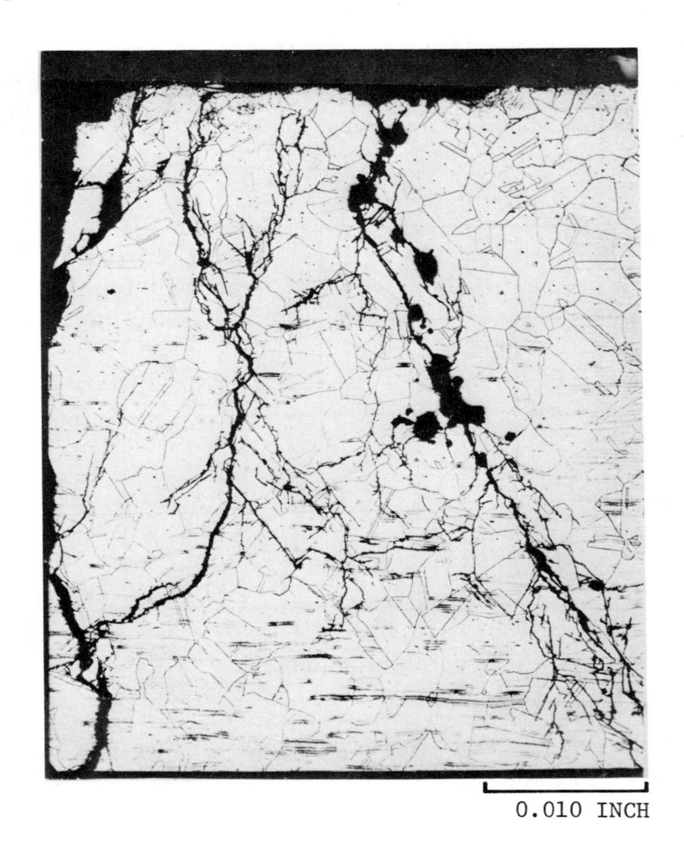

FIG. 7. Microstructure of Type 304 stainless steel that experienced stress-corrosion cracking (Steigerwald, (24)). Copyright American Society for Metals, 1975.

Time Dependent (Cyclic) Stress

A second serious problem area is that corrosion fatigue has all the insidious features of normal fatigue behavior but is aggravated by the presence of the chemical environment. The metallurgical problem is that of evaluating corrosion fatigue in environments of coal gasification, i.e., finding the endurance limit (if it exists) when the state of stress is cyclic. The area is under investigation by ERDA (19).

2.4 Metal Fabrication Problems

Industrial Capacity Limitations

Domestic fabrication capacity is limited when thick-walled (greater than four inches) pressure vessels are needed (25). It is inadequate to handle growth rates that are being projected for coal gasification and nuclear power (1). This problem is one of industrial capacity rather than technology, and causes production and construction delays. It illustrates an important point, namely that the production of gaseous energy from coal is constrained as much by institutional and financial constraints as by materials technology. Technology is still important and improvements are desirable; however, the constraints seen by the private sector center on (i) availability of risk capital, (ii) availability of government-insured loans, (iii) assurances of price stability, and (iv) evidence of stable government policy concerning coal reserves on federal land.

An alternative route, no longer viable, is to use foreign industrial capacity to manufacture pressure vessels. This route would appear closed because representatives of the Lurgi process have stated that their capabilities are completely committed and are suggesting that their overflow business be taken care of in America (26).

Industrial Capability Limitations

There is another aspect to the fabrication problem. Advanced concepts (27) call for gasification units with wall thicknesses up to 12 inches. Current domestic practice is limited to 8 to 9 inches. Only one domestic company can handle that thickness. If the 12-inch wall thickness should become a realistic requirement, it would pose a significant problem. This thickness in finished product means the original value must be 15 to 20 inches. The largest openings in forging mill heat treating furnaces is 40 inches. With these restrictions, it appears difficult to achieve the 4:1 reductions deemed necessary in current practice (for metallurgical uniformity) and still have 12-inch wall thicknesses.

Another capability problem involves the question of field fabrication of very large vessels. Large vessels are attractive because they offer a promise of economy-of-scale. Field fabrication is needed because 1800-ton vessels cannot be transported by rail, where transport is limited to loads of 250 to 400 tons (depending on the exact route) and to loads of 13-feet diameter.

Field fabrication is necessary and has been practiced in the construction industry, but never on vessels that have all the dimensions of the conceptual vessels; therefore, it is necessary that field fabrication techniques be developed and qualified for thick pressure vessels.

2.5 *Role of Coatings and Claddings*

An ever-present problem, inspection, always is associated with the use of coatings (27). Both alonized (aluminum) and chromized coatings (chromium) are heavily employed in complex internal parts of gasifiers. Both are difficult to inspect after preparation in complex and large structures, and techniques are needed to prepare coatings and inspect them on parts that are field fabricated.

2.6 *Materials/Design Interface*

An important area demanding attention exists in the interface between the materials scientist and the design engineer.

Current practice in design of pressure vessels is to use a stress criterion (25 percent of yield strength) coupled to a generous corrosion allowance and this is coupled to a larger aging factor. The result is that, in the opinion of many, the overall method is a brute force approach. In contrast, aircraft machinery is routinely operated at much closer to the limit of performance, particularly in high temperature areas such as turbines. The reason is straightforward: the aircraft design engineer has at his disposal a large body of materials performance data, ranging all the way from traceability data on individual metal heats to final and detailed inspection data on finished engine parts. Fabrication procedures and inspection techniques are both detailed and thorough and are based on the requirements that the aircraft meet rigid and demanding vehicle performance criteria. A comparable state of affairs does not exist today in coal gasification. There has been no need for it. However, as the technical community plans for the years 1985-2000, it is clear that several distinct benefits can accrue from a total metal characterization program:

- A data base will be established whereby metals can be used at conditions much closer to their limiting cases.
- Alternatively, if the processing conditions are not changed*, the data base will provide a basis for making gasification reactors (i) thinner, or (ii) from less expensive steel, or (iii) by simpler welding techniques.

---

*For example, the temperature of a methanation bed must not exceed ~900°F. because if it does, the methane that will have been produced from CO and H_2 will decompose (crack) back to H_2 and carbon.

- The data base can provide a basis for revising plant maintenance schedules and reduce the frequency and nature of scheduled but unnecessary maintenance shut-downs.

The interface between the materials and design engineering functions is critical in order to effect substantial savings of both investment capital and natural resources over the next decade.

3.0 CERAMICS

Of all coal conversion processes, high-Btu gasification (Fig. 1) is potentially the biggest user of refractories, for the reason that in high-Btu gasifiers loss of sensible heat must be avoided. It is presently the consensus of gasification process engineers and designers that high-Btu gasifiers will be refractory lined.

Refractory usage in low-Btu gasifiers will follow the same patterns as high-Btu gasifiers. Most low-Btu gasifiers will be built as a part of combined cycle power plants, where there is a large demand for process steam. Thus, it may be advantageous to use tube wall heat exchangers as a partial replacement for refractory lined walls, as is done in the Koppers-Totzek gasifier. Water cooling of relatively thin refractory lined walls is also a possibility as the hot water or steam can be used elsewhere in the power plant. Thus, low-Btu gasifiers will not be discussed further as all statements on high-Btu gasifiers also will apply, to a lesser degree, to low-Btu gasifiers.

3.1 Need for Ceramics in Coal Gasification

Durable exotic metals, such as platinum, molybdenum, tungsten, tantalum, etc., are available but are very expensive (upwards of $100/lb). Many ceramic materials are available as liners that protect the more common metals (structural steels) and thereby obviate the need for exotic metals. Ceramic lining materials probably will consist of rather conventional materials similar to those presently produced by the ceramics industry. Suggested candidate ceramic materials are compiled in Table 2 (29,30).

Gasifier reactors (Fig. 1), burners or regenerators will require refractories.

A conceptual engineering study (28-31) determined the most likely amounts of refractories (Table 3). Each commercial plant will require approximately 1500-3000 tons of

TABLE 2

CANDIDATE CERAMIC MATERIALS

TYPE	CANDIDATES
CASTABLE	DENSE, HIGH PURITY, 94% ALUMINA DENSE, HIGH PURITY, 60% ALUMINA INSULATING LOW IRON, 47% ALUMINA INSULATING BUBBLE ALUMINA (97%) LIGHTWEIGHT (53% ALUMINA)
BRICK	DENSE MULLITE (70% ALUMINA) DENSE PHOSPHATE BONDED (85% ALUMINA) DENSE 90% ALUMINA DENSE 99% ALUMINA INSULATING LOW IRON, HIGH ALUMINA INSULATING 97% ALUMINA FUSED CAST ALUMINA (99%)
MORTAR	HIGH ALUMINA-CHEMICAL BOND HIGH ALUMINA, AIR SETTING
PLASTICS	CHEMICAL BOND - 60% ALUMINA CHEMICAL BOND - 90% ALUMINA
BONDED CARBIDES	SILICON CARBIDE

TABLE 3

ESTIMATED NEEDS FOR REFRACTORIES BY ADVANCED COAL GASIFICATION TECHNOLOGIES

PROCESS	UNIT	DIAMETER (FT)	HEIGHT (FT)	WEIGHT IN TONS		UNITS PER PLANT
				METAL	REFRACTORIES	
HYGAS	GASIFIER	24 (I.D.)	220	1700	1070	2
AGGLOMERATING ASH	GASIFIER	62	110	900	1425	3
	BURNER	30	90	300	525	3
CO_2 ACCEPTOR	GASIFIER	59 (I.D.)	100	1500	1250	1
	REGENERATOR	44 (I.D.)	50	800	490	1
SYNTHANE	GASIFIER	31 (I.D.)	75	1800	500	3
BI-GAS	GASIFIER	STAGE I-9 (I.D.) STAGE II-12 (I.D.)	185	400	825	1

refractories. Some energy plans call for the construction of 36 such plants by 1990. This will require 54,000 to 108,000 tons of refractories for first installation.

Assuming that a reline will be needed every five years, the repeat business could be about seven gasifier relines per year, requiring 10,000-20,000 tons of refractories. In addition to this, some yearly maintenance may be required on all gasifiers. This business is estimated at about 300 tons/gasifier or about 10,000 tons/year total. Thus, total amount of refractories required for coal gasification could be 20,000-30,000 tons/year, in addition to new installations.

3.2 Basic Process Conditions and Possible Failure Mechanisms

Basic process conditions have been described in Section 1. Most of the potential failure and wear mechanisms are neither new nor unique to the coal gasification process, but have been experienced to a lesser or greater extent in various other processes. However, some of them may well be more pronounced in coal gasification processes. Therefore, each anticipated failure mechanism will be discussed in some detail.

3.2.1 *Gaseous Corrosion*

Hydrogen

The ability of hydrogen to reduce SiO_2 or silicate minerals has been well documented, especially by Crowley (32,33). The effect of silica loss on the strength of the refractories involved is less well documented. Crowley reports a noticeable loss in strength for high fired superduty brick, but not for a calcium aluminate bonded castable using a crushed fireclay brick aggregate. Fortunately, the magnitude of reduction by hydrogen is decreased in the presence of steam (Fig. 8).

Since dry ash gasifiers operate at temperatures below the softening point of coal ash, i.e., below 2000°F, and the gasifier atmosphere contains usually less than 40% H_2 and

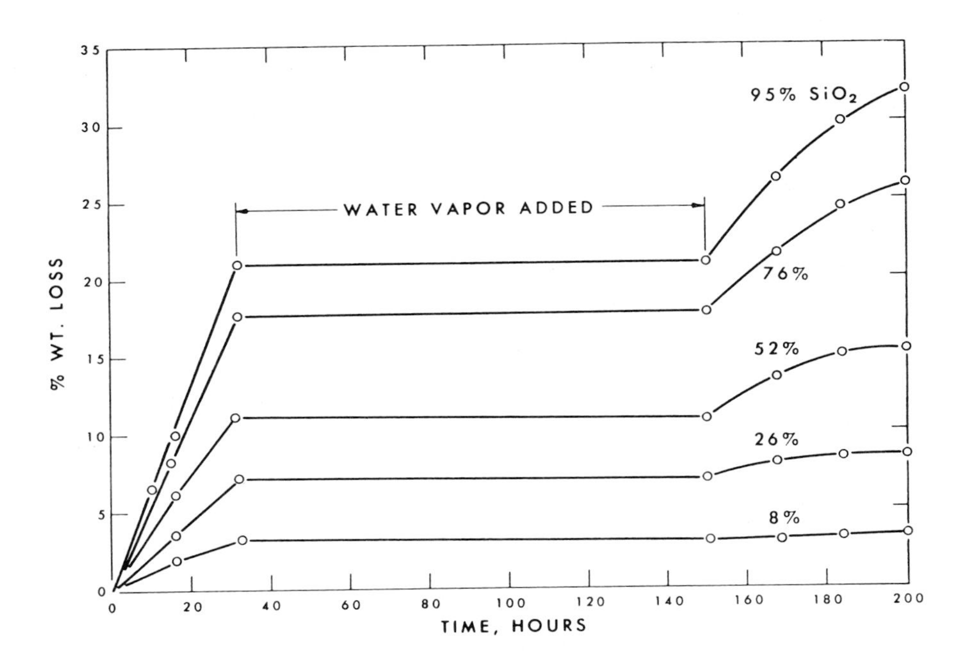

FIG. 8. *Effect of Steam Addition on Silica Reduction of Refractories by Hydrogen.*

considerable quantities of steam, it may be that hydrogen reduction of refractory oxides, especially SiO_2, will not be a major mechanism. In slagging gasifiers the opposite may be true.

Steam

Weight losses caused by leaching with steam at high temperature (34) also increase with increasing silica content of the refractory, at least at temperatures below 2000°F. At temperatures above 2000°F, fireclay brick appears more sensitive to steam leaching than silica brick, probably because of the high amount of glassy phase present in fireclay brick.

Weight losses increase with increasing steam pressure. The effect of steam corrosion on the strength of various refractories is less well known. Budnikov (35) reports a 50% loss of strength when high alumina refractories are exposed to steam at 590°C and 1420 psi. This observation has recently been confirmed by workers at the NBS for a 95% Al_2O_3 calcium aluminate bonded castable but not for a 50-60% Al_2O_3 castable with a similar bond (36). Work at the University of Missouri at Rolla at lower temperatures and pressures indicates an increase in strength when a dense 95% Al_2O_3 castable is exposed to a steam-N_2 atmosphere but a decrease when exposed to a steam-CO atmosphere (37). Further work will be necessary to determine under which conditions rapid deterioration can occur under actual service conditions. A failure analysis of a spalling event in the COGAS gasifier indicates the possibility of some weakening of the refractory behind the hot face, but no clean-cut conclusions could be drawn.

Carbon Monoxide and Carbon Dioxide (38,39)

The disintegration of refractories by carbon monoxide occurs at relatively low temperature, about 500°C. Disintegration of the refractory is caused by the catalytic reduction of CO to carbon and the deposition of carbon in the pores of the refractory in the presence of free iron or iron oxide. Resistance to carbon monoxide disintegration can be built in by firing the brick to a high enough temperature to ensure the formation of iron silicates. Through the use of high fired fireclay brick, carbon monoxide disintegration can be prevented easily. It is therefore not considered a major problem at this time.

Reactions of the oxides of carbon with the calcium aluminate bond of refractory concretes also are a definite possibility, especially below the decomposition temperature of $CaCO_3$, and should be studied in more detail.

3.2.2 *Alkali Salts*

The action of alkali vapors on refractories is probably the major cause of lining wear in blast furnaces and has been studied extensively there.

What is not known at this point is whether the conditions in various coal gasification processes are reducing enough to cause the reduction of alkali compounds. The probability of alkali reduction is not only dependent on the gas composition and pressure, but also on the temperature and the type of alkali compounds present in the coal ash. It is easily conceivable that alkali attack will be more probable when gasifying North Dakota lignite containing 8-10% Na_2O than Eastern bituminous coals containing only 1-3% alkalies. This problem will be the subject of an ERDA-FER sponsored study in the near future.

3.3 *Corrosion/Erosion by Slag*

Coal slags are quite variable. It is therefore quite conceivable that refractory linings for slagging gasifiers must be tailor-made for each gasifier, depending on the ash composition of the coal to be used.

Good resistance to any type of slag usually depends on chemical compatibility and low permeability and porosity. Fused cast refractories generally do well in slag tests regardless of the slag composition. Among the more conventional refractories, chemical compatibility is more important a consideration, especially the ability to form a stable reaction layer between the refractory and the slag. In many cases this layer consists of spinels, both on high alumina and basic brick (Figs. 9 and 10). Our experience with the relatively acid slags (39,40) with chemical compositions similar to coal ashes is that alumina-chrome and magnesite-chrome brick withstand slag erosion much better than high alumina brick even when high purity raw materials are used (Fig. 11).

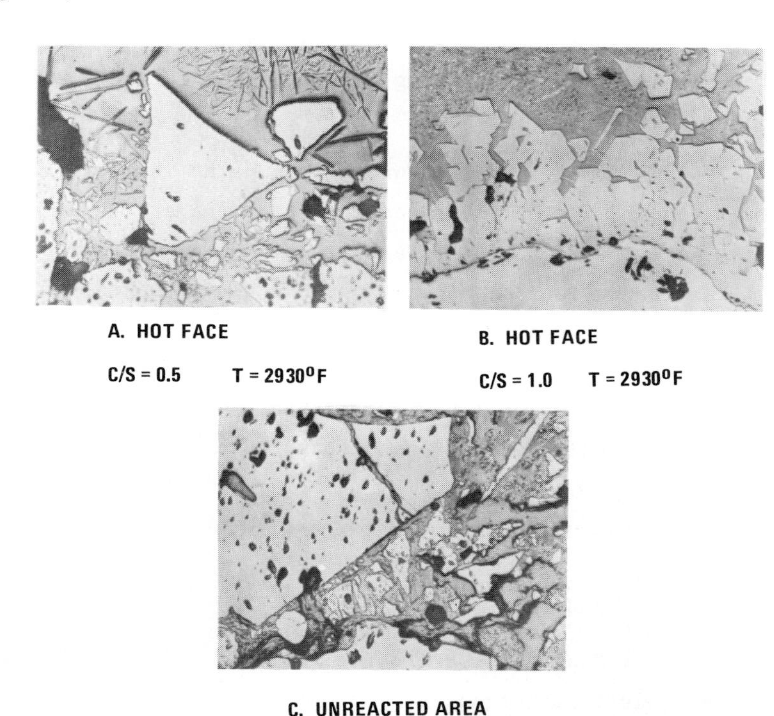

FIG. 9. Spinel Formation in Slagged 90% Al_2O_3 Brick. The C/S Ratio Refers to Calcia/Silica. ∿50 X magnification

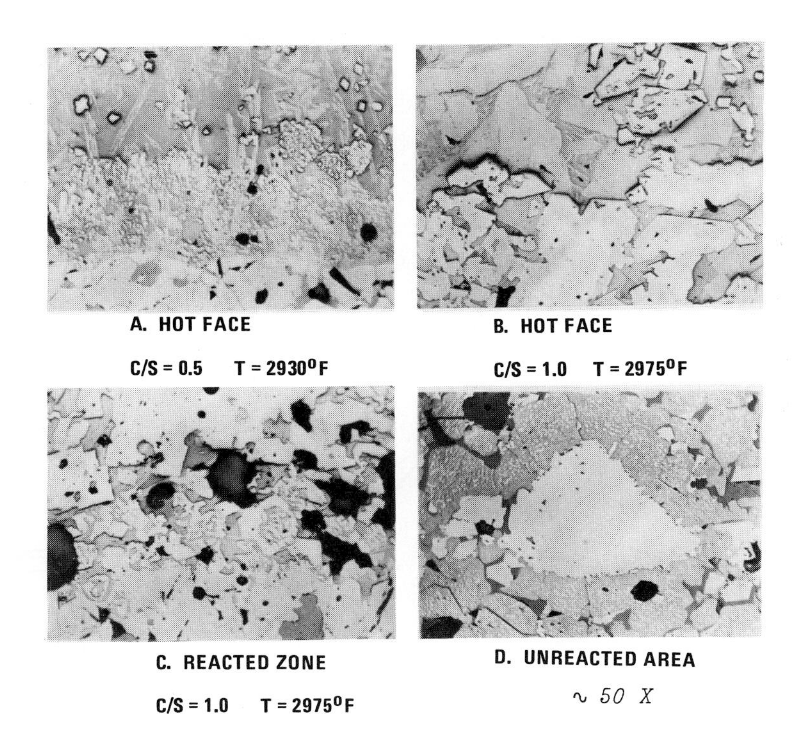

FIG. 10. Spinel Formation in Slagged 60% MgO Basic Brick.

When one looks beyond the short-term results of most slag tests, it is evident that all conventional refractories eventually will dissolve in a complex slag containing SiO_2, Al_2O_3, CaO, FeO, alkalies and other oxides, since the eutectic temperature of such a complex mixture of silicates is very low.

Since service life counted in years rather than months is needed, the possibility of developing refractory compositions, which will dissolve slowly enough to last several years, appears very remote. Koenig's theory on refractory stability in the blast furnace bosh appears therefore attractive for use in the design of slagging gasifier linings, with some modifications (41,42).

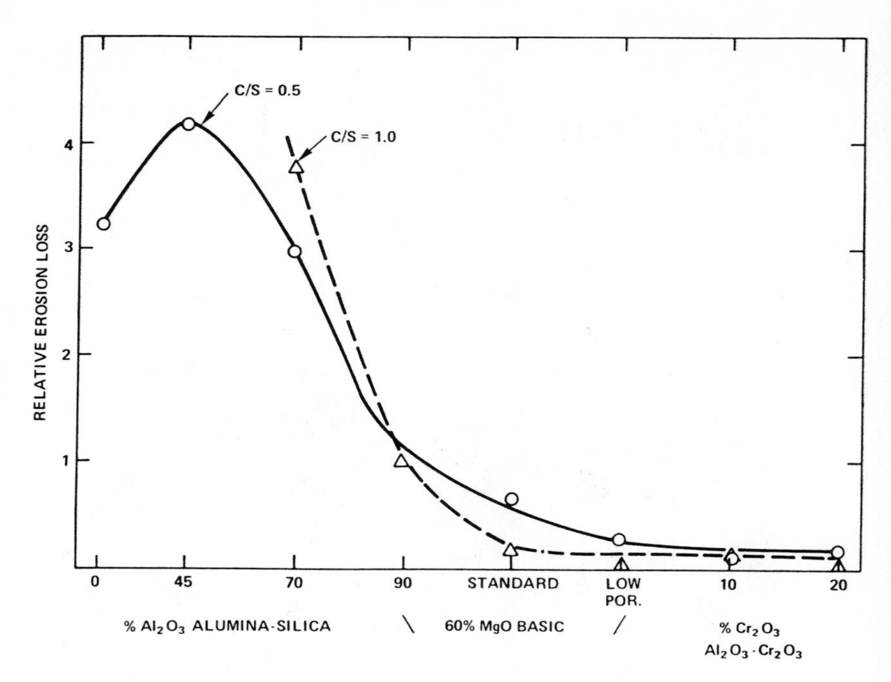

FIG. 11. Resistance of Refractories to Acid Slag.

Briefly, Koenig's theory states that the residual thickness of the refractory lining in the blast furnace bosh is only dependent on the thermal conductivity of the refractory and the lowest temperature at which the refractory will react with any component in the blast furnace atmosphere, be it liquid, gaseous or solid. A long service life is believed to be dependent on a thick residual refractory lining. In the blast furnace, the stability of the lining and the resulting trouble free service of the blast furnace is the only consideration. Loss of heat, i.e. loss of process efficiency, is not considered significant. In coal gasification plants, especially oxygen blown high Btu plants, prevention of excess heat loss must be obtained without sacrificing too many Btu's. Therefore, a refractory with a relatively high T_c (minimum reaction temperature) and low thermal conductivity is needed. The properties of the frozen slag and semi-solid slag layers are also important. The ideal refractory should have the ability to form a thick stable slag layer of low thermal conductivity (equation 5).

$$X = K\left[\left(\frac{T_c - T_w}{T_f - T_s} \cdot \frac{1}{\alpha_1}\right) - \left(\frac{1}{\alpha_2} + \frac{S_p}{K_p} + \frac{S_s}{K_s}\right)\right] \qquad (5)$$

X	equilibrium brick thickness, in.
K	thermal conductivity of brick, Btu-in./ft^2· h ·^{o}F.
T_c	minimum temperature of chemical reaction, ^{o}F.
T_w	cooling water temperature = 212°F.
T_f	furnace temperature = 2192°F.
T_s	melting point of slag = 2012°F.
α_1	heat transfer number, furnace to slag = 41 Btu/ft^2· h ·^{o}F.
α_2	heat transfer number, steel to cooling water = 820 Btu/ft^2· h ·^{o}F.
S_p	thickness of ram material behind lining = 1.97 in.
K_p	thermal conductivity of ram material = 80.6 Btu-in/ft^2· h ·^{o}F.
S_s	thickness of bosh steel = 1.97 in.
K_s	thermal conductivity of bosh steel = 322.4 Btu-in/ft^2· h ·^{o}F.

3.4 *Erosion and Abrasion by High Velocity Particle-Gas Streams*

This type of wear is relatively easy to simulate in laboratory tests, but the fundamental mechanisms leading to loss of material are rather complex and not very well understood. Thus, to make meaningful laboratory studies the test parameters must be very well defined and measured. In general, the higher the strength of the refractory the better its abrasion resistance. However, this relation is not linear; therefore, actual erosion testing is needed.

Most erosion/abrasion studies until now have been made at room temperature after various firing treatments. Recent experiments at the NBS Laboratory (43) indicate that for 95% alumina concretes the erosion loss is less at 1000°C than when measured at room temperature.

The combined impact of simultaneous corrosion and erosion has not been studied extensively in the laboratory and needs further work. A direct correlation between the decrease in strength, due to the corrosive action by the furnace atmosphere and decreasing resistance to erosion, may be expected to exist as refractories are porous and permeable and thus open to gaseous corrosion throughout their cross section.

3.5 *Thermo-mechanical Failures Causing Hot Spots at the Shell*

This type of failure is the most common and also the most insidious failure in many petrochemical processes working at high pressures and with high thermal conductivity gases (44-46). It is safe to extrapolate that this type of failure will be even more prevalent in coal gasification process vessels.

Precise data on the stresses and strains occurring in monolithic refractory lining during heat-up do not exist. The causes of cracking and mechanical deterioration are only intuitively understood. A systematic approach intended to reduce, and if possible eliminate, cracking is lacking. A careful study of the mechanical behavior of gasifier linings, with the objective of reducing stresses to a level at which crack growth is negligibly slow, therefore appears urgently required.

3.6 *Acidic Condensation*

Acid condensation at the shell will occur when the furnace atmosphere contains large quantities of acid such as SO_2, HCl, etc., and the steel shell is below the dew point of the acid. At this point it is not known if acid corrosion of refractories and the shell will be a real problem. More work is needed to provide exact data for design purposes.

3.7 Future Materials and Research Requirements

The foregoing discussion has included various possible difficulties which ceramics in coal gasifiers may face. From this, it appears that some problems probably will occur with existing materials, but that radically new materials and design concepts will most likely not be needed. What is needed, however, is a better definition of the process conditions and failure mechanisms in coal gasification processes to aid refractory selection and optimization.

For gasifier process vessels, this means better definition and analysis of the gas composition itself. A better understanding is needed of the mechanical behavior of refractory wall systems under the impact of thermal and mechanically induced stresses and strains, in much the same way as stress/strain analyses are carried out for complex metal process vessels and components.

Finally, it is well known that performance of any materials system is only as good as the quality of workmanship used during its manufacture, installation and operation. It therefore becomes necessary to utilize strict quality assurance procedures during manufacture and installation and to monitor refractory wear during operation.

4.0 CATALYTIC MATERIALS

Coal gasification, in common with many modern industrial processes, uses a variety of catalytic materials. They are almost always solids, characteristically possessing very high surface areas. They can be metals or metal oxides or sometimes sulfides. Metal catalysts are generally members of Group VIII of the periodic table, e.g., platinum or nickel, or Group IB, e.g., copper. Some of the oxide catalysts are of the insulation type which can act as solid acids, e.g., silica-alumina. However, most oxide catalysts are non-stoichiometric solids, frequently of transition metal elements, e.g., chromium, molybdenum or zinc. The electronic structure of these non-stoichiometric oxides which leads to their semiconductivity properties also is in general responsible for their catalytic activity.

Catalysts generally have surface areas in the range of 50 to 500 square meters per gram and consist of small particles whose diameter is between 10 and 100 Å. Catalytic activity is usually proportional to surface area, which accounts for the importance of high surface area in catalytic applications. If particles 10 Å in diameter grow to 100 Å in diameter, the surface area decreases to one tenth and almost all catalytic activity lost. Surface area can be lost by "sintering" which occurs at a much lower temperature than the melting point. The presence of certain gases can greatly accelerate loss of surface area. Specifically, the presence of steam even at low partial pressures accelerates greatly the sintering of oxide solids.

Many phenomena in catalytic and metallic corrosion are similar. For example, chlorine-containing gases cause both catalyst poisoning and metallic stress-corrosion cracking. In both cases, chlorine preferentially attacks "edge" atoms which in metals are at the grain boundaries and which in catalysts are the active sites.

In addition to sintering, catalysts lose activity by poisoning, especially by sulfur compounds, and by physical blocking by coke formation. Thus, a large part of the technology of keeping catalysts in an active and selective condition consists of avoiding conditions which lead to one of the above.

Since the particle size of active catalysts is so small, special techniques are required to examine their physical state, such as electron microscopy, x-ray line broadening, chemisorption, LEED, etc. (47-49).

4.1 *Catalysts for Gasification Reactions*

The conversion of coal to pipeline gas, i.e., methane, consists of a series of chemical reactions carried out in a series of process steps (Fig. 1). There are three steps where catalysts are significant, namely gasification, shift reaction and methanation, indicated in Table 4.

TABLE 4

CATALYSTS FOR GASIFICATION REACTIONS

LOCATION	GASIFIER	SHIFT CONVERTER	METHANATOR
REACTION	$C+H_2O \rightarrow CO+H_2$	$CO+H_2O \rightarrow CO_2+H_2$	$CO+3H_2 \rightarrow CH_4+H_2O$
CATALYST	K_2O, Na_2O	IRON OXIDE-CHROMIA Cu-Zn, Cr_2O_3 CoO, MoO_3/SPINEL	Ni

Gasification

The reaction of coal and steam to form CO and H_2 (equation 1) is essential to the gasification process. This endothermic reaction is relatively slow and, in fact, only proceeds at a rate of industrial interest at temperatures of about 1800^oF (1000^oC) and above, to about 3000^oF (1600^oC). It has been found that alkali and alkaline earth oxides and carbonates can act as catalysts, speeding up the carbon-steam reaction so that they proceed sufficiently rapidly at about $1200\text{-}1400^oF$. These alkaline elements are naturally present in some coal substances, notably in lignites. Therefore, the practicality of recycle of ash and deliberate addition of potassium oxide (including recovery) is under study. The ability to use significantly lower temperature in the gasifier could drastically improve the materials problems in the gasifier.

Shift Reaction

The shift reaction (equation 4) is carried out to increase the hydrogen content of the synthesis gas to bring the H_2:CO ratio to 3:1 needed for the methanation reaction. Prior to 1963, the "standard" shift catalyst was a relatively sulfur-resistant catalyst of 70-85% iron oxide promoted by 5-15% chromia. It was operated at about 650°F (343°C). It was the practice to use two catalyst beds with interbed cooling and CO_2 removal.

The shift conversion processing steps were greatly simplified when catalyst manufacturers made low-temperature shift catalysts available to the industry in 1963. This is a copper-zinc based catalyst, sometimes containing chromia, which is active at temperatures as low as 375°F. It is used in the 375-450°F (200-232°C) range where the equilibrium of the shift reaction is so favorable to carbon dioxide and hydrogen formation that the carbon monoxide content can be reduced to the 0.3-0.5 mole percent range without intermediate removal of CO_2. The catalyst is more sensitive to poisons, particularly sulfur and chloride than the high-temperature shift catalyst.

Recently, shift catalysts have been described which are very active so that they can operate at low temperatures, are active in their sulfided form, and are resistant to loss of activity and hardness, even if contacted inadvertently with liquid water. This new catalyst is a cobalt/molybdenum catalyst supported on spinel, which can be operated successfully below 600°F (316°C).

Shift catalyst work has been directed to the commercial manufacture of hydrogen. However, the improvements made also demonstrate the opportunity for developing improved catalysts for coal gasification fuels - namely, those active at lower temperatures, able to withstand sulfur, and capable of avoiding coke formation.

Methanation (50)

The methanation reaction (equation 3) is highly exothermic. The heat of reaction is minus 50,353 calories at 127°C (260°F). Catalytic methanation has been known for 70 years and has been utilized extensively in removing small amounts of CO from hydrogen-containing gases. The equilibrium is unfavorable at high temperatures, so a temperature over 825°F (440°C) is avoided.

The selectivity and conversion efficiency of different catalysts to H_2 + CO mixtures varies (Fig. 12). Based on conversion data, the metals can be arranged in order of decreasing activity as Ru, Rh, Re, Pt, Pd, and Os. However, their selectivity to methane is different. Hydrocarbons other than methane are formed.

The problem is that methanation has been used in gas streams that contain only about 2 mole percent oxides of carbon. The streams in coal gasification techniques may contain 25 mole percent oxides of carbon. Heat removal is a major engineering/materials problem. To overcome this problem, recently, Forney and associates have described a very interesting catalyst termed the tube-wall reactor (51). This catalyst is prepared by flame-spraying a nickel-aluminum alloy on tubes of two-inch diameter so as to form a Raney-type catalyst adherent to the tube wall. Having a heat transfer material such as Dowtherm on one side of the tube makes for a more practical engineering system. The

SYNTHANE pilot plant is being designed with this form of a tube-wall reactor.

Work on rate of quenching of the 50:50 Ni/Al alloy has shown that the amount of $NiAl_3$ is greater the faster the cooling rate. The Ni_2Al_3 phase which is to be avoided is shown in Fig. 13, labeled A. The $NiAl_3$ phase is lableed B, and

CATALYST	Pt	Pd	Ru	Rh	Os	Re	Mo
CAT No	L1138	L1140	L1141	L1139	L1163	L1154	L6137
RUN No	Z360	Z362	Z363	Z361	Z408	Z394	Z353
TEMP °C	513	499	222	441	600	451	420
SVH	297	318	303	302	295	281	288
CO_2 – FREE CONTRACTION, %	38.2	24.8	67.1	64.8	57.5	64.4	44.2
CONVERSION %	56.4	35.5	79.7	86.7	70.0	85.5	56.4
USAGE RATIO H_2/CO	1.8	2.2	2.4	2.4	2.3	2.4	1.8

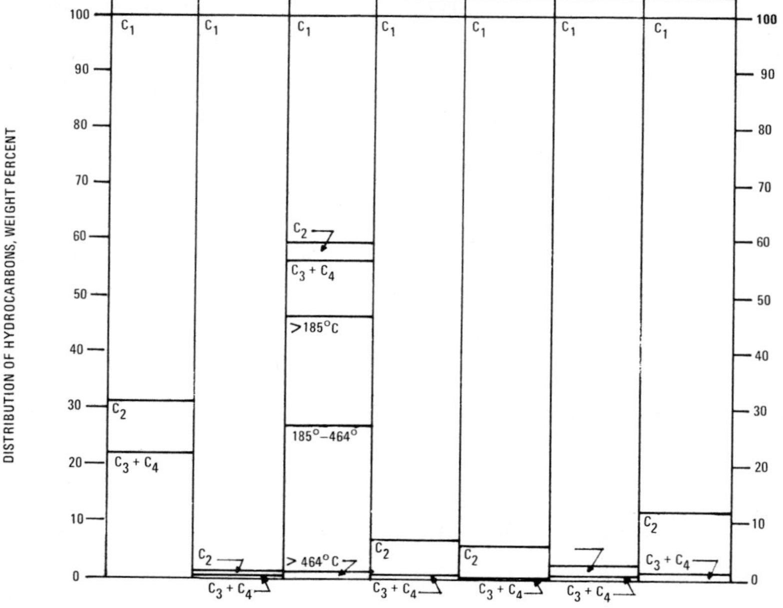

FIG. 12. Methanation Conversion Data for Metal Catalysts. Courtesy Marcel Dekker Publishers, 1973.

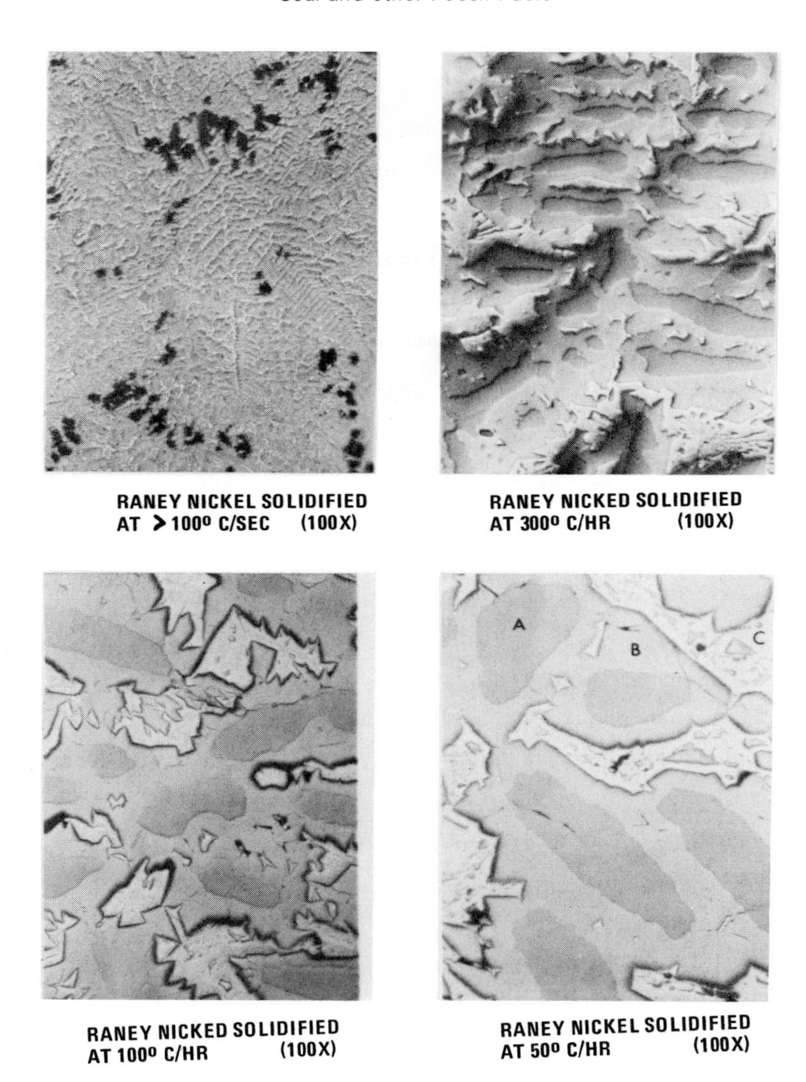

FIG. 13. Microstructures of 50/50 Ni/Al Alloys Under Different Cooling Rates.

C is the eutectic which contains very small islands of $NiAl_3$. The relationship between grain size and catalytic performance is being tested.

4.2 *Future*

Catalytic materials - how to fabricate better ones, and how to maintain them at optimum effectiveness - offer fertile fields for research with great rewards.

Improved gasification catalysts, by lower process temperatures, could alter equipment requirements and improve the economics of pipeline gas manufacture.

A combination shift and methanation catalyst seems possible.

Methanation catalysts are needed which are much less sensitive to sulfur and are able to withstand higher temperatures without sintering and loss of activity. Also needed are catalytic reactor systems capable of removing heat without costly equipment or high gas recycle rates. While at present high selectivity to methane is desired, future catalysts may require a selectivity that provides for simultaneous production of high hydrocarbons valuable for both energy and chemical applications.

5.0 ACKNOWLEDGMENT

The authors acknowledge the assistance of Dr. J. R. Ogren of TRW, Inc., in the preparation of this manuscript.

6.0 REFERENCES

1. Project Independence Blueprint, Synthetic Fuels from Coal, Interagency Task Force Report (1974).

2. H. Perry, Scientific American, 1230, (3), 19 (1974).

3. H. Storch, N. Golumbic, R. B. Anderson, Fischer-Tropsch and Related Synthesis, John Wiley & Sons, Inc., New York (1974).

4. Clean Energy from Coal Technology, Department of Interior, Office of Coal Research, U.S. Government Printing Office - 0-527-141 (1973).

5. L. A. Heredy, A. E. Kostyo, M. E. Neuworth, Coal Science, R. F. Gould, ed., American Chemical Society Publications, Washington, D. C. (1966).

6. W. W. Bodle, K. C. Vyas, The Oil and Gas Journal, 73, (August 26, 1974).

7. L. K. Mudge, G. F. Schiefelbein, C. T. Li, R. H. Moore, The Gasification of Coal, Battelle Energy Program Report (1974).

8. J. Huebler, Heating/Piping/Air Conditioning, 149, (January 1973).

9. G. A. Mills, Coal Utilization R&D Program, Office of Coal Research, Government Institute, Inc., Washington, D. C.

10. Energy Research Program of the U.S. Department of Interior - FY 1976, Office of Research and Development (1975).

11. H. Perry, Chemical Engineering, 88, (July 22, 1974).

12. A Survey of R&D Projects Directed Toward the Conversion of Coal to Gaseous and Liquid Fuels, I-A-15, Institute of Gas Technology (1972).

13. Materials and Components in Fossil Energy Applications, No. 1, ERDA Newsletter, U.S. Energy Research and Development Administration, ERDA-Fossil Energy, Washington, D. C. 20036 (July, 1975).

14. H. E. Frankel, Improving the Reliability of Fossil Energy Plants by Means of a Failure Analysis and Reporting System, 7th Annual Synthetic Pipeline Gas Symposium, Chicago, Illinois, October 1975, available from American Gas Association, Arlington, Virginia.

15. Materials Problems and Research Opportunities in Coal Conversion, Vol. 1, Conclusion and Recommendations from a Workshop, April 16-18, 1974, available from Corrosion Center, Ohio State University.

16. Materials Technology in the Near Term Energy Program, National Academy of Sciences (1974).

17. Report of the Committee on Materials (COMAT), Interagency Task Force Report, available from ERDA Conservation, Washington, D. C. 20036 (December, 1975).

18. G. E. Dieter, Mechanical Metallurgy, McGraw-Hill, New York (1961).

19. A. O. Schaefer, C. H. Samans, M. A. Howes, S. Bhattacharyya, E. R. Bangs, V. C. Hill, F. C. Chang, A Program to Discover Materials Suitable for Service Under Hostile Conditions Obtaining in Equipment for the Gasification of Coal and Other Solid Fuels, Annual Progress Report for 1975 on ERDA Contract E(49-18)-1784, Report FE-1784-12 (January 20, 1976).

20. F. A. Shunk, Constitution of Binary Alloys, Second Supplement, 274, McGraw-Hill, New York (1969).

21. R. F. Hochman, Metal Dusting: Catastrophic Deterioration of Metals and Alloys in Carbonaceous Gases, 6th International Congress on Metallic Corrosion, Sydney, Australia, December 9, 1975, available from National Association of Corrosion Engineers, Houston, Texas (1975).

22. ASM Metals Handbook, 8th ed., Vol. 1, 603, Properties and Selection of Metals (1961).

23. Commodity Data Summaries - 1975 (includes 1975 mineral industry data), available from U.S. Bureau of Mines, Dept. of Interior (1975).

24. R. F. Steigerwald, Metallurgical Engineering Quarterly 15, (4), (1975).

25. R. L. Brooks, Availability of Pressure Vessel Steels for Coal Gasification Plants, 7th Annual Synthetic Pipeline Gas Symposium, Chicago, Illinois, October 1975, available from American Gas Association, Arlington, Virginia.

26. P. F. H. Rudolph, R. K. Herbert, Conversion of Coal to High Value Products, University of Pittsburgh Symposium on Coal Gasification, Liquefaction and Utilization, August 5-7, 1975, available from Lurgi Mineralötechnik GmbH, Coal Technology D-6, Frankfurt (Main), Bockenheimer Landstrasse 42, P.O.B. 119181.

27. S. J. Dapkunas, Evaluation of Hot Corrosion Behavior of Thermal Barrier Coatings, Proc. 1974 Gas Turbine Materials in the Marine Environment Conference, J. W. Fairbanks and I. Machlin, eds., MCIC-75-27 (July 1974).

28. R. D. Howell, Mechanical Design Considerations in Commercial Scale Coal Gasification Plants, 6th Annual Synthetic Pipeline Gas Symposium, Chicago, Illinois (October, 1974).

29. A. J. McNab, Chemical Engineering Prog. 71, 11, 51 (November, 1975).

30. A. M. Hall, Materials Engineering, 16 (July, 1974).

31. J. J. Bollin, Commercial Concept Designs, Proc. 5th Synthetic Pipeline Gas Symposium, Chicago, Illinois (1973).

32. M. S. Crowley, Bulletin American Ceramics Society, 48, (7), 679 (1967).

33. M. S. Crowley, Bulletin American Ceramics Society, 49, (5), 527 (1970).

34. M. S. Crowley, Hydrogen-Steam Reaction in Refractories, Proc. American Ceramics Society, Chicago, Illinois (April, 1974).

35. P. Budnikov, F. Charitonov, Silicate Technology, 18, (8), 243 (1967).

36. E. R. Fuller, Jr., D. R. Robins, Effect of Elevated Steam Pressure on the Strength of Refractory Castables, Paper 14-R-75F, American Ceramics Society, Refr, Div. Mgt., Bedford Springs (October, 1975).

37. F. Gac, Reactions in Refractory Castables Exposed to High Pressure Steam-N_2 and Steam-CO Atmospheres, M.S. thesis, University of Missouri, Rolla (January, 1975).

38. Van Vlack, Journal American Ceramics Society, 31 (8), 22 (1948).

39. R. B. Snow, Bulletin American Ceramics Society, 48, 11 (1969).

40. R. L. Walker, J. V. O. Gorman, Mineral Matter and Trace Elements in U.S. Coals, R&D Report No. 61, Coal Research Section, College of Earth and Mineral Sciences, Pennsylvania State University (July 15, 1975).

41. R. H. Herron, K. A. Baab, Bulletin American Ceramics Society, 54 (7), 654 (1975).

42. G. Koenig, Stahl und Eisen, 91 (2), 63 (1971).

43. S. M. Wiederhorn, High Temperature Erosive Wear of Calcium Aluminate Bonded Castable Refractories, Paper 13-R-75F, American Ceramics Society, Refr. Div. Mtg., Bedford Springs, (October, 1975).

44. M. S. Crowley, J. F. Wygant, Bulletin American Ceramics Society, 52 (11), 828 (1973).

45. L. G. Huggett, Lining of Secondary Reformers in Materials Technology in Steam Reforming, ed. F. Edeheanu, Pergammon Press, London (1966).

46. M. S. Crowley, Refractory Problems in Coal Gasification Reactors, Paper, American Ceramics Society, Refr. Div. Mtg., Washington, D. C. (April, 1975).

47. G. A. Mills, S. Weller, E. B. Cornelius, The State of Platinum in a Reforming Catalyst, 2221, 2nd International Congress on Catalysis, ed. Tecknip, Paris (1961).

48. J. H. Sinfelt, Annual Review of Materials Science, 2, 641 (1972).

49. R. Van Herdevelt, A. Van Montfoort, Surface Science, 4, 396 (1966).

50. G. A. Mills, F. W. Steffgen, Catalytic Reviews, 8 (2), 159 (1973).

51. A. J. Forney, W. P. Haynes, The Synthane Coal-to-Gas Process, Preprints, Div. Fuels Chem., American Chemical Society (September, 1971).

Chapter 19

NEW CATALYTIC MATERIALS FOR THE CONVERSION OF COAL

M. Boudart

Department of Chemical Engineering
Stanford University
Stanford, California 94305

James A. Cusumano and Ricardo B. Levy

Catalytica Associates, Inc.
2 Palo Alto Square
Palo Alto, California 94304

I. INTRODUCTION

Inorganic chemistry encompasses the discovery, preparation and characterization of new inorganic and organometallic materials with unusual catalytic properties. At present, there is a significant need for new catalytic materials with unprecedented activity, selectivity and long life. This review of inorganic catalytic materials attempts to identify those classes of compounds which have the greatest potential for the discovery of new catalysts. Applications to the catalytic conversion of coal to chemicals and clean burning fuels are emphasized.

Inorganic chemistry has contributed to catalysis in two primary ways. In the broadest sense, the synthesis and characterization of new materials has provided an enormous wealth of interesting compounds with new binary, ternary and higher stoichiometries. Secondly, and more specifically, the solid state chemistry of materials has been used with increasing frequency over the last decade to solve problems such as thermal and chemical stability of catalysts, and the stabilization of oxidation states with desired catalytic properties.

The present discussion surveys some of the most important families of compounds, their properties of interest to catalysis and some of their applications to the catalytic conversion of coal.

II. COMPOUNDS OF INTEREST FOR THE CATALYTIC CONVERSION OF COAL

A look at the materials studied in solid state and inorganic chemistry is, in effect, a look at the periodic table of the elements. The combinations are innumerable. The wealth of materials is expanded even further due to the ability of many elements, and especially the transition metals, to exhibit a large number of valence states. Nevertheless, in spite of the vast number of compounds, there are groups of materials which exhibit common properties and belong to distinct chemical families. The most important families are shown in Table 1. Of these, only a few have been considered for catalytic applications. Furthermore, among the groups that are used as catalysts (oxides and metals, in particular), there are still large numbers of compositions that have not been explored.

Before a new composition can be considered as a catalyst for coal conversion, there are two important constraints which must be met: thermal and chemical stability. Thermal stability is a problem in several highly exothermic coal conversion reactions such as the synthesis of methanol and hydrocarbons from CO and H_2. It is also a general problem during catalyst regeneration, in particular when oxidative regeneration is used to remove carbonaceous deposits. The latter is a severe problem in coal liquefaction and many of the upgrading reactions for coal liquids. The chemical stability of a catalyst is a problem because of a number of reactive compounds that are present in the coal conversion environment, in particular sulfur, carbon and oxygen. In liquefaction, for example, sulfur is present at high concentration ~ 1-5%. This severely limits the

TABLE 1: SCOPE OF INORGANIC CHEMISTRY PERTINENT TO COAL CONVERSION CATALYSIS

MATERIALS	EXAMPLES	ADVANTAGES	CONSTRAINTS (1)	AREAS OF INSUFFICIENT DATA	COMMENTS
OXIDES	Simple: Al_2O_3, CoO, Cr_2O_3 Complex: $MgAl_2O_4$, $FeTiO_3$, $CuCrO_2$, $BaTiO_3$	Thermally & chemically stable - use as supports HDS ability with partial sulfidation Versatility with structures and stoichiometries Oxidative regeneration possible	Many oxides unstable in H_2S	Thermochemical data on oxysulfides Sulfur sensitivity of complex oxides	Complex oxides such as perovskites offer extensive opportunity for optimizing valence states for activity and sulfur tolerance
SOLID ACIDS	$TiO_2 \cdot SiO_2$, $ZrO_2 \cdot SiO_2$, Zeolites, Natural clays (montmorillonite), Super acids (SbF_5/HF), Supported halides ($AlCl_3$/Al_2O_3)	Extensive range of cracking activity Facilitates S/N removal Thermally stable Controlled physical properties Oxidative regeneration possible	Poisoning effects for some acids by S, N compounds	Activation of S, N in heterocyclics to facilitate removal Nature of acid in liquefaction environment	Although some solid acids are poisoned by sulfur & nitrogen compounds, many appear to function at liquefaction conditions (e.g. $ZnCl_2$, $SnCl_2$) even though the original acid may have a different structure during use.
SOLID BASES	CaO, MgO, $NaNH_2$, BaO, K_2O, Na_2O	Not poisoned by H_2O, N and S compounds Thermally stable Controlled physical properties Oxidative regeneration possible	Alkyl rearrangements and cracking may not occur through base-catalyzed mechanism	Cracking and rearrangement reactions of hydrocarbons over bases Reactions over basic mixed oxide	Very little catalytic work has been done with solid bases for reactions which are relevant to coal liquefaction
SULFIDES	Simple: VS, V_5S_8, MoS_2 Complex: CoV_2S_4, $SrZrS_3$, $Al_{0.5}Mo_2S_4$	Proven HDS & hydrogenation activity Versatility in complex sulfide formation Stable in H_2S Thermally stable		Thermochemical data on simple and complex sulfides	Defect chemistry may be crucial for activity: thus need to study small particles Variable stoichiometry may allow control of catalytic properties
CARBIDES NITRIDES	WC, Pt_3ZnC Fe_2N, Ni_3N_2, Ni_3AlN	Show some cracking, isomerization, and hydrogenation activity Thermally stable	Most carbides and nitrides unstable in H_2S Unstable in oxidative regeneration	Kinetics of sulfidation Thermochemistry of partial sulfidation	Stability in H_2S is a thermodynamic evaluation, no data available on kinetics of sulfidation Partial sulfidation may occur; could lead to an active catalyst
BORIDES SILICIDES PHOSPHIDES	Ni_3B, FeB, ZrB_{12} TiSi, RhSi, $TaSi_2$ MnP, NiP, TaP_2	Thermal stability Possible chemical stability in H_2S (Group VIII) Show some hydrogenation-hydrogenolysis activity (borides)	Probably unstable in oxidative regeneration	Limited thermochemical data Limited catalytic activity data	Optimism based on possible stability in H_2S; has to be corroborated. Need to be synthesized in high surface area
METALS & MULTIMETALLICS	NiCu, PtAu, NiW, RuCu, Cu_3Au, $ZrPt_3$	Preparable in high surface area Hydrogenation/hydrogenolysis activity	Most form bulk sulfides during operation, should be studied as sulfides	Kinetic and thermodynamic data for interaction of S with highly dispersed metals and multi-metallic systems	Technology for synthesis of highly dispersed metals may be applicable to sulfides.
MOLTEN SALTS	$ZnCl_2$, Na_2CO_3	Cracking activity Effective contacting with coal Good heat transfer	High catalyst/coal ratios Handling & containment problems (corrosion) Eutectic & compound formation with coal ash Complex regeneration, poor catalyst recovery	Model reactions in melts (hydrogenation, HDS, HDN, etc.) Catalytic data for melts other than $ZnCl_2$	
HOMOGENEOUS CATALYSTS	$Co_2(CO)_8$, $RhCOCl_2$	Catalyst/coal contacting very effective Possibly tolerant of S, N compounds if catalyst doesn't degrade Controlled hydrogenation activity and selectivity (minimize H_2 consumption)	Many materials thermally unstable at conditions of operation Catalyst recovery problems	Effects of S and N compounds on catalysis HDS and HDN activity New techniques for catalyst recovery	
ANCHORED CATALYSTS	SiO_2/-$OSi(C_2H_4)(Ph_2P)RhCl(CO)_2$	Catalytic activity and selectivity of homogeneous catalysts retained	Thermal stability poor Chemical stability not clear Unlikely to survive coal liquefaction environment		

(1) These constraints have to be overcome before these materials can be considered promising.

type of materials that can be used as catalysts. In $CO-H_2$ synthesis reactions, the sulfur contamination has to be reduced below 1 ppm for currently known catalysts to operate with reasonable activity and life. The presence of carbon is an important factor in methanation, where carbide formation leads to deactivation and degradation of the catalyst. Finally, the presence of oxygen during carbon removal of spent liquefaction and upgrading reactions imposes an additional constraint on the chemical stability of new potential catalytic materials.

In the following discussion, the properties of a number of important families of compounds are examined, with emphasis on thermal and chemical stability and some of the other properties and compositions of interest. A more detailed discussion may be found elsewhere (1).

A. *Oxides*

Metallic oxides constitute a very large class of compounds. They are important catalytic materials and are used in a number of industrial processes, some of which are illustrated in Table 2.

TABLE 2

SOME REACTIONS CATALYZED BY OXIDES

Reaction	Catalyst
Partial or a Complete Oxidation	$CoO-MoO_3$
Oxidative Dehydrogenation	$Bi_2O_3-MoO_3$
Cracking	Aluminosilicates (including transitional, acid treated)
Reforming	Al_2O_3
Polymerization	MoO_3/Al_2O_3
Water Gas Shift	$CuO + ZnO-Al_2O_3$ $Fe_3O_4-Cr_2O_3$
Methanol Synthesis	$CuO-ZnO-Al_2O_3$

The extensive and varied chemistry of oxides leads to a broad range of thermal and chemical stabilities and a number of interesting and unusual structures and compositions. Some of these are discussed below.

1. *Thermal Stability: Volatibility and Sintering*

In general, oxides exhibit high thermal stability at the temperatures and environments of interest for catalysis. Exceptions are the oxides of Os or Ru, which are volatile at low temperatures. This eliminates the use of these metals in oxidizing environments. Under most conditions, however, oxides maintain high specific surface areas up to temperatures greater than 800 K. Some, if stabilized by the incorporation of small amounts of certain cations, can be made resistant to sintering at higher temperatures. For example, Al_2O_3 can be stabilized with Cs^+, Ce^{+4} or La^{+4} so as to maintain a surface area of 20-50 m^2/g up to 1500 K (2).

Because of this high thermal stability, many oxides (such as Al_2O_3 and SiO_2) are routinely used as catalyst supports. However, there are limitations. Steam, for example, leads to crystal growth and sintering at elevated temperatures, often due to chemical vapor transport (3). Thus, materials such as MoO_3 show increased volatility at high temperatures in the presence of water vapor. This phenomenon frequently puts a constraint on the use of steam in certain processes. It can also lead to catalyst degradation during oxidative regeneration if excessive partial pressures of water are allowed to form. Such catalyst loss can sometimes be minimized by the use of appropriate supports, e.g. Al_2O_3 to stabilize MoO_3.

2. *Chemical Stability*

The chemical stability of oxides varies across the periodic table. This is particularly evident when one examines the behavior of various oxides in an H_2S environment, especially at elevated temperatures (1). Typical examples are shown in Table 3. Three degrees of stability can be distinguished. The most stable oxides include those of Al, Si, Be and Mg (1). The oxides of the rare earth elements as well as those of Group IV (Ti, Zr, Hf) show intermediate stability. All other transition metal oxides are thermodynamically unstable in H_2S. It should be stressed, however, that the criteria of stability depend on the severity of the environment (e.g.

TABLE 3

FREE ENERGY OF FORMATION OF SULFIDES FROM OXIDES AT 700 K[(a)]

Group	Reaction	$\Delta G^o_{700\ K}$ kcal g-atom^{-1}
IIIB	$La_2O_3 \rightarrow La_2S_3$	-20
	$La_2O_3 \rightarrow LaS_2$	-8
IVB	$TiO \rightarrow TiS$	+42
	$TiO \rightarrow TiS_2$	-42
	$TiO_2 \rightarrow TiS$	+92
	$TiO_2 \rightarrow TiS_2$	-50[(b)]
VIB	$MoO_2 \rightarrow MoS_2$	-59
	$MoO_3 \rightarrow MoS_2$	-138
VIIB	$MnO \rightarrow MnS$	-42
	$MnO \rightarrow MnS_2$	-25
	$Mn_3O_4 \rightarrow MnS$	-42
	$Mn_3O_4 \rightarrow MnS_2$	-8
VIIIB	$CoO \rightarrow Co_9S_8$	-75
	$CoO \rightarrow CoS_2$	-71
	$Co_3O_4 \rightarrow Co_9S_8$	-92
	$Co_3O_4 \rightarrow CoS_2$	-89
IB	$CuO \rightarrow CuS$	-54
	$Cu_2O \rightarrow CuS$	-92
IIIA	$Al_2O_3 \rightarrow Al_2S_3$	+105

(a) Details of the calculations in this and subsequent tables can be found in references 1 and 4.
(b) Expected reaction at 700 K

temperature, pressure, H_2 partial pressure). Thus, there is a direct relationship between the free energy change and the pressure of H_2S. For the above classification, the high H_2S concentrations expected in coal liquefaction ($\sim$ 1%) were used. However, the relative ranking remains the same for lower concentrations. An important point is that support materials such as oxides of Al and Si are stable even in the most severe H_2S environments.

The same conclusions can be drawn in a reducing environment. A look at the free energy changes for the reduction of different transition metal oxides to the zero valent metal shows that reducibility increases as one progresses to the right of the periodic table (1). This trend persists through the Group VIII metals. Thus, the reducibility trend in the first series is Fe $\ll$ Co $<$ Ni (5).

3. *Other General Properties of Catalytic Interest*

As illustrated in Table 2, oxides have been used extensively in catalysis. Properties of particular interest are surface acidity, which can be varied over several orders of magnitude (6-8), and the ability of oxides to exchange lattice oxygen with gas phase O_2 (3). The former is used in the petroleum industry extensively for cracking, reforming, isomerization and other reactions which require an acid function (1), while the latter is a property that relates to the ability of oxides to catalyze the complete oxidation of hydrocarbons (10).

As can be seen from the examples in Table 1, complex oxides offer a wealth of compositions and structures. To illustrate the variety in these structures, it is of interest to examine the difference between spinels and perovskites. In spinels, the basic structural unit is a close packed network of oxygen ions. The metal ions reside in the tetrahedral and octahedral interstices formed by this network. The type of site occupied depends roughly on the size of the cation (1). In general, spinels are thermally very stable and exhibit a range of compositions that depend on the extent of the occupation of interstitial sites. Typical examples are $CrAl_2O_4$ (ruby) and $MgAl_2O_4$ (the mineral spinel). In perovskites, on the other hand, one of the metal cations of the complex oxide is too large to fit in the interstitial sites. This cation becomes part of the close packed network, with a metal to oxygen concentration ratio of 1:3, while the other smaller cations again reside in octahedral oxygen interstices. The interesting feature of this structure is the ability to exist with less than stoichiometric amounts of the large cation, leaving a rather open structural network. One extreme case is WO_3, where all the sites available to large cations such as Li, Na, K and Rb are empty. The structure is nevertheless stable, and even admits smaller ions such as H^+. This ability to exist in a number of stoichiometries allows for practically continuous variations in oxidation states, and therefore continuous changes in the electronic properties of the solid.

The large number of compositions and structures result in complex oxides with a variety of electronic and chemical properties. These are likely to be reflected in changes in surface properties which may prove to be of catalytic interest. In many cases, metals are located in unique and unusual environments. Certain oxides, for example, exhibit smaller metal-metal distances than encountered in the pure element itself, e.g. $BaRuO_3$ with a Ru-Ru distance of 2.55 Å (11), or clusters of several transition metals e.g. $Mg_2Mo_3O_8$ with Mo_3 clusters (12). The catalytic properties of such interesting compositions have not been studied in detail. However, they have been applied in certain cases to solve specific problems. The $BaRuO_3$ system provides an example of such an application. It has been known for some time that ruthenium is one of the most effective NO_x reduction catalysts, in particular because of minimum NH_3 formation (13). However, previously it could not be used as an automotive emissions control catalyst because the oxidizing conditions encountered in the exhaust led to volatilization of the ruthenium as RuO_4. Recently, a procedure was developed to stabilize ruthenium under such conditions (14). Here the ruthenium is supported on a thermally stabilized support such as Al_2O_3 in combination with an alkaline-earth or rare-earth oxide. Under oxidizing conditions, the elements form the above mentioned ruthenate. Under reducing conditions, the stable state is a two phase system consisting of Ru metal and the alkali or rare-earth metal oxide. The solid state chemistry of the ruthenium-alkaline earth system, therefore, provides a solution to the problem of thermal stability of ruthenium. Under oxidizing conditions, a stable, non-volatile complex perovskite is formed. Under reducing conditions, the zero valent metal is "regenerated" and available to perform its catalytic function. This example illustrates the use of inorganic chemistry in the development of new catalystic materials.

4. *Zeolites*

One family of oxides that deserves special attention in this review of inorganic chemistry is that of zeolites. These materials have been used in many catalytic reactions of importance to coal conversion including upgrading of coal liquids and, more recently, the conversion of methanol to high octane gasoline (79, 80). Their synthesis and structure has been studied extensively over the last decade. Of special interest are their structure, acidic properties, potential thermal stability and the indication of considerable metal catalyst-zeolite interaction.

Zeolites are members of a large class of compounds called aluminosilicates. They consist of SiO_4 tetrahedra that share corners. Silicon is partially substituted by Al ions with positive ions such as Na^+ or K^+ to compensate for the charge difference between Al^{3+} and Si^{4+}. The charge compensating cations can be exchanged for many others. This property plus the fact that the structures have cavities (cages) and channels of varying sizes accessible to even rather large molecules, make these materials of great catalytic interest. Naturally occurring zeolites display fibrous structures like edingtonite, $Ba(Al_2Si_3O_{10})\cdot 4H_2O$, lamellar structures such as phillipnite, $(K, Na)_5\ Si_{11}\ Al_5O_{32}\cdot 10H_2O$, and zeolites with 3-dimensional polyhedral cavities such as chabazite, $Ca(Al_2Si_4)O_{12}\cdot 6H_2O$ (15). The zeolites with large accessible cages have found greatest application in catalysis. They can be synthesized with controlled cavity and channel sizes, thereby providing a considerable variation in physical properties. In general, the more open zeolite structures are prepared from sodium rather than potassium-containing gells (16). This is a direct function of the size of the hydrated ion, since the tetrahedral building blocks of the zeolite are believed to surround this hydrated ion during synthesis.

The <u>acidity</u> of zeolites is a property of great interest to catalysis. It is largely responsible for the extensive use of these materials since their introduction as fluid cracking catalysts in 1962 (17). The acidity depends on a number of factors that are related to the inorganic chemistry of these compounds. It is primarily controlled by the types of ions present in the structure which compensate the charge difference between Al^{3+} and Si^{4+}. Zeolites with univalent cations (e.g. Na^{+}, K^{+}) have low acidity. Ion exchange of these materials with ions such as Ca^{+2}, Mg^{+2}, Sr^{+2}, and Zn^{+2} leads to an increase in the acidity, with a distribution concentrated towards strong acid sites (18, 19).

<u>Thermal stability</u> is another important parameter of zeolites which is a function of composition. For example, zeolites with substituted germanium ions are considerably less stable than those with silicon ions. (20). On the other hand, it has been shown that Si-Al zeolites which are deficient in aluminum show increased thermal stability and can tolerate high temperatures and severe hydrothermal conditions (21, 22). Some zeolites have been synthesized which can endure temperatures approaching 1300 K without any appreciable loss of crystallinity. This property is important for catalytic application which require frequent oxidative regeneration, e.g. catalytic cracking and hydrocracking. Regeneration is highly exothermic and high local temperatures can cause catalyst degradation. The thermal stability of many zeolites makes these materials excellent catalysts or catalyst supports for such applications.

Finally, there is a property of zeolites which has been reported recently and may have interesting catalytic consequences: the presence of <u>metal catalyst-zeolite interactions</u> in the case of metallic clusters in zeolites. First reported for Pt in a Y zeolite (23), and confirmed for Rh and Pd, also in a Y zeolite (24), the small metal clusters in the zeolite cages are found to be electron deficient. This is concluded from ESR spectra (24) as well as the catalytic behavior of the system (23, 25). In the case of platinum -Y zeolites (23), the activity of platinum for reactions of hydrogenation and hydrogenolysis is more typical of that of iridium, platinum's neighbor in the periodic table. This ability to modify activity by varying the electronic interaction with the zeolite support is an important aspect of these materials. In fact, as pointed out by Dalla Betta and Boudart (23), it has possible implications for the sulfur resistance of these catalysts in addition to the observed changes in catalytic activity.

B. *Sulfides*

Formation of transition metal sulfides is particularly important in processes directed towards synthetic fuels. Many reactions are carried out in high concentrations of H_2S, and therefore require catalysts that survive such environments. This is particularly true for hydrodesulfurization. In other coal conversion reactions, the problem of sulfur tolerance is also severe. It is likely that many catalysts in their working state are totally or partially sulfided. The study of sulfides is therefore an important aspect of a general review of potential catalytic materials.

In spite of their importance, only few sulfides have been tested as catalysts. They have been used primarily for hydrogenation reactions, in particular when sulfur-containing compounds are present. In general, the catalysts tested have been mixed sulfides of Ni, Co, W, and Mo (26). Only limited information is available on the structure of these materials at reaction conditions. However, a detailed study of the Ni-W sulfide system by Voorhoeve et al. (27) has shown that the interaction between added nickel and the WS_2 matrix has an important effect on the catalytic activity of these materials in reactions of hydrogenation. This suggests that complex sulfides should be explored more extensively for catalytic applications, especially in view of the growing interest in these materials in the solid state chemistry literature (28, 29, 30). For the present review, the discussion of sulfides is conveniently divided into those aspects concerning thermal ability, chemical stability and examples of typical compositions.

1. *Thermal Stability*

The examples shown in Table 1 illustrate the large number of stoichiometries possible for each transition metal sulfide. The behavior of vanadium is typical of a large number of transition metals. In addition to various compositions, metal sulfides often exhibit extensive homogeneity ranges (e.g. $VS \rightarrow V_5S_8$). The thermal stability of a particular composition is a sensitive function of sulfur pressure. A discussion of the thermal properties of sulfides is therefore closely related to the chemical environment of the system.

The thermochemistry of a number of metal sulfides was studied as early as the 1930's. For vanadium, for example, Biltz and co-workers investigated a broad range of compositions at various sulfur pressures (31). These studies were followed by extensive work of French workers who followed the decomposition of vanadium sulfides gravimetrically (32, 33). More data is lacking, especially at the low sulfur concentrations of interest in most catalytic systems. In spite of this, the general evidence is that the lower sulfides such as CoS and VS are stable to high temperatures. For example, VS has been reported to be stable up to 1800 K in vacuum (34). By contrast, stoichiometric compounds such as VS_4 decompose at temperatures as low as 600 K even in the presence of high sulfur pressures (32, 33). This low stability explains the decomposition of VS_4 catalysts tested for hydrogenation and desulfurization (35), and illustrates the importance of the study of the thermochemistry of these compounds if they are considered for catalytic applications.

2. *Chemical Stability*

The stability of sulfides in H_2S is a function of pressure and temperature. As discussed extensively in reference 1, in the case of H_2S the important parameter is the equivalent (or "virtual") sulfur pressure which is in equilibrium with an H_2S/H_2 mixture. This parameter is also important in determining the possibility of sulfide formation for other materials. For metals, for example, the minimum H_2S concentration for bulk sulfide formation can be calculated from free energy data (1). A few examples are shown in Table 4. These data show that the sulfiding tendency of transition metals varies widely. In severe environments such as encountered in coal liquefaction, thermodynamics predicts that all transition metals can sulfide. Therefore, if the reaction proceeds, it probably does so in the presence of a metal sulfide. At lower H_2S concentrations, however, this will not be the case. At conditions of methanation ($H_2S < 1$ ppm), sulfide catalysts such as NiS will be unstable. However, considerable interaction of H_2S with the surface can still occur with possible effect on activity and selectivity of the catalysts.

TABLE 4

MINIMUM CONCENTRATION OF H_2S (AS PERCENT H_2S IN H_2) REQUIRED FOR BULK SULFIDATION; T = 700 K

Stoichiometry	Concentration, %
CrS	4×10^{-4} (4 ppm)
MnS	2×10^{-11}
FeS	7×10^{-2} (700 ppm)
CoS	3×10^{-2}
NiS	2×10^{-1}
CuS	73
RuS	1×10^{-6} (10 ppb)
OsS	2×10^{-2}
PtS	1.3

As indicated earlier, stability in the presence of oxygen is also of concern with coal liquefaction catalysts. The above remarks of sulfide stability apply to a specific H_2S/H_2 environment. Several sulfides (such as MnS or VS) are stable in very low H_2S pressures or, in the limit, in vacuum. However, most sulfides can oxidize in the presence of O_2 (1). This is an important aspect of the chemical stability of these materials because many catalytic processes use oxidative regeneration procedures. In such circumstances, the balance between the oxide and sulfide becomes important and the possibility exists of partial sulfidation - namely the formation of oxysulfides. There is little information on these materials, which deserve more research, in particular when oxidative regeneration of the catalytic material is considered. The oxidation of both bulk and supported sulfides should be studied. The latter is especially important because the chemical behavior of supported systems may be different from that of the bulk, as has been suggested for cobalt molybdate catalysts supported on Al_2O_3 (36).

3. *Compositions of Interest*

Like oxides, there are a number of complex sulfides that take advantage of the extensive chemistry of transition metals. Many of the properties and structures of these compounds have been studied recently. The layered structure of a number of disulfides leads to a large number of complex sulfides that resemble intercalation compounds, such as FeV_2S_4, $TiCr_2S_4$, $CuCrS_2$ (37, 38). Compounds that have the perovskite structure $MM'S_3$ (M = Sr, Ca, Pb; M' = Zr, Ti) have also been synthesized and studied (39). Complex sulfides that incorporate a rare earth metal (such as La_4NiS_7 and $La_2Fe_2S_5$) are another group of sulfides which has received recent attention (40). Finally, sulfospinels have been investigated (41, 42), and exhibit the same non-stoichiometry shown by oxides. $Al_xMo_2S_4$, for example, has been prepared with values of x as low as 0.5. The latter stoichiometry exhibits unusual molybdenum clusters similar to that found in some of the complex oxides discussed earlier. Many of these contain cations which are known to exhibit catalytic activity in a sulfide matrix, such as Mo^{4+} and Ni^{2+}. They are therefore interesting candidates for the catalytic conversion of coal in the presence of sulfur.

C. *Carbides and Nitrides*

As outlined in detail in a recent review (4), carbides and nitrides of transition metals show similarities in many of their chemical and physical properties. The structures of these materials, in particular, are closely related. Because of the large difference in size between transition metals and C or N (radius ratios greater than 1.7), the non-metal occupies interstitial sites formed by close packing of transition metal atoms. The presence of the non-metal in these sites substantially modifies properties of the parent transition metal including the thermal and chemical stability.

1. *Thermal Stability*

Carbides and nitrides of transition metals are known to be refractory compounds which are stable at very high temperatures. Thus, several carbides and nitrides melt or decompose above 3000 K and TaC has the highest melting point known for any material (about 4250 K) (43). This stability of bulk carbides and nitrides is likely to persist when these materials are prepared in fine dispersion, since carbides are known to resist sintering. In fact, metallurgists have had to add binders such as cobalt to achieve dense "cemented" carbides (using powder metallurgical procedures) for cutting tools (44). Carbides and nitrides are therefore an interesting class of materials in terms of their potential application to such exothermic reactions as methanation and CO-H_2 syntheses where sintering can be a severe problem.

2. *Chemical Stability*

The resistance of carbides such as WC to corrosion is well know, and is one of the reasons why these compounds have been tested as battery or fuel cell electrodes (45). In the presence of oxygen, however, carbides and nitrides will oxidize according to thermodynamic data (1). Kinetically, of course, this oxidation does not take place at low temperatures. However, at temperatures of catalyst regeneration, oxidation may be a problem in the use of these materials as catalysts. If this were the case, the catalyst re-synthesis (for example of the oxide with NH_3 or a hydrocarbon to form the respective nitride or carbide) could be a problem. With sulfides (e.g. sulfidation of cobalt molybdate) such re-synthesis occurs readily.

In the presence of sulfur containing compounds, thermodynamic calculations indicate that most carbides are expected to sulfide (4). Several examples are shown in Table 5. The sulfur level at which this occurs, and the rate of sulfidation, have yet to be determined experimentally. The same is true, in general, for nitrides. However, the most stable nitrides, such as TiN, show a positive free energy of sulfide formation, and are therefore expected to survive even in severe sulfiding environments (4).

It is interesting to note, with respect to the sulfur reactivity of these compounds, that formation of a strong refractory material does not necessarily improve sulfur tolerance. Tungsten, for example, reacts with H_2S according to the equation:

$$W + 2H_2S \longrightarrow WS_2 + 2H_2$$

The standard free energy change for this reaction at 700 K is -69.9 kJ mol^{-1} (1). For the carbide, on the other hand, the reaction

$$WC + 2H_2S \longrightarrow WS_2 + CH_4$$

has a standard free energy change of -104.7 kJ mol^{-1} (1). The main reason for the increased tendency of WC to sulfide in the presence of H_2S in this illustration is the formation of CH_4, which is very stable.

3. *Structures and Compositions of Interest*

With the exception of several Group VIII noble metals (Ru, Rh, Pd, Ir, and Pt), all transition metals form a number of carbide and nitride stoichiometries (43). They follow the empirical rules postulated by Hägg in 1931: if the radius ratio of non-metal to metal is less than 0.59, the compounds formed have a simple, interstitial structure (1). In addition to such binary compounds, a number of very interesting multimetallic carbides and nitrides have been synthesized. Of particular interest are the Nowotny octahedral phases (43). They contain two or more transition metals, and display complex structures, many of them bearing a relationship to the oxide structures discussed earlier. Some examples are shown in Table 1. Particularly interesting are the carbide and nitride perovskites such as Pt_3ZnC (1). These compounds are likely to show interesting modifications of the bulk properties of the parent metals which in turn should alter the surface chemistry of these materials.

TABLE 5

FREE ENERGY CHANGE FOR THE REACTION OF CARBIDES AND NITRIDES WITH H_2S

Compound	Group	Reaction		$\Delta G^o_{700\ K}$ kJ g-atom^{-1}
Carbides	IVB	TiC	TiS	-33
		TiC	TiS_2	-117
	VIB	WC	WS_2	-105
Nitrides	IVB	TiN	TiS	+92
		TiN	TiS_2	+8
		ZrN	ZrS_2	-134
	VB	TaN	TaS_2	-29

In addition to the interesting carbides and nitrides mentioned above, carbonitrides and oxide-carbides are formed by many transition metals (46). This presents new opportunities for the synthesis of novel compounds with a broad range of physical and chemical properties. The synthesis of these materials in high surface areas is a key challenge to their application in catalysis. In general, the attention of materials scientists has been focused on the preparation of dense materials (47). However, the methods of powder metallurgy and some of the recently developed techniques for catalyst preparation are likely to have a bearing on the synthesis problems discussed above.

D. *Borides, Silicides and Phosphides*

Carbides and nitrides were discussed together because of the similarities in structural and physicochemical properties of these materials. For this same reason, it is convenient to treat borides, silicides and phosphides as a group. These compounds are formed by most transition metals. The occurrence of these materials is, in fact, more extensive than that of carbides and nitrides since even Group VIII noble metals form a number of compounds with B, Si, and P. However, they have not been studied in detail, especially with respect to thermochemical properties. High thermal and chemical stability, unusual mechanical strength, and stable structures are among the more interesting properties of these materials which could be pertinent to catalytic applications.

1. *Thermal Stability*

Borides and silicides have thermal stabilities that are comparable to the carbides and nitrides discussed above (e.g. TiB_2 melts at 3253 K). Although little work has been reported for phosphides, they are also expected to form compounds with melting points above 2500 K (48). However, while the stability of carbides and nitrides varies quite sharply across the periodic table (decreasing as one approaches Group VIII), silicides and borides do not show such large variations. As indicated above, the latter elements form stable binary phases even with the noble metals of Group VIII. This presents the possibility of modifying the catalytic properties of the Group VIII metals, which are of primary importance in catalysis.

2. *Chemical Stability*

A comparison of the enthalpies of formation of borides, silicides, and phosphides with those of oxides clearly shows that the former materials tend to oxidize in the presence of O_2. In practice, these compounds, and in particular the silicides, have shown very high resistance to oxygen attack (49). They are used as heating elements because of this property. However, as pointed out by Searcy (49), this chemical resistance is likely to be the result of a thin passivation layer of SiO_2 which protects the compound from bulk oxidation very much the way aluminum metal is protected from further oxidation by a thin film of Al_2O_3. If the metal silicide is the important catalytic material, this coating would therefore lead to a decline in activity.

An interesting catalytic property of this group of materials becomes evident upon examination of the thermodynamics of sulfidation. As expressed earlier, the thermochemistry of these compounds has not been studied very extensively. For borides, in particular, there are only limited data for certain compositions of the Group IV metals (50). Data for silicides are more widespread, but vary by as much as 30% among different laboratories (49). In spite of this uncertainty, a definite trend is apparent (1) as seen from Table 6: from left to right across the Periodic Table, silicides become more resistant to sulfidation. Group VIII silicides, in fact, are expected to be sulfur tolerant even in high H_2S concentrations. A similar trend is observed for phosphides, and is expected for borides in view of the similarity in other chemical properties of these three materials (1). This has obvious implications for catalysis, especially for reactions such as methanation for

TABLE 6

FREE ENERGY CHANGE FOR THE REACTION OF BORIDES, SILICIDES AND PHOSPHIDES WITH H_2S

Compound	Group	Reaction	$\Delta G^o_{700\ K}$ kJ g-atom^{-1}
Borides	IV B	$TiB \rightarrow TiS$	-46
		$TiB \rightarrow TiS_2$	-130
		$TiB_2 \rightarrow TiS_2$	-50
Silicides	IV B	$TiSi \rightarrow TiS$	-75
		$TiSi \rightarrow TiS_2$	-159
	VII B	$MnSi \rightarrow MnS$	-50
		$MnSi \rightarrow MnS_2$	-4
	VIII	$FeSi \rightarrow FeS$	+38
		$FeSi \rightarrow FeS_2$	+25
		$OsSi \rightarrow OsS$	+13
		$OsSi \rightarrow OsS_2$	+33
		$CoSi \rightarrow CoS_2$	+59
		$IrSi \rightarrow Ir_2S_3$	+46
		$IrSi \rightarrow IrS_2$	+50
		$NiSi \rightarrow NiS$	+50
		$PtSi \rightarrow PtS$	+38
		$PtSi \rightarrow PtS_2$	+67
Phosphides	IV B	$TiP \rightarrow TiS_2$	-13
	VII B	$MnP \rightarrow MnS$	-50
		$MnP \rightarrow MnS_2$	0
	VIII	$FeP \rightarrow FeS$	+88
		$FeP \rightarrow FeS_2$	+75

which currently used Group VIII metals are extremely sulfur sensitive.

3. *Structure and Stoichiometry*

The common feature that is most evident for borides, phosphides and silicides is structure. The size of phosphorous and silicon is too large to form interstitial compounds of the type formed by C and N. Boron is on the borderline of the Hägg rule (36) stated earlier for interstitial carbides and nitrides, and in general behaves more like Si or P than C or N. The structures formed are characterized by a wide range of stoichiometries, and the formation of non-metal networks at high non-metal to metal ratios. Some examples are shown in Table 1.

In addition to simple compounds, a number of complex structures are formed. This is particularly the case for boron, which forms series of compounds such as MoCoB, WFeB, WCoB. Perovskite-like structures have also been reported. They exhibit interesting properties such as metal cluster formation in the case of $(Co_{13})(Co_8)Hf_2B_6$ (51). In addition to these multi-metallic compounds, the formation of complex compounds with mixed non-metals is also of great interest. They are formed primarily by borides, as illustrated by the formation of numerous borocarbides such as Mo_2BC and ScB_2C_2 (52). The latter compounds, also formed by the 4f elements (lanthanide series), display close metal-metal distances and interesting B-C networks (52). They are likely to have unusual chemical and physical properties which have not been explored to date. These properties may possibly give rise to unusual catalytic properties in certain reactions.

E. *Alloys*

Alloys have been the subject of study as catalytic materials for some time. Recent theoretical work on alloys dates back to concepts proposed by Douden (53, 54) and Schwab (55), and was directed at understanding the relationship between catalytic activity and the electronic structure of metals as expressed by bands and Brillouin zones. Over the last decade, the early ideas have been refined considerably. For example, the rigid band picture of the electronic structure of alloys (53-55), has been replaced by more realistic models such as that of the virtual bound state (56, 57). The development of photoelectron spectroscopic techniques has led to a better understanding of the surface and catalytic properties of alloys related to surface composition as determined, for example, by Auger electron spectroscopy. Finally, while the early work dealt mostly with massive alloys (wire, foils, powders), Sinfelt and co-workers have now shown how to obtain supported alloys with a high dispersion of the metals (58). These aspects have been reviewed in detail elsewhere (1, 59). The following discussion gives a brief treatment of thermal and chemical stability and is followed by a few examples of catalytic applications. For more details, the reader is referred to references 1 and 59. Also, for simplicity, only binary alloys are considered, although many of the basic concepts discussed are applicable to multi-component complex systems.

1. *Thermal Stability: Sintering and Surface Composition*

The effects of temperature on a supported binary alloy catalyst depends on a number of factors including initial metal dispersion (crystallite size), nature of the support and the chemical environment. In general, high initial dispersion favors decreased sintering rates (60, 61). Metal support interactions can be used to minimize sintering (62) and in certain chemical environments to eliminate volatilization of the metal component from the support, as discussed earlier in this review. Similarly, metals are also affected by halogens. Under some conditions, a halogen such as chlorine facilitates dispersion of the metal, however, under others crystallite growth can occur.

Another aspect of supported alloys is the change in surface composition with crystallite size, as expected from the thermodynamics of dispersed alloy systems (64-68). It is an effect that is generally found, even in relatively miscible systems such as NiCu (69, 70) and NiAu (65). Equilibrium surface composition of alloys depends on temperature and reflects the thermodynamic driving force for the metal with the lowest surface energy to concentrate on the surface. Recent work has also shown that the chemical environment can have a marked effect on surface composition. Thus, the presence of oxygen over a clean Ni-Au surface which is rich in gold leads to a surface exhibiting nickel exclusively (71). Similar studies by Sachtler and co-workers (72, 73) with Pd-Ag surfaces show that chemisorbed CO brings more Pd to the surface than is found in the absence of CO.

Each of the above phenomena which alter surface composition or metal surface area must be assessed in considering the use of supported alloys for high temperature applications or in chemically reactive environments.

2. *Chemical Stability*

By far the two major demands on a catalyst in environments involved in processing coal are compatability with sulfur and stability with respect to oxidative regeneration (1, 59).

With respect to sulfur, as discussed earlier, all metals can sulfide in an environment such as for the direct liquefaction of coal to boiler fuels. Alloying is not expected to have a significant effect on the thermodynamic driving force for sulfidation under these conditions (1). However, at lower levels of sulfur (e.g. 10-1000 ppm), alloy formation may alter the poisoning effects of sulfur although there are few data in the literature to support this contention (74). The example of the resistance of the very stable intermetallic compound $ZrPt_3$ to sulfidation has been discussed in detail elsewhere (1, 59). Here it was shown that increased exothermicity for the heat of formation of the alloy leads to a greater thermodynamic resistance to bulk sulfidation. Indeed, intermetallic compounds of the Brewer-Engel-type (59) may hold promise for applications in this respect, provided these materials are used in environments where the sulfur levels are kept below about 1000 ppm H_2S (59).

Problems are also caused by the oxidative regeneration which is required for carbon burn-off from catalysts. Sintering due to high local surface temperatures promotes crystallite growth and is a problem for most metals. However, for alloy catalysts, an additional factor is important. This involves changes in surface composition and phase segregation or separation with increased crystallite size (64-68). For some bimetallic systems (e.g. PtCo, PtPd, RuCu), there is an increase in thermal stability with respect to sintering (75, 76) over that observed for the pure metals. However, this increased stability is not necessarily maintained in reactive oxidative environments. For example, cycling between the metallic and oxide state in some bimetallic catalysts has been reported to give rise to the formation of catalysts with a different distribution of surface composition than the original catalyst (77). Such changes must be taken into account when studying these catalysts under such conditions.

3. *Other Catalytic Implications of Alloy Formation*

From the above discussion, it is clear that alloy catalysts have potential applications for a number of coal processing steps which do not involve high sulfur concentrations. The development of these materials over the last decade has led to catalysts with controlled activity and selectivity, increased thermal resistance and activity maintenance, and a potential for improved sulfur resistance at low levels.

Some of these applications involve the upgrading of coal liquids to clean-burning fuels and a number of synthesis reactions (e.g. methanation) starting from carbon monoxide-hydrogen mixtures. Of particular interest, is the ability of certain alloy catalysts (e.g. $PtFe/Al_2O_3$) to show enhanced tolerance to carbon deposition (78) or to minimize this deposition (79). Both effects lead to increased activity maintenance, an increase that has been attributed for some systems such as Pt-Au, Pt-Sr and Pt-Cu, to a disruption of contiguous absorption residues on metal surfaces (80).

In upgrading coal liquids to clean fuels such as gasoline or turbine fuels, maximum liquid yields are important. The use of alloy catalysts to maintain hydrogenation activity while controlling hydrogenolysis reactions which form light hydrocarbon gases could be important. Studies by Sachtler (81) and Sinfelt (82) with Group VIII-IB alloys such as Ni-Cu are relevant in this respect. These workers have shown that the addition of a Group IB metal to a Group VIII metal, in general, promotes or maintains hydrogenation activity, but decreases carbon-carbon bond hydrogenolysis by several orders of magnitude.

Only recently has some work been reported concerning the synthesis of fuels from CO-H_2 mixtures with alloy catalysts (74). In this respect, Vannice has shown that the binding energy of H_2 and CO to the metal surface can have a strong effect on the synthesis reaction (83, 84). It is known (85) that alloy formation can affect this binding energy, and therefore, it would be expected that this might lead to controlled product selectivity. As previously mentioned, increased tolerance to sulfur and enhanced thermal stability with respect to sintering are also expected for some alloy systems. This could be important for CO-H_2 reactions such as methanation which are very exothermic and very sensitive to sulfur poisoning.

Alloy catalysts represent an important segment in the development of heterogeneous catalysis. Studies of these materials have played a major role in increasing the understanding of the electronic and geometric factors in catalysis by metals. The necessity for differentiating between surface and bulk compositions in such work cannot be over-emphasized. Failure to do so may seriously affect the conclusions concerning the catalytic properties of these systems. However, advances in catalyst preparation (59) and the development of improved characterization techniques for surface analysis, including selective chemisorption, Auger electron spectroscopy and photoelectron spectroscopy will contribute to the increased understanding of these materials and their applications to areas such as the catalytic conversion of coal.

F. *Organometallic Compounds*

The wide scope and versatility of organometallic chemistry is exemplified by the many volumes which have been written on this subject (86-88). Organometallic complexes have been used for catalytic reactions for many years. The hydroformylation of olefins to make alcohols and the carbonylation of methanol to acetic acid using cobalt and rhodium catalysts, respectively, are two commercial examples. Therefore, in the context of the present discussion, these types of compounds do not represent a novel class of materials. However, organometallics deserve consideration since they have been of interest for coal liquefaction for some time. There are three reasons for this interest. First, as homogeneous catalysts they maximize contacting with coal during the liquefaction process, and therefore, in principle should be more effective than heterogeneous catalysts. Second, organometallic complexes tend to be highly active, specific, and selective catalysts, and thus one would hope to minimize unwanted reactions and unnecessary hydrogen consumption. Third, it has been suggested, without much experimental evidence, that this class of materials may be more poison-resistant than heterogeneous systems. Also, there are two recent advances in organometallic chemistry of interest to catalysis, namely the synthesis of polynuclear cluster complexes and the anchoring of complexes to heterogeneous surfaces. The former promises to extend the ranges of catalytic properties for organometallic complexes and to give a greater degree of control on catalyst selectivity and specificity. The latter has been developed as a mechanism for obviating the separation problems which have been a primary constraint for the commercial development of organometallic catalysts. These two advances will be discussed in turn.

1. *Polynuclear Organometallic Complexes*

The primary interest in the synthesis of organometallic complexes with metal clusters is the possibility of new catalytic properties not exhibited by the mono-nuclear species. As pointed out in a recent review by Norton (89), these complexes provide the opportunity to take advantage of several unique features of heterogeneous catalysts in a well defined and reproducible organometallic framework: the possibility of bonding with several metal atoms, thus leading to selectivity control; the migration of species on the catalyst cluster "surface"; and, finally, the added flexibility in achieving modifications of catalytic properties due to metal-metal interactions.

In the last few years, a number of cluster compounds have been synthesized. They range from small clusters of Pt and Ir compounds such as $Pt_3[CN(C_4H_9)]_6$ (90) and $Ir_3(CO)_{12}$ (91) to large cluster compounds of gold, $Au_{11}[P(C_6H_5)_3]_7I_3$ (92), and rhodium, $Rh_{13}(CO)_{24}H_3$ (93). Several interesting aspects of these cluster compounds have been studied. The mobility of some of the ligands is substantial. In $(\pi\text{-}C_5H_5)_2Rh_2(CO)_3$, for example, there are two types of carbonyl groups: bridging and terminal. Carbon 13 NMR, which can in principle distinguish between these two types of CO species, cannot resolve them in the case of this di-rhodium complex because of rapid rearrangement of the carbonyl groups (94). As pointed out by Norton (89), this mobility parallels the fast surface diffusion of some reactive chemisorbed species on heterogeneous materials. There are also clear indications of metal-metal interactions in these clusters (89). Recent reports of the preparation of multimetallic clusters in organometallic complexes (95, 96), therefore, opens up the application of these materials to many of the interesting phenomena that have motivated the development of the supported alloy catalysts described in the preceding section. However, there are preliminary indications that clusters exhibit unusual catalytic chemistry. This is exemplified by the work of Muetterties et al. (96) who observed cyclization of acetylene to benzene and butadiene to cyclo-octadienes at room temperature in the presence of $Ni_4(CNC(CH_3)_3)_7$. No such reaction has been observed with the monomer which does not contain the Ni cluster network. Other recent studies have reported the synthesis of organometallic clusters which catalyze the formation of methane (97), methanol (98), and glycols (99) from CO-H_2 mixtures, and there are indications (100) that certain base-promoted organometallic complexes, e.g. $Fe(CO)_5$, catalyze the synthesis of methanol and ethylene from carbon monoxide and water.

2. *Anchoring of Homogeneous Catalysts*

There are several limitations to the application of homogeneous catalysts to coal conversion (1). The most important ones are thermal and chemical instability and poor catalyst recovery. The former is a serious problem in liquefaction (in particular because of the high H_2S levels). For CO-H_2 syntheses, on the other hand, the combination of high

CO pressures and low S concentrations may be compatible environments for certain organometallic complexes. The problem of separation of the catalyst from the product, however, is a general problem regardless of the type of process being considered. The problem is extreme, of course, when expensive metals are used in the organometallic complex. Unless recovery of rhodium is greater than 99%, for example, use of this metal in chemical processes becomes uneconomic (101). Even with less expensive metals, recovery and separation are severe obstacles.

Over the last 5 to 10 years there has been an increasing effort to solve the separation problem by the reaction of the organometallic complex with a solid containing a ligand that is capable of binding to the complex. Both organic and inorganic supports have been studied, in particular polystyrene and silicagel. As discussed in several recent reviews, the resulting "anchored" complex retains many of the important features of the homogeneous species (102, 103). The carbonyl stretching frequencies in carbonyl complexes, for example, are only slightly shifted (102), and the catalytic activity and selectivity patterns of many reactions remain the same (103). However, this area is still in the early stages of development, and many problems, such as the rigidity of the polymeric supports, have to be solved.

3. *Potential Applications*

The most promising application for organometallic catalysts is the synthesis of chemicals and clean fuels from synthesis gas. These reactions are highly exothermic, and the possibility of liquid phase catalysis offers a very efficient means of temperature control. Also, organometallic complexes can frequently be designed to be highly selective catalysts. There have been suggestions that homogeneous catalysts can be sulfur tolerant (104), however, there is no strong evidence to this effect in the literature except for a report by Soviet workers for the hydrogenolysis of carbon-sulfur bonds using organometallic complexes (105). Furthermore, it should be emphasized that the above references to the reduction of CO by H_2 (97-100) to form chemicals and fuels are recent and unconfirmed. Further work is required.

It should be recognized that while anchoring solves the catalyst separation problem, it defeats the contacting advantage one obtains with homogeneous catalysts. Also, for gas phase reactions over anchored systems, the effective heat transfer capabilities of the corresponding liquid phase system is lost, although this would not be the case for a continuous liquid phase system. Therefore, the advantages of anchored catalysts are primarily ease of catalyst separation and recovery and the possibility of unusual reactivity due to the effect of the ligands.

G. *Molten Salts*

A major problem in direct catalytic liquefaction of coal is the contact between solid catalyst, gas, and liquid phases. Homogeneous catalysis has been explored with limited success in solving this problem. An alternative method is to disperse coal in a catalytic molten salt. The physicochemical (106, 107) and catalytic (108) properties of molten salts have been detailed elsewhere. In this section, a brief review of the properties and possible relevance of molten salts to catalysis is presented.

Upon melting, as a result of coulombic interactions each ion in the melt is surrounded by a number of counter ions (109). The original long range order in the solid is destroyed. Molten salts can, however, be best described by a defect solid model known as the "hole" model. Experimental data seem to support the "hole" model (108) which attributes an increase in volume from solid to fused states to the empty volumes in the salts.

The species present in melts vary for different salts. A melt of BeF_2, for example, exhibits polymeric chains while $ZnCl_2$ forms two dimensional cross-linked layers.

Fused salts have a high capacity for solvating many materials. Gases often dissolve either by reacting or simply filling the void spaces in the melt ("hole" model). Containment of molten salts is a problem because they dissolve many other inorganic salts as well as most refractory materials. In alkali metal hydroxide melts, the presence of oxygen or water leads to the formation of peroxides which dissolve both noble metals and ceramics. In many instances, fused salts also have the ability to dissolve the parent metal. The metal is highly dispersed throughout the medium, and produces metal-like properties such as increases in electrical conductivity (110). Water readily dissolves in many molten salts, especially halides. As an example, $ZnCl_2$ retains some H_2O even at 1300 K (106).

The possible compositions of melts are extensive. Variations in composition give rise to systems with different melting points and chemical properties ranging from strongly oxidizing to strongly reducing, including a wide range of acidities. This variation offers much flexibility in controlling chemical reaction conditions.

In chemical reactions involving molten salts, the melt can be considered either as a reactant, a solvent, or as a catalyst. Molten salts are especially interesting in catalysis for several reasons: possibility of dispersing solid reactants for better contacting, high thermal conductivity for heat removal, continuous regeneration or exposure of fresh catalyst surface, and high polarization forces in the melts which may affect catalytic activity and selectivity.

Only a few areas have been studied with respect to melts. Chlorination and oxidative chlorination of hydrocarbons such as methane, ethylene, and benzene have been explored using $NaCl/AlCl_3$, $NaCl/AlCl_3/FeCl_3$, and $CuCl_2$ melts (107). Other work has involved Friedel-Crafts catalytic melts such as $AlCl_3$ and $SbCl_3$ (111). Recently, Shell Oil Company has described the use of zinc halide melts as hydroconversion catalysts for heavy petroleum fractions (112, 113). One aspect which has not been explored for catalysts and may have interesting possibilities is the use of melts in which the zero-valent metal has been dispersed. Such melts could provide the possibility of a multi-functional system, i.e. a cracking and hydrogenation function in the same melt.

Two processes using molten salts have been used to produce synthetic fuels. The first involves coal gasification using alkali metal carbonate or oxide melts at high temperatures such as the use of Na_2CO_3 in the Kellogg Process (114). The second uses $ZnCl_2$ and other Lewis acid halides in the Consol Process of Consolidation Coal Company (115). The latter process uses molten salts for hydrocracking, desulfurization, and denitrogenation of coal and coal extract to low sulfur distillate fuel oil or high-octane gasoline. The process, however, presents several problems, including high catalyst to coal ratios, deactivation of the catalyst due to formation of ZnS, ZnO, $ZnCl_2 \cdot (NH_3)_4$ complexes, build up of carbon residues and the necessity of moving large volumes of corrosive materials due to large catalyst to coal ratios. It is therefore evident that while the properties of molten salts appear very interesting for catalysis, especially due to increased contact effectiveness between catalysts and reactants, there are severe problems that have to be solved before they can be readily used. Efforts in this direction are underway (116).

III. CONCLUSIONS

Inorganic chemistry will permit us to discover new materials with improved catalytic properties. It affects all of the catalytic steps in coal conversion, as illustrated by the many new concepts and materials discussed above and summarized in Table 1.

The catalytic chemist is faced with two challenges. First, he has to choose among the literally thousands of compositions and structures available to him through the inorganic chemistry literature. Second, he has to devise new methods for the synthesis of the materials in high enough specific surface areas to be attractive for catalytic applications.

For the first challenge, thermodynamics provides an initial guideline to the choice of materials that will withstand a particular environment. This has been discussed briefly in the above sections.

For the second challenge, many of the advances made over the last decade in the area of catalyst synthesis should provide some basic guidelines for the synthesis of new materials with high specific surface areas.

ACKNOWLEDGMENTS

The authors would like to thank the Electric Research Institute (Contract RP415-1) and the Energy Research and Development Administration (Contract E(49-18)-2017) for their sponsorship of part of this work.

REFERENCES

1. M. Boudart, J. A. Cusumano, and R. B. Levy, New Catalytic Materials for the Liquefaction of Coal, Report RP-415-1, Electric Power Research Institute, October 30, 1975.

2. R. Gauguin, M. Granlier, and D. Papee, in Catalysts for the Control of Automotive Pollutants (James E. McEnvoy, Ed.), Advances in Chemistry Series No. 143, p. 147. ACS, Washington, D. C., 1975.

3. H. Schafer, Chemical Transport Reactions. Academic Press, Inc., New York, 1964.

4. R. B. Levy, in Advanced Materials in Catalysis (J. J. Burton and R. L. Garten, Eds.). Academic Press, New York, 1976.

5. R. B. Anderson, in Catalysis (P. H. Emmett, Ed.), vol. 4, p. 1. Reinhold Publishing Corp., New York, 1956.

6. K. Tanabe, Solid Acids and Bases. Academic Press, New York, 1970.

7. J. B. Donnet, Bull. Soc. Chim. France, 3353 (1970).

8. L. Forni, Catal. Rev. 8, 65 (1973).

9. G. K. Boreskov, in Advances in Catalysis (D. D. Eley, H. Pines, P. B. Weisz, Eds.), vol. 15, p. 285. Academic Press, New York, 1964.

10. J. Haber, Int. Chem. Eng. 15, 21 (1975).

11. J. G. Dickson, L. Katt, and R. Ward, J. Amer. Chem. Soc. 83, 3026 (1961).

12. S. J. Tauster, J. Catal., 26, 487 (1972).

13. M. Shelef, and H. S. Gandhi, Ind. Eng. Chem., Prod. Res. Dev., 393 (1972).

14. M. Shelef, and H. S. Gandhi, Plat. Metals Rev., 18, 2 (1974).

15. A. F. Wells, Structural Inorganic Chemistry, p. 828. Oxford University Press, London, 1975.

16. D. W. Breck, J. Chem. Ed., 41, 678 (1964).

17. S. C. Eastwood, R. D. Drew, and F. D. Hartzell, Oil and Gas J., 60, 152 (1962).

18. A. E. Hirschler, J. Catal., 2, 428 (1963).

19. H. Otnouma, Y. Arai, and H. Ukihashi, Bull. Chem. Soc. Japan, 42, 2449 (1969).

20. L. Lerat, G. Poncelet, M. L. Dubru, and J. J. Fripiat, J. Catal., 37, 396 (1975).

21. G. T. Kerr, J. Phys. Chem., 73, 2780 (1969).

22. G. T. Kerr, J. Phys. Chem., 71, 4155 (1967).

23. R. A. Dalla Betta, and M. Boudart, in Catalysis (J. Hightower, Ed.), p. 1329. North Holland Publishing Co., Amsterdam/London, 1972.

24. T. Gallezot, A. Alarcon-Diaz, J. A. Dalmon, J. R. Renouprez, and B. Imelik, J. Catal., 39, 334 (1975).

25. J. Datka, P. Gallezot, B. Imelik, J. Massardier, and M. Primet, Presentation at the 6th International Congress on Catalysis, London, July 1976.

26. O. Weisser, and S. Landa, Sulphide Catalysts, Their Properties and Applications. Pergamon Press, Oxford, 1973.

27. R. J. H. Voorhoeve, and J. C. M. Stuiver, J. Catal., 23, 228 (1971).

28. J. M. Vandenberg, and D. Brasen, J. Solid State Chem., 14, 203 (1975).

29. W. P. F. A. M. Omloo, J. C. Bommerson, H. H. Heikens, H. Risseloda, M. B. Vellinga, C. F. van Bruggen, C. Haas, and T. Jellinek, Phys. Stat. Sol., (a) 5, 349 (1971).

30. G. Collin, and J. Flahaut, J. Solid State Chem., 9, 352 (1974).

31. W. Biltz, and A. Kocher, Z. Anorg. Allgem. Chem., 241, 324 (1939).

32. J. Tudo, and G. Tridot, Compt. Rend. Acad. Sci. Paris, 258 6437 (1964).

33. J. Tudo, and G. Tridot, Compt. Rend. Acad. Sci. Paris, 257, 3602 (1963).

34. F. Fransen, and S. Westman, Acta. Chem. Scand., 17, 2353 (1963).

35. W. T. Gleim, U. S. Patent No. 3,694,352 (1973).

36. G. C. A. Schuit, and C. Gates, AIChE J., 19, 417 (1973).

37. M. Chevreton, and A. Sapet, Compt. Rend. Acad. Sci. Paris, 261, 928 (1965).

38. B. Van Laar, and D. J. W. Ijdo, J. Solid State Chem., 3, 590 (1971).

39. S. Yamoka, J. Amer. Ceram. Soc., 111 (1972).

40. G. Collin, and J. Flahaut, NBS Special Publication, 364, 645 (1972).

41. H. Barz, Mater. Res. Bull., 8, 983 (1973).

42. D. Brasen, J. M. Vandenberg, M. Robbins, R. H. Willens, W. A. Reed, R. C. Sherwood, and X. J. Pinder, J. Solid State Chem., 13, 298 (1975).

43. L. E. Toth, Transition Metal Carbides and Nitrides. Academic Press, New York, 1971.

44. Ibid. p. 9.

45. H. Bohm, Electrochim. Acta, 15, 1273 (1970).

46. A. F. Wells, Structural Inorganic Chemistry, p. 761. Oxford University Press, London, 1975.

47. T. W. Barbee, Jr., private communication.

48. B. Aronsson, T. Lundstrom, and S. Rundquist, Borides, Silicides, and Phosphides. Wiley, New York, 1965.

49. A. W. Searcy, in Chemical and Mechanical Behavior of Inorganic Materials (A. W. Searcy, D. V. Ragone, and U. Colombo, Eds.), p. 1. Wiley Interscience, 1970.

50. H. Nowotny, in MTP International Review of Science (F. R. S. Emelens and L. E. J. Roberts, Eds.), vol. 10, p. 184. Butterworths, London, 1972.

51. Ibid. p. 180.

52. A. F. Wells, Structural Inorganic Chemistry, p. 845. Oxford University Press, London, 1975.

53. D. A. Douden, J. Chem. Soc., 242 (1950).

54. D. A. Douden, and P. Reynolds, Discuss. Faraday Soc., 8, 184 (1950).

55. G. M. Schwab, Discuss. Faraday Soc., 8, 166 (1950).

56. S. Hufner, G. K. Wertheim, R. L. Cohen, and J. H. Wernick, Phys. Rev. Lett., 28, 488 (1972).

57. D. H. Seib, and W. E. Spicer, Phys. Rev. Lett., 20, 1441 (1968).

58. J. H. Sinfelt, J. Catal. 29, 308 (1973).

59. M. Boudart, J. A. Cusumano, and R. B. Levy, Scientific Resources Relevant to the Catalytic Problems in the Conversion of Coal, Parts 1 & 2, ERDA contract no. E(49-18)-2017, March 10, 1976.

60. L. F. Norris, and G. Parravano, in Reactivity of Solids (J. W. Mitchell et al., Eds.), p. 149. Wiley Interscience, New York, 1969.

61. J. C. Schlatter, "Sintering of Supported Metals", presented at 4th International Conference on Sintering and Related Phenomena, Univ. of Notre Dame, May 26-28, 1975.

62. A. A. Slinkin, and E. A. Fedorovskaya, Russ. Chem. Rev., 40, 860 (1971).

63. Ya. I. Ivashentsev, and R. I. Timonova, Russ J. Inorganic Chem., 14, 9 (1969).

64. F. L. Williams and D. Nason, Surf. Sci., 45, 377 (1974).

65. F. L. Williams, and M. Boudart, J. Catal., 30, 438 (1973).

66. J. J. Burton, E. Hyman, and D. Fedak, J. Catal., 37, 106 (1975).

67. D. F. Ollis, J. Catal., 23, 131 (1971).

68. D. W. Hoffman, J. Catal., 27, 374 (1972).

69. W. M. H. Sachtler, and R. Jongepier, J. Catal., 4, 665 (1965).

70. D. T. Quinto, V. S. Sundaram, and W. D. Robertson, Surface Sci., 28, 504 (1971).

71. F. L. Williams, PhD Thesis, Stanford University, 1972.

72. R. Bouwman, R., G. J. M. Lippits, and W. M. H. Sachtler, J. Catal., 25, 350 (1972).

73. R. Bouwman, and G. J. M. Lippits, J. Catal., 26, 63 (1972).

74. C. H. Bartholomew, Quarterly Tech. Prog. Rept. (April 22-July 22, 1975), ERDA Contract No. E(49-18)-1790, August 6, 1975.

75. J. W. Myers, and F. A. Prange, U. S. Patent No. 2,911,357 (1959).

76. J. H. Sinfelt, and A. E. Barnett, U. S. Patent No. 3,567,625 (1971).

77. S. D. Robertson, S. C. Kloet, and W. M. H. Sachtler, J. Catal., 39, 234 (1975).

78. E. F. Schwarzenbek, Advan. Chem. Ser., 103, 94 (1971).

79. R. R. Cecil, W. S. Kmak, J. H. Sinfelt, and C. W. Chambers, presented at Advances in Reforming during 37th mid-year meeting of API Div. Ref., New York, May 1972.

80. J. K. A. Clark, Chem. Rev., 75, 291 (1975).

81. V. Ponec, and W. M. H. Sachtler, J. Catal., 24, 250 (1972).

82. J. H. Sinfelt, J. L. Carter, and D. J. C. Yates, J. Catal., 24, 283 (1972).

83. M. A. Vannice, J. Catal., 37, 449 (1975).

84. M. A. Vannice, J. Catal., 37, 462 (1975).

85. K. Christmann, and G. Ertl, Surface Sci., 33, 254 (1972).

86. D. Forster, and J. F. Roth, Eds., Homogeneous Catalysis-II, Advances in Chemistry Series No. 132. American Chemical Society, Washington, D. C., 1974.

87. G. N. Schrauzer, Ed., Transition Metals in Homogeneous Catalysis. Marcel Dekker, Inc., New York, 1971.

88. J. Chatt, and J. Halpern, in Catalysis, Progress in Research (F. Basolo and R. L. Burwell, Jr., Eds.). Plenum Press, New York, 1973.

89. J. R. Norton, "Catalysis by Polynuclear Complexes: A Bridge Between Homogeneous and Heterogeneous Catalysis", presented at American Chemical Society Petroleum Chemistry Award Symposium, New York, April 1976.

90. "New Routes to Novel Organic Compounds", C & E News, p. 26, January 20, 1975.

91. K. J. Karel, and J. R. Norton, J. Amer. Chem. Soc., 96, 6812 (1974).

92. R. Cariati, and L. Naldini, Inorg. Chim. Acta., 5, 172 (1971).

93. V. G. Albans, A. Ceriotti, P. Chini, G. Ciani, S. Martinengo, and W. M. Anker, Chem. Commun., 859 (1975).

94. A. J. Canty, B. F. G. Johnson, J. Lewis, and J. R. Norton, Chem. Commun., 79 (1973).

95. G. C. Smith, J. P. Chojnacki, S. R. Dasgupta, K. Iwatate, and K. Watters, Inorg. Chem., 14, 1419 (1975).

96. V. W. Day, R. O. Day, J. S. Kristoff, F. J. Hirsedorn, and E. L. Muetterties, J. Amer. Chem. Soc., 97, 2571 (1975).

97. J. R. Norton, Amer. Chem. Soc., Div. Petrol. Chem., Prepr. 21, 343 (1976).

98. J. Bercaw, J. Chatt, in C & E News, March 1, 1976, p. 14.

99. Belgian Patent 792,086 (1972).

100. C & E News, April 26, 1976, p. 29.

101. J. F. Roth, J. H. Craddock, A. Hershman, and F. E. Paulik, Chem. Tech., 600 (1971).

102. J. G. Allum, R. D. Hancock, I. V. Howell, S. McKenzie, R. C. Pitkenthly, and P. J. Robinson, J. Org. Chem., 87, 203 (1975).

103. J. C. Bayer, Jr., Catal. Rev., 10, 17 (1974).

104. "NSF Workshop on Fundamental Research in Homogeneous Catalysis as Related to U. S. Energy Problems", held at Stanford University, December 4-6, 1974.

105. A. S. Berenblyum, Izv. Akad. Nauk SSSR, Ser. Khim., 11, 2650 (1973).

106. H. Bloom, The Chemistry of Molten Salts. W. A. Benjamin, Inc., New York, 1967.

107. W. Sundermeyer, Angew. Chem. Int. Ed., 4, 222 (1965).

108. C. N. Kenney, Catal. Rev., Sci. and Eng., 11, 197 (1975).

109. M. I. Temkin, Acta Physicochimica URSS, 20, 411 (1945).

110. H. R. Bronstein, and M. A. Bredig., J. Amer. Chem. Soc., 80, 2077 (1958).

111. H. A. Cheney, U. S. Patent No. 2,342,073 (1944).

112. D. E. Hardesty, and T. A. Rodgers, U. S. Patent No. 3,677,932 (1972).

113. D. O. Geymer, U. S. Patent No. 3,844,928 (1974).

114. A. E. Cover, W. C. Schreiner, and G. T. Skaperdas, Chem. Eng. Prog., 69, 31 (1973).

115. Consolidation Coal Company, R & D Report No. 39, OCR Contract No. 14-01-0001-310, Vols. I, II, III (1969).

116. ERDA Contract No. 1743, Continental Oil Company, Library, Pennsylvania.

6

VI. SUPERCONDUCTING MATERIALS

In considering materials problems in each of the aspects of energy; its generation, transmission and storage, superconducting components are important because they can influence the overall efficiencies of three major energy systems: magnets for nuclear fusion reactors, inductors for energy storage, and superconducting power transmission lines.

The objectives of current research in superconductors are twofold: to understand the basics of the phenomena so that improvements in the critical field, critical current and higher transition temperature can be realized in present superconducting materials and secondly to be able to predict new and superior performing compounds.

The progress which has been made in this field since its discovery in 1911 by Kamerlingh-Onnes and the nature of the research currently engaging solid state physicists are described in the two chapters of this Section. It is enlightening to follow the dramatic changes in our conception of the phenomena since the time it was first considered an esoteric property of materials. Today, after discovering that most metallic elements are superconducting, this property is now regarded as a normal attribute of metals, while those elements which do not display superconductivity are accepted as abnormal.

The interesting, often exciting research into superconducting composites and new ternary compounds and the role lattice instabilities; crystal symmetry; electron densities of state; electron-phonon coupling; and electron mean-free-path play in producing superconductivity at higher critical fields, current densities and transition temperatures are detailed in the following two chapters.

Chapter 20

SUPERCONDUCTING MATERIALS

T. H. Geballe

Department of Applied Physics
Stanford University
Stanford, California 94305

I. INTRODUCTION

Kamerlingh-Onnes envisioned a superconducting technology soon after he discovered superconductivity. In his 1913 Nobel Laureate speech, he said, "Thus, a bright lead wire which had been dipped in liquid helium and which remained superconducting up to a current density of 420 A/mm^2 melted at a current density of 940 A/mm^2." He goes on, "The great question still not solved is whether this first impulse precedes from bad sections of the wire or is also produced in pure, evenly crystallized metal. If the potential phenomena are to be attributed to bad spots, we shall learn to eliminate them. The potential phenomena may, however, depend on the nature of the metal in that the resting vibrators of the now superconducting metal are not made to move until the electron wind which blows past them has reached a certain strength."

He faced the materials problem then and that is what we face now. After six decades of research, we have a vast array of known superconductors to draw on. We also have an understanding of the basic physics underlying their performance in magnetic fields and with current flowing. The question is really not, 'Are superconducting technologies feasible?' - they are; but 'How competitive are they in the marketplace? In most cases the superconducting component is part of larger technology that is not yet proved, but becomes easier as the superconducting component is improved. The largest scale potential application of superconductors are all concerned with energy — its generation, transmission, and storage. Other fields, such as transportation and particularly instrumentation and signal processing have superconducting futures of their own, but we will not be concerned with them in this chapter.

There are many reasons for continuing vigorous research programs on superconducting materials. Most important is the possibility of increasing the transition temperature (T_c). We will hear more of this in the next paper, also. There has been a slow, steady increase in the superconducting transition temperatures since Kamerlingh-Onnes' day; finally to the present-day limit of $23^{\circ}K$, and there is no reason to believe that it has saturated. It is important, and perhaps crucial, for energy-related superconducting technology that the potential increases of the transition temperature be brought to fruition. This is true for several reasons.

(1) The applications of superconductivity are all large scale. They become competitive with conventional modes

of operation above some threshold. That threshold may be so large as to be of no practical use on this planet. The higher the superconducting operating temperature, the lower the threshold, and, of course, the operating temperature depends heavily upon T_c. Even when the superconducting mode is competitive, there is the universal problem of introducing new technology, particularly when it is large scale and not familiar. By analogy, if the aircraft industry had had to produce a jumbo jet rather than the "Spirit of St. Louis" to demonstrate the competitiveness of heavier-than-air flight, I might have had no choice but to come to Albuquerque on the Santa Fe Chief. (But it might have been levitated by superconducting magnets!)

(2) Higher transition temperatures can remove the almost total dependence of today's technology upon the availability of an adequate supply of helium for refrigeration. There is enough helium in storage (having been separated from natural gas) to satisfy the level of present-day usage for probably fifty years. The accumulative needs of the three major energy-related uses, namely, superconducting power transmission lines, magnets for nuclear fusion, and inductors for energy storage, are potentially large. It would be ironic if all three were succeed in every way except to be limited by the cost of helium. The situation is discussed in depth in recent reports by Hammel (1) and by Laverick (2). Each has made reasonable, although somewhat different, assumptions about the future. There does exist sufficient helium to get started, but our generation is shortchanging future ones. One would think that our generation could be farsighted enough to not throw away the helium while we are using up all the natural gas. By 1990, most of the resources of helium will be exhausted because most of the natural gas will be exhausted. From then on, the demand on helium will exceed supply. I believe the consumers of the natural gas should pay for returning the helium back underground. For more complete details, I refer you to those reports. The fact remains that we have enough to get started and

go ahead for the next 25 years but that long-range problems are being created by present-day loose practice.

(3) Operation at higher temperatures will obviously cut down on refrigeration costs. It will also improve the stability of the superconductor in dynamic operation when heat is evolved. The thermal stability is improved at higher temperatures because the heat capacity of the superconductor is mainly due to the lattice and increases as T^3.

The properties of superconductors with which we have to be concerned if we are going to use them in addition to T_c include the critical magnetic field H_c and the critical current J_c. Values for three materials are plotted in Fig. 1 in the parameters space which these properties define. H_c

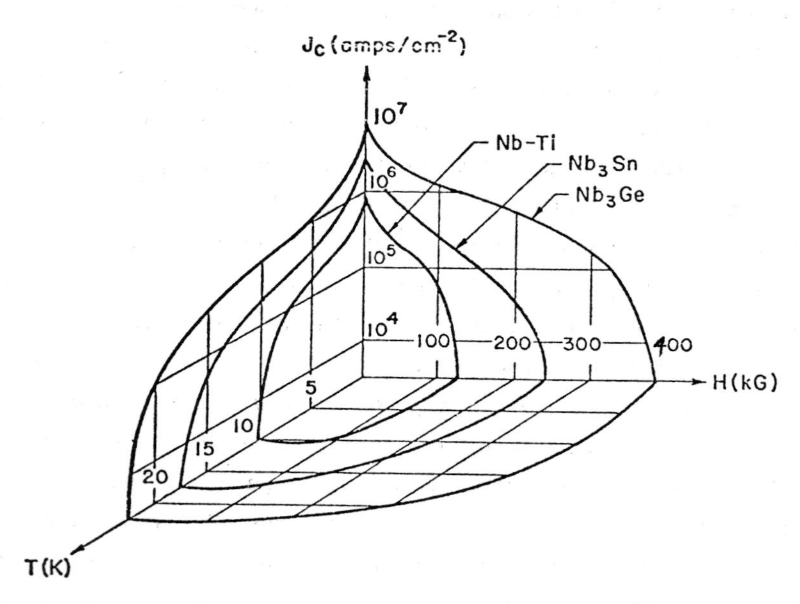

FIG. 1. Parameter Space for the Practical Superconductors Nb-Ti and Nb_3Sn and the Potentially Important Nb_3Ge [After Gavaler, et al., IEEE Trans. Mag. MAG-11, 192 (1975)]

is the magnetic field above which the superconductors cannot exist; and the critical current, J_c , is the maximum current that can be put through the superconductor. It is clear insofar as intrinsic superconducting properties are concerned that Nb_3Ge is to be preferred. However, NbTi is presently the preferred commercial material because it is ductile and much easier to manufacture. The pure superconductor doesn't exist by itself in a vacuum. It exists as part of a wire with other materials around it and has to perform as an integrated composite. Before considering the more practical problems associated with composite wires or cable, it is necessary to understand the intrinsic properties in more detail.

II. THE CRITICAL TEMPERATURE (T_C)

What does the superconducting transition temperature look like? In zero magnetic field it is a second-order transition, as shown for a good single crystal of Nb_3Sn in Fig. 2. The discontinuity in heat capacity at 17.8°K , which you see here, is the transition. Below that, the heat capacity falls off rapidly. The superconductor has a lower ground state than the normal metal by an amount which is $\sim 4\ k_BT_c$, where k_B is Boltzman's constant. Something like twenty years ago, Bernd Matthias first synthesized Nb_3Sn by melting some tin over some powdered niobium. We found its T_c to be 18.05° . For that particular compound the highest T_c possible was achieved in the very first samples made because nature was cooperative. In other cases such as "Nb_3Ge," it is only a metastable Nb_3Ge phase, prepared with difficulty, which has the high T_c .

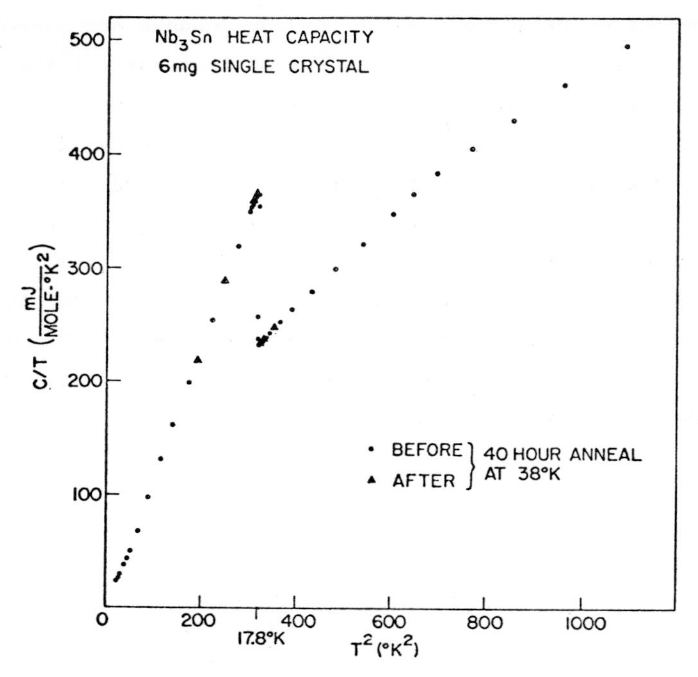

FIG. 2. Superconducting Transition of a Homogeneous Superconductor as Evidenced by a Sharp Discontinuity in Heat Capacity [*After Harper, et al, J. of Less-Common Metals* 43, *5-11 (1975)*]

III. THE CRITICAL FIELD (H_C, H_{C1}, H_{C2}

I am only going to make a few introductory remarks. Excellent texts which review the different critical fields are easily available (3). There is, first of all, the thermodynamic critical field, H_c , whose square is proportional to the superconducting state condensation energy. The type-II superconductors we will be dealing with have an H_{c1} below which all magnetic flux is excluded, and an H_{c2} above which the sample reverts to the normal (nonsuperconducting) state. Between H_{c1} and H_{c2} , the superconductor is in a mixed state. Fluxoids exist which are in quanta of magnetism defined by vortices of supercurrent.

H_{c1} and H_{c2} usually are materials-related extrinsic parameters which depend very much upon the mean free path for electrical conductivity. The shorter the mean free path, the higher H_{c2} and the lower H_{c1}. It is necessary but not sufficient to have a high H_{c2} in order to operate a solenoid in high fields. Nature can be kind, because the "dirtier" materials can have the better performance insofar as H and J are concerned. Insofar as T_c is concerned, for a lot of materials, it doesn't matter whether the mean free path is long or short.

IV. THE CRITICAL CURRENT (J_C)

The critical current which can be carried by a material which is in a field above H_{c1} depends on the "pinning" of the quantized flux lines. As a result J_c is material dependent and can vary greatly with processing procedures. Pinning forces in the superconductor resist the Lorentz force resulting from the flow of the current in a magnetic field, and when sufficiently strong, keep the fluxoids immobile. If the fluxoids move, then there is dissipation and heat is generated. Pinning centers are local regions where there are energy minima in which the fluxoids can sit. They can be inclusions of a second phase or normal material, grain boundaries, dislocation tangles, or simply a surface, etc. The cold-drawing of ductile wire introduces pinning centers, and thus is compatible with the manufacture of NbTi wire referred to above. Actually, J_c is very intimately dependent upon the metallurgical treatment. For further discussion of the intrinsic properties, there are some excellent texts and review articles (4).

V. PRESENT-DAY COMPOSITE SUPERCONDUCTORS

Let us now consider what happens when the superconductor is fabricated into a usable wire which together with insulation and mechanical reinforcement can be used to manufacture a solenoid. Copper or aluminum is added to increase stability, and temporarily to provide a shunt path for the current. The physics of the interaction between the superconductor and the normal metal is understood on a microscopic level at least on a qualitative basis. Performance of any given system, even though based upon semiquantitative calculations, must be checked. Microscopic interactions of the defects with the fluxoids are important in determining the pinning. There is a lot of work that remains to be done in order to obtain the order of magnitude improvement in J_c which is theoretically possible.

Figure 3 shows the history of evolution of NbTi

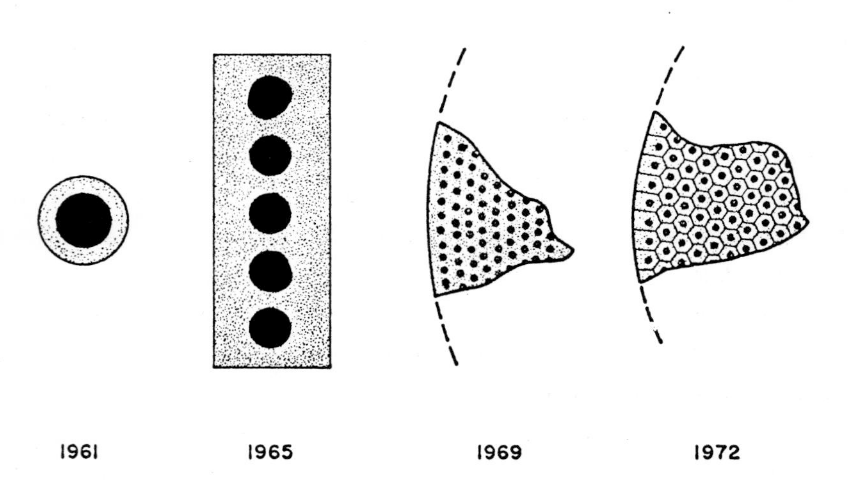

FIG. 3. Evolution of Ductile Alloy Wire.

the currently available workhorse material for large-scale uses such as motors, generators, containment magnets, energy storage devices, and magnets for high energy physics experiments. In 1961 superconducting magnets were made simply from a core of NbZr surrounded by a shroud of copper. They weren't satisfactory because of instabilities. The fluxoids would break loose or "jump"; they would then generate heat and drive the superconductor normal, sometimes in an abrupt or catastrophic manner. The Kamerlingh-Onnes metal-vaporization experiment of 1913 was reenacted.

By 1965 enough copper was added to make reliable bubble chamber magnets. The copper could carry the current by itself without heating up very much. The composite was cryostatically stable in the sense that one could dispense with the superconductor altogether and operate a cryogenic magnet in liquid helium, although without the superconductor, unconscionable amounts of helium would be boiled away. It took a lot of space and copper (filling factor of 20 Cu:1 NbZr). If there were an instability in the superconductor which persisted for only a short time, the current could be shunted through the copper without any problem. So these 1965 versions were very satisfactory and are still in use in the big bubble chambers where volume is not much of a problem. However, for rotating machinery where size is important, such as airborne generators, in particular, the amount of copper must be reduced.

By 1969 copper NbTi composites were made with hundreds, or even thousands, of micron-sized filaments of NbTi imbedded in copper with a reasonable filling factor (2 Cu: 1 NbTi). Flux jumping in one little filament doesn't propagate to the next because the electromagnetic propagation is damped by the intervening copper. As an extra refinement, the filaments are twisted about their central axes to reduce the size of eddy current loops in the copper between neighboring filaments.

Finally, in 1972 barriers of high resistivity

alloys (such as cupro-nickel) were added resulting in a mixed matrix which reduced eddy current losses even further.

The problems remaining today are concerned with establishing good quality control so that engineering and manufacturing can be done reliably. There are also reasons for improving the mechanical stability of magnets since it is believed that mechanical instability limits the performance of pulsed magnets.

VI. COMPOUND SUPERCONDUCTORS

In order to operate at higher temperature and fields than can be achieved with ductile alloys, one must turn to intermetallic compounds — in particular, those with the A15 structure such as Nb_3Sn. High temperature superconductors have been known in the A15 family since they were first discovered by Hulm and Hardy (5), and by Matthias and co-workers (6) more than twenty years ago. The only known superconductors with T_c's above 18°K — namely, Nb_3Sn, Nb_3Ge, $Nb_3A\ell$, Nb_3Ga, and $Nb_3(A\ell_{0.8}Ge_{0.2})$ — are all members of the A15 (or β-W) family.

The structure is shown in Fig. 4. The stoichiometric formula is A_3B, where B is nontransition metal such as silicon, germanium, gallium, or aluminum, or tin on the body-centered cubic lattice. On each face there are equispaced niobium atoms which form continuous chains through the system. The three chains do not intersect. There are a number of anomalies in the electrical and lattice properties of these structures (7) which have been extensively studied, not only because they are interesting in their own right but because of the belief that the anomalies might be related to the enhancement of T_c.

Lattice anomalies are commonly found - strongly

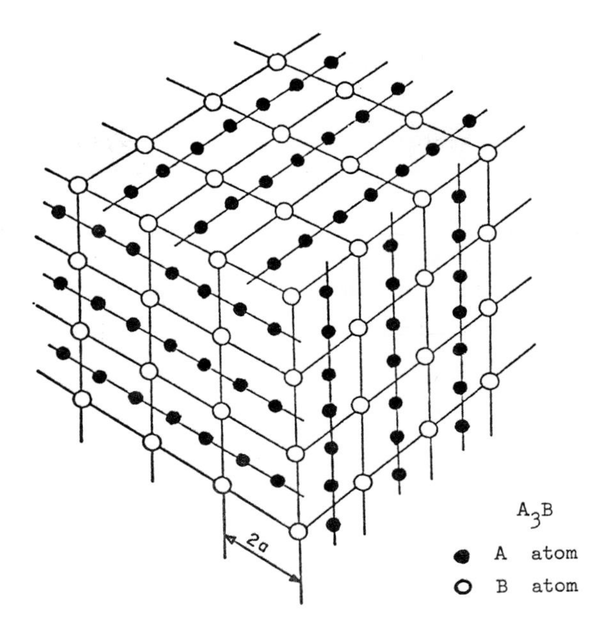

FIG. 4. A15 Structure

temperature-dependent elastic constants and ultrasonic attenuation coefficients (7). Certain modes in the acoustical spectrum soften with temperature over a wide range through the Brillouin zone. A cubic-to-tetragonal transformation occurs at low temperatures. There are also anomalies that can be associated with the electronic degrees of freedom. Strongly temperature-dependent magnetic susceptibility, and Knight shifts were found as well as unusual ρ vs T curves. These can all be explained by motion of the Fermi level as thermal energy $0 < T < 300$ became comparable to the energy scale of sharp structure in the electronic density of states.

Table I shows some of the important A15 compounds that people have studied. Starting with Nb_3Ge at 23° .. Does it form the exact stoichiometric ratio? No. Does it have any anomalies in the lattice properties? No one has found any, but the samples may not be perfect enough. Examination of Nb_3Ge

TABLE I

A15 (β-W) STRUCTURE
T_C LATTICE AND ELECTRONIC ANOMALIES

Compound	T_c	Exactly 3:1	Lattice Anomalies	Electronic Anomalies	Cubic Tetragonal
Nb_3Ge	-23	No	-	No?	No?
Nb_3Ga	20+	No	-	-	-
$Nb_3A\ell$	18.8	No	Yes(?)	No	No
Nb_3Sn	18.1	Yes	Yes	Yes	Yes
V_3Si	17.1	Yes	Yes	Yes	Yes
V_3Ge	6.1	Yes	Slight	No	No

produced by chemical vapor deposition below room temperature shows nothing unusual (8).

Recently, we searched via optical absorption for the sharp features in density of states. Careful studies by O'Connor (9) have failed to detect any evidence. It is risky to conclude the nonexistence of structure from its nondetection; nevertheless, I am becoming convinced that the strongly temperature-dependent electronic properties are more likely due to the Fermi surface itself changing with temperature. This could be due to the formation of something akin to the charge density wave state found in the transition metal dichalcogenides (10). Gor'kov (11) has developed such a theory based upon the breaking of symmetry at the X points of the Brillouin zone as the temperature is lowered.

I think the fact that the three highest T_c materials do not occur in stoichiometric compounds is significant. (If you try to make Nb_3Ge , you get the A15 structure with the composition of approximately Nb_4Ge , plus the B-rich phase, Nb_5Ge_3 .

A lot of work being done today is trying to force Nb_4Ge to form as stoichiometric Nb_3Ge using techniques that promote the formation of a metastable phase. Quenching from the vapor is the favorite. It also seems necessary to have site order, i.e., the A atoms on the A sites and the B atoms on the B sites. Data of Sweedler and Cox at Brookhaven (Fig. 5) shows what happens

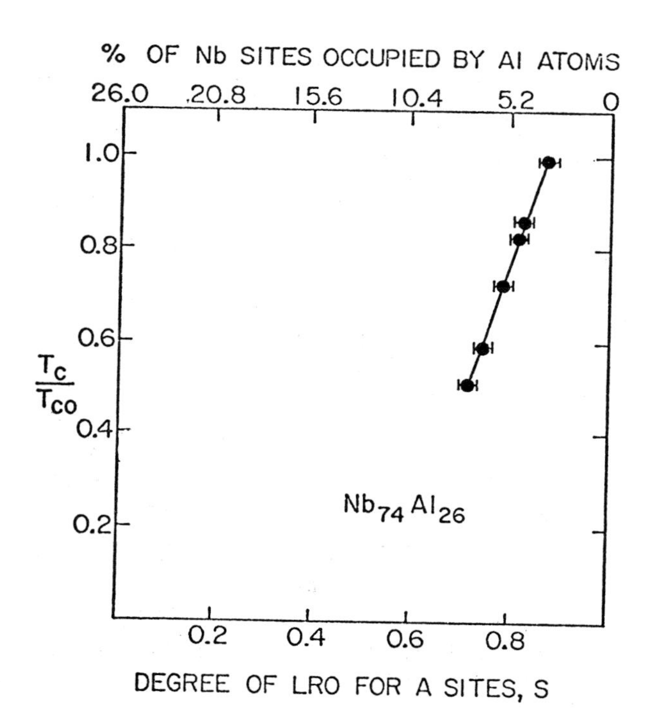

FIG. 5. The Fractional Reduction in Critical Temperature for Irradiated Nb_3Al Vs. Degree of Long-Range Order [After Sweedler and Cox, Phys. Rev. (in press)]

to Nb_3Al when the atoms are exchanged. Upon annealing the disordered Nb_3Al to 700° , T_c is brought back to its original value. The necessity for order could be detrimental in nuclear fusion containment magnets. The magnet will be subject to a large fluence of neutrons, although probably an order of magnitude less in twenty years than would be expected to reduce T_c appreciably. But the experimental uncertainties are still large so that the margin of safety is not overwhelming. So far, the measurements have been made only at room temperature.

The high T_c metastable A15 compounds have smeared out transitions into the superconducting state. Figure 6

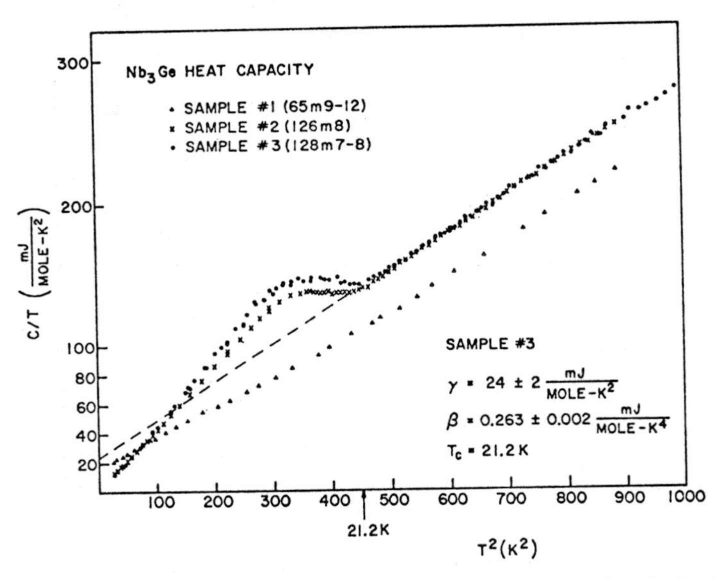

FIG. 6. Heat Capacity of Chemical-Vapor-Deposited Nb_3Ge [After Harper, et al., J. of Less-Common Metals 43, 5-11 (1975)]

shows the data for some of Newkirk's Nb_3Ge, as measured by Harper (8). It shows an onset of the transition at 21.2°. The smeared transition is probably caused by a distribution of metastable material. The intercept indicates a surprisingly low density of states (see Fig. 9). There is no reason to believe that we are at the maximum T_c for Nb_3Ge, $Nb_3A\ell$, or Nb_3Ga. In Fig. 7

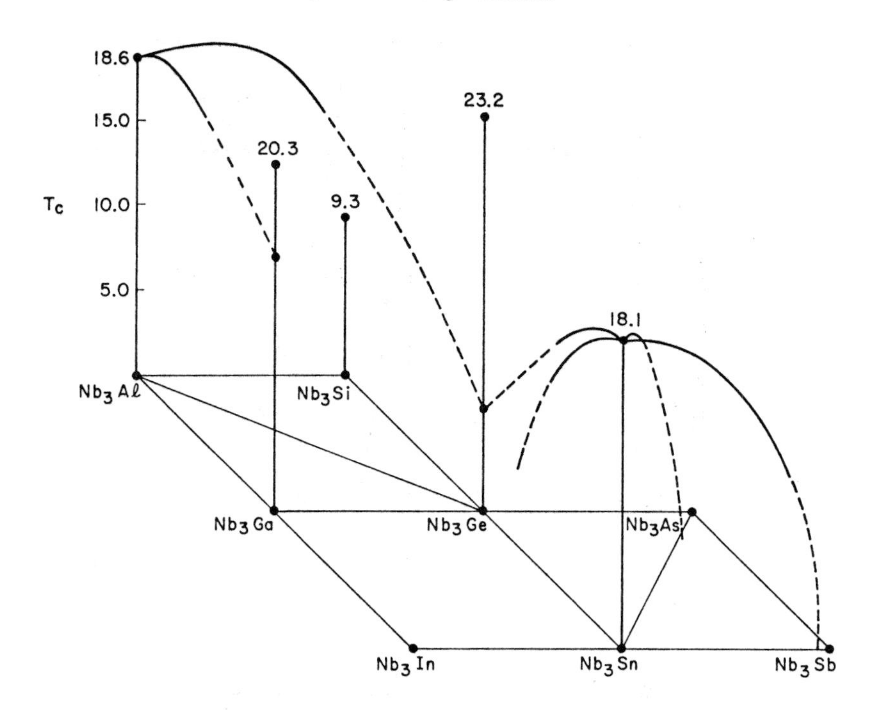

FIG. 7. Niobium-Based, High-T_c Binary A15 Compounds and the T_c's of Some of Their Ternary Solid Solutions

is part of the Periodic Table which shows the possibilities for two- and three-component Nb-based systems. It seems likely that the compounds that do not naturally occur on stoichiometry; namely, the Al , Ga , Ge , and Si compound, are candidates for producing higher T_c's than presently can be obtained if conditions for quenching the metastable phases can be improved.

Work at Stanford has been under way with Nb_3Ge using the electron-beam codeposition technique developed by Hammond (12). Figure 8 is the phase diagram obtained by Hallak (13) for the temperature range below 1000°C . It is significant

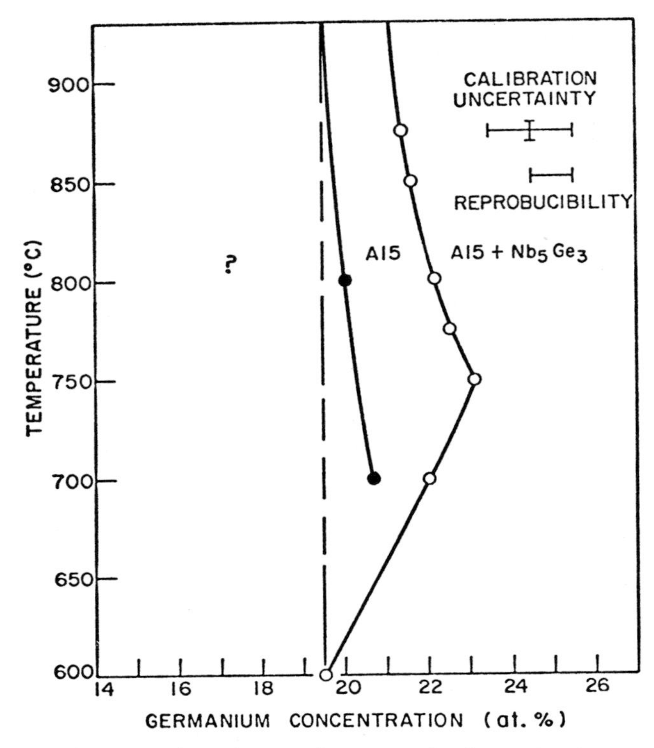

FIG. 8. Portion of the Phase Diagram of Nb-Ge System as a Function of Oxygen Partial Pressure Present During the Sample Preparation. 0 - Oxygen Partial Pressure ~ 2 x 10^{-6} Torr; 0 - Oxygen Partial Pressure ~ 7 x 10^{-8} Torr

that small concentrations of oxygen in the vacuum system during deposition seem to affect the growth kinetics in a temperature-dependent way. Using oxygen partial pressures ~ 10^{-6} Torr, it is possible to improve the stoichiometry and obtain films with high T_c's. It would see probable that further work on $Nb_3A\ell$ and Nb_3Ga could also result in higher T_c's being produced.

Nb_3Si is different. The Si is too small relative to Nb to form the phase at all, even off stoichiometry, except with great difficulty (12). A favorite game during the last couple of years has been predicting how high the T_c for Nb_3Si

should be. There are lots of different ways of extrapolating; some are very sophisticated, but they are no better than those based simply on the Periodic Table. I can extend the SnGe line (Fig. 8) and predict that Nb_3Si is 28°, or I could take another curve and predict something else. I don't think there is any point in getting excited about predictions. Even the most sophisticated theory is likely to have no more insight than is obtainable from the Periodic Table.

Recent work in the NbAℓ system has been carried out along the 1100°C isotherm and the phase boundary of the A15 phase established at 21 at% Aℓ (14). The authors extrapolate to 26°K for the T_c of the stoichiometric compounds.

In the simplest form of the BCS theory of superconductivity, the transition temperature is given roughly by $T_c = \langle\theta\rangle \exp - 1/G$, where θ is the Debye temperature and G is the coupling constant between the electrons and phonons. McMillan (15) has formulated the strong-coupling theory of Eliashberg in a way that permits us to extract the attractive part of the electron-electron interaction which gives rise to the superconductivity from the coupling constant. Figure 9 is generated from McMillan's formulation using experimental heat capacity data. The A15 structures end up here with very high densities of states and moderately attractive interactions. The solid curve is drawn arbitrarily with $N(0)\,V_{ph} = 1$. It suggests that there is a stability boundary regardless of whether a low density of states material or high density of states material is considered. Attempts to cross the boundary result in some other (unwanted) structure being formed or in a departure from stoichiometry which causes lower density of states and a return to the boundary. How do you get around this? Well, that hopefully will be covered in the next chapter!

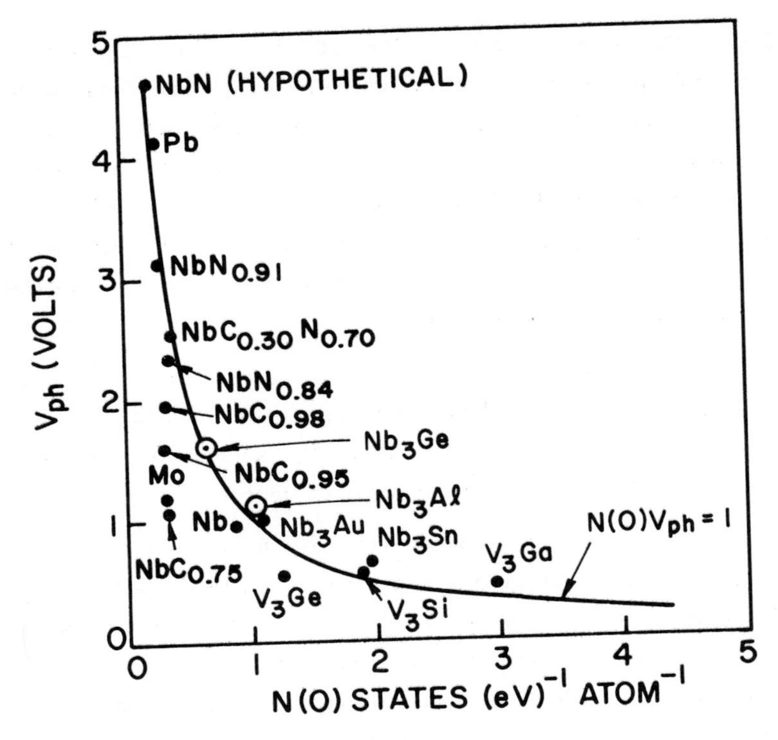

FIG. 9. Attractive Phonon Interaction V_{ph} vs Density of States N(O). Nb_3Ge and $Nb_3A\ell$ have Lower N(O) than Expected Suggesting the "Ideal" Metastable Phase is not Present

VII. SUPERCONDUCTING POWER TRANSMISSION LINES

I would like to spend the remaining section on superconducting power transmission lines, a subject we have been active in at Stanford, along with larger groups at Brookhaven, Los Alamos, and Linde. Such lines become feasible when large blocs of power must be transmitted. The losses must be sufficiently low so that the refrigeration requirement is not excessive. For a dc line the requirement is less severe since there are no dc losses and ac losses enter only as a ripple. For ac lines research has shown that if care is taken with the surface, the losses can be kept to $\sim 10\ \mu W/cm^2$ at operating conditions — an entirely acceptable figure.

Materials for the ac line include Nb , and Nb_3Sn . The greater difficulty of fabricating the Nb_3Sn line tends to be compensated for by the higher temperature of operation and the increased critical current density obtainable. The latter is useful for handling fault currents, i.e., currents that can be ten times the normal load for short periods of time. Figure 10 shows the cross section of a multilayered configuration in which it has been shown that large current densities that are found for thin films ($\sim 8 \times 10^6\ A/cm^2$) can be achieved in coatings thick enough to make practical lines. The losses remain low. The prospects for making satisfactory composites able to operate in the required high current - low loss regime are good. Of course, the superconductor is only one component of the whole complex power transmission system.

VIII. CONCLUSIONS

Thus, I conclude as I started, by emphasizing that the superconducting technology does not stand alone — it usually is a component of a much bigger energy-related system. The emergence of large-scle superconductivity depends upon factors far beyond the solution of the technical superconducting problems. However, the better the superconducting performance, the better the overall prospects, because there are always tradeoffs whereby being able to demand more of the superconducting component will enable tough requirements in other areas to be relaxed. It is

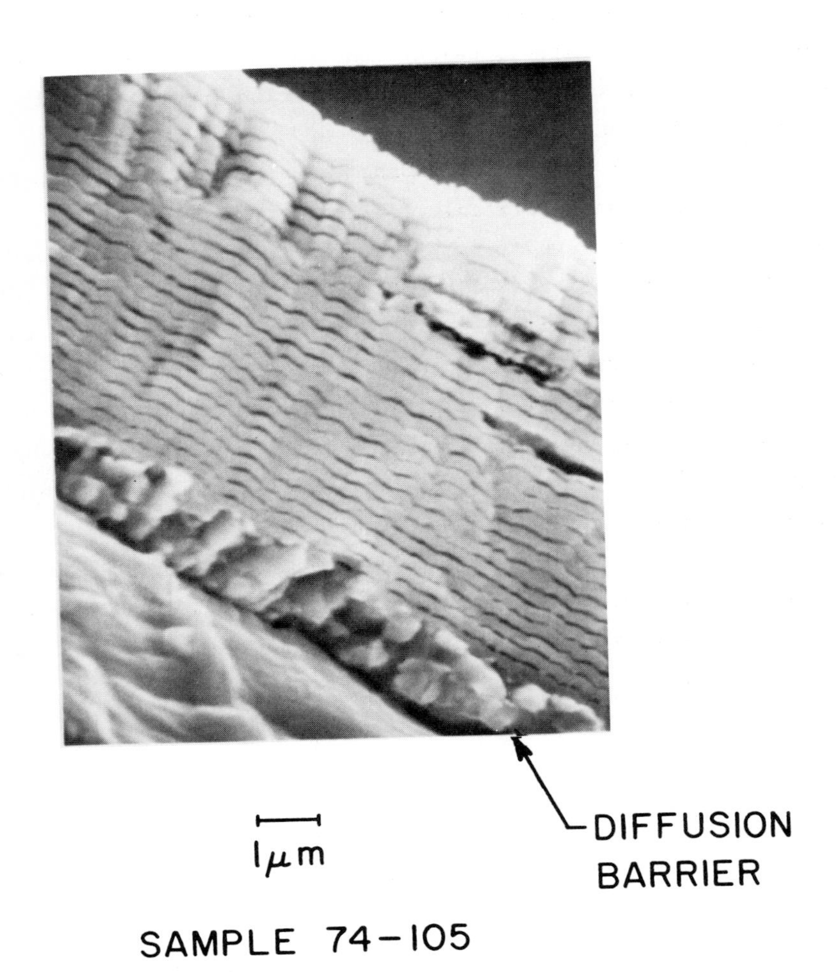

FIG. 10. SEM Micrograph of a Sample Containing 25 Layers of 1890 Å Thick Nb_3Sn Separated by 600 Å Barriers of Nb 5% Sn Alloy. An Initial 3000 Å Thick Layer of Nb_3Sn that was Deposited on the Hastelloy B Substrate as a Ni Diffusion Barrier can also be seen.

almost certain that further research can result in increased critical current densities which are still at least an order of magnitude below their theoretical intrinsic limit. It also seems possible that increased critical fields can be achieved in ductile alloys and thus the performance of NbTi-like composites can be improved. By careful work in experimentally exploring phase diagrams, particularly regions of metastability, we should be able to raise T_c moderately. We must then learn how to put these better superconductors into composites and learn more about the physics of their operation. It will then be possible to design and build reliable and economic competitors in today's technology. Finally, we always have the hope of developing new compounds. I am happy to leave that for my colleague to answer in the next chapter.

REFERENCES

1. R. C. Seamans, "A Report to the President and to the Congress of the United State," ERDA-13, April 11, 1975.

2. C. Laverick, "Helium — Its Storage and Use if Future Years," Report to the National Science Foundation, 1975.

3. R. D. Parks (ed.), Superconductivity (Marcel Dekker, Inc., New York, 1969), Vol. II; M. Tinkham, Introduction to Superconductivity (McGraw-Hill, Inc., New York, 1975); G. D. Cody and G. W. Webb, CRC Critical Reviews in Solid State Sciences, December 1973, p. 27.

4. A. M. Campbell and J. E. Evetts, Adv. Phys. 21, 199 (1972).

5. G. F. Hardy and J. K. Hulm, Phys. Rev. 93, 1004 (1964).

6. B. T. Matthias, T. H. Geballe, S. Geller, and E. Corenzwit, Phys. Rev. 95, 1435 (1954).

7. L. R. Testardi, in Physical Acoustics, edited by W. P. Mason and R. N. Thurston (Academic Press, New York, 1973), Vol. 10, p. 193; L. R. Testardi, Rev. Mod. Phys. 47, 637 (1975); and M. Weger and I. B. Goldberg, in Solid State Physics, edited by H. Ehrenreich, F. Seitz, and D. Turnbull (Academic Press, New York, 1973), Vol. 28, p. 1.

8. J.M.E. Harper, T. H. Geballe, L. R. Newkirk, and F. A. Valencia, J. of the Less-Common Metals, 43 (1975) 5-11.

9. M. C. O'Connor, Ph.D. Dissertation, Stanford University, 1976 (unpublished).

10. J. A. Wilson, F. J. DiSalvo, and S. Mahajan, Adv. in Phys. 24, 117 (1975).

11. L. P. Gor'kov, Sov. Phys.-JETP 38, 830 (1974).

12. R. H. Hammond, IEEE Trans. Mag. MAG-11, 201 (1975).

13. A. B. Hallak, R. H. Hammond, and T. H. Geballe, to appear in Applied Physics Letters.

14. L. Kokot, R. Horyn, and N. Iliew, J. Less-Common Metals 44, 215 (1976).

15. W. L. McMillan, Phys. Rev. 167, 331 (1968).

Chapter 21

SUPERCONDUCTIVITY

Bernd T. Matthias

Institute for Pure and Applied Physical Sciences*
University of California, San Diego, La Jolla, California 92093

and

Bell Laboratories
Murray Hill, New Jersey 07974

I. INTRODUCTION

From 1921 until 1933, only one ferroelectric compound was known and thus ferroelectricity was considered one of the rarest collective phenomenon. Finally, a large number of new ferroelectrics were found and it became evident that ferroelectricity was a rather general phenomenon. This suggested that the same approach be utilized in superconductivity. At that time, there were very few superconducting compounds known aside from some of the elements. Since then, the number of superconducting elements has

*Research in La Jolla sponsored by the Air Force Office of Scientific Research, Air Force Systems Command, USAF, under AFOSR Contract #AFOSR/F-44620-72-C-0017.

not only more than doubled, but the number of superconducting compounds and alloys now reaches into the thousands. Similarly, after the discovery of a great many new superconducting elements and compounds, superconductivity has gradually emerged as the normal behavior of metals at very low temperatures.

Although superconductivity was discovered in 1911 nevertheless, the transition temperature has been raised only slowly, although steadily. In 1911 it started at 4°K and the highest transition temperature known today is at ~23°K for the compound Nb_3Ge.

Why does it rise so slowly? One of the reasons is that during the last 25 years, all new superconductors with high transition temperatures have been discovered in only four laboratories throughout the world: Bell Laboratories, La Jolla, Los Alamos and Westinghouse.

At present, one recognizes one feature which is common to all superconductors — they are all metals. Contrary to what one might have heard or read or have been told, there are no superconductors today which are not real metals. To be quite specific, this author's work in the area of ferroelectricity demonstrated that ammonium sulfate was ferroelectric after almost half a century of speculations by many people. It was then realized that probably many more compounds were ferroelectric or anti-ferroelectric, and to a large extent, at the present time this has turned out to be true.

It may be illustrative to point out how the superconductivity of the element molybdenum was discovered. A lot of people had worked on molybdenum; in fact a paper had reported that superconductivity did not exist in molybdenum, down to 40 millidegrees. However, Geballe, Corenzwit and the present author found its superconductivity by just boiling molybdenum for a short period of time in the arc furnace. At present, molybdenum is superconducting at .98°K. What had been overlooked earlier was the purity of the molybdenum. Now, clearly by the year 1963 or 1964, one might have expected that before the absence of superconductivity was reported down to a temperature of 40 millidegrees, one might have looked a little bit closer at the purity of the metal.

Since the time that research in superconducting compounds was started 25 years ago, we have also more than doubled the number of superconducting elements. Sometimes it was not just simply improving the purity of the element but in addition one had to subject materials to pressure in order to make them metallic. For example, phosphorous became superconducting at the respectable temperature of 5°K under a pressure of 80 kilobars. From all this, one gradually comes to the conclusion that at low temperatures everything will be superconducting unless it is magnetic or ferroelectric.

Stated more directly: superconductivity is not an esoteric property of materials. Most metallic elements today are superconducting and since it is the majority that defines normalcy, superconductivity is now the normal property, and the so called "normal" (i.e., not superconducting) metals are anomalous because they are now in the minority. It is all really a matter of definition.

In 1960 everybody was assured that everything in superconductivity was clearly defined and that there were few unsolved problems since the isotope effect had been predicted and found. At that time T. Geballe and this author decided to measure the isotope effect of ruthenium which had not been measured before. Ruthenium is superconducting below .5 degrees. It was measured and then there was no isotope effect found, whatsoever. This was astonishing since a large segment of the scientific community had believed that the regular isotope effect occurred throughout the entire periodic system. When it was demonstrated that ruthenium had no isotope effect at all, there was a certain amount of astonishment, malaise and embarrassment. However, recovery soon set in, and after two years the lack of an isotope effect was finally predicted. Several years ago M. Fowler and H. Hill, at Los Alamos, looked into the isotope effect of uranium since, if the present author's picture of superconductivity was correct, uranium should show the opposite isotope effect. Indeed, the transition temperature of α-uranium under pressure was found to be approximately proportional to the square of the mass, and not to the inverse square root of the mass. Because of these findings, the theory of transition temperatures is, at present, of not much help to us. On the contrary, it may be a hindrance to the search for high temperature superconductors. Due to the enormous success of the theory in other directions such as pairing, tunneling and ultrasonic attenuation, all efforts or almost all efforts towards finding high temperature superconductors are being directed along definite theoretical lines. Unfortunately, without a single exception, these approaches have all failed.

In like manner, efforts towards getting to higher transition temperatures have been impeded by errors due to the experimentalists as well.

For example, in 1942, Justi had found a new superconductor, niobium nitride, at 15°K. There was great joy in Germany at the time and it was thought that the transition temperature could quickly be raised to room temperature. Soon Justi reported superconducting nuclei at room temperature. What was called nuclei in 1942 is what is now called fluctuations in 1976. Needless to say, what was wrong was that Justi forgot to stir the hydrogen bath. Another example occurred about five or six years later when Ogg reported superconductivity of metal-liquid ammonia solutions. This superconductivity involved very high transition temperatures, namely in the range of 200°K or so. What had gone wrong with Ogg's experiment was that he precipitated metal bridges of sodium and had misinterpreted their high conductivity for superconductivity which he realized and admitted shortly thereafter. Justi was more difficult to convince. He persisted that his was a real effect. His persistance lasted for 12 years until he was forced to present his experiment at a public meeting of the German Physical Societies where it failed, of course. Following these errors a great number of minor papers reported very high temperature superconductivity which, subsequently, were all proven not to be factual.

This is indeed unfortunate since there is no doubt that if one could raise the transition temperature to 30 or 60°K or even better, room temperature, it would change our whole technology, drastically.

The research effort, of searching for metals with high transition temperatures is between 5 and 10% of the total monetary and manpower effort currently directed by theoretical considerations. Of course ever since the theory came into existence, the theorists direct and lead the experimentalists. As a consequence, most of the effort today directed toward superconductivity is done so without any hope of applicability in the future. This is the reason why all progress towards real high transition temperatures is so slow — too few people and little money. This has been with us for many decades. But, let me come back to the errors of experimentalists.

The latest example relating to erroneous high temperature superconducting materials pertains to work on tetra-thiofulvalenium - tetracyanoquinodimethane single crystals. In three simultaneous press releases, superconducting fluctuations at 58°K were announced. The normal conductivity of TTF-TCNQ had been known and reported much earlier. In the present instance a very simple mistake had been made: on three crystals out of seventy the electrodes were painted on poorly. Consequently, when vanishing resistivity was noticed, it was due to the fact that there was actually no current to speak of. TTF-TCNQ is a highly antisotropic crystal and hence has a highly anisotropic conductivity. This has all since been shown by Gordon Thomas.(1) In addition, in a recent Physical Review there is a paper - which in solid state physics is an entirely novel event - in which 30 authors from many different laboratories together published the finding that in TTF-TCNQ there are no superconducting fluctuations or otherwise. "The high values reported for a few crystals are incorrect."(2) In reality TTF-TCNQ is an impure antiferroelectric. It has been known for many years that ferroelectrics and antiferroelectrics in the impure state have an enormously high conductivity at the Curie point (3) which drops as the temperature decreases. For instance, if barium titanate contains small amounts of lanthanum, scandium, or samarium oxide as an impurity to induce better conductivity, these values change by five orders of magnitude at the Curie point - a truly staggering effect. This effect is true not only for barium titanate but for ice at its Curie point, for ammonium sulfate - in fact, all ferroelectrics display this same phenomenon. For that reason, when the data for TTF-TCNQ were published, this author predicted that it would be either a ferroelectric or antiferroelectric, one can never be quite sure in advance. However, when the dielectric constant was finally measured and found to be 5,000 the veracity of this argument became self evident.

II. *EXPERIMENTAL EFFORTS TO OBTAIN HIGHER TRANSITION TEMPERATURE SUPERCONDUCTORS*

In order to achieve higher temperature superconductors one has to be cognizant of changes taking place at all temperatures in the metal. To illustrate this, see Fig. 1, in which those elements which have been found to be superconducting are shown in the periodic system. One can see immediately that the earlier statement that the majority of all metallic elements are superconducting is verified and that superconductivity in itself is a general phenomenon. Now, how does one achieve higher temperatures?

From the beginning it was realized that the only way to get to higher temperatures experimentally would be through

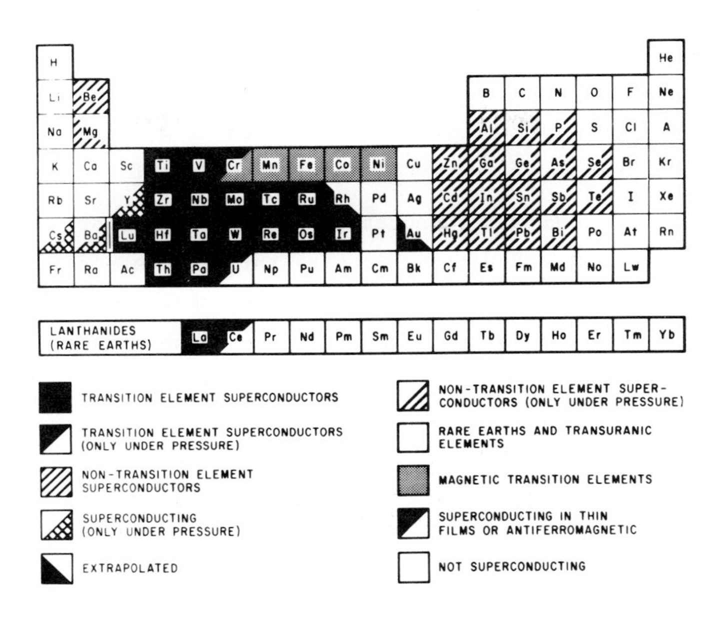

FIG. 1. The Periodic System of the Elements

a very primitive and empirical approach. In other words, one had to look at so many superconductors that a rule would be immediately apparent and obvious. At that time there was no doubt that one could get to higher temperatures. Unfortunately, superconductivity is closely related to the melting point of elemental metals, which, like superconductivity is just not understood. In fact, the melting point is more poorly understood than superconductivity as far as our ability to make predictions are concerned. The melting point is the oldest collective phenomenon known to mankind, yet even today, with all our sophistication in science, it seems no one can predict melting points.

For quite awhile measured melting points of compounds, as a function of time, went up until in 1929 - a bad year like everything else - it suddenly stopped going up. That was when

Agte and Alterthun (4) found the melting point of hafnium tantalum carbide near 4,200C. Hard as it may seem, ever since that time no one has found a material with a higher melting point. That's remarkable because if one could get to a melting point of only 5,000C our whole technology would be radically different.

In superconductivity we are more fortunate because one can make superconductors which are ternary systems. In binary systems, the melting point is sometimes at a maximum, however, in ternary systems it is always lower. Since we have more or less exhausted all binary systems in superconductivity, we have now found that ternary systems have great possibilities and that in some cases have even much higher critical temperatures. Therefore, the hope today for obtaining higher transition temperatures resides in looking at the superconductivity of ternary metallic compounds and alloys.

Unfortunately, nature, not just our scientific environment, has also raised enormous obstacles to our obtaining higher superconducting temperatures. It has been pointed out already that some of the high temperature superconductors are intrinsically unstable. As a matter of fact, it is this author's feeling that all high transition temperature superconductors are unstable, without exception. When experimenters sometimes cannot find the instabilities, it is simply due to the laboratory difficulties involved. For every crystal investigated, the introduction of a low temperature and low symmetry modification, which is thought to be its equilibrium state, has invariably resulted in a lower (5) superconducting temperature. Figure 2 shows several superconducting transition temperatures and the depression of the transition temperatures once the low temperature modification is induced. While no mathematician would be willing to fit a curve at this stage of the game, one could guess that the maximum transition temperature achievable will be somewhere between 25 and 30°K. Paul Stein has extrapolated from the time of the discovery of superconductors how long it will take to get to 25-30°K. See Figure 3.

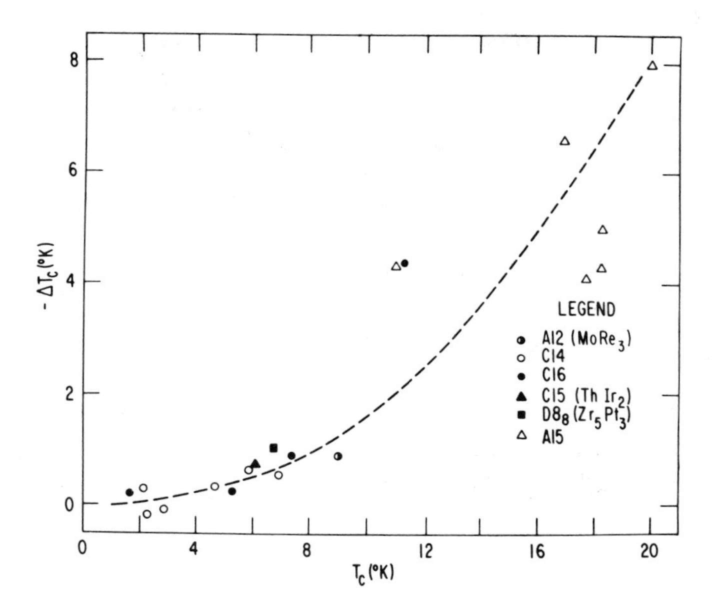

FIG. 2. Depression of the Superconducting Transition Temperature for the Low Temperature Modification

Not only must these instabilities be overcome, but one must also overcome anisotropies because cubic crystals, isotropic crystals or high symmetry crystals are the ones which have the highest transition temperatures. When the symmetry is lowered, the transition temperature will also go down. That is the reason why all two-dimensional crystals never get much above 5 or 6 degrees, and why pseudo one-dimensional crystals, such as $(SN)_x$ do not get above .5°K. One can immediately see the importance of cubic symmetry in Figure 4 (6). This is the beta-wolfram crystallographic morphology. Nb_3Si usually crystallizes in the tetragonal form which has a c/a_o ratio of exactly 2. A great deal

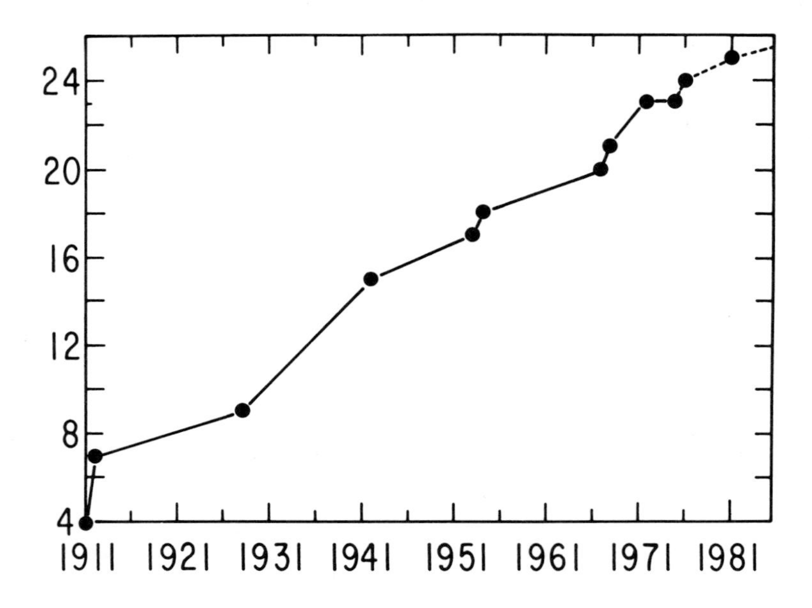

FIG. 3. Superconducting Transition Temperature as a Function of Time

of experimental work has been expended in attempts to change this structure into cubic beta wolfram. However, even though they are so closely related, with a_o being exactly what it should be in beta-wolfram, success, as yet, has not been achieved. It would seem that nature has an aversion to cubic Nb_3Si. Now it is true, some groups have claimed to have made "Nb_3Si." But it turns out to be about the most nonstoichiometric and poorest compound there is. While silicon is smaller than germanium yet the lattice constant of the reported Nb_3Si is a good deal larger than that of Nb_3Ge. If one could make it properly, if one could overcome the visicitudes of nature, one would easily raise the superconducting transition temperature somewhat above 25°K. Alas, this has yet to be accomplished.

We seem to be fighting against nature. However,

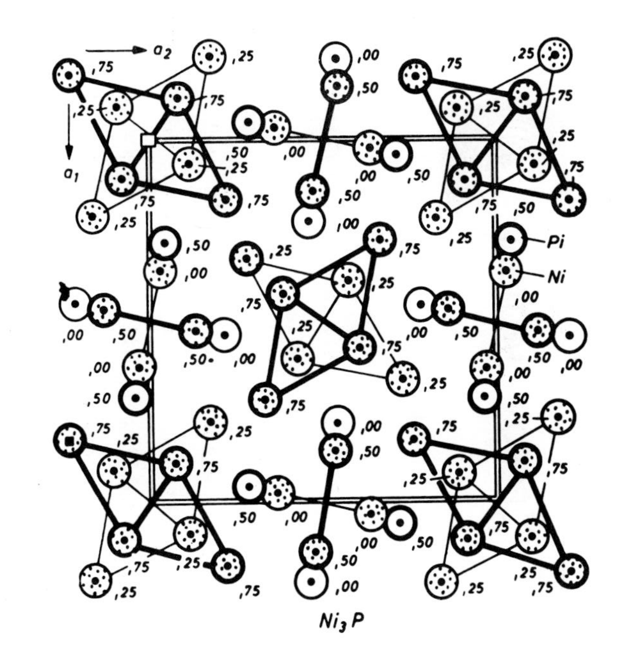

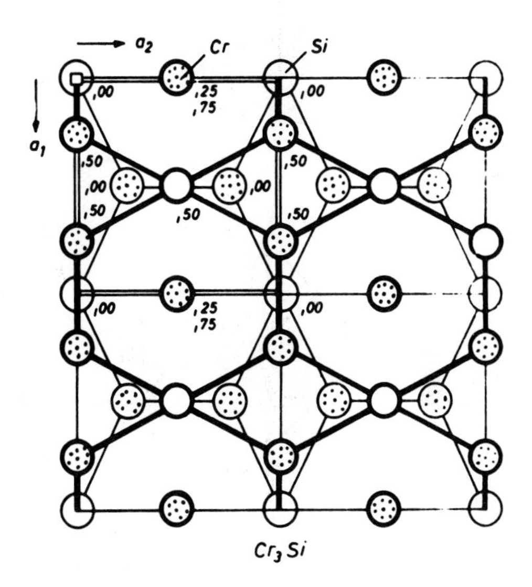

FIG. 4. A_3B: *Tetragonal vs. Cubic Symmetries*

it has been demonstrated that if one has a cubic metallic crystal, it will usually be superconducting if one addresses the problem correctly.

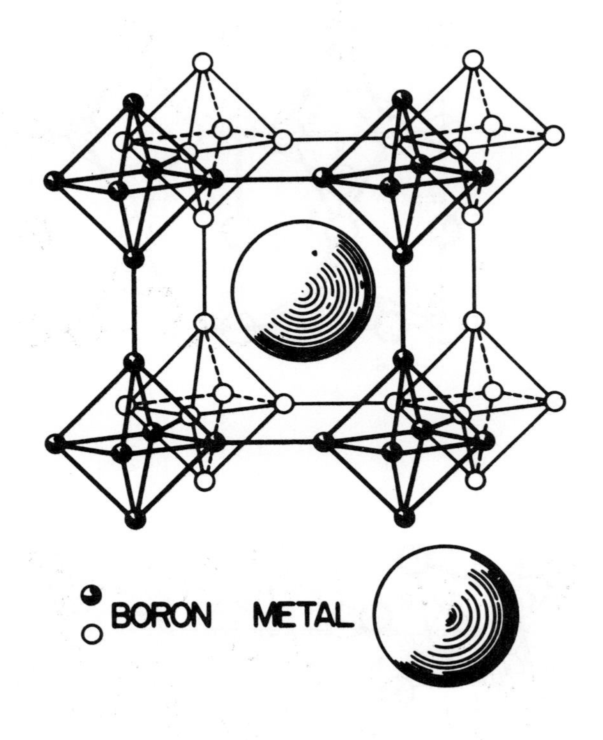

FIG. 5. Cubic Hexaboride Compounds

Take, for example, the XB_6 compounds all of which have been found to be superconducting. It really doesn't matter what the metallic atom X is. As long as the compound has cubic symmetry and is metallic, it will either be superconducting or magnetic. As another example, consider the B_{12} compounds. Here, again, the actual chemistry of the compound doesn't matter; provided crystallographically they are cubic and metallic they are always found to be superconducting. Unfortunately, these are not

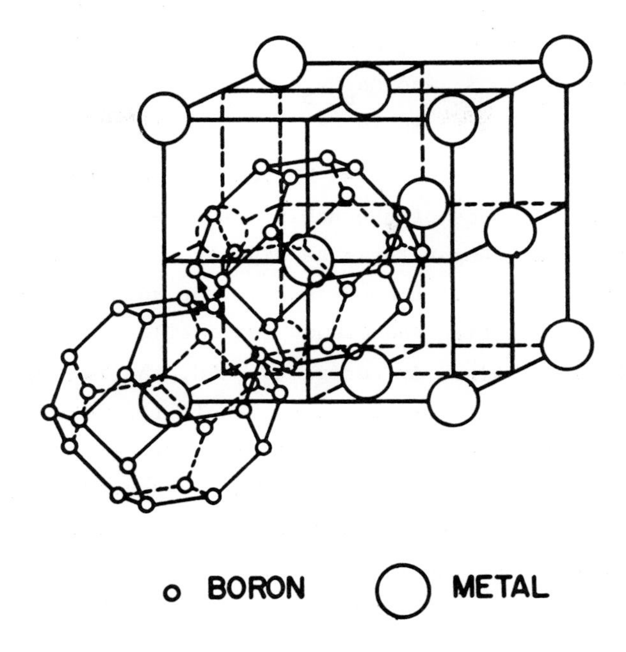

FIG. 6. Cubic Dodecaboride Compounds

very high temperature superconductors and sadly, we were never able to get the transition temperature above 7 degrees.

III. FUTURE HIGH TRANSITION TEMPERATURE SUPERCONDUCTORS

It may be instructive to speculate here about all those structures which show promise for future high transition temperature superconductivity. In the first instance are those structures of the sodium chloride form, such as NbN and NbC, which were discovered by Justi many years ago. Then there is the beta wolfram structure which Hardy and Hulm first found for V_3Si in 1953. Following this came the plutonium sesquioxide structure, which Giorgi, Szklarz and Krikorian found in 1968 through the synthesis of $(Y,Th)_2C_3$.(7) More recently, the first ternary phases were found which gave transition temperatures up to 16 degrees. Some examples of these are the cubic spinel compound $LiTi_2O_4$ (8) and the rhombohedral XMo_6S_8 and Mo_6Se_8 (9) phases where X includes many elements from Mg to Ag to La to Pb and many more.

It should also be pointed out that during the last few years hydrides of the noble metal palladium have become superconducting at temperatures as high as 17°K. (10) The reason for this high transition temperature is not quite understood, but it is a very interesting phenomenon indicating, once again, the role of instabilities. Since in contrast to Pd, the metals Cu, Ag and Au have little affinity for hydrogen, their lattices become quite unstable and thus the transition temperatures will rise. The effect may also be due to metallic hydrogen which, according to some speculation, has been predicted to be superconducting at room temperature. This is unlikely, and if it is ever made, it will probably be superconducting in the range between 15 and 20°K.

Instabilities can sometimes be overcome either by rapidly quenching the crystal lattice or by imposing high hydrostatic pressures. Instabilities can always be enhanced by uniaxial pressure, whereas hydrostatic pressures tend to keep the lattice deformation symmetrical. Consequently, at Bell Labs when instabilities were observed at lower temperatures in the lanthanum sulfides, it was immediately suggested that the application of large hydrostatic pressures would directly effect these instabilities and their superconducting transition temperatures. While both expectations were verified, the polymorphic transition which occurs more or less in parallel with the superconducting transition temperature, was different from what had been expected. Since the crystal is less isotropic below the phase transition than above it, this seemed to be a contradiction to our assumption. After all, the higher the transition temperature, the higher the anisotropy at low temperatures. But something had to happen to make them more isotropic as their transition temperature increased. What happened under pressure at these high temperatures is that an increasingly smaller phase change occurs. Consequently, with high hydrostatic pressures one would expect to achieve higher transition temperatures. This proved to be correct. Lanthanum sulfide rose from 8 degrees to 12 degrees and lanthanum selenide shows more or less the same behavior (see Figure 7a).

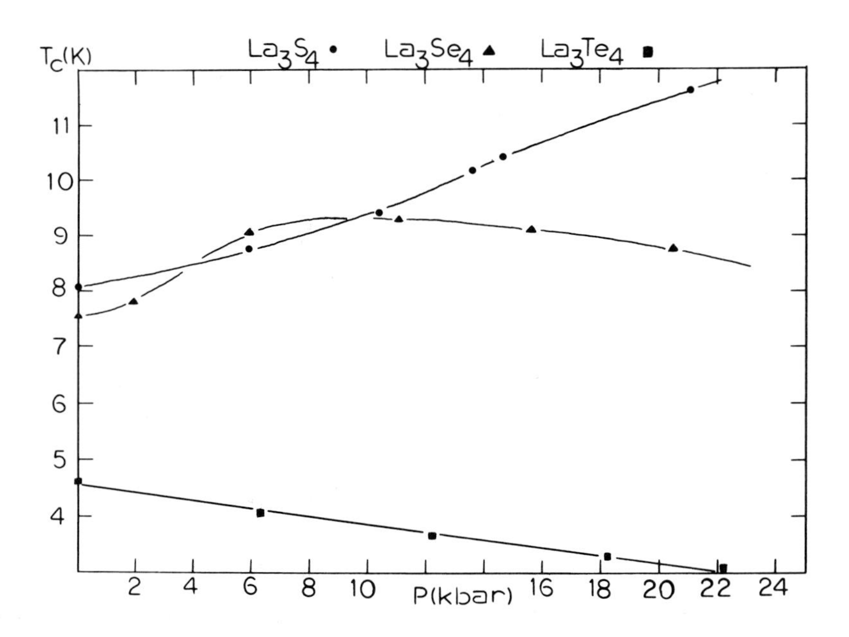

FIG. 7a. Superconducting Transition Temperature of Lanthanum Chalcogenides

This is a beautiful demonstration which shows how instabilities correlate closely with the corresponding superconductivity. Instabilities are the real, and major obstacle to high temperature superconductivity. Unfortunately, as pointed out before, they are very difficult to overcome.

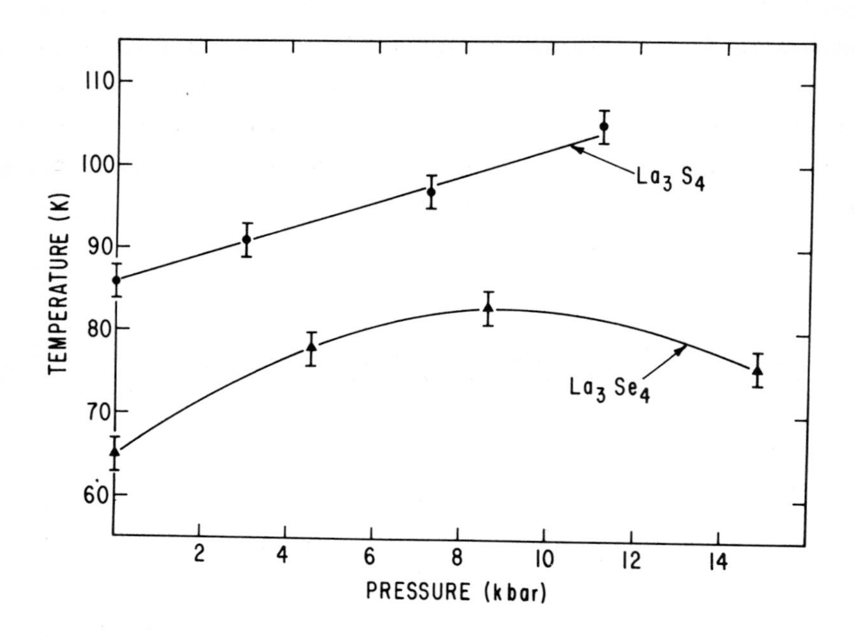

FIG. 7b. Polymorphic Transition Temperatures of Lanthanum Chalcogenides

IV. TERNARY COMPOUNDS

At present, ternary compounds seem to offer attractive superconducting possibilities. Of these the most promising system is that of the molybdenum sulfides. Experimenting with this system is very exciting because this type of behavior has never been encountered before and one can see through its use the possibility of attaining higher transition temperatures. Mo_3Se_4 is superconducting somewhat above 6°K, while, as stated before, all compounds of $X_{0.5}Mo_3S_4$, with X covering the range from Cu to Pb, are also superconducting. The transition temperatures range from 2-3°K when substituting divalent elements but up to 16°K for the four valent ones.

The crystal structure of these compounds is also

very interesting. The Mo-S or Mo-Se cubes are slightly canted and are arranged in such a way so that there are three channels of voids running through the crystal in a three-dimensional arrangement. This three dimensionality - even though it is only voids - is of primary importance, since the additional X elements are located here. The systematic occurrence of superconductivity in this system is a very unusual one. The total number of valence electrons, which is always very close to 6, hardly varies throughout this system and yet the transition temperatures cover a very wide range. It has been demonstrated that the only elements from which X can be chosen in order to get superconductivity, are those which are both immiscible with Mo and form no compounds with Mo in binary systems. This can be seen particularly well when X is a trivalent metal. Al, Ga and In react with Mo in binary systems and consequently do not form these phases. The trivalent transition elements, from Sc to Lu all form the rhombohedral phase, being inert with Mo in binary systems. This tendency is very strong, so much so that even many of the magnetic rare earths such as Pr, Nd or Gd form superconductors with temperatures as high as 9°K. This phenomenon is radically new and will lead to the first truly magnetic superconductivity of single compounds, where superconductivity will <u>not</u> occur <u>without</u> the presence of a strongly magnetic element.

V. CONCLUSIONS

It was once thought that all high temperature superconductivity could be characterized by the number of valence electrons - the electrons outside of the filled shell, per atom. That is to say, the more electrons there are, the better is the possibility for superconductivity. As indicated for ternary compounds, this idea doesn't seem to be correct and yet in the past it was so very applicable for structures such as beta wolfram. After all, using the criteria of the number of valence electrons

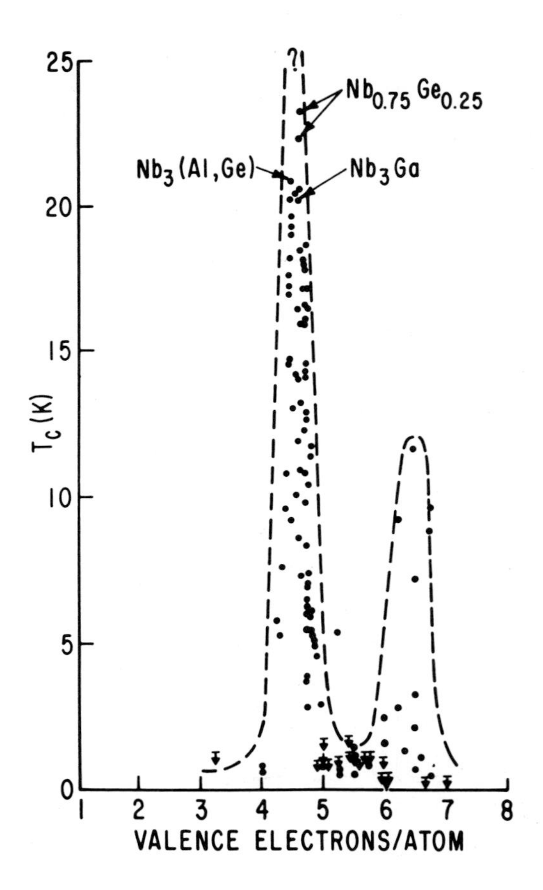

FIG. 8. Superconducting Transition Temperature in the β-W Structure as a Function of Valence Electrons/Atom

made it possible even 10 years ago for us to predict higher transition temperatures for Nb_3Ge. (11) It was already then quite obvious from this criteron that the transition temperature of Nb_3Ge should be much higher than 17 or 18°K. One conclusion becomes evident. Namely that the number of electrons is crucial as long as one is dealing with a binary system; however, with regard to these new Mo sulfides and selenides, the number of valence electrons is totally irrelevant. The reason for this change with valency electrons from binary to ternary systems is not understood at present.

The ternary systems are attractive from another viewpoint: many of the binary systems which exist, have been tried. But, when the number of ternary systems is considered there is, at present, no real limit. Consequently, today the most promising way to get to higher transition temperatures seems to be just to read the literature, and look for new ternary systems.

Eventually superconducting transition temperatures are expected to reach the vicinity of 25°K. This would be a truly significant achievement because hydrogen will be used without the need to pump on it.

The exciting possibility to explore a situation where there are many unanswered questions and to discover new, high transition temperature superconductors will attract, I hope, many new people to this field which should help accelerate progress toward this goal.

REFERENCES

1. D. E. Schafer, F. Wudl, G. A. Thomas, J. P. Ferraris and D. O. Cowan, Sol. St. Commun. 14, 347 (1974).

2. G. A. Thomas, D. E. Schafer, F. Wudl, P. M. Horn, D. Rimai, J. W. Cook, D. A. Glocker, M. J. Skove, C. W. Chu, R. P. Groff, J. L. Gillson, R. C. Wheland, L. R. Melby, M. B. Salamon, R. A. Craven, G. De Pasquali, A. N. Bloch, D. O. Cowan, V. V. Walatka, R. E. Pyle, R. Gemmer, T. O. Poehler, G. R. Johnson, M. G. Miles, J. D. Wilson, J. P. Ferraris, T. F. Finnegan, R. J. Warmack, V. F. Raaen and D. Jerome, Phys. Rev. 13B 5105 (1976).

3. V. M. Guervich, Electric Conductivity of Ferroelectrics, Izdatel' stvo Komiteta Standartov, Mer i Izmeritel'nykh Priborov pri Sovete Ministrov SSSR (translated from Russian, 1971).

4. C. Agte and H. Alterthum. Zschr. J. Tech. Physik 11, 182 (1930).

5. S. L. McCarthy, Jrl. of Low Temp. Physics 4, 669 (1971); A. C. Lawson and R. N. Shelton, Ferroelectrics (in press).

6. W. Rossteutscher and K. Schubert, Z. Metallk. 56, 820 (1965).

7. A. L. Giorgi, E. G. Szklarz, M. C. Krupka, T. C. Wallace and N. J. Krikorian, J. Less-Common Metals 14, 247 (1968).

8. D. C. Johnston, H. Prakash, W. H. Zachariasen and R. Viswanathan, Mat. Res. Bull. 8, 777 (1973).

9. R. N. Shelton, R. W. McCallum and H. Adrian, Phys. Letters 56A, 213 (1976).

10. W. Buckel and B. Stritzker, Physics Letters 43A, 403 (1973).

11. B. T. Matthias, T. H. Geballe, R. H. Willens, E. Corenzwit and G. W. Hull, Jr., Phys. Rev. 139, A1501 (1965).

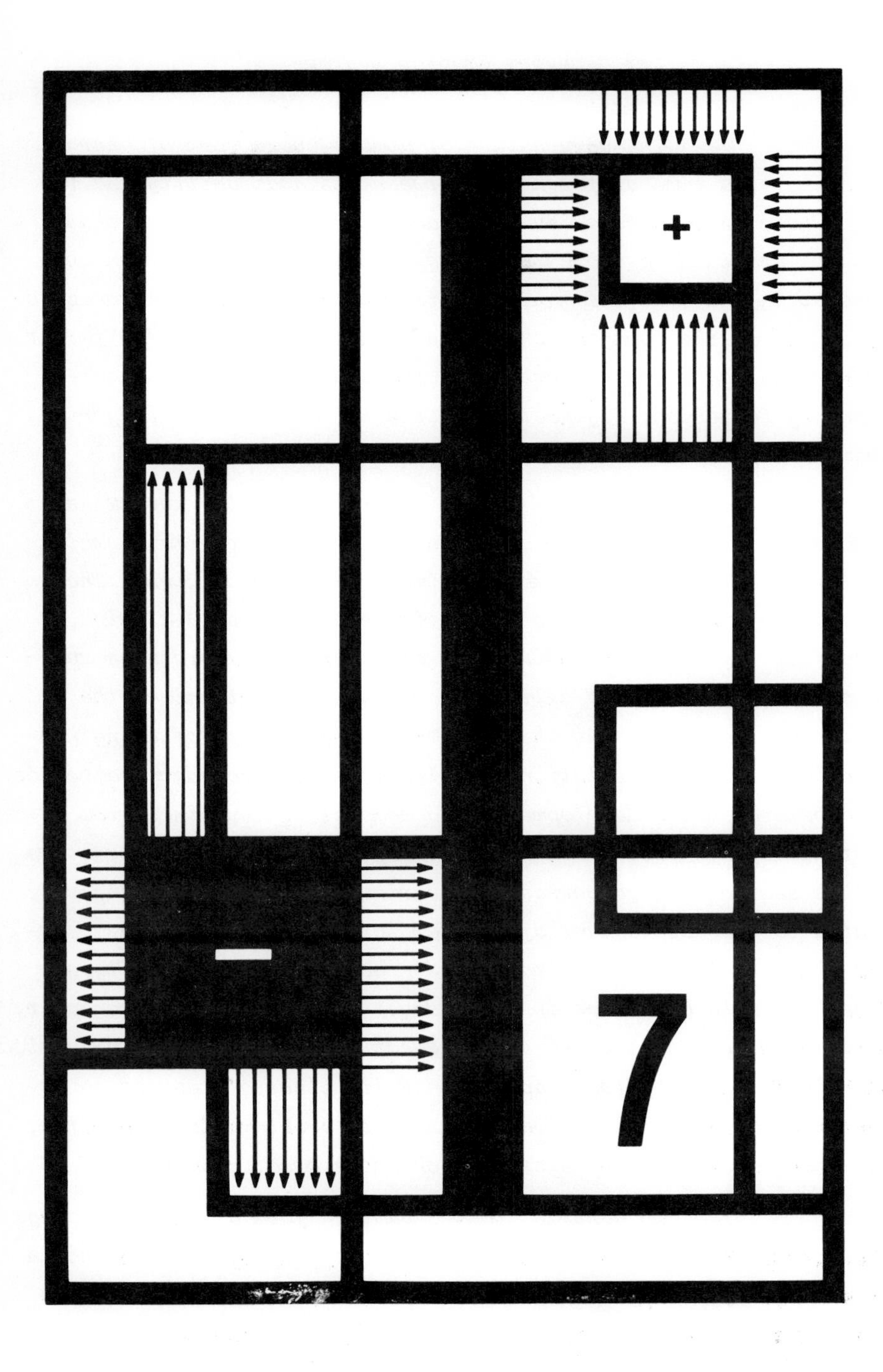
+
7

VII. ENERGY STORAGE DEVICES

The very nature of almost every form of energy necessitates a time differential between its availability and its use. The consequences of such a phase difference requires that means be provided to store the energy between its production and consumption and that the manifestation the storage system takes is predicated on the form the energy is produced in. Three different kinds of energy storage systems are treated in this Section: Secondary Batteries, Flywheels, and the Enthalpy of Metal Hydride Decomposition.

Because of recent materials development, we seem to be on the verge of major advances in battery performance which should directly impact a number of important technologies. The discovery of the solid ionic conducting electrolyte, β*-alumina, whose fast transport of sodium ions at moderate temperatures approaches the behavior of a liquid has truly revolutionized the electrochemical cell. At the same time, the concept of using non-stoichiometric solid solutions as electrodes in secondary batteries has accelerated the development of a completely contained, low maintenance, all solid component battery. The first five chapters of this section are concerned with various materials challenges related to the development of more efficient batteries. The overview chapter discusses the materials problems of the major types of rechargeable batteries in use today. In the succeeding chapter, problems in the development of new electrode materials are presented which is followed by two chapters on solid electrolytes and a chapter which is illustrative of materials problems in a specific, high temperature battery system which shows great promise.*

In contrast to the battery form of energy storage, materials problems encountered in the use of superflywheels and in the storage of energy through metal hydride decomposition are the subject matter for the last two chapters of this section.

Chapter 22
MATERIALS PROBLEMS IN RECHARGEABLE BATTERIES

Elton J. Cairns

Research Laboratories, General Motors Corporation
Warren, Michigan 48090

I. INTRODUCTION

The selection of the most appropriate materials for use in a specific rechargeable battery is a complex process. Both technical and economic factors exercise a strong influence on the ultimate choice of materials. During the last decade, the work on advanced batteries has shifted from a position in which economics was of minor importance (NASA programs) to one in which economics is a major factor (off-peak energy storage, electric vehicles). This shift has caused the problem of materials selection to become significantly more difficult. In this chapter, major emphasis will be placed upon the technical aspects of materials problems for rechargeable batteries. The discussion to follow deals with both the active materials, which are intended to take part in chemical and electrochemical reactions, and the inactive materials, which are intended to remain inert.

II. *GENERAL GUIDELINES FOR THE SELECTION OF ACTIVE MATERIALS*

The selection of active materials for rechargeable batteries is carried out with the objective of meeting certain goals for the intended applications. For automobile propulsion, it is important that the battery have high specific energy (W·h/kg) for maximum range between recharges, with an acceptable battery weight; it must have high specific power (W/kg) for good acceleration and top speed; it must have an acceptably low initial cost ($/kW·h of energy storage capability). The same general goals are suitable for many other applications, but high specific energy and high specific power are less necessary for off-peak energy storage, where cost and lifetime are of overriding importance.

The four types of goals indicated above carry certain implications which influence the choices of active mater ials to be used in batteries. The goals and their implications are shown in Table I. Some simple mass and energy considerations are useful in focusing attention on the better candidate electrochemical couples for applications which require high specific energy. The maximum amount of energy which might be obtained from an electrochemical reaction under ideal conditions (and at constant temperature and pressure) is the Gibbs free energy of reaction (ΔG), which can be calculated with the aid of

TABLE I

GENERAL GOALS FOR BATTERIES

Goal	Implication
High Specific Energy	High cell voltage; low equivalent weight; high utilization of active material.
High Specific Power	High cell voltage, rapid reactions; rapid mass transport; low internal resistance.
Long Life	Reversible reactions; low corrosion rates; negligible rates of side reactions.
Low Cost	Inexpensive, plentiful materials; simple manufacturing processes; high efficiency; little or no maintenance.

thermodynamic tables (1). The minimum weight which might be associated with a battery is that of the reactants (or products) themselves (with no allowance for structural materials). Therefore, the maximum specific energy theoretically available, called the theoretical specific energy is:

$$\text{Theor. Specific Energy} = \frac{-\Delta G}{\Sigma MW} = \frac{nFE}{\Sigma MW} \qquad (1)$$

where n is the number of electrochemical equivalents involved in the cell reaction, F is the Faraday constant (96487 coul/equiv.), E is the average emf of the cell reaction, and ΣMW is the summation of the molecular weights (times the number of moles) of the reactants (or products).

It is clear from Equation 1 that the theoretical specific energy is maximized by having a high emf and a low summation of reactant weights. A high emf is obtained by selecting negative electrode reactants having a low electronegativity (those elements located at the left-hand side of the periodic chart), and positive electrode reactants with high electronegativities (located at the right-hand side of the periodic chart). The elements of low atomic weight are located at the top of the periodic chart, so the elements of first choice for negative electrode reactants include H_2, Li, Na, Be, Mg, Ca, and a few others. As compromises are made regarding various properties, the list of reactants for negative electrodes grows to include Fe, Zn, Cd, and even Pb. The list of candidate elements for positive electrode reactants includes those from Group VI, (O_2, S) and Group VII (Cl_2, Br_2) and compounds containing these elements (e.g., PbO_2, NiOOH, FeS_2 Ag_2O_2, MnO_2). These elements and compounds provide for a long list of possible cell reactions, but the list is shortened considerably when the problem of providing an electrolyte appropriate for both the negative and the positive electrode reactant of a proposed couple is taken into account.

The couples which might be considered for use in major energy storage systems (off-peak energy storage or vehicle propulsion) are listed in Table II. Some candidate electrode reactants have been eliminated for reasons of cost and availability (Ag_2O_2, Cd, HgO) or for technical reasons relating to electrochemical performance (Mg, Ca, MnO_2, Be). Note that lithium and sodium are used only with nonaqueous electrolytes because they react with water (they also react with air). Ambient-temperature nonaqueous systems are not shown because they have inadequate ability to be recharged repeatedly. Equation 1 has been applied to the systems of Table II, with the results shown in Figure 1, where the theoretical specific energy is plotted against the equivalent weight of the reactants. Lines of constant voltage are included in Figure 1 to aid in

TABLE II

CANDIDATE SYSTEMS FOR MAJOR ENERGY STORAGE

AMBIENT TEMPERATURE SYSTEMS	
Acid Electrolyte	Alkaline Electrolyte
$Pb/H_2SO_4/PbO_2$, $Zn/ZnCl_2/Cl_2$	Fe/KOH/NiOOH, Fe/KOH/Air Zn/KOH/NiOOH, Zn/KOH/Air
HIGH TEMPERATURE SYSTEMS (300-500°C)	
Molten Salt Electrolyte	Solid Electrolyte
$Li/LiCl\text{-}KCl/FeS_2$, $Li/LiCl\text{-}KCl/FeS$ $LiAl/LiCl\text{-}KCl/FeS_2$, $LiAl/LiCl\text{-}KCl/FeS$ $Li_4Si/LiCl\text{-}KCl/FeS_2$, $Li_4Si/LiCl\text{-}KCl/FeS$	$Na/Na_2O{\cdot}XAl_2O_3/S$ Na/Na_2O glass/S $Na/Na_2O{\cdot}XAl_2O_3/SbCl_3$ in $NaAlCl_4$

selecting the higher-voltage systems. The cells with aqueous electrolytes tend to be located toward the bottom of Figure 1, whereas those that contain alkali metals and operate at high temperatures tend to be higher on the theoretical specific energy scale, with Li/S and Li/Cl_2 being highest.

The active materials which look most attractive in Figure 1 (high theoretical specific energy and high voltage) are also those that present the greatest problems with regard to the inactive materials from which the cell containers, current collectors, and separators are to be made. The negative electrodes, such as lithium, sodium, and zinc represent strongly reducing conditions; the positive electrode materials, such as Cl_2, S, O_2, PbO_2 and FeS_2 represent strongly oxidizing conditions. In many cases, compromises must be made in cell voltage in order to ease the difficulties of inactive material selection.

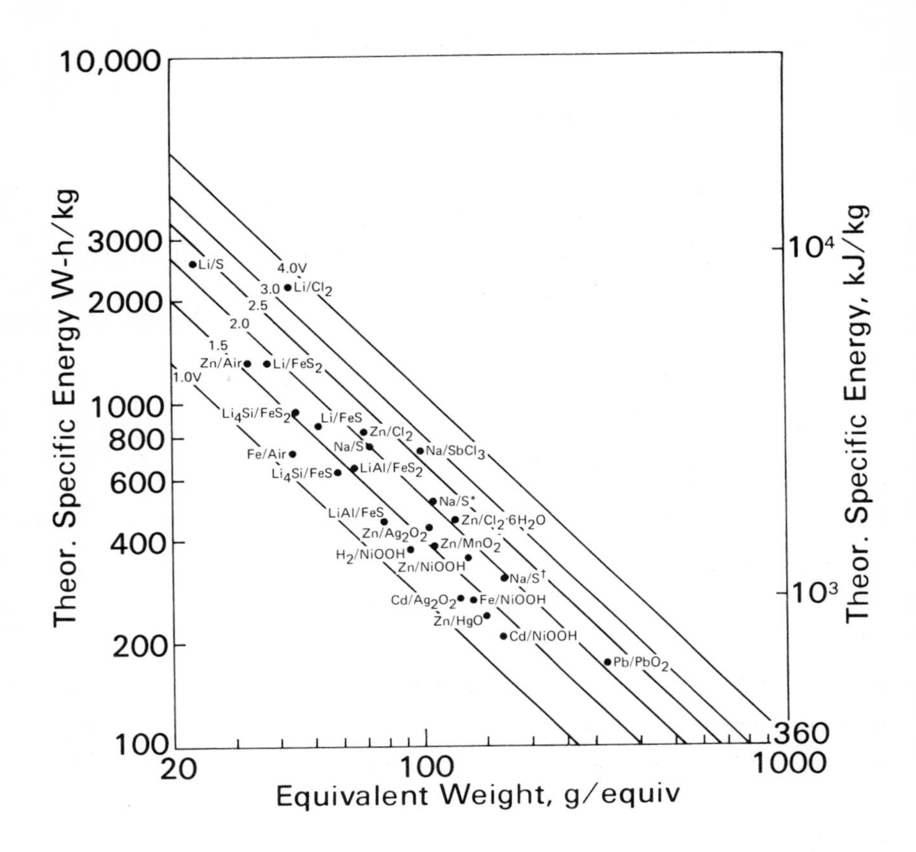

*FIG. 1. Theoretical Specific Energy of a Number of Electrochemical Couples. * Corresponds to $2Na + 5.2S \rightarrow Na_2S_{5.2}$, Operation in the Two-phase Region Only. † Corresponds to $4.4Na + 3Na_2S_{5.2} \rightarrow 5.2Na_2S_3$, Operation in the Single-phase Region Only.*

In the sections to follow, the status of each of the cells shown in Table II will be reviewed, and the problems, especially those relating to materials, will be discussed.

III. AMBIENT TEMPERATURE CELLS WITH ACID ELECTROLYTES

A. The $Pb/H_2SO_4/PbO_2$ Cell

The lead-acid cell is made in larger quantity than any other rechargeable cell, in spite of its low specific energy. The important features of low cost, good cycle life and availability of materials make it attractive for many applications. The status of cell performance, cost and life for acid-electrolyte cells, shown in Table III, indicate that recent improvements have been made in the specific energy, now up to 40 W·h/kg, or 23% of the theoretical value for commercial batteries with claims of up to 60 W·h/kg (34% of theoretical) for experimental batteries (5) of undisclosed, but short, cycle life.

Maintenance-free cells have been developed which contain all of the electrolyte necessary for the life of the cell. Specially-designed vents prevent electrolyte loss, but allow the gases (H_2, O_2) formed during recharge to escape. The development of this cell has been made possible by the great reduction in gassing provided by the replacement of antimony ($\sim$4 w/o) with calcium as an alloying agent in the positive electrode current collector (which is mostly lead). Additional improvement to decrease the gassing rate sufficiently to allow truly sealed operation is important for improved life and higher specific energy (since less electrolyte would be required). Further materials research on reduction of the corrosion rate of the lead alloys used as positive electrode current collectors is expected to lead to longer-lived cells. Cohesion of the PbO_2 to itself and adhesion of PbO_2 to the current collector continue to be problems. The adoption of porous polyvinyl chloride separators has resulted in a decrease of cell internal resistance, but further resistance reduction is required for vehicle applications.

TABLE III

AMBIENT TEMPERATURE CELLS WITH ACID ELECTROLYTES

$Pb + PbO_2 + 2H_2SO_4 \rightarrow 2PbSO_4 + 2H_2O$ E = 2.095 V; 175 W·h/kg Theor.		Ref.	$Zn + Cl_2 \rightarrow ZnCl_2$ E = 2.12 V; 825 W·h/kg Theor.**		Ref.
STATUS			STATUS		
Specific Energy	22-40 W·h/kg @ 10 W/kg	2	Specific Energy	66+ W·h/kg @ 3-4 W/kg	6
Specific Power	50-100 W/kg @ 10 W·h/kg	4	Specific Power	60 W/kg for seconds	6
Cycle Life	300+ @ 10 W/kg, 60% DOD*	2,4	Cycle Life	10-30	6,7
Cost	\$35-50/kW·h	3,4	Cost	>\$100/kW·h	
RECENT WORK			RECENT WORK		
Replace Sb with Ca in positive current collector		2	Precious metal on Ti for Cl_2 electrode		9
Maintenance-free cells		2	Additives for Zn deposition		6,7,8
Use $4PbO \cdot PbSO_4$ instead of $PbO + Pb_3O_4$ in positive		2	System components		9
PVC separators		2			
PROBLEMS			PROBLEMS		
Sealing of cells (gassing)			Precious metals		
Corrosion of positive current collector			Bulky		
Cohesion and adhesion of PbO_2			Complex		
High internal resistance			Low specific power		
Heavy			Cycle life not demonstrated		
			Gaskets		

* Depth of discharge as percent of nominal (or rated) capacity

** 460.6 W·h/kg for $Zn/Cl_2 \cdot 6H_2O$

B. *The $Zn/ZnCl_2/Cl_2$ Cell*

The zinc/chlorine cell is in the early stages of development. Each of the individual electrodes has been charged and discharged a large number of times, and some cells and batteries have been built and operated (6,7,8). The cell voltage of 2.12 V and theoretical specific energy of 825 W·h/kg for the Zn/Cl_2 cell are very attractive, but since chlorine gas is dangerous, and difficult to handle conveniently, solid $Cl_2 \cdot 6H_2O$ (at a few degrees centigrade) has been used as the stored form of Cl_2. This offers safety and storage advantages, but the weight of the water causes a reduction in the theoretical specific energy to 460.6 W·h/kg. A specific energy of 66 W·h/kg (14% of theoretical) has been demonstrated, and it is likely that a somewhat higher specific energy can be realized as the system is developed further.

A simplified schematic diagram of the system is shown in Figure 2. Because of the storage of $Cl_2 \cdot 6H_2O$ in a separate area, the diagram resembles that of a fuel cell, and the system complexity adds to the weight, making it more difficult to achieve high specific power. The heat exchanger must resist acid chloride-chlorine attack, and requires the use of special materials, such as titanium. The separators, gaskets, pumps, and seals must also resist attack of the $ZnCl_2$ electrolyte containing dissolved chlorine. For the Cl_2 electrode, the early work made use of porous carbon and porous graphite (6-8), but more recent reports indicate that precious-metal-coated titanium is currently used (9), probably adding to the cost. The zinc electrode in $ZnCl_2$ electrolyte operates very reliably (6-8), but requires a proprietary additive in the electrolyte to prevent zinc dendrite formation (8).

It is clear from the schematic diagram and the discussion above that this system requires the development of not

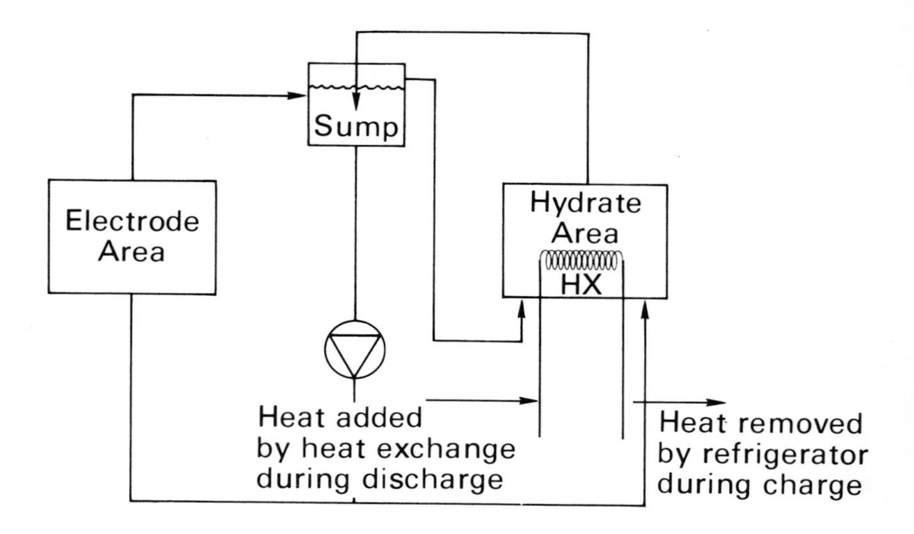

FIG. 2. Simplified Schematic Diagram of the $Zn/ZnCl_2/Cl_2 \cdot 6H_2O$ Systems (6-8).

only a reliable, long-lived battery module containing inexpensive corrosion-resistant materials, including the chlorine reduction catalyst, but the development of various system components. These components include heat exchangers, a refrigerated hydrate formation and storage system, pumps, a flow and pressure control system and an overall control system. Both electronically conducting and insulating materials resistant to acid chlorine-chloride attack play a key role in the further development of the Zn/Cl_2 battery.

IV. AMBIENT TEMPERATURE CELLS WITH ALKALINE ELECTROLYTES

A. The Fe/KOH/NiOOH Cell

This cell dates back to the turn of the century, when Jungner and Edison independently received patents on very similar Fe/NiOOH cells. Even at that time, these cells delivered up to 22 W·h/kg, a remarkable specific energy. As shown in Table IV, the modern versions of this cell deliver 30-44 W·h/kg (11-16% of theoretical). From the beginning, the Fe/NiOOH cell has been characterized by a very long life (some cells have been in daily use for as long as 60 years), and rugged mechanical and electrochemical characteristics. Recently, improvements have been made in the construction of the electrodes for this cell, so that specific energies near 45 W·h/kg may be obtained.

The iron electrode has been rather inefficient during recharge, evolving hydrogen at a significant rate. Mercury has been added to this electrode to mitigate this problem (10), but overall energy efficiencies remain near 60%, a low value compared to typical values of 80% for other systems. The low efficiency causes problems regarding temperature control, due to excessive heat evolution. The hydrogen evolution prevents sealing of these cells, which would allow them to be maintenance-free. Further improvements in the active material of the iron electrode are needed to raise the efficiency, to reduce the rate of self-discharge on open circuit, and to improve active material utilization at low temperatures (below 0°C).

TABLE IV

AMBIENT TEMPERATURE CELLS WITH ALKALINE ELECTROLYTES AND IRON ELECTRODES

$Fe + 2NiOOH + 2H_2O \rightarrow Fe(OH)_2 + 2Ni(OH)_2$

E = 1.37 V; 267 W·h/kg Theor.

		Ref.
STATUS		
Specific Energy	30-44 W·h/kg @ 10 W/kg	10
Specific Power	25-100 W/kg @ 10 W·h/kg	10
Cycle Life	500-2000 @ 10-25 W/kg, 60% DOD	10
Cost	>$100/kW·h	10
RECENT WORK		
Improved Fe and NiOOH electrodes with higher specific energy		11,12
PROBLEMS		
H_2 evolution during recharge; can't be sealed		10
Heat evolution		
Low efficiency ∿60%		10
Capacity loss at low temperature		10
High rate of self-discharge, with H_2 evolution		

$Fe + \frac{1}{2}O_2 + H_2O + Fe(OH)_2$

E = 1.2 V; 720 W·h/kg Theor.

		Ref.
STATUS		
Specific Energy	90 W·h/kg @ 10 W/kg	13
Specific Power	30 W/kg @ 40 W·h/kg	13,14
Cycle Life	200 @ ∿20 W/kg	13,14
Cost	>$100/kW·h	13,14
RECENT WORK		
Improved Fe and air electrodes		13,14
Systems of several kilowatts built for vehicles		13
Additives to reduce H_2 gassing		13
PROBLEMS		
Maintenance of water balance		13,14
H_2 evolution during charge		
Self-discharge		13
Low specific power		
High internal resistance		

B. *The Fe/KOH/Air Cell*

The iron/air cell makes use of an iron electrode which is very similar to that used in the iron/nickel oxide cell discussed above. The use of an air electrode raises the theoretical specific energy to the high value of 720 W·h/kg. In practice, 90 W·h/kg has been achieved (13% of theoretical), a rather impressive value compared to those of the other systems discussed above. Part of the price for the high specific energy is the system complexity associated with the operation of the air electrode, and the problem of CO_2 from the air converting the KOH of the electrolyte to K_2CO_3. The accumulation of high concentrations of K_2CO_3 must be avoided because the iron electrode will dissolve anodically in a carbonate electrolyte (14).

The iron electrode presents the same general materials problems in the Fe/air cell as it does in the Fe/NiOOH cell. Hydrogen evolution during recharge and during open-circuit is a major problem. The iron electrode is also sensitive to impurities, such as aluminum and silicon, which could be introduced in make-up water. Some additives have been used to reduce the rate of hydrogen evolution (13).

The air electrodes used in Fe/air cells have rather complex structures (13,14), making them expensive. The cycle lives of the dual-layer nickel electrodes have been reported to be more than 1000 cycles (13).

The overall charge-discharge efficiency of the Fe/air cell suffers from the current inefficiency of the Fe electrode (H_2 evolution) and the voltage inefficiency of the air electrode (low oxygen exchange current density, i.e., slow kinetics), so that the charge-discharge efficiency is only 35% (14) vs. about 80% for many other rechargeable cells.

The remaining materials problems are mostly those of the active materials -- better iron electrodes with no hydrogen evolution, and better air electrodes with higher reaction rates for oxygen reduction and evolution.

C. *The Zn/KOH/NiOOH Cell*

This cell is in some ways similar to the Fe/NiOOH cell, but has a higher theoretical specific energy (373 vs. 267 W·h/kg), and a higher demonstrated specific energy (up to 66 W·h/kg, or 18% of theoretical). Many of the problems with the Fe electrode (such as rapid hydrogen evolution) are avoided by using the Zn electrode, but the cycle life of the Zn electrode is not as long as desired.

The zinc electrode suffers from dendrite formation and a redistribution of zinc from the edges of the electrode toward the center, as the cell is cycled repeatedly (15,19), resulting in a loss of capacity and eventual cell failure. Dendrite formation and the rate of shape change are influenced by a number of factors (19), including the separator. Recent work with inorganic separators has indicated that the cycle life of the zinc electrode is extended to 200 and more cycles from about 100 cycles with cellulosic separators. Further work in this promising materials area seems justified.

The use of inorganic separators permits the cells to be sealed, which offers the advantage of maintenance-free operation, maintenance of state-of-charge balance between the positive and negative electrodes, and minimum electrolyte weight in the cell. Sealed cells with inorganic separators have demonstrated 100-200 cycles at 60% DOD, 4-h rate (16). This is a significant improvement, but cycle lives in excess of 300 cycles at more than 60% DOD are needed.

Nonsintered NiOOH electrodes have been developed (15) which contain only 40% of the amount of nickel used in conventional sintered NiOOH electrodes, resulting in a weight and cost advantage. Also, the processing time for electrode preparation is reduced by a large factor (10 to 50). All NiOOH electrodes exhibit some inefficiency during recharge, resulting in oxygen evolution. In sealed cells, this oxygen is recombined with zinc by use of a fuel-cell type of oxygen reduction electrode (16). If oxygen evolution could be prevented by improvement of the NiOOH electrode, the cell could be simplified, and the efficiency would be improved by about 10%. The status of this cell is summarized in Table V.

D. *The Zn/KOH/Air Cell*

The zinc/air cell (Table V) has long attracted attention because of its high theoretical specific energy, 1310 W·h/kg, rivaled only by some of the high-temperature cells (see Fig. 1), although only less than 10% of the theoretical specific energy has been obtained. This cell is of higher current efficiency than the Fe/air cell, because the zinc electrode is very efficient, but still is only in the range of 50%, largely because of the low rate of the oxygen electrode reactions. The cycle lives of both electrodes of this cell are modest (100-200 cycles).

High specific energies look most achievable in the simplest systems, however problems with zinc deposition on recharge have caused attention to be directed to zinc/air cells with a circulating zinc-electrolyte slurry (5,20), which adds to the total electrolyte weight in the system. If the system with a zinc slurry is designed for high specific energy, then the specific power is likely to be low, and vice-versa. At present, the spent slurry (ZnO in KOH) is recharged outside of the cell, an added complication.

In general, the materials of construction problems for this system are not particularly severe. The main

TABLE V

AMBIENT TEMPERATURE CELLS WITH ALKALINE ELECTROLYTES AND ZINC ELECTRODES

$Zn + 2NiOOH + H_2O \rightarrow ZnO + 2Ni(OH)_2$ E = 1.74 V; 373 W·h/kg Theor.			$Zn + \frac{1}{2}O_2 \rightarrow ZnO$ E = 1.6 V; 1310 W·h/kg Theor.		
		Ref.			Ref.
STATUS			STATUS		
Specific Energy	50-66 W·h/kg @ 30 W/kg	15,16	Specific Energy	120 W·h/kg @ 70 W/kg	5,20
Specific Power	200-300 W/kg @ 35 W·h/kg	18	Specific Power	140 W/kg @ 70 W·h/kg	20
Cycle Life	100-200 @ 25-50 W/kg 60% DOD	15,16,17	Cycle Life	<100	5
Cost	>$100/kW·h		Cost	>$100/kW·h	
RECENT WORK			RECENT WORK		
Inorganic separators (e.g., K_2TiO_3, ZrO_2, others		15,16,17	Improved circulating zinc electrodes		20
			Improved air electrodes		20
Sealed cells		15,16	Multicell batteries built (∿1 kW)		20
Nonsintered electrodes		15,16			
PROBLEMS			PROBLEMS		
Sealing of cells - O_2 evolution and recombination			Rechargeable air electrodes		20
			Control of zinc in slurry		20
Shape change of zinc electrode			Wetting		
Separators			Carbonation		
			Short life		

problems lie in the active materials: recharging of the zinc electrode, robust, effective catalysts for the air electrode, and good control of the gas/solid/liquid contact zone in the air electrode.

V. HIGH TEMPERATURE CELLS WITH MOLTEN SALT ELECTROLYTES THE LITHIUM/METAL SULFIDE CELLS

For the purposes of this discussion, all of the cells listed under molten salt electrolyte in Table II have been grouped together, because the materials problems are very similar for all members of this group. These cells, Figure 3, make use of lithium (27) or a high-lithium-content alloy as the negative electrode (22-24), a molten salt mixture, such as LiCl-KCl as the electrolyte, and a metal sulfide, such as FeS_2 or FeS, as the positive electrode. The operating temperature is 375-475°C. As indicated in Table VI, light-weight LiAl/FeS_2 cells of the design shown in Figure 3 have demonstrated specific energies up to 150 W·h/kg (vs. 650 W·h/kg theor., see Fig. 1) and specific powers up to 80 W/kg (22). The cycle lives of single cells have been in excess of 300 deep cycles, and such cells have operated continuously for more than 6000 h. These performance data show the promise that lithium/metal sulfide cells hold: very high specific energy and specific power, with low-cost active materials.

The major problems for lithium/metal sulfide cells are ones of corrosion. Lithium attacks all but a few of the candidate electronic insulators needed for use as separators and feedthrough components. Table VII shows a listing of the best ceramics identified to date for use with lithium at 400°C in lithium/metal sulfide cells. Additional materials work is

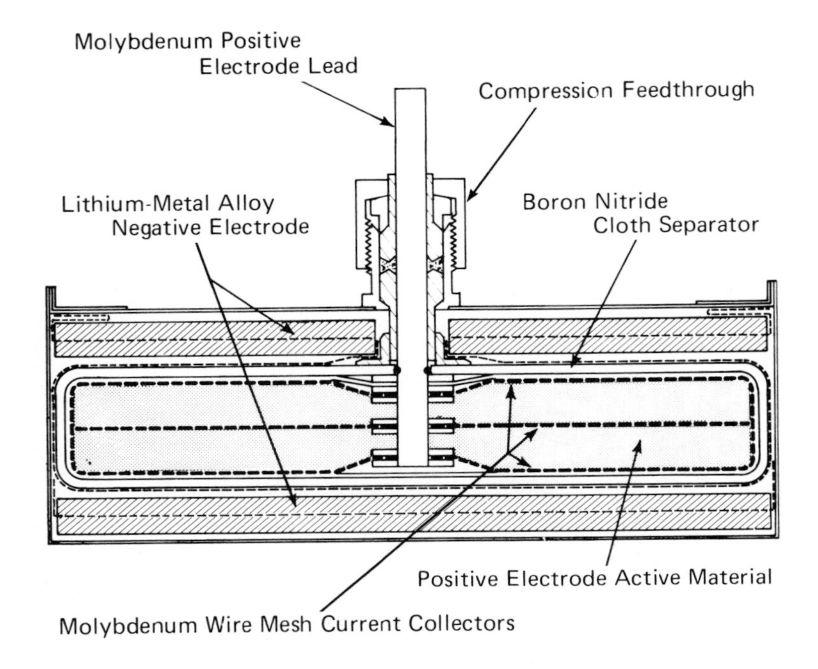

FIG. 3. Schematic Cross Section of a Lithium Aluminum/Iron Sulfide Cell, which Operates at 375-475°C.

needed to provide thin (∿1 mm) strong, flexible, corrosion resistant separators. Materials are needed for bonding the ceramics of Table VII to electronic conductors that are stable to the attack of sulfur or metal sulfides (such as FeS_2). Table VIII shows the corrosion rates of some electronic conductors in lithium/metal sulfide cells (21). Only molybdenum, tungsten, carbon, graphite, TiN, and FeB have acceptably low corrosion rates in contact with FeS and FeS_2 in cells. Graphite is commonly used in conjunction with molybdenum as a current collector for the positive electrodes. A wider choice of materials of construction for these cells is clearly needed.

An important materials research opportunity exists for the development of a solid lithium-ion conducting

TABLE VI

HIGH TEMPERATURE CELLS

$4Li + FeS_2 \rightarrow 2Li_2S + Fe$			$2Na + 3S \rightarrow Na_2S_3$		
E = 2.1 - 1.6 V; 1320 W·h/kg Theor.			E = 2.08 - 1.75 V; 758 W·h/kg Theor.		
		Ref.			Ref.
STATUS ($LiAl/FeS_2$)			STATUS		
Specific Energy	150 W·h/kg @ 15 W/kg	21,22	Specific Energy	80 W·h/kg @ 15 W/kg	28
Specific Power	80 W/kg @ 80 W·h/kg	22	Specific Power	154 W/kg peak	28
Cycle Life	>300 @ 100% DOD	21,22	Cycle Life	(∿100 @ 0.1 A/cm^2)*	31
Lifetime	>6000 h	21	Lifetime	(3000 h)*	31
Cost	>$100/kW·h		Cost	>$100/kW·h	
RECENT WORK			RECENT WORK		
Compact, lightweight cells		21,22	Operation in two-phase region		29,30
Simplified construction		21	New materials as coatings		29
Improved separators		21,22	New electrode designs		29,30
Improved feedthroughs and insulators		25,26	New seals		31
Lithium-silicon alloys		23,24			
PROBLEMS			PROBLEMS		
Corrosion of metals by FeS_2			Corrosion of conductors by polysulfide		
Current collection from FeS_2			Corrosion of seal materials		
Seals			Cracking of β-Al_2O_3		
Corrosion of separators by lithium			Flaking of β-Al_2O_3 (impurities)		
Feedthrough development			Short life in two-phase region		
			Difficult to recharge		

* Estimated for operation from S to Na_2S_3 from incomplete information in the literature.

TABLE VII

BEST CERAMIC MATERIALS FOR Li/FeS_2 CELLS

FEEDTHROUGHS		SEPARATORS	
AlN	- Molten Li, 400°C, 10,000 h	BN	- Cloth, paper, purified. Molten Li, 400°C, 2500 h
BN	- Purified - Molten Li, 400°C, 1300 h	ZrO_2	- Cloth. In contact with positive only
$CaZrO_3$	- Stoichiometric, high purity. Promising	Y_2O_3	- Cloth. Not very strong.
BeO	- High purity, hot pressed, molten Li, 400°C, ∿1000 h		

TABLE VIII

COMPATIBILITY OF PROSPECTIVE POSITIVE ELECTRODE MATERIALS IN LITHIUM/IRON SULFIDE CELLS

Material	Positive Electrode	Results of Corrosion Tests or Cell Tests
Mo	FeS_2, FeS	Little or no attack on properly prepared material
W	FeS_2	
C, Graphite	FeS_2, FeS	
TiN, FeB Coating on Fe	FeS_2	
Fe	FeS	moderate attack or dissolution
Ni	FeS	
Fe	FeS_2	Severe attack or dissolution
Ni	FeS_2	
Cu	FeS	
Nichrome	FeS_2	
Nb	FeS_2	

electrolyte, resistant to lithium and sulfur attack. This would simplify cell construction and allow the use of the Li/S couple, which has a theoretical specific energy of 2600 W·h/kg, higher than that of any other system under consideration.

VI. HIGH TEMPERATURE CELLS WITH SOLID ELECTROLYTES THE SODIUM/SULFUR CELLS

Sodium/sulfur cells make use of a solid electrolyte which conducts sodium ions. This electrolyte, called beta alumina, which has a composition of $Na_2O \cdot XAl_2O_3$ where X is in the range 8 to 11, is at once a great advantage, in that it provides for simple cell design, and a source of problems. This cell operates at 300-400°C, and usually has a tubular form, as shown in Figure 4 (31). Sodium is contained in the tubular electrolyte, and moves as ions into the outer region where sulfur is held in a porous carbon felt. The sodium ions react with the sulfur and electrons to produce sodium polysulfides, with the overall reaction shown in Table VI.

Sodium/sulfur cells have demonstrated specific energies as high as 80 W·h/kg, and specific powers up to 154 W/kg. These cells, unfortunately had short lifetimes. It has proven difficult to operate over the full composition region implied by the equation of Table VI, and work continues on the design of the graphite current collector with the goal of optimizing the mass transport of the immiscible sulfur and polysulfide phases present (29,30).

One of the major life-limiting factors is failure of the beta-alumina electrolyte. The most common failure mode was sodium penetration during recharge, causing cracking of the electrolyte. This problem has been minimized by developing

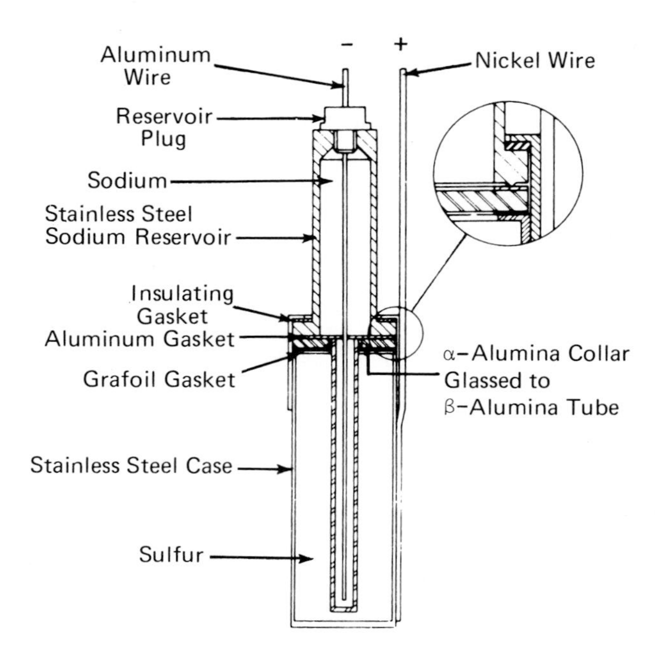

FIG. 4. Schematic Cross Section of a Na/S Cell Showing Detail of the Seal (31).

very dense beta alumina, and modifying its composition. Some impurities, such as potassium, are harmful and cause cracking by entering the beta-alumina lattice. Silica impurity forms a coating over the beta alumina, blocking the electrode reaction. Metallic impurities, such as Fe, Mn, Cr, form sulfide deposits, interfering with mass transport. Cell lifetime has increased as these problems have been brought under control.

The major remaining materials problem for the Na/S cell is the selection of a case material which is electronically conducting and resists sulfur-polysulfide corrosion at 300-400°C. Table IX (29) shows a recent summary of the materials tested in sodium polysulfide. Some conductive oxides, such as Cr_2O_3 may be useful as coatings on stainless steel cell cases. Work in this materials area continues.

TABLE IX

COMPILATION OF MATERIALS SCREENED IN SODIUM TETRASULFIDE AT 400°C

Group I - Extensively Corroded Materials
TaB_2, ZrC, VN, VC, NbB_2, $ZrSi_2$, $TiSi_2$, CrB_2, ZrN, CrC, $CaTiO_3$ (+ 0.3% Fe_2O_3), anodized Ti, Ti
Group II - Corrosion Protection from Sulfide or Oxide Film Formation
Zr, $CrSi_2$, TiC, ZrB_2, TiN, Si, TaC, TaN, AISI 446 stainless steel, Inconel 600 and 601, Mo
Group III - Intrinsically Corrosion Resistant Materials
Cr_2O_3, MoS_2, ZrO_2, $La_{0.84}Sr_{0.16}CrO_3$, TiO_2 (single crystal and polycrystalline doped with Ta). $SrTiO_3$, $CaTiO_3$ (+ 3.0% Fe_2O_3). Polyphenylene (Pph), polyphenylene-graphite composites, oxidized stainless steels, NiO

VII. FUTURE DIRECTIONS

From the foregoing can be extracted a list of critical materials problems which require additional effort. These are: corrosion resistant current collectors for PbO_2 electrodes; iron electrodes which do not evolve hydrogen; rechargeable oxygen electrodes; zinc electrodes of long cycle life; oxidation-resistant separators for zinc cells; lithium-ion-conducting solid electrolytes; lithium-resistant separators (thin, lightweight); sulfur compounds with small ΔG_f° and small MW; sulfur-resistant electronic conductors; and sulfur and polysulfide resistant seals.

REFERENCES

1. JANAF Thermochemical Tables, 2nd ed., U.S. Dept. of Commerce, Wash., D.C., NSRDSNBS37, June, 1971.

2. A. C. Simon and S. M. Caulder, in Proc. of the Symp. and Workshop on Advanced Battery Research and Design, Argonne Nat'l Lab., Argonne, Ill., March, 1976, ANL 76-8.

3. D. L. Douglas, in Proc. Symp. on Power Systems for Electric Vehicles, Columbia Univ., New York, Apr. 1967.

4. N. J. Maskalick, J. T. Brown, and G. A. Monito, in Proc. 10th IECEC, IEEE, New York, 1975.

5. N. P. Yao, in Proc. of the Symp. and Workshop on Advanced Battery Research and Design, Argonne Nat'l Lab., Argonne, Illinois, March 1976, ANL 76-8.

6. P. C. Symons, Preprint No. 730253, Soc. Automotive Engr's, New York, 1973.

7. P. C. Symons and H. K. Bjorkman, Jr., Presented at AIChE Meeting, Detroit, Mich., June, 1973.

8. P. C. Symons, Presented at Society for Electrochemistry Meeting, Brighton, England, Dec., 1973.

9. P. C. Symons, in Proc. 3rd Internat. Electric Vehicle Symp., Wash., D.C., Feb., 1974.

10. S. U. Falk and A. J. Salkind, Alkaline Storage Batteries, John Wiley & Sons, New York, 1969.

11. E. Buzzelli, The Electrochemical Society Meeting, Dallas, Oct., 1975.

12. E. Buzzellii, Private communication, 1976.

13. O. Lindström, in Power Sources 5, D. H. Collins, ed., Academic Press, New York, 1975.

14. H. Cnobloch, D. Groppel, D. Kühl, W. Nippe, and G. Siemsen, in Power Sources 5, D. H. Collins, ed., Academic Press, New York, 1975.

15. E. J. Cairns and J. McBreen, Industrial Research, June, 1975, p. 56.

16. A. Charkey, in Proc. 10th IECEC, IEEE, New York, 1975, p. 1126.

17. R. G. Gunther, presented at The Electrochemical Society Meeting, Wash., D.C., May, 1976, Abstract No. 2.

18. F. P. Kober and A. Charkey, in Power Sources 3, D. H. Collins, Ed., Oriel Press, Newcastle, England, 1971.

19. J. McBreen, J. Electrochem. Soc., 119, 1620 (1972).

20. A. J. Appleby, J. P. Pompon, and M. Jacquier, in Proc. 10th IECEC, IEEE, New York, 1975, p. 811.

21. E. J. Cairns and J. S. Dunning, in Proc. of the Symp. and Workshop on Advanced Battery Research and Design, Argonne Nat'l Lab., Argonne, Ill., March, 1976, ANL 76-8.

22. W. J. Walsh, et al., in Proc. 9th IECEC, ASME, New York, 1974.

23. L. R. McCoy and S. Lai, in Proc. of the Symp. and Workshop on Advanced Battery Research and Design, Argonne Nat'l Lab., Argonne, Ill., March, 1976, ANL 76-8.

24. R. A. Sharma and R. N. Seefurth, General Motors Research Publication, GMR-2116, March, 1976; J. Electrochem. Soc., submitted.

25. E. J. Cairns and R. A. Murie, in Corrosion Problems in Energy Conversion and Generation, C. S. Tedmon, Jr., ed., The Electrochem. Soc., Princeton, N. J., 1974.

26. R. A. Sharma, R. A. Murie, and E. J. Cairns, J. Electrochem. Soc., in press; Extended Abstracts of The Electrochem. Soc. Meeting, Wash., D.C., May, 1976.

27. E. J. Cairns, et al., Development of High-Energy Batteries for Electric Vehicles, Progress Report for the Period July 1970-June 1971, Argonne Nat'l Lab. Report No. ANL 7888, Dec. 1971.

28. R. W. Minck, in Proc. 7th IECEC, Amer. Chem. Soc., Wash., D.C., 1972.

29. S. A. Weiner, "Research on Electrodes and Electrolyte for the Ford Sodium-Sulfur Battery," Semiannual Report for the Period June 30, 1975 to Dec. 30, 1975, Ford Motor Co., Dearborn, Mich., Jan., 1976.

30. R. W. Minck, in Proc. of the Symp. and Workshop on Advanced Battery Research and Design, Argonne Nat'l Lab., Argonne, Ill., March, 1976, ANL 76-8.

31. J. L. Sudworth, in Proc. 10th IECEC, IEEE, New York, 1975.

Chapter 23

SOLID SOLUTION ELECTRODES

B.C.H. Steele

Department of Metallurgy and Materials Science
Imperial College
London, S.W.7. England

I. INTRODUCTION

Selected non-stoichiometric compounds can function as a host lattice for the incorporation of many elements such as hydrogen, lithium, sodium, oxygen, which can be employed as the electro-active species in a variety of electrochemical devices. Providing the kinetics of dissolution into the solid compounds are rapid and providing that relatively large quantities of the solute can be incorporated, then these non-stoichiometric solids can be used as solid solution electrodes (S.S.E.) in appropriate electrochemical cells. The characteristics of a solid solution electrode (S.S.E.) are somewhat different from those normally associated with traditional electrode materials and these differences can be illustrated with reference to the following two electrochemical cells:

$$\underset{(\Delta\overline{G}_{Na})}{Na\ (l)}\ /\ Na-\beta\text{-}Al_2O_3\ /\ \underset{(\Delta\overline{G}''_{Na})}{S\ (l)}$$

$$Na\ (s)\ /\ \underset{(Na^+)}{\text{organic electrolyte}}\ /\ TiS_2\ (s)$$

The emf developed by all these cells is given by the relationship,

$$E = \frac{\Delta\overline{G}''_{Na} - \Delta\overline{G}_{Na}}{F} = \frac{\Delta\overline{G}''_{Na}}{F} \qquad (1)$$

as the activity of sodium at the left hand electrode is unity. The factors governing $\Delta\overline{G}''_{Na}$ can be shown on the relevant free energy composition diagrams for these systems (Figs. 1 and 2). As the electro-active species (Na^+) is transported through the cell the ratio Na/electrode component will gradually increase and the relevant value for $\Delta\overline{G}''_{Na}$ is given by the intercept of the tangent to the free energy composition curve at the appropriate composition. For the first cell the transport of sodium to the right hand electrode results in the formation of two immiscible liquids (Fig. 1) which establishes a constant value for $\Delta\overline{G}''_{Na}$ between the compositions $Na_{0.01}S$ (point a) and $Na_{0.29}S$ (point b). The introduction of more sodium will result in a steady decrease of cell voltage until the composition (c) is attained which is in equilibrium with the solid phase Na_2S_2 at (d). The tangent c - d now establishes a constant value for the open circuit voltage of the cell. As liquid phases are necessary to permit large current densities, the operation of the cell is thus usually restricted to the composition limit of $Na_{0.67}S$ (Na_2S_3), which corresponds to a maximum theoretical specific energy density of approximately 500 WH/Kg.

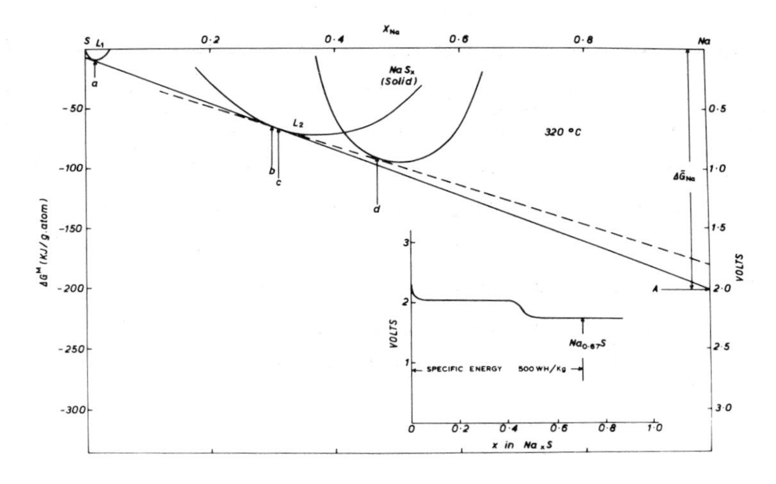

FIG. 1. Schematic Free Energy Composition Diagram for Na-S System

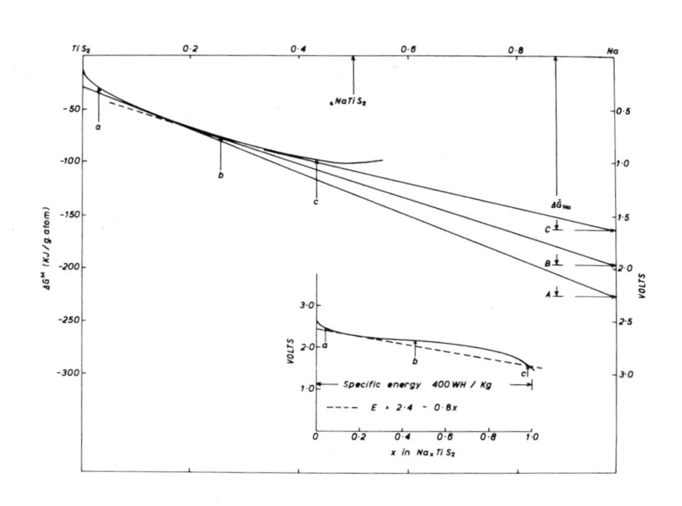

FIG. 2. Schematic Free Energy Composition Diagram for Pseudo Binary Na - TiS_2 System

An alternative approach is provided by the application of solid solution electrodes and the thermodynamic data for the system $Na_x\ TiS_2$ shown in Fig. 2 provides a suitable example. In this system the integral ΔG_M curve can be represented approximately by a shallow curve and so the slope of the tangents and thus the associated $\Delta\overline{G}''_{Na}$ values change with composition. For this system the incorporation of 1 gm atom of sodium ($TiS_2 \rightarrow Na_xTiS_2$) is accompanied by a voltage change of 2.4 - 1.4. Assuming that $\Delta\overline{G}''_{Na}$ can be represented by a linear function of x, i.e., $\Delta\overline{G}''_{Na}$ = -240,000 + 80,000 x (J), then the theoretical energy is given by

$$\int_{x=0}^{x=1} \Delta\overline{G}''_{Na} \, dx \qquad (2)$$

and thus the associated theoretical specific energy density is 400 WH/Kg. This value is comparable to that calculated for the Na/S battery. This type of thermodynamic analysis does suggest therefore, that S.S.E.'s could be incorporated into viable secondary batteries provided that the rate of mass transport within the non-stoichiometric solid is high enough for the required current densities.

II. SPECIFICATIONS FOR SOLID SOLUTION ELECTRODES

The desirable properties of solid solution electrodes have been discussed by Steele (1) and are summarized below.

(i) Large range of homogeneity.

(ii) Partial molar free energy ($\Delta\overline{G}$) of the electro-active species should be relatively constant over a wide range of composition.

(iii) Good electronic conductor.

(iv) Rapid chemical diffusion of the electro-active species within the non-stoichiometric electrode.

(v) No significant interaction with the electrolyte phase.

(vi) Low weight materials, economical to fabricate.

III. MEASUREMENTS OF THERMODYNAMIC AND TRANSPORT PROPERTIES AS SOLID SOLUTION ELECTRODES

Solid solution electrodes have often been used in high temperature electrochemical cells incorporating solid electrolytes, and reviews (2, 3, 4, 5) are available describing the measurement of the relevant thermodynamic and kinetic properties. Coulometric titrations enable the partial molar free energy ($\Delta\overline{G}$) of the electro-active species in the S.S.E. to be measured as a function of composition. Potentiostatic, galvanostatic and other electro-chemical relaxation techniques (6, 7) can then be employed to investigate how the chemical diffusion coefficient ($\tilde{D}$) varies as a function of composition. If appropriate, the chemical diffusion data can be supplemented by measurements of ionic conductivity using blocking electrodes, and by determination of self diffusion coefficients (D^*) using radioactive tracers, and other techniques such as N.M.R. (8). The various diffusion coefficients should be consistent with relationships of the type,

$$\tilde{D}_i = D_i^* \frac{d \ln a_i}{d \ln x_i} \tag{3}$$

and so a check on the measurements is often possible. The electronic properties of the S.S.E. can also be determined by well established four-point techniques, and it is often possible to combine these measurements with investigations into the e.m.f. composition relationship using the Van der Pauw method (9). Finally the performance of the S.S.E. has to be assessed in the appropriate electrochemical cell.

IV. PROPERTIES OF SELECTED SOLID SOLUTION ELECTRODES

A. Incorporation of Oxygen

The non-stoichiometric fluorite phases, UO_{2+x}, CeO_{2-x} were among the first S.S.E. electrodes to be investigated. The e.m.f. composition data was first reported (10) in 1962 and the associated chemical diffusion data measured in 1970 (11) using solid state cells of the following type at elevated temperatures (800 - 1000°C),

$$Cu, Cu_2O \;/\; Zr_{0.85}Ca_{0.15}O_{1.85} \;/\; UO_{2+x}$$

the relevant data are collected together in Fig. 3 and attention is drawn to the large values of $\tilde{D}_o$ and their variation with composition. At these high temperatures there was sufficient plastic flow to ensure good interfacial contacts and the rate limiting step was usually the ionic flux through the solid electrolyte. While these data are of importance for the fabrication of ceramic nuclear fuel elements they are not applicable to the problem of electrochemical energy storage.

More relevant are the investigations of Kudo et al (12). Examination of their paper indicates that the perovskite solid solutions $Nd_{1-x}Sr_xCoO_{3-x}$ can behave as solid solution electrodes in alkaline solutions at 25°C. A typical voltage composition curve obtained under dynamic conditions is depicted in Fig. 4, and the effective diffusion coefficients were in the range 10^{-11} - 10^{-14} cm^2/s. The low oxygen capacity of the system and the relatively low diffusion coefficients reported make this non-stoichiometric oxide unsatisfactory for an electrochemical storage electrode.

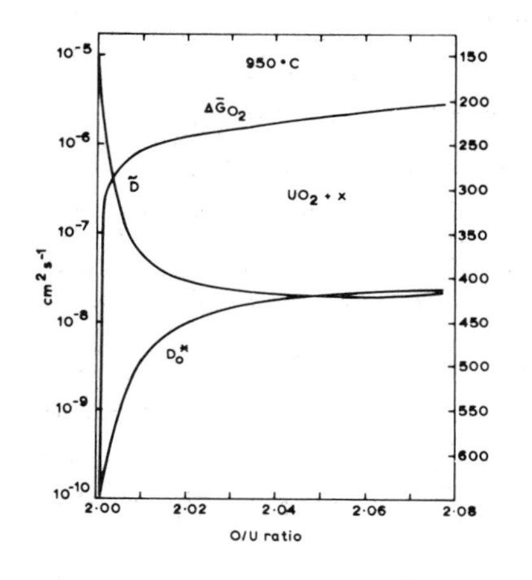

FIG. 3. Thermodynamic and Transport Properties of Oxygen in UO_{2+x} as a Function of Composition

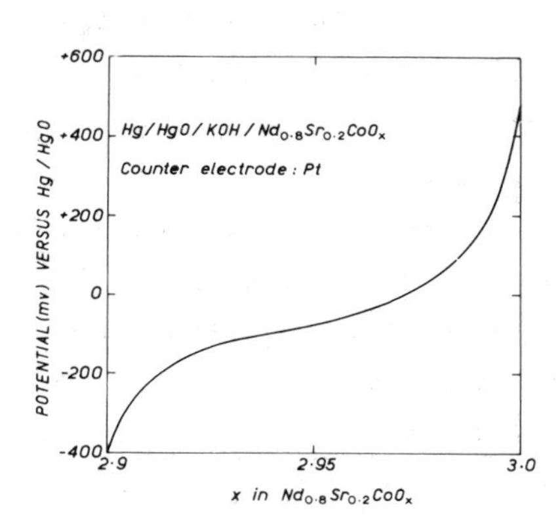

FIG. 4. Potential Composition Curve for $Nd_{0.8}Sr_{0.2}CoO_x$

This type of behavior however, has obvious implications for the proposed application (13) of this material as an oxygen electrode.

B. *Incorporation of Hydrogen*

It is well known that large quantities of hydrogen can dissolve in certain metals and alloys and often the hydrogen diffusion coefficient is of the appropriate magnitude (10^{-7} - 10^{-6} cm^2/s) for the system to be considered as a viable solid solution electrode. Many of the simple binary metal hydrides are too stable thermodynamically to be used in electrochemical systems incorporating aqueous electrolytes. However, using nickel as an alloying element it is possible to produce ternary hydride systems in which $\Delta\overline{G}_{H_2}$ has values appropriate for use in aqueous electrolytes. La Ni_5H_x for example dissolves 6 atoms of hydrogen per mole $LaNi_5$ at 25°C and the corresponding hydrogen pressure is only 2.5 atm. Moreover NMR investigations (8) suggest a high proton mobility. This material and other rare earth metal alloys are being investigated as hydrogen storage electrodes (14) but detailed results remain to be published. Information, however, is available for another system, $Ti_2NH_{2.5}$ which has been successfully employed as an electrode (15). In the presence of small quantities of the second phase TiNiH approximately two atoms of hydrogen per mole Ti_2Ni can be rapidly and reversibly incorporated into the electrode. The O.C.V. composition curve for this electrode is shown in Fig. 5 which also includes for comparison the emf composition curve for palladium hydride which also exhibits a high proton diffusivity (5 x 10^{-7} cm^2/sec) at room temperature (6).

Protons can also dissolve in non-stoichiometric compounds such as oxides but very little quantitative data are available (16). It is known for example that protons can be incorporated into MnO_{2-x} (17), and Gabano (18) has measured a value of D_H in MnO_{2-x} of 10^{-19} cm^2/s at 25°C. Higher values (10^{-10} - 10^{-11} cm^2/s) have been reported (19) for proton diffusion coefficients in nickel hydroxide.

C. *Incorporation of Copper and Silver*

The use of Ag_2S as a solid solution electrode has been exploited for many years particularly by Wagner and Rickert (4) to provide many elegant experiments which illustrate the various measurement techniques possible in solid state electrochemistry. The high temperature form of Ag_2S can be deformed relatively easy by small stresses and so excellent solid/solid interfaces (e.g., Ag_2I/Ag_2S or $RbAg_4I_5/Ag_2S$) can be prepared. This facility when combined with the remarkably high $\tilde{D}_{Ag}$ (10^{-2} - 10^{-1} cm^2/s) ensures that very high currents can be passed across this solid/solid interface. However, although the transport

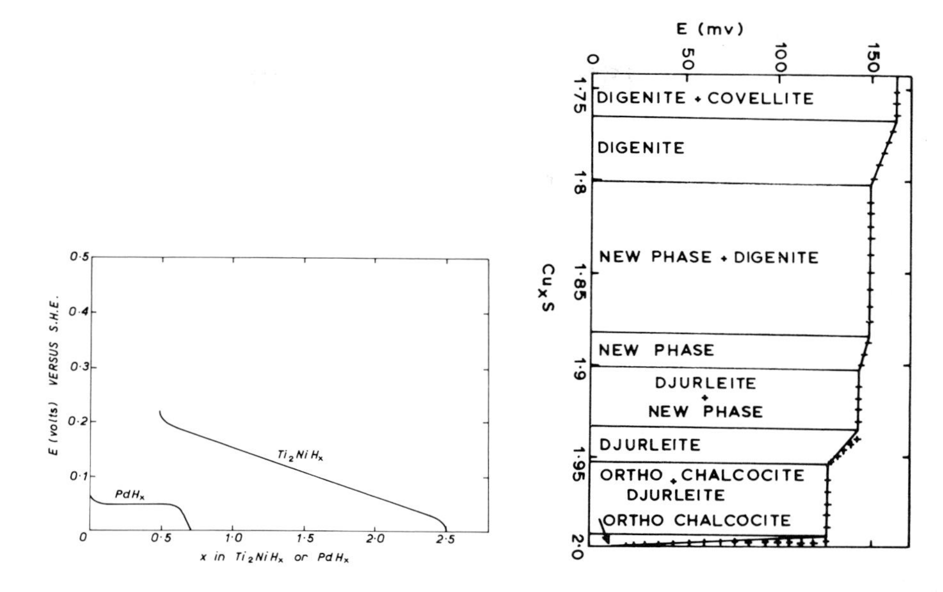

FIG. 5. Potential Composition Curve for $Ti_2Ni\ H_x$ and Pd_x

FIG. 6. Potential Composition Curve for Cu - S System at 25°C

properties of Ag_2S are excellent it has a poor storage capacity because of the limited range of stoichiometry in Ag_2S ($Ag_{2.000}S$ - $Ag_{2.002}$). Similar comments apply to other silver electrodes, e.g., Ag_2Se, and Ag - S - Se - P - O solid solutions (20).

The copper chalcogenides can also exhibit high electronic and ionic conductivities. Moreover, the ranges of composition available for the incorporation of copper is usually higher than in the corresponding silver compounds and of course copper has the added advantage of being lighter and cheaper. The thermodynamics of the Cu - S system have been studied at ambient temperatures by Rickert (21) using aqueous electrolytes and the measured emf composition curve is shown in Fig. 6. A notable feature is the relatively flat potential composition curve over much of the range studied. Also, using an electrochemical technique, Etienne (22) has reported a value for the copper diffusion coefficient in Cu_2S of approximately 10^{-10} cm^2/s. This low value would prevent Cu_2S being used as an S.S.E. at ambient temperatures but at higher temperatures the copper mobility is certainly sufficient.

D. *Incorporation of Alkali Ions*

(i) Transition metal bronzes with one dimensional channels. The alkali metal tungsten and vanadium bronzes (e.g., Na_xWO_3 and $Na_xV_2O_5$) have structures which contain channels of various diameters in which the alkali ions reside. The channels in the tetragonal and hexagonal structures should be large enough to permit translational motion of the alkali ions and so these materials have been investigated as possible S.S.E. as they also exhibit high electronic conductivities. Thermodynamic and kinetic measurements made with cells of the type:

Na/Propylene carbonate (NaI)/Na_xWO_3

or

Na/beta alumina/Na_xWO_3;

quickly revealed (23) that although the ΔG_{Na} values were very promising (see Fig. 7), the chemical diffusion coefficients were very low ($\sim 10^{-15}$ cm^2/s) which prevented their application as S.S.E. It appears that transport along the tunnels can be impeded by the presence of impurity ions, or more likely crystallographic imperfections exist which effectively block the tunnels. High resolution electron micrographs (24) of other tetragonal bronze structures reveal complex distorted arrangements of the WO_6 octahedra in contrast to the simple idealized structure originally proposed for these materials. It is relevant to note that disappointing results have also been obtained for ionic transport in other structures having relatively larger tunnels such as hollandite ($K_{1.6}Mg_{0.8}Ti_{7.2}O_{16}$).

(ii) Graphite and Fluorographite. It is well

known that graphite can intercalate alkali metals and emf data (25) for the potassium graphite system are included in Fig. 7. The maximum potassium content is represented by the formula C_8K, and attention is drawn to the high activity of potassium in this phase. In contrast, the intercalation compound C_8CrO_3, prepared by Armand (26) behaves as a very deep electron sink and consequently lithium atoms incorporated into this structure have a very low activity (~ 3.9V). The range of stoichiometry in this compound, $Li_x C_8CrO_3$ appears to be very large and the lithium diffusion coefficient approaches 10^{-6} cm^2/s. The principal problem with this type of compound is the additional incorporation of solvated species into the compound when the material is used as an electrode in electrochemical cells incorporating organic electrolytes such as propylene carbonate. It should also be noted that the fluorographite (CF) n electrode probably behaves as a solid solution electrode during the initial stage of its reaction with lithium. This suggestion (27) could explain why the voltage developed by Li CF_x cells is considerably less than that calculated for the Li-LiF couple and also why the discharge characteristics are markedly influenced by the organic electrolyte used in the cells (28).

(iii) Di-chalcogenides. The metal di-chalcogenides (MX_2) of Groups IV, V, VI adopt a layer structure (CdI_2) in which slabs of XMX layers are held together by Van der Waals bonding. Recent work (29, 30, 31) has confirmed that alkali metals can be rapidly and reversibly intercalated within the Van der Waals layer at ambient temperatures and emf composition results (31) for the incorporation of sodium into single crystals of Ti_xS_2 are depicted in Fig. 8. The emf results shown were obtained after 3-4 cycles and were slightly dependent upon the Ti/S ratio in the original crystals. The Ti/S ratio, however, had a profound influence upon the alkali metal chemical diffusion coefficients as measured by electrochemical techniques. The excess titanium residing in the normally vacant Van der Waals layer must develop strong bonds between the adjacent sulphur layers as well as blocking sites for the transport of the alkali metal.

Lithium can also be rapidly and reversibly electro-intercalated into TiS_2 (31). It is interesting to note that the voltage differential for the incorporation of 1 g atom of lithium is only 0.5 V (2.4 - 1.9V) and thus the theoretical specific energy density (500 WH/Kg) is even more favorable than that calculated for the Na - TiS_2 system (400 WH/Kg). It is interesting to note that TiS_2 has recently been successfully employed (32) as a reversible solid solution cathode in the high temperature cell LiAl/KCl - LiCl/TiS_2. Niobium diselenide has also been used (29) as a solid solution electrode for lithium in similar cells using organic electrolytes.

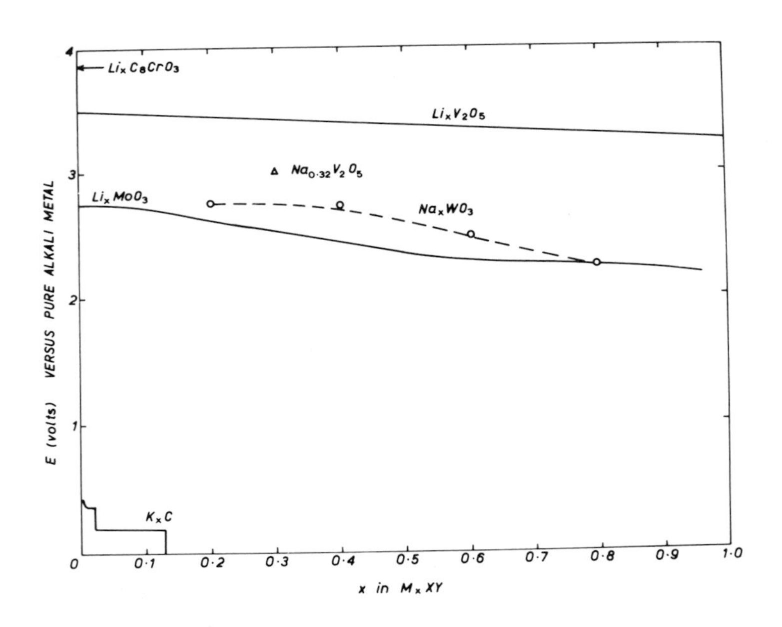

FIG. 7. Potential Composition Curves for Alkali Metal Incorporated in Selected Non-Stoichiometric Electrode Materials

The transition metal trichalcogenides MX_3 also intercalate lithium (29, 33) to form compounds of the type Li_3MX_3. At present, detailed emf and transport data for these compounds are not available but certain of these components appear to be able to function as cathodes in secondary cells (29) using propylene carbonate as the electrolyte.

(iv) Beta ferrites. Potassium beta ferrite ($K_{Hx}Fe_{11}O_{17}$) is one of the isomorphous iron analogues of beta alumina and exhibits electronic as well as ionic conductivity. Additional potassium ions can be incorporated into the structure and is charge compensated by the formation of Fe^{2+}. Emf composition relationships have been obtained using either propylene carbonate (25 - 60°C) or beta alumina (50 - 300°C) as the electrolyte phase (9). At this temperature the current density (100 uA/cm^2) used in the coulometric titrations was limited by transport of the potassium ions across the ceramic electrolyte/electrolyte interfaces. These interfaces are difficult to prepare because the components do not exhibit the plastic deformation which is characteristic of, for example, AgI/Ag_2S interfaces. Although this system is interesting insomuch as one can investigate the ionic conductivity as a function of charge carrier density it is not of course suitable for a S.S.E. in a high energy density battery.

(v) Other oxides. A variety of non-stoichiometric oxides and sulphides have been employed as cathodic depolarizers in organic electrolyte primary batteries usually as additions to graphite and other positive plate materials. There have been reports in which certain of these materials, e.g., MoO_3, MnO_2 V_2O_5 (e.g. 34) have been the only constituent present in the electrode and typical results are shown in Fig. 8. The conventional electrochemists' interpretation of these results is to assume a cathode reaction of the type:

$$MoO_3 + 2Li \rightarrow MoO_2 + Li_2O. \quad (4)$$

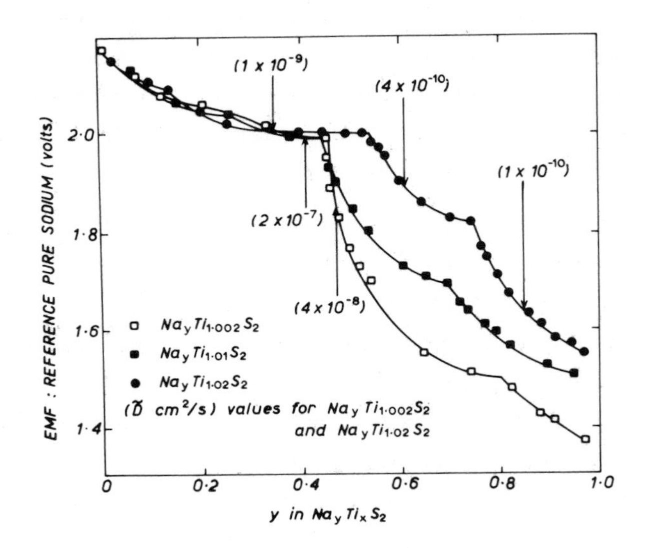

FIG. 8. Potential Composition Curves for Na - Ti_xS_2 System

However, thermodynamic data indicate that the maximum voltage to be expected for reactions of this type is about 2V. A more plausible explanation is to assume that lithium is dissolving in the MoO_3 host lattice and the cell voltage reflects the activity of lithium in this non-stoichiometric phase. The transport of lithium in these compounds appears to be sufficiently high for their application as S.S.E. although there is little information about the ability of these materials to tolerate repeated cycling. It is also worth mentioning that V_2O_5 and MoO_3 may be regarded as layered structures which obviously favor the rapid transport of lithium.

V. DISCUSSION

The concept of using S.S.E. in secondary

batteries is relatively new and very little data are available to assess their future development. The favorable properties of hydrogen in selected alloys and of lithium in titanium disulphide are already being exploited in the development of new battery systems, and it appears likely that other technologically interesting materials will be prepared. It is appropriate therefore to conclude with a few comments about the theoretical models relevant to the specifications suggested in section II for S.S.E.

A. *Electronic Conductivity*

It is impossible to predict, a priori, the band structures of complex oxides and sulphides. The approach of Goodenough (35) based on chemical bonding and crystallography however, does provide a framework which is useful for the preliminary selection of materials likely to exhibit high electronic conductivities.

B. *Ionic Transport*

The factors controlling fast ion transport have been discussed in the Belgirate Conference (1) and many reviews are available (36, 37, 38). Recent experience (39, 40) suggests that it is very difficult to prepare new solid electrolytes with conductivities comparable to that exhibited by B-Al_2O_3. The restrictions are somewhat relaxed for S.S.E. with the possibility of introducing covalent and metallic type bonding and it should be easier to prepare compounds exhibiting both high ionic and electronic conductivity. It should be noted that the possibility of varying the density of ionic charge carriers in S.S.E. provides an elegant method of testing many theoretical models for fast ion transport.

C. *Free Energy Composition Relationships*

The magnitude of $\Delta\overline{G}$ of the dissolved electroactive species and its variation with composition is not easy to predict because of lack of knowledge of both the relevant electronic levels and the ion-ion interactions. A variety of statistical thermodynamic models have been proposed and recent reviews are available (41).

D. *Interfacial Problems*

The detailed mechanism of the injection of ions from a liquid or solid electrolyte into a solid solution electrode and vice-versa remains obscure and it certainly requires systematic investigation in which the environment of the electro-active species is varied both in the electrolyte and electrode phase. Recent relevant reviews on the topic include (42, 43, 44).

Finally it should be noted that the present survey has been principally concerned with the application of S.S.E. in secondary batteries. Other applications include the use of S.S.E. in electro-chromic passive light display devices. Cells of the type Ag/Rb Ag_4I_5/MoO_3; SnO_2, can be used to reversibly inject ions (H^+, Ag^+, Li^+) into thin films of MoO_3 or WO_3 electrodes. The color centers produced can be used as a display unit (45, 46) requiring only a small energy input (cf. liquid crystals).

REFERENCES

1. B.C.H. Steele, Fast Ion Transport in Solids, Ed. W. Van Gool (North Holland, Amsterdam, 1973), p.103
2. B.C.H. Steele, Heterogeneous Kinetics at Elevated Temperatures, ed. G.R. Belton and W.L. Worrell (Plenum Press New York, 1970) p.135
3. Physics of Electrolytes, Vols 1 and 2, ed. J. Hladik, Academic Press (New York 1972)
4. Einfuhrung in die Elektrochemie fester Stoffe, H. Rickert (Springer Verlag, Berlin, 1973)
5. The Chemistry of Imperfect Crystals, F.A. Kroger (North Holland, Amsterdam, 1974) Vol. 3.
6. H. Zuchner and N. Boes, Ber. d. Bun. Gesellschaft 76, 1972, 783.
7. W.F. Chu, H. Rickert, and W. Weppner, Fast Ion Transport in Solids ed. W. Van Gool (North Holland, Amsterdam, 1973) p.181
8. T.K. Holstead, J. Solid State Chem. 11, 1974, 114
9. G.J. Dudley, B.C.H. Steele and A.T. Howe, J. Solid State Chem to be published.
10. T.L. Markin and R.J. Bones U.K.A.E.A. Report A.E.R.E. R 4042 1962
11. B.C.H. Steele and C.C. Riccardi, Proc. Int. Symp. on Metallogical Chemistry, ed. O. Kubaschewski (H.M.S.O. London, 1972) p.123.
12. T. Kudo, H. Obayashi, and T. Gejo, J. Electrochem. Soc. 122, 1975, 159
13. U.S. Patent 3, 804, 674 (16/4/74)
14. P.A. Boter, Abs. 8 Symp. on Novel Electrode Materials Sept. 25/26th Brighton U.K.
15. M.A. Gutjar, H. Buchner, K.D. Beccu, and H. Saufferer, Power Sources 4 ed. D.H. Collins, (Oriel Press, Newcastle, U.K., 1973)
16. L. Glasser, Chem. Reviews 75, 1975, 21
17. J. McBreen, Electrochimica Acta, 20, 1975, 221
18. J.P. Gabano, J. Seguret, and J.F. Laurent, J. Electrochem. Soc. 117, 1970, 147.
19. D.M. MacArthur, J. Electrochem Soc. 117, 1970, 729
20. T. Takahashi, E. Nomura, and O Yamamoto, J. Appl. Electrochem, 3, 1973, 23

21. H.J. Mathiess and H. Rickert, Z. Physik. Chemie. N.F. 79, 1972, 315
22. A. Etienne, J. Electrochem. Soc. 117, 1970, 870
23. B.C.H. Steele, Mass Transport Phenomena in Ceramics, ed. A.R. Cooper and A.H. Heuer (Plenum Press, New York, 1975) p.269
24. S. Iijima and J.G. Allpress, Acta Cryst. A30, 1974, 22
25. S. Aronson, F.J. Salzano, and D. Bellafiore, J. Chem. Phys. 49, 1968, 434
26. M.B. Armand, Fast Ion Transport in Solids, ed. W. Van Gool, (North Holland, Amsterdam, 1973) p.665
27. M.S. Whittingham, J. Electrochem. Soc. 122, 1975, 526
28. R.G. Gunther, Power Sources 5, ed. D.H. Collins, (Academic Press, London, 1975) p.729
29. J. Broadhead and F.A. Trumbore, Power Sources 5, ed. D.H. Collins, (Academic Press, London, 1975) p661
30 M.S. Whittingham and F.R. Gamble, Mat. Res. Bull, 10, 1975, 363
31. D.A. Winn and B.C.H. Steele, Mat. Res. Bull, 11, 1976 to be published.
32. D. Inman and Y.E.M. Mariker, Abs. 2, Symp on Novel Electrode Materials, Sept. 25/26th, Brighton, U.K.
33. R.R. Chianelli and M.B. Dines, Inorg. Chem.14, 1975, 2417
34. F.W. Dampier, J. Electrochem. Soc. 121, 1974, 656
35 J.B. Goodenough, Prog. in Solid State Chem. Vol. 5 ed. H. Reiss (Pergamon Press, Oxford, U.K. 1971) p.145
36 W. Van Gool, Am. Rev. of Materials Science, 4, 1974, 311
37 B.C.H. Steele and G.J. Dudley, M.T.P. International Rev. of Science, Series Two, Inorganic Chemistry Vol. 10, (Butterworths, London, 1975) p.181
38 M.S. Whittingham, Electrochimica Acta 20, 1975, 575
39 J. Singer, H. Kautz, W. Fielder and J. Fordyce, NASA Technical Memo, TMX-71753, (May 1975)
40 W.L. Roth and O. Muller, NASA Technical Memo, CR-134610 (April 1974)
41. B.E.F. Fender, M.T.P. International Review of Science Series One, Inorganic Chemistry Vol. 10, (Butterworths London, 1972) p.243
42. D.O. Raleigh, Electroanalytical Chem. Vol. 6. Ed. A.J. Bard (M Dekker, New York 1973)
43. A.D. Franklin, J. AM. Ceram. Soc. 58, 1975, 465
44 Electrode Processes in Solid State Ionics, ed. M. Kleitz, (Reidel, Dordrecht, to be published.
45. I.F. Chang, B.L. Gilbert, and T.I. Sun, J. Electrochem. Soc. 122, 1975, 955

Chapter 24

SOLID ELECTROLYTES

Robert A. Huggins

Department of Materials Science
and Engineering
Stanford University
Stanford, California 94305

I. INTRODUCTION

In recent years there has been rapidly accelerating interest in a group of materials which, although solid, exhibit many of the properties normally associated with liquid electrolytes. Because of the wide range of potential uses of such materials, exploratory efforts have been undertaken in a number of laboratories aimed at the discovery of new materials which exhibit this type of behavior, and there also has been greatly increased interest in the acquisition of better understanding of the physical phenomena involved in the rapid translation of ions through such solids, this property being the outstanding characteristic of this class of materials.

After a very brief discussion of the nature of an electrolyte and some of the applicational motivation of the study of such materials, this paper will discuss the special characteristics of the structure of solid electrolytes which enables unusually rapid ionic motion, giving particular attention to the material family called "beta alumina".

II. GENERAL CONSIDERATIONS RELATING TO ELECTROCHEMICAL CELLS

A generalized electrochemical cell is shown in Fig. 1. Two phases with different values of chemical potential for some species, i, are separated from each other by a material

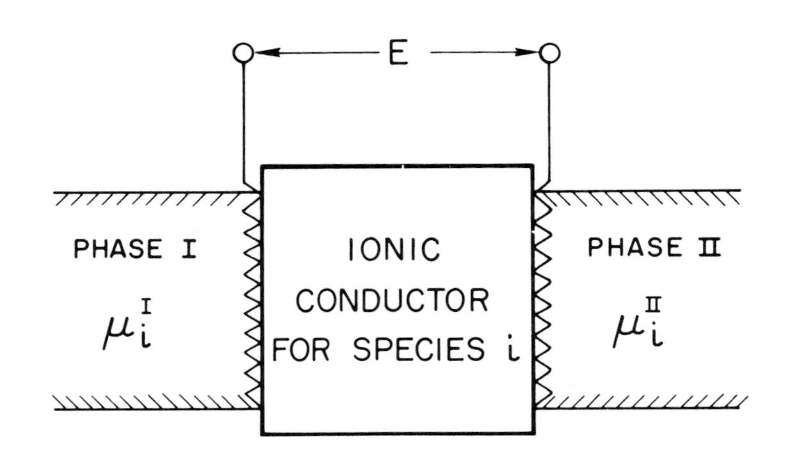

FIG. 1. Schematic Representation of Simple Electrochemical Cell

which is an ionic conductor for species i. Electrical contacts are present at the two interfaces. If equilibrium is established, and the material in the center is an electronic insulator, an electric potential difference, E, should in principal be measurable between the electrodes at the two interfaces in accordance with the Nernst equation if no current is drawn. This simple system acts as an electrochemical transducer, in which a chemical difference applies a driving force for the transport of species i from one side to the other, and under equilibrium (open circuit) conditions, this chemical driving force is exactly balanced by an electrical driving force of equal magnitude in the opposite direction, so that the resultant force acting upon the charged species within the electrolyte is zero.

Such a simple transducer system can act as a battery in which chemical energy is transformed into electrical energy. It can also be used as a sensor whereby electrical measurements are used to deduce the difference in chemical quantities. It can be used as a fuel cell, or as a pump in which chemical species are caused to be transported from one side to the other by the application of an electrical potential difference.

One can visualize the mechanism whereby this electrochemical cell works in a rather simple way. A difference in chemical potential means that there will be a driving force tending to cause chemical species to move from one side to the other. In order that immense electrostatic potentials are not generated, there must be no overall transport of electric charge. This means that either neutral species must move or electrically neutral combinations of species must move from one side of the cell to the other.

One can think of the ionic conductor as acting as a simple filter at whose surface neutral chemical species are divided into their ionic and electronic constituent parts. The ionic parts (ions) can move readily through the ionic conductor, but the electronic parts (electrons or holes) cannot. If the electrodes upon the two interfaces are connected so that there is an electrical path for the electronic species to move from one side of the cell to the other, the transport of neutral combinations is allowed if the charge fluxes of the ionic species through the electrolyte and the electronic species through the external electrical circuit are equal.

III. COMMENTS ON APPLICATIONS

One can, of course, make use of the electronic current in this external circuit to do work, in which case this simple system will be acting as a fuel cell or a "primary" battery. If it is reversible, and species can be pumped back in the other direction by the application of a reverse electric potential across the electrodes, this system then would operate as a "secondary" battery.

It is also quite obvious that if one applies an external potential difference between the electrodes at the two interfaces, a force will be exerted upon the charged species in the electrolyte, causing them to tend to move from one side to the other, so that this arrangement serves as a simple chemical pump.

This simple system can be used for a variety of different purposes, depending upon how it is manipulated and what materials are used to construct it. The critical element is the electrolyte, which must be a relatively good conductor for ionic species and at the same time must not allow passage of electrons or holes. If the electrolyte were to allow both ionic and electronic species to move through it, the electrodes would be effectively shorted out and the externally measurable electric potential difference drastically reduced.

Table 1 includes a list of various types of

TABLE 1

EXAMPLES OF THE APPLICATION OF SOLID STATE ELECTROCHEMICAL TECHNIQUES

Static Emf Measurements

- Free energy of formation of binary and ternary compounds
- Free energy changes accompanying various cell reactions
- Thermodynamics of binary phases
- Determination of limits of stoichiometry of compounds
- Phase diagram determination
- Thermodynamics of phase transformations
- Effective mass of electrons or holes in semiconductors
- Solubility of gases in liquids

Time-dependent Emf Measurements

- Phase boundary migration kinetics
- Supersaturation required for nucleation within and upon solids and liquids

Combinations of Emf and Current Measurements

- Diffusion in liquid and solid metals and mixed conductors
- Transport of both ionic and electronic species across phase boundaries
- Kinetics of condensation and vaporization processes
- Oxidation and reduction reactions on solid surfaces
- Thermodynamics of gaseous species
- Studies of the mechanism of catalysis
- Ionic and electronic partial conductivities in mixed conductors
- Structure of electrode-electrolyte interfaces

Technological Applications

- Batteries
- Fuel cells
- Catalysts
- Electrochromic display elements
- Variable resistors
- Thermoelectric devices
- Memory elements
- Solute valence control in semiconductors and ionic solids
- Purification of liquids and gases
- Measurement and control of liquid and gas compositions

---

applications for such electrochemical transducer systems. From this list one can see that there is a large number of potential uses for both scientific and technological purposes.

Mention will be made here of only one such use, which is attracting a great deal of attention at the present time. This is the utilization of electrochemical cells for battery systems. This is obviously a very old story, as we all know that electrochemical batteries have been in actual use for many decades. There is a new wind blowing through this area at the present time, however, and we seem to be on the verge of major advances in the performance of battery systems which should, in turn, have a large influence upon a number of important technologies. These include electric vehicles and large scale energy storage related to utility load leveling and potential solar energy conversion.

The possibility of major advances in battery systems can be seen by consideration of Fig. 2, in which the maximum theoretical specific energy (the energy that can be stored per unit weight) is plotted versus equivalent weight for a number of possible electrochemical couples. The data used in this figure, which was presented some time ago by Cairns and Shimotake (1), can be calculated from thermodynamic information and represent the maximum theoretical values that could be obtained from such systems assuming 100% energy efficiency, and that the total weight of the battery system is made up only of the reactants themselves. That is, no allowance is made for the weight of the container, the electrolyte, connectors, and all of the other paraphernalia that is normally present in a battery system. Most practical batteries have values of specific energy that are between 20 and 35% of the maximum theoretical calculation.

It can be seen that there is a wide disparity between the theoretical maxima for various systems. The common

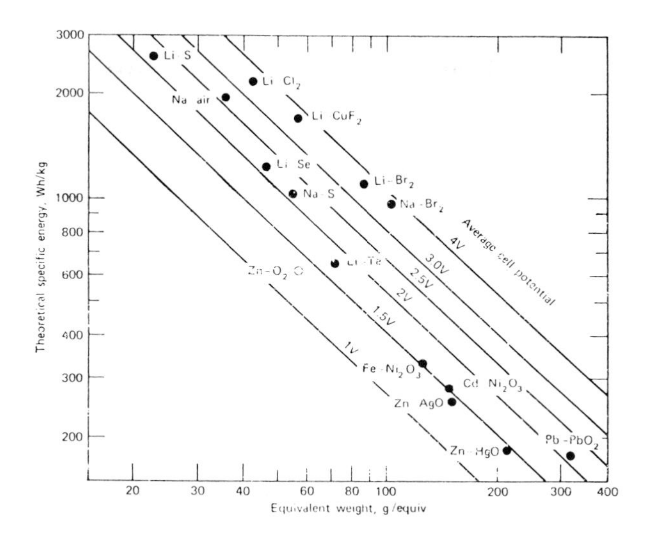

FIG. 2. Maximum Theoretical Specific Energy Versus Equivalent Weight for a Number of Possible Electrochemical Couples (1)

lead-acid cell is near the bottom of the figure, and it is quite apparent that if one could work with some of the systems whose data lie near the top of the figure, a great deal more energy could be stored per unit weight than is the case with current battery systems. It is also evident that these more advantageous chemical combinations tend to involve alkali metals on the one hand, and very oxidizing species, such as oxygen or sulfur, on the other. It should be readily apparent that this is not an easy technological task, particularly as it is not feasible to use an aqueous electrolyte with such high energy cells, for the reactive constituents would immediately decompose water.

This, of course, has been known for a long time. What is new is that just about a decade ago workers at the Ford Motor Company announced the invention of a radically different type of electrochemical cell is based upon the use of a solid, rather than a liquid, electrolyte. This solid, known as "beta alumina", was shown to have some truly remarkable properties. It is stable in the presence of both sodium and sulfur, and despite being a solid with a melting point of over 2000°C, behaves as a sodium ion-conducting electrolyte. The sodium ions move through its crystal structure very readily, almost as fast as though it were a liquid. On the other hand, it is a hard, white ceramic which is electronically insulating.

The discovery of this material and recognition of its special properties led to the introduction of the concept of an inverted type of electrochemical cell, with liquid electrodes and a solid electrolyte. Although we shall not dwell upon them here, this type of design has several very important potential advantages with respect to current density and rechargeability. It also opens the door to the consideration of the use of a number of the high specific energy electrochemical systems, if solid electrolytes can be found with sufficiently high values of ionic conductivity for one of the active ingredients and sufficient stability under the necessarily severe oxidizing and reducing conditions found in these high energy systems.

IV. GENERAL CHARACTERISTICS OF USEFUL SOLID ELECTROLYTES

As mentioned in the previous section, an important parameter related to the use of an electrolyte in a battery system is its ionic conductivity. If the conductivity is low, there will be a large internal resistance and such a cell will have a limited power capability, as well as a low efficiency during recycling.

The most striking feature of the beta alumina material was its extremely high ionic conductivity at moderate temperatures. The recognition of this has led to a great deal of interest in ionic conductors, and the wide range of possible technological applications is causing a great deal of excitement in the research and development communities at the present time.

Ionic conductivity in solids has been known for many years and the study of the motion of ions and ionic defects in crystalline solids has long been a subject in good standing in the solid state research community. It is now evident, however, that materials such as beta alumina are drastically different from the more conventional ionic conductors which have received most of the research attention in the past. This is illustrated in Fig. 3 in which the conductivity-temperature product is plotted versus the reciprocal of temperature for several materials which are representative of different classes of behavior. KCl is an example of materials with reasonably dense crystal structures in which ionic motion occurs by transport of vacancies. AgCl is an example of a group of materials in which ionic transport occurs but the motion of interstitial (Ag in this case) species. The conductivity is typically much higher in the

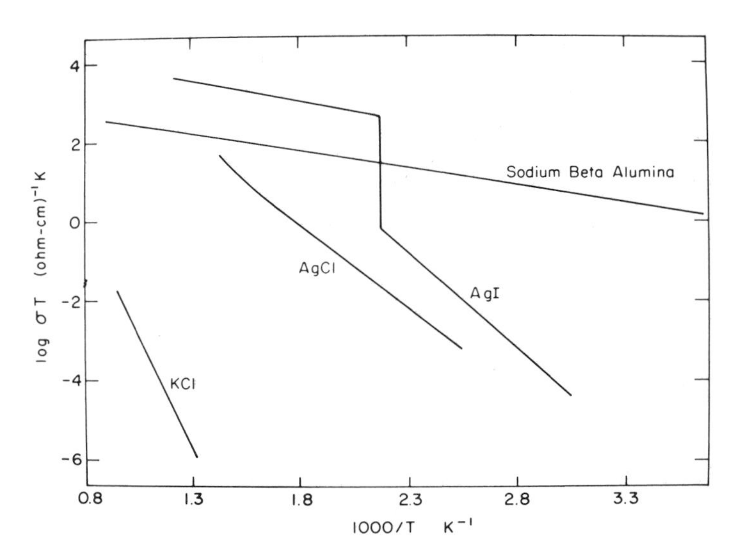

FIG. 3. Temperature-Dependence of the Ionic Conductivity of Several Representative Materials

materials with interstitial motion than it is in materials in which ions move by the vacancy transport mechanism. Also, one can see that the temperature dependence, which is characterized by the activation enthalpy, is considerably greater in the latter case. AgI is one of a moderate number of materials which were found some years ago to behave at lower temperatures in a manner similar to that of the interstitial conductors. At an elevated temperature (146°C), a phase transformation takes place and the high temperature phase has an unusually large ionic conductivity which is not very dependent on temperature. In fact, the ionic conductivity of solid silver iodide is extremely large, and its variation with temperature near the melting point is illustrated in Fig. 4. It can be seen that the ionic conductivity is actually greater in the solid just below the melting point than it is in the liquid form of the same material. That is, the conductivity actually decreases when silver iodide melts, and the structure changes from that of a relatively rigid crystal to that of a liquid.

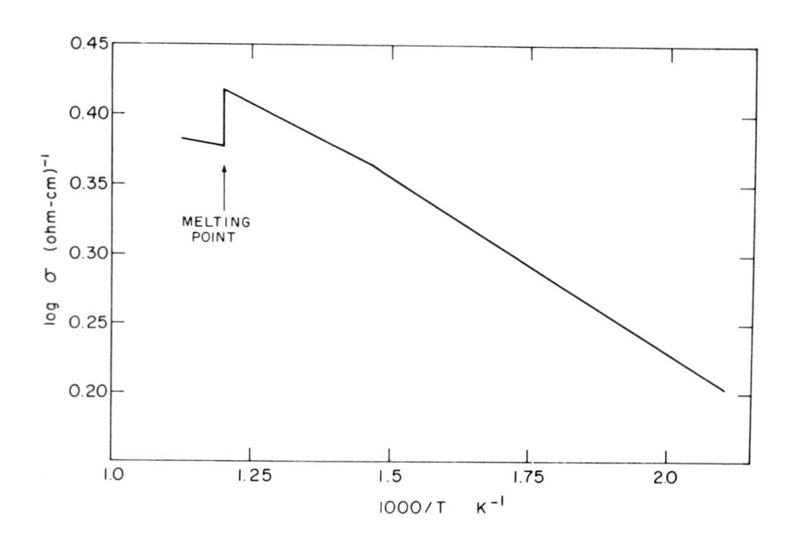

FIG. 4. Temperature-Dependence of the Ionic Conductivity of AgI Near the Melting Point

Turning back to Fig. 3, it is seen that the conductivity of sodium beta alumina is somewhat lower than that of the high temperature phase of AgI, but they have about the same temperature dependence. A very important difference, however, is that the high conductivity of the beta alumina persists to unusually low temperatures, rather than decreasing suddenly due to the onset of a phase transformation. This indicates that the physical phenomena involved in ionic motion in the high temperature phase of AgI and in the beta alumina are similar, and distinctly different from those involved in the conductivity process in the other two groups of materials whose mechanisms involve conventional interstitial or vacancy motion.

While only a few materials have been used here for purposes of illustration, there have been several reviews in the last few years which have compiled information about many other materials in each of these several groups (2-5).

The importance of this difference in mechanism can be illustrated by considering the values of ionic conductivity of these various materials at room temperature. If one extrapolates the data for NaCl it is found that its conductivity at 25°C is about 14 orders of magnitude lower than that of the beta alumina, an immense difference.

The unusual behavior of these materials is directly related to the details of their crystallographic structure. They have unusual crystal structures which are characterized by the existence of atomically-sized tunnels permeating a relatively rigid skeleton. These tunnels are dilutely populated by the mobile species, which reside in relatively shallow potential wells, and thus can rather readily move through them.

V. BETA ALUMINA

Fig. 5 is a schematic representation of the

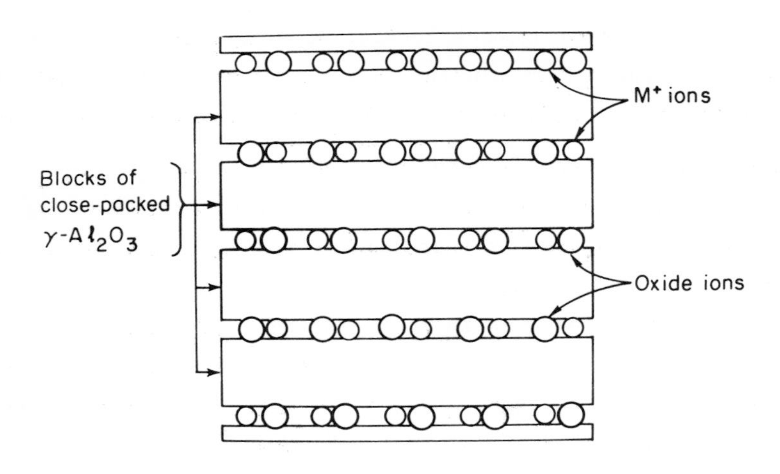

FIG. 5. Schematic Representation of the Structure of Beta Alumina

crystallographic structure of beta alumina, which has the nominal formula $MAl_{11}O_{17}$. The structure can be described in terms of rather densely packed blocks of material in which the oxygen atoms are arranged in a face centered cubic array, and the aluminum ions occupy interstitial sites equivalent to those found in the spinel structure. These spinel-like blocks are four close-packed oxygen layers thick, and are separated by "bridging layers" which are only two-thirds full. These layers contain one-third as many oxygen ions as the normal close-packed oxygen layers and an equal number of alkali metal (sodium) ions. The other third of the sites which would be occupied if this were a normal close-packed layer, are empty. The oxygen ions in these bridging layers are held rigidly in place by the existence of tetrahedrally coordinated aluminum ions both above and below them. On the other hand, the sodium ions can be thought of as existing in a set of crystallographic tunnels which are only half full. It is through these dilutely occupied tunnels, which form an interconnected hexagonal array in the plane of the bridging layer, that the sodium ions move with unusual mobility. Thus, this special crystallographic structure causes this material to have an extremely anisotropic ionic conductivity, and the particular nature of the tunnels which exist within the bridging layers makes it possible for the sodium ions to be extremely mobile, leading to the unusually high values of ionic conductivity in this sandwich-type structure.

Measurements have been made of the transport properties of sodium beta alumina over an extremely wide range of temperature by the use of three independent but related techniques, radiotracer diffusion, ionic conductivity, and dielectric loss measurements. The radiotracer measurements were made at the Ford Motor Company (6), as were the dielectric loss measurements (7), whereas the conductivity measurements were made at Stanford (8).

The results of these measurements are shown in Fig. 6. It can be seen that they correspond exceedingly well, despite the fact that they were made on different samples and in

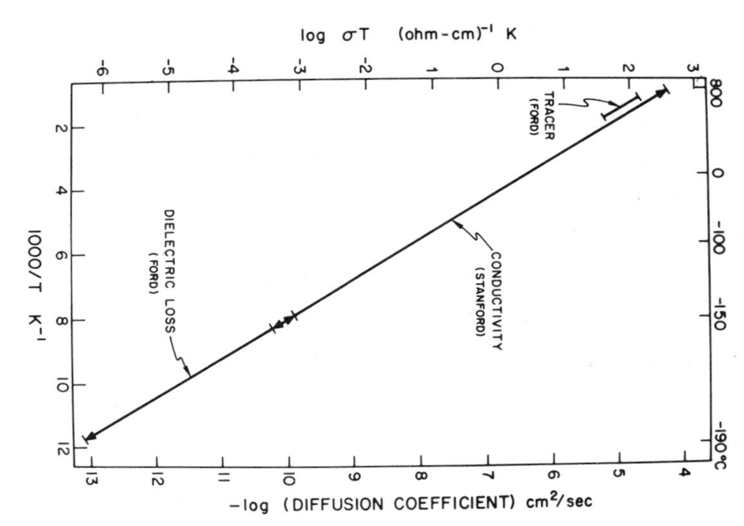

FIG. 6. Temperature-Dependence of the Ionic Conductivity and the Self-Diffusion Coefficient of Sodium Beta Alumina, as Evaluated by the Use of Three Different Techniques

different laboratories. These measurements extend over an unusually wide range of temperature for an ionic transport process, and the consistency between the results of the three measurement techniques as well as the linearity of this behavior over such a wide span of temperature indicate that a single physical process is being observed in all three cases. They also give a considerable degree of confidence in the value of the activation enthalpy and preexponential constant which can be derived from the data by utilizing the Arrhenius expression

$$\sigma = (\sigma_o/T)\ \exp(-\Delta H/RT)$$

It should be pointed out that the disparity between the radiotracer diffusion results and those obtained by direct conductivity measurements is exactly what is expected, since correlation effects, expressed as the Haven Ratio, must be taken into account in the interpretation of radiotracer data. Calculations were made (9) of the Haven Ratio, assuming that the sodium ion motion occurs by the interstitialcy mechanism through the tunnels in the bridging layer. The calculated value, 0.6, agrees very well with that which was found experimentally.

One of the characteristics of this type of crystal structure is that it is possible to perform solid state ion exchange, in which one species of mobile cation can be exchanged for another by interdiffusion between the beta alumina and an external phase, such as a molten salt. This occurs quite readily at moderate temperatures where the diffusion coefficient of the cationic species is very high, whereas the rest of the structure is relatively rigid. It has been shown experimentally (6) that a number of different cations can be introduced into this structure in this manner.

By making use of the ion exchange method as well as the technique of "electrochemical pumping" (10) ionic conductivity measurements have been made on single crystals of beta alumina containing a variety of different cations (11). The results are shown in Fig. 7. One of the interesting features of these data is the fact that there is a most favorable ionic size for rapid diffusion (high conductivity), as both the smallest and the largest ions have lower mobility. The explanation for this in terms of a relatively simple model will be discussed in a later section of this paper.

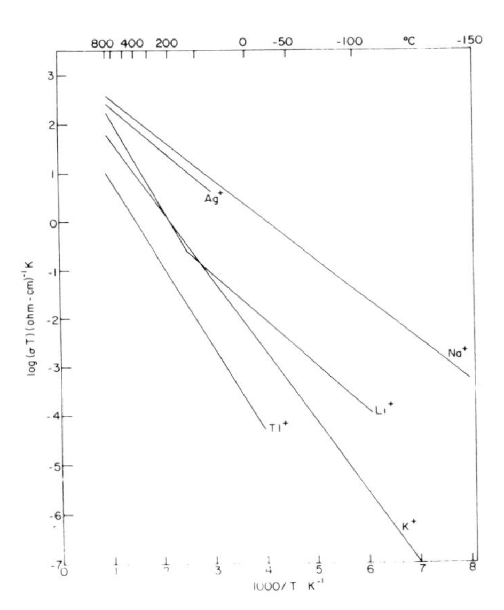

FIG. 7. Temperature-Dependence of the Ionic Conductivity of Single Crystal Beta Alumina Containing Several Different Mobile Cations

Another important characteristic of a material to be used as a solid electrolyte is the ratio of electronic to ionic conductivity. The fraction of the total charge current carried by electronic species is defined as the electronic transference number t_e. It is desirable of course that t_e be as small as possible for a good electrolyte.

Measurements have been made for the case of silver beta alumina (9). It was found that the fraction of the current carried by electronic species is very small and decreases as the temperature is lowered, becoming essentially unmeasurable at temperatures below about 500°C. Thus, the beta alumina family exhibits two unusually good properties, a very high value of ionic conductivity, and also an unusually low electronic conductivity.

Data relating to the conductivity parameters of various ions in single crystal beta alumina are presented in Table 2. Also included are the experimentally obtained values of the Haven Ratio, and it is seen that they are essentially identical to the value of 0.6 which was calculated theoretically. It might also be pointed out that the room temperature conductivity of sodium beta alumina is about $10^{-2}(\text{ohm cm})^{-1}$. This is only about 1 order of magnitude lower than that found in common aqueous electrolytes at the same temperature.

VI. MECHANISM OF FAST IONIC CONDUCTION

One of the clues to the physical mechanism for this very rapid transport in solids was furnished by very careful x-ray diffraction work which can indicate the actual locations of the mobile ions on a time-averaged basis. The results of such work in the case of high chalcocite (Cu_2S) are presented in Fig. 8, in the form of an electron density map (12). The contours indicate the average electron density, and thus the ionic occupation in this structure at various places. It is seen that although the sulfur ions have well defined locations, the copper ions are distributed rather uniformly along paths which extend between what would normally be considered crystallographic sites.

From this kind of information it is reasonably obvious that in such materials part of the structure (the sulfur sublattice in this case) is reasonably normal, whereas ions of

TABLE 2

IONIC CONDUCTIVITY OF SINGLE CRYSTAL BETA ALUMINA CONTAINING DIFFERENT MOBILE IONS

Ion	Temperature Range (°C)	σ_o $((ohm\text{-}cm)^{-1}K)$	Activation Enthalpy (kJ/Mole)	Conductivity at 25°C $(ohm\text{-}cm)^{-1}$	Haven Ratio (D_T/D_σ)
Ag	25 → 800	1.6×10^3	16.6	6.7×10^{-3}	0.61
Na	-150 → 820	2.4×10^3	15.8	1.4×10^{-2}	0.61
K	- 70 → 820	1.5×10^3	28.4	6.5×10^{-5}	-
Tl	- 20 → 800	6.8×10^2	34.3	2.2×10^{-6}	0.58
Li	180 → 800	9.7×10^3	35.8	-	1.0
Li	-100 → 180	5.4×10^1	18.0	1.3×10^{-4}	-

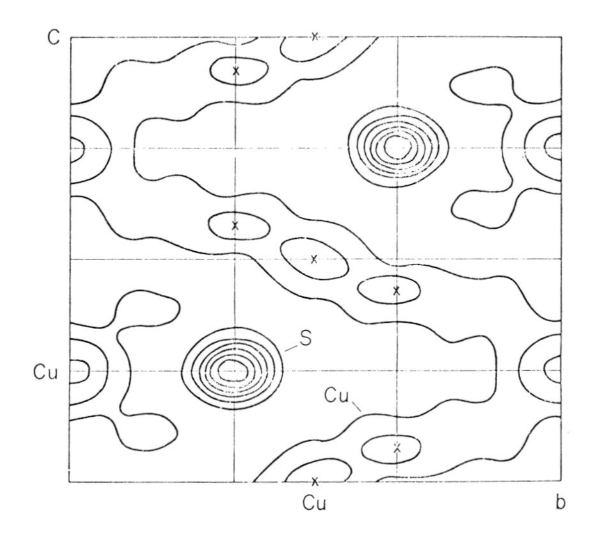

FIG. 8. Electron Density Section ρ(0 x y) Through the Orthohexagonal Cell of High Chalcocite (12)

the other constituent (copper) do not sit upon well defined lattice sites, occasionally jumping from one site to the next. Instead, they flow rather continually through the interstices of the other rigid sublattice. One way of describing such a structure is in terms of a "molten" sublattice permeating a framework defined by the other ions present.

This "molten sublattice" model of fast ion conductors is consistent with thermodynamic data. As it has been pointed out (13) the entropy change upon melting such materials is considerably lower than that for normal materials, indicating that part of the structure has behaved thermally as though it were already molten before reaching the melting point of the structure as a whole.

The prototype for fast ion conductors is the high temperature (alpha) phase of silver iodide. The crystal structure of this phase consists of a body centered cubic array of iodine ions. The silver ions move through the iodine lattice, and there has been a long standing quandary concerning the precise location of the silver ions, since they do not give rise to the type of x-ray diffraction pattern normally expected of a well defined structure.

From considerations of the symmetry of the crystal, it can be shown that there are sites with two-fold coordination, sites with three-fold coordination, and sites with four-fold coordination upon which the silver ions might reside.

A calculation was undertaken a few years ago to try to help understand this problem. Since this is a cubic structure, symmetry considerations allow a great deal of simplification, and one can treat the silver ions as both residing and moving within a set of orthogonal tunnels parallel to the cube edge direction. If one looks along one such tunnel within a structure, it is found that silver ions move between a sequence of pairs of iodines with alternating north-south and east-west orientations. This geometry is shown in Fig. 9.

Calculations were made (14) using the general method of Born and Mayer (15) in which the total interaction energy between a mobile ion, such as silver, and the surrounding, relatively static lattice ions (such as iodine) is expressed as the sum of three terms, relating to electrostatic point charge interactions, dipolar polarization interactions, and overlap repulsion between adjacent closed shell ions.

By use of computer techniques, the total interaction energy between a mobile ionic species arbitrarily placed at any position within the crystal structure and the total lattice of fixed ions can be calculated.

By this method the total energy was found as a function of the cation (e.g., silver) position within the tunnels

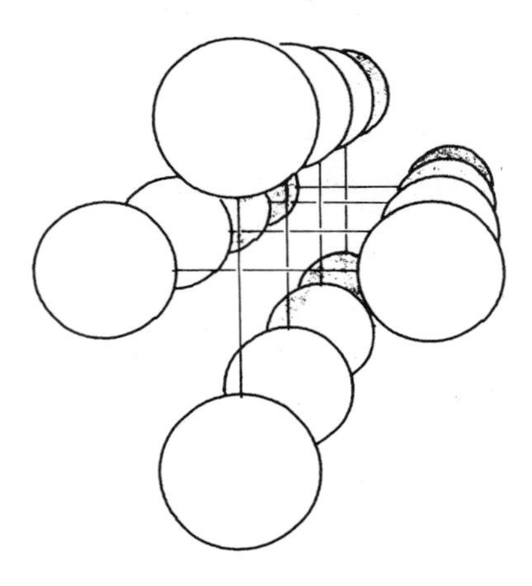

FIG. 9. Arrangement of Anions Along Tunnel Through bcc AgI Structure

which run through the body centered cubic anion lattice, assuming a specific value of cation radius. From this information the minimum energy path through the tunnel could be deduced, and it was found that the location of this path was strongly dependent upon the radius of the cation. The minimum energy paths for small cations tends to deviate in the direction of the nearby anions, whereas the minimum energy paths for larger cations, where the repulsion rather than polarization energy terms are more important, tend to deviate away from the anion pairs as the cation progresses through the tunnel. After obtaining information about the minimum energy (preferred) path through the tunnel for ions of different sizes, it was possible to calculate the energy variation along the path for ions of any given size. The results of this calculation are shown in Fig. 10. It can be seen that both

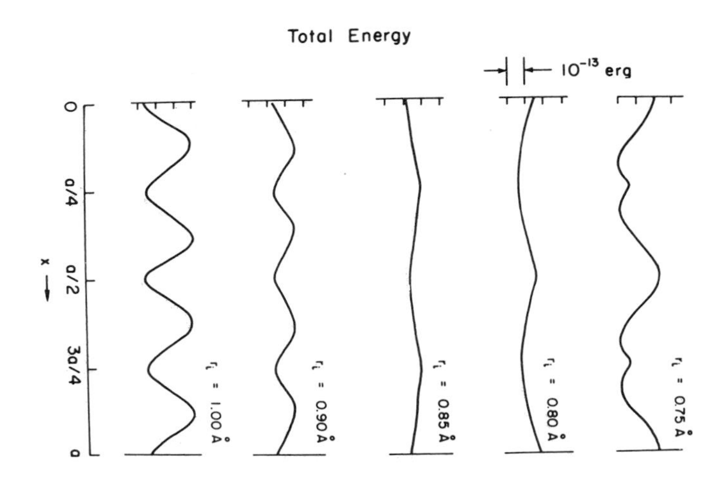

FIG. 10. Variation of Potential Energy Along Minimum Energy Paths for Cations of Various Sizes Within Unit Cell of bcc AgI Structure

quite small and quite large ions have relatively large variations in potential energy along their preferred paths, whereas ions of intermediate size, where the repulsive and attractive energy terms are more evenly matched, tend to have much flatter potential energy profiles. From these data, one can deduce activation enthalpies for ions of different size as shown in Fig. 11.

This theoretical result predicts that for any given crystal structure there will be an optimum ionic size with a minimum value of activation enthalpy for motion, and thus maximum mobility. This, indeed, has been found to be the case for the materials of the beta alumina family (11). In addition a very interesting experiment was performed (16) to evaluate the influence of pressure upon the mobility of ions of different size in the beta alumina structure. It was found that hydrostatic pressure, which reduces the size of the tunnel through which the mobile ion moves, enhances the mobility of lithium ions, which are quite small. On the other hand, the mobility of potassium,

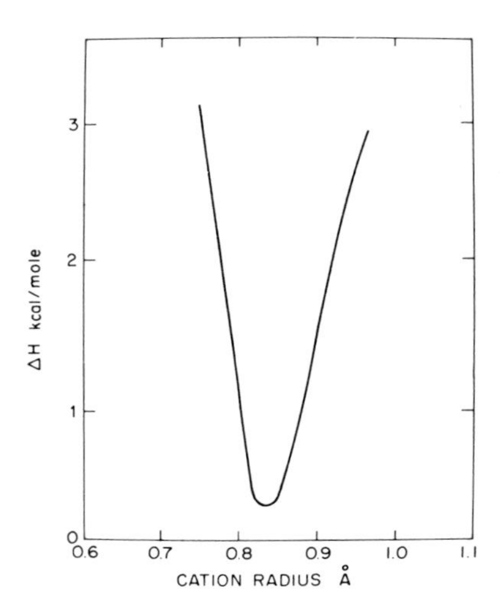

FIG. 11. Calculated Variation of Activation Enthalpy with Cation Radius in the bcc AgI Structure

a relatively large ion, is reduced, as expected from this model. The mobility of sodium ions, which are near the optimum size, was not influenced significantly. More recent experiments in Japan have shown that the mobility of sodium is decreased somewhat under the application of significantly higher pressure.

VII. CLOSING REMARKS

Thus it seems that materials exhibiting fast ionic conduction are characterized by highly mobile species which behave in a manner similar to that of liquids upon their sub-lattice. These ions move through crystallographic tunnels formed by the static skeleton of the other sublattice. Furthermore, the preferred motional path typically deviates from the center line between positions normally considered to be crystallographic sites, depending upon the radius of the mobile ion, as does the potential variation along the path. The result is that for the same crystal structure, one expects an optimum ionic size for maximum mobility.

During the last decade, this field has moved very rapidly from the first recognition that one may use the unusual properties of a few fast ionic conductors for a variety of scientific measurement purposes (17-19), to the present situation in which a much greater number of materials are now recognized as having the characteristic of fast ionic conductors, and technological as well as scientific applications are burgeoning. There is a considerable amount of scientific effort going into the development of the understanding of the unusual properties of this group of materials, and much attention is being given to the search for new materials of this type. In addition, practical applications in a number of areas are moving forward rapidly. The most visible and heavily funded, of course, being the development of advanced battery systems based upon the use of solid electrolytes.

VIII. ACKNOWLEDGMENT

Work in this area at Stanford is supported by the Defense Advanced Research Projects Agency and monitored by the Office of Naval Research under Grant Number N00014-75-C-1056.

IX. REFERENCES

1. E. J. Cairns and H. Shimotake, Science, 164, 1347 (1969).

2. J. Hladik, Ed., Physics of Electrolytes, Vols. 1 and 2, Academic Press (1972).

3. R. A. Huggins, in Diffusion in Solids: Recent Developments, Eds. A. S. Nowick and J. J. Burton, Academic Press (1975) p. 445.

4. R. A. Huggins, in Advances in Electrochemistry and Electrochemical Engineering, Vol. 10, Eds. C. W. Tobias and H. Gerischer, John Wiley and Sons (1976).

5. W. van Gool, Ed. Fast Ion Transport in Solids, North Holland, Amsterdam (1973).

6. Y. F. Y. Yao and J. T. Kummer, J. Inorg. Nucl. Chem. 29, 2453 (1967).

7. R. H. Radzilowski, Y. F. Yao, and J. T. Kummer, J. Appl. Phys., 40, 4716 (1969).

8. M. S. Whittingham and R. A. Huggins, J. Chem. Phys., 54, 414 (1971).

9. M. S. Whittingham and R. A. Huggins, J. Electrochem. Soc., 118, 1 (1971).

10. M. S. Whittingham, R. W. Helliwell, and R. A. Huggins, U. S. Gov. Res. Develop. Rep. 69, 158 (1969).

11. M. S. Whittingham and R. A. Huggins, in Solid State Chemistry, Eds. R. S. Roth and S. J. Schneider, Nat. Bur. Stand. U. S. Spec. Publ. 364, Washington, D.C. (1972) p. 139.

12. M. J. Buerger and B. J. Wuensch, Science, 141, 276 (1963).

13. M. O'Keeffe and B. G. Hyde, Phil. Mag., 33, 219 (1976).

14. W. H. Flygare and R. A. Huggins, J. Phys. Chem. Solids, 34, 1199 (1973).

15. M. Born and J. E. Mayer, Z. Phys., 75, 1 (1932).

16. R. H. Radzilowski and J. T. Kummer, J. Electrochem. Soc, 118, 714 (1971).

17. C. Wagner, J. Chem. Phys., 21, 1819 (1953).

18. K. Kiukkola and C. Wagner, J. Electrochem. Soc., 104, 308 (1957).

19. K. Kiukkola and C. Wagner, J. Electrochem. Soc., 104, 379 (1957).

Chapter 25

SOME ASPECTS OF SOLID ELECTROLYTES

W. van Gool
Department of Inorganic Chemistry
State University Utrecht, Netherlands

I. ENERGY STORAGE SYSTEMS

Time differences between the production or availability of energy and its use are common in the total energy system. There are many reasons for these phase differences and a more complete analysis would fill completely the allocated time.

One result of such an analysis is that we cannot expect to find one general energy storage system for bridging the time between availability and use. Optimizing the storage systems has to be done with respect to the critical factors. Depending upon the systems under consideration one or more of the following factors might be critical: time needed for charging or discharging, the time during which the storage is required, the amount of the stored energy, the type of energy available before storage and the type required at a certain moment, the energy density - in weight or in volume - of the storage system, life time of the storage system, possibility of replacement, the materials needed to produce the storage system, and - predominating in many applications - the price of the system. Here it is not the right place to analyze all possible storage systems with respect to their potential use in different situations.

It will not require a large imagination to accept that electrochemical batteries are not only important in a number of present applications, but that some other future options exist, for example electric vehicles, peak sharing in electricity production, storage in combination with the use of wind or solar energy, and space applications. The use of solid electrolytes in batteries or even complete solid state batteries is a part of present day research directed to an enlargement of the number of useful storage systems.

In the following sections we first analyze in a general way the implications of the use of solid electrolytes in batteries. Next we look into the problem of finding new solid electrolytes as a materials engineering problem.

II. GENERAL ASPECTS OF BATTERIES

Most of the well-known storage batteries are of the following type:

+ S / L / S -

This means that the electrodes are solid phases. They are separated by the liquid electrolyte L. Basic to all electrochemical batteries is that the potential chemical reaction between the electrode materials is separated into an ionic reaction and a transport of electrons. The two processes are separated in physical space: the ions go through the electrolyte, the electrons follow a path outside the battery. The lead-acid battery is a well-known example of the electrochemical batteries. Furthermore, it is an example of rechargeable batteries: by applying a voltage from an outside source the battery can be recharged. These batteries are called secondary, in contrast to the one-shot or primary batteries. These can be used only once and are a waste product after use. Tremendous amounts of primary batteries, such as the Leclanche Cell, are used in radios, toys, apparatus, etc.

The reactants do not need to be solids. In the zinc-air cell, for example, oxygen from the air is used to oxidize the zinc. The battery is of the type

S / L / (S) G

A solid electrode is also needed on the gas side in order to permit the electrons to go around.

All these facts are well-known and they do not need further amplification. More important is the analysis of the function of the parts of the batteries: both the electrodes and the electrolyte phase.

The electrodes generally contain the chemical mass which has to react in order to produce work (the delivered electrical energy). An exception is the air electrode in the zinc-air battery, which actually is a semi-fuel cell rather than a battery. Thermodynamics tells us that the maximum amount of available electrical energy is equal to the Gibbs free energy change, ΔG, of the reaction. This data is available from tables, and we have some idea about the maximum energy density (stored energy per unit of weight or volume). Reactive couples reach values well above 1000 Wh/kg. These extreme values cannot be obtained in real batteries. Some reasons are very obvious. We want, for example, conducting electrodes and many of the very reactive materials (F_2, Cl_2, liquid sulphur) do not fulfill this condition. Furthermore, the reaction is an oxidation-reduction process. We prefer to have electronically conducting materials before and after the discharging. Not many couples are satisfactory in this respect. Note that even the frequently used lead acid batteries do not fulfill this condition. Basically the reaction is

$$Pb + PbO_2 + 2H_2SO_4 \rightarrow 2\ PbSO_4 + 2\ H_2O$$

The lead sulphate formed during the reaction is certainly not electronically conducting. Much ingeneous work has been done in constructing electrodes such as to minimize the influence of the insulating $PbSO_4$.

This was just one factor determining the effectiveness of a battery. Several other factors, such as shelf-life, cycle-life, etc., are decisive to the question of whether a certain battery is interesting or not.

The meaning of this analysis is obvious. Although in theoretical discussions much stress is laid on the energy density of the system, other factors like morphological compatibility of oxidized and reduced phases, their conductivity, etc., are often more important. We will return to this evaluation later on.

The function of the electrolyte phase can be defined uniquely: it must be a barrier for electrons, but it must give an easy passage to the relevant ions. Here we find an important indication for the use of solid electrolytes. In the common construction with liquids as an electrolyte phase, a distance of say a few millimeters must be maintained between the electrodes in order to prevent an electronic short circuit. This space must be filled with liquid electrolyte, which lowers the energy density of the battery, requires materials increasing the investment, etc. When the functions of the electrolyte phase can be fulfilled by a solid layer of, say, 1 micron, an improvement could be made in the battery system. In this way we are confronted with solid electrolytes as a possibility to improve battery systems.

III. GENERAL ASPECTS OF SOLID ELECTROLYTES

Once the function of the solid electrolyte phase has been recognized some other options must be considered. Substitution of the liquid phase by a solid electrolyte phase leads to the all solid state battery

S / S / S

There are some essential problems in this construction, to which we will return more extensively later on. One of these problems concerns the interfacial contact: the transfer of ions through the solid-solid interface is geometrically more difficult than between an L/S interface. For this reason it might be worthwhile to investigate the possibilities of an

L / S / S

or an

L / S / L

battery. In the last situation the construction is reversed with respect to the common types of batteries. Obviously, other problems are introduced: the container must be resistant to both an electropositive and an electronegative liquid, which limits the choice of applicable materials. Furthermore, the solid electrolyte phase has to fulfill the same condition.

Considering this aspect with respect to life

time of the battery, material problems remain predominating, although they are different from the problems which we do know already for a long time for the known types of batteries. It is the expectation that new problems are easier to solve than known problems that stimulates the interest in solid electrolytes. It must be admitted, however, that history does generally not confirm this expectation.

It is worthwhile here to remember the fact that solid electrolytes are useful in many other applications, for example in fuel cells and thermodynamic measurements. These applications are discussed elsewhere (1, 2). Even though solid electrolytes may not lead to large industrial applications in battery systems, it is worthwhile to investigate them in connection with these other applications and in relation to the extension of our knowledge of materials science.

Returning to the battery systems, we are facing two conflicting themes:

1) In order to get a high energy density (per weight or per volume) we are inclined to search for very reactive couples of materials (say Li and F_2, Na and S). The theoretical energy density of these systems is well above 1000 Wh/kg.

2) Since these materials have to be used in battery systems we have to find solid electrolytes, solid electrodes and container materials which are stable with respect to both reactive materials. Safety, in the form of preventing a rupture of containers of a solid interphase, is a different way of saying that the materials must be stabile in contact with both a very electropositive and a very electronegative component.

It is important to realize that whenever we start with very reactive couples of materials, we have to make a compromise in the design of the battery, in which we loose much of the original high energy density. It is not unusual that this loss decreases the energy density to 10-25% of the value of the pure reactants. It might be quite well possible that starting with a less extreme combination of reactive materials ultimately a higher energy density of the total system is obtained. Since 100 Wh/kg is an interesting target for several applications, we need only 250 Wh/kg if we reserve 60% of the weight for the non-storage parts of the battery.

Estimates have shown that 250 Wh/kg might be too high for the type battery in which at least one electrode is a solid electrolyte. This requires that the migrating ion enters and leaves the solid electrode. However, energy densities typical for lead acid batteries are in the range of what might be obtained by the indicated processes.

IV. SOLID ELECTROLYTES: A MATERIALS PROBLEM

Surveying the presently available data (2, 3), it is obvious that the number of applicable solid electrolytes is very limited. Since much interest exists for high current density applications at or slightly above ambient temperature the choice is limited to the diffusion of Ag^+ or Na^+, the last ion requiring 200-300 C in β-Al_2O_3 in order to reach the conductivity range necessary for high-duty batteries. It must be stressed that the prospects for an all solid-state battery applicable in electric vehicles are not very good at the moment: the use of silver as diffusing ion excludes large scale applications for economic reasons. However, batteries of the type L / S / L are under development for heavy electrical traction and the type L / S / S cannot be ruled out completely for future applications in electrical vehicles. Furthermore, there are other applications in which a long shelf-life together with low current-densities during use are relevant. This has stimulated the industrial development of all-solid state batteries.

Depending upon the application under consideration

one or more of the following problems might be worthwhile to pursue:

(a) the development of - electronically insulating - solid electrolytes with a high conductivity for specified ions at or slightly above ambient temperature. A good proton conductor - for example - will be interesting for the development of fuel cells. In other applications special interests exist in solid electrolytes with F^-, Cl^-, O^{2-}, Li^+ or Na^+ as mobile ion.

(b) the development of mixed conductors (electron - and ion conductivity). These materials are important as solid electrode phases in the S / •• / •• / type batteries.

(c) the development of a system with a solid electrolyte and at least one solid electrode phase. Here the diffusing ion must be the same in both phases. Special problems do occur with respect to interfacial compatibility.

(d) the development of the aforementioned materials and systems with readily available and inexpensive materials.

Since the information about the presently known solid electrolytes has been reviewed extensively, we concentrate the remaining part of this contribution upon the development of solid electrolytes as a materials engineering problem.

V. *CONDITIONS FOR OPTIMIZED ION CONDUCTION*

According to the preceding sections, the goal of the research is very well fixed: find materials with a conductivity of 0.1 $\Omega^{-1}cm^{-1}$ or better at ambient temperature in which the charge is carried for at least 99% by a specified ion.

The history of materials science showed many problems of this type. The solution was often sought by the empirical approach: long series of materials were scanned and many materials problems were solved. Obviously, one always tried to limit the number of experiments by combining theory - good and wrong - knowledge, and intuition.

Due to the large increase of knowledge about chemical bonding, structures, etc., materials science comes to the point where we ask: can we construct the materials we want? At this moment the number of successful approaches along this line is very limited. Solid electrolytes might be a good example to find out whether we do know enough from basic science to answer this question.

Obviously, in order to synthesize a solid electrolyte or to select a good solid electrolyte from the known materials, we need to know how good ion conductivity is obtained. We need a theory. However, not every theory is useful. A very detailed theory deriving the properties of a good ion conductor from more or less fundamental assumptions, is not necessarily the best one in the search for or the construction of new materials. The theory we need in materials engineering must include the essential conditions of the investigated property, but should not depend upon the detailed characteristics of one specially known material. In the present problem a detailed interpretation of the properties of β-Al_2O_3 or α-AgI does not need to be helpful in constructing materials with a high mobility of F^-, O^{2-}, etc.

The generalized conditions for high ion mobility have been described elsewhere (4). We summarize the relevant aspects.

(a) The fraction of the ions ready to move at a certain moment must be high, say, for example, 1-10% of the ions present. This property is characteristic: in the normal defect-type diffusion mechanism diffusion is determined by small amounts of defects, for example 0.0001-0.1%. Thus solid electrolytes can have conductivities which are a factor 10 to 10^5 higher than in the defect type diffusion. This has led to the suggestion to describe the high ion conduction as "optimized ion conductivity" rather than super ion conduction.

(b) after each individual movement of an ion, the situation remains equivalent: the same fraction of ions must be in a diffusion-ready situation.

(c) the transition state for the moving ion must have a low activation energy.

(d) the diffusion path for a specified ion should be continuous and two- or three-dimensionally branched. One-dimensional paths as occur in tunnel structures seem to be less favorable.

The first two conditions can be fulfilled by having more than one site for the mobile ion, for example three equivalent sites for two ions or two sites for one ion. A rather straightforward interpretation of this condition leads to a domain model: adjacing domains have the mobile ion in a different but symmetrically and energetically equivalent position (5).

The question has been raised why a domain model is more probable than a statistical distribution. In the latter case the two positions are occupied in an ad-hoc way by half the amount of ions. The difference between the two descriptions is rather small. The assumption is that an ion in position A which is surrounded completely or largely by ions in position B will also switch to the B position, and vice versa. This assumption will hold in many cases, since an ion in position A surrounded by ions in positions B, is a defect situation. The energy will decrease by converting B into A. High temperatures will sustain a larger disorder, but the configuration entropy leads generally to an energy small compared to enthalpy increase due to the defect situation. Figure 1 demonstrates this principle two-dimensionally. Figure 1A was obtained by filling 504 cells either in A or in B position by using a table of random numbers. Figure 1B is obtained by changing the position when more than a half (3 or 4 for non-edge elements) of the direct neighbor have a different position.

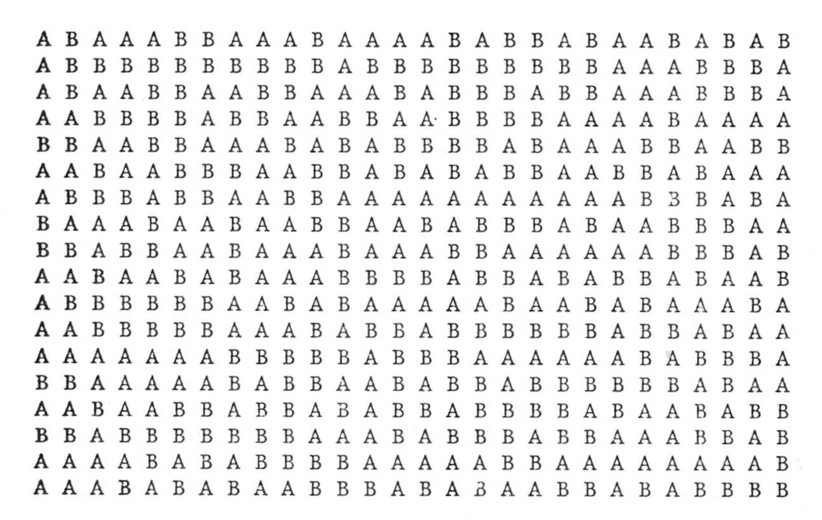

FIG. 1A. Statistical Occupation of 504 Unit Cells with Two Equivalent Positions

```
A A A B B B B B B B B B B B B B B B B B B A A A B B B B
A A A B B B B B B B B B B B B B B B B B B A A A B B B B
A A A B B B B B B B B B B B B B B B B B B A A A B B B B
A A A B B B A A A A A B B B B B B B B A A A A A A A B B
A A A A B B A A A A A B B B B B B B B A A A B B A A A A
A A A A A A A A A A B B A A A A A A A A A A B B B A A A
B B B A A A A A A A B B A A A A A A A A A A B B B A A A
B B B A A A A A A A B B A A A A A A A A A A A B B B A A
B B B B A A A A A A A B A A A A B B A A A A A B B B A A
B B B B A A A A A A A B B B B B B B A A A A B B A A A A
B B B B B B B A A A A B B A A B B B A A A A B B A A A A
B B B B B B B A A A B B B A A B B B B B A A B B B A A A
A A A A A A A B B B B B B B B B B B B B A A B B B B A A
A A A A A A A B B B B A A B B B B B B B B B B B B B A A
A A A A A B B B B B A A A B B B B B B B B B A A B B B B
A A A A B B B B B B A A A A A B B B B B B A A A B B B B
A A A A B B B B B B B B A A A A A B B B A A A A B B B B
A A A A B B B B B B B B A A A A A B B B A A A A B B B B
```

FIG. 1B. Domains Obtained from Fig. 1A by Procedure Described in the Paper

This procedure is not quite unique. When a choice had to be made, statistical procedures were used.

The last generalized condition is derived from the properties of well-known solid electrolytes. The two-dimensional case is well illustrated by β-Al_2O_3, whereas $|(CH_3)_4N|_2Ag_{13}I_{15}$ is a good example of a material with a three-dimensional path for the Ag-ions (6).

The movement of the boundary between the domains represents a mass transport. It seems probable that other diffusion mechanisms do occur. One of the well-known solid electrolytes, viz. α-AgI, has been presented as a simple example of the aforementioned principles: there are a lot of positions available for each Ag^+-ion and this should be responsible for the high mobility of the Ag^+-ions. A more detailed investigation of the local aspects of diffusion-ready situations leads to many difficulties. A reevaluation of the available experimental data leads to doubts about the simplicity of this case. It cannot be excluded that the diffusion of the Ag-ions is correlated with certain phonon modes of the iodine lattice. Local configurations are a tetragonal deformation of the (x-ray averaged) bcc-structure. For diffusion the instantaneous configuration is more important than the time - and place - averaged information of the X-ray analysis.

In the case of sulphates, molybdates, nitrates, one may doubt whether concerted rotations of tetrahedral or plane triangular groups are responsible for the creation of "equivalent" positions of the moving ions (7). In these cases the equivalent positions need not to be present in the original - low temperature - configuration. See Figure 2 for an illustration of this principle.

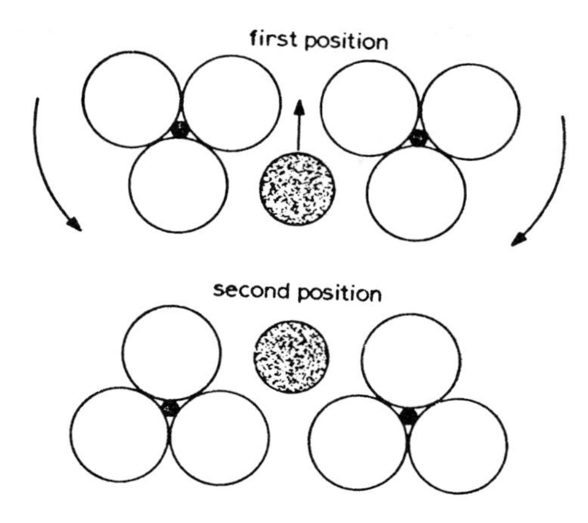

FIG. 2. Creation of Equivalent Sites by Coordinated Rotation of Planar Groups

VI. SEARCH FOR NEW MATERIALS

Several methods have been proposed in order to assist in the selection of possibly interesting new solid electrolytes. Conductivity measurement remains the most straightforward method. This method is very time consuming and the automatization of the conductivity measurements was a condition for a more extensive application (8).

There remains a need for methods which help in selecting promising materials. Experimentally NMR line narrowing with increasing temperature has been used (9). Theoretically, a high transition entropy for a solid-solid phase transition might be an indication for ionic conductivity in the high temperature phase (10).

The best approach seems to be the use of the conditions formulated in the last section. Some structures fulfill the conditions of having, say, two equivalent sites for one ion, although no statistical occupation of both sites is reported. This is true for example in the sphalerite structure. The ionic conductivity does not reach interesting values: obviously the condition for a low transition energy is not fulfilled.

A more conclusive indication is obtained when X-ray analysis leads to a fractional occupation of equivalent sites. Heating of the material often leads to phase transitions with hysteresis effects in the cooling curves. Li_4SiO_4 is such a material (11). The ion conductivity reaches interesting values at high temperatures (12). Solid solutions with other oxides have been prepared, some of which have good ion conductivities. Secondary properties - which are of importance when the materials are applied as solid electrolytes - are disappointing.

Following the approach sketched above, it remains attractive to investigate certain silicates and phosphates for good Li^+ or Na^+ conduction. It must be remarked that the number of cases in which we have available all relevant information - phase transition temperature and entropy, structures, fractional occupation - is limited. There are even principal reasons why this information cannot be complete for good ion conductors.

It is in agreement with the idea of a directed material development that we concentrate upon finding or - perhaps - constructing a conductor for some specified ion. For this reason we will not dwell upon the detailed properties of well known materials. It is pointed out that we still need stable and inexpensive materials which have a good conductivity for Na^+, Li^+, H^+ and O^{2-} at ambient temperature, or slightly higher.

We conclude with some remarks about the possibilities for good conductivity for H^+ and O^{2-}.

Apart from many other aspects, it can be stated that the great advantage of the proton for fast conduction - viz. its smallness - is also its disadvantage: it polarizes most of the negative - ions and it sticks locally. The best chances are thus in the range of materials with low polarizabilities like fluorides, hydrazines, etc. Detailed results, especially obtained with KHF_2, have been published (13-15).

For oxygen migration another problem occurs: the O^{2-} ion is so big, that the creation of a two- or three dimensional continuous path requires very large ions. These will generally be complex. The known solid electrolytes with good O^{2-} migrations are based upon another principle. In materials like doped zirconia the defect type diffusion mechanism is brought into the range of optimized conduction by increasing the defect chemistry massively. Oxygen diffusion might be one of the examples in which we have to engineer favorable structures.

In summary it has been illustrated how we try to develop a directed approach to materials science problems: we abstract from some well-known examples generalizing concepts which are useful in selecting - and later perhaps constructing - the required materials.

REFERENCES

1. W. van Gool, Ed. "Fast Ion Transport in Solids," North-Holland, 1973.

2. J. Hladik, Ed. "Physics of Electrolytes," Academic Press, 1972.

3. W. van Gool, Annual Rev. of Materials Sc., 4, 311, 1974.

4. W. van Gool, J. Solid State Chem., 7, 55, 1973

5. W. van Gool, P.H. Bottelberghs, J. Solid State Chem., 7, 59, 1973.

6. S. Geller, M.D. Lind, J. Chem. Phys., 52, 5854, 1970.

7. W. van Gool in "Mass Transport Phenomena in Ceramics," A.R. Cooper and A.H. Heuer, Eds., Plenum Press, 1975, p 139.

8. P.H. Bottelberghs et al., J. Appl. Electrochem., 5, 165, 1975.

9. M.S. Whittingham, in [1], p 429.

10. W. van Gool in "Phase Transitions 1973," L.E. Cross, Ed., Pergamon Press, 1973, p 373.

11. H. Vollenklee et al., Monatsh. fur Chem., 99, 1360, 1968.

12. A.R. West, J. Appl. Electrochem., 3, 327, 1973.

13. J. Bruinink, J. Appl. Electrochem., 2, 239, 1972.

14. J. Bruinink, G.H.J. Broers, J. Phys. Chem. Solids, 33, 1713, 1972.

15. J. Bruinink, B. Kosmeyer, J. Phys. Chem. Solids, 34, 897, 1973.

Chapter 26

MATERIALS FOR HIGH-TEMPERATURE Li-Al/FeS_x SECONDARY BATTERIES*

James E. Battles

Chemical Engineering Division
Argonne National Laboratory
Argonne, Illinois 60439

I. INTRODUCTION

Lithium-aluminum/iron sulfide secondary batteries of high specific energy and high specific power are being developed at Argonne National Laboratory (1) for two major

*Work performed under the auspices of the U.S. Energy Research and Development Administration.

applications: 1) electric-powered automobiles and 2) load-leveling, energy-storage devices for electric utility systems. Recent efforts have concentrated on development of cells employing negative electrodes of solid lithium-aluminum alloy and positive electrodes of iron sulfide, either FeS or FeS_2. The negative electrodes are normally prepared by the electrochemical reaction of lithium with a porous aluminum structure or by a pyrometallurgical preparation of β-LiAl which is then used to fabricate electrodes. The positive electrodes are prepared by any one of a number of procedures which include a) dry loading a metal sulfide powder into a macroporous metal or carbon structure, b) preparing a paste of positive electrode mix, and c) hot- or cold-pressing a mixture of metal sulfides and electrolyte. Both electrodes may be characterized as packed particle beds in an electrolyte matrix which is molten at cell operating temperatures (400-450°C). The melting point of the LiCl-KCl eutectic electrolyte is 352°C.

Lithium has been a very popular electrode among electrochemists because of its promise of yielding the highest specific energy of any alkali metal (2). However, the utilization of both molten lithium and molten sulfur electrodes in a high-temperature cell employing a molten electrolyte poses severe engineering problems. The basic problem is one of containment and separation of the three liquid phases. To circumvent the problems associated with molten electrodes, the solid β-LiAl intermetallic compound was substituted for lithium, and iron sulfide (FeS_2 or FeS) was substituted for molten sulfur. The disadvantages of these substitutions are the decreases that occur in cell voltage, specific energy, and specific power. Although these sacrifices are undesirable, the more important consideration is the fact that practical cells can be fabricated and operated with long lifetimes and still meet performance goals as a result of substituting solid electrodes.

A simplified Li-Al/FeS_2 cell is shown in Fig. 1. The overall reaction that takes place upon discharge can be written as

$$2Li(Al) + FeS \rightarrow Li_2S + Fe + 2Al \quad (1a)$$

$$\text{or} \quad 4Li(Al) + FeS_2 \rightarrow 2Li_2S + Fe + 4Al \quad (1b)$$

where the composition of the cathode at complete discharge of a stoichiometric cell would be Li_2S and Fe. The Li_2S product dissociates on recharging. The half-cell reactions corresponding to the overall cell reaction are shown in Fig. 1. At the anode, lithium metal from β-LiAl is oxidized to lithium cations, which migrate through the molten lithium-halide electrolyte, and electrons, which pass through the electrical circuit, doing work as they go. At the cathode, lithium cations from the electrolyte react with electrons from the electrical circuit and sulfur (FeS_2) to form the lithium sulfide, Li_2S. The reverse of these processes occurs on recharging the cell.

Because the active cell materials (lithium, lithium-aluminum and iron sulfides) are very reactive at elevated temperatures, the choices of materials for application as insulators, separators, positive- and negative-electrode current collectors, and cell housings are severely limited. Thus, an extensive corrosion-testing program (3,4) of both metals and ceramics is being conducted to enable suitable materials to be selected for constructing the various cell components, with emphasis on those that are lightweight and low cost. Corrosion tests of materials are conducted according to the intended application, *i.e.*, ceramics for insulators are tested in lithium or Li-Al + LiCl-KCl electrolyte, and metals for current collectors are tested in the respective electrode environments. The final evaluation of all materials is based on performance in operating cells.

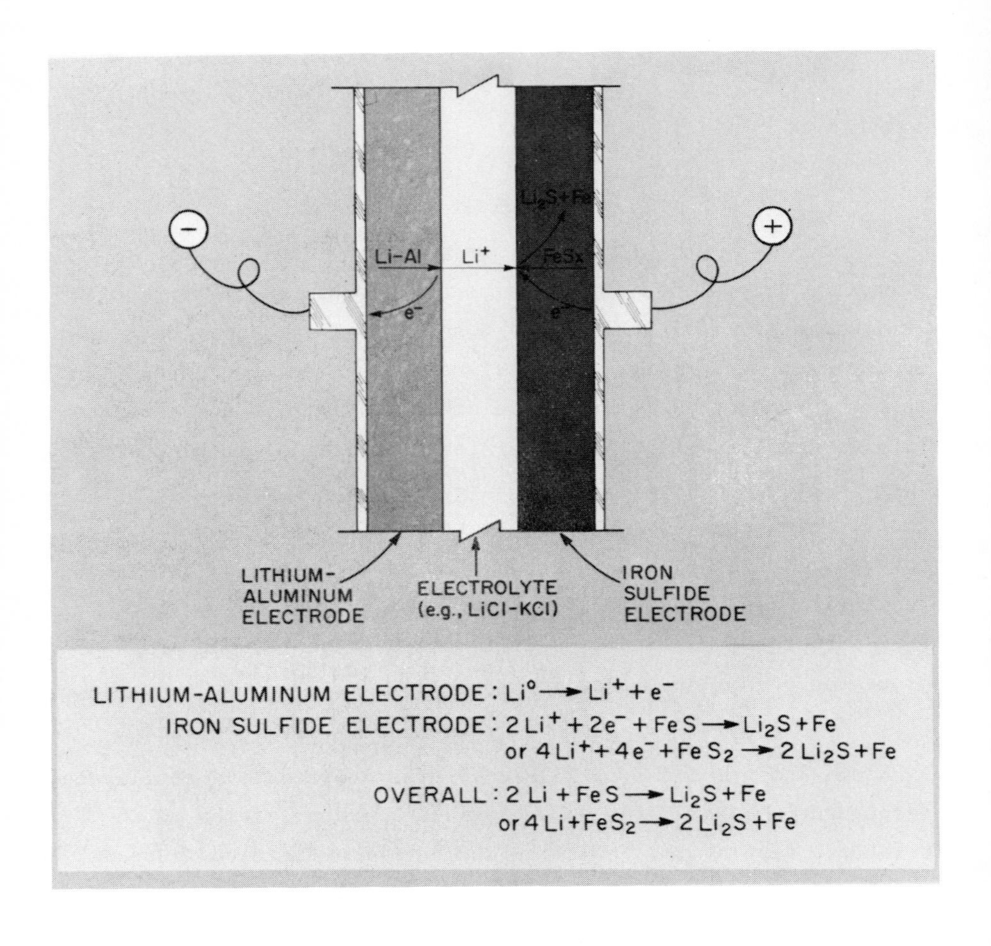

FIG. 1. Representation of the Lithium-aluminum/ Iron Sulfide Cell Reactions.

Ceramics are utilized in cell construction as electrical insulators and as electrode separators. The most stringent requirements placed upon these components are resistance to attack by lithium and lack of conductivity at elevated temperature. Preliminary selection of candidate materials can be readily made by examination of the thermodynamics of various oxide and nitride ceramics. The Gibbs free energies of formation of the more stable oxides and nitrides are listed in Tables I and II, respectively, and are arranged in order of decreasing stability. Accordingly, the oxides appearing above Li_2O, such as CaO, ThO_2, BeO, and Y_2O_3, are more stable and, hence, should be resistant to attack by lithium. Likewise, the nitrides appearing above Li_3N, such as AlN, TiN, BN, and Si_3N_4, are more stable. Although the activity of lithium is reduced in the two phase α-Al + β-LiAl system ($\sim$0.3 V), the reactivity of lithium is not significantly decreased.

TABLE I

FREE ENERGY OF FORMATION OF SELECTED METALS OXIDES AT 600, 700, AND 800 K

	FREE ENERGY OF FORMATION*, $-\Delta G_f^\circ$ [kcal/g-atom O]		
Compound	600K	700 K	800 K
Y_2O_3	137.71	135.42	133.14
CaO	136.72	134.28	131.85
ThO_2	(133)	(131)	(129)
BeO	129.0	126.65	124.31
$Y_3Al_5O_{12}$	-	-	-
MgO	128.19	125.62	123.06
Li_2O	124.82	121.47	118.12
α-Al_2O_3	118.46	115.97	113.49
δ-Al_2O_3	117.11	114.66	112.23
ZrO_2	117.34	115.08	112.84
$CaZrO_3$	-	-	-
TiO_2 (Rutile)	99.66	97.49	95.33
SiO_2	95.77	93.62	91.48

*Reference, JANAF THERMOCHEMICAL TABLES, 2nd Ed. Dow Chemical Co. (1971).

TABLE II

FREE ENERGY OF FORMATION OF SELECTED METAL NITRIDES AT 400 AND 500°C

Compound	FREE ENERGY OF FORMATION*, $-\Delta G_f^o$ (kcal/g-atom N)	
	400°C	500°C
ZrN	72.0	69.3
TiN	65.5	63.3
AlN	58.4	56.6
BN	45.9	43.2
Si N	31.2	28.7
Li_3N**	16.8	13.5

*Reference, JANAF THERMOCHEMICAL TABLES, 2nd Ed., Dow Chemical Co. (1971).

**R. M. Yonco, E. Veleckis and V. A. Maroni, J. Nucl. Mater., 57, 317-324 (1975).

Materials for application as current collectors in the positive electrode are exposed to metal sulfides (e.g., FeS, FeS_2, Li_2S and Li_xFeS_y) and LiCl-KCl eutectic salt. The corrosiveness of this environment is compounded by the fact that most metals react electrochemically with the electrolyte at potentials less than the normal charge cutoff voltages for cells having sulfur or FeS_2 positive electrodes. The Gibbs free energies of formation of various metal sulfides are listed in order of their increasing stability in Table III. Unlike the data for the oxides and nitrides, the values for the free energy of formation for the sulfides indicate that none of the low cost metals Fe, Ni, Cu, Cr, and Co, are likely to be compatible with the positive electrode environment. The cell potentials for various Li-Al/LiCl-KCL/M cells at 450°C are listed in Table IV, where M represents the more common construction materials. The

TABLE III

FREE ENERGY OF FORMATION OF SELECTED METAL SULFIDES AT 600, 700 AND 800°K

Compound	FREE ENERGY OF FORMATION*, $-\Delta G_f^\circ$ [kcal/g-atom S]		
	600°K	700°K	800°K
Ag_2S	15.69	15.06	14.51
CoS_2	20.31	18.39	15.99
FeS_2	21.50	19.14	16.80
NiS	23.92	23.34	21.40
Cu_2S	27.05	26.30	25.63
$CoS_{0.89}$	28.20	26.17	24.13
FeS	28.61	27.18	25.80
WS_2	32.51	30.27	28.06
MoS_2	34.54	32.32	30.12
CrS	37.36	35.78	34.21
Al_2S_3	59.60	57.60	55.17
Li_2S	102.8	102.4	101.9

*Reference, K. C. Mills, Thermodynamic Data for Inorganic Sulphides, Selenides, and Tellurides, Butterworth & Co. Ltd., London(1974).

probable reactions are shown in the third column. Typical charge cutoff potentials are ∿2.1 and 1.8 V for Li-Al/FeS_2 and Li-Al/FeS cells, respectively. A comparison of the data in Table IV shows that Nb, Ni, Mo and W would show an insignificant rate of reaction at the required charge potential for FeS_2 cells. Carbon is also a satisfactory material. For FeS cells, the selection of materials is much less restricted because of the lower charge cutoff potential required. Thus, materials that are unsatisfactory for application in FeS_2 cells may be completely suitable for cells that employ FeS.

TABLE IV

POTENTIALS OF Li-Al/LiCl-KCl/M CELLS AT 450°C*

Material	Potential, V	Probable Reaction
Cr	1.58	$Cr° \rightarrow Cr^{2+}$
Nb	2.1**	$Nb° \rightarrow Nb^{4+}$
Mo	2.40	$Mo° \rightarrow Mo^{3+}$
Mo-30W	2.4**	$Mo° \rightarrow Mo^{3+}$
W	3.36	$Cl^{-} \rightarrow Cl_2$
Fe-26Cr	1.8**	$Fe° \rightarrow Fe^{2+}$
Fe	1.83	$Fe° \rightarrow Fe^{2+}$
Cu	2.05	$Cu° \rightarrow Cu^{+}$
Ni	2.21	$Ni° \rightarrow Ni^{2+}$
Co	2.01	$Co° \rightarrow Co^{2+}$
Al	1.34	$Al° \rightarrow Al^{3+}$

*J. A. Plambeck, J. Chem. Eng. Data, 12 (1), 77 (1967).

**E. C. Gay *et al.*, Proc. 8th IECEC p. 96 (1973).

Although thermodynamic properties are a useful guide in selecting candidate materials (both metals and ceramics) for cell applications, corrosion testing in the intended environment is required for confirmation because of unknown kinetic factors and the effects of impurities in both the materials and the environment. Therefore, the final evaluation is based on the materials behavior during operation.

II. EXPERIMENTAL PROCEDURES

The procedures used for determining the compatibility of various materials included 1) static immersion tests, 2) planned-interval corrosion tests (described below), and 3) postoperative examination of cells. The corrosive media included lithium, LiCl-KCl eutectic, mixtures of Li-Al + LiCl-KCl*, FeS + LiCl-KCl,* and FeS_2+LiCl-KCl.* All test samples were carefully weighed and the surface areas were accurately determined prior to immersion in the selected corrosive medium. Test temperatures were 400 and 500°C. Helium-atmosphere gloveboxes were used in the preparation for all tests to prevent reaction of the corrosive media with the environment; the corrosion tests were conducted in furnace wells attached directly to the gloveboxes. Closed quartz crucibles were used in the compatibility studies that employed LiCl-KCl eutectic and mixtures of LiCl-KCl and FeS_2 (or FeS). Lithium and lithium-aluminum compatibility tests were conducted in Type 304 stainless steel containers. After immersion in one of the corrosive solutions for the prescribed time, the samples were removed and the adhering lithium or LiCl-KCl eutectic was dissolved in methanol or water. Each sample was then weighed to determine the weight change. Weight loss was converted to an annual corrosion rate reported in micrometers (or mils) per year penetration according to the formula (5)

$$\text{Corrosion rate (mils/yr)} = \frac{534\ W}{DAT}$$

where, W is the weight loss (mg), and DAT is the specific gravity, area ($in.^2$), and test time (hr). Weight gains were taken as an indication of reaction product formation, for example, an adherent film. All samples were examined for localized attack (pitting) and for the presence of a surface reaction product. The formation of a conductive film on ceramic materials precludes their use as insulators, whereas the formation of nonconductive films on metals eliminates them from application as electrode

*Each mixture contained 50 vol % LiCl-KCl eutectic.

current collectors. Test specimens were sectioned and examined metallographically for less obvious corrosion effects such as intergranular attack. Additional examinations such as X-ray diffraction, and ion and electron microprobe analyses were conducted to identify reaction products.

Wachter and Treseder (6) developed a planned-interval test procedure for evaluating the effect of time on the corrosion of materials and also on the corrosiveness of the environment. These tests produce data not only on the accumulated effects of corrosion at several times under a given set of conditions, but also on the initial rate of corrosion of fresh metal, the more or less instantaneous corrosion rate of metal after long exposure, and the initial corrosion rate of fresh metal during the same period of time as the latter. Thus, the results from these tests can be used to evaluate the long-term corrosion behavior of materials without the need for conducting a long-term corrosion test. The time intervals in the planned-interval test may be arbitrarily selected as one-day or longer periods. The time interval selected is generally determined by the expected corrosion rate of the material being tested. An outline of the test procedure and information obtained is given in Table V.

III. RESULTS AND DISCUSSION

In the conceptual design of lithium-aluminum/iron sulfide cells, ceramic materials are used for electrode separators and feedthrough insulators, whereas metallic materials are used for the cell housings, retainers, and for electrode current collectors. Because lithium is highly reactive toward ceramics, insulator materials are principally evaluated for compatibility with lithium. The corrosive conditions are considered moderate for metallic parts in contact with and at the negative

TABLE V

PLANNED-INTERVAL CORROSION TEST[a]

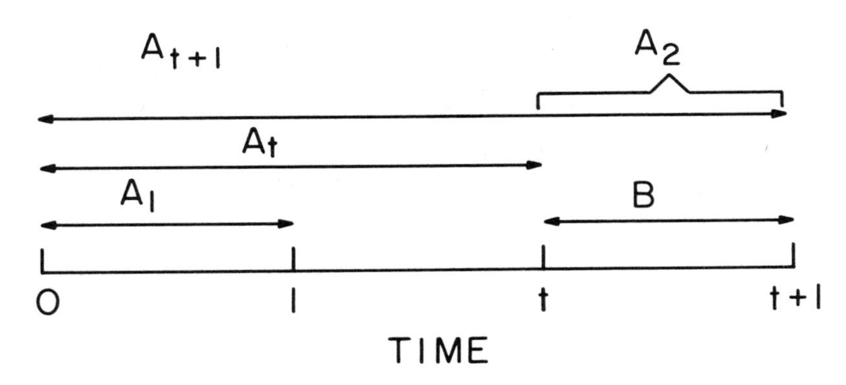

Occurrences during Corrosion Test			
Liquid Corrosiveness	Criteria	Metal Corrodibility	Criteria
Unchanged	$A_1 = B$	Unchanged	$A_2 = B$
Decreased	$B < A_1$	Decreased	$A_2 < B$
Increased	$A_1 < B$	Increased	$B < A_2$

electrode potential. These components are normally exposed only to LiCl-KCl eutectic and the lithium-aluminum alloys. On the other hand, materials used as current collectors in the positive electrode are exposed to metal sulfides (FeS, FeS_2, Li_2S, and Li_xFeS_y) and LiCl-KCl eutectic. The corrosiveness of this environment is compounded by the fact that most metals react electrochemically with the electrolyte at potentials less than the normal charge cutoff potential for FeS_2 cells. Because the corrosive conditions and compatibility criteria are different, the corrosion studies have dealt with each area separately and, hence, the results are presented in a like manner.

A. *Corrosion of Ceramic Materials*

Static compatibility tests have been conducted on various ceramic materials in both lithium and Li-Al + LiCl-KCl. The results for some of the materials that have been evaluated in this program are summarized in Table VI. The materials tested were generally greater than 90% of theoretical density; most were prepared by dry pressing or isostatic pressing followed by sintering, but some test samples were also prepared by hot-pressing. The fibrous materials tested as potential electrode separators were obtained from commercial suppliers. The test temperatures and durations are listed in Table VI. Analysis of the test results indicated that failure resulted from one of the following causes: 1) thermodynamic instability (SiO_2), 2) impurity content (BeO, MgO, BN), and 3) formation of a conductive surface layer ($CaZrO_3$, ZrO_2). Similar type studies have been conducted at other laboratories (7,8).

For materials that are thermodynamically stable, such as, BeO, MgO, AlN, and BN, tests results have shown that their corrosion resistance is an extremely sensitive function of impurity content and the location of impurities in the specimen. In most instances, materials failure have been directly attributable to reaction of lithium with impurities segregated at grain boundaries. In the case of BeO, failure was attributed primarily to the presence of SiO_2 in grain boundaries. The silica is selectively attacked by lithium, which results in rapid disintegration of the sample. On the other hand, high-purity beryllia (essentially free of SiO_2) has demonstrated excellent compatability with lithium. The effects of impurities in the grain boundaries are further illustrated by comparing the results of single-crystal and hot-pressed, translucent MgO (3). The single-

TABLE VI

COMPATIBILITY OF CERAMIC MATERIALS WITH LITHIUM AT 400°C

Material	Results
$CaZrO_3$, sintered or hot-pressed	Weight gain (<2%); discolored; conductive surface layer; intergranular reaction; no mechanical deterioration (200-4200 hr).
Y_2O_3, sintered or hot-pressed	Slight discoloration and weight loss in 200-2550 hr; nonconductive.
ZrO_2 (CaO stabilized), sintered	Severe discoloration (black); conductive; mechanical failure.
Al_2O_3, sintered	Severe discoloration (black); conductive; disintegrated.
MgO, single crystal	Slight weight loss; no apparent reaction; nonconductive
MgO, hot-pressed	Severe weight loss (>10%); slight discoloration; nonconductive.
$MgAl_2O_3$, hot-pressed	Complete disintegration to a powder.
SiAlON (Si_3N_4 + Al_2O_3)	Sample severely cracked.
Si_3N_4, hot-pressed	Sample severely cracked.
BN, hot-pressed, pre-treatment at 1700°C in N_2	Slight discoloration; nonconductive (1300 hr).
BeO, high purity	No apparent reaction.
SiC, hot-pressed	Disintegrated.
SiC·AlN, hot-pressed	Circumferential surface fracture, nonconductive.
SiC·2 AlN, hot-pressed	Sample fragmented, nonconductive.

crystal sample showed no apparent evidence of attack, and the weight loss was less than one percent after 300 hr at 400°C. However, the polycrystalline (and less pure) MgO was visibly attacked and showed a weight loss in excess of 10 percent. The cracking phenomenon found to occur in Si_3N_4 and the SiAlON (solid solution of Si_3N_4-Al_2O_3) is not the result of thermal shock or thermal stress loading. There is evidence to indicate that crack propagation exhibits preferred orientation and is probably associated with lithium attack of impurities along certain crystallographic planes and grain boundaries.

As noted earlier, a reaction which results in the formation of a conductive surface layer precludes utilization of a material as an insulator. This type of reaction has been observed with both $CaZrO_3$ (3) and ZrO_2 (stabilized with either CaO or Y_2O_3). Figure 2 shows the appearance of the conductive surface layer formed on $CaZrO_3$. The morphology of the reaction layer is the same whether the $CaZrO_3$ is exposed to lithium or the mixture of Li-Al and LiCl-KCl eutectic. The reaction product layer exhibits high electrical conductivity, and is responsible for the precipitous drop in the apparent resistivity of the bulk insulator. Although the precise identity of the layer is not known, electron microprobe analysis has shown that it contains both calcium and zirconium and that zirconium depletion occurs only at the outermost surfaces. Unlike calcium zirconate, zirconia exhibits severe cracking, in addition to the formation of a conductive surface layer, when exposed to lithium. In spite of this behavior, zirconia fabric is successfully utilized as a particulate retainer in the positive electrode.

The conflicting results shown in Table VI for boron nitride exposed to lithium has been attributed to the reaction of lithium with B_2O_3 impurities in the BN. Commercial boron nitride, which normally contains about 4 wt % oxygen (as B_2O_3), forms a black conductive surface layer, whereas material

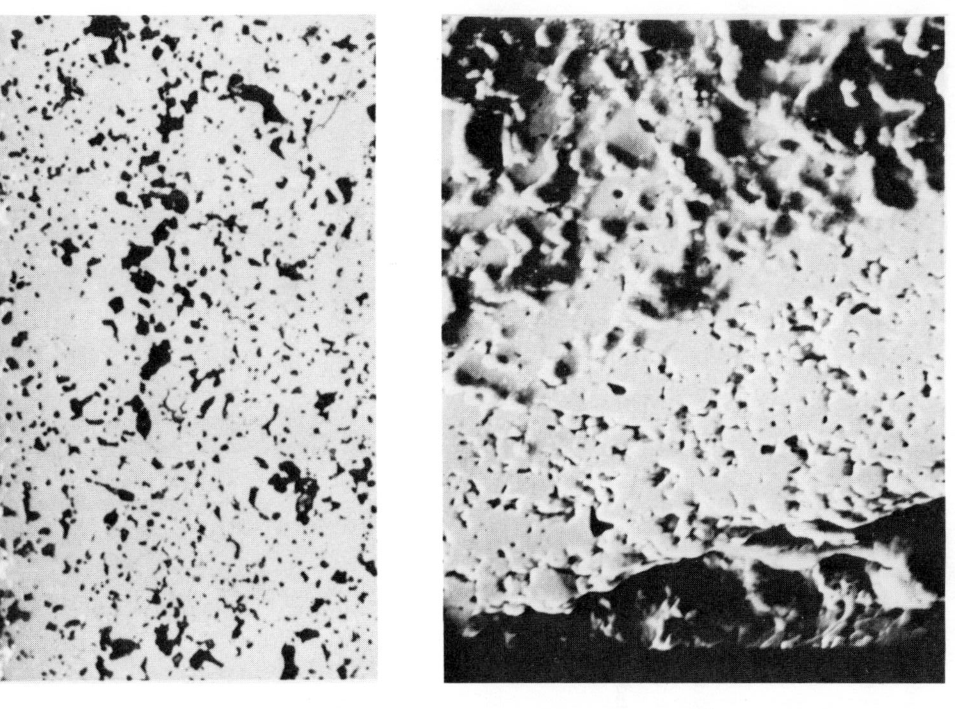

(a) OPTICAL MICROGRAPH AT 275X

(b) SCANNING ELECTRON MICROGRAPH AT 800X

FIG. 2. Micrographs Showing the Conductive Surface Layer Formed on $CaZrO_3$ exposed to Lithium

that is heat-treated at 1700°C in flowing nitrogen for removal of B_2O_3 shows good compatibility with lithium. Also, pretreated BN insulators and fabric separators have been used successfully in cells with solid Li-Al negative electrodes for more than 6400 hr (1).

The use of a solid Li-Al alloy in an electrolyte matrix, rather than molten lithium, as the negative electrode material reduces the lithium activity and mobility and, hence, should decrease the extent of lithium attack. However, compatibility studies of most of the materials listed in Table VII

TABLE VII

*COMPATIBILITY OF CERAMIC MATERIALS WITH Li-Al ALLOYS AT 400°C**

Material	Results
BeO, High-Purity	No apparent reaction
Y_2O_3	No apparent reaction
$CaZrO_3$	Slight weight gain; discolored and conductive
Al_2O_3, High-Purity	Slight reaction; nonconductive
ZrO_2 fabric	Discolored and conductive
SiO_2 fabric	Severe reaction
BN fabric	No apparent reaction
SiC, Hot-Pressed	Sample fracture; small surface cracks visible
SiC·AlN, Hot-Pressed	Minor surface attack, nonconductive
SiC·2 AlN	Minor surface attack; nonconductive

*50 vol % Li-Al in LiCl-KCl eutectic salt.

using a 50 vol % mixture of Li-Al and LiCl-KCl eutectic salt yielded results that are in qualitative agreement with those obtained in lithium tests. These results confirm that the lithium activity in this alloy is not sufficiently reduced to alter the thermodynamic relationships between these materials. The reaction rates are moderately reduced by the solid-solid reaction (as opposed to liquid-solid) and by the reduced contact area. These results have also been substantiated by in-cell tests with Li-Al negative electrodes. Materials such as ZrO_2, SiO_2, Al_2O_3, and $CaZrO_3$ that become conductive or react extensively in lithium compatibility tests show the same behavior in in-cell tests with Li-Al negative electrodes. From these results and thermodynamic considerations, it is apparent that only BeO, Y_2O_3, BN, AlN, and possibly high-purity MgO are suitable for cell application. Of these materials only BN and Y_2O_3 are readily available in the form of fibers for the fabrication of electrode separators.

B. *Electrode Separator*

A porous electrode separator which is stable in the cell environment, maintains electrical separation between the electrodes, allows ionic transport, and permits close electrode spacing is required for a compact cell. Boron nitride fabric has been used successfully in test cells for more than 6400 hr, but its present cost of $\$515/ft^2$ and projected mass-production cost of about $\$15/ft^2$ precludes its use in low-cost commercial cells. This cost could be reduced through the use of papers which require less fiber per square foot of separator or by using a less expensive fiber. Since no low-cost fibers that are expected to be stable in the cell environment are currently available, efforts have been concentrated on the development of paper or felt-like separators.

The only suitable materials that are presently available are boron nitride (9-12) and yttria (13). However, these fibers are not good paper formers, and binders must be used to produce papers with the strength necessary for handling and cell assembly. Earlier evaluations of paper separators in a test cell showed that organic and inorganic (colloidal SiO_2 and Al_2O_3) binders were not stable in the cell. When the binders failed, the papers lost their integrity and the electrodes came into contact. We are attempting to solve this problem through design changes in the cell, and through the development of a boron nitride binder which is stable in the cell.

Fabrication of the boron nitride bonded paper* involves a process of blending about 10 wt % B_2O_3 fibers with the BN fibers in an appropriate solution to obtain a uniform distribution. The resulting mat is dried and then heated at a temperature adequate to melt the B_2O_3 fibers (550-700°C). The B_2O_3 is then converted to BN through the normal nitriding process (14). The scanning electron micrograph in Fig. 3 shows the general structure obtained in this process. From this photograph, it is apparent that bonding was achieved only in a limited number of sites and that the B_2O_3 did not readily wet the BN fibers. Although the paper is quite weak, it has sufficient strength to permit handling and is very flexible. Cell test results with this separator are very encouraging and work is in progress to improve the bonding in this paper.

Another approach in the development of electrode separators has been the utilization of the excellent sheet-forming properties of chrysotile asbestos as a binder for BN and Y_2O_3 papers (15). The scanning electron micrograph in Fig. 4 shows the typical appearance of BN fibers bonded with asbestos fibers. The incorporation of chrysotile asbestos in small amounts (5-10 wt %) markedly improved the physical properties of the sheets. Although asbestos is not compatible with the cell environment, a yttria paper containing 10 wt % asbestos fiber as a binder has been tested in a cell for more than 1000 hr with good results.

*Development contract with the Carborundum Company.

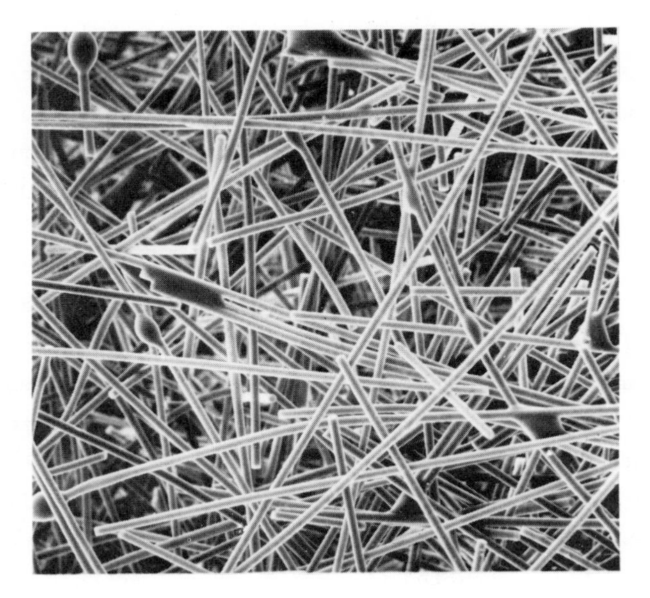

FIG. 3. Scanning Electron Micrograph of BN Fibers Bonded with BN (original magnification 300 X).

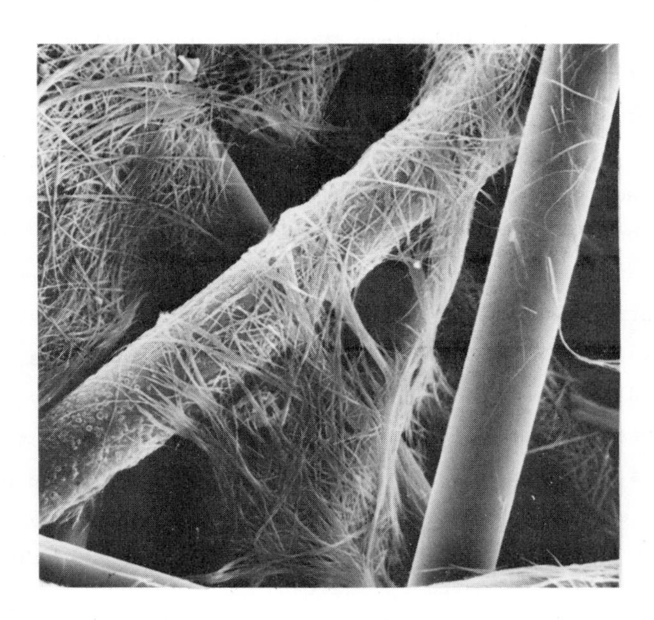

FIG. 4. Scanning Electron Micrograph of BN Fibers Bonded with Asbestos (original magnification 3000 X).

The development of low-cost, thin, porous electrode separators that are mechanically and chemically compatible with the cell environment is essential to the successful development of the lithium-aluminum/metal sulfide battery. Also, from the above discussion, it is apparent that much development remains to be done in this area.

C. *Materials for Negative Electrode Applications*

The corrosiveness of lithium toward a variety of materials over a wide range of temperatures has been the subject of numerous studies (3, 16-20). DeVries (16) tested a large number of iron-base alloys and miscellaneous metals for corrosion resistance in impure lithium at 315 and 480°C. The effects of extended exposure to lithium on microstructure, tensile strength, and stress corrosion were evaluated in these tests. DeStefano and Litman (17) conducted studies which showed that impurities such as oxygen, nitrogen, or carbon in either the refractory metal or the alkali metal had a gross effect on the compatibility of refractory metals with alkali metals. These and similar studies (18-20) have shown that the presence of oxygen in the refractory metal leads to severe penetration of the refractory metal by the alkali metal. Both intergranular and transgranular attack can occur in niobium and tantalum (18) when the oxygen concentrations exceed certain minimum values. In general, the lithium compatibility studies of the more common construction materials, *e.g.*, low-carbon steel, nickel, and stainless steels, indicate that no serious problems are encountered at temperatures less than 500°C. Similarly, corrosion tests of the planned-interval type have shown that the stainless steels and low-carbon steels are compatible with dry LiCl-KCl eutectic salts at 400 and 500°C (3). The austenitic stainless steels were tested in the sensitized and solution-annealed conditions with comparable results.

Postoperative examinations of several engineering-scale cells, which were operated at or above 500°C for at least part of their lifetime, have shown that extensive reaction occurred between the metal housing, metal current collector, and the Li-Al alloy negative electrode to form an Fe-Al intermetallic compound. A photomicrograph and electron microprobe scanning images of the reaction zone are shown in Fig. 5. The

PHOTOMICROGRAPH 300X AS-POLISHED

SPECIMEN CURRENT IMAGE

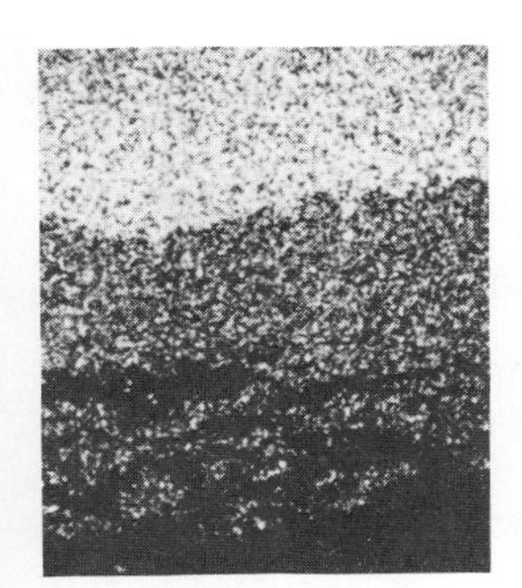

Fe K_α SCANNING IMAGE

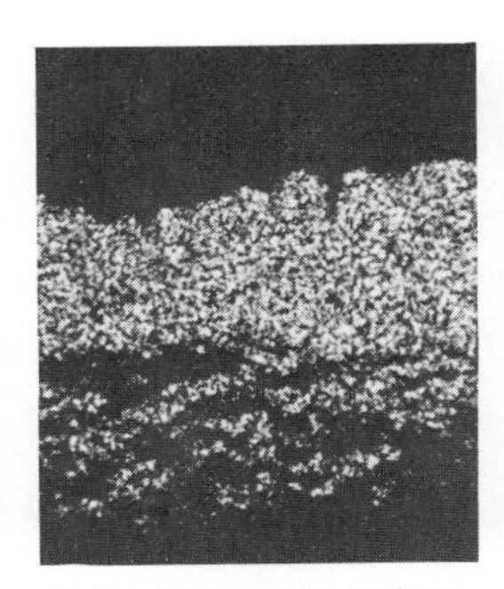

Al K_α SCANNING IMAGE

Mg K_α SCANNING IMAGE

FIG. 5. Photomicrograph and Electron Microprobe Scanning Images of Li-Al Reaction with Iron (original magnifications 300 X and 500 X).

electron microprobe analysis showed that the porous, outer reaction-product layer contained iron, aluminum, and magnesium (the latter from the 5056 aluminum alloy used to prepare the Li-Al alloy). The inner reaction layer showed only the presence of iron and aluminum and was identified as the intermetallic compound $FeAl_2$. The brittleness of the reaction product is apparent from the presence of numerous cracks. Corrosion studies were initiated in an effort to fully characterize this reaction. In the initial tests, samples of Armco electromagnet iron, AISI-1008 steel, nickel, and Types 304 and 316 stainless steel were immersed in an equal-volume mixture of Li-Al alloy and LiCl-KCl eutectic at 400 and 500°C for exposure times of 200, 400, and 600 hr. The Li-Al alloy used in the 400 and 500°C tests contained 47 and 45 at. % Li, respectively. Post-test examination showed no evidence of reaction, thereby confirming that the Li-Al alloy does not react with these materials under static conditions.

The nonreactivity demonstrated in the tests described above indicated that other conditions were necessary to account for the reaction observed in the postoperative cell examinations; thus, other mechanisms were considered. Near the end of the discharge cycle, the Li-Al alloy is depleted of lithium and is in physical contact with the metallic housing materials; moreover, electrochemical contact is provided by the LiCl-KCl electrolyte. These conditions satisfy the requirements for a galvanic couple, and the reaction rate would be expected to accelerate as the temperature is increased.

Galvanic couples of pure aluminum and various iron- and nickel-base alloys were assembled to test this hypothesis. The couples were inserted into quartz crucibles filled with LiCl-KCl eutectic salt for periods of 9 to 10 days at both 400 and 500°C. The results of these tests are shown in Table VIII. The weight loss of the aluminum anode was matched by a similar weight gain by the cathodic metals, and the weight change was one to two orders of magnitude larger at the higher temperature, as shown in columns 2 and 3 of Table VIII. The compositions of the reaction products as determined from electron microprobe analyses are listed in the last column.

TABLE VIII
REACTION OF DISSIMILAR METAL COUPLES IN LiCl-KCl ELECTROLYTE

Couple	Rate of Aluminum Pick-Up by the Cathodic Member, (mg/cm^2-yr)				Reaction Product
	400°C-APL	500°C APL	400°C Lithcoa**	500°C Lithcoa	500°C-APL
Fe/Al	8,30	95,514	2 -	3,14	$FeAl_2$
304 SS/Al	2,8	55,44	1 -	8 -	-
316 SS/Al	1	224	-	4	Inner layer $Fe(Cr)Al_2$ Outer layer $Fe(Ni,Cr)Al_2$
Ni/Al	5	337	-	11	$NiAl_3$, Ni_2Al_3
Ni-20 Cr/Al	3	105	-	27	-
E-Brite/Al	3	41	-	-	$Fe(Cr)Al_2$
1008/Al	55	90	-	-	-

*LiCl-KCl eutectic salt supplied by Anderson Physics Lab.
**LiCl-KCl eutectic salt supplied by Lithcoa.

The initial tests in this study were conducted using polarographic grade LiCl-KCl eutectic salt from the Anderson Physics Laboratory (APL). During the course of these experiments, electrolyte from the Lithium Corporation of America (Lithcoa) was substituted for the APL material. This salt is less pure and considerably less expensive; however, it has been successfully utilized in engineering cells. The results obtained with the Lithcoa salt differed considerably from the initial results, particularly at 500°C, as shown in columns 4 and 5. Since impurities appeared to be the most likely cause for the reduced reaction rate, cyclic voltammetry studies were conducted to characterize these impurities. The results showed much higher levels of OH^-, $O^=$, and possibly $CO_3^=$ anions in the Lithcoa salt (21). The obvious conclusion from these results is that one or more of these anions is involved in the formation of a passivating film on the aluminum anode surface that impedes further reactions. With the assumption that a reaction rate of 20 mg/cm^2-yr is acceptable, the results in Table VIII show that none of the materials would be suitable in cells operated at 500°C with APL salt. Based on the test results using Lithcoa salt, these same materials would be acceptable at 500°C provided the passivating impurities are not consumed in other cell reactions.

D. *Materials for Positive Electrode Applications*

Current collectors in the positive electrode are exposed to metal sulfides (FeS, FeS_2, Li_2S and Li_xFeS_y) and LiCl-KCl eutectic, a condition which results in rapid attack on most commonly used metals, *e.g.*, iron, nickel, copper and alloys of these elements. The corrosiveness of this environment is compounded by the fact that most metals react electrochemically with the electrolyte at potentials less than the normal charge cutoff potential of approximately 2.1 V for cells employing FeS_2 positive electrodes (see Table IV). For FeS positive electrodes, the conditions are less severe because of the reduced sulfur activity and the lower charge cutoff potential (∿1.8 V). Because of the importance of developing current collectors, a number of metals were subjected to corrosion tests in both FeS and FeS_2 environments at 400 and 500°C. The test specimens (3.9 to 4.5 cm^2 surface area) were immersed in molten, equal-volume mixtures of both FeS + LiCl-KCl and FeS_2 + LiCl-KCl and removed for examination after 500 and 1000 hr. The results for representative materials tested are summarized in Tables IX and X, which present both the corrosion rates and the metallographic observations for the FeS and FeS_2 environments, respectively.

Based on an acceptable corrosion rate of 75 μm/yr or less, a survey of the results in Table IX indicates that the following materials are suitable for application in FeS cells up to 500°C: molybdenum, niobium, nickel, Hastelloys B and

TABLE IX

RESULTS OF CORROSION TESTS IN FeS+LiCl-KCl at 500°C

Material	Corrosion Rate [μm/yr]*		Comments
	500 hr	1000 hr	
Armco iron	480	340	General attack, with intergranular penetration 1 to grains deep.
Nickel	+58	+39	Iron-nickel sulfide reaction layer, intergranular attack.
Molybdenum	9	12	Minor surface attack.
Niobium	+18	+30	Iron-niobium reaction layer.
Hastelloy B	+9	+12	Weakly adherent reaction layer consists of an internal molybdenum sulfide band and a plated iron layer. Localized areas of intergranular attack.
Hastelloy C	+26	+9	Similar to Hastelloy B but no apparent intergranular attack.
Fe-6Mo	110	130	Intergranular attack resulted in grain fallout.
Fe-6Mo-7Ni	25	79	Surface attack after 500 hr with intergranular penetration after 1000 hr.
Fe-6Mo-20Ni	74	90	Internal sulfidation of molybdenum resulted in a band of porosity 30 μm deep after 1000 hr.
Fe-10Mo-20Ni	17	+14	Localized areas of intergranular attack on 500-hr sample, but a reaction layer also developed on areas of the 1000-hr sample.
Fe-15Mo-20Ni	32	26	Intergranular attack resulted in grain fallout.
Fe-15Mo-30Ni	59	21	Intergranular attack was more severe for the 500-hr sample.

*Numerical values preceded by + indicate the rate of formation of a reaction layer in μm/yr.

C, Fe-20Ni-10Mo, Fe-20Ni-15Mo, and Fe-30Ni-15Mo. Comparison of the test results at 500 and 1000 hr shows that the corrosion rate for some materials decreases with time, while that of others increases. The mechanisms involved in the decreasing reaction rates are not well understood at this time, but are believed to be associated with a reduction in sulfur activity and scale formation. The photomicrograph and electron scanning images shown in Fig. 6 are an example of sulfide scale formation on Type 316 Stainless Steel. Although this sample was removed from an FeS cell, the

PHOTOMICROGRAPH 200X AS-POLISHED

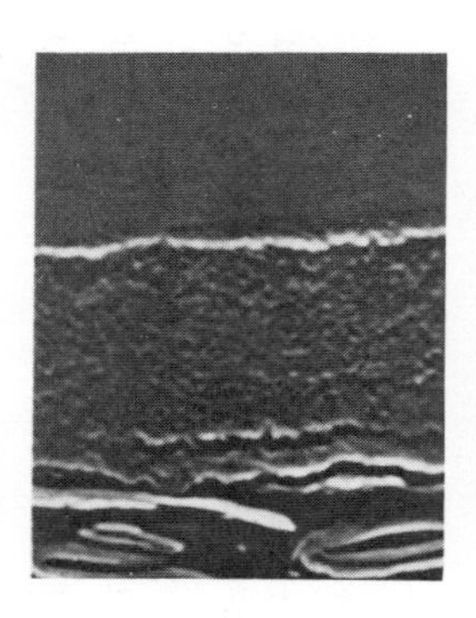

SPECIMEN CURRENT IMAGE

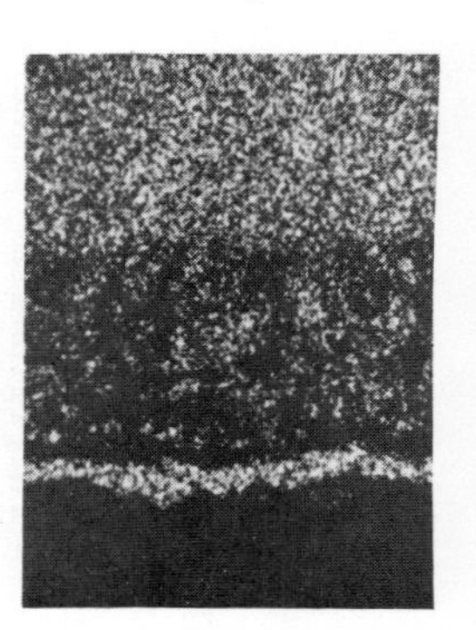

Fe K_α SCANNING IMAGE

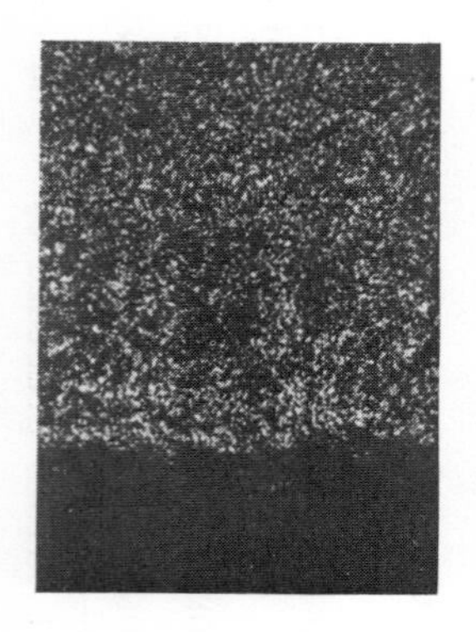

Cr K_α SCANNING IMAGE

Ni K_α SCANNING IMAGE

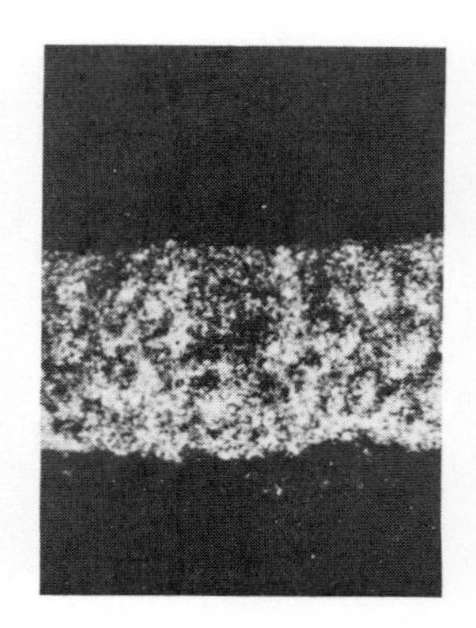

S K_α SCANNING IMAGE

FIG. 6. Photomicrograph and Electron Microprobe Scanning Images of Sulfide Reaction with Type 316 SS.

morphology of the sulfide scale is representative of that observed in corrosion tests where sulfide layers were formed. In some cases, the sulfide layer was so fragile that it was partially or completely lost in the water dissolution treatment used to remove the adhering electrolyte. As shown in Fig. 6, a thin layer of essentially pure iron is present on the outermost surface of the sulfide layer. This layer of iron apparently was formed as the FeS reacted with the alloy to form the sulfide scale; the iron freed by this reaction plated out behind the advancing sulfide scale. Sulfidation proceeds by the diffusion of the metal ion through the sulfide layer according to the dissociative mechanism proposed by Mrowec (22) for the formation of monophase scales. The presence of fissures in the thin iron layer (and the sulfide scale) provides additional access for the FeS reactant for the continued sulfidation. The formation of such sulfide layers is undesirable because the electrical conductivity is reduced. However, because of the tendency for the sulfide layers to spall, the reduction in conductivity has not substantially affected cell operation.

The iron-nickel-molybdenum alloys were prepared at ANL in an attempt to capitalize on the resistance of molybdenum to sulfide attack. Furthermore, these alloys have a lower cost potential and are considerably easier to fabricate into useful current collector forms. Since molybdenum has limited solubility in iron ($\sim$6 wt.%), nickel was added to the iron-molybdenum alloy to increase the solubility of molybdenum, with the aim of obtaining an alloy that combines the corrosion properties of molybdenum with the fabricability of a ductile material. Although the test results show that these alloys are significantly better than iron, they do not appear to offer any advantage over the commercially available alloys, Hastelloys B and C, for applications in FeS cells.

The results of the corrosion tests of various

metals in the mixture of FeS_2 and LiCl-KCl at 400 and 500°C are listed in Table X. As expected, the corrosion rates for all of

TABLE X

RESULTS OF COMPATIBILITY TESTS IN FeS_2+LiCl-KCl AT 400 AND 500°C

Metal	Corrosion Rate (mm/yr)		Remarks
	400°C	500°C	
Molybdenum	0.001	+0.011	Weakly adherent MoS_2 layer formed at 500°C
Niobium	0.72	>4.5	Completely reacted after 500 hr at 500°C
Hastelloy B	0.10	4.0	Minor intergranular attack
Hastelloy C	0.24	3.3	Minor Intergranular attack
Fe-6Mo	0.77	>22	Extensive reaction after 500 hr at 500°C
Fe-20Ni-6Mo	-	>22	Completely reacted
Fe-20Ni-10Mo	-	>22	Completely reacted
Fe-20Ni-15Mo	-	>22	Completely reacted
Fe-30Ni-15Mo	-	11	Porous internal structure and outer sulfide layer

the metals were much higher than those exposed to the FeS environment. Molybdenum was the only material that showed an acceptably low reaction rate at both temperatures.* The electron microprobe scanning images of the sulfide scale formed on molybdenum exposed to the FeS_2 environment for 1000 hr at 500°C is shown in Fig. 7. Similar sulfide layers have been observed on molybdenum current collectors from FeS_2 cells operated near 500°C.

*Recrystallized tungsten appears resistant to sulfide attack, but is even more difficult to fabricate and more costly than molybdenum. Also, carbon is not attacked by the sulfides and has been used in some cells.

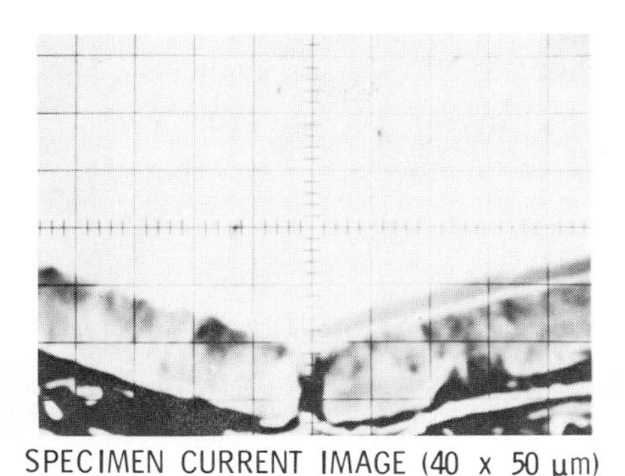

SPECIMEN CURRENT IMAGE (40 x 50 μm)

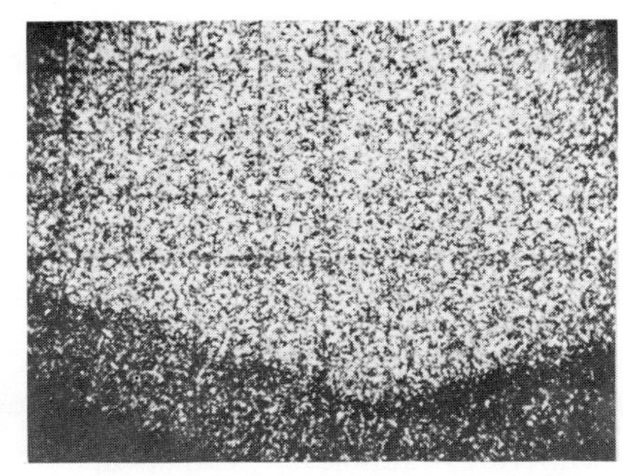

MOLYBDENUM SCANNING IMAGE

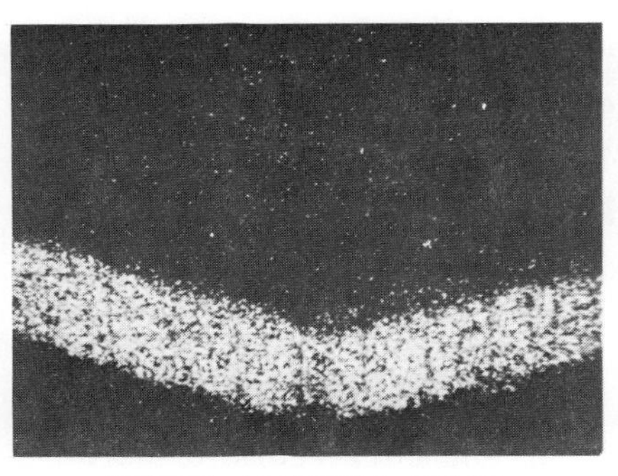

SULFUR SCANNING IMAGE

IRON SCANNING IMAGE

FIG. 7. Electron Microprobe Scanning Images of Molybdenum After Reaction with FeS_2/LiCl-KCl for 1000 hr at 500°C

The sulfide has been identified as MoS_2 by X-ray diffraction analysis. No evidence of MoS_2 has been observed on samples tested at 400°C in either FeS or FeS_2 environments. The sulfidation mechanism for molybdenum appears to be the same as that discussed above for Type 316 stainless steel. Although the sulfide layer appeared quite uniform, it was not particularly adherent and showed evidence of spalling; thus, the sulfide layer cannot be classified as protective.

At 500°C, all of the other metals exhibited

reaction rates that were much higher than those observed at 400°C; for example, the corrosion rates of the two Hastelloys were one to two orders of magnitude higher at 500°C. In general, the morphology of the sulfide scales was similar to that for the Hastelloys and the iron-base alloys (4). Two distinct reaction zones were observed. The outer zone consisted of three sulfide layers that had apparently formed initially at the original interface of the sample; the intermost of the three layers was a mixture of iron-nickel sulfides and molybdenum, the middle layer was molybdenum sulfide, and the outer layer was a mixture of weakly adherent iron and nickel sulfides. The inner zone, adjacent to the unreacted base metal, was adherent, porous and depleted of both iron and nickel, and was free of sulfur. This zone was apparently formed by the preferential outward diffusion of iron and nickel through the sulfide scale. Additional studies are being conducted to characterize the reaction mechanisms and kinetics involved in the sulfidation of metals in mixtures of FeS + LiCl-KCl and FeS_2 + LiCl-KCl.

The discussion above has dealt with the evaluation and selection of metals for current collectors in both positive and negative electrodes. At this time, it appears appropriate to discuss the structure or shape of these current collectors since these factors can have a pronounced effect on the performance of the electrodes. Typical current collectors have included steel wool, wire screens, Feltmetal[R] plaques and expanded mesh fabricated from metals such as low carbon steel, nickel, stainless steel, and molybdenum. Except for molybdenum expanded mesh, all of these structures and metals have been used as current collectors in negative electrodes and in FeS positive electrodes. Molybdenum expanded mesh has been the normal current collector* for FeS_2 positive electrodes. These structures have been moderately efficient as current collectors but are difficult

*Carbon structures and molybdenum honeycombs have been used in a few cells.

to work with. Figure 8 shows a scanning electron micrograph of a low-density, macroporous metal structure*. This type of cellular

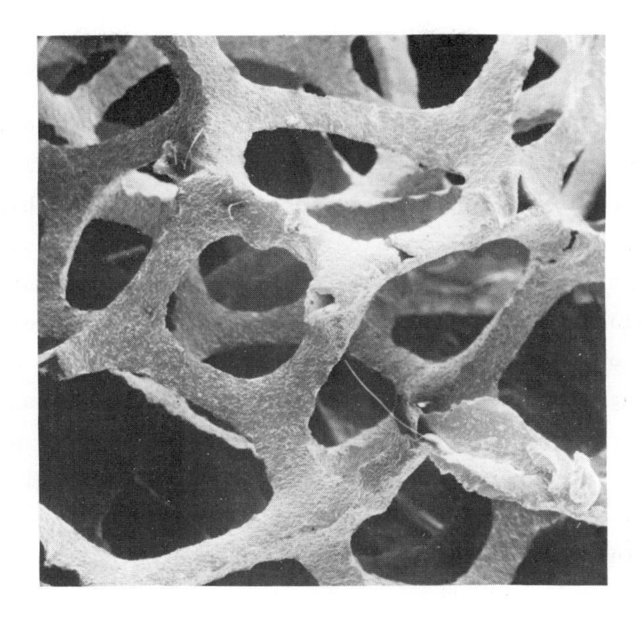

FIG. 8. Scanning Electron Micrograph of Macroporous Metal Current Collector (original magnification 26 X)

structure appears ideal for current collectors for two reasons: 1) this particular form adds rigidity and dimensional stability to the particle-like electrode, and 2) the uniform and continuous network provides more efficient current collection throughout the electrode. Disadvantages include difficulties in obtaining adequate loadings of active materials to meet specific energy design goals, and the present high cost of the material. Efforts are under way to resolve these problems and to develop similar structures using alloy compositions that are suitable for

*Foametal^R is a product of Hogen Industries, Willoughby, Ohio. A similar material, Retimet^R, is a product of Dunlop, Ltd. England.

application in FeS_2 positive electrodes. In the latter case, preliminary results have been encouraging.

CONCLUSIONS

Results have been reported on the extensive materials evaluation program for the selection of suitable low-cost materials for application in high-temperature, high-energy lithium-aluminum/metal sulfide batteries being developed at Argonne National Laboratory. Ceramic and metallic materials have been evaluated using static corrosion tests in various environments according to their intended application, *i.e.*, ceramics for feedthrough insulators and electrode separators, metals for negative electrode current collectors and cell hardware, and metals for positive electrode current collectors. The final evaluation of all materials has been based on their behavior in operating cells. In general, the results of the static corrosion tests have correlated well with the behavior observed in postoperative cell examinations of these materials. Analysis of the results leads to the following conclusions regarding materials for application in lithium-aluminum/metal sulfide batteries.

1. Lithium compatibility studies indicate that the only ceramic materials that are compatible with molten lithium (or Li-Al alloy) are those which are thermodynamically more stable than the corresponding lithium compounds. Corrosion resistance of these materials is an extremely sensitive function of impurity content and location of impurities. The following materials have been identified as suitable insulator and separator materials: Y_2O_3, BeO, MgO, BN and AlN. Of these materials only Y_2O_3 and BN are commercially available as fibers suitable for the fabrication of electrode separators.

2. Corrosion studies have shown that low-cost, iron-base alloys (low carbon steels and stainless steels) are

suitable construction materials for negative electrode current collectors and cell housings at the negative potential. At elevated temperature (∿500°C), a galvanic reaction can occur between these materials and the Li-Al alloy, depleted of lithium (a fully discharged condition), resulting in a brittle intermetallic compound of the general type $FeAl_2$.

3. For FeS positive electrodes, no problems have been encountered in the identification of suitable current collector materials. Alloys containing molybdenum, such as Hastelloys B and C and iron-nickel-molybdenum, have shown acceptable resistance to sulfide attack at both 400 and 500°C.

4. Molybdenum is the only metal that has shown good compatibility with FeS_2 cells at both 400 and 500°C. At 400°C, one or more of the molybdenum alloys listed above may be a suitable current collector material. Although molybdenum is a difficult material to fabricate, expanded mesh current collectors have been prepared and successfully used in FeS_2 cells.

ACKNOWLEDGMENT

The author is grateful to L. Burris, D. S. Webster, and P. A. Nelson of the Chemical Engineering Division of ANL for providing administrative support. The author wishes to acknowledge the contributions of the members of the Materials Group: F. C. Mrazek, J. A. Smaga, J. P. Mathers, K. M. Myles, J. L. Settle and T. W. Olszanski. The editorial assistance of G. M. Kesser is gratefully acknowledged.

REFERENCES

1.(a) P. A. Nelson et al., "ANL High-Energy Batteries for Electric Vehicles," in Proc. 3rd Int. Electric Vehicle Symp., Washington, D. C. (1974).

(b) H. Shimotake et al., "Lithium/Sulfur Cells and Their Potential for Vehicle Propulsion," in Proc. 1st Int. Electric Vehicle Symp., p. 392, Electric Vehicle Council, New York (1969).

(c) M. L. Kyle et al.,"Lithium/Sulfur Batteries for Electric Vehicle Propulsion," in Proc. 1971 IECEC, p. 80, SAE, New York (1971).

(d) P. A. Nelson et al., Development of High-Specific-Energy Batteries for Electric Vehicles, Progress Report for the Period February 1973-July 1973, USAEC Report ANL-8039, Argonne National Laboratory (December 1973).

(e) P. A. Nelson et al., High-Performance Batteries for Off-Peak Energy Storage, Progress Report for the Period January-June 1973, USAEC Report ANL-8038, Argonne National Laboratory (1974).

(f) P. A. Nelson et al., High-Performance Batteries for Off-Peak Energy Storage and Electric-Vehicle Propulsion: Progress Report for the Period July-December 1975, ANL-76-9, Argonne National Laboratory (in preparation); Progress Report for the Period January-June 1975, ANL-75-36 (in press); Progress Report for the Period July-December 1974, ANL-75-1 (1975); Progress Report for the Period January-June 1974, ANL-8109 (1974).

2. E. J. Cairns and R. K. Steunenberg, "High-Temperature Batteries," in Progress in High Temperature Physics and Chemistry, C. A. Rouse, Ed., p. 63, Pergamon Press, New York (1973).

3. J. E. Battles, F. C. Mrazek, W. D. Tuohig, and K. M. Myles, "Materials Corrosion in Molten-Salt Lithium/Sulfur Cell," Corrosion Problems in Energy Conversion and Generation, pp. 20-31, C. S. Tedmon, Jr. (ed), The Electrochemical Society, (1974).

4. K. M. Myles et al., "Materials Development in the Lithium-Aluminum/Iron Sulfide Battery Program at Argonne National Laboratory," Proc. Symp. and Workshop on Advanced Battery Research and Design. USERDA Report ANL-76-8, p. B-50 (March 1976).

5. Mars G. Fontana and Norbert D. Greene, Corrosion Engineering, p. 8, McGraw-Hill Book Co., New York (1967).

6. A. Wachter and R. S. Treseder, Chem. Eng. Progr. 43, 315 (1947).

7. D. L. Beals and W. H. Mapes, Separator Materials for the Lithium-Chlorine Battery, Research and Development Technical Report ECOM-3105 (March 1969).

8. F. C. Arrance and M. J. Plizga, Stable Inorganic Matrix Materials for High Temperature Batteries, Technical Report ECOM-0456-F (August 1968).

9. J. Economy and R. V. Anderson, Textile Res. J., 36, 994 (1966).

10. J. Economy and R. V. Anderson, J. Polymer Sci., Part C, 19, 283 (1967).

11. J. Economy., Research/Development, June, 1967.

12. J. Economy, W. D. Smith, and R. Y. Lin, App. Polymer Symp. 21, 131 (1973).

13. B. H. Hamling, Zircar Products, Inc. N.Y., private communication.

14. R. Hamilton, Carborundum Company, Niagara Falls, N.Y., private communication.

15. R. D. Walker, Jr., "Separators for Lithium Alloy-Iron Sulfide Fused Salt Batteries," Proc. Symposium and Workshop on Advanced Battery Research and Design. USERDA Report ANL-76-8, p. B-41 (March 1976).

16. Gerrit DeVries, in Corrosion by Liquid Metals, J. E. Draley and J. R. Weeks, Eds., p. 251, Met. Soc. AIME, Plenum Press, New York (1970).

17. J. R. DiStefano and A. P. Litman, Corrosion, 20, 392t (1964).

18. R. L. Klueh, in Corrosion by Liquid Metals, J. E. Draley and J. R. Weeks, Eds., p. 177, Met. Soc. AIME, Plenum Press, New York (1970); R. L. Klueh, Met. Trans., 5, 875 (1974).

19. R. W. Harrison, Ibid, p. 217.

20. J. R. DiStefano, Corrosion of Refractory Metals by Lithium, USAEC Report ORNL-3551 (April 1966).

21. C. Melendres, Argonne National Laboratory, Argonne, Ill., private communication.

22. Stanislaw Mrowec, "Mechanism of High Temperature Metallic Corrosion by Sulfur Vapors," High-Temperature Metallic Corrosion of Sulfur and its Compounds, Z. A. Foroulis, Ed., p. 55, The Electrochemical Society (1970).

Chapter 27

THE MATERIALS SCIENCE OF THE SUPERFLYWHEEL

David W. Rabenhorst

The Johns Hopkins University
Applied Physics Laboratory
Laurel, Maryland 20810

I. INTRODUCTION

Nurbei Gulia in his recent book on flywheel technology (Ref. 1) gives a most detailed historical documentation on kinetic energy storage, from the Uruk potter's wheel of 4000 years ago to the recent flywheel-powered buses in Europe, the Soviet Union and the United States. Much of this early history was originally detailed by Robert Clerk (Ref. 2) in 1963, who described the applicability of the flywheel to a number of vehicles, including British race cars, automobiles, switching engines and the American flywheel-powered Howell torpedo of 1885.

One of the earliest applications of the flywheel as a practical electric power supply for electricity apparently was a Soviet wind power generating machine, which was built in 1931 and is said to be still in operation (Ref. 1).

Most of these historical applications used steel flywheels. They illustrated the many advantages of flywheel energy storage, even though the energy storage per pound was much less than that of contemporary batteries, and performance was very poor according to present technology. The advantages included:

1. Rapid charge/discharge capability;
2. Unlimited depth of discharge and number of cycles;
3. Simplicity, no maintenance, and infinite shelf life;
4. Flexibility, i.e., input/output can be electrical (AC or DC), hydraulic, or mechanical, or any combination; and
5. Reliability — Performance not degraded by environmental conditions.

But these early flywheels also had a number of very important disadvantages, and these disadvantages inhibited flywheel development for 4000 years. Clerk described these disadvantages as:

1. Relatively poor energy storage per pound;
2. Poor efficiency, i.e., short rundown time, and
3. The ever-present hazard of catastrophic failure.

Although modern technology has provided an order of magnitude improvement in steel flywheel energy density and efficiency, the hazard problem has gotten proportionately worse. Consequently, modern steel flywheels typically are derated to a small fraction of their theoretical maximum performance, which is generally inversely proportional to the size of the flywheel. For example, a small modern steel flywheel theoretically can store as much as 26 watt-hours per pound (Wh/lb), but practical considerations limit its rating to about 12 Wh/lb (Ref. 3). On the other hand, a steel flywheel currently used for plasma physics experiments at the Applied Physics Laboratory weighs 7700 pounds and is rated at about 0.25 Wh/lb.

The flywheel is being introduced in an increasing number of applications throughout the world, including small road and non-road vehicles, buses, a variety of electrical power supplies, hoists, aircraft catapults, trains, and earth-moving vehicles.

A 20,000-horsepower, electric-powered-dragline earth-moving machine currently in use in Australia (Ref. 4) uses a flywheel to reduce its peak power demands on the local transmission line within acceptable limits. In New York City an experimental subway train is being equipped with a flywheel braking system to recover a large portion of the braking energy usually lost in heat. This would virtually eliminate overheating of the subway tubes while, at the same time, providing an annual saving of tens of millions of dollars through the use of the recovered energy for acceleration of the trains. This program is being supported by the New York Metropolitan Transit Authority and the Federal Department of Transportation.

Flywheels are also being applied to urban buses. During a 17-year period ending about 1948, a fleet of flywheel-powered buses operated in Switzerland. Because of the flywheel performance limitations of that time (about 3 Wh/lb), those buses were limited to the range between bus stops. Today flywheels are being incorporated in buses for reasons other than primary power. In the Soviet Union, flywheel braking systems have proven that a bus can be operated more efficiently, with less maintenance, and at greater route speeds (Ref. 1). An experimental bus program in the city of San Francisco will demonstrate that the present electric trolley-buses will be able to operate for several miles on route extensions without trolley lines (Ref. 5).

Because of its capability of delivering energy at very high rates, the flywheel is ideally suited for the aircraft catapult, and hardware programs are under way in this country as well as in England and the Soviet Union. The Soviets have also demonstrated successful flywheel installations in wind power-generating machines, experimental trucks, switching locomotives, large road scrapers, and a variety of electric power supplies, including an electric welding unit (Ref. 1).

Modern flywheels are also used throughout the world in a variety of electric power supplies. Occasionally they are used to provide extremely high power peaks that would not otherwise be practical, such as for the plasma physics experiments in the United States, France, Germany, Japan and other countries. On the other hand, countless flywheels are used throughout the world to provide uninterrupted emergency power for various lengths of time for computers and other sensitive equipment, which require uninterrupted power in order to avoid very expensive consequences.

In the past, utilities have found pumped-hydro storage an economical means for utilizing some off-peak energy to pump water to an elevated reservoir for storage. Recovering the potential energy of this stored water during peak periods has allowed minimum use of fossil fired gas turbines with their high operating costs. However, suitable pumped hydro sites are scarce and are also meeting with environmental opposition.

In the future many homes and industrial facilities will make use of on-site power generation systems, all of which demand effective energy storage in order to be practical. The solar home or plant must have adequate energy storage, since there will be no sunlight for more than half of the time. If wind generators are employed, there will be a rather obvious need for energy storage, both to supply power in calm periods and to provide a steady output in the face of a rapidly changing energy input.

The flywheel represents additional potential savings over other types of energy storage being considered, because it can have lower initial cost, longer operating life or higher efficiency. The flywheel is also superior to all other storage methods for urban locations, considering safety, environmental impact, land use, aesthetic and other factors of community acceptability. There are essentially no restrictions on location in the urban environment.

But, a conventional steel flywheel is not cost-effective for many of the applications described in the foregoing. However, it is generally agreed that flywheels which can make effective use of various available filamentary super materials can be quite cost effective, provided that the incremental manufacturing cost can be held to a fraction of the material cost. These superflywheels, as they are called, make use of unidirectional composite or filament materials having many times the strength-to-density ratio of the best available steel.

A concept developed at The Johns Hopkins University Applied Physics Laboratory, which utilizes a large number of thin rods made of unidirectional composite material is illustrated in Fig. 1, and is described in detail in Ref. 6. This cir-

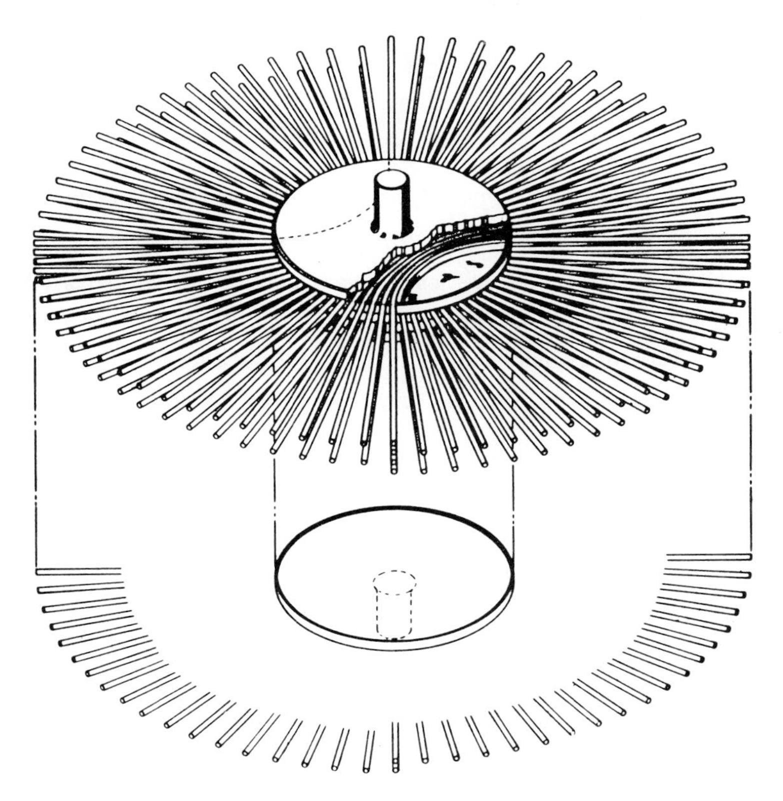

FIG. 1. Laminated Circular Brush Rotor.

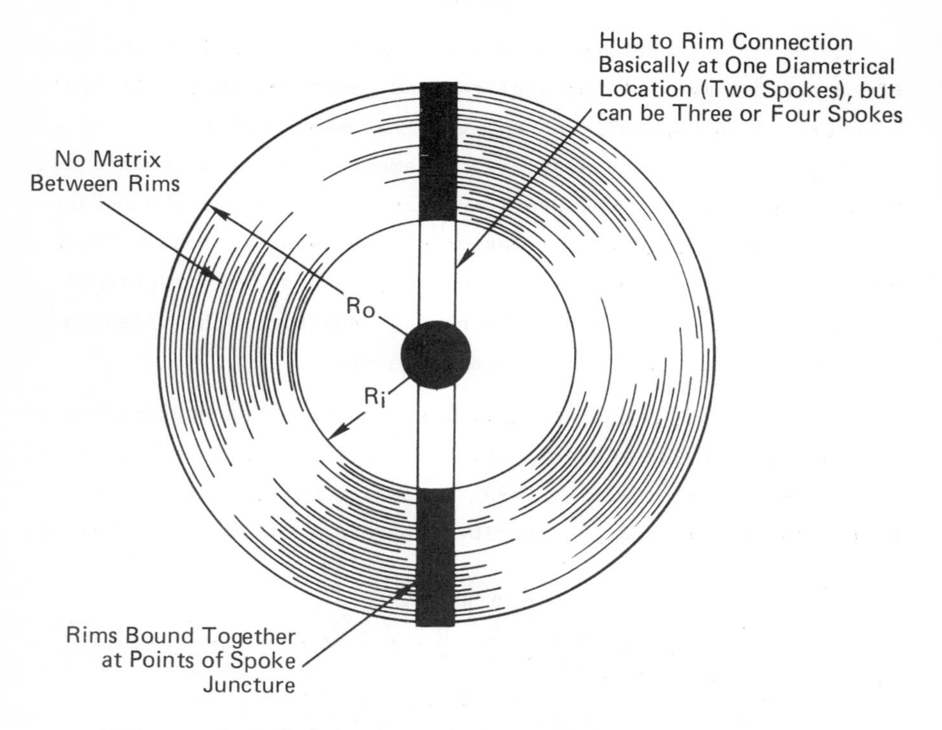

FIG. 2. Multirim Superflywheel Basic Configuration.

cular brush configuration has the advantage of intrinsic simplicity and safety, but its principal drawbacks are its relatively low volume efficiency and somewhat higher rpm requirement, relative to other possible flywheel types.

The second applicable filamentary flywheel is the multi-ring type, which does not have these disadvantages and provides some additional performance in energy per unit of volume. It is illustrated in Fig. 2 and is described in detail in Ref. 7. The comparison of theoretical performance of the circular brush and the multi-ring flywheel is shown in the table in Fig. 3.

The performance, or figure of merit, of the multi-ring Superflywheel is a function of its inner radius in proportion to its outside radius. Performance in terms of energy/weight is high when this ratio is maximum (one infinitely thin

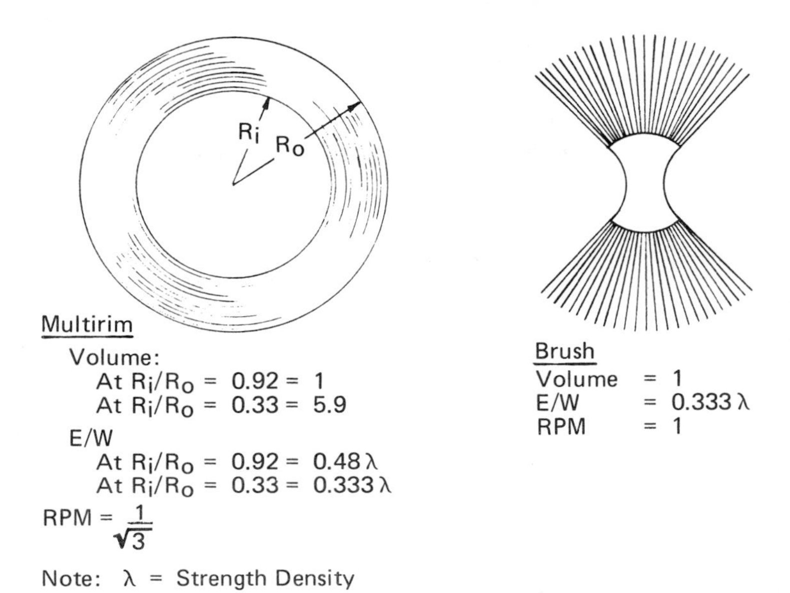

FIG. 3. Theoretical Performance Comparison Multirim vs Brush Superflywheel.

ring) and low when the inside radius is zero, or when the flywheel is a solid disc of thin rings. The performance in terms of energy/volume is maximum when the disc made by the rings is a solid disc, and the inside radius is zero, even though the energy/weight is minimum at this condition. These relationships are illustrated in Fig. 4.

It will be seen that both of these basic configurations can be configured to make effective use of bare filament, matrix-less materials with quite significant advantages in performance and cost. The strength-to-density of several bare filament materials is generally more than twice that of the filaments in a unidirectional composite material. This results in a corresponding increase in the performance of flywheels making effective use of this property.

Before considering the details of these configurations it will be of interest to discuss the current state-of-art of composite flywheel technology.

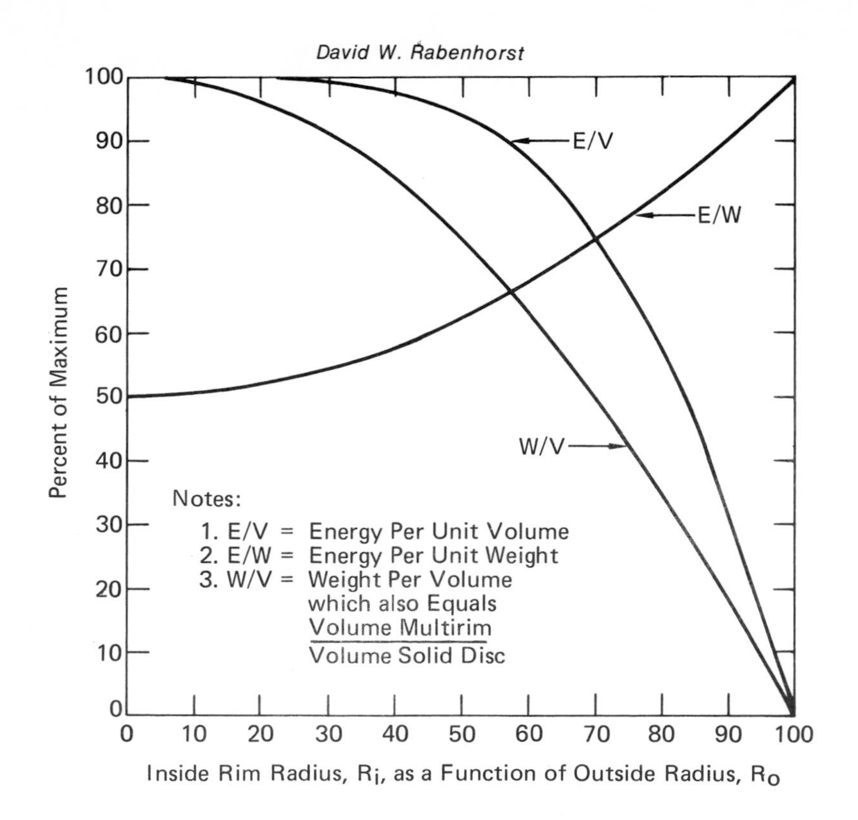

FIG. 4. Multirim Superflywheel Performance vs Inside Radius.

II. COMPOSITE FLYWHEEL STATE-OF-THE-ART

The initial flywheel applications being proposed for funding by the Federal Government and others include the following:

1. Subway accel-decel system
2. Electric Trolley bus range extension
3. Flywheel/internal combustion engine hybrid car

4. Mine shuttle car
5. Switching engine
6. The Urban Car
7. Aircraft Landing Gear Power Boost
8. Helicopter High Speed Hoist
9. High powered mobile equipment
10. Utilities peaking.

It will be noted that most of these applications involve steel flywheels having rather nominal performance. The flywheels in the experimental New York Subway System, for example, have an operating capability of only 5.5 watt hours per pound.

Some agencies have taken steps to investigate anisotropic flywheels with improved performance over the conventional steel flywheel. These studies cover the range of 30 to 40 watt hours per pound; however, some difficulties are anticipated in meeting these requirements with the composite flywheel configurations currently being investigated, as discussed in the following.

Previous attempts to develop multi-ring flywheels in the form of filament wound disc and ring structures have been only partially successful. The principal reason that they have not been completely successful is that differential radial stresses caused experimental rotors to break up in concentric rings at only a fraction of the rotational speed that would have resulted in the filaments being stressed to their design limit. The differential radial stresses are caused by the fact that the stress difference between two thin spinning rings is a function of the square of their spin radii. Thus, if one ring, such as the outer winding of a wound structure, has twice the radius of another ring, such as the inner radius of a wound structure, then the stress in the outer ring would be four times the stress in the inner ring. If this were the only factor involved, the outer ring would stretch approximately four times that of the inner ring, since the strain (stretch) of the typical materials being considered is directly proportional to the applied stress. However, the amount of actual stretch is also a function of the length (circumference) of the rings which varies directly in proportion to the radius. Therefore the total stretch of the rings will be proportional to the cube of the radius.

It is also characteristic of these unidirectional materials that they have only a few percent as great a stress capability in the transverse direction normal to the fibers, as they have in the direction of the fibers. This transverse stress capability is generally insufficient to accommodate the radial stress loads in a filament wound flywheel rotor disc.

In theory, if a specific ring in a multi-ring configuration is thin enough, then its differential radial stress load can be held to a level within the transverse stress capability of the composite material used in its construction. The remaining problem is then to arrive at an acceptable multi-ring flywheel configuration which will allow satisfactory tieing of the rings together in one contiguous structure, yet be able to accommodate the different expansion of the rings.

One possible method of accomplishing this is being pursued by the Electric Power Research Institute under a contract to the Brobeck Associates of Berkeley, California. This flywheel is the invention of Richard and Stephen Post. It is illustrated in Fig. 5, and is described in detail in Refs. 8 and 9. The flywheel is divided into a number of thin rings, which, in turn, are held together by elastomer rings. The elastomer allows one adjacent ring to expand from rotation independently with respect to the other adjacent ring. In order to keep this differential expansion within manageable limits, that is, within the capabilities of usable elastomers, it is necessary that the ratio of ring modulus of elasticity to ring density be increased

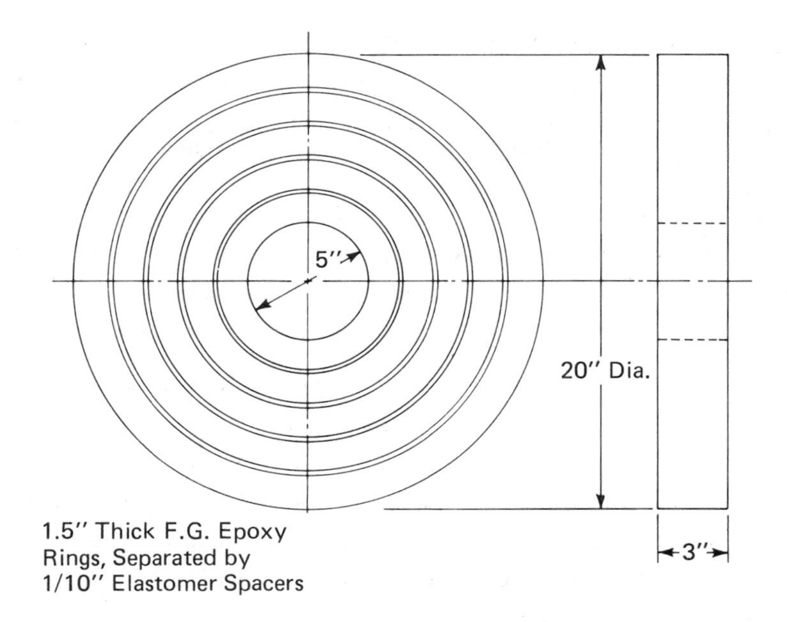

FIG. 5. Post-Brobeck Superflywheel.

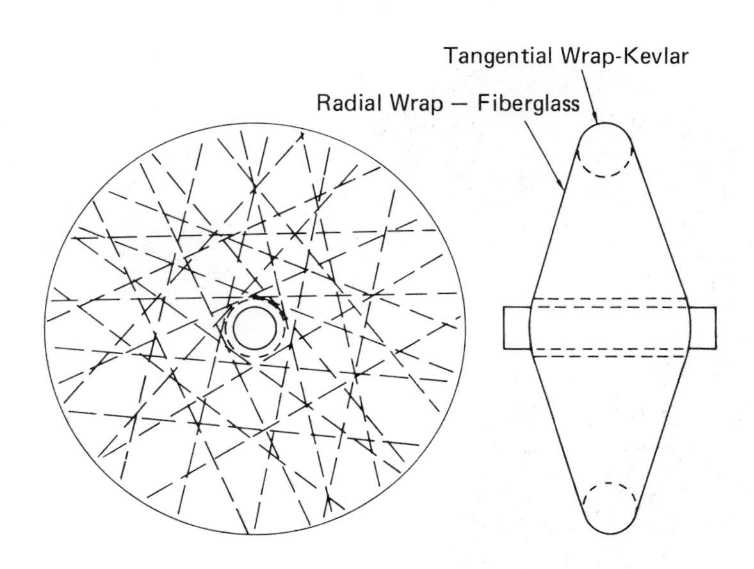

FIG. 6. Rockwell Composite Flywheel.

in a special relationship as the diameter of each successive concentric ring is increased. This means, of course, that each ring must be made of a different material, in order to obtain the desired modulus to density ratio.

The Electric Power Research Institute recently announced that this flywheel configuration is not currently cost-competitive with other energy storage methods being considered, and has abandoned further support of this program (Ref. 10).

A second approach is being investigated by Rockwell International under the sponsorship of the Energy Research and Development Administration. The major features of this configuration are illustrated in Fig. 6. While the performance claimed for it is somewhat better than that of the Post-Brobeck flywheel, its manufacturing cost is yet to be established. It is probably of the same order of magnitude as the one shown in Fig. 7, since it is quite similar. This latter type is manufactured by Aerospatiale (France) under contract with the (U.S.) Comsat Corp. Performance is nominal and cost is relatively high.

FIG. 7. Aerospatiale Kevlar Wound Flywheel.

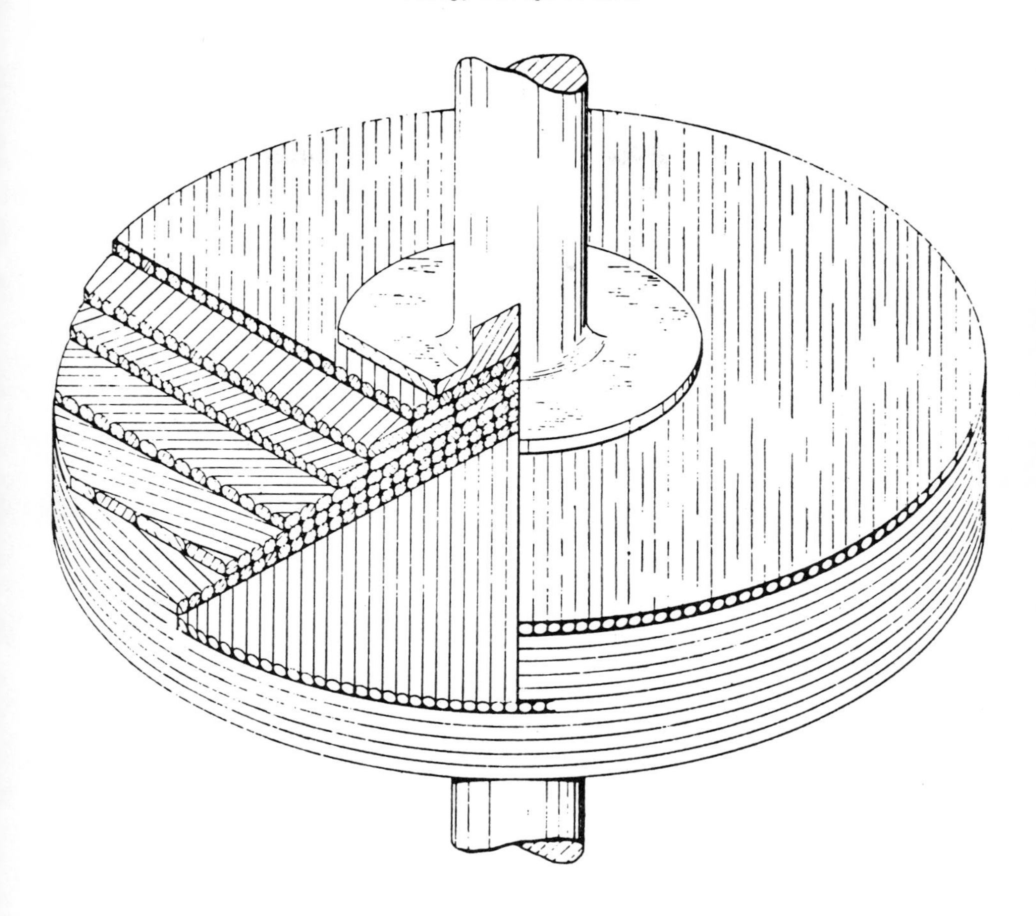

FIG. 8. Pseudo-Isotropic Configuration.

Also, the Applied Physics Laboratory is developing flywheels in this intermediate performance category (30-50 watt-hours per pound) under a grant from the National Science Foundation. In addition to the thin rod and thin ring types mentioned previously, this program emphasizes the configuration shown in Fig. 8. While the energy per unit weight is moderate ($\approx$ 30 watt hours per pound) its volume effectiveness and low manufacturing cost appear to be its main attributes.

III. THE APL SPIN TEST PROGRAM

The Applied Physics Laboratory has conducted several hundred spin tests of a wide variety of potential anisotropic flywheel materials over the past five years. These led to a number of preliminary flywheel spin tests of promising superflywheel configurations. Most of these tests were conducted without the benefit of outside financial support; however, the most recent tests were conducted in anticipation of the Laboratory receipt of a grant for this work from the National Science Foundation. The NSF funded program began on March 1, 1976.

A typical spin test consists of spinning the test article at increasing rotational speed until failure occurs. It is then a relatively simple matter to calculate the ultimate tensile strength and intrinsic energy density from the rotational speed at failure. The facility used for these tests is illustrated in Fig. 9. This equipment, which was built at the Applied

FIG. 9. APL Spin Test Facility.

Physics Laboratory, can accommodate test articles up to a diameter of 32 inches and weighing about five pounds. Rotational speeds up to 50,000 rpm can be handled with this equipment, and it can be upgraded to about 100,000 rpm by the addition of another timing belt and associated pulleys.

Samples of typical materials test articles are shown in Figs. 10, 11, and 12. The materials evaluation

FIG. 10. Graphite/Epoxy Rod Test.

FIG. 11. Glass Rod Test

tests have generally confirmed predictions that materials are currently available to satisfy the three basic flywheel applications performance categories. These are the very high performance/high cost; the high performance/low cost; and the very low cost flywheels.

FIG. 12. Superflywheel Test Rotor Fiberglass Wire

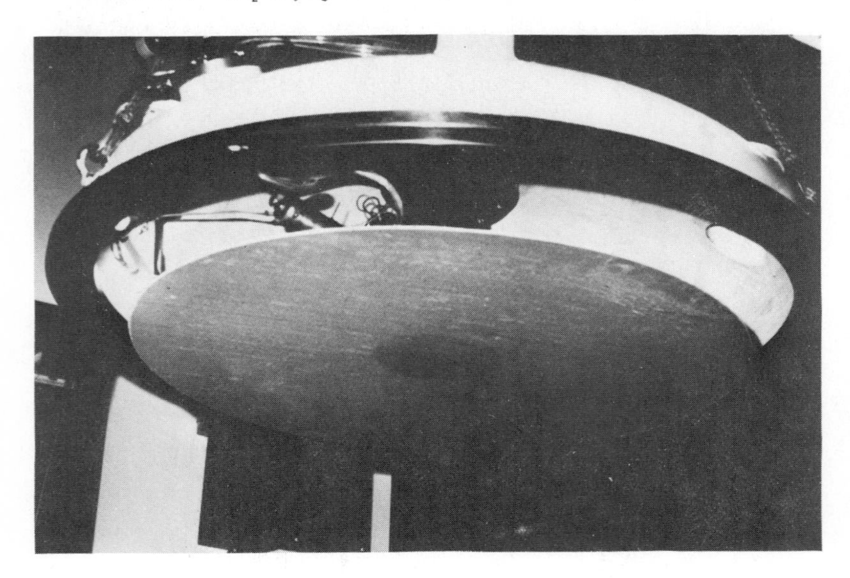

FIG. 13. 3M Scotch-Ply Superflywheel Test Rotor

Examples of some of the flywheel rotors tested are illustrated in Figs. 13 through 16. These represent a fairly good cross section of the utilization of the basis material configurations which are applicable to the anisotropic flywheel.

FIG. 14. Wood Cross-Ply Test Rotor

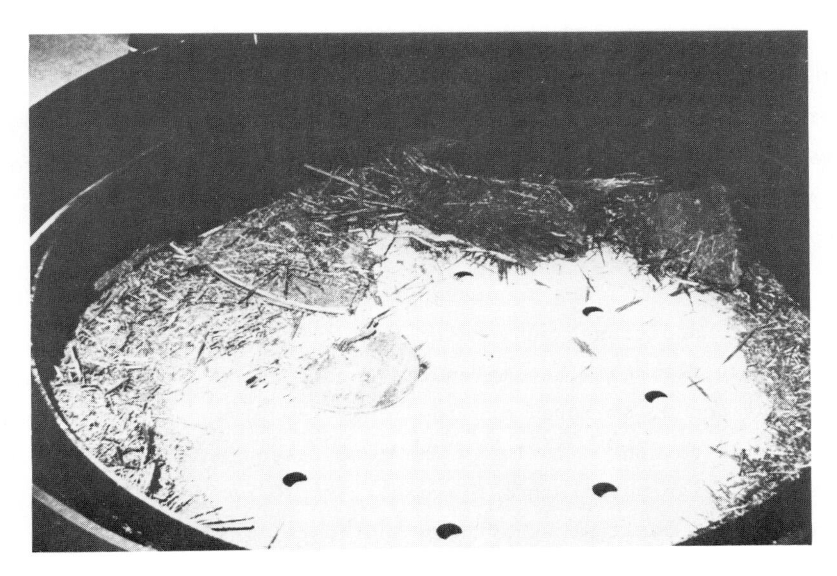

FIG. 15. Debris from Wood Test Rotor

The first type is illustrated in the fiberglass wire coil rotor in Fig. 12, which was manufactured by the Condex Corp. of Camden, N.J. This fiberglass wire is 0.045 inch in diameter, and has an advertised tensile strength of about 400,000 psi.

The second rotor configuration illustrated is that shown in Fig. 13. This is fabricated from standard E-glass cross-ply sheets produced by the 3M company.

Failure of these composite rotors usually results in most of the rotor mass being reduced to dust-like particles, as shown in Fig. 15.

Corresponding pictures of a very low cost wooden cross-ply rotor are illustrated in Figs. 14 and 15.

Figure 16 shows the basic configuration of one of many so-called bare filament rotors that are being tested at the Applied Physics Laboratory. The principal rotor material illustrated in this configuration is steel wire having a tensile strength of 350,000 psi. Such wire has an intrinsic energy density of about 20 watt hours per pound in this bare filament configuration. At about 50¢ per pound, this material is an excellent candidate for a very low cost flywheel. More importantly, however, much of the bare filament rotor technology derived from these inexpensive tests is applicable to future bare filament rotors having very high performance.

Tests conducted during the subject program, together with additional subsequent tests, indicate the feasibility of building flywheels having energy densities of around 100 watt hours per pound, or more, provided that low flywheel cost is not a governing design criterion. The tests also confirmed that moderately high performance flywheels, at 30 to 40 watt hours per pound, can be built at acceptably low costs. They also indicated that very low cost/low performance flywheels are feasible for

FIG. 16. Bare Filament Test Rotor Steel Wire

those applications where flywheel weight and volume are not critical.

REFERENCES

1. Gulia, N. V., USSR Book, "Inertial Energy Accumulators," Voronezh University Press, Voronezh, 1973.

2. Clerk, R. C., "The Utilization of Flywheel Energy," ASME Paper No. 711A, June 1963.

3. Gilbert, R. R., et al., "Flywheel Drive Systems Study Final Report," Lockheed Missiles and Space Company, Inc., Report No. LMSC-D246393, July 31, 1972.

4. Kilgore, L. A., and D. C. Washburn, Jr., "Energy Storage at Site Permits Use of Large Excavators on Small Power Systems," Westinghouse Engineer, November 1970, Vol. 30, No. 6.

5. Lawson, L. J., "Kinetic Energy Storage for Mass Transportation," Mechanical Engineering, September 1974.

6. U.S. Patent No. 3,737,694, "Fanned Circular Filament Rotor," June 5, 1973, D. W. Rabenhorst.

7. APL/JHU Report TG 1240, "The Multirim Superflywheel," D. W. Rabenhorst, August 1974.

8. U.S. Patent No. 3,683,216, "Inertial Energy Storage Apparatus and System for Utilizing the Same" August 8, 1972, Richard F. Post.

9. U.S. Patent No. 3,741,034, "Inertial Energy Storage Apparatus," June 26, 1973, Stephen F. Post.

10. Industrial Research Magazine, March 1976, p. 20-22, "Multiring Flywheel Study Stops Spinning."

11. APL/JHU Report CP 011, "Heat-Engine/Mechanical-Energy-Storage Hybrid Propulsion Systems for Vehicles-Final Report," March 1972, G. L. Dugger et al.

Chapter 28

METAL HYDRIDES FOR ENERGY STORAGE

G. G. Libowitz

Materials Research Center
Allied Chemical Corporation
Morristown, New Jersey 07960

I. PROPERTIES OF METAL HYDRIDES

Metal hydrides which may be used for energy storage are solid compounds which are formed by direct combination of a metal or an alloy with hydrogen according to the following reaction:

$$M + \frac{x}{2}H_2 \rightleftarrows MH_x \quad (1)$$

This is a spontaneous exothermic reaction which can be easily reversed by the application of heat. If the formation reaction is carried out isothermally, it can be represented by the pressure-composition isotherm in

Fig. 1. Hydrogen first dissolves in the metal alloy

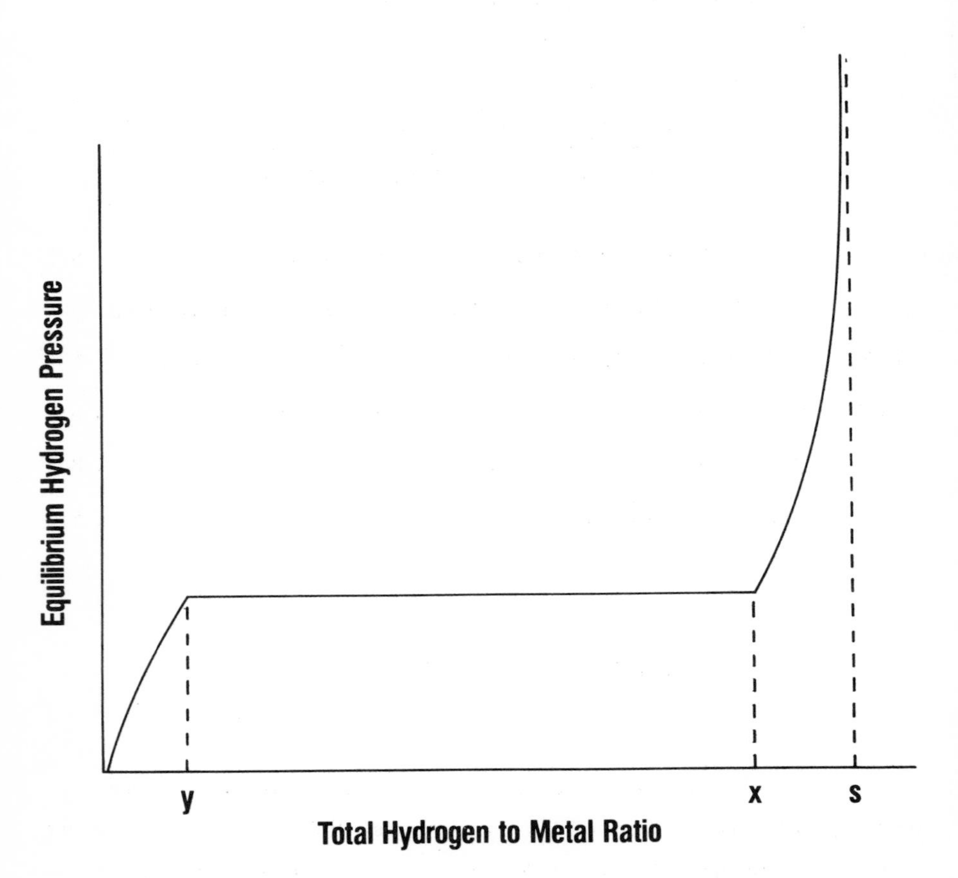

FIG. 1. Typical pressure-composition isotherm for a metal-hydrogen system.

to form a solid solution. When the metal becomes saturated with hydrogen at point y, a hydride phase is formed. With further addition of hydrogen, more metal is converted to hydride and the pressure remains constant across the two-phase region. This plateau pressure is referred to as the dissociation pressure of the hydride at the temperature of the isotherm. At point x, the metal phase has been completely converted to hydride phase (nonstoichiometric), and as the hydride approaches its stoichiometric composition, s, the hydrogen pressure increases. The reverse process occurs on de-hydriding.

The ease of reversibility of Eq(1) is one major factor in the use of metal hydrides for energy storage. The other is the high density of hydrogen atoms in metal hydrides which permit this group of compounds to store hydrogen more efficiently with respect to volume than liquid, or even solid, hydrogen as illustrated in Table 1. In the case of titanium

TABLE 1

HYDROGEN DENSITIES IN SOME HYDROGEN-CONTAINING COMPOUNDS

Compound	Number of Hydrogen Atoms per $cm^3 \times 10^{-22}$
Liquid hydrogen (20°K)	4.2
Solid hydrogen (4.2°K)	5.3
LiH	5.9
TiH_2	9.2
ZrH_2	7.3
YH_2	5.7
UH_3	8.2
$FeTiH_{1.7}$	6.0
$LaNi_5H_{6.7}$	6.1

hydride, the hydrogen density is more than twice that in liquid hydrogen. This hydride has the largest known packing efficiency of hydrogen.

In this chapter, the properties required of metal hydrides in their application for energy storage will be reviewed, and the development of new alloy hydrides will be discussed.

II. TYPES OF ENERGY STORAGE WITH METAL HYDRIDES

Metal hydrides may be used to store energy in four different forms: (1) hydrogen which is used directly as a fuel (2) electrochemical energy in solid solution electrodes (3) thermal energy and (4) magnetic energy in superconducting coils.

A. STORAGE OF FUEL HYDROGEN

Because of its versatility as a fuel, pollution-free combustion products, ease of transportation and storage, and abundant source (water), hydrogen is seriously being considered (1) as a replacement for fossil fuels. Although hydrogen can be stored as liquid hydrogen, the volume efficiency of metal hydrides is greater, as shown in Table 1. Additional advantages of metal hydride over liquid hydrogen storage is the elimination of heavy walled containers and insulation, no loss of hydrogen by evaporation, and increased safety considerations. The storage pressure of hydrogen using metal hydrides is usually close to atmospheric. Also, if a leak develops in the storage container, the escaping hydrogen causes the temperature to drop and the loss of hydrogen ceases automatically, since the dissociation of a hydride is an endothermic reaction.

The desired properties of a metal hydride which can be used for the storage of hydrogen are listed in Table 2. The advantage of item 1 is obvious.

TABLE 2

PROPERTIES OF METAL HYDRIDES FOR HYDROGEN STORAGE

1. High hydrogen retentive capacity (i.e. high H/M ratio)
2. Low temperature of dissociation ($\leq$ 100°C)
3. High rates of hydrogen uptake and discharge
4. Low heats of formation
5. Stable with respect to oxygen and moisture
6. Low cost of alloy
7. Light weight

Low temperatures of dissociation (item 2) are needed in order for the hydrogen to be easily recoverable. For example, if hydrogen is stored as a fuel in an automobile, either the radiator hot water or the hot exhaust gases can be used to dissociate the storage hydride. Relatively high rates of metal hydride formation and dissociation are desirable so that the times necessary to recover the hydrogen and recharge the hydride are not excessively long.

On formation of the hydride, the heat of formation must be dissipated. Consequently, hydrides with low enthalpies of formation would be preferable.

A low enthalpy of formation also is advantageous in order to keep the energy requirements low when recovering the hydrogen.

The hydrides also should be relatively stable with respect to oxygen and moisture although charging and discharging of hydrogen probably will be carried out in a closed system, it would be desirable if the hydride did not completely deteriorate in the event of a leak.

The last two requirements in Table 2, low cost and light weight, can be partially offset by using hydrides with very high hydrogen retentive capacities (item 1) since such hydrides would require less metal or alloy, thus lowering the total cost and weight.

B. *Electrochemical Storage*

The subject of "Solid Solution Electrodes" is discussed by B.C.H. Steele in Chapter 23 of this book. For electrochemical storage with metal hydrides, a hydride-forming alloy is used as the cathode in a battery. During the charging cycle the following reaction would occur (in an alkaline solution:

$$\frac{1}{x}M + H_2O + e^- \rightarrow \frac{1}{x}MH_x + OH^-$$

The reverse reaction takes place during discharge. Gutjahr et al(2) have demonstrated this concept utilizing the intermetallic compound hydrides TiNiH and $Ti_2NiH_{2.5}$(3).

The requirements of a hydride for this application are nearly the same as those shown in Table 2, except that item 5 becomes extremely important for operation in aqueous solutions. Item 4 is no longer valid since ΔH is now included in the free energy of the electrochemical reaction, and item 3 now refers to rates of electrochemical charging and discharging.

C. *Thermal Storage*

Libowitz(4,5) has proposed that the enthalpies of formation of metal hydrides be utilized to store thermal energy. In this concept solar energy (or waste heat) provides the heat necessary to dissociate a metal hydride. The evolved hydrogen is stored either as compressed gas or in a less stable secondary hydride. In order to recover the stored heat, the hydrogen is permitted to re-combine with the dehydrided metal, and the enthalpy of reaction is used to provide heat. This concept also may be employed in solar air conditioning whereby the endothermic dissociation reaction is used to cool warm air(5).

The metal hydrides needed for these applications should meet all the requirements listed in Table 2, except for item 4 where just the reverse is required. Since the thermal storage capacity of a metal hydride is directly proportional to the absolute value of its enthalpy of formation, ΔH, it is advantageous to have as high a value of ΔH as possible without adversely affecting other properties of the hydride.

D. *Magnetic Storage in Superconductors*

Large quantities of energy for electrical use may be stored as magnetic energy in superconducting inductors(6). It has been shown recently that for palladium(7), thorium(8) and some niobium-based alloys (9), the formation of the hydride significantly increases the superconducting transition temperature, T_c, over that of the un-hydrided metal or alloy. By appropriate alloying of Pd with copper prior to formation of the hydride, a T_c value as high as 16.6°K has been obtained(10). These results suggest the possibility that new alloy hydride superconductors can be developed which would permit efficient storage of magnetic energy. However, the properties required of such superconducting hydrides and the concepts governing their behavior(11) are entirely different from those hydrides which will be utilized in the other three types of storage mentioned above. Hence, a discussion of the development of hydride superconductors will not be included in this Chapter.

III. THE NATURE OF METAL HYDRIDES

In order to predict alloy compositions which would form new hydrides having appropriate properties for the applications discussed above, the nature of the reaction between metals and hydrogen must be understood. All metals and alloys dissolve hydrogen to a greater or lesser extent; however, it is only those metals, which form a definite hydride phase which is of interest here. The formation of a hydride phase is denoted by the constant pressure plateau (or change in slope in multicomponent systems) in the isotherm illustrated in Fig. 1, as well as by the changes in structure of the metal sub-lattice with hydride formation as shown in Table 3. All known binary

TABLE 3

STRUCTURE OF METALS AND CORRESPONDING HYDRIDES

Metal	Metal Structure	Hydride	Hydride Structure
Alkali, metals	b.c.c.	LiH - CsH	f.c.c.
Ca, Sr	f.c.c.	CaH_2, SrH_2	orthorhombic
Ba	b.c.c.	BaH_2	orthorhombic
Mg	h.c.p.	MgH_2	b.c.t. rutile
Sc	h.c.p.	ScH_2	f.c.c.
La, Pr, Nd	d.h.c.p.	LaH_2 - NdH_2	f.c.c.
Y, Gd-Tm, Lu	h.c.p.	YH_2, GdH_2-TmH_2, LuH_2	f.c.c.
		YH_3, GdH_3-TmH_3, LaH_3	hexagonal
Ce	f.c.c. (a_o=5.16Å)	CeH_2	f.c.c. (a_o = 5.58Å)
Sm	rhombohedral	SmH_2, SmH_3	f.c.c., hexagonal
Eu	b.c.c.	EuH_2	orthorhombic
Yb	f.c.c.	YbH_2, YbH_{3-x}	orthorhombic, f.c.c.
Ac	f.c.c. (a_o= 5.31Å)	AcH_2	f.c.c. (a_o = 5.67Å)
Th	f.c.c.	ThH_2	f.c. tetragonal
		Th_4H_{15}	b.c.c. (complex)
Pa	b.c. tetragonal	PaH_3	β-W (cubic)
U	orthorhombic	β-UH_3	β-W (cubic)
		α-UH_3	b.c.c.
Np	orthorhombic	NpH_2	f.c.c.
		NpH_3	hexagonal

TABLE 3
(Cont'd.)

Metal	Metal Structure	Hydride	Hydride Structure
Cm,Bk		CmH_2, BkH_2	f.c.c.
Pu	monoclinic	PuH_2	f.c.c.
		PuH_3	hexagonal
Am	f.c.c. (a_o= 4.89Å)	AmH_2	f.c.c. (a_o=5.35Å)
		AmH_3	hexagonal
Ti,Zr,Hf	h.c.p.	TiH_2, ZrH_2, HfH_2	f.c.c., f.c.t.
V	b.c.c.	VH, VH_2	b.c.t., f.c.c.
Nb	b.c.c.	NbH, NbH_2	orthorhombic, f.c.c.
Ta	b.c.c.	TaH	orthorhombic
Pd	f.c.c. (a_o= 3.89Å)	PdH	f.c.c. (a_o=4.03Å)

Data on the amercium hydrides, CmH_2, and BkH_2 are given in refs(12), (13), and (14), respectively. More detailed information on the structures of the other hydrides listed in Table 3 (including references) are given in ref.(15).

hydrides formed by direct reaction of metal and hydrogen are listed in Table 3. In those cases where there is no change in metal sub-lattice structure (Ce,Ac,Am, Pd), there is a discontinuous change in lattice parameter on hydride formation.

The alkali and alkaline earth metal hydrides are predominantly ionic. However, the nature of the chemical bonding in the transition metal hydrides is still not completely understood although there has been a great deal of progress recently. For many years, there were two opposing views of the bonding in transition metal hydrides; the protonic model and the anionic model(16). The protonic model assumes that the hydrogen acts as a metal and essentially forms an alloy with the transition metal. In this case, the electrons of the hydrogen occupy the d-band of the transition metal so that the hydrogen atoms exist as screened protons in the metal sub-lattice. The opposite view is that hydrogen may be considered the first member of the halogen series needing one electron to complete its outer shell. Therefore, the hydrogen would take electrons from the metal to form hydride anions and transition metal cations to form an essentially ionic compound. The protonic model or alloy model was supported by the fact that most transition metal hydrides are metallic conductors; in some cases the electronic conductivity of the metal hydride is higher than that of the corresponding un-hydrided metal(16). On the other hand, the trihydrides of the rare earths become semiconductors and their electronic properties can best be explained by the ionic model(17). Theoretical and experimental studies of hydrides utilizing such techniques as Mössbauer spectroscopy, magnetic susceptibilities, nuclear magnetic resonance, and positron annihilation have been interpreted(18) in favor of one or the other of these two models.

On the basis of energy band calculations, Switendick(19) has suggested a model which would explain both the alloy-like and ionic behavior of transition metal hydrides. The model can be explained in terms of the following simplified qualitative discussion. As hydrogen is added to a metal some of the energy states become strongly perturbed or new hydrogen states are formed. If the perturbed or new hydrogen states fall below the Fermi energy, they will be occupied by the electrons from the hydrogen atoms in preference to existing unperturbed metallic states, and the hydride will appear to be anionic. On the other hand, if the hydrogen states are already filled or if they appear above unfilled metallic states, the metallic states will be preferentially occupied by the electrons and the hydrogen will appear to "donate" its electrons to the metal energy bands; hence the protonic model. In the rare earth trihydrides, the separation between the highest filled state and the lowest unfilled state becomes high enough (~1eV) so that the hydride becomes a semiconductor. According to Switendick's calculations of the relative positions of the electronic energy states, the hydrides of palladium, and the Group V metals, niobium, vanadium, and tantalum, should appear to fit the "protonic" model, while the rare earth hydrides appear to be "anionic". A review(18) of many investigations on metal hydrides reveals that indeed the experimental data appear to support the existence of anionic hydrogen in the rare earth hydrides and more metallic behavior for the other hydrides.

IV. DEVELOPMENT OF NEW ALLOY HYDRIDES

A. Approach

None of the binary metal hydrides shown in Table 3 meet all (or most) of the requirements listed in Table 1. In most cases, the temperatures required to dissociate the hydride in order to recover the hydrogen are too high. In order to develop new hydrides which will be suitable for the applications discussed in section II, it is necessary to find alloy systems which would form stable hydride phases with the required properties.

Although energy band calculations have been very valuable in explaining the bonding in hydrides as well as the relative stabilities of various hydrides, at present, they cannot be used for predicting new hydride phases. Because such calculations are very complex, certain simplifying assumptions must be made, the most important of which is that the metal sub-lattice does not change its crystal structure on formation of the hydride. As can be seen in Table 3, this is not the case. The energy band calculations were actually carried out by starting with the structure of the metal sub-lattice as it appears in the hydride phase and computing the effect of adding hydrogen and its electrons to this metal sub-lattice. Consequently, the structure of the hydride phase must be known before such calculations can be performed.

There are two approaches which can be taken in the development of new alloy hydrides having the required properties for energy storage; the syntheses of new intermetallic compound hydrides and the modification of the properties of known hydrides by appropriate alloying.

B. *Intermetallic Compound Hydrides*

In general, the properties of intermetallic compound hydrides appear to have a little or no resemblance to the properties of the constituent metal hydrides, and therefore, such hydrides may be viewed as pseudo-binary metal hydrides. Table 4 lists some

TABLE 4

HYDRIDES OF INTERMETALLIC COMPOUNDS

Intermetallic Compound Hydride	Constituent Hydrides	Ref.
$LaNi_5H_{6.7}$	LaH_3	(20)
$PrCo_5H_{3.6}$	PrH_3	(21)
$DyCo_3H_5$	DyH_3	(22)
$ErCo_3H_5$	ErH_3	(22)
$ZrNiH_3$	ZrH_2	(23)
Zr_7NiH_{17}	ZrH_2	(24)
$ThCoH_4$	ThH_2 or Th_4H_{15}	(25)
$Th_7Fe_3H_{28}$	ThH_2 or Th_4H_{15}	(26)

examples of known intermetallic compound hydrides. $LaNi_5$ and $PrCo_5$ are representative of a large group of rare earth-nickel (or cobalt) intermetallics which easily form hydrides having relatively high hydrogen contents. Since cobalt and iron do not form hydrides, and nickel can form a hydride only with difficulty at extremely high hydrogen pressures(27) or by indirect techniques(28), the hydrogen-to-metal ratios of the intermetallic compound hydrides shown in Table 4 are higher than would be expected on the basis of the constituent metal hydride formulas.

The fact that the properties of intermetallic compound hydrides cannot be predicted from properties of constituent metal hydrides is further illustrated by the data in Table 5. The structure of

TABLE 5

PROPERTIES OF ZrH_2 AND $ZrNiH_3$

	ZrH_2(15)	$ZrNiH_3$	
Structure	Tetragonal (distorted fluorite)	Orthorhombic	(29)
Dissoc. Press at 250°C	3×10^{-11} torr	200 torr	(23)
Zr-H distance	2.09Å	1.96Å	(29)

the intermetallic compound hydride, $ZrNiH_3$ is quite different from that of ZrH_2. In addition, although the Zr-H distance is less in $ZrNiH_3$ indicating stronger bonds in the intermetallic compound hydride, the dissociation pressure of $ZrNiH_3$ is higher than ZrH_2 by a factor of almost 10^{13} which shows that the stability of $ZrNiH_3$ is considerably less than ZrH_2. Thus, it is clear that the properties of intermetallic compound hydrides cannot be extrapolated in any simple manner from a knowledge of the properties of the constituent hydrides.

Since there are an extremely large number of known and possible intermetallic compounds, it would be very desirable to be able to predict which of these compounds would form hydrides as well as the approximate properties of such new hydrides. An attempt to do this by Miedema and coworkers (30,31) has resulted in a "Rule of Reversed Stability" which qualitatively states that the greater the stability of an intermetallic compound, the less stable the corresponding hydride of that intermetallic compound relative to other compounds within a given series. The enthalpies of formation of the rare earth-nickel(cobalt) hydrides appear to agree(30) with the rule. A simple explanation for the rule can be given in terms of the intermetallic compounds $LaNi_5$ and $LaCo_5$. As hydrogen is introduced into the compounds to produce the hydride, lanthanum-to-hydrogen bonds are formed at the expense of the lanthanum-to-nickel (or cobalt) bonds. If the La-Ni bond is stronger than the La-Co bond, the strength of the La-H bond in $LaNi_5H_x$ will be less than that of the La-H bond in $LaCo_5H_x$ and consequently the former hydride will be less stable.

The Rule of Reversed Stability may be stated(31) more quantitatively as follows:

$$\Delta H(MN_aH_{b+c}) = \Delta H(MH_b)+\Delta H(N_aH_c)-\Delta H(MN_a) \qquad (2)$$

where ΔH represents enthalpy of formation and MN_a is an intermetallic compound, MN_aH_{b+c} its hydride, and MH_b and N_aH_c the constituent metal hydrides. Eq(2) is applicable when M is a strong hydride-forming metal and $a \geq 1$.

Unfortunately the Rule of Reversed Stability only has limited applicability at the present time for several reasons; (a) its use in calculating relative stabilities is limited to a particular structural or compositional series, (b) the thermodynamic properties of many intermetallic compounds are unknown and must be calculated from semi-empirical principles(32); (c) if N is a non-hydride forming metal, the extrapolation of ΔH at high hydrogen concentrations from dilute solution solubility data is of questionable validity (d) in some cases, there is a degree of arbitrariness in choosing the constituent hydrides.

In a recent paper, Buschow et al(31) showed a rough correlation between the stabilities of a group of thorium-based intermetallic compound hydrides and the corresponding values of ΔH as calculated from Eq(2). The treatment was subject to the limitations mentioned above and the compound Th_2Fe_{17} may be used as an example. The calculated value of ΔH for the intermetallic compound was -25 kcal/mole(32), while the experimentally measured value is -56.6 kcal/mole(33). For this system, Buschow et al wrote Eq(2) as follows:

$$\Delta H(Th_2Fe_{17}H_8) = 2\Delta H(ThH_2)+\Delta H(Fe_{17}H_4)-\Delta H(Th_2Fe_{17}) \quad (3)$$

The hydride $Fe_{17}H_4$ does not exist, and the value of +16 kcal/mole used for this hypothetical hydride was apparently extrapolated from dilute solution data. Using a value of -35 kcal/mole for ΔH of ThH_2(15), a value of about +3 kcal/mole was obtained for ΔH of $Th_2Fe_{17}H_8$. The experimental studies indicated that it would take hydrogen pressures well in excess of 26 atm to produce this hydride. Therefore it appears that the calculated positive value of ΔH is in agreement with the observed instability of this hydride. However, thorium also forms a higher hydride Th_4H_{15}, with ΔH ≃ 210 kcal/mole of hydride(15) and Eq(2) also may be written as follows:

$$\Delta H(Th_2Fe_{17}H_8) = \tfrac{1}{2}\Delta H(Th_4H_{15})+\Delta H(Fe_{17}H_{0.5})-\Delta H(Th_2Fe_{17}) \quad (4)$$

Eq(4) yields a value of -46 kcal/mole for the enthalpy of formation of the hydride which is quite different from the value of +3 kcal/mole and would indicate a quite stable hydride. Whether Eqs(3) or (4) should be chosen appears to be quite arbitrary and therefore for this reason, and the ones mentioned above, the Rule of Reversed Stability does not appear to be of much value in predicting the stability of unknown intermetallic compound hydrides. It should be mentioned, however, that the Rule of Reversed Stability can be very useful in predicting alloying elements necessary to modify the properties of known hydrides. This will be discussed in more detail in Section IV C.

Clearly, there is need for further investigations relating the energy band and crystal structures of intermetallic compounds with their behavior towards hydrogen in order to find correlations which would be useful in developing new alloy hydrides. Along this line, some preliminary studies(34) on metallic glasses indicate that in some cases the effect of crystal structure may be more important than electronic structure with regard to hydride formation.

C. *Alloy Modifications of Hydrides*

It may be possible to improve the properties of known hydrides relative to the required characteristics listed in Table 2 by adding appropriate amounts of an alloying element. Some general approaches to modifying properties in this manner will be discussed.

The rare earth dihydrides all have the

fluorite structure (except Eu and Yb) illustrated in Fig. 2. The metal atoms form a f.c.c. sub-lattice

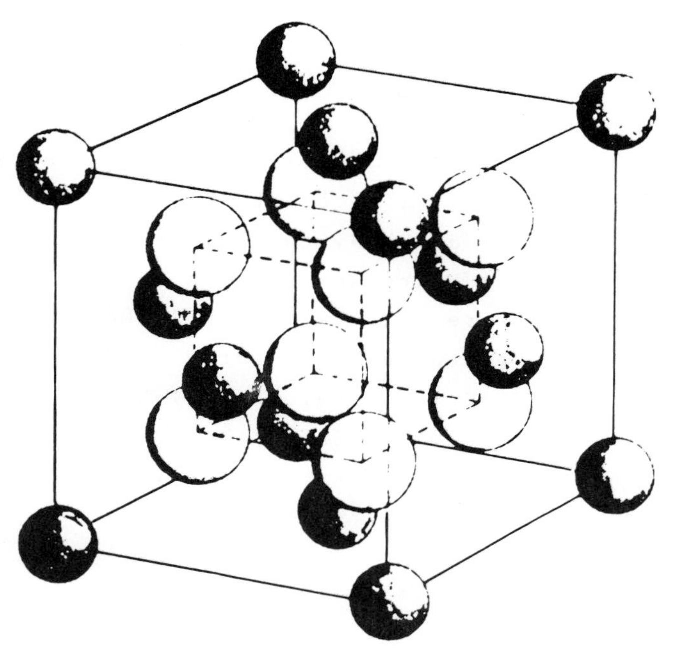

FIG. 2. The structure of some metal dihydrides.

(dark spheres) and the hydrogen atom (light spheres) occupy the tetrahedrally coordinated sites. As can be seen in Table 3, many other metal hydrides have this structure (or a slight tetragonal distortion of the structure) including the Group IV metal hydrides and the dihydrides of Nb and V. Additional hydrogen can be added to the rare earth dihydrides (and some actinide dihydrides having the same structure). These additional hydrogen atoms enter the octahedral sites of the metal sub-lattice (center and edges of the cube) and in some cases these positions are completely filled to form a cubic trihydride, MH_3. Although the Group IV metal hydrides, TiH_2, ZrH_2, and HfH_2, have basically the same structure, no additional hydrogen can be added. This may be explained(19) by the fact that the octahedral-to-tetrahedral hydrogen distance in the fluorite structure must be greater than about 2.30Å in order for a trihydride to be stable. The early rare earths which form cubic trihydrides all have H(oct)-H(tet) distances greater than 2.30Å. On the other hand, the distance between octahedral and tetrahedral interstices in the Group IV metal hydrides is less than 2.18A. The question then may be raised: Is it possible to add an alloying metal having a larger metallic radius than the Group IV metals which could enter the lattice substitutionally so as to expand it to the point where additional hydrogen atoms could occupy octahedral as well as tetrahedral sites? This would have to be done without significant changes in the electronic band structure of the parent metal. If an appropriate combination of alloying elements could be found, it may be possible to significantly increase the hydrogen density in some hydrides.

Many hydrides, such as those of the rare earth metals, have high hydrogen contents, but as mentioned in Section IV A, the dissociation temperatures are much too high. For example, the rare earth hydrides have dissociation temperatures in the vicinity of 1000°C while the desired dissociation temperatures, as listed in Table 2, should be closer to 100°C.

The fact that alloying elements can increase the dissociation pressure of a hydride (and therefore decrease its dissociation temperature) was demonstrated several years ago by Reilly and Wiswall (35). Fig. 3 shows the effects of small amounts of Si on the dissociation pressure of VH_2. The isotherms are similar to the one described in Fig. 1. Increasing the concentration of Si increases the dissociation (plateau) pressure and it appears that up to 0.40 mole% of Si can be added without significantly affecting the maximum hydrogen content of the dihydride (as indicated by the rapidly rising portion of the isotherms). At higher Si contents the dissociation pressure continues to increase but the maximum hydrogen-to-metal ratio decreases.

In general, the dissociation pressure of a metal hydride, P_{H_2}, is a function of its enthalpy, ΔH, and entropy, ΔS, of formation as shown by the integrated form of the Van Hoff equation for Eq(1):

$$\ln P_{H_2} = (2/x)[(\Delta H/RT)-(\Delta S/R)] \quad (5)$$

The value of the enthalpy of formation is, generally, an indication of the bond strength in a compound. Since reaction (1) is an exothermic reaction, ΔH is negative, and the stronger the M-H bond in the hydride, the higher the negative value of ΔH. Therefore, in order to increase the dissociation pressure of a metal hydride, alloying elements should be introduced which will weaken the bonding and make ΔH less negative.

The effect of changing overall bond

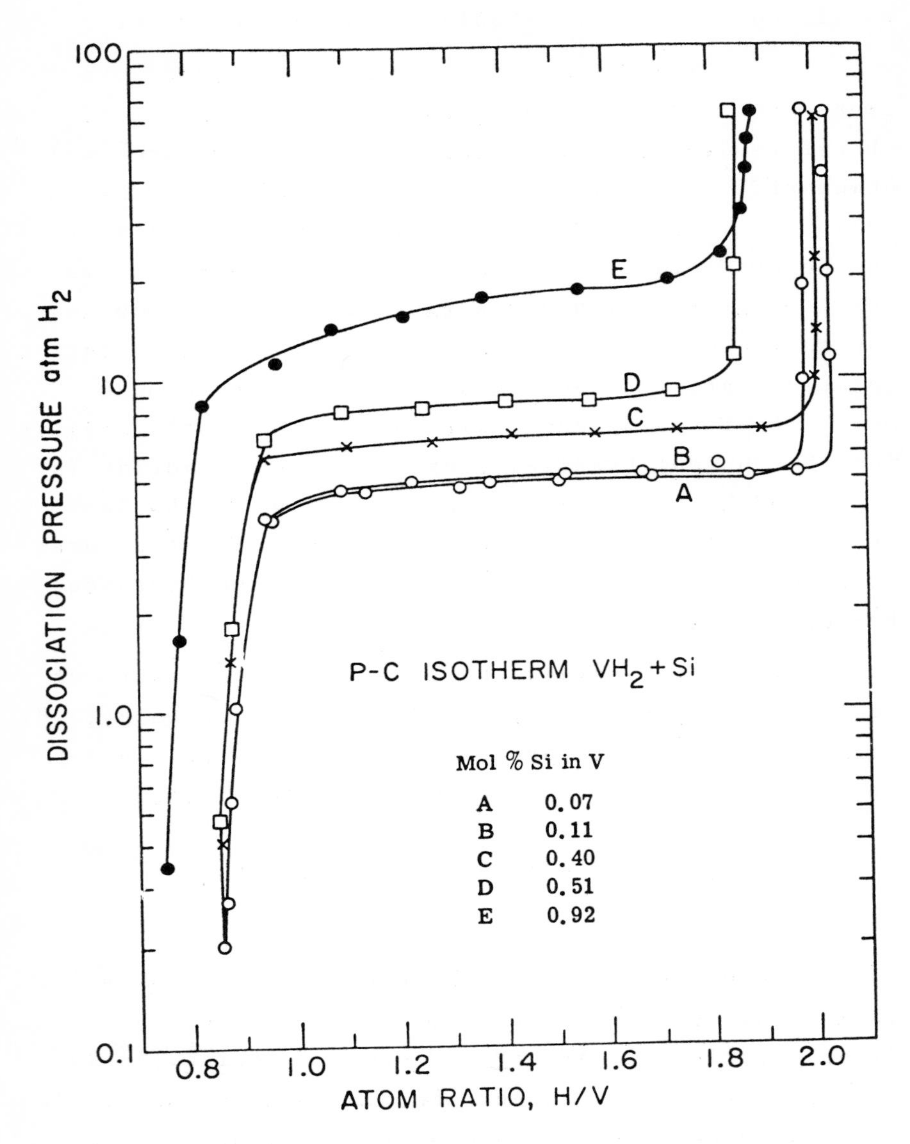

FIG. 3. Effect of silicon on the dissociation pressure of VH_2 (taken from Ref. 35).

strength on dissociation pressure may be seen in Fig. 4 which shows some isotherms for the hydride of

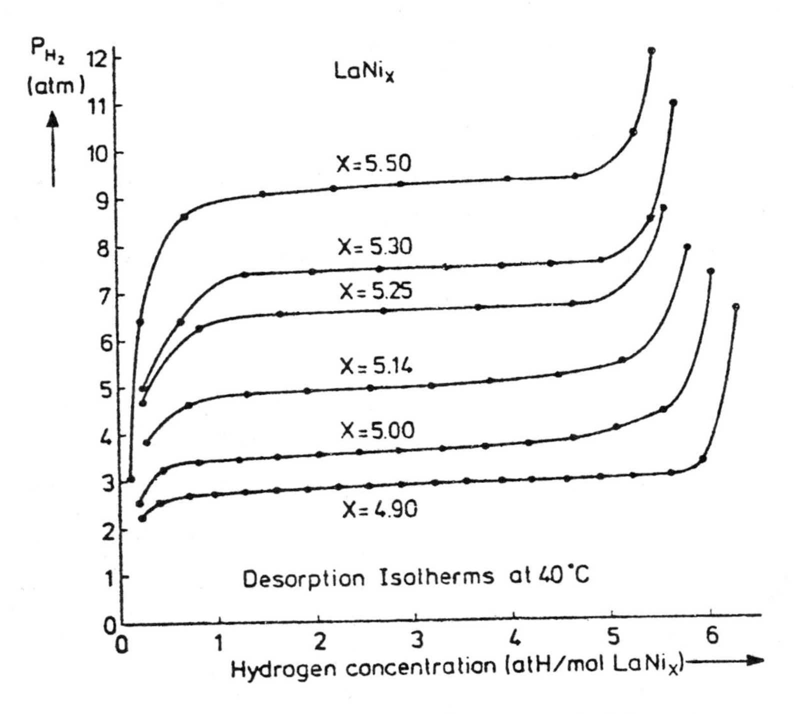

FIG. 4. Effect of nonstoichiometry on the dissociation pressure of lanthanum pentanickel hydride [taken from K.H.J. Buschow and H.H. VanMal, J. Less-Common Metals 29, 203 (1972)].

$LaNi_5$. The hydride of the stoichiometric intermetallic compound has a dissociation pressure of about 3.5 atm at 40°C. It can be seen that as Ni is substituted for La in the intermetallic compound (increasing x) the dissociation pressure increases. Since La forms a strong hydride and Ni forms a normally unstable hydride, the La-H bond would be much stronger than the Ni-H bond in this compound. Consequently, the result of substituting nickel for La would be to weaken the overall bonding of the intermetallic compound hydride, thereby making ΔH less negative and increasing P_{H_2} according to Eq(5). Predictably, the reverse effect is obtained when La is substituted for Ni (x = 4.90 in Fig. 4).

It was mentioned earlier that the Rule of Reversed Stability could be useful in modifying the properties of intermetallic compound hydrides. This is illustrated in Fig. 5 which shows the effects of

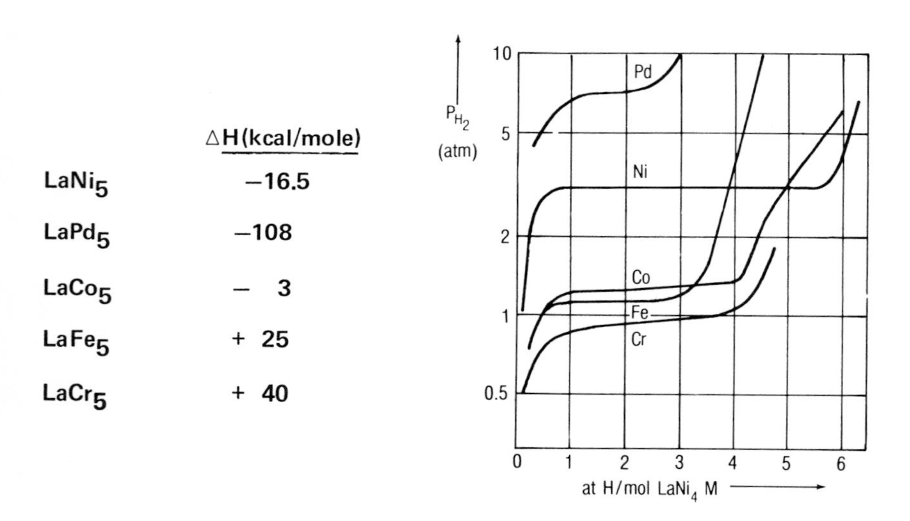

FIG. 5. Effect of metal substitutions on the dissociation pressure of lanthanum pentanickel hydride (taken from Ref. 30).

substituting 20% of the nickel in $LaNi_5$ by another alloying metal. As seen from the values of ΔH, $LaPd_5$ is more stable than $LaNi_5$. Therefore, if Pd is substituted for some of the Ni in $LaNi_5$ the resulting hydride should be less stable as suggested by the Rule of Reversed Stability. That this is the case is illustrated by the increased dissociation pressure of the Pd substituted hydride. Conversely, the substitution of metals which form less stable intermetallic compounds with La lead to lower dissociation pressures (e.g. Co,Fe,Cr).

Metal hydrides which may be used as thermal storage materials require a very large negative value for ΔH (i.e. the absolute value of ΔH should be large as pointed out in Section II C). According to Eq(5), this may result in a dissociation pressure of the hydride which would be too low to permit rapid mass transfer of hydrogen. As seen from Eq(5), the pressure also may be increased by decreasing the value of ΔS, either by decreasing the entropy of the hydride or by increasing the entropy of the alloy. One way of accomplishing the latter is to use disordered alloys.

V. CONCLUSION

Four ways of using metal hydrides for energy storage were discussed and methods of developing new hydrides for these applications were suggested. The methods mentioned were illustrative cases, particularly with respect to modification of properties of existing hydrides, and it should be pointed out that there are many other approaches which may be taken. However, it is important to emphasize that there is need for a great deal of further materials research on fundamental theories related to the formation and stabilities of alloys and intermetallic compounds and their interactions with hydrogen.

REFERENCES

1. W.E. Winsche, K.C. Hoffman, F.J. Salzano, Science, 180, 1325 (1973); D.P. Gregory, Sci. Am., 228 #1, 13 (1973); see also the Proceedings of the Hydrogen Economy Miami Energy Conference, 1974 and the First World Hydrogen Energy Conference, 1976, both published by the School of Engineering and Environmental Design, Univ. of Miami, Coral Gables, Fla.

2. M.A. Gutjahr, H. Buchner, K.D. Beccu, and H. Säufferer, Paper 40(a) at the Brighton(U.K.) Power Sources Symposium, 1972.

3. H. Buchner, M.A. Gutjahr, K.D. Beccu, and H. Säufferer, Zeit. Metallk. 63, 497 (1972).

4. G.G. Libowitz, Proc. Ninth Intersoc. Energy Conf. Paper No. 749025, pp. 322-325, (1974).

5. G.G. Libowitz and Z. Blank, paper presented at Centennial Meeting of the Am. Chem. Soc., April 1976; to be published in Advances in Chemistry Series.

6. R.W. Boom and H.A. Peterson, IEEE Trans.-Magnetics, MAG 8, 701 (1972).

7. T. Skoskiewicz, Phys. Stat. Sol.(a)11, K123 (1972).

8. C.B. Satterthwaite and I.L. Toepke, Phys. Rev. Lett. 25, 741 (1970); C.B. Satterthwaite and D.T. Peterson, J. Less-Common Metals 26, 361 (1972).

9. C.G. Robbins and J. Muller, J. Less-Common Metals 42, 19 (1975).

10. B. Stritzker, Z. Physik 268, 261 (1974).

11. D.A. Papaconstantopoulos and B.M. Klein, Phys. Rev. Lett. 35, 110 (1975); B.N. Ganguly, Z. Physik B 22, R.J. Miller and C.B. Satterthwaite, Phys. Rev. Lett. 34, 144 (1975).

12. W.M. Olson and R.N.R. Mulford, J. Phys. Chem. 70, 2935 (1966).

13. B.M. Bansal and D. DAmien, Inorg. Nucl. Chem. Letters 6, 603 (1970).

14. J.A. Fahey, J.R. Peterson, and R.D. Baybarz, Inorg. Nucl. Chem. Letters 8, 101 (1972).

15. W.M. Mueller, J.P. Blackledge, and G.G. Libowitz, Metal Hydrides, Academic Press, New York (1968).

16. G.G. Libowitz, The Solid State Chemistry of Binary Metal Hydrides, W.A. Benjamin, Inc., New York (1965).

17. G.G. Libowitz, Ber. Bunsenges. Physik. Chem. 76, 837 (1972).

18. G.G. Libowitz, MTP Internatl. Rev. Sci., Inorg. Chem. Ser. 1, Vol. 10, Solid State Chemistry, L.E.J. Roberts, Ed., Butterworths Ltd. London (1972) pp. 79-116.

19. A.C. Switendick, Solid State Commun. 8, 1463 (1970) Int. J. Quant. Chem. 5, 459 (1971); Proc. Hydrogen Econ. Miami Energy Conf., T.N. Verizoglu, Ed, Univ. of Miami, Coral Gables, Fla. (1974) p. S6-1.

20. J.H.N. von Vucht, F.A. Kuijpers, and H.C.A.M. Bruning, Philips Res. Repts. 25, 133 (1970).

21. F.A. Kuijpers and B.O. Loopstra, J. Phys. Chem. Solids 35, 301 (1974).

22. T. Takeshita, W.E. Wallace, and R.S. Craig, Inorg. Chem. 13, 2283 (1974).

23. G.G. Libowitz, H.F. Hayes, and T.R.P. Gibb, J. Phys. Chem. 62, 76 (1958).

24. I.R. Tannenbaum, W.L. Korst, J.S. Mohl, and G.G. Libowitz, U.S.A.E.C. Report No. NAA-SR-7132 (1963).

25. W.L. Korst, U.S.A.E.C. Report No. NAA-SR-6881 (1962).

26. K.H.J. Buschow, H.H. Van Mal, and A.R. Miedema, J. Less-Common Metals 42, 163 (1975).

27. B. Baranowski, K. Bochenska, and S. Majchrzak, Rocz. Chem. 41, 2071 (1967).

28. B. Baranowski and M. Smialkowski, J. Phys. Chem. Solids 12, 206 (1959); B. Baranowski, Rocz. Chem. 38, 1019 (1964).

29. S.W. Peterson, V.N. Sadana, and W.L. Korst, J. Phys. (Paris) 25, 451 (1964).

30. H.H. Van Mal, K.H.J. Buschow, and A.R. Miedema, J. Less-Common Metals 35, 65 (1974).

31. K.H.J. Buschow, H.H. Van Mal, and A.R. Miedema, J. Less-Common Metals 42, 163 (1975).

32. A.R. Miedema, J. Less-Common Metals 32, 117 (1973); 41, 238 (1975); 46, 67 (1976).

33. W.H. Skelton, N.J. Magnani, and J.F. Smith, Met. Trans. 4, 917 (1973).

34. A.J. Maeland and G.G. Libowitz - unpublished work.

35. J.J. Reilly and R.H. Wiswall, Proc. Internat. Meeting on Hydrogen in Metals, Kernforschungsanlage Julich, Germany, (1972) p. 38.

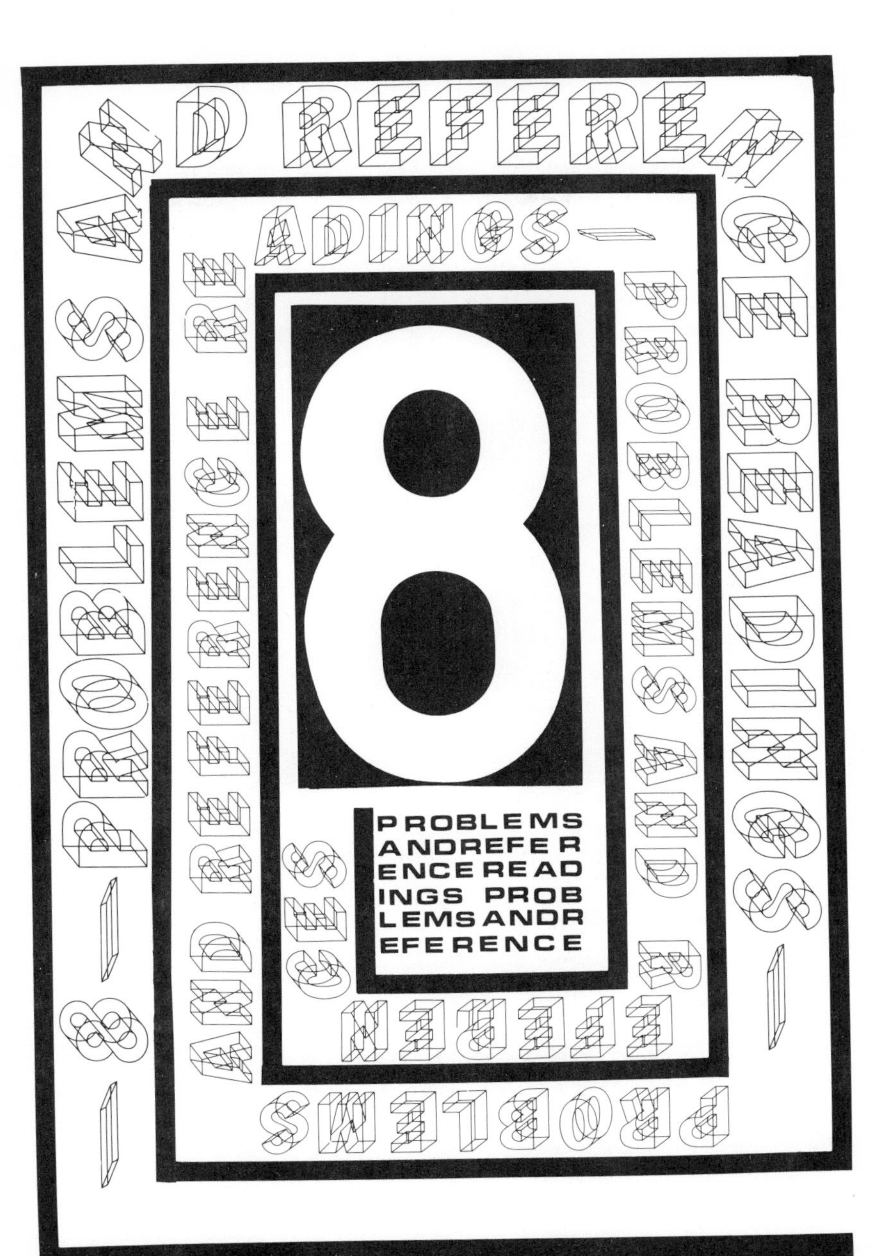
8
PROBLEMS
ANDREFER
ENCEREAD
INGS PROB
LEMSANDR
EFERENCE

Chapter 29

PROBLEMS AND REFERENCE READINGS

L.E. Murr

Department of Metallurgical and Materials Engineering
New Mexico Institute of Mining and Technology
Socorro, New Mexico, U.S.A.

Prob. 1 Discuss, in historical perspective, the utilization of energy sources in the United States since 1900 and describe what you consider materials limiting aspects of this utilization; or the generation, storage, and distribution of energy. The total electrical energy generation in billions of kilowatt hours is estimated to have been 7×10^2 in 1962, 10^3 in 1966, 1.05×10^3 in 1970 and 1.15×10^3 in 1974. Is there a correlation between electrical energy generation and the supply of (and use of) copper during this period? What might happen if alternative conductors are not developed in the future?

REFERENCES

L.V. Azaroff, Energy, Environment, and Materials, JOM, 28, 25 (1976).

National Materials Policy, Proc. of Joint Meeting of the NAS-NAE, National Acad. Sci., Washington, D.C. (1975).

Prob. 2 A novel, proposed coal liquefaction process is designed to liquify coal in an underground seam in-situ by placing a pipe string within the seam in order to inject solvent and/or hydrogen and to recover the liquefied coal. If the coal seam is located 2000 ft. below the surface, and the liquefaction process is carried out at a pressure of 4000 psi and a temperature of roughly 400°C, compare the use of type 304-L and 314 stainless steel in this application. Outline the materials considerations in arriving at your conclusions and basing your comparisons. What will be the pipe size if the inside diameter is specified to be 2 in.? If the coal to be liquefied contains pyrite inclusions of a size and distribution illustrated in Fig. P.2, what additional materials problems must be considered in the pipe string design and evaluation? Describe briefly how the coal, if brought to the surface, might be treated to eliminate or reduce its sulfur content if the sulfur is contained in pyrite particles as shown in Fig. P.2.

REFERENCES

W.E. Berry, Corrosion in Nuclear Applications, J. Wiley and Sons, New York (1971).

Corrosion Fatigue: Chemistry, Mechanics and Microstructure, (Int. Corrosion Conf. Series; NACE2), National Assoc. of Corrosion Engr., Houston (1972).

Corrosion Problems of the Petroleum Industry, MacMillan, New York (1960).

M.G. Fontana and N.D. Greene, McGraw-Hill, New York (1967).

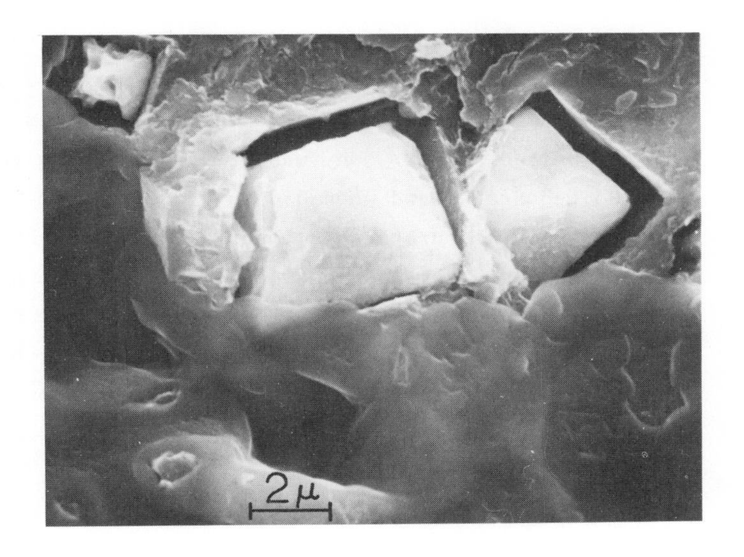

FIG. P.2. Scanning electron micrograph of pyrite crystals in Carthage, New Mexico coal.

M.G. Fontana and R.W. Staehle (Eds.) Advances in Corrosion Science and Technology, Plenum Press, New York (1970).

J. Huebler, "Gasification and Liquefaction of Coal", Mining Congress Journal, October, 16 (1975).

D.W. vanKrevelen, Coal Typology, Chemistry, Physics, Constitution, Elsevier, New York (1961).

W.D. Robertson, (Ed.) Stress Corrosion Cracking and Embrittlement, J. Wiley and Sons, New York (1956).

M.P. Seah, "Interface Adsorption, Embrittlement, and Fracture in Metallurgy: A Review", Surface Sci., 53, 168 (1975).

H.H. Uhlig, Corrosion and Corrosion Control; An Introduction to Corrosion Science and Engineering, 2nd ed., J. Wiley and Sons, New York (1971).

W.S. Wise, Solvent Treatment of Coal, Mills and Boon, London (1971).

Prob. 3 Most conventional heat resisting and corrosion-resistant alloys are fundamentally unstable, and contain chromium in varying amounts to control their useful metallurgical properties. Discuss the production of chromium (Cr metallurgy) and describe sources of Cr in the United States. Why was Cr production halted in the United States and when did this occur? Because all Cr for alloying purposes is imported, it would be advantageous to develop an alloy or alloys with appropriate substitutions for Cr. Describe possible schemes to investigate an elemental alternative to Cr. Describe basically what effect Cr additions to Ni have on the mechanical properties of the alloy, i.e. how is cross-slip of dislocations affected?

REFERENCES

R.W. Cahn, (Ed.) Physical Metallurgy, Elsevier, New York (1970).

Ductile Chromium and Its Alloys, ASM, Metals Park, Ohio (1957).

A.G. Guy, Introduction to Materials Science, McGraw-Hill, New York (1972).

A. Prince, Alloy Phase Equilibria, Elsevier, New York (1966).

A.H. Sully, Chromium, Academic Press, New York (1954).

Prob. 4 Briefly describe the Lurgi process for coal gasification and the Synthoil process for coal liquefaction, and note especially the role of catalysts in the associated reactions. Identify the catalyst or catalysts and discuss their structure and properties. Are these catalysts reforming (dispersion) catalysts? Consider that experimental evidence has shown that the catalytic properties of large single crystals of platinum are essentially the same as platinum clusters (of roughly 10 Å diameter) in the Al_2O_3 - Pt dispersion catalyst. What might this indicate about the structure of the Pt clusters in the dispersion catalyst? Show some sketches and defend your conclusions. Discuss possible techniques for directly observing the structure of Pt clusters in a dispersion catalyst.

REFERENCES

J. Huebler, "Gasification and Liquefaction of Coal", Mining Congress Journal, October, 16 (1975).

J.J. McCarroll, "Surface Physics and Catalysis", Surface Sci., 53, 297 (1975).

L.E. Murr, Interfacial Phenomena in Metals and Alloys, Addison-Wesley, Reading, Mass. (1975).

M. Pettre, Catalysis and Catalysts (translated by D. Antin), Dover Publications, New York (1963).

G.N. Schrauzer (Ed.) Transition Metals in Homogeneous Catalysis, Marcel Dekker, New York (1971).

Prob. 5 Consider four experimental flywheels to be 0.5 meter in diameter and 10 cm. thick. One is cast of solid nickel, the other a nickel eutectic alloy (Ni-Al), the third is an aluminum-silicon carbide fiber reinforced matrix with the fibers oriented along the diameters through the flywheel, and the fourth is a solid spun graphite composite with multi-directional fiber fabrication. Calculate the maximum speed to which each flywheel can be accelerated. Show all assumptions you make. (Note that the SiC fiber content in the aluminum flywheel is 5 percent by volume.) What will be the stored energy for each flywheel at maximum speed? Discuss the effect of other flywheel geometries on energy storage capabilities of these materials.

REFERENCES

G.S. Holister and C. Thomas, Fibre Reinforced Materials, Elsevier, New York (1966).

A. Kelly, Strong Solids, Clarendon Press, Oxford (1973).

A.P. Levitt (Ed.) Whisker Technology, Wiley-Interscience, New York (1970).

N.J. Parratt, Fibre-Reinforced Materials Technology, Van Nostrand-Reinhold Co., London (1972).

D.H. Pletta and D. Frederick, Engineering Mechanics: Statics and Dynamics, The Ronald Press Co., New York (1964).

Prob. 6 Compare the solubilities of hydrogen in Fe-base systems with V, Ta, Nb, Ti, and Zr. Which of these are capable of forming hydride phases with sufficient ease to make them good energy storage units? Are such systems effective in hydrogen storage? How much hydrogen can be stored in 1 cc of Zr for example? How does this compare with conventional storage schemes?

REFERENCES

Hydrogen in Metals, ASM, Metals Park, Ohio (1974).

B. Lustman and F. Kerze, Jr. (Eds.) The Metallurgy of Zirconium, McGraw-Hill, New York (1955).

A.G. Quarrell (Ed.) Niobium, Tantalum, Molybdenum, and Tungsten, Elsevier, New York (1961).

F.T. Sisco and E. Epremian (Eds.) Columbium and Tantalum, J. Wiley and Sons, Inc., New York (1963).

Prob. 7 A key to efficient, high performance batteries is the development of efficient solid electrolytes. These electrolytes presently constitute a general group characterized as layered structures having basic features similar to AgI or CdI for example. Efficient ion conductivity in such materials is explained by high ion mobilities in the channel structures of the associated crystal regimes. In actual fabrications of electrolyte materials, it is not possible to utilize (or fabricate) single crystals, and the linear variation of conductivity with temperature is observed to decrease slightly (by ~ factor of 3) from that observed in the ideal, single crystal electrolyte. Make a sketch of such an ideal single crystal electrolyte structure (for example sodium beta-alumina). Now make a sketch showing a polycrystalline electrolyte and explain a possible reason for an ionic conductivity loss by a factor of 2-3. What can you say about the structure of grain boundaries in such polycrystalline-electrolytes with regard to ionic mobility? On the basis of this response, make a sketch of a grain boundary in a layered electrolyte of the sodium beta-alumina type.

REFERENCES

H.C. Brinkhoff, J. Phys. Chem. Solids, 35, 1225 (1974).

J.A.V. Butter (Ed.) Electrical Phenomena at Interfaces, in Chemistry, Physics, and Biology, Methuen Publishers, London (1951).

L.E. Murr, Interfacial Phenomena in Metals and Alloys, Addison-Wesley Publishing Co., Reading, Mass. (1975).

H. Sato, J. Chem. Phys., 55, 677 (1971).

W. VanGool (Ed.) Solid-State Batteries and Devices, American Elsevier, New York (1973).

M.S. Whittingham and R.A. Huggins in Solid State Chemistry, Proc. 5th Materials Res. Symp., Special Publication No. 364, National Bureau of Stds., Washington, D.C. (1972) p. 139.

Prob. 8 Make a sketch of a battery utilizing a solid-solution electrolyte. Describe the operation of the battery you sketch. Describe the solid-solution structure (showing the crystal lattice). Discuss the basic difference between solid electrolytes and liquid electrolytes with respect to interfacial problems. Make a sketch of the important battery interfaces and depict the details of the associated electrochemical-interfacial phenomena.

REFERENCES

J.A.V. Butter (Ed.) Electrical Phenomena at Interfaces in Chemistry, Physics, and Biology, Methuen Publishers, London (1951).

I. Fried, The Chemistry of Electrode Processes, Academic-Press, New York (1973).

E. Gileadi, E. Kirowa-Eisner and J. Penciner, Interfacial Electrochemistry: An Experimental Approach, Addison-Wesley, Reading, Mass. (1975).

J. Fouletier, P. Fabry and M. Kleitz, J. Electrochem. Soc., 123, 204 (1976).

W. VanGool (Ed.) Solid-State Batteries and Devices, American Elsevier Publishing Co., New York (1973).

L.M. Foster and J.E. Scardefield, J. Electrochem. Soc., 123, 141 (1976).

Prob. 9 Sketch (to scale) a sphalerite unit cell. Now extend this structure to show in three and two dimensions a channel structure (open lattice regime) through which ions might diffuse as in a battery electrolyte. Could you use sphalerite as an electrolyte material? Considering the scaled lattice array you have sketched, name several ionic specie which could rapidly diffuse through this structure. What does this tell you about the necessary conditions for a good electrolyte material? List three important features of a solid electrolyte as a good battery material.

REFERENCES

R.C. Evans, Crystal Chemistry, Cambridge Univ. Press, Cambridge (1966).

J.B. Goodenough, H. Y-P. Hong and J.A. Kafalas, Mater. Res. Bull., 11, 203 (1976).

F.G. Mills and S.P. Mitoff, J. Electrochem. Soc., 122, 457 (1975).

L. Pauling, Nature of the Chemical Bond, Cornell Univ. Press, Ithaca, New York (1960).

W. VanGool (Ed.) Solid-State Batteries and Devices, American Elsevier, New York (1973).

Prob. 10 As you might recall from Prob. 7, one key to efficient ionic conduction is the ease of diffusion of some particular ion. This is dependent to a large extent, upon the diffusional paths. Name an electrolyte which is characterized by a 1-dimensional transport path mechanism, one which is characterized by a 2-dimensional transport path mechanism and one characteristic of 3-dimensional diffusion paths. Make a sketch of each. Now, reconsidering your response to Prob. 7, discuss the performance of a polycrystalline electrolyte. Would you expect any one of the transport features to be predominant in thin film solid-state batteries having electrolyte thicknesses of less than 1 micron? Illustrate your response.

REFERENCES

G. Belonger, C. Lamarre and A.K. Vijh, J. Electrochem. Soc., 122, 47 (1975).

L.C. DeJonge, J. Mater. Sci., 11, 206 (1976).

T. Katan and H.F. Bauman, J. Electrochem. Soc., 122, 77 (1975).

J.T. Kummer, Prog. Sol. State Chem., 7, 141 (1972).

H. Sato, J. Chem. Phys., 55, 677 (1971).

H.L. Tuller and A.J. Norwick, J. Electrochem. Soc., 122, 225 (1975).

Prob. 11 With reference to the electromotive series, explain why zinc has been the most popular material for anodes in batteries. Show the reactions for two different zinc-anode batteries. What other metals can be used as effective anode materials in batteries? How does the zinc-nickel oxy-hydroxide cell differ from the Lalande cell? Show the reactions of each cell. Would you expect that corrosion of zinc would be as severe in each of the cells and why?

REFERENCES

J.W. Diggle and A. Damjanovic, J. Electrochem. Soc., 119, 1649 (1972).

A.B. Garrett, Batteries of Today, Research Press, Inc., Dayton, Ohio (1957).

J.L. Weininger and C.R. Morelock, J. Electrochem. Soc., 122, 1161 (1975).

Prob. 12 Consider the high-temperature (450°C) reaction

$$2Li + FeS \longrightarrow Li_2S + Fe.$$

Prove that the lithium sulfide formed is a highly stable sulfide. How does its energy of formation compare with FeS or MoS_2? What would the corresponding reaction be in the case of FeS_2? Sulfide reactions are, to a large extent, limiting features of battery design of various types, and an important aspect of materials problems in battery development. Are there any metallic or other elements which do not react with sulfur? Show the complete phase diagram for the Ni-S system. Can a battery be made for this system? Defend your conclusions.

REFERENCES

E.J. Cairns, G.H. Kucera and P.T. Cunningham, J. Electrochem. Soc., 120, 595 (1973).

J.A. Chitty and W.W. Smeltzer, J. Electrochem. Soc., 120, 1362 (1973).

M. Hansen and K. Anderko, Constitution of Binary Alloys (2nd Edition), McGraw-Hill Book Co., New York (1958).

S.D. James, J. Electrochem. Soc., 122, 921 (1975).

B.N. Popov and H.A. Laitinen, J. Electrochem. Soc., 120, 1346 (1973).

K.P. Staudhammer and L.E. Murr, Atlas of Binary Alloys, Marcel Dekker, New York (1973).

Prob. 13 The temperature variation of the critical field for a superconductor is expressed approximately by

$$H_c(T) = H_c(0)[1 - (\frac{T}{T_c})^2],$$

where $H_c(0)$ is the critical field at T = 0°K. Make a plot of $H_c(T)$ versus T and compare this with a plot of critical current versus critical field. Since it is known that crystal defects influence J_c (critical current) and T_c (critical temperature), show using appropriate sketches, how these parameters would be affected for beryllium in going from the microstructure in Fig. P.13(a) to those shown in (b) and (c) below. Describe the mechanism responsible for the changes you depict. How, in the fabrication of superconducting transmission lines, could these principles be utilized in achieving higher current densities?

REFERENCES

J.F. Bussiere, M. Garber, and M. Suenaja, J. Appl. Phys., 45, 4611 (1974).

G.P. DeGennes, Superconductivity in Metals and Alloys, W.A. Benjamin, New York (1966).

T.H. Geballe, Scientific American, 225, 22 (October, 1971).

J.M.E. Harper, T.H. Geballe, L.R. New Kirk and F.A. Valencia, J. Less Common Metals, 43, 5 (1975).

V.C. Newhouse (Ed.) Applied Superconductivity, Academic Press, New York (1975).

Prob. 14 Make separate plots of melting temperature versus superconducting transition temperature for elements which crystallize in cubic and hexagonal structures. What can you deduce about the effect of melting temperature and crystal structure on the superconducting transition temperature? Consider the same features for at least 6 binary alloys having a cubic structure and compare the effects of melting temperature on superconducting transition temperature. If, a difference exists between the single element plot and the binary plot of T_m versus T_c for cubic structures, and if, ideally, the alloying of a third element to produce ternary superconductors has a similar

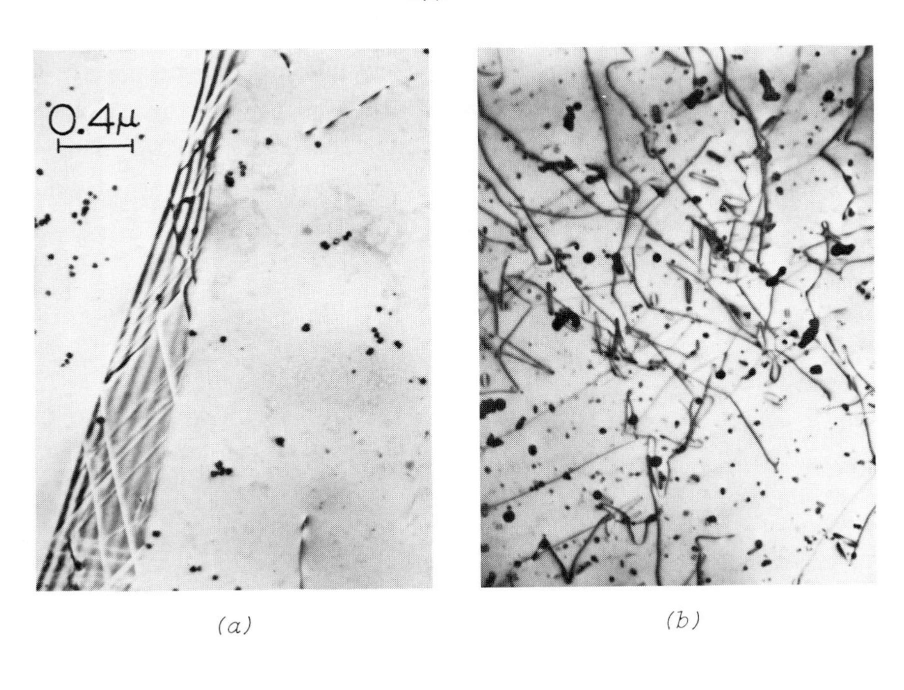

(a) (b)

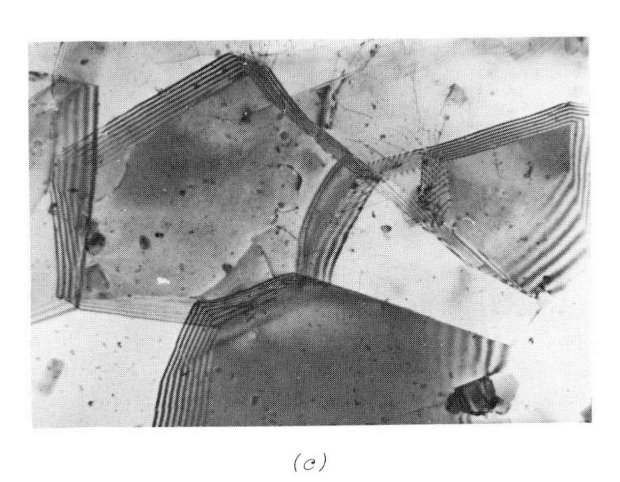

(c)

FIG. P.13. Electron micrographs of beryllium thin foils prepared from annealed (a), annealed and shock-loaded (b), and rolled and annealed sheet (c).

incremental change in superconducting transition temperature, make some predictions concerning the possibility of attaining T_c values above 25°K. Show from experimental data that variations in compositions of elements in ternary alloys affects T_c.

REFERENCES

M. Drys, J. Less Common Metals, 44, 229 (1976).

M. Drys and N. Iliew, J. Less Common Metals, 44, 235 (1976).

R. Horyn, J. Less Common Metals, 44, 221 (1976).

L. Lokot, R. Horyn and N. Iliew, J. Less Common Metals, 44, 215 (1976).

B.T. Matthias, Phys. Rev., 97, 74 (1954).

B.T. Matthias, J. Less Common Metals, 43, 1 (1975).

K.P. Staudhammer and L.E. Murr, Atlas of Binary Alloys: A Periodic Index, Marcel Dekker, New York (1973).

Prob. 15 Show the formation of a thin-film CdTe-CdS photovoltaic junction with appropriate electrical contacts, i.e. show how it could be fabricated. Is the junction homogeneous or heterogeneous? Describe the difference between these junction types. What will be the difference in performance of the device if the contacts are Cu or Ni? Make a complete energy-level diagram for the device utilizing Cu, then Ni contacts and demonstrate whether the contacts are ohmic or rectifying in each case. Will the efficiency of the device vary with contact metal? How will aluminum contacts perform as compared with Cu or Ni? (Note that the work functions for Cu, Ni, and Al are 5.30, 4.61, and 3.81 eV respectively.)

REFERENCES

P.D. Ankrum, Semiconductor Electronics, Prentice-Hall, Englewood Cliffs, New Jersey (1971).

R.H. Bube, Photoconductivity of Solids, J. Wiley and Sons, New York (1960).

A.L. Fahrenbruch and R.H. Bube, J. Appl. Phys., 45, 1264 (1975).

L.P. Hunter, Introduction to Semiconductor Phenomona and Devices, Addison-Wesley, Reading, Mass. (1966).

S. Larach (Ed.) Photoelectronic Materials and Devices, Van Nostrand, Princeton, New Jersey (1965).

J.L. Moll, Physics of Semiconductors, McGraw-Hill Book Co., New York (1964).

B. Pellegrini, Solid-State Electronics, 17, 217 (1974).

Prob. 16 As in most junction devices, the concept of an interface, both electronic and structural in nature, may be an efficiency limiting phenomena in the operation of photovoltaic heterojunction systems. Show a relationship between semiconductor junction photovoltaic efficiency and the associated lattice mismatch for heterojunction systems. What conclusions can be made from a plot of this data? Assume the junction is made along [001] in all heterojunctions. Show your assumptions and mismatch calculations. Show also, in a series of simple sketches, the differences in the interface structures based on accommodation of the mismatch by the formation of misfit dislocations.

REFERENCES

R.H. Bube, Photoconductivity of Solids, J. Wiley and Sons, New York (1960).

E.G. Bylander, Materials for Semiconductor Functions, Hayden Book Co., New York (1971).

S. Larach (Ed.) Photoelectronic Materials and Devices, Van Nostrand, Princeton, New Jersey (1965).

L.E. Murr, Interfacial Phenomena in Metals and Alloys, Addison-Wesley, Reading, Mass. (1975).

P. Perfetti, F. Cerrina, C. Coluzza, and G. Margaritondo, J. Appl. Phys., 45, 972 (1974).

C. Wu and R.H. Bube, J. Appl. Phys., 45, 648 (1974).

Prob. 17 With reference to Prob. 13, discuss the effect of radiation damage on the magnet coils of a proposed fusion reactor. In particular, discuss the effects of high neutron fluxes upon superconductivity and normal conductivity in the magnet coils. Will the effect be complimentary or non-complimentary?

REFERENCES

R.J. Arsenault (Ed.) Proc. 1973 Intl. Conf. on Defects and Defect Clusters in B.C.C. Metals and Their Alloys, National Bureau of Standards, Gaithersburg, Maryland (1973).

G.W. Cullen and G.D. Cody, J. Appl. Phys., 44, 2838 (1973).

D. Dew-Hughes, Phil. Mag., 30, 293 (1974).

D. Dew-Hughes, Rep. Prog. Phys., 34, 821 (1971).

S. Foner and B.B. Schwartz (Eds.) Superconducting Machines and Devices, Plenum Press, New York (1974).

A.D. McInturff and G.G. Chase, J. Appl. Phys., 44, 2378 (1973).

A. Seeger, D. Schumacher, W. Schilling and J. Diehl (Eds.) Vacancies and Interstitials in Metals, Wiley-Interscience, New York (1970).

S.J. Thompson and P.E.J. Flewitt, J. Less Common Metals, 40, 269 (1975).

Prob. 18 Liquid lithium has been suggested as a prime candidate for coolant applications in fusion reactors. What are the specific advantages of lithium as a coolant? What other functions can lithium perform in such a reactor environment? Discuss the materials problems associated with the containment of lithium. Indicate metals or alloys which might be candidates for coolant channel materials. Are the materials you suggest also generally considered reactor-compatible materials? Discuss your response.

REFERENCES

H.M. Finniston and J.P. Howe (Eds.) Metallurgy and Fuels, Vol. 1, Pergamon Press, Ltd., London (1956).

M.D. Fiske and W.W. Havens, Jr. (Eds.) Physics and the Energy Problem - 1974, American Institute of Physics, New York (1974).

C.A. Hampel (Ed.) Rare Metals Handbook, Reinhold Publishing Corp., New York (1954).

C.J. Smithells, Metals Reference Book, 2 vols., (Third Edition), Butterworths, London (1962).

J.H. Stang, E.M. Simons, J.A. DeMastry and J.M. Genco, Compatibility of Liquid and Vapor Alkali Metals with Construction Materials, DMIC Report 227, Defense Metals Information Center, Battelle Memorial Institute, Columbus, Ohio, April 15 (1966).

K.P. Staudhammer and L.E. Murr, Atlas of Binary Alloys: A Periodic Index, Marcel Dekker, Inc., New York (1973).

Prob. 19 On considering materials requirements for the first wall of the proposed fusion reactors, particularly TOKAMAK, it is noted that candidate materials must have a high thermal conductivity, high melting point, high strength, low sputtering rates, low diffusion rates and solubilities of hydrogen and helium, and they must be able to be welded and fabricated for high vacuum applications and cheap (or reasonably cheap). Current first-wall candidate materials include Nb, V,

V-Zr, Nb-V alloys, several stainless steels including 304 and 316, and several Ni-base alloys and beryllium. Make a table of five specific metals or alloys mentioned above, indicating all of the properties cited, and discuss your candidate for a first wall material on the basis of the best projected performance at 1000°C.

REFERENCES

M.D. Fiske and W.W. Havens, Jr. (Eds.) Physics and the Energy Problems - 1974, American Institute of Physics, New York (1974).

G.K. Kulcinski, P.G. Duran and M.A. Abdon, American Soc. for Testing and Materials Special Tech. Publication STP-570, ASTM, Philadelphia (1975).

T. Lyman (Ed.) Metals Handbook, Properties and Selection of Metals, Vol. I; American Soc. Metals, Metals Park, Ohio (1961).

J. Smithells, Metals Reference Book, Vol. I and II, Third Edition, Butterworths, London (1962).

V.F. Zackay (Ed.) High Strength Materials, J. Wiley, New York (1965).

Prob. 20 Suppose type 304 stainless steel is chosen as the first-wall material. Describe the possible effects of hydrogen and helium diffusion into the first wall at 1000°C. If the first wall thickness is 1 cm, and the temperature is maintained at 1000°C continuously, how long might it take to grow grains large enough to produce direct grain boundary channels from surface to surface? Could grain growth be a serious problem in the first wall? Why? Discuss your answer.

REFERENCES

G.K. Kulcinski, P.G. Doran and M.A. Abdou, American Society for Testing and Materials Special Tech. Publication, STP-570, ASTM,

Philadelphia (1975).

A.G. Guy, Introduction to Materials Science, McGraw-Hill Book Co., New York (1972).

P.G. Shewmon, Diffusion in Solids, McGraw-Hill Book Co., New York (1963).

R.J. Walter, R.P. Jewett and W.T. Chandler, Mater. Sci. Engr., 5, 98 (1969/70).

Prob. 21 In the proposed Theta-Pinch fusion reactor, the insulator first wall is expected to differ in its response to the radiation fluxes when compared with metal or alloy first wall materials in the TOKAMAK design. Discuss the major differences. Figure P.21 below shows a CaF_2 single-crystal section which has been exposed to high-dose electron irradiation in an electron microscope. Compare the defect lattice with those lattices observed for metals such as Ni, Mo, or Nb following high-dose neutron irradiation. If the lattice in Fig. P.21 arises by color-center aggregates, how does it differ from the neutron-irradiation induced void lattices in Nb or Mo? Could the defect lattice in Fig. P.21 be a colloid lattice? How would it differ from a color-center aggregate lattice? Would the void-like lattice in the CaF_2 shown in Fig. P.21 cause swelling? How much?

REFERENCES

J.L. Brimhall and G.L. Kulcinski, Rad. Effects, 20, 25 (1973).

G.L. Kulcinski, J.L. Brimhall and H.E. Kissinger, J. Nuclear Mater., 40, 166 (1971).

G.L. Kulcinski, P.G. Duran and M.A. Abdou, American Society for Testing and Materials Special Tech. Publication, STP-570, ASTM, Philadelphia (1975).

L.E. Murr, Phys. Stat. Sol.(a), 22, 239 (1974).

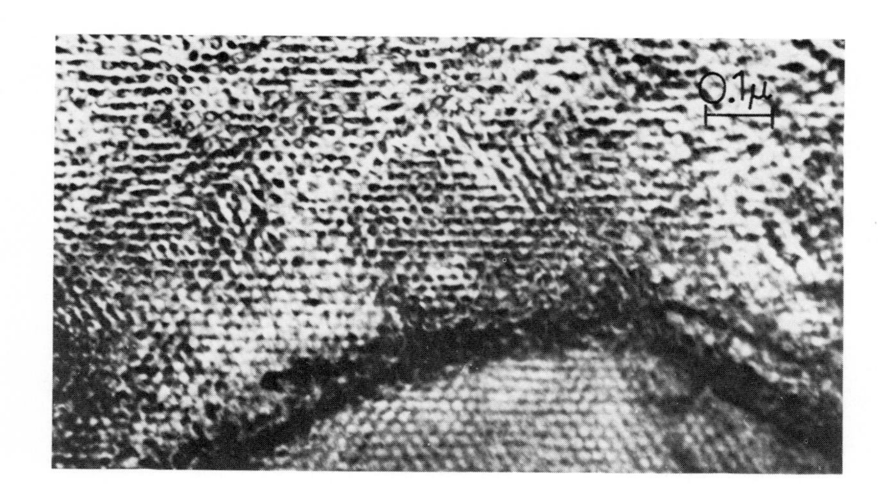

FIG. P.21. Electron micrograph of defect-aggregate lattice in electron-irradiated CaF_2. The print surface is parallel to (111).

L.E. Murr in, Diffraction Studies of Real Atoms and Real Crystals, Proc. Int. Crystallography Conf., Australian Acad. Sci., Canberra (1974). p. 167.

A.W. Searcy, D.V. Ragone, and V. Colombo (Eds.) Chemical and Mechanical Behavior of Inorganic Materials, Wiley-Interscience, New York (1970).

Prob. 22 What will be the effect of dislocations on hydrogen embrittlement of a first wall of 304 stainless steel in a TOKAMAK if the temperature remains somewhat constant at 1000°C? What will happen to the grain boundaries? That is, will the grain boundaries retain hydrogen and add to brittle fracture by inducing intergranular fracture? If, at this temperature, the material swells by 3% and then diffuses to the surface to form a continuous array of blisters which just touch one another, what will be the blister size if the individual blisters are considered to be hemispherical?

REFERENCES

R.J. Arsenault (Ed.) Proc. 1973 Intl. Conf. on Defects and Defect Clusters in B.C.C. Metals and Their Alloys, National Bureau of Standards, Gaithersburg, Maryland (1973).

Hydrogen in Metals, ASM, Metals Park, Ohio (1974).

G.L. Kulcinski, P.G. Duran and M.A. Abdou, American Society for Testing and Materials Special Tech. Publication, STP-570, ASTM, Philadelphia (1975).

R.J. Walter, R.P. Jewett, and W.T. Chandler, Mater. Sci. Engr., 5, 98 (1969/70).

Prob. 23 It is currently recognized that thin film optics is the weakest link in developing high energy fusion laser systems. In this regard, experimental evidence suggests that laser damage thresholds are generally higher in R.F. sputtered films when compared with vacuum vapor-deposited thin films of the same material. If the difference in laser damage threshold is due to differences in the degree of crystallinity, describe an experiment to verify and quantify this contention utilizing silver films vapor-deposited onto copper-substrate-mirror surfaces. Describe how the grain size could be systematically altered assuming that the deposited Ag film structure on Cu would be the same as that upon NaCl.

REFERENCES

Laser Damage in Optical Materials, NBS Special Publication 435, National Bureau of Standards, Washington, D.C. (1975).

L.E. Murr and M.C. Inman, Phil. Mag., 14, 135 (1966).

L.E. Murr and W.R. Bitler, Mater. Res. Bull., 2, 787 (1967).

J.F. Ready, Effects of High-Power Laser Radiation, Academic Press, New York (1970).

Prob. 24 In. Fig. P.24 are shown several examples of silver films vapor-deposited onto single crystal NaCl surfaces and subsequently observed in the transmission electron microscope. If we assume that the average damage threshold occurring for 1.06 μ short-pulse laser radiation is estimated to be approximately 2, 6, 13, and 27 MW/cm^2 for the films in Fig. P.24 (a) - (d), respectively; make a plot of damage threshold versus a materials property of the silver films. What can you say about this response? Is this possible or probable in actual laser systems? Discuss your response. Calculate the thickness of the films in Fig. P.24 if all films are assumed to have the same thickness. What fraction of the radiation wave length does this film thickness represent?

REFERENCES

J.R. Bettis, R.A. House, A.H. Guenther, and R. Austin, "The Importance of Refractive Index, Number Density, and Surface Roughness in the Laser-Induced Damage of Thin Films and Bare Surfaces", in Laser Induced Damage in Optical Materials, NBS Special Publication 435, National Bureau of Standards, Washington, D.D. (1975).

R.A. House, J.R. Bettis, A.H. Guenther, and R. Austin, "Correlation of Laser-Induced Damage with Surface Structure and Preparation Technique of Several Optical Glasses at 1.06 μ", Ibid. (1975).

L.E. Murr and M.C. Inman, Phil. Mag., 14, 135 (1966).

L.E. Murr, Electron Optical Applications in Materials Science, McGraw-Hill Book Co., New York (1970).

J.F. Ready, Effects of High-Power Laser Radiation, Academic Press, New York (1971).

W.A. Szilva, M.S. Thesis, New Mexico Institute of Mining and Technology, Socorro (1976).

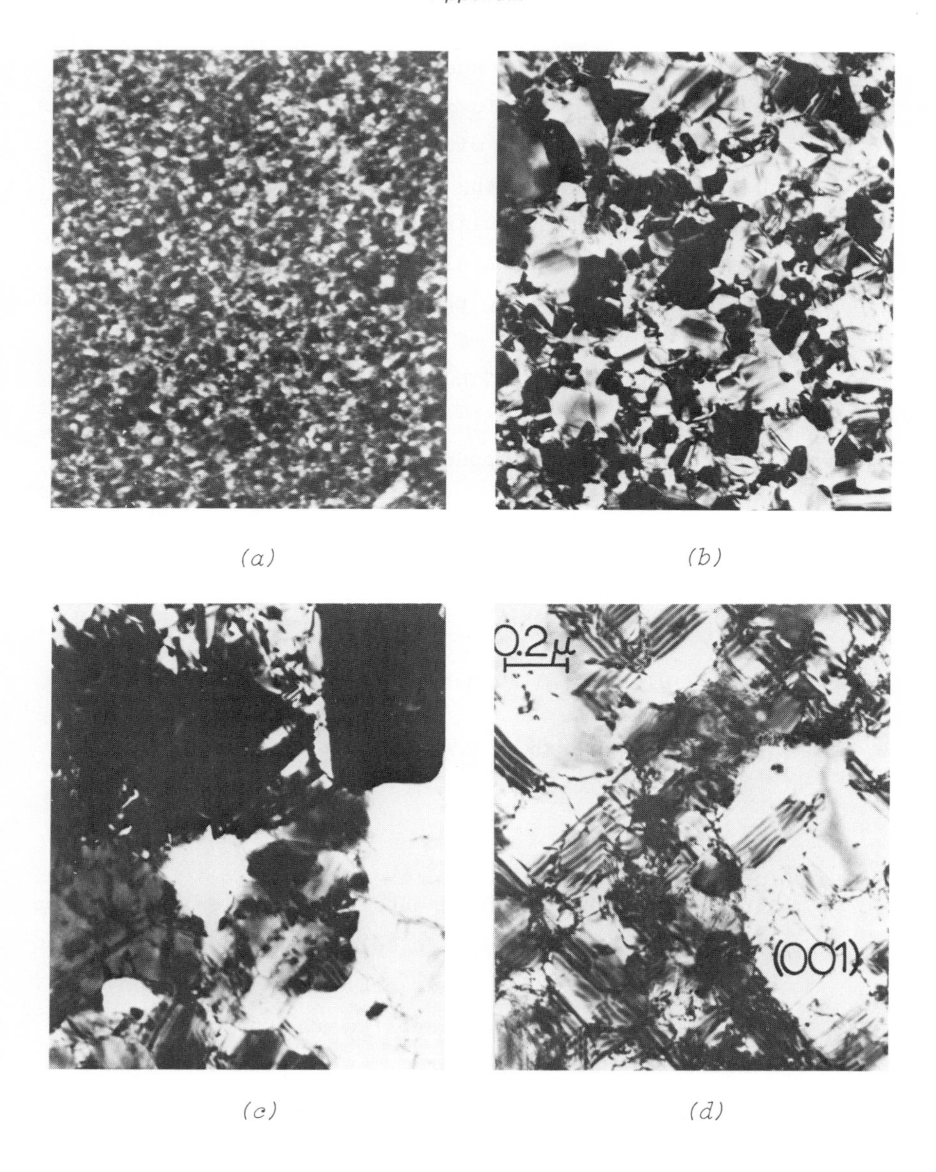

(a) (b)

(c) (d)

FIG. P.24. Bright-field transmission electron micrographs of vapor-deposited silver films. The films were vapor-deposited at substrate temperatures of (a) 25°C, (b) 100°C, (c) 200°C, and (d) 250°C. The operating reflection governing contrast in (d) was [200] at an alcelerating voltage of 200KV.

Prob. 25 Show how tritium can cause embrittlement in metals and alloys which compose fusion reactor critical components. Support your response with appropriate equations and sketches. Show also the breeding reactions for the solid ceramics Li_2O, $LiAlO_2$, and Li_3N. Describe the crystal structure of $LiAlO_2$ and discuss how it would change in the breeding reaction

REFERENCES

E.A. Evans, Tritium and Its Compounds, Van Nostrand, Princeton, New Jersey (1966).

M.D. Fiske and W.W. Havens, Jr. (Eds.) Physics and the Energy Problems - 1974, American Institute of Physics, New York (1974).

Nuclear Science and Technology for Ceramists: Symposium Proc., National Bureau of Standards (NBS Publication 285), Washington, D.C. (1967).

V.F. Zackay (Ed.) High Strength Materials, J. Wiley and Sons, New York (1975).

See also appropriate handbooks, e.g. Handbook of Chemistry and Physics (latest edition).

Prob. 26 Tin oxide (SnO_2) has been used as an optical coating on glass for use in flat-plate solar collectors. Describe the advantages of SnO_2 over MgF_2 as a coating material. Could these materials be applied as transparent, conducting coatings for other solar collector sandwich devices such as photovoltaic devices? What advantages would they have? If SnO_2 is applied as a coating to a flat-plate collector and the absorption-reflection properties measured at 25°C, how would these properties be altered at an operating temperature of 200°C?

REFERENCES

J.A. Duffie and W.A. Beckman, Solar Energy Thermal Processes, J. Wiley and Sons, New York (1974).

R.C. Weast (Ed.) Handbook of Chemistry and Physics, Edition 54, Chemical Rubber Co., Cleveland, Ohio (1976).

J.R. Williams, Solar Energy: Technology and Applications, Ann Arbor Science Publishers, Ann Arbor, Michigan (1974).

Prob. 27 Compare the heat (or thermal energy) which can be stored on melting 1 ft^3 of NaCl, NaOH, $Na_2H_2P_2O_7 \cdot 6H_2O$, and Fe. Discuss the applications of these materials as solar-thermal storage materials. Discuss also the specific container problems for each of these materials. In materials such as $Na_2H_2P_2O_7 \cdot 6H_2O$ and NaOH, it is usually necessary to utilize a nucleating agent to promote cyclic storage functions. What properties must such an agent possess? Describe the difference between epitaxial and pseudomorphic nucleation and growth.

REFERENCES

J.A. Duffie and W.A. Beckman, Solar Energy Thermal Processes, J. Wiley and Sons, New York (1974).

M. Telkes, ASHRAE Journal, September, 40 (1974).

R.C. Weast (Ed.) Handbook of Chemistry and Physics, Edition 54, Chemical Rubber Co., Cleveland, Ohio (1976).

J.R. Williams, Solar Energy: Technology and Applications, Ann Arbor Science Publishers, Ann Arbor, Michigan (1974).

Prob. 28 Because of the necessity to operate magnetohydrodynamic (MHD) generators at temperatures in the 1200°C temperature range, a stringent materials problem arises, particularly with regard to maintaining strength at temperature. It has been suggested that thoria-dispersed (TD) nichrome might be suitable for such applications. Figure P.28 illustrates the TD-nichrome microstructure by comparison with commercial nichrome.

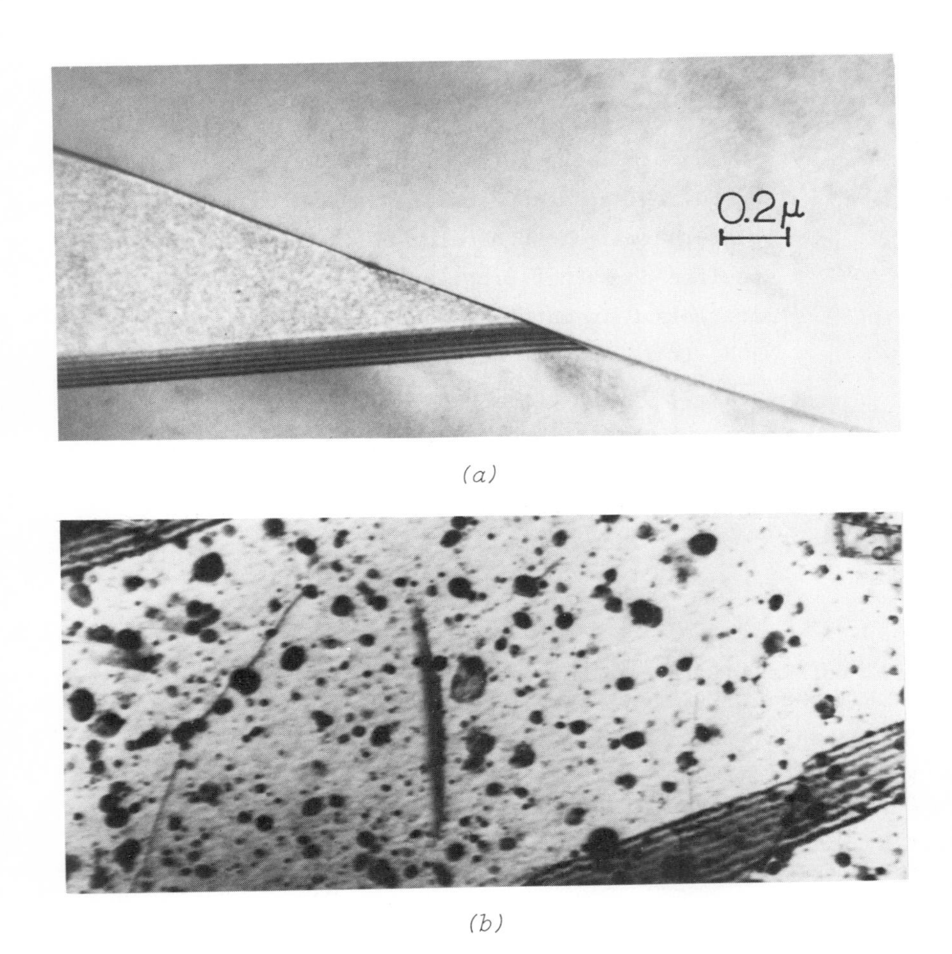

FIG. P.28. Transmission electron micrographs of (a) nichrome (80 Ni / 20 Cr) and (b) TD-nichrome (2 vol. percent ThO_2, 20 percent Cr). The thoria particles having a distribution of sizes are shown dispersed in the nichrome matrix in (b).

Discuss the mechanism of high-temperature strengthening in TD-nichrome. What would be the effect of shock deforming TD-nichrome sheet before use in MHD applications? Would this be feasible in terms of fabrication technology? Would it be practical in terms of the proposed MDH high temperature applications? Qualify your response.

REFERENCES

A.G. Guy, Introduction to Materials Science, McGraw-Hill Book Co., Inc., New York (1972).

L.E. Murr, H.R. Vydyanath and J.V. Foltz, Met. Trans., 1, 3215 (1970).

L.E. Murr and H.R. Vydyanath, Acta Met., 18, 1047 (1970).

R.J. Rosa, Magnetohydrodynamic Energy Conversion, McGraw-Hill Book Co., Inc., New York (1968).

G.W. Sutton, Engineering Magnetohydrodynamics, McGraw-Hill Book Co., Inc., New York (1965).

Prob. 29 There are essentially four prominent methods or mechanisms known for strengthening alloy systems for use at high temperatures (> 500°C) in environments involving turbine devices such as jet engines of various types and magnetohydrodynamic systems. Discuss these mechanisms and give an appropriate example of each. Describe an alloy system in which it might be possible to utilize all of these prominent strengthening mechanisms. A recent technique for producing fine oxide dispersions in systems such as Ni-Co-Mo or Ni-Cr-Al involves mechanical alloying by grinding metal powders. Describe the possible applications of this alloying technique in achieving high-strength, high-temperature materials.

REFERENCES

J.S. Benjamin, Scientific American, 234, 40, May (1976).

C.C. Law and A.F. Giamei, Met. Trans., 7A, 5 (1976).

Magnetohydrodynamics: Power Generation and Theory (TID-3356), ERDA Bibliography Series, National Information Service, U.S. Dept. of Commerce, Springfield, Virginia, Nov. (1975).

P.R. Sahm and M.O. Speidel (Eds.) High Temperature Materials in Gas Turbines, Elsevier, Amsterdam (1974).

V.F. Zackey (Ed.) High Strength Materials, J. Wiley and Sons, New York (1975).

Prob. 30 A considerable effort has been devoted to understanding or controlling the mechanical properties of iridium as a cladding material for nuclear fuel beads in fission (fast-breeder) reactor applications. The major problem seems to hinge upon the uncharacteristic brittle fracture of iridium, particularly at the grain boundaries in highly purified metal. Current thought on this problem centers upon either segregation of oxygen or some other impurity (K or P) at the interface, or the occurrence of special grain boundary structures which are conducive to intergranular fracture. Describe an experiment or experiments to differentiate between grain boundary structural or segregation effects. Show how grain boundary structure can be effective in promoting the nucleation of intergranular cracks.

REFERENCES

J.M. Galbraith and L.E. Murr, J. Mater. Sci., 10, 2025 (1975).

L.E. Murr and O.T. Inal, Metals Engr. Quarterly, 12, 29 (1972).

L.E. Murr, Interfacial Phenomena in Metals and Alloys, Addison-Wesley Publishing Co., Inc., Reading, Mass. (1975).

INDEX

B

C

D

G

H

S

U

A 6
B 7
C 8
D 9
E 0
F 1
G 2
H 3
I 4
J 5